CAMBRIDGE MONOGRAPHS ON MATHEMATICAL PHYSICS

General editors: P.V. Landshoff, W.H. McCrea, D.W. Sciama, S. Weinberg

SPINORS AND SPACE-TIME

Volume 2: Spinor and twistor methods in space–time geometry

SPINORS AND SPACE–TIME

Volume 2

Spinor and twistor methods in space–time geometry

ROGER PENROSE

Rouse Ball Professor of Mathematics, University of Oxford

WOLFGANG RINDLER

Professor of Physics, University of Texas at Dallas

CAMBRIDGE
UNIVERSITY PRESS

PUBLISHED BY THE PRESS SYNDICATE OF THE UNIVERSITY OF CAMBRIDGE
The Pitt Building, Trumpington Street, Cambridge, United Kingdom

CAMBRIDGE UNIVERSITY PRESS
The Edinburgh Building, Cambridge CB2 2RU, UK http://www.cup.cam.ac.uk
40 West 20th Street, New York, NY 10011–4211, USA http://www.cup.org
10 Stamford Road, Oakleigh, Melbourne 3166, Australia

First published 1986
First paperback edition (with corrections) 1988
Reprinted 1989, 1993, 1999

Printed in the United Kingdom at the University Press, Cambridge

A catalogue record for this book is available from the British Library

ISBN 0 521 25267 9 hardback
ISBN 0 521 34786 6 paperback

Contents

Preface

This is a companion volume to our introductory work *Spinors and space–time*, Volume 1: *two-spinor calculus and relativistic fields*. There we attempted to demonstrate something of the power, utility and elegance of 2-spinor techniques in the study of space–time structure and physical fields, and to advocate the viewpoint that spinors may lie closer to the heart of (even macroscopic) physical laws than the vectors and tensors of the standard formalism. Here we carry these ideas further and discuss some important new areas of application. We introduce the theory of twistors and show how it sheds light on a number of important physical questions, one of the most noteworthy being the structure of energy–momentum/angular momentum of gravitating systems. The illumination that twistor theory brings to the discussion of such physical problems should lend further support to the viewpoint of an underlying spinorial structure in basic physical laws.

Those who have some familiarity with the standard 2-spinor formalism should be able to read this volume as an independent work. All necessary background material has been collected together in an introductory section which summarizes the relevant contents of Volume 1. There are many references to this earlier volume in the present work for the results and notations that are needed, but these are all explicitly provided in the summary, the numbering of results being the same as that of Volume 1 so that they can be located in either place without ambiguity. Detailed derivations are not repeated, however, and the reader who wishes to pursue these results *per se* would be well advised to refer back to the earlier volume.

The main topics introduced and discussed in this volume are twistor theory and related matters such as massless fields and the geometry of light rays, energy–momentum and angular momentum (from an unusual but, we hope, illuminating point of view), and the conformal structure of infinity. In addition we provide, in Chapter 8, a detailed classification of space–time curvature by 2-spinor means. Though we are mainly concerned with the classification of the Weyl (or empty-space Riemann) tensor, we also give a comprehensive treatment of the Ricci tensor, showing how this arises as a particular case of the classification of symmetric spinors generally. Chapter 8 can be read independently of the other chapters in this volume, though there is some interrelation between its results and those of the other chapters.

Some remarks about the role of twistors in this volume are in order. Twistor theory can be viewed in two ways. On the one hand it may be seen simply as providing new mathematical techniques for the solution of problems within standard physical theory. On the other, it may be viewed as suggesting an alternative framework for the basis of all physics, characterized by the relegation of the concept of event (space–time point) from a primary to a secondary role. In this volume, twistors are treated very much in the former rather than in the latter way. This is not a work on twistor physics as such, though it can serve as a fairly comprehensive introduction to the subject for those who are interested in following up some of its more detailed, sophisticated, or speculative ideas.

We are concerned here mainly with the way that the mathematics of twistor theory interrelates with that of 2-component spinors. Spinor-field descriptions of twistors are presented in more detail than have been given hitherto. Many applications of twistor methods are given, such as: a proof that the vanishing of the Weyl tensor implies that the space–time is locally Minkowskian; the Kerr theorem for generating shear-free congruences in Minkowski space (including a generalization that applies in curved space–times); massless free-field contour integrals with basic sheaf cohomology ideas; the Ward construction for general self-dual Yang–Mills fields (though, to our regret, the original gravitational analogue of this construction is beyond the scope of this work – for we have felt it necessary to limit our discussion essentially to results with direct application to real, i.e. Lorentzian, space–times); conformal Killing vectors; Killing spinors; cosmological models; and the Grgin phenomenon for massless fields. An elaborate 'twistorial' discussion of energy–momentum and angular momentum in linearized gravitation is given, and this is generalized in our presentation of some recent developments which use corresponding twistor-type techniques in full general relativity and which provide a suggestive 'quasi-local' definition of mass. This leads us to a discussion of energy–momentum and angular momentum at null infinity and their behaviour under the Bondi–Metzner–Sachs (BMS) group – the group describing the asymptotic symmetries of an asymptotically flat space–time. We also give details of the full Bondi–Sachs mass-loss formula and present a version of Witten's argument for demonstrating mass positivity in general relativity as it applies at null infinity. None of this material on energy–

momentum/angular momentum has appeared before in book form and it illustrates a significant new area in which spinor (and twistor) arguments can find powerful physical application. Much of .he detailed spinor discussion of conformal infinity is presented here for the first time.

In an appendix we show how 2-spinors, Dirac 4-spinors, twistors and spin-weighted functions fit in with the general scheme of *n*-dimensional spinors.

As with Volume 1, there are many to whom we owe thanks, for assistance of a varied nature, either direct or indirect. Most particularly we are grateful to Michael Atiyah, Toby Bailey, Nick Buchdahl, Judith Daniels, Mike Eastwood, Robert Geroch, Denny Hill, Lane Hughston, Ben Jeffryes, Ron Kelly, Roy Kerr, Ted Newman, Zoltán Perjés, Ivor Robinson, Niall Ross, Ray Sachs, Engelbert Schücking, William Shaw, Iz Singer, Paul Sommers, George Sparling, Paul Tod, Helmuth Urbantke, Ronnie Wells and Nick Woodhouse. Again, our special thanks goes to Dennis Sciama for his encouragement and faith in this project for well over twenty years, and also to Tsou Sheung Tsun who has again provided invaluable help with the references.

Roger Penrose

1985 Wolfgang Rindler

Summary of Volume 1

We provide here a connected account summarizing all the material of Volume 1 (*Two-spinor calculus and relativistic fields*) needed for the present volume. The equation numbering (and occasional section numbering) is precisely that of Volume 1, though sometimes the equations appear a little out of strict numerical order.

The light cone in Minkowski space

We use standard (*restricted*, i.e. right-handed and isochronous) Minkowski coordinates

$$x^0 = T, \quad x^1 = X, \quad x^2 = Y, \quad x^3 = Z$$

for (affine) Minkowski space $\mathbb{M}$ [*or* Minkowski vector space $\mathbb{V}$]. Points P, Q with coordinates $P^{\mathbf{a}} = (P^0, P^1, P^2, P^3)$, $Q^{\mathbf{a}} = (Q^0, Q^1, Q^2, Q^3)$, in $\mathbb{M}$, have invariant (squared) separation

$$\Phi(P, Q) = (Q^0 - P^0)^2 - (Q^1 - P^1)^2 - (Q^2 - P^2)^2 - (Q^3 - P^3)^2 \tag{1.1.22}$$

The *light cone* of P [*or* correspondingly the *null cone* in $\mathbb{V}$ when P is the origin] is the set of $Q \in \mathbb{M}$ [*or* $Q \in \mathbb{V}$] for which $\Phi(P, Q) = 0$. The light rays through P are the straight lines (generators) of the cone. The light rays through (say) the origin O can be parametrized by $\zeta \in \mathbb{C} \cup \{\infty\}$, where

$$\zeta = \frac{X + \mathrm{i}Y}{T - Z} = \frac{T + Z}{X - \mathrm{i}Y} \tag{1.2.19}$$

and, in terms of standard spherical polar coordinates on the sphere S^+, whose equation is

$$T = 1, \quad X^2 + Y^2 + Z^2 = 1,$$

we have

$$\zeta = \mathrm{e}^{\mathrm{i}\phi} \cot \frac{\theta}{2}. \tag{1.2.10}$$

S^+ is sometimes referred to as the *anti-celestial sphere*. The light rays through O put it in one-to-one correspondence with the sphere S^- (given by $T = -1$, $X^2+Y^2+Z^2=1$): the *celestial sphere* of the observer at O with time-axis $X = Y = Z = 0$. The two are related by an antipodal map.

Under a (restricted) Lorentz transformation ζ transforms as

$$\zeta \to \tilde{\zeta} = \frac{\alpha\zeta + \beta}{\gamma\zeta + \delta} \tag{1.2.17}$$

where the coefficients α, β, γ, δ constitute the *spin-matrix*

$$\begin{pmatrix} \alpha & \beta \\ \gamma & \delta \end{pmatrix} \quad (\alpha\delta - \beta\gamma = 1; \quad \alpha, \beta, \gamma, \delta \in \mathbb{C}).$$

This provides a (general) *conformal* motion of S^+, where we think of the points of S^+ as simply labelling the generators of the future light cone of P. The matrix acts on $\binom{\xi}{\eta}$, the components $\kappa^0 = \xi$, $\kappa^1 = \eta$ of a *spin-vector* $\boldsymbol{\kappa}$, where

$$\zeta = \xi/\eta = \kappa^0/\kappa^1.$$

The null vector ('flagpole') corresponding to this spin-vector has coordinates

$$T = \frac{1}{\sqrt{2}}(\xi\bar{\xi} + \eta\bar{\eta}), \quad X = \frac{1}{\sqrt{2}}(\xi\bar{\eta} + \eta\bar{\xi}),$$

$$Y = \frac{1}{\mathrm{i}\sqrt{2}}(\xi\bar{\eta} - \eta\bar{\xi}), \quad Z = \frac{1}{\sqrt{2}}(\xi\bar{\xi} - \eta\bar{\eta}), \tag{1.2.15}$$

(*cf.* also (3.1.31), (3.2.2)). *Unitary* spin-matrices describe rotations, while an active *boost* in the Z-direction of velocity v is described by

$$\begin{pmatrix} \tilde{\xi} \\ \tilde{\eta} \end{pmatrix} = \pm \begin{pmatrix} w^{\frac{1}{2}} & 0 \\ 0 & w^{-\frac{1}{2}} \end{pmatrix} \begin{pmatrix} \xi \\ \eta \end{pmatrix}, \tag{1.2.37}$$

with

$$w = \left(\frac{1+v}{1-v}\right)^{\frac{1}{2}}. \tag{1.2.36}$$

The Lorentz invariant cross-ratio of four null directions (or points of S^+) with complex parameters $\zeta_1, \zeta_2, \zeta_3, \zeta_4$ is

$$\chi = \{\zeta_1, \zeta_2, \zeta_3, \zeta_4\} = \frac{(\zeta_1 - \zeta_2)(\zeta_3 - \zeta_4)}{(\zeta_1 - \zeta_4)(\zeta_3 - \zeta_2)}, \tag{1.3.9}$$

i.e.

$$\chi = \frac{(\xi_1\eta_2 - \xi_2\eta_1)(\xi_3\eta_4 - \xi_4\eta_3)}{(\xi_1\eta_4 - \xi_4\eta_1)(\xi_3\eta_2 - \xi_2\eta_3)}, \tag{1.3.10}$$

where $\zeta_j = \xi_j/\eta_j$. If the null directions are permuted, χ gets replaced by one of

$$\chi,\ 1-\chi,\ \frac{1}{\chi},\ \frac{1}{1-\chi},\ \frac{\chi-1}{\chi},\ \frac{\chi}{\chi-1}. \tag{1.3.12}$$

The condition for the four null directions to lie in a hyperplane, i.e. for the corresponding points of S^+ to be concyclic, is that χ be real. The *harmonic* case $\chi = -1$ (or 2 or $\frac{1}{2}$) occurs when the four points can be transformed to the vertices of a square; and the *equianharmonic* case $\chi = -e^{\pm 2i\pi/3}$, to the vertices of a regular tetrahedron.

§1.5 Spinorial objects and spin structure

A non-zero spin-vector $\boldsymbol{\kappa}$ may be represented, up to a sign, as a null flag (i.e. a future-null vector with an oriented null 2-plane through it, cf. (3.2.2), (3.2.9); compare also Payne 1952) or as a pair of infinitesimally separated future-null directions. However, the *sign* of $\boldsymbol{\kappa}$ cannot be given a local geometrical interpretation since a continuous rotation through 2π (about any axis) sends $\boldsymbol{\kappa}$ into its negative, whereas it must restore any local geometrical structure (in the ordinary sense) to its original state. Thus we widen the concept of 'local geometry' to include *spinorial objects*, like $\boldsymbol{\kappa}$, which are restored to their original states when rotated continuously through 4π but not through 2π. For spinorial objects to exist on a manifold $\mathcal{M}$, a restriction on its topology is needed, and then $\mathcal{M}$ is said to have *spin structure*. For spin-vectors to be defined consistently on a space–time manifold $\mathcal{M}$, we need to consider continuous motions of a null flag in which the point P at which it is situated is allowed to move continuously as well as rotations of the flag at P being allowed. Such a motion is called a *flag path*. For consistency $\mathcal{M}$ must, in addition to having spin structure, be space- and time-orientable – in which case $\mathcal{M}$ is said to have *spinor structure*. A (non-zero) spin-vector at point $P \in \mathcal{M}$ is an element of the (universal) double cover of the space of null flags at P; the (non-zero) spin-vector bundle of $\mathcal{M}$ is a (not necessarily universal) double cover of the null-flag bundle of $\mathcal{M}$. If $\mathcal{M}$ is not simply-connected it may possess inequivalent possible spin-vector bundles, i.e. different spinor structures. One two-fold ambiguity arises from each independent loop in $\mathcal{M}$ no odd multiple of which is continuously deformable to a point.

§2.2 The abstract-index formalism for tensor algebra

Tensor (or spinor) indices in *standard lightface italic type* are simply *abstract markers* and do not take numerical values or denote components in some basis frame. For example, the symbol V^a denotes the actual vector V paired with the marker a (an element of some pre-assigned labelling set $\mathscr{L}$); and V^b, V^{a_0}, etc., though representing the same geometrical vector V, are distinct elements of the abstract-index algebra because the vector is, in each case, paired with a different element, b, a_0, etc. of $\mathscr{L}$. Similarly, the spin-vector $\boldsymbol{\kappa}$ may be represented by various distinct elements κ^A, κ^{B_3}, etc., of the algebra, or by their complex conjugates $\bar{\kappa}^{A'}$, $\bar{\kappa}^{B'_3}$, etc. This notation allows commutativity (e.g. $U^aV^b = V^bU^a$, but $\neq V^aU^b$) in the expression for tensor product. Indices in **bold upright** type* are *numerical* and label *components* with respect to some chosen basis (e.g., world-tensor component indices $\mathbf{a}$, $\mathbf{b}$,... ranging over 0, 1, 2, 3 and spinor component indices $\mathbf{A}$, $\mathbf{B}$,... or $\mathbf{A}'$, $\mathbf{B}'$,... ranging over 0, 1 or 0′, 1′, respectively).

Denoting a *basis* (in the n-dimensional case) by $\delta_{\boldsymbol{\alpha}}^{\alpha} = \delta_{\mathbf{1}}^{\alpha}, \ldots \delta_{\mathbf{n}}^{\alpha}$, with dual basis $\delta_{\alpha}^{\boldsymbol{\alpha}} = \delta_{\alpha}^{\mathbf{1}}, \ldots, \delta_{\alpha}^{\mathbf{n}}$, we can express the *components* of a tensor $A_{\lambda\ldots\nu}^{\alpha\ldots\gamma}$ as

$$A_{\boldsymbol{\lambda}\ldots\boldsymbol{\nu}}^{\boldsymbol{\alpha}\ldots\boldsymbol{\gamma}} = A_{\lambda\ldots\nu}^{\alpha\ldots\gamma}\delta_{\alpha}^{\boldsymbol{\alpha}}\cdots\delta_{\gamma}^{\boldsymbol{\gamma}}\delta_{\boldsymbol{\lambda}}^{\lambda}\cdots\delta_{\boldsymbol{\nu}}^{\nu} \qquad (2.3.13)$$

where, inversely, the tensor itself can expressed in terms of its components by

$$A_{\lambda\ldots\nu}^{\alpha\ldots\gamma} = A_{\boldsymbol{\lambda}\ldots\boldsymbol{\nu}}^{\boldsymbol{\alpha}\ldots\boldsymbol{\gamma}}\delta_{\boldsymbol{\alpha}}^{\alpha}\cdots\delta_{\boldsymbol{\gamma}}^{\gamma}\delta_{\lambda}^{\boldsymbol{\lambda}}\cdots\delta_{\nu}^{\boldsymbol{\nu}}. \qquad (2.3.14)$$

Different abstract index letters (which may be a mixture of upper and lower ones, if need be) may be 'clumped together' in the form of a single *composite* (abstract) *index* denoted, in general situations, by a capital script letter. Composite indices are useful for general statements about tensor (spinor) systems. The symbol $\mathfrak{S}^{\mathscr{A}}$ (or $\mathfrak{S}_{PQ_0}^{aB}$, for example, when the constituents of $\mathscr{A}$ are written explicitly) denotes the system ($\mathfrak{S}$-module) of tensors (spinors) whose abstract-index structure is that denoted by $\mathscr{A}$ in the upper position (the constituent indices being in either position). The scalars constitute a commutative ring $\mathfrak{S}$ with identity (normally C^∞ complex scalar fields on the space–time manifold $\mathscr{M}$). For real fields we use $\mathfrak{T}$ in place of $\mathfrak{S}$. The dual of $\mathfrak{S}^{\mathscr{A}}$ is $\mathfrak{S}_{\mathscr{A}}$ and, more generally, we have (for a totally reflexive system)

(2.2.38) PROPOSITION

The set of all $\mathfrak{S}$-multilinear maps from $\mathfrak{S}^{\mathscr{A}} \times \mathfrak{S}^{\mathscr{B}} \times \cdots \times \mathfrak{S}^{\mathscr{D}}$ to $\mathfrak{S}^{\mathscr{X}}$ may be identified with $\mathfrak{S}_{\mathscr{A}\mathscr{B}\ldots\mathscr{D}}^{\mathscr{X}}$, where the maps are achieved by means of contracted product.

* In handwritten calculations it is convenient to distinguish such indices by underlining them with a wavy line.

Accordingly, any element $T^{\mathscr{X}}_{\mathscr{A}\mathscr{B}\ldots\mathscr{D}} \in \mathfrak{S}^{\mathscr{X}}_{\mathscr{A}\mathscr{B}\ldots\mathscr{D}}$ may be regarded as providing a map

$$T^{\mathscr{X}}_{\mathscr{A}\mathscr{B}\ldots\mathscr{D}}: \mathfrak{S}^{\mathscr{A}} \times \mathfrak{S}^{\mathscr{B}} \times \cdots \times \mathfrak{S}^{\mathscr{D}} \to \mathfrak{S}^{\mathscr{X}}$$

in addition to all the various other corresponding maps obtained by grouping the indices in different ways. Distinct script index letters generally indicate distinct clumpings of constituent tensor (or spinor) indices. If similar clumpings are required then suffixes are used on the composite indices, e.g. $\mathscr{A}_1, \mathscr{A}_2, \ldots$.

For spinors or tensors at a point P of a manifold $\mathscr{M}$ we may write, explicitly, $\mathfrak{S}^{\mathscr{A}}[P]$, and correspondingly $\mathfrak{S}^{\mathscr{A}}[\mathscr{U}]$ for the restrictions of the elements of $\mathfrak{S}^{\mathscr{A}}$ to a subset $\mathscr{U}$ of $\mathscr{M}$.

Tensor–spinor abstract index correspondence

World-tensor indices are regarded *as* composite indices, each simply standing for a clumped pair of corresponding spinor indices, one primed and one unprimed. We adopt the standard convention

$$a = AA', \quad b = BB', \quad c = CC', \quad \ldots, \quad z = ZZ',$$
$$a_0 = A_0A'_0, \quad \ldots, \quad a_1 = A_1A'_1, \quad \ldots \tag{3.1.2}$$

Primed spinor indices are the complex conjugates of their corresponding unprimed ones (in the sense $\overline{\phi^{AB'}} = \bar{\phi}^{A'B}$ etc.). Spinor indices are raised and lowered with the anti-symmetrical ε-spinors:

$$\psi^{\mathscr{C}A} = \varepsilon^{AB}\psi^{\mathscr{C}}{}_B = -\psi^{\mathscr{C}}{}_B\varepsilon^{BA} \tag{2.5.14}$$

$$\psi^{\mathscr{C}}{}_B = \psi^{\mathscr{C}A}\varepsilon_{AB} = -\varepsilon_{BA}\psi^{\mathscr{C}A} \tag{2.5.15}$$

(and similarly for primed indices, with $\varepsilon^{A'B'} := \bar{\varepsilon}^{A'B'} = \overline{\varepsilon^{AB}}$, $\varepsilon_{A'B'} := \bar{\varepsilon}_{A'B'} = \overline{\varepsilon_{AB}}$), this being consistent with the normal world-tensor index raising and lowering (using g_{ab} and g^{ab}), where

$$g_{ab} = \varepsilon_{AB}\varepsilon_{A'B'}, \quad g_a{}^b = \varepsilon_A{}^B\varepsilon_{A'}{}^{B'}, \quad g^{ab} = \varepsilon^{AB}\varepsilon^{A'B'}. \tag{3.1.9}$$

(Note that $g_a{}^b = g^b{}_a$, $\varepsilon_A{}^B = -\varepsilon^B{}_A$ and $\varepsilon_{A'}{}^{B'} = -\varepsilon^{B'}{}_{A'}$ are 'Kronecker delta' symbols.) While we must be careful to preserve the ordering of the unprimed indices on a spinor symbol, and also of the primed ones, the *relative* ordering between primed and unprimed indices is immaterial; for example $\psi^a{}_b{}^Q$ is each of the following except the last:

$$\psi^{AA'}{}_{B'B}{}^Q = \psi^{A'A}{}_B{}^Q{}_{B'} = \psi^A{}_B{}^{QA'}{}_{B'} = \psi^{AA'Q}_{BB'} \neq \psi^{AA'Q}{}_{BB'}. \tag{2.5.33}$$

The ε-spinors satisfy the identities

$$\varepsilon_{AB}\varepsilon_{CD} + \varepsilon_{BC}\varepsilon_{AD} + \varepsilon_{CA}\varepsilon_{BD} = 0, \tag{2.5.21}$$

their raised versions such as

$$\varepsilon_{AB}\varepsilon_C{}^D + \varepsilon_{BC}\varepsilon_A{}^D + \varepsilon_{CA}\varepsilon_B{}^D = 0, \tag{2.5.20}$$

$$\varepsilon_A{}^C\varepsilon_B{}^D - \varepsilon_B{}^C\varepsilon_A{}^D = \varepsilon_{AB}\varepsilon^{CD}, \tag{2.5.22}$$

and also the complex conjugates of these relations. We deduce

$$\phi_{\mathscr{D}AB} - \phi_{\mathscr{D}BA} = \phi_{\mathscr{D}C}{}^C\varepsilon_{AB}, \tag{2.5.23}$$

so for $\phi_{\mathscr{D}AB}$ *skew* in A, B,

$$\phi_{\mathscr{D}AB} = \tfrac{1}{2}\phi_{\mathscr{D}C}{}^C\varepsilon_{AB}. \tag{2.5.24}$$

Bases and components

The standard symbols for a *spinor basis* for $\mathfrak{S}^A$, or *dyad*, are o^A, ι^A, where $o^{A'} := \bar{o}^{A'} (= \overline{o^A})$ and $\iota^{A'} := \bar{\iota}^{A'} (= \overline{\iota^A})$. These need not be normalized to unity, and we define

$$\chi = \varepsilon_{AB}o^A\iota^B = o_A\iota^A. \tag{2.5.46}$$

When $\chi = 1$, the dyad is a *spin-frame* and we have

$$\varepsilon^{AB} = o^A\iota^B - \iota^Ao^B, \quad \varepsilon_{AB} = o_A\iota_B - \iota_Ao_B, \quad \varepsilon_A{}^B = o_A\iota^B - \iota_Ao^B. \tag{2.5.54}$$

The condition for a general dyad, on the other hand, is simply $\chi \neq 0$, since

(2.5.56) PROPOSITION

The condition $\alpha_A\beta^A = 0$ at a point is necessary and sufficient for α_A, β_A to be scalar multiples of each other at that point.

Associated with any spin-frame is a *null tetrad* l^a, n^a, m^a, $\bar{m}^a$ defined by

$$\begin{aligned} l^a &= o^Ao^{A'}, & n^a &= \iota^A\iota^{A'}, \\ m^a &= o^A\iota^{A'}, & \bar{m}^a &= \iota^Ao^{A'}, \end{aligned} \tag{3.1.14}$$

and satisfying

$$l^al_a = n^an_a = m^am_a = \bar{m}^a\bar{m}_a = 0, \tag{3.1.15}$$

$$l^an_a = 1, \quad m^a\bar{m}_a = -1, \tag{3.1.16}$$

$$l^am_a = l^a\bar{m}_a = n^am_a = n^a\bar{m}_a = 0, \tag{3.1.17}$$

$$l^a = \bar{l}^a, \quad n^a = \bar{n}^a, \tag{3.1.18}$$

and also a standard restricted Minkowski tetrad

$$t^a = \frac{1}{\sqrt{2}}(l^a + n^a) = \frac{1}{\sqrt{2}}(o^Ao^{A'} + \iota^A\iota^{A'}),$$

$$x^a = \frac{1}{\sqrt{2}}(m^a + \bar{m}^a) = \frac{1}{\sqrt{2}}(o^A\iota^{A'} + \iota^A o^{A'}),$$

$$y^a = \frac{\mathrm{i}}{\sqrt{2}}(m^a - \bar{m}^a) = \frac{\mathrm{i}}{\sqrt{2}}(o^A\iota^{A'} - \iota^A o^{A'}),$$

$$z^a = \frac{1}{\sqrt{2}}(l^a - n^a) = \frac{1}{\sqrt{2}}(o^A o^{A'} - \iota^A\iota^{A'}), \tag{3.1.20}$$

where reciprocally

$$l^a = \frac{1}{\sqrt{2}}(t^a + z^a) = o^A o^{A'},$$

$$n^a = \frac{1}{\sqrt{2}}(t^a - z^a) = \iota^A\iota^{A'},$$

$$m^a = \frac{1}{\sqrt{2}}(x^a - \mathrm{i}y^a) = o^A\iota^{A'},$$

$$\bar{m}^a = \frac{1}{\sqrt{2}}(x^a + \mathrm{i}y^a) = \iota^A o^{A'}. \tag{3.1.21}$$

See Fig. 1–17 for the geometrical relation between o^A, ι^A and t^a, x^a, y^a, z^a. The components of a vector K^a in the Minkowski tetrad are related to the

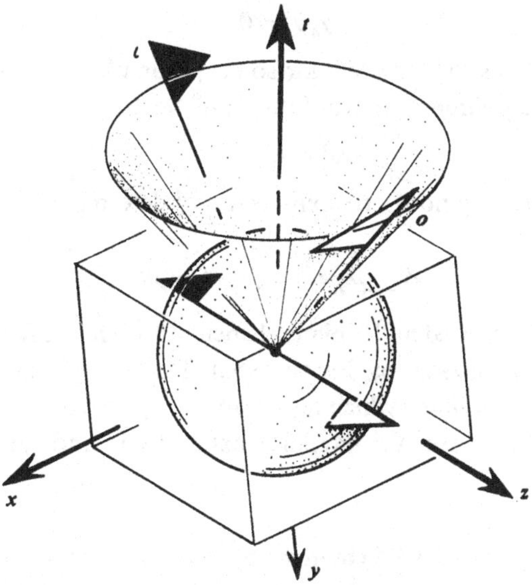

Fig. 1-17. The standard relation between a spin-frame ***o***, ***ι*** and a (restricted) Minkowski tetrad ***t***, ***x***, ***y***, ***z***.

spin-frame components of $K^{AA'}$ by

$$\frac{1}{\sqrt{2}}\begin{pmatrix} K^0+K^3 & K^1+\mathrm{i}K^2 \\ K^1-\mathrm{i}K^2 & K^0-K^3 \end{pmatrix} = \begin{pmatrix} K^{00'} & K^{01'} \\ K^{10'} & K^{11'} \end{pmatrix}. \tag{3.1.31}$$

To express the corresponding relations between Minkowski tetrad components and spin-frame components for a general tensor it is often convenient to use the translation symbols:

$$g_0{}^{\mathbf{AB'}} = \frac{1}{\sqrt{2}}\begin{pmatrix} 1 & 0 \\ 0 & 1 \end{pmatrix} = g_{\mathbf{AB'}}{}^0, \quad g_1{}^{\mathbf{AB'}} = \frac{1}{\sqrt{2}}\begin{pmatrix} 0 & 1 \\ 1 & 0 \end{pmatrix} = g_{\mathbf{AB'}}{}^1,$$

$$g_2{}^{\mathbf{AB'}} = \frac{1}{\sqrt{2}}\begin{pmatrix} 0 & \mathrm{i} \\ -\mathrm{i} & 0 \end{pmatrix} = -g_{\mathbf{AB'}}{}^2, \quad g_3{}^{\mathbf{AB'}} = \frac{1}{\sqrt{2}}\begin{pmatrix} 1 & 0 \\ 0 & -1 \end{pmatrix} = g_{\mathbf{AB'}}{}^3. \tag{3.1.49}$$

Null vectors and null flags

A complex null vector χ^a, in the normal sense $\chi^a\chi_a = 0$, has the spinor form

$$\chi^a = \kappa^A \xi^{A'} \tag{3.2.6}$$

where we can take $\xi^{A'} = \bar{\kappa}^{A'}$ [*or* $-\bar{\kappa}^{A'}$] whenever χ^a is real and future-null [past-null] (or zero). Two complex null vectors χ^a and ψ^a, with $\chi^a \neq 0$ given by (3.2.6), are orthogonal in the sense

$$\chi_a \psi^a = 0 \tag{3.2.22}$$

if and only if $\psi^a = \kappa^A \eta^{A'}$ or $\tau^A \xi^{A'}$ for some $\eta^{A'}$ or τ^A.

In addition to defining the real future-null vector

$$K^a = \kappa^A \bar{\kappa}^{A'}, \tag{3.2.2}$$

called its *flagpole*, any non-zero spin-vector κ^A determines the real simple null bivector

$$P^{ab} = \kappa^A \kappa^B \varepsilon^{A'B'} + \varepsilon^{AB} \bar{\kappa}^{A'} \bar{\kappa}^{B'} \tag{3.2.9}$$

which defines an oriented null 2-plane element (through the flagpole) called the *flag plane* (or an 'axe' by Payne 1952). The flagpole and flag plane together define the *null flag* which represents κ^A up to sign. Multiplying κ^A by $re^{\mathrm{i}\theta}$ (r, θ real) increases the flagpole's extent by r^2 and rotates the flag plane by 2θ.

§3.3 Symmetry operations

We adopt the standard convention that round brackets (parentheses) around indices denote that the symmetric part is being taken and square brackets, the antisymmetric part. Vertical bars are used at either side of a set of indices to be omitted from an (anti-) symmetrization. If a pair of brackets appears nested within another, then the inside operation is to be performed first and the resulting indices (if not blocked off by vertical bars) then partake again of the larger (anti-) symmetry operation. (Though such a notation is, strictly speaking, redundant it can be very useful in the formulation of a proof, e.g. see (4.3.17).) While the bracket notation normally applies just to individual tensors or spinors, it may also be used to denote *subspaces* of $\mathfrak{S}_{\mathscr{A}}$ for which (anti-) symmetry relations are imposed. Thus, for example:

$$\chi_{\mathscr{A}}{}^{\alpha}{}_{\lambda\mu}{}^{\beta}{}_{\nu}{}^{\mathscr{B}_1\mathscr{B}_2} \in \mathfrak{S}^{[\alpha\beta](\mathscr{B}_1\mathscr{B}_2)}_{\mathscr{A}(\lambda\mu\nu)} \quad \textit{iff} \quad \chi_{\mathscr{A}}{}^{\alpha}{}_{\lambda\mu}{}^{\beta}{}_{\nu}{}^{\mathscr{B}_1\mathscr{B}_2} = \chi_{\mathscr{A}}{}^{[\alpha}{}_{(\lambda\mu}{}^{\beta]}{}_{\nu)}{}^{(\mathscr{B}_1\mathscr{B}_2)}. \quad (3.3.14)$$

Various simple results concerning symmetries are useful. For example (with Greek letters denoting n-dimensional abstract tensor indices):

(3.2.22) PROPOSITION

$$\phi_{\mathscr{A}\alpha\beta\ldots\delta}X^{\alpha}X^{\beta}\ldots X^{\delta} = 0 \textit{ for all } X^{\alpha}\in\mathfrak{S}^{\alpha} \textit{ iff } \phi_{\mathscr{A}(\alpha\ldots\delta)} = 0.$$

Hence

the function $\phi_{\mathscr{A}}(X) \equiv \phi_{\mathscr{A}\alpha\ldots\delta}X^{\alpha}\ldots X^{\delta}$ *serves to define the tensor* $\phi_{\mathscr{A}(\alpha\ldots\delta)}$ *uniquely.* (3.3.23)

Combinations of symmetries and anti-symmetries can be applied to produce a tensor with a *Young tableau* symmetry and which is, accordingly, (pointwise) *irreducible* under the group of general linear transformations (about that point). Some relevant properties are discussed in the extended 'footnote' on p. 143 of Volume 1. In particular, a tensor $\phi_{\alpha\ldots\lambda}$ with tableau symmetry – of the version in which groups of non-increasing lengths of *symmetric* (rather than anti-symmetric) indices are exhibited – may be represented by a polynomial $P(X^{\alpha},\ldots,Z^{\alpha})$, homogeneous of non-increasing degrees in $X^{\alpha},\ldots,Z^{\alpha}$ separately. (Here (3.3.23) is being applied to *each* symmetric group of indices of $\phi_{\ldots.}$.) The *tableau property* (i.e. 'hidden saturation by anti-symmetries') may then be expressed as:

$$X^{\alpha}\partial P/\partial Y^{\alpha} = 0,\ldots, X^{\alpha}\partial P/\partial Z^{\alpha} = 0,\ldots, Y^{\alpha}\partial P/\partial Z^{\alpha} = 0$$

(earlier variable contracted with derivative with respect to later variable) in which case P can be expressed as a function just of the skew products like $X^{[\alpha}Y^{\beta}Z^{\gamma]}$, $X^{[\alpha}Y^{\beta]}$, X^{α} (consecutive variables starting with X^{α}). Equivalently, the tableau property may be expressed as *any further symmetrization of* $\phi_{\alpha\ldots\lambda}$ *which includes one entire symmetric group of indices together with one index from a later group, vanishes.* For such a $\phi_{\alpha\ldots\lambda}$, if the number of groups of indices exceeds n (i.e. if the number of variable vectors $X^\alpha,\ldots,Z^\alpha$ in $P(X^\alpha,\ldots,Z^\alpha)$ exceeds n), where n is the dimension of space, then $\phi_{\alpha\ldots\lambda}=0$. This is because of the presence of a 'hidden anti-symmetry' of length greater than n in $\phi_{\alpha\ldots\lambda}$.

For spinor indices $n=2$, so anti-symmetrizations of length greater than two always yield zero. Moreover, in accordance with (2.5.24), one of length two can always be 'split off' as an ε-spinor. Accordingly, only totally symmetric spinors need be considered. With both primed and unprimed indices being allowed to be present, the *irreducible spinors* (pointwise irreducible under the restricted Lorentz group, or, more correctly under the spin group $SL(2,\mathbb{C})$), turn out to be those which, if all indices are in the lower position, are *totally symmetric* in all unprimed indices and also in all primed indices. The number of independent components of such a spinor is given by:

(3.3.62) PROPOSITION

If $\varphi_{A\ldots CP'\ldots R'}$ *is symmetric of valence* $\left[\begin{smallmatrix}0&0\\p&q\end{smallmatrix}\right]$, *then it has* $(p+1)(q+1)$ *independent (complex) components.*

Here the *valence* symbol $\left[\begin{smallmatrix}r&s\\p&q\end{smallmatrix}\right]$ is being used, for a spinor with r upper unprimed, s upper primed, p lower unprimed and q lower primed indices. In the cases when $r\neq 0$ or $s\neq 0$, irreducibility is expressed as: not only symmetry in all four groups of indices but also as the vanishing of all contractions between upper and lower indices (implying total symmetry when all indices are lowered). The component count in (3.3.62) would then become $(p+r+1)\,(q+s+1)$.

Tensor–spinor translation of expressions; duals

For a spinor $\phi_{A\ldots FA'\ldots F'}$ of valence $\left[\begin{smallmatrix}0&0\\r&r\end{smallmatrix}\right]$, *irreducibility*

$$\phi_{ab\ldots f}=\phi_{AB\ldots FA'B'\ldots F'} = \phi_{(AB\ldots F)(A'B'\ldots F')} \tag{3.3.58}$$

can be expressed *tensorially* as

$$\phi_{ab\ldots f}=\phi_{(ab\ldots f)} \tag{3.3.59}$$

$$\phi^a{}_{ac\ldots f}=0 \tag{3.3.60}$$

and the number of independent components is $(r+1)^2$.

World-vector space being four-dimensional, there is an *alternating tensor* e_{abcd} with $e_{abcd}=e_{[abcd]}$, where we take $e_{0123}=1$ in a standard Minkowski tetrad (*cf.* (3.1.20)). In spinor terms

$$e_{abcd} := \mathrm{i}\varepsilon_{AC}\varepsilon_{BD}\varepsilon_{A'D'}\varepsilon_{B'C'} - \mathrm{i}\varepsilon_{AD}\varepsilon_{BC}\varepsilon_{A'C'}\varepsilon_{B'D'}. \tag{3.3.31}$$

For a *symmetric* tensor T_{ab}, the *trace-free part*

$$S_{ab} = T_{ab} - \tfrac{1}{4}T_c{}^c g_{ab} \tag{3.4.9}$$

can be expressed as

$$S_{ABA'B'} = T_{(AB)(A'B')} = T_{(AB)A'B'} = T_{AB(A'B')}. \tag{3.4.5}$$

If symmetry is not assumed for T_{ab}, its *symmetric* trace-free part is expressed by the second expression (3.4.5). For symmetric T_{ab}, the *trace-reversed* tensor $\hat{T}_{ab}$ has simple spinorial expressions

$$T_{ab} - \tfrac{1}{2}T_c{}^c g_{ab} = \hat{T}_{ABA'B'} = T_{BAA'B'} = T_{ABB'A'}. \tag{3.4.13}$$

Any (complex) *anti-symmetric* F_{ab} can be expressed (uniquely) as

$$F_{ab} = F_{AA'BB'} = \phi_{AB}\varepsilon_{A'B'} + \varepsilon_{AB}\psi_{A'B'}, \tag{3.4.17}$$

where ϕ_{AB} and $\psi_{A'B'}$ are *symmetric*. For *real* F_{ab} we have

$$F_{ab} = \phi_{AB}\varepsilon_{A'B'} + \varepsilon_{AB}\bar{\phi}_{A'B'}. \tag{3.4.20}$$

Defining the *dual* of F_{ab} as

$$^*F_{ab} := \tfrac{1}{2}e_{abcd}F^{cd} = \tfrac{1}{2}e_{ab}{}^{cd}F_{cd} \tag{3.4.21}$$

we find

$$^*F_{ab} = {}^*F_{ABA'B'} = -\mathrm{i}\phi_{AB}\varepsilon_{A'B'} + \mathrm{i}\varepsilon_{AB}\psi_{A'B'} \tag{3.4.22}$$

where $\psi_{A'B'} = \bar{\phi}_{A'B'}$ if F_{ab} is real, and

$$^*F_{ABA'B'} = \mathrm{i}F_{ABB'A'} = -\mathrm{i}F_{BAA'B'}. \tag{3.4.23}$$

We can dualize also when further indices are present: if $G_{ab\mathscr{A}} = G_{[ab]\mathscr{A}}$, then

$$^*G_{ab\mathscr{A}} := \tfrac{1}{2}e_{ab}{}^{cd}G_{cd\mathscr{A}} \tag{3.4.25}$$

with spinor expressions similar to (3.4.22) and (3.4.23). These show that

$$^{**}G_{ab\mathscr{A}} = -G_{ab\mathscr{A}}.$$

We note that

$$^*G_{[abc]\mathscr{B}} = 0 \Leftrightarrow {}^*G^{ab}{}_{a\mathscr{B}} = 0. \tag{3.4.26}$$

Duals for other numbers of skew indices can also be defined:

$$^\dagger J_{abc\mathscr{A}} = e_{abc}{}^d J_{d\mathscr{A}} \tag{3.4.29}$$

$$^\ddagger K_{a\mathscr{B}} = \tfrac{1}{6}e_a{}^{bcd}K_{bcd\mathscr{B}} \tag{3.4.30}$$

where $K_{abc\mathscr{B}} = K_{[abc]\mathscr{B}}$. Then

$$^\dagger J_{abc\mathscr{A}} = \mathrm{i}\varepsilon_{AC}\varepsilon_{B'C'}J_{BA'\mathscr{A}} - \mathrm{i}\varepsilon_{BC}\varepsilon_{A'C'}J_{AB'\mathscr{A}} \tag{3.4.33}$$

$$^\ddagger K_{a\mathscr{B}} = \frac{\mathrm{i}}{3}K_{AB'BA'}{}^{BB'}{}_{\mathscr{B}} \tag{3.4.34}$$

and

$$^{\dagger\dagger}J_{a\mathscr{A}} = J_{a\mathscr{A}}, \quad ^{\dagger\dagger}K_{abc\mathscr{B}} = K_{abc\mathscr{B}} \tag{3.4.31}$$

In the case of two indices, if F_{ab} is skew and *complex*, we say that it is respectively (i) *anti-self-dual*, or (ii) *self-dual*, if

$$\text{(i)}\ {}^*F_{ab} = -\mathrm{i}F_{ab}, \quad \text{or} \quad \text{(ii)}\ {}^*F_{ab} = \mathrm{i}F_{ab}. \tag{3.4.35}$$

Generally, for F_{ab} given by (3.4.17), we define its *anti-self-dual* part

$$^-F_{ab} := \tfrac{1}{2}(F_{ab} + \mathrm{i}\ {}^*F_{ab}) = \phi_{AB}\varepsilon_{A'B'}. \tag{3.4.38}$$

and *self-dual part*

$$^+F_{ab} := \tfrac{1}{2}(F_{ab} - \mathrm{i}\ {}^*F_{ab}) = \varepsilon_{AB}\psi_{A'B'} \tag{3.4.39}$$

so that ${}^{*\pm}F_{ab} = \pm\mathrm{i}^{\pm}F_{ab}$ and

$$F_{ab} = {}^-F_{ab} + {}^+F_{ab}. \tag{3.4.40}$$

The conditions (3.4.35) for anti-self-duality and self-duality can be restated as (i) $\psi_{A'B'} = 0$, (ii) $\phi_{AB} = 0$, respectively, i.e.

$$\text{(i)}\ F_{ab} = \phi_{AB}\varepsilon_{A'B'}, \quad \text{(ii)}\ F_{ab} = \varepsilon_{AB}\psi_{A'B'}. \tag{3.4.41}$$

Combining results for the spinor expressions for symmetric and anti-symmetric parts we obtain the following tensor expression for the interchange of two spinor indices

$$H_{BAA'B'} = \tfrac{1}{2}(H_{ab} + H_{ba} - H_c{}^c g_{ab} + \mathrm{i}e_{abcd}H^{cd}) \tag{3.4.53}$$

$$H_{ABB'A'} = \tfrac{1}{2}(H_{ab} + H_{ba} - H_c{}^c g_{ab} - \mathrm{i}e_{abcd}H^{cd}) \tag{3.4.54}$$

where $H_{ab} = H_{ABA'B'}$ is general (and may have other indices as well). Note that the skew part of H_{ab} is

$$H_{[ab]} = H_{[AB](A'B')} + H_{(AB)[A'B']} = \tfrac{1}{2}\varepsilon_{AB}H_C{}^C{}_{(A'B')} + \tfrac{1}{2}\varepsilon_{A'B'}H_{(AB)C'}{}^{C'} \tag{3.4.55}$$

with a similar expression for $H_{(ab)}$. The results of this subsection clearly also hold if additional indices are present.

Some properties of tensors and spinors at a point

We now state a number of simple properties which hold for tensors or spinors at any *one point* (so that $\mathfrak{S}$ is now a *division* ring) and which do not necessarily hold for tensor or spinor fields.

(3.5.15) PROPOSITION

$\psi_{(\mathscr{A}_1\ldots\mathscr{A}_r}{}^{\mathscr{B}}\phi_{\mathscr{A}_{r+1}\ldots\mathscr{A}_{r+s})}{}^{\mathscr{C}} = 0$ *implies either* $\psi_{(\mathscr{A}_1\ldots\mathscr{A}_r)}{}^{\mathscr{B}} = 0$ *or* $\phi_{(\mathscr{A}_1\ldots\mathscr{A}_s)}{}^{\mathscr{C}} = 0.$

(3.5.18) PROPOSITION

If $\phi_{AB\ldots L}=\phi_{(AB\ldots L)}\neq 0$, *then*

$$\phi_{AB\ldots L}=\alpha_{(A}\beta_{B}\ldots\lambda_{L)}$$

for some $\alpha_A, \beta_A, \ldots, \lambda_A \in \mathfrak{S}_A$. *Furthermore, this decomposition is unique up to proportionality or reordering of the factors.*

This is the *canonical decomposition* of $\phi_{A\ldots L}$, and we call α_A, $\beta_A\ldots$ (and their non-zero multiples) the *principal spinors of* $\phi_{A\ldots L}$. The flagpoles of these spinors are its *principal null vectors*, and the associated null directions are its *principal null directions* (PNDs). If $\xi^A\neq 0$, then

$$\phi_{AB\ldots L}\xi^A\xi^B\ldots\xi^L=0. \tag{3.5.22}$$

if and only if ξ^A is a principal spinor of $\phi_{A\ldots L}$. We say that $\phi_{A\ldots L}$ is *null* if all its PNDs coincide. If α_A is an exactly *k-fold* principal spinor, then

$$\phi_{AB\ldots DE\ldots L}\alpha^E\ldots\alpha^L=\kappa\alpha_A\alpha_B\ldots\alpha_D \tag{3.5.24}$$

with $\kappa\neq 0$, the number of αs on the right being k. Moreover,

(3.5.26) PROPOSITION

A necessary and sufficient condition that $\xi_A\neq 0$ *be a k-fold principal spinor of the non-vanishing symmetric spinor* $\phi_{AB\ldots L}$ *is that*

$$\phi_{A\ldots GH\ldots L}\xi^H\ldots\xi^L$$

should vanish if $n-k+1$ ξ*s are transvected with* $\phi_{A\ldots L}$ *but not if only* $n-k$ ξ*s are transvected with* $\phi_{A\ldots L}$.

(3.5.27) PROPOSITION

If $\xi^A\neq 0$, $\phi^{*}_{A\ldots G}=\phi^{*}_{(A\ldots G)}$, *and*

$$\xi^A\ldots\xi^C\xi^D\phi^{*}_{A\ldots CDE\ldots G}=0,$$

then there exists a $\psi^{*}_{A\ldots C}$ *such that*

$$\phi^{*}_{A\ldots CDE\ldots G}=\psi^{*}_{(A\ldots C}\xi_D\xi_E\ldots\xi_{G)}.$$

As a particular case, we have:

$$\textit{If } \lambda_B\neq 0,\textit{ then } \psi_{\mathscr{A}B}\lambda^B=0 \textit{ implies } \psi_{\mathscr{A}B}=\chi_{\mathscr{A}}\lambda_B \textit{ for some } \chi_{\mathscr{A}}. \tag{3.5.17}$$

Note that if $\varphi_{AB}=\alpha_{(A}\beta_{B)}$ then

$$\varphi_{AB}\varphi^{AB}=-\tfrac{1}{2}(\alpha_A\beta^A)^2, \tag{3.5.29}$$

so the vanishing of $\varphi_{AB}\varphi^{AB}$ is a criterion for the nullity of φ_{AB}. Note also that $\det(\varphi_{AB}) = \frac{1}{2}\varphi_{AB}\varphi^{AB}$.

An *anti*-symmetric tensor (in n dimensions) which is a skew product of vectors is called *simple*. One criterion for simplicity is:

(3.5.30) PROPOSITION

If $F_{\alpha\beta\ldots\rho}$ is skew in all its indices, then

$$F_{[\alpha\beta\ldots\rho}F_{\sigma]\tau\ldots\omega} = 0 \Leftrightarrow F_{\alpha\beta\ldots\rho} = a_{[\alpha}b_{\beta}\ldots r_{\rho]}$$

for some $a_\alpha, b_\beta, \ldots, r_\rho$.

Another way of stating this criterion is

$$*F^{\delta\ldots\rho\sigma}F_{\sigma\tau\ldots\omega} = 0. \qquad (3.5.32)$$

Here the asterisk denotes n-dimensional dual. Consequently,

(3.5.34) PROPOSITION

$F_{\alpha\ldots\gamma}$ is simple if and only if its dual $*F^{\delta\ldots\sigma}$ is simple.

(3.5.35) PROPOSITION

In four dimensions the bivector F_{ab} is simple if and only if any of the following conditions holds:

$$\text{(i) } F_{[ab}F_{cd]} = 0, \quad \text{(ii) } F_{ab}{*F^{ab}} = 0, \quad \text{(iii) } \det(F_{ab}) = 0.$$

Active Lorentz and spin transformations

A Lorentz transformation

$$V^b \to L_a{}^b V^a$$

which sends world-vectors to world-vectors (at a point) is given by one of the following spinor expressions

$$L_{AA'}{}^{BB'} = \pm\, \theta^B_{A'}\bar{\theta}_A{}^{B'} \qquad (3.6.14)$$

$$L_{AA'}{}^{BB'} = \pm\, \phi_A{}^B\bar{\phi}_{A'}{}^{B'}. \qquad (3.6.15)$$

where we can take

$$\det(\theta^B_{A'}) = 1, \quad \det(\phi_A{}^B) = 1. \qquad (3.6.16)$$

The upper [lower] signs in (3.6.14) and (3.6.15) refer to the orthochronous [non-orthochronous] transformations, improper for (3.6.14) and proper for (3.6.15). In the improper case the determinant condition (3.6.16) can be rewritten

$$\theta^a\theta_a = 2, \tag{3.6.22}$$

and in the proper case

$$\phi_{AB}\phi^{AB} = 2. \tag{3.6.30}$$

A restricted (i.e. proper isochronous) Lorentz transformation other than the identity leaves invariant precisely two (possibly coincident) null directions. These null directions are in fact the PNDs of $\phi_{(AB)}$. When the PNDs coincide (i.e. when $\phi_{(AB)}$ is null) we call the Lorentz transformation a *null rotation.* When the PNDs are distinct we can use them as the flagpole directions of a spin-frame, with respect to which $\phi_{\mathbf{A}}{}^{\mathbf{B}}$ is then *diagonal.* In the case of a null rotation, we can choose the flagpole of ι^A in the repeated PND (the flagpole *vector* being now invariant) and we have

$$\phi_{\mathbf{A}}{}^{\mathbf{B}} = \begin{pmatrix} 1 & -\xi \\ 0 & 1 \end{pmatrix}. \tag{3.6.47}$$

Covariant derivative

We use the symbol ∇, indexed as appropriate, (e.g. ∇_α, ∇_a, etc.) to denote *covariant derivative* on a (say n-dimensional) manifold $\mathscr{M}$, with respect to some connection (not necessarily the standard Christoffel one). When acting on scalars ∇_α is the ordinary *gradient* operator and it relates the abstract-index version V^α of a vector field to its differential operator version V by

$$V = V^\alpha\nabla_\alpha \tag{4.1.40}$$

(on scalars). Generally, ∇_α operators do not commute, and we define

$$\Delta_{\alpha\beta} := \nabla_\alpha\nabla_\beta - \nabla_\beta\nabla_\alpha = 2\nabla_{[\alpha}\nabla_{\beta]}. \tag{4.2.14}$$

The *torsion* $T_{\alpha\beta}{}^\gamma$ is then defined by

$$\Delta_{\alpha\beta}f = T_{\alpha\beta}{}^\gamma\nabla_\gamma f \tag{4.2.22}$$

and the *curvature* $R_{\alpha\beta\gamma}{}^\delta$ by

$$(\nabla_\alpha\nabla_\beta - \nabla_\beta\nabla_\alpha - T_{\alpha\beta}{}^\gamma\nabla_\gamma)V^\delta = R_{\alpha\beta\gamma}{}^\delta V^\gamma \tag{4.2.31}$$

(for an arbitrary scalar f and vector V^γ).

If $T_{\alpha\beta}{}^{\gamma} = 0$ (i.e. ∇_α is *torsion-free*), this is simply $\Delta_{\alpha\beta} V^\delta = R_{\alpha\beta\gamma}{}^{\delta} V^\gamma$. Then we can write the abstract index version W^α of the *Lie bracket*

$$\boldsymbol{W} = [\boldsymbol{U}, \boldsymbol{V}] := \boldsymbol{U} \circ \boldsymbol{V} - \boldsymbol{V} \circ \boldsymbol{U} \tag{4.3.26}$$

as

$$U^\alpha \nabla_\alpha V^\beta - V^\alpha \nabla_\alpha U^\beta. \tag{4.3.2}$$

More generally, the *Lie derivative* of a tensor field with respect to $\boldsymbol{V}$ is

$$\begin{aligned} \underset{\boldsymbol{V}}{\pounds} H^{\alpha\ldots\gamma}_{\lambda\ldots\nu} := {} & V^\rho \nabla_\rho H^{\alpha\ldots\gamma}_{\lambda\ldots\nu} - H^{\alpha_0\ldots\gamma}_{\lambda\ldots\nu} \nabla_{\alpha_0} V^\alpha - \cdots - H^{\alpha\ldots\gamma_0}_{\lambda\ldots\nu} \nabla_{\gamma_0} V^\gamma \\ & + H^{\alpha\ldots\gamma}_{\lambda_0\ldots\nu} \nabla_\lambda V^{\lambda_0} + \cdots + H^{\alpha\ldots\gamma}_{\lambda\ldots\nu_0} \nabla_\nu V^{\nu_0}. \end{aligned} \tag{4.3.3}$$

Lie brackets and Lie derivatives are independent of the choice of (torsion-free) ∇_α. Thus they can be evaluated in components by replacing ∇_α by $\partial/\partial x^\alpha$.

We adopt a convention for *differential forms* that the particular index letters $\iota_1, \iota_2, \ldots$ (occurring in an antisymmetrical set) may be omitted from a kernel symbol, that symbol being now presented in **bold** type, so

$$\boldsymbol{A} := A_{\iota_1\iota_2\ldots\iota_p} \tag{4.3.10}$$

denotes a p-form, where $A_{\iota_1\iota_2\ldots\iota_p} = A_{[\iota_1\iota_2\ldots\iota_p]}$. Additional index letters may be present also, where $\iota_1, \ldots, \iota_p$ are considered as occurring 'first':

$$\boldsymbol{B}_{\mathscr{Q}} = B_{\iota_1\iota_2\ldots\iota_p\mathscr{Q}}.$$

This is in the general n-dimensional case; for space–time we use i_r or $I_r I_r'$ in place of ι_r. The *exterior product* of $\boldsymbol{A}$ with $\boldsymbol{C}$ is

$$\boldsymbol{A} \wedge \boldsymbol{C} := A_{[\iota_1\ldots\iota_p} C_{\iota_{p+1}\ldots\iota_{p+q}]} \tag{4.3.13}$$

and the *exterior derivative* of $\boldsymbol{A}$ (provided ∇_α is torsion-free) is

$$\mathrm{d}\boldsymbol{A} := \nabla_{[\iota_1} A_{\iota_2\ldots\iota_{p+1}]} \tag{4.3.14}$$

(where other indices may be present as well). Provided there are no other indices, $d^2 = 0$:

$$\mathrm{d}(\mathrm{d}\boldsymbol{A}) = 0. \tag{4.3.15)(viii}$$

Equation (4.3.14) holds equally well if ∇_α is replaced (locally) by any other torsion-free derivative, say by ∂_α whose torsion and curvature *both* vanish. So a simple proof of (4.3.15) (viii) is

$$\mathrm{d}(\mathrm{d}\boldsymbol{A}) = \partial_{[.}\partial_{[.}A_{\ldots]]} = \partial_{[.}\partial_{.}A_{\ldots]} = \partial_{[[.}\partial_{.]}A_{\ldots]} = 0. \tag{4.3.17}$$

We sometimes take advantage of the notational convention that $\mathrm{d}\phi\psi$ stands for $(\mathrm{d}\phi)\psi$ and not $\mathrm{d}(\phi\psi)$, with the corresponding rule for other differential operator symbols (e.g. ∇_α, ð, etc.).

If x^α are local coordinates on $\mathscr{M}$ then (locally)

$$A = A_{\alpha_1 \ldots \alpha_p} \mathrm{d}x^{\alpha_1} \wedge \cdots \wedge \mathrm{d}x^{\alpha_p} \tag{4.3.20}$$

where components are taken in the associated *coordinate basis* defined by

$$\delta^{\alpha}_{\boldsymbol{\alpha}} = \nabla_{\boldsymbol{\alpha}} x^{\alpha}. \tag{4.2.55}$$

(Note that $\mathrm{d}x^\alpha = \nabla_{\iota_1} x^\alpha$.) The integral of a p-form A over an oriented p-dimensional surface $\mathscr{P}$ is

$$\int_{\mathscr{P}} A = \int_{\mathscr{P}} A_{\alpha_1 \ldots \alpha_p} \mathrm{d}x^{\alpha_1} \wedge \cdots \wedge \mathrm{d}x^{\alpha_p}$$

$$= p! \int_{\mathscr{P}} A_{1 \ldots p} \mathrm{d}x^1 \wedge \cdots \wedge \mathrm{d}x^p \tag{4.3.24}$$

whenever $\mathscr{P}$ is defined (locally) by $x^{p+1} = \cdots = x^n = 0$. The *fundamental theorem of exterior calculus* states

$$\int_{\mathscr{Q}} \mathrm{d}A = \oint_{\partial \mathscr{Q}} A, \tag{4.3.25}$$

where $\mathscr{Q}$ is a *compact* $(p+1)$-surface with boundary $\partial\mathscr{Q}$.

Sometimes the notation

$$\underset{X}{\nabla} := X^\alpha \nabla_\alpha \tag{4.3.31}$$

for *directional covariant derivative* is useful. Then the expression (4.2.31) for the curvature becomes

$$(\underset{X}{\nabla}\underset{Y}{\nabla} - \underset{Y}{\nabla}\underset{X}{\nabla} - \underset{[X,Y]}{\nabla})Z^\delta = R_{\alpha\beta\gamma}{}^{\delta} X^\alpha Y^\beta Z^\gamma, \tag{4.3.33}$$

the third term on the left now being needed even in the absence of torsion.

For a *space-time* $\mathscr{M}$ we use $a, b, \ldots$ in place of $\alpha, \beta, \ldots$ where we can rewrite a as AA', b as BB', etc. in accordance with our spinor rules. Thus $\nabla_{AA'} = \nabla_a$, $\nabla_{BB'} = \nabla_b$, etc., and corresponding operators $\nabla_{AB'}$, $\nabla_{A}^{A'}$, etc. can be defined, these being extended to act on spinorial quantities (uniquely, where we demand $\nabla_a \varepsilon_{BC} = 0$). Thus we can express, for example, the Dirac–Weyl equation for a (massless) neutrino as

$$\nabla_{AA'} \phi^A = 0. \tag{4.4.61}$$

Spin-coefficients

Choosing a *dyad* o^A, ι^A i.e. a basis $\varepsilon_{\mathbf{A}}{}^A$, and dual $\varepsilon_A{}^{\mathbf{A}}$, so that

$$\varepsilon_0{}^A = o^A, \quad \varepsilon_1{}^A = \iota^A; \quad \varepsilon_A{}^0 = -\chi^{-1}\iota_A, \quad \varepsilon_A{}^1 = \chi^{-1} o_A,$$

(*cf.* (2.5.46)), where we need not normalize to $\chi = 1$, we define the *spin-coefficients* to be the sixteen quantities

$$\gamma_{\mathbf{AA'C}}{}^{\mathbf{B}} := \varepsilon_A{}^{\mathbf{B}} \nabla_{\mathbf{AA'}} \varepsilon_{\mathbf{C}}{}^A = -\varepsilon_{\mathbf{C}}{}^A \nabla_{\mathbf{AA'}} \varepsilon_A{}^{\mathbf{B}} \tag{4.5.2}$$

for which a standard notation is

$\gamma_{\mathbf{AA'B}}{}^{\mathbf{C}} =$

$\mathbf{AA'}$ \ $\mathbf{C}$	0	1	0	1
$\mathbf{B}$	0	0	1	1
00′	ε	$-\kappa$	$-\tau'$	γ'
10′	α	$-\rho$	$-\sigma'$	β'
01′	β	$-\sigma$	$-\rho'$	α'
11′	γ	$-\tau$	$-\kappa'$	ε'

Then, for $\gamma_{\mathbf{AA'BC}}$, we have

$$\begin{array}{|cccc|}\hline \kappa & \varepsilon & \gamma' & \tau' \\ \rho & \alpha & \beta' & \sigma' \\ \sigma & \beta & \alpha' & \rho' \\ \tau & \gamma & \varepsilon' & \kappa' \\ \hline \end{array} = \chi^{-1} \times \begin{array}{|c|c|c|c|}\hline o^A D o_A & \iota^A D o_A & -o^A D \iota_A & -\iota^A D \iota_A \\ o^A \delta' o_A & \iota^A \delta' o_A & -o^A \delta' \iota_A & -\iota^A \delta' \iota_A \\ o^A \delta o_A & \iota^A \delta o_A & -o^A \delta \iota_A & -\iota^A \delta \iota_A \\ o^A D' o_A & \iota^A D' o_A & -o^A D' \iota_A & -\iota^A D' \iota_A \\ \hline \end{array} \tag{4.5.21}$$

$$= \chi^{-1}\bar{\chi}^{-1} \times \begin{array}{|cccc|}\hline m^a D l_a & \tfrac{1}{2}(n^a D l_a + m^a D \bar{m}_a + \bar{\chi} D \chi) & \tfrac{1}{2}(l^a D n_a + \bar{m}^a D m_a + \bar{\chi} D \chi) & \bar{m}^a D n_a \\ m^a \delta' l_a & \tfrac{1}{2}(n^a \delta' l_a + m^a \delta' \bar{m}_a + \bar{\chi} \delta' \chi) & \tfrac{1}{2}(l^a \delta' n_a + \bar{m}^a \delta' m_a + \bar{\chi} \delta' \chi) & \bar{m}^a \delta' n_a \\ m^a \delta l_a & \tfrac{1}{2}(n^a \delta l_a + m^a \delta \bar{m}_a + \bar{\chi} \delta \chi) & \tfrac{1}{2}(l^a \delta n_a + \bar{m}^a \delta m_a + \bar{\chi} \delta \chi) & \bar{m}^a \delta n_a \\ m^a D' l_a & \tfrac{1}{2}(n^a D' l_a + m^a D' \bar{m}_a + \bar{\chi} D' \chi) & \tfrac{1}{2}(l^a D' n_a + \bar{m}^a D' m_a + \bar{\chi} D' \chi) & \bar{m}^a D' n_a \\ \hline \end{array} \tag{4.5.22}$$

where

$$\begin{aligned} D &:= \nabla_{00'} = o^A o^{A'} \nabla_{AA'} = l^a \nabla_a = \bar{D} \\ \delta &:= \nabla_{01'} = o^A \iota^{A'} \nabla_{AA'} = m^a \nabla_a = \bar{\delta}' \\ \delta' &:= \nabla_{10'} = \iota^A o^{A'} \nabla_{AA'} = \bar{m}^a \nabla_a = \bar{\delta} \\ D' &:= \nabla_{11'} = \iota^A \iota^{A'} \nabla_{AA'} = n^a \nabla_a = \bar{D}'. \end{aligned} \tag{4.5.23}$$

If, as is most usual, we normalize to the case $\chi = 1$ of a *spin-frame*, then

$$\gamma_{\mathbf{AA'BC}} = \gamma_{\mathbf{AA'CB}}, \tag{4.5.5}$$

and if we introduce the frequently employed symbols π, λ, μ, ν, we have

$\gamma_{AA'BC} =$

AA' \ BC	00	10 or 01	11
00'	κ	$\varepsilon = -\gamma'$	$\pi = -\tau'$
10'	ρ	$\alpha = -\beta'$	$\lambda = -\sigma'$
01'	σ	β	$\mu = -\rho'$
11'	τ	γ	$\nu = -\kappa'$

(4.5.29)

The primed letters are preferred here, these and (4.5.23) being used consistently with the general *priming operation* which effects the interchange

$$o^A \mapsto i\iota^A, \iota^A \mapsto io^A, o^{A'} \mapsto -i\iota^{A'}, \iota^{A} \mapsto -io^{A'}. \tag{4.5.17}$$

Spinor form of the curvature

The Riemann curvature tensor has the spinor expression

$$R_{abcd} = \mathrm{X}_{ABCD}\varepsilon_{A'B'}\varepsilon_{C'D'} + \Phi_{ABC'D'}\varepsilon_{A'B'}\varepsilon_{CD} + \bar{\Phi}_{A'B'CD}\varepsilon_{AB}\varepsilon_{C'D'} + \bar{\mathrm{X}}_{A'B'C'D'}\varepsilon_{AB}\varepsilon_{CD}, \tag{4.6.1}$$

where

$$\mathrm{X}_{ABCD} = \mathrm{X}_{(AB)(CD)}, \quad \Phi_{ABC'D'} = \Phi_{(AB)(C'D')} \tag{4.6.3}$$

and

$$\mathrm{X}_{ABCD} = \mathrm{X}_{CDAB}, \quad \bar{\Phi}_{ABC'D'} = \Phi_{ABC'D'}. \tag{4.6.4}$$

Having *two* pairs of skew indices, R_{abcd} may be dualized on the second pair (R^*_{abcd}), the first ($^*R_{abcd}$), or both ($^*R^*_{abcd}$). The effect on the spinor expression (4.6.1) of applying these operations is given by

$$\begin{aligned} R_{abcd} &= \mathrm{X} + \Phi + \bar{\Phi} + \bar{\mathrm{X}} \\ R^*{}_{abcd} &= -\mathrm{i}\mathrm{X} + \mathrm{i}\Phi - \mathrm{i}\bar{\Phi} + \mathrm{i}\bar{\mathrm{X}} \\ {}^*R_{abcd} &= -\mathrm{i}\mathrm{X} - \mathrm{i}\Phi + \mathrm{i}\bar{\Phi} + \mathrm{i}\bar{\mathrm{X}} \\ {}^*R^*{}_{abcd} &= -\mathrm{X} + \Phi + \bar{\Phi} - \bar{\mathrm{X}} \end{aligned} \tag{4.6.11}$$

(where indices and ε-spinors have been suppressed on the right).

We define

$$\Lambda := \tfrac{1}{6}\mathrm{X}_{AB}{}^{AB} \tag{4.6.18}$$

so that

$$\mathrm{X}_{ABC}{}^{B} = 3\Lambda\varepsilon_{AC} \tag{4.6.19}$$

and find

$$\Lambda = \bar{\Lambda} \tag{4.6.17}$$

as an expression of the cyclic identity $R_{a[bcd]} = 0$. We remark here that the double dual ${}^*R^*_{abcd}$ satisfies this also:

$${}^*R^*{}_{a[bcd]} = 0. \tag{4.6.9}$$

Defining*

$$R_{ab} = R_{acb}{}^{c}, \quad R = R_{a}{}^{a},$$

we further find

$$R = 24\Lambda \tag{4.6.22}$$

and

$$R_{ab} = 6\Lambda g_{ab} - 2\Phi_{ab}. \tag{4.6.21}$$

Einstein's field equations (with cosmological constant λ and gravitational constant G)**

$$R_{ab} - \tfrac{1}{2}Rg_{ab} + \lambda g_{ab} = -8\pi G T_{ab}$$

become, in terms of these quantities,

$$\Phi_{ab} = 4\pi G(T_{ab} - \tfrac{1}{4}T_q{}^q g_{ab}), \quad \Lambda = \tfrac{1}{3}\pi G T_q{}^q + \tfrac{1}{6}\lambda. \tag{4.6.32}$$

The *symmetric part* of X_{ABCD}

$$\Psi_{ABCD} := \mathrm{X}_{(ABCD)} = \mathrm{X}_{A(BCD)} \tag{4.6.35}$$

is referred to as the *gravitational spinor* or the *Weyl (conformal) spinor*. We have

$$\mathrm{X}_{ABCD} = \Psi_{ABCD} + \Lambda(\varepsilon_{AC}\varepsilon_{BD} + \varepsilon_{AD}\varepsilon_{BC}), \tag{4.6.34}$$

and the full Riemann tensor becomes

$$\begin{aligned} R_{abcd} = {} & \Psi_{ABCD}\varepsilon_{A'B'}\varepsilon_{C'D'} + \bar{\Psi}_{A'B'C'D'}\varepsilon_{AB}\varepsilon_{CD} \\ & + \Phi_{ABC'D'}\varepsilon_{A'B'}\varepsilon_{CD} + \bar{\Phi}_{A'B'CD}\varepsilon_{AB}\varepsilon_{C'D'} \\ & + 2\Lambda(\varepsilon_{AC}\varepsilon_{BD}\varepsilon_{A'C'}\varepsilon_{B'D'} - \varepsilon_{AD}\varepsilon_{BC}\varepsilon_{A'D'}\varepsilon_{B'C'}). \end{aligned} \tag{4.6.38}$$

The terms involving Ψ_{ABCD} alone yield the *Weyl (conformal) tensor*

$$C_{abcd} := \Psi_{ABCD}\varepsilon_{A'B'}\varepsilon_{C'D'} + \bar{\Psi}_{A'B'C'D'}\varepsilon_{AB}\varepsilon_{CD} \tag{4.6.41}$$

whose tensor expression is given by

$$C_{ab}{}^{cd} = R_{ab}{}^{cd} - 2R_{[a}{}^{[c}g_{b]}{}^{d]} + \tfrac{1}{3}Rg_{[a}{}^{c}g_{b]}{}^{d}. \tag{4.8.2}$$

* The sign conventions we have adopted, when applied to a *positive-definite* space, give a Ricci tensor and scalar curvature with a sign *opposite* to that which is usual in the pure mathematical literature.

** Two small notational changes from Volume 1 are the use of G rather than γ, for the gravitational constant, and the lower case 'c' rather than '$\mathscr{C}$' as the subscript for the conformally invariant ð and þ (*cf.* (5.6.33) below).

Its anti-self-dual and self-dual parts are, respectively,

$$^{-}C_{abcd} := \Psi_{ABCD}\varepsilon_{A'B'}\varepsilon_{C'D'} \tag{4.6.42}$$

and

$$^{+}C_{abcd} := \bar{\Psi}_{A'B'C'D'}\varepsilon_{AB}\varepsilon_{CD} \tag{4.6.43}$$

and the same results are obtained if we dualize instead on the right. By analogy with a corresponding operation on the Maxwell field, we define *duality rotations*

$$\begin{aligned} C_{abcd} \mapsto {}^{(\theta)}C_{abcd} &= C_{abcd}\cos\theta + {}^{*}C_{abcd}\sin\theta \\ &= \mathrm{e}^{-\mathrm{i}\theta}\ {}^{-}C_{abcd} + \mathrm{e}^{\mathrm{i}\theta}\ {}^{+}C_{abcd} \end{aligned} \tag{4.8.15}$$

which correspond to

$$\Psi_{ABCD} \mapsto \mathrm{e}^{-\mathrm{i}\theta}\Psi_{ABCD}. \tag{4.8.16}$$

Spinor Ricci identities

The decomposition (3.4.20), applied to the derivative commutator (4.2.14), gives

$$\Delta_{ab} = 2\nabla_{[a}\nabla_{b]} = \varepsilon_{A'B'}\square_{AB} + \varepsilon_{AB}\square_{A'B'}, \tag{4.9.1}$$

where

$$\square_{AB} = \nabla_{X'(A}\nabla_{B)}^{X'}, \quad \square_{A'B'} = \nabla_{X(A'}\nabla_{B')}^{X}. \tag{4.9.2}$$

Operating on a spin-vector κ^C, (4.9.1) yields

$$\Delta_{ab}\kappa^C = \{\varepsilon_{A'B'}\mathrm{X}_{ABE}{}^{C} + \varepsilon_{AB}\Phi_{A'B'E}{}^{C}\}\kappa^E, \tag{4.9.7}$$

which splits into

$$\square_{AB}\kappa^C = \mathrm{X}_{ABE}{}^{C}\kappa^E, \quad \square_{A'B'}\kappa^C = \Phi_{A'B'E}{}^{C}\kappa^E. \tag{4.9.8}$$

Similarly, on κ_C it yields

$$\square_{AB}\kappa_C = -\mathrm{X}_{ABC}{}^{E}\kappa_E, \quad \square_{A'B'}\kappa_C = -\Phi_{A'B'C}{}^{E}\kappa_E. \tag{4.9.11}$$

To deal with primed indices, we take the complex conjugates of (4.9.8) and (4.9.11). For a spinor with several indices, these relations combine together, to give, e.g.

$$\begin{aligned} \square_{AB}\theta^{C}{}_{D}{}^{E'}{}_{F'} &= \mathrm{X}_{ABQ}{}^{C}\theta^{Q}{}_{D}{}^{E'}{}_{F'} - \mathrm{X}_{ABD}{}^{Q}\theta^{C}{}_{Q}{}^{E'}{}_{F'} \\ &\quad + \Phi_{ABQ'}{}^{E'}\theta^{C}{}_{D}{}^{Q'}{}_{F'} - \Phi_{ABF'}{}^{Q'}\theta^{C}{}_{D}{}^{E'}{}_{Q'} \end{aligned} \tag{4.9.13}$$

$$\begin{aligned} \square_{A'B'}\phi^{C'}{}_{D'}{}^{E}{}_{F} &= \bar{\mathrm{X}}_{A'B'Q'}{}^{C'}\phi^{Q'}{}_{D'}{}^{E}{}_{F} - \bar{\mathrm{X}}_{A'B'D'}{}^{Q'}\phi^{C'}{}_{Q'}{}^{E}{}_{F} \\ &\quad + \Phi_{A'B'Q}{}^{E}\phi^{C'}{}_{D'}{}^{Q}{}_{F} - \Phi_{A'B'F}{}^{Q}\phi^{C'}{}_{D'}{}^{E}{}_{Q}. \end{aligned} \tag{4.9.14}$$

Relations involving Ψ_{ABCD} and Λ directly are sometimes useful, e.g.

$$\square_{AB}\kappa_C = -\Psi_{ABC}{}^{D}\kappa_D - \Lambda(\varepsilon_{AC}\kappa_B + \varepsilon_{BC}\kappa_A), \tag{4.9.15}$$

whence

$$\square_{(AB}\kappa_{C)} = -\Psi_{ABC}{}^{D}\kappa_D, \tag{4.9.16}$$

$$\square_{AB}\kappa^B = -3\Lambda\kappa_A. \tag{4.9.17}$$

Spinor Bianchi identity

The Bianchi identity

$$\nabla_{[a}R_{bc]de} = 0 \tag{4.10.1}$$

translates to

$$\nabla^A_{B'}\mathrm{X}_{ABCD} = \nabla^{A'}_{B}\Phi_{CDA'B'}, \tag{4.10.3}$$

i.e. to

$$\nabla^A_{B'}\Psi_{ABCD} = \nabla^{A'}_{(B}\Phi_{CD)A'B'} \tag{4.10.7}$$

$$\nabla^{CA'}\Phi_{CDA'B'} + 3\nabla_{DB'}\Lambda = 0. \tag{4.10.8}$$

When Einstein's vacuum equations (with cosmological term)

$$\Phi_{ABC'D'} = 0, \quad \Lambda = \tfrac{1}{6}\lambda, \tag{4.10.10.}$$

hold, (4.10.7) reduces to the form of a 'massless free-field equation':

$$\nabla^{AA'}\Psi_{ABCD} = 0. \tag{4.10.9}$$

With matter present, Einstein's field equations (4.6.32) lead to a 'source term' on the right:

$$\nabla^A_{B'}\Psi_{ABCD} = 4\pi G\nabla^{A'}_{(B}T_{CD)A'B'}. \tag{4.10.12}$$

Spinor curvature components

In conjunction with the spin-coefficients (4.5.16), we introduce curvature components

$$\Psi_0 := \chi^{-1}\bar{\chi}\Psi_{0000}, \quad \Psi_1 := \chi^{-1}\bar{\chi}\Psi_{0001}, \quad \Psi_2 := \chi^{-1}\bar{\chi}\Psi_{0011},$$
$$\Psi_3 := \chi^{-1}\bar{\chi}\Psi_{0111}, \quad \Psi_4 := \chi^{-1}\bar{\chi}\Psi_{1111}, \tag{4.11.6}$$

$$\Pi := \chi\bar{\chi}\Lambda \tag{4.11.7}$$

$$\begin{aligned} &\Phi_{00} := \Phi_{000'0'} \quad \Phi_{01} := \Phi_{000'1'} \quad \Phi_{02} := \Phi_{001'1'} \\ &\Phi_{10} := \Phi_{010'0'} \quad \Phi_{11} := \Phi_{010'1'} \quad \Phi_{12} := \Phi_{011'1'} \\ &\Phi_{20} := \Phi_{110'0'} \quad \Phi_{21} := \Phi_{110'1'} \quad \Phi_{22} := \Phi_{111'1'}. \end{aligned} \tag{4.11.8}$$

the dyad being not necessarily normalized. (Note: $\Phi_{010'0'} = \Phi_{ABC'D'}o^A\iota^B o^{C'}o^{D'}$, etc.). In terms of the corresponding null tetrad we have

$$\begin{aligned} &\Psi_0 = \chi^{-1}\bar{\chi}^{-1}C_{abcd}l^a m^b l^c m^d, \quad \Psi_1 = \chi^{-1}\bar{\chi}^{-1}C_{abcd}l^a m^b l^c n^d \\ &\Psi_2 = \chi^{-1}\bar{\chi}^{-1}C_{abcd}l^a m^b \bar{m}^c n^d, \quad \Psi_3 = \chi^{-1}\bar{\chi}^{-1}C_{abcd}l^a n^b \bar{m}^c n^d \\ &\Psi_4 = \chi^{-1}\bar{\chi}^{-1}C_{abcd}\bar{m}^a n^b \bar{m}^c n^d \end{aligned} \tag{4.11.9}$$

and

$$\begin{array}{lll}\Phi_{00} = -\frac{1}{2}R_{ab}l^a l^b & \Phi_{01} = -\frac{1}{2}R_{ab}l^a m^b & \Phi_{02} = -\frac{1}{2}R_{ab}m^a m^b \\ \Phi_{10} = -\frac{1}{2}R_{ab}l^a \bar{m}^b & \Phi_{11} = -\frac{1}{2}R_{ab}l^a n^a + 3\Pi & \Phi_{12} = -\frac{1}{2}R_{ab}m^a n^b \\ \Phi_{20} = -\frac{1}{2}R_{ab}\bar{m}^a \bar{m}^b & \Phi_{21} = -\frac{1}{2}R_{ab}\bar{m}^a n^b & \Phi_{22} = -\frac{1}{2}R_{ab}n^a n^b.\end{array} \quad (4.11.10)$$

Compacted spin-coefficient formalism

A (normally scalar) quantity η which scales as

$$\eta \mapsto \lambda^{r'}\bar{\lambda}^{t'}\mu^r\bar{\mu}^t\eta \quad (4.12.9)$$

under

$$o^A \mapsto \lambda o^A, \quad \iota^A \mapsto \mu\iota^A \quad (4.12.2)$$

is said to be a *weighted* quantity of *type* $\{r', r; t', t\}$. Note that χ has type $\{1, 1; 0, 0\}$:

$$\chi \mapsto \lambda\mu\chi. \quad (4.12.3)$$

So if we wish to preserve the normalization $\chi = 1$, only the two numbers

$$p = r' - r, \quad q = t' - t \quad (4.12.10)$$

are defined and as we say, instead, that η has type $\{p, q\}$ or, equivalently, *spin-weight* $\frac{1}{2}(p - q)$ and *boost-weight* $\frac{1}{2}(p + q)$. We sometimes refer to such a scalar η simply as an $\{r', r; t', t\}$*-scalar* or a $\{p, q\}$*-scalar*.

For a scalar (or tensor, or spinor) of type $\{r', r; t', t\}$ we define

$$\begin{aligned} þ\eta &= (D - r'\varepsilon - r\gamma' - t'\bar{\varepsilon} - t\bar{\gamma}')\eta, \\ ð\eta &= (\delta - r'\beta - r\alpha' - t'\bar{\alpha} - t\bar{\beta}')\eta, \\ ð'\eta &= (\delta' - r'\alpha - r\beta' - t'\bar{\beta} - t\bar{\alpha}')\eta, \\ þ'\eta &= (D' - r'\gamma - r\varepsilon' - t'\bar{\gamma} - t\bar{\varepsilon}')\eta, \end{aligned} \quad (4.12.15)$$

and find that the operators þ, ð, ð′, þ′ are *weighted* (in the sense of sending weighted quantities to weighted quantities) and have the following types

$$þ:\{1,0;1,0\}, \quad ð:\{1,0;0,1\}, \quad ð':\{0,1;1,0\}, \quad þ':\{0,1;0,1\}. \quad (4.12.17)$$

The spin-coefficients appearing on the right in (4.12.15) are precisely those which are *not* weighted. The others all have types, namely

$$\begin{array}{llll}\kappa:\{2,-1;1,0\}, & \sigma:\{2,-1;0,1\}, & \rho:\{1,0;1,0\}, & \tau:\{1,0;0,1\} \\ \kappa':\{-1,2;0,1\}, & \sigma':\{-1,2;1,0\}, & \rho':\{0,1;0,1\}, & \tau':\{0,1;1,0\}.\end{array} \quad (4.12.13)$$

The curvature components are weighted scalars of types

$$\Psi_r:\{3-r, r-1; 1, 1\}, \quad \Pi:\{1,1;1,1\}, \quad \Phi_{rt}:\{2-r, r; 2-t, t\} \quad (4.12.25)$$

and for a general symmetric spinor of valence $[{}_{r'+r}^{\;0}\;\; {}_{t'+t}^{\;0}]$, the various components

$$\xi_{r,t} = \xi_{A\ldots D\ldots G'\ldots K'\ldots}\underbrace{o^A\ldots}_{r'}\underbrace{\iota^D\ldots}_{r}\underbrace{o^{G'}\ldots}_{t'}\underbrace{\iota^{K'}\ldots}_{t} \tag{4.12.26}$$

have respective types $\{r', r; t', t\}$. The components of the derivatives of $\xi_{A\ldots M'}$, are, in this formalism

$$\begin{aligned}
(o^A\ldots\iota^{K'}\ldots)D\xi_{A\ldots K'\ldots} &= \text{þ}\xi_{r,t} + r'\kappa\xi_{r+1,t} + r\tau'\xi_{r-1,t} \\
&\quad + t'\bar{\kappa}\xi_{r,t+1} + t\bar{\tau}'\xi_{r,t-1} \\
(o^A\ldots\iota^{K'}\ldots)\delta\xi_{A\ldots K'\ldots} &= \eth\xi_{r,t} + r'\sigma\xi_{r+1,t} + r\rho'\xi_{r-1,t} \\
&\quad + t'\bar{\rho}\xi_{r,t+1} + t\bar{\sigma}'\xi_{r,t-1} \\
(o^A\ldots\iota^{K'}\ldots)\delta'\xi_{A\ldots K'\ldots} &= \eth'\xi_{r,t} + r'\rho\xi_{r+1,t} + r\sigma'\xi_{r-1,t} \\
&\quad + t'\bar{\sigma}\xi_{r,t+1} + t\bar{\rho}'\xi_{r,t-1} \\
(o^A\ldots\iota^{K'}\ldots)D'\xi_{A\ldots K'\ldots} &= \text{þ}'\xi_{r,t} + r'\tau\xi_{r+1,t} + r\kappa'\xi_{r-1,t} \\
&\quad + t'\bar{\tau}\xi_{r,t+1} + t\bar{\kappa}'\xi_{r,t-1}.
\end{aligned} \tag{4.12.27}$$

The following relations are also useful:

$$\begin{array}{llll}
\text{þ}o^A = -\kappa\iota^A, & \text{þ}\iota^A = -\tau' o^A, & \text{þ}o^{A'} = -\bar{\kappa}\iota^{A'}, & \text{þ}\iota^{A'} = -\bar{\tau}' o^{A'} \\
\eth o^A = -\sigma\iota^A, & \eth\iota^A = -\rho' o^A, & \eth o^{A'} = -\bar{\rho}\iota^{A'}, & \eth\iota^{A'} = -\bar{\sigma}' o^{A'} \\
\eth' o^A = -\rho\iota^A, & \eth'\iota^A = -\sigma' o^A, & \eth' o^{A'} = -\bar{\sigma}\iota^{A'}, & \eth'\iota^{A'} = -\bar{\rho}' o^{A'} \\
\text{þ}' o^A = -\tau\iota^A, & \text{þ}'\iota^A = -\kappa' o^A, & \text{þ}' o^{A'} = -\bar{\tau}\iota^{A'}, & \text{þ}'\iota^{A'} = -\bar{\kappa}' o^{A'}.
\end{array} \tag{4.12.28}$$

Furthermore

$$\text{þ}\chi = 0, \quad \eth\chi = 0, \quad \eth'\chi = 0, \quad \text{þ}'\chi = 0. \tag{4.12.23}$$

The prime notation is being used consistently here to denote the operation (4.5.17). Then for η of type $\{r', r; t', t\}$ we have

$$(\eta')' = (-1)^{r'+t'-r-t}\eta \tag{4.12.24}$$

and

$$(\bar{\eta})' = \overline{\eta'}.$$

Moreover the curvature components (4.11.6), (4.11.8) are replaced according to

$$\begin{aligned}
&\Psi_r \mapsto \Psi_s : 0 \leftrightarrow 4,\ 1 \leftrightarrow 3,\ 2 \leftrightarrow 2 \\
&\Phi_{rs} \mapsto \Phi_{tu} : 0 \leftrightarrow 2,\ 1 \leftrightarrow 1.
\end{aligned} \tag{4.11.13}$$

Compacted spin-coefficient equations

The use of the operators (4.12.15) casts the differential relations satisfied by the (weighted) spin-coefficients (and curvature quantitities) into the following relatively simple form

$$
\begin{aligned}
&þ\rho - ð'\kappa = \rho^2 + \sigma\bar{\sigma} - \bar{\kappa}\tau - \tau'\kappa + \Phi_{00} && (a)\\
&þ\sigma - ð\kappa = (\rho + \bar{\rho})\sigma - (\tau + \bar{\tau}')\kappa + \Psi_0 && (b)\\
&þ\tau - þ'\kappa = (\tau - \bar{\tau}')\rho + (\bar{\tau} - \tau')\sigma + \Psi_1 + \Phi_{01} && (c)\\
&ð\rho - ð'\sigma = (\rho - \bar{\rho})\tau + (\bar{\rho}' - \rho')\kappa - \Psi_1 + \Phi_{01} && (d)\\
&ð\tau - þ'\sigma = -\rho'\sigma - \bar{\sigma}'\rho + \tau^2 + \kappa\bar{\kappa}' + \Phi_{02} && (e)\\
&þ'\rho - ð'\tau = \rho\bar{\rho}' + \sigma\sigma' - \tau\bar{\tau}' - \kappa\kappa' - \Psi_2 - 2\Pi && (f) \quad (4.12.32)
\end{aligned}
$$

together with their primed (and complex conjugated) versions. The unweighted spin-coefficients do not appear explicitly, but their differential relations are subsumed into the following *commutator equations*, as applied to an $\{r', r; t', t\}$-scalar:

$$
\begin{aligned}
þþ' - þ'þ = {} & (\bar{\tau} - \tau')ð + (\tau - \bar{\tau}')ð' - p(\kappa\kappa' - \tau\tau' + \Psi_2 + \Phi_{11} - \Pi)\\
& - q(\bar{\kappa}\bar{\kappa}' - \bar{\tau}\bar{\tau}' + \bar{\Psi}_2 + \Phi_{11} - \Pi) \qquad (4.12.33)
\end{aligned}
$$

$$
\begin{aligned}
þð - ðþ = {} & \bar{\rho}ð + \sigma ð' - \bar{\tau}'þ - \kappa þ' - p(\rho'\kappa - \tau'\sigma + \Psi_1)\\
& - q(\bar{\sigma}'\bar{\kappa} - \bar{\rho}\bar{\tau}' + \Phi_{01}) \qquad (4.12.34)
\end{aligned}
$$

$$
\begin{aligned}
ðð' - ð'ð = {} & (\bar{\rho}' - \rho')þ + (\rho - \bar{\rho})þ' + p(\rho\rho' - \sigma\sigma' + \Psi_2 - \Phi_{11} - \Pi)\\
& - q(\bar{\rho}\bar{\rho}' - \bar{\sigma}\bar{\sigma}' + \bar{\Psi}_2 - \Phi_{11} - \Pi), \qquad (4.12.35)
\end{aligned}
$$

together with their primed versions (for which $p \mapsto -p$ and $q \mapsto -q$), their complex conjugate versions (for which $p \mapsto q$ and $q \mapsto p$) and their primed complex conjugated versions (for which $p \mapsto -q$ and $q \mapsto -p$). (We assume p and q are real.) The Bianchi identity becomes

$$
\begin{aligned}
& þ\Psi_1 - ð'\Psi_0 - þ\Phi_{01} + ð\Phi_{00}\\
& = -\tau'\Psi_0 + 4\rho\Psi_1 - 3\kappa\Psi_2\\
& \quad + \bar{\tau}'\Phi_{00} - 2\bar{\rho}\Phi_{01} - 2\sigma\Phi_{10} + 2\kappa\Phi_{11} + \bar{\kappa}\Phi_{02}, \qquad (4.12.36)
\end{aligned}
$$

$$
\begin{aligned}
& þ\Psi_2 - ð'\Psi_1 - ð'\Phi_{01} + þ'\Phi_{00} + 2þ\Pi\\
& = \sigma'\Psi_0 - 2\tau'\Psi_1 + 3\rho\Psi_2 - 2\kappa\Psi_3\\
& \quad + \bar{\rho}'\Phi_{00} - 2\bar{\tau}\Phi_{01} - 2\tau\Phi_{10} + 2\rho\Phi_{11} + \bar{\sigma}\Phi_{02}, \qquad (4.12.37)
\end{aligned}
$$

$$
\begin{aligned}
& þ\Psi_3 - ð'\Psi_2 - þ\Phi_{21} + ð\Phi_{20} - 2ð'\Pi\\
& = 2\sigma'\Psi_1 - 3\tau'\Psi_2 + 2\rho\Psi_3 - \kappa\Psi_4\\
& \quad - 2\rho'\Phi_{10} + 2\tau'\Phi_{11} + \bar{\tau}'\Phi_{20} - 2\bar{\rho}\Phi_{21} + \bar{\kappa}\Phi_{22}, \qquad (4.12.38)
\end{aligned}
$$

$$
\begin{aligned}
& þ\Psi_4 - ð'\Psi_3 - ð'\Phi_{21} + þ'\Phi_{20}\\
& = +3\sigma'\Psi_2 - 4\tau'\Psi_3 + \rho\Psi_4\\
& \quad - 2\kappa'\Phi_{10} + 2\sigma'\Phi_{11} + \bar{\rho}'\Phi_{20} - 2\bar{\tau}\Phi_{21} + \bar{\sigma}\Phi_{22} \qquad (4.12.39)
\end{aligned}
$$

$$\text{þ}\Phi_{11}+\text{þ}'\Phi_{00}-\eth\Phi_{10}-\eth'\Phi_{01}+3\text{þ}\Pi$$
$$=(\rho'+\bar{\rho}')\Phi_{00}+2(\rho+\bar{\rho})\Phi_{11}-(\tau'+2\bar{\tau})\Phi_{01}$$
$$-(2\tau+\bar{\tau}')\Phi_{10}-\bar{\kappa}\Phi_{12}-\kappa\Phi_{21}+\sigma\Phi_{20}+\bar{\sigma}\Phi_{02} \quad (4.12.40)$$

$$\text{þ}\Phi_{12}+\text{þ}'\Phi_{01}-\eth\Phi_{11}-\eth'\Phi_{02}+3\eth\Pi$$
$$=(\rho'+2\bar{\rho}')\Phi_{01}+(2\rho+\bar{\rho})\Phi_{12}-(\tau'+\bar{\tau})\Phi_{02}$$
$$-2(\tau+\bar{\tau}')\Phi_{11}-\bar{\kappa}'\Phi_{00}-\kappa\Phi_{22}+\sigma\Phi_{21}+\bar{\sigma}'\Phi_{10}. \quad (4.12.41)$$

The *zero rest-mass* (or *massless*) *free-field equation*

$$\nabla^{AA'}\phi_{AB\ldots L}=0, \quad (4.12.42)$$

with $\phi_{A\ldots L}=\phi_{(A\ldots L)}$ takes the form

$$\text{þ}\phi_r-\eth'\phi_{r-1}=(r-1)\sigma'\phi_{n-2}-r\tau'\phi_{r-1}+(n-r+1)\rho\phi_r$$
$$-(n-r)\kappa\phi_{r+1}, \quad (r=1,\ldots,n), \quad (4.12.44)$$

together with the primed versions, where the components have types

$$\phi_r=\phi_{\underbrace{0\ldots0}_{n-r}\underbrace{1\ldots1}_{r}}=\mathrm{i}^{-n}\phi'_{n-r}:\{n-r,r;0,0\}, \quad (r=0,\ldots,n). \quad (4.12.43)$$

The *twistor equation* $\nabla_{A'}^{(A}\omega^{B)}=0$ (with ω^0 of type $\{0,1;0,0\}$ and ω^1 of type $\{1,0;0,0\}$) becomes

$$\kappa\omega^0=\text{þ}\omega^1, \sigma\omega^0=\eth\omega^1, \eth'\omega^0=\sigma'\omega^1, \text{þ}'\omega^0=\kappa'\omega^1,$$
$$\text{þ}\omega^0+\rho\omega^0=\eth'\omega^1+\tau'\omega^1, \eth\omega^0+\tau\omega^0=\text{þ}'\omega^1+\rho'\omega^1. \quad (4.12.46)$$

Geometry of spacelike 2-surfaces

We set up a spin-frame ($\chi=1$) at each point of an oriented spacelike 2-surface $\mathscr{S}$ in a space–time $\mathscr{M}$, such that the flagpoles of o^A and ι^A are orthogonal to $\mathscr{S}$ at each point, with the spatial projection of the o^A flagpole pointing in the positive direction away from $\mathscr{S}$. We use the compacted formalism, so the choice of flag planes for o^A and ι^A is immaterial. Hence, no role is played by what could otherwise be topological restrictions on the smooth choice of these flag planes. We smoothly extend the spin-frames into a neighbourhood of $\mathscr{S}$ in $\mathscr{M}$ and find:

(4.14.2) PROPOSITION

If the null vectors l^a and n^a are orthogonal to a spacelike 2-surface $\mathscr{S}$, then ρ and ρ' are both real at $\mathscr{S}$.

The *projection operator* into $\mathscr{S}$ (at $\mathscr{S}$) is

$$S_a{}^b=-m_a\bar{m}^b-\bar{m}_am^b \quad (4.14.6)$$

and S_{ab} then acts as the negative definite metric tensor intrinsic to $\mathscr{S}$. The *complex curvature* of $\mathscr{S}$

$$K = \sigma\sigma' - \Psi_2 - \rho\rho' + \Phi_{11} + \Lambda \tag{4.14.20}$$

involves an imaginary part which is an extrinsic quantity but

(4.14.21) PROPOSITION

$K + \bar{K}$ is the Gaussian curvature of $\mathscr{S}$.

We note that K is the sum of two parts

$$\sigma\sigma' - \Psi_2 \quad \text{and} \quad \Phi_{11} + \Lambda - \rho\rho', \tag{4.14.41}$$

the first of which turns out to have simple conformal properties, and the second of which is real. From the Gauss–Bonnet theorem we deduce that if $\mathscr{S}$ is a closed surface of genus g (with $g = 0$ for a topological sphere), then

$$\oint_{\mathscr{S}} K\mathscr{S} = 2\pi(1-g) \tag{4.14.44}$$

and the conformally invariant quantity

$$\oint_{\mathscr{S}} (\sigma\sigma' - \Psi_2)\mathscr{S} \quad \textit{is real} \tag{4.14.45}$$

where $\mathscr{S}$ is the surface-area 2-form on $\mathscr{S}$, given by

$$\mathscr{S} = \mathrm{i}\bar{\boldsymbol{m}} \wedge \boldsymbol{m} \tag{4.14.65}$$

where $\boldsymbol{m} = m_{i_1}$, $\bar{\boldsymbol{m}} = \bar{m}_{i_1}$.

We note that if

$$\boldsymbol{\beta} = \beta_{ab}\mathrm{d}x^a \wedge \mathrm{d}x^b \tag{4.14.52}$$

then its restriction to $\mathscr{S}$ is

$$\tfrac{1}{2}\mathrm{i}\beta := \beta_{01'10'} = -\beta_{10'01'} = \beta_{ab}m^a\bar{m}^b \tag{4.14.53}$$

whence

$$\int_\Gamma \boldsymbol{\beta} = \int_\Gamma \beta\mathscr{S} \tag{4.14.66}$$

for a region $\Gamma \subset \mathscr{S}$.

We can (locally) introduce a (type $\{0,0\}$) *holomorphic coordinate* ξ into $\mathscr{S}$, characterized by the property

$$\eth'\xi = 0$$

(i.e. $\delta'\xi = 0$), and we can define the $\{1,-1\}$-scalar P on $\mathscr{S}$ by

$$\eth = P\frac{\partial}{\partial\xi} \text{ (on type } \{0,0\} \text{ scalars).} \tag{4.14.27}$$

Then we find

$$\bar{m} = -P^{-1}\,\mathrm{d}\xi, \quad m = -\bar{P}^{-1}\,\mathrm{d}\bar{\xi} \tag{4.14.29}$$

and, for any $\{s, -s\}$-scalar η,

$$ð\eta = P\bar{P}^{-s}\frac{\partial}{\partial\xi}(\bar{P}^{s}\eta). \tag{4.14.34}$$

If $\mathscr{S}$ is a closed surface,

$$\oint_{\mathscr{S}} ð'\alpha\,\mathscr{S} = 0, \oint_{\mathscr{S}} ð\tilde{\alpha}\mathscr{S} = 0 \tag{4.14.70}$$

where α has type $\{1, -1\}$ and $\tilde{\alpha}$ type $\{-1, 1\}$. Hence

$$\oint_{\mathscr{S}} \chi ð'\eta\mathscr{S} = -\oint_{\mathscr{S}} \eta ð'\chi\mathscr{S},$$

$$\oint_{\mathscr{S}} \tilde{\chi}ð\tilde{\eta}\mathscr{S} = -\oint_{\mathscr{S}} \tilde{\eta}ð\tilde{\chi}\mathscr{S}, \tag{4.14.71}$$

where the types of χ, η add up to $\{1, -1\}$ and those of $\tilde{\chi}$, $\tilde{\eta}$ to $\{-1, 1\}$.

On a null hypersurface

We consider a compact 3-region Σ of a null hypersurface $\mathscr{N}$, where Σ is smoothly bounded by two spacelike 2-surfaces:

$$\partial\Sigma = \mathscr{S}' - \mathscr{S}. \tag{4.14.73}$$

We choose the orientations so that the flagpole of ι^A points in the null direction in $\mathscr{N}$ (i.e. normal to $\mathscr{N}$, since $\mathscr{N}$ is null). We take u to be a smoothly increasing parameter on each generator of $\mathscr{N}$, scaled in relation to n^a according to

$$n^a\nabla_a u = U \quad (\neq 0) \tag{4.14.88}$$

so U is a $\{-1, 1\}$-scalar. For a (null) 'volume' element on $\mathscr{N}$ we take

$$\mathscr{N} := \mathrm{i}\bar{m} \wedge m \wedge l = U^{-1}\mathscr{S} \wedge \mathrm{d}u. \tag{4.14.89}$$

The fundamental theorem of exterior calculus (4.3.25), as applied to Σ, then takes the form

$$\int_{\Sigma} \{(þ' - 2\rho')\mu_0 - (ð - \tau)\mu_1\}\mathscr{N} = \oint_{\mathscr{S}'} \mu_0\mathscr{S} - \oint_{\mathscr{S}} \mu_0\mathscr{S}, \tag{4.14.92}$$

where μ_0 has type $\{0, 0\}$ and μ_1 type $\{-2, 0\}$.

§4.15 Functions on a metric sphere

In the study of (spin-weighted) spherical harmonics it is convenient to take $\mathscr{S}$ to be a *metric sphere of radius R* in $\mathbb{M}$, arising as the intersection of the future light cone $\mathscr{L}$ of a point L, and the past light cone $\mathscr{N}$ of a point N. A variable point Q of $\mathscr{S}$ has position vector vl^a relative to L and $-un^a$ relative to N. Then v is a $\{-1,-1\}$-scalar and u is a $\{1,1\}$-scalar, the *spin-frame* (o^A, ι^A) varying with Q in a way consistent with our previous discussion. We are interested in $\{p,q\}$-scalars defined on $\mathscr{S}$, particularly in relation to properties which are invariant under proper *rotations* of $\mathscr{S}$ (restricted Poincaré motions of $\mathbb{M}$ leaving both L and N fixed), and also properties invariant under *conformal motions* of $\mathscr{S}$. The latter are induced by restricted Poincaré motions of $\mathbb{M}$ leaving L fixed, where $\mathscr{S}$ is now identified as the space of generators of $\mathscr{L}$; or alternatively those leaving N fixed, where $\mathscr{S}$ is now identified as the space of generators of $\mathscr{N}$. We call $s=\frac{1}{2}(p-q)$ the spin-weight as before but now the boost-weight $b=\frac{1}{2}(p+q)$ is re-interpreted either as $b=w$ (in the case when L is held fixed) or as $b=-w$ (in the case when N is held fixed), w being the *conformal weight.*

We find

$$ðu=0,\quad ðv=0,\quad ð'u=0,\quad ð'v=0, \tag{4.15.28}$$

$$(ðð'-ð'ð)f=-sR^{-2}f, \tag{4.15.29}$$

where f has spin-weight s, and

$$ð^a ð'^b f = ð'^b ð^a f \tag{4.15.36}$$

whenever $b-a=2s$. Moreover, if g is defined by*

$$g=(vð')^{p+1}f \tag{4.15.30}$$

where f is of type $\{p,q\}$, with $p\geqslant 0$, then the relation between f and g is invariant under restricted Lorentz transformations about L; as is the relation between f and h given by

$$h=(vð)^{q+1}f \tag{4.15.32}$$

if $q\geqslant 0$. The conformally invariant operation in (4.15.32) is, in effect, $ð^{w-s+1}$ acting on a scalar of spin-weight s and conformal weight $w\geqslant -s$, while for (4.15.30) it is $(ð')^{w+s+1}$ where $w\geqslant s$.

We define a *spin-weighted spherical harmonic*** to be a $\{p,q\}$-function

* The factor v arising here has significance only in providing the relevant quantities with a consistent boost-weight. In practical calculations we may, if desired, set $v=1$.

** See Newman and Penrose (1966); essentially the same objects. somewhat confusingly called 'monopole harmonics', have been described by Wu and Yang (1976), *cf.* Dray (1985), and date back to earlier work: Fierz (1944), Dirac (1931) and Tamm (1931); see also the reference to Gegenbauer functions in Chandrasekhar (1983).

on $\mathscr{S}$ which is an eigenfunction h of $ð'ð$:

$$ð'ðh = -(j+s+1)(j-s)\tfrac{1}{2}R^{-2}h \tag{4.15.54}$$

where s is the spin-weight of h, and j is an integer or half-integer (the 'total spin'), satisfying

$$|s| \leqslant j.$$

Spin-weighted spherical harmonics can be characterized in another way as the components with respect to o^A, ι^A of a *constant* spinor

$$f_{\underbrace{A\ldots D}_{p}\underbrace{H'\ldots H'}_{q}} \in \mathbb{S}_{(A\ldots D)(E'\ldots H')} \tag{4.15.42}$$

for which

$$f_{A\ldots DE'\ldots H'}T^{E'}_{E}\ldots T^{H'}_{H} \in \mathbb{S}_{(A\ldots H)}$$

T^a being a timelike vector in the direction LN. Here the notation $\mathbb{S}_{\mathscr{A}}$ is being used for the subsystem (vector space over $\mathbb{C}$) of $\mathfrak{S}_{\mathscr{A}}$ of *constant* spinor fields, and the bracket notation of (3.3.14) is also being adopted.

The following table is useful:

j \ s	$-\frac{5}{2}$	-2	$-\frac{3}{2}$	-1	$-\frac{1}{2}$	0	$\frac{1}{2}$	1	$\frac{3}{2}$	2	$\frac{5}{2}\cdots$
$j=0$						1					
$\frac{1}{2}$					2		2				
1	$\overleftarrow{ð'}$			3		3		3		$\overrightarrow{ð}$	
$\frac{3}{2}$			4		4		4		4		
2		5		5		5		5		5	
$\frac{5}{2}$	6		6		6		6		6		6
$\vdots$	$\cdot^{\cdot^{\cdot}}$										$\ddots$

(4.15.60)

The numbers in the triangular array (extending indefinitely downwards) are the complex *dimensions* of the various spaces of these harmonics; $ð$ carries us one s-unit to the right and yields zero if and only if this moves us off the array; the action of $ð'$ is similar but in the other direction. In particular we have

(4.15.59) PROPOSITION

If f, defined on $\mathscr{S}$, has negative [positive] spin-weight then $ðf=0$ [or $ð'f=0$] implies $f=0$.

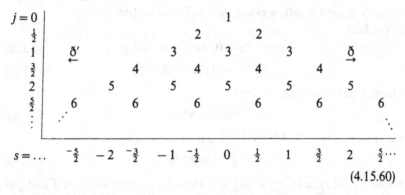

We note that if $s=j=0$ (the apex of the triangle), then the harmonic is

a *constant*. Hence if h has $s=0$, *each of the equations* $\eth h=0$, $\eth' h=0$, $\eth'\eth h=0$, $\eth\eth' h=0$ *implies that h is constant on* $\mathscr{S}$.

The table (4.15.60) is also helpful in the context of conformal transformations of $\mathscr{S}$. If we fix attention on the point (s,j), then we find that, for conformal weight $w=j$, the spaces represented by the set of points of the s-column above and including that point will *together* form a $(j+s+1)(j-s+1)$-dimensional space which is transformed to itself under conformal motions of $\mathscr{S}$. (This is the space of components of the spinors (4.15.42), with $p=j+s$, $q=j-s$.) If, on the other hand, we take the conformal weight to be $w=-j-2$ (the dual situation), then we find that the property of a quantity of spin-weight s that it contain *no* contribution from this space is conformally invariant.

If we choose a standard stereographic coordinate ζ on $\mathscr{S}$, in accordance with (1.2.10), we find that it is *anti*-holomorphic and so we can take

$$\xi=\bar{\zeta}. \tag{4.15.115}$$

Then we find, making the specialization $P>0$, that

$$P=\frac{\zeta\bar{\zeta}+1}{R\sqrt{2}} \tag{4.15.116}$$

and, for η of type $\{s,-s\}$:

$$\eth\eta=\frac{1}{R\sqrt{2}}((\zeta\bar{\zeta}+1)^{1-s}\frac{\partial}{\partial\bar{\zeta}}((\zeta\bar{\zeta}+1)^{s}\eta),$$

$$\eth'\eta=\frac{1}{R\sqrt{2}}((\zeta\bar{\zeta}+1)^{1+s}\frac{\partial}{\partial\zeta}((\zeta\bar{\zeta}+1)^{-s}\eta). \tag{4.15.117}$$

We remark that for $j=\frac{1}{2}$ the spin-weighted spherical harmonics (in the ζ-system) are linear combinations of

$$\frac{1}{\sqrt{\zeta\bar{\zeta}+1}} \quad\text{and}\quad \frac{\zeta}{\sqrt{\zeta\bar{\zeta}+1}} \quad (\text{where } s=-\tfrac{1}{2})$$

(multiples of ${}_{-\frac{1}{2}}Y_{\frac{1}{2},-\frac{1}{2}}$ and ${}_{-\frac{1}{2}}Y_{\frac{1}{2},\frac{1}{2}}$respectively) or of

$$\frac{\bar{\zeta}}{\sqrt{\zeta\bar{\zeta}+1}} \quad\text{and}\quad \frac{1}{\sqrt{\zeta\bar{\zeta}+1}} \quad (\text{when } s=\tfrac{1}{2}).$$

(multiples of ${}_{\frac{1}{2}}Y_{\frac{1}{2},-\frac{1}{2}}$ and ${}_{\frac{1}{2}}Y_{\frac{1}{2},\frac{1}{2}}$). If $j=1$ and $s=0$ they are linear combinations of

$$\frac{\zeta}{\zeta\bar{\zeta}+1}, \quad \frac{\zeta\bar{\zeta}-1}{\zeta\bar{\zeta}+1} \quad\text{and}\quad \frac{\bar{\zeta}}{\zeta\bar{\zeta}+1}.$$

(multiples of ${}_{0}Y_{-1,-1}$, ${}_{0}Y_{-1,0}$ and ${}_{0}Y_{-1,1}$).

Derivatives of charged fields

In the presence of charged fields, the symbol ∇_a denotes the appropriate (covariant) derivative operator which involves the electromagnetic field as well as the space–time curvature. Thus for a scalar field ψ, in $\mathcal{M}$, of charge e we have

$$\mathrm{i}\Delta_{ab}\psi = eF_{ab}\psi \tag{5.1.30}$$

where $\underset{e}{\Delta_{ab}}$ is as in (4.2.14) and where F_{ab} is the Maxwell field tensor. More generally, if $\psi^{\mathscr{A}} \in \underset{e}{\mathfrak{S}}^{\mathscr{A}}$ (where $\underset{e}{\mathfrak{S}}^{\mathscr{A}}$ denotes the module of spinor fields of charge e and index type $\mathscr{A}$), then

(5.1.31) PROPOSITION

$\Delta_{ab}\psi^{\mathscr{A}}$ differs from the result of a commutator acting on an uncharged $\psi^{\mathscr{A}}$ simply by the additional term $-\mathrm{i}eF_{ab}\psi^{\mathscr{A}}$.

In particular,

$$\Delta_{ab}\psi_c = -R_{abc}{}^{d}\psi_d - \mathrm{i}eF_{ab}\psi_c. \tag{5.1.34}$$

In accordance with (3.4.20), F_{ab} has a spinor expression

$$F_{ab} = \varphi_{AB}\varepsilon_{A'B'} + \varepsilon_{AB}\bar{\varphi}_{A'B'}, \tag{5.1.39}$$

which defines the *electromagnetic spinor* $\varphi_{AB} = \varphi_{(AB)}$. When $\square_{AB}$ (cf. (4.9.2)) is applied to a field $\psi^{\mathscr{A}}$ of charge e, the result differs from that applied to an uncharged $\psi^{\mathscr{A}}$ simply by the additional term $-\mathrm{i}e\varphi_{AB}\psi^{\mathscr{A}}$, and correspondingly for $\square_{A'B'}$, e.g.

$$\square_{AB}\psi^D = X_{ABC}{}^{D}\psi^C - \mathrm{i}e\varphi_{AB}\psi^D \tag{5.1.44}$$

$$\square_{A'B'}\psi^D = \Phi_{A'B'C}{}^{D}\psi^C - \mathrm{i}e\bar{\varphi}_{A'B'}\psi^D. \tag{5.1.45}$$

A *potential* Φ_a, for which

$$F_{ab} = \nabla_a\Phi_b - \nabla_b\Phi_a \tag{5.1.37}$$

relates to φ_{AB} according to

$$\varphi_{AB} = \nabla_{A'(A}\Phi_{B)}^{A'}. \tag{5.1.46}$$

If the Lorenz gauge condition

$$\nabla^a\Phi_a = 0 \tag{5.1.47}$$

holds, then (5.1.46) simplifies to

$$\varphi_{AB} = \nabla_{A'A}\Phi_B^{A'}. \tag{5.1.49}$$

With a source present

$$\nabla_a F^{ab} = 4\pi J^b. \tag{5.1.38}$$

The current J^a is real (and uncharged) and subject to

$$\nabla_a J^a = 0. \tag{5.1.54}$$

In spinor terms, (5.1.38) becomes

$$\nabla^{A'B}\varphi^A{}_B = 2\pi J^{AA'}. \tag{5.1.52}$$

When $J^a = 0$ this becomes another example (along with (4.4.61) and (4.10.9)) of the massless free-field equation (4.12.42), namely

$$\nabla^{AA'}\varphi_{AB} = 0. \tag{5.1.57}$$

The relation between the components $\varphi_{\mathbf{AB}}$ and those of the Maxwell field is, explicitly,

$$\begin{aligned}
\varphi_{00} &= \tfrac{1}{2}(F_{31} + F_{01} - \mathrm{i}F_{32} - \mathrm{i}F_{02}) = \tfrac{1}{2}(C_1 - \mathrm{i}C_2)\\
\varphi_{01} &= \tfrac{1}{2}(-F_{03} - \mathrm{i}F_{12}) = -\tfrac{1}{2}C_3\\
\varphi_{11} &= \tfrac{1}{2}(F_{31} - F_{01} + \mathrm{i}F_{32} - \mathrm{i}F_{02}) = -\tfrac{1}{2}(C_1 + \mathrm{i}C_2),
\end{aligned} \tag{5.1.59}$$

where the complex 3-vector $\boldsymbol{C}$ is related to the electric 3-vector $\boldsymbol{E}$ and the magnetic 3-vector $\boldsymbol{B}$ by

$$\boldsymbol{C} = \boldsymbol{E} - \mathrm{i}\boldsymbol{B}. \tag{5.1.60}$$

The well-known scalar invariants

$$P = B^2 - E^2, \quad Q = 2\boldsymbol{E}\cdot\boldsymbol{B} \tag{5.1.67}$$

arise spinorially as

$$K := \varphi_{AB}\varphi^{AB} = \tfrac{1}{2}{}^-F_{ab}{}^-F^{ab} = \tfrac{1}{2}F_{ab}{}^-F^{ab} = P + \mathrm{i}Q. \tag{5.1.68}$$

Sometimes we allow *complex* Maxwell fields, and then the $\bar{\varphi}_{A'B'}$ in (5.1.39) must be replaced by an independent quantity $\tilde{\varphi}_{A'B'}$. Accordingly

$$\tilde{K} := \tilde{\varphi}_{A'B'}\tilde{\varphi}^{A'B'} = \tfrac{1}{2}F_{ab}{}^+F^{ab} = P - \mathrm{i}Q \tag{5.1.69}$$

is independent of K, and we have, inversely

$$P = (K + \tilde{K})/2, \quad Q = (K - \tilde{K})/2\mathrm{i}. \tag{5.1.70}$$

A real field is purely electric [*or* purely magnetic] (i.e. Lorentz transformable to one for which $\boldsymbol{B} = 0$ [*or* $\boldsymbol{E} = 0$]) if $K < 0$ [*or* $K > 0$]. It is *simple* whenever K is real.

The electromagnetic energy tensor has the simple spinor form

$$T_{ab} = \frac{1}{2\pi}\varphi_{AB}\bar{\varphi}_{A'B'} \tag{5.2.4}$$

and if it acts as the only source of the gravitational field Einstein's equations (4.6.32) become

$$\Phi_{ABA'B'} = 2G\varphi_{AB}\bar{\varphi}_{A'B'}, \quad \Lambda = \tfrac{1}{6}\lambda \tag{5.2.6}$$

and the Bianchi identity (4.10.7) becomes

$$\nabla^A_{B'}\Psi_{ABCD} = 2G\bar{\varphi}_{A'B'}\nabla^{A'}_B\varphi_{CD}. \tag{5.2.7}$$

Yang–Mills fields

We use capital Greek letters for *bundle indices*. In particular, a Yang–Mills charged (YM-charged) field λ^{Φ} is so labelled. The symbol ∇_a now extends, when applied to such fields, to become a *bundle connection*, subject to

$$\nabla_a : \mathfrak{T}^{\Phi} \to \mathfrak{T}^{\Phi}_a \quad [or\ \mathfrak{S}^{\Phi} \to \mathfrak{S}^{\Phi}_a], \tag{5.4.17}$$

$$\nabla_a(\lambda^{\Phi} + \mu^{\Phi}) = \nabla_a\lambda^{\Phi} + \nabla_a\mu^{\Phi}, \tag{5.4.18}$$

$$\nabla_a(f\lambda^{\Phi}) = \lambda^{\Phi}\nabla_a f + f\nabla_a\lambda^{\Phi}, \quad f\in\mathfrak{T} \quad [or\ \mathfrak{S}], \tag{5.4.19}$$

where $\mathfrak{T}^{\Phi}$, in the case of a real vector bundle, or $\mathfrak{S}^{\Phi}$, in the case of a complex vector bundle, denotes the module of *cross-sections* of the bundle (paired with the abstract index Φ). The *bundle curvature* $K_{ab\Omega}{}^{\Phi}$ (in the torsion-free case; compare (4.2.31)) is defined by

$$K_{ab\Omega}{}^{\Phi}\lambda^{\Omega} = (\nabla_a\nabla_b - \nabla_b\nabla_a)\lambda^{\Phi}. \tag{5.4.23}$$

In the case of Yang–Mills (YM) fields we frequently assume a unitary (Hermitian) structure for the bundle (i.e. complex conjugation interchanges upper and lower capital Greek indices). In place of $K_{ab\Omega}{}^{\Phi}$ we use the (in that case Hermitian) $F_{ab\Omega}{}^{\Phi} = \mathrm{i}K_{ab\Omega}{}^{\Phi}$ which satisfies, e.g.

$$\Delta_{ab}\beta_{\Psi} = \mathrm{i}\beta_{\Theta}F_{ab\Psi}{}^{\Theta}. \tag{5.5.30}$$

Its bundle components are related to a YM potential $\Phi_{a\boldsymbol{\Theta}}{}^{\boldsymbol{\Psi}}$ (a matrix of covectors) by

$$\tfrac{1}{2}F_{ab\boldsymbol{\Theta}}{}^{\boldsymbol{\Psi}} = \nabla_{[a}\Phi_{b]\boldsymbol{\Theta}}{}^{\boldsymbol{\Psi}} - \mathrm{i}\Phi_{\boldsymbol{\Lambda}}{}^{\boldsymbol{\Psi}}{}_{[a}\Phi_{b]\boldsymbol{\Theta}}{}^{\boldsymbol{\Lambda}} \tag{5.5.28}$$

(allowing space–time indices to move across the YM component indices). Gauge transformations effect

$$\Phi_{a\boldsymbol{\Psi}}{}^{\boldsymbol{\Theta}} \mapsto \hat{\Phi}_{a\hat{\boldsymbol{\Psi}}}{}^{\hat{\boldsymbol{\Theta}}} = \Phi_{a\boldsymbol{\Psi}}{}^{\boldsymbol{\Theta}} r_{\boldsymbol{\Theta}}{}^{\hat{\boldsymbol{\Theta}}} q_{\hat{\boldsymbol{\Psi}}}{}^{\boldsymbol{\Psi}} + i r_{\boldsymbol{\Psi}}{}^{\hat{\boldsymbol{\Theta}}}\nabla_a q_{\hat{\boldsymbol{\Psi}}}{}^{\boldsymbol{\Psi}} \tag{5.5.25}$$

and

$$F_{ab\boldsymbol{\Theta}}{}^{\boldsymbol{\Psi}} \mapsto F_{ab\hat{\boldsymbol{\Theta}}}{}^{\hat{\boldsymbol{\Psi}}} = F_{ab\boldsymbol{\Theta}}{}^{\boldsymbol{\Psi}} r_{\boldsymbol{\Psi}}{}^{\hat{\boldsymbol{\Psi}}} q_{\hat{\boldsymbol{\Theta}}}{}^{\boldsymbol{\Theta}}, \tag{5.5.29}$$

but preserve the abstract curvature quantity $F_{ab\Theta}{}^{\Psi}$ (the YM *field*). The matrices $(r_{\boldsymbol{\Psi}}{}^{\hat{\boldsymbol{\Psi}}})$ and $(q_{\hat{\boldsymbol{\Psi}}}{}^{\boldsymbol{\Psi}})$ are inverses of one another (and also Hermitian conjugates in the unitary case).

In spinor terms we have a decomposition

$$F_{ab\Theta}{}^{\Psi} = \varphi_{AB\Theta}{}^{\Psi}\varepsilon_{A'B'} + \varepsilon_{AB}\chi_{A'B'\Theta}{}^{\Psi}, \tag{5.5.36}$$

where

$$\varphi_{AB\Theta}{}^{\Psi} = \varphi_{(AB)\Theta}{}^{\Psi} = \tfrac{1}{2}F_{ABC'}{}^{C'}{}_{\Theta}{}^{\Psi} \qquad (5.5.37)$$

$$\chi_{A'B'\Theta}{}^{\Psi} = \chi_{(A'B')\Theta}{}^{\Psi} = \tfrac{1}{2}F_{C}{}^{C}{}_{A'B'}{}^{\Psi}$$

(these being complex conjugates of one another in the unitary case). We have spinor 'Ricci identities'

$$\Box_{AB}\mu^{\Psi} = -\mathrm{i}\mu^{\Theta}\varphi_{AB\Theta}{}^{\Psi}, \quad \Box_{A'B'}\mu^{\Psi} = -\mathrm{i}\mu^{\Theta}\chi_{A'B'\Theta}{}^{\Psi},$$

$$\Box_{AB}\lambda_{\Psi} = \mathrm{i}\lambda_{\Theta}\varphi_{AB\Psi}{}^{\Theta}, \quad \Box_{A'B'}\lambda_{\Psi} = \mathrm{i}\lambda_{\Theta}\chi_{A'B'\Psi}{}^{\Theta}, \qquad (5.5.40)$$

and the expressions in terms of potentials

$$\varphi_{AB\Theta}{}^{\Psi} = \nabla_{A'(A}\Phi^{A'}_{B)\Theta}{}^{\Psi} - \mathrm{i}\Phi_{\Lambda}{}^{\Psi}{}_{A'(A}\Phi^{A'}_{B)\Theta}{}^{\Lambda},$$

$$\chi_{A'B'\Theta}{}^{\Psi} = \nabla_{A(A'}\Phi^{A}_{B')\Theta}{}^{\Psi} - \mathrm{i}\Phi_{\Lambda}{}^{\Psi}{}_{A(A'}\Phi^{A}_{B')\Theta}{}^{\Lambda}. \qquad (5.5.41)$$

The YM ('source-free') field equations

$$\nabla^{a}F_{ab\Theta}{}^{\Psi} = 0 \qquad (5.5.35)$$

together with the identity

$$\nabla_{[a}F_{bc]\Theta}{}^{\Psi} = 0 \qquad (5.5.34)$$

take the spinor form

$$\nabla^{B}_{A'}\varphi_{AB\Theta}{}^{\Psi} = 0 = \nabla^{B'}_{A}\chi_{A'B'\Theta}{}^{\Psi}. \qquad (5.5.44)$$

The *self-dual* and *anti-self-dual* parts of a YM field are, respectively

$$^{+}F_{ab\Theta}{}^{\Psi} = \varepsilon_{AB}\chi_{A'B'\Theta}{}^{\Psi} \qquad (5.5.49)$$

$$^{-}F_{ab\Theta}{}^{\Psi} = \varphi_{AB\Theta}{}^{\Psi}\varepsilon_{A'B'}. \qquad (5.5.50)$$

The YM field is said to be self-dual [anti-self-dual] if its anti-self-dual [self-dual] part vanishes—and then (5.5.35) follows from (5.5.34).

Conformal rescalings

We accompany a rescaling

$$g_{ab} \mapsto \hat{g}_{ab} = \Omega^{2}g_{ab} \qquad (5.6.1)$$

of the space–time metric, by the 'geometrically natural'

$$\varepsilon_{AB} \mapsto \hat{\varepsilon}_{AB} = \Omega\varepsilon_{AB} \qquad (5.6.2)$$

(together with its complex conjugate), where Ω is a nowhere vanishing (normally positive) real scalar field. ('Hatted' quantities have indices raised and lowered with 'hatted' gs and εs.) Defining

$$\Upsilon_{a} = \Omega^{-1}\nabla_{a}\Omega = \nabla_{a}\log\Omega \qquad (5.6.14)$$

we find that covariant derivative transforms as follows

$$\hat{\nabla}_{AA'}\chi_{B\ldots F'\ldots}^{P\ldots S'\ldots} = \nabla_{AA'}\chi_{B\ldots F'\ldots}^{P\ldots S'\ldots} - \Upsilon_{BA'}\chi_{A\ldots F'\ldots}^{P\ldots S'\ldots} - \cdots - \Upsilon_{AF'}\chi_{B\ldots A'\ldots}^{P\ldots S'\ldots} - \cdots$$
$$+ \varepsilon_A{}^P\Upsilon_{XA'}\chi_{B\ldots F'\ldots}^{X\ldots S'\ldots} + \cdots + \varepsilon_{A'}{}^{S'}\Upsilon_{AX'}\chi_{B\ldots F'\ldots}^{P\ldots X'\ldots} + \cdots. \quad (5.6.15)$$

Various different scalings may be applied to a dyad when accompanying (5.6.1), (5.6.2), so we write

$$\hat{o}^A = \Omega^{w_0}o^A, \quad \hat{\iota}^A = \Omega^{w_1}\iota^A, \quad \hat{\chi} = \Omega^{w_0+w_1+1}\chi. \qquad (5.6.2)$$

Setting

$$\omega = \log\Omega \qquad (5.6.23)$$

and

$$\Sigma = \Omega^{w_0 - w_1},$$

we find that the spin-coefficients transform as

$$\begin{bmatrix} \hat{\kappa} & \hat{\varepsilon} & \hat{\gamma}' & \hat{\tau}' \\ \hat{\rho} & \hat{\alpha} & \hat{\beta}' & \hat{\sigma}' \\ \hat{\sigma} & \hat{\beta} & \hat{\alpha}' & \hat{\rho}' \\ \hat{\tau} & \hat{\gamma} & \hat{\varepsilon}' & \hat{\kappa}' \end{bmatrix} = \qquad (5.6.25)$$

$$\Omega^{w_0+w_1} \times \begin{bmatrix} \kappa\Sigma^2 & [\varepsilon + (w_0+1)\mathrm{D}\omega]\Sigma & (\gamma' + w_1\mathrm{D}\omega)\Sigma & \tau' - \delta'\omega \\ (\rho - \mathrm{D}\omega)\Sigma & \alpha + w_0\delta'\omega & \beta' + (w_1+1)\delta'\omega & \sigma'\Sigma^{-1} \\ \sigma\Sigma & \beta + (w_0+1)\delta\omega & \alpha' + w_1\delta\omega & (\rho' - \mathrm{D}'\omega)\Sigma^{-1} \\ \tau - \delta\omega & (\gamma + w_0\mathrm{D}'\omega)\Sigma^{-1} & [\varepsilon' + (w_1+1)\mathrm{D}'\omega]\Sigma^{-1} & \kappa'\Sigma^{-2} \end{bmatrix}$$

In particular,

$$\kappa, \sigma, \kappa', \sigma' \textit{ are all conformal densities} \qquad (5.6.28)$$

(of respective weights $3w_0 - w_1$, $2w_0$, $3w_1 - w_0$, $2w_1$), and also

$$\tau - \bar{\tau}' \textit{ and the imaginary parts of } \rho, \rho', \varepsilon, \varepsilon', \gamma, \gamma'$$
$$\textit{are all conformal densities} \qquad (5.6.29)$$

(of respective weights $w_0 + w_1$, $2w_0$, $2w_1$, $2w_0$, $2w_1$, $2w_0$, $2w_1$), where we say that η has *conformal weight* w if η changes to

$$\hat{\eta} = \Omega^w\eta \qquad (5.6.32)$$

under (5.6.1), (5.6.2), (5.6.22). Suppose that η is also of type $\{r', r; t', t\}$ (i.e. undergoes (4.12.9) under (4.12.2)). Then we define the following operators by their effect on such η:

$$\text{þ}_c = \text{þ} + [w - r'(w_0+1) - rw_1 - t'(w_0+1) - tw_1]\rho$$
$$\text{þ}'_c = \text{þ}' + [w - r'w_0 - r(w_1+1) - t'w_0 - t(w_1+1)]\rho'$$
$$\eth_c = \eth + [w - r'(w_0+1) - rw_1 - t'w_0 - t(w_1+1)]\tau$$
$$\eth'_c = \eth' + [w - r'w_0 - r(w_1+1) - t'(w_0+1) - tw_1]\tau'. \qquad (5.6.33)$$

These enjoy a conformal invariance:

$$\begin{aligned} \text{þ}_c\eta &= \hat{\text{þ}}_c\hat{\eta} = \Omega^{w+2w_0}\text{þ}_c\eta \\ \text{þ}'_c\eta &= \hat{\text{þ}}'_c\hat{\eta} = \Omega^{w+2w_1}\text{þ}'_c\eta \\ \eth_c\eta &= \hat{\eth}_c\hat{\eta} = \Omega^{w+w_0+w_1}\hat{\eth}_c\eta \\ \eth'_c\eta &= \hat{\eth}'_c\hat{\eta} = \Omega^{w+w_0+w_1}\eth_c\eta. \end{aligned} \tag{5.6.34}$$

Using these operators we can simplify the appearance of the compacted equations (4.12.44) for the massless free-field equations to

$$\begin{aligned} \text{þ}_c\phi_r - \eth'_c\phi_{r-1} &= (r-1)\sigma'\phi_{r-2} - (n-r)\kappa\phi_{r+1}, \\ \eth_c\phi_r - \text{þ}'_c\phi_{r-1} &= (r-1)\kappa'\phi_{r-2} - (n-r)\sigma\phi_{r+1}, \end{aligned} \tag{5.6.37}$$

of the compacted equations (4.12.46) for the twistor equation to

$$\begin{aligned} \eth'_c\omega^0 &= \sigma'\omega', \quad \eth_c\omega^1 = \sigma\omega^0, \\ \text{þ}'_c\omega^0 &= \kappa'\omega^1, \quad \text{þ}_c\omega^1 = \kappa\omega^0, \\ \eth_c\omega^0 &= \text{þ}'_c\omega^1, \quad \text{þ}_c\omega^0 = \eth'_c\omega^1, \end{aligned} \tag{5.6.38}$$

and also of the equation (4.14.92), for the null hypersurface version of the fundamental theorem of exterior calculus, to

$$\oint_\Sigma \{\text{þ}'_c\mu_0 - \eth_c\mu_1\}\mathcal{N} = \oint_{\mathcal{S}'}\mu_0\mathcal{S} - \oint_{\mathcal{S}}\mu_0\mathcal{S}. \tag{5.6.40}$$

Massless free fields

We have already given the field equation (4.12.42) for a massless free field $\phi_{AB\ldots L} = \phi_{(AB\ldots L)}$ of spin $\frac{1}{2}n$ (n indices). The complex conjugate form

$$\nabla^{AA'}\theta_{A'B'\ldots L'} = 0, \quad \theta_{A'B'\ldots L'} = \theta_{(A'B'\ldots L')} \tag{5.7.3}$$

also describes a massless free field of spin $\frac{1}{2}n$, but there is the distinction that if we are considering *positive frequency* wave functions, then (5.7.3) describes *right-handed* massless particles (helicity $+\frac{1}{2}n\hbar$) whereas (4.12.42) describes left-handed ones (helicity $-\frac{1}{2}n\hbar$). We must remark, also, that in general these equations are satisfactory only in (conformally) flat space–time since the calculation

$$\begin{aligned} 0 = \nabla^B_{A'}\nabla^{AA'}\phi_{ABC\ldots L} &= \nabla^{(B}_{A'}\nabla^{A)A'}\phi_{ABC\ldots L} \\ &= \square^{AB}\phi_{ABC\ldots L} \\ &= -ie\varphi^{AB}\phi_{ABC\ldots L} - \mathrm{X}^{ABM}{}_A\phi_{MBC\ldots L} - \mathrm{X}^{ABM}{}_B\phi_{AMC\ldots L} \\ &\quad - \mathrm{X}^{ABM}{}_C\phi_{ABM\ldots L} - \cdots - \mathrm{X}^{ABM}{}_L\phi_{ABC\ldots M} \end{aligned} \tag{5.8.1}$$

yields the Fierz–Pauli–Buchdahl–Plebański constraint

$$(n-2)\phi_{ABM(C\ldots K}\Psi_{L)}{}^{ABM} = -ie\varphi^{AB}\phi_{ABC\ldots L}. \tag{5.8.2}$$

The case $n = 4$ in $\mathbb{M}$ is of considerable significance since it corresponds to the weak-field limit of general relativity. We take the space–time metric to be a smoothly varying function $g_{ab}(u)$ of a parameter u, where $g_{ab} = g_{ab}(0)$ is the Minkowski metric, with

$$g_{ab}(u) = g_{ab} + uh_{ab} + O(u^2).$$

We find

$$K_{abcd} := \lim_{u\to 0}(u^{-1}R_{abcd}(u)) = 2\nabla_{[a}\nabla_{|[c}h_{d]|b]} \tag{5.7.4}$$

for the weak-field curvature, the weak-field limit of Einstein's equations being

$$K_{abc}{}^{b} - \tfrac{1}{2}g_{ac}K_{bd}{}^{bd} = -8\pi G E_{ac} \tag{5.7.6}$$

(where E_{ab} is the weak-field energy–momentum tensor). K_{abcd} has the symmetries of the Riemann tensor, and in the absence of sources enjoys a spinor description like (4.6.41):

$$K_{abcd} = \phi_{ABCD}\varepsilon_{A'B'}\varepsilon_{C'D'} + \bar{\phi}_{A'B'C'D'}\varepsilon_{AB}\varepsilon_{CD}. \tag{5.7.8}$$

In the absence of sources, ϕ_{ABCD} satisfies the massless free-field equation (4.12.42) (*cf.* (4.10.9)) by virtue of the linearized Bianchi identity

$$\nabla_{[a}K_{bc]de} = 0. \tag{5.7.9}$$

In terms of the linearized metric h_{ab} we have

$$\phi_{ABCD} = \tfrac{1}{2}\nabla^{A'}_{(A}\nabla^{B'}_{B}h_{CD)A'B'}. \tag{5.7.12}$$

Behaviour under conformal rescalings

We can preserve the massless free-field equation (4.12.42) under conformal rescaling, for each n (compare McLennan 1956), by taking

$$\hat{\phi}_{AB\ldots L} = \Omega^{-1}\phi_{AB\ldots L} \tag{5.7.17}$$

since then (5.6.15) implies

$$\hat{\nabla}^{AA'}\hat{\phi}_{AB\ldots L} = \Omega^{-3}\nabla^{AA'}\phi_{AB\ldots L}. \tag{5.7.20}$$

Another conformally invariant equation is the vanishing divergence relation

$$\nabla^a T_{ab} = 0 \tag{5.9.1}$$

on a *trace-free* symmetric tensor $T_{ab} = T_{(AB)(A'B')}$, since if

$$\hat{T}_{ab} = \Omega^{-2}T_{ab}, \tag{5.9.2}$$

then by (5.6.15) we obtain

$$\hat{\nabla}^a\hat{T}_{ab} = \Omega^{-4}\nabla^a T_{ab}.$$

We have seen that the electromagnetic energy–momentum tensor (5.2.4) is of this type. So also is that for the Dirac–Weyl equation (4.4.61) (with ν_A for ϕ_A):

$$T_{ab} = k(\mathrm{i}\nu_{(A}\nabla_{B)A'}\bar{\nu}_{B'} - \mathrm{i}\bar{\nu}_{(A'}\nabla_{B')A}\nu_B) \tag{5.8.3}$$

(k some real constant). The verification of (5.9.1) for (5.8.3) in a curved space–time is facilitated by use of the following identity, needed later:

$$\nabla^A_{A'}\nabla_{B'(A}\xi_{B)} = \nabla_{B'B}\nabla^A_{A'}\xi_A - \tfrac{1}{2}\nabla_{A'B}\nabla^A_{B'}\xi_A + 3\Lambda\varepsilon_{A'B'}\xi_B - \Phi_{ABA'B'}\xi^A. \tag{5.8.4}$$

The charge-current conservation law (5.1.54) is also conformally invariant, with

$$\hat{J}_a = \Omega^{-2} J_a. \tag{5.9.3}$$

This follows directly from the fact that (5.1.54) simply asserts that the exterior derivative of ${}^\dagger\boldsymbol{J}$ vanishes, where (*cf.* (3.4.29))

$${}^\dagger\boldsymbol{J} = {}^\dagger J_{i_1 i_2 i_3} = e_{i_1 i_2 i_3 a} J^a, \tag{5.9.5}$$

since

$$\mathrm{d}{}^\dagger\boldsymbol{J} = \nabla_{[i_1}{}^\dagger J_{i_2 i_3 i_4]} = -\tfrac{1}{4}(\nabla_a J^a) e_{i_1 i_2 i_3 i_4}. \tag{5.9.6}$$

Note that

$$\hat{e}_{abcd} = \Omega^4 e_{abcd} \tag{5.9.7}$$

so that (5.9.3) is compatible with ${}^\dagger\hat{\boldsymbol{J}} = {}^\dagger\boldsymbol{J}$. The natural scaling for $\boldsymbol{F} = F_{i_1 i_2}$ as a 2-form, namely $\hat{\boldsymbol{F}} = \boldsymbol{F}$, is also compatible with the Maxwell equations with *source* (5.1.38),

$$\mathrm{d}\boldsymbol{F} = 0, \quad \mathrm{d}\,{}^*\boldsymbol{F} = \frac{4\pi}{3}{}^\dagger\boldsymbol{J}, \tag{5.9.13}$$

so (5.1.52) is also conformally invariant with this, the standard massles field scaling (5.7.17):

$$\hat{\varphi}_{AB} = \Omega^{-1}\varphi_{AB}, \quad \text{i.e.,}\ \hat{\varphi}^{AB} = \Omega^{-3}\varphi^{AB}. \tag{5.9.8}$$

Various other field equations

The *wave equation*

$$\Box\phi := \nabla^{AA'}\nabla_{AA'}\phi = 0 \tag{5.10.6}$$

can be restated as the spinor property

$$\nabla^{A'}_{[A}\nabla^{B'}_{B]}\phi = 0, \tag{5.10.7}$$

i.e. as the fact that $\nabla_{AA'}\nabla_{BB'}\phi$ is a symmetric spinor (in the sense of Proposition (3.3.62)). In $\mathbb{M}$, any massless free field (including ϕ subject to (5.10.6)) has the property that *any* number of uncontracted derivatives will always yield *symmetric* spinors. Let such a differentiated spinor be $\theta^{P' \ldots S'}_{A \ldots E}$

(with appropriate index positioning). Then

$$\nabla^{AA'}\theta^{P'\ldots S'}_{AB\ldots E}=0,\quad \nabla_{PP'}\theta^{P'Q'\ldots S'}_{A\ldots E}=0. \tag{5.10.9}$$

Conversely (see (3.3.14)):

(5.10.10) PROPOSITION

If (5.10.9) *holds, then* $\theta^{P'\ldots S'}_{A\ldots E}\in\mathfrak{S}^{(P'\ldots S')}_{(A\ldots E)}$ *is an* rth *derivative of some massless free field.*

The *Dirac equation* for the electron (spin $\frac{1}{2}$) may be written in the 2-spinor form

$$\nabla^{AA'}\psi_A=\mu\chi^{A'},\quad \nabla_{AA'}\chi^{A'}=-\mu\psi_A, \tag{5.10.15}$$

where $\hbar\mu\sqrt{2}$ is the mass. (The difference in choice of index positioning between ψ_A and $\chi^{A'}$ has no significance.) The *Schrödinger–Klein–Gordon* equation (spin 0) is

$$(\Box+2\mu^2)\theta=0. \tag{5.10.20}$$

The (Dirac) equation for higher spin (in $\mathbb{M}$, free of electromagnetic field) is the coupled pair

$$\nabla^{AP'}\psi^{Q'\ldots T'}_{AB\ldots D}=\mu\chi^{P'Q'\ldots T'}_{B\ldots D},\ \nabla_{AP'}\chi^{P'Q'\ldots T'}_{B\ldots D}=-\mu\psi^{Q'\ldots T'}_{AB\ldots D} \tag{5.10.35}$$

the spin being one half the total number of indices of each spinor, these spinors being assumed symmetric (with the mass as before). This symmetry implies that the 'subsidiary conditions'

$$\nabla^A_{Q'}\psi^{Q'R'\ldots T'}_{AB\ldots D}=0,\quad \nabla^B_{P'}\chi^{P'Q'\ldots T'}_{B\ldots D}=0 \tag{5.10.36}$$

hold. Moreover, each of $\psi^{\cdots}_{\cdots}$, $\chi^{\cdots}_{\cdots}$ satisfies (5.10.20).

Null hypersurface data

A system of spinor fields, subject to field equations and derivative commutator equations, is said to form an *exact set* if all the symmetrized derivatives can be specified arbitrarily at any one point and if all unsymmetrized derivatives can be expressed in terms of the symmetrized ones. The standard 'consistent' systems (e.g. the Einstein–Maxwell–Dirac system) can all be put into the form of exact sets. Let $\mathcal{N}$ be a null hypersurface, ξ^A being defined over $\mathcal{N}$ with flagpole pointing along the generators of $\mathcal{N}$. If $\psi_{A\ldots EP'\ldots S'}$ is a field belonging to the given system, then its *null-datum at a point* $X\in\mathcal{N}$ is

$$\psi=\xi^A\ldots\xi^E\bar{\xi}^{P'}\ldots\bar{\xi}^{S'}[\psi_{A\ldots EP'\ldots S'}]_X. \tag{5.11.11}$$

The exact-set condition asserts that, for a light cone (in an analytic space-time $\mathscr{M}$), the null-data, on $\mathscr{N}$, for all the various fields of the system, form a complete irredundant (constraint-free) initial data set.

The simplest type of exact set is a single massless free field $\phi_{A\ldots L}$ in $\mathbb{M}$. Let $\mathscr{N}$ be a general (but suitable) null hypersurface in $\mathbb{M}$. Then the *generalized Kirchhoff–d'Adhémar formula*

$$\phi_{AB\ldots L}(P) = \frac{1}{2\pi}\oint_{\mathscr{S}} \eta_A \eta_B \ldots \eta_L \frac{\text{þ}_c \phi}{r}\mathscr{S} \tag{5.12.6}$$

can be used to express the field at a point $P \in \mathbb{M}$ in terms of the null-datum ϕ at the points Q on the intersection, $\mathscr{S}$, of the light cone $\mathscr{C}$ of P with $\mathscr{N}$. As Q varies, the flagpoles of η^A point along the generators of $\mathscr{N}$, and r is defined by

$$(\overrightarrow{QP})^a = r\eta^A \bar{\eta}^{A'}. \tag{5.12.2}$$

The operator þ_c, is given by (5.6.33) (1), where the spin-frame is

$$o^A = \xi^A, \quad \iota^A = -\eta^A \tag{5.12.7}$$

(so that $\xi_A \eta^A = 1$), and we have, explicitly here,

$$\text{þ}_c = \text{þ} - (n+1)\rho = D - n\varepsilon - (n+1)\rho. \tag{5.12.8}$$

$\mathscr{S}$ is the 2-form element of surface area of $\mathscr{S}$ at Q, as before.

To prove (5.12.6) we fix $\mathscr{C}$ and vary $\mathscr{S}$, showing first that the RHS is independent of the cross-section $\mathscr{S}$ of $\mathscr{C}$ that is chosen, and then taking the limit as $\mathscr{S}$ shrinks down to P. The independence of cross-section depends upon (4.14.92) (complex conjugated, and with $\mathscr{C}$ in place of $\mathscr{N}$) and on the vanishing of A, as defined in

$$A := (\text{þ}' - 2\bar{\rho}')\mu_{0'} - (\eth' - \bar{\tau})\mu_1 = 0. \tag{5.12.17}$$

Here

$$\mu_{0'} = \omega^n r^{-1} \text{þ}_c \phi \tag{5.12.16}$$

and

$$\mu_{1'} = \omega^n r^{-1}(\eth\phi - (n+1)\tau\phi + n\sigma\phi_1) \tag{5.12.18}$$

with $\phi_1 = \phi_{10\ldots 0}$ and $\omega = \omega^A \eta_A$, where ω^A is constant in $\mathbb{M}$. If desired, we can replace $\omega^n r^{-1}$ by a $\{-n-1, -1\}$-scalar Γ, subject (when $\rho' = 0$) to

$$\text{þ}_c'\Gamma = 0 = \eth_c'\Gamma; \quad \text{i.e., } \text{þ}'\Gamma = 0, \quad \eth'\Gamma = 0. \tag{5.12.32}$$

This concludes our summary of the relevant portions of Volume 1. That volume was concerned mainly with the setting up of the basic 2-spinor formalism and geometry, and with providing spinor descriptions of the most familiar physical fields. In what follows we shall enter more thoroughly into certain important questions of space–time geometry where spinor techniques have proved particularly valuable. We shall treat asymptotics and energy–momentum in general relativity, the detailed classification of space–time curvature and the geometry of null rays. As a major part of our development we shall also need to provide a basic introduction to the powerful techniques of twistor theory. That is where we begin our discussion.

6
Twistors

6.1 The twistor equation and its solution space

At various places in Volume 1 we stressed the fact that the two-component spinor calculus is a very specific calculus for studying the structure of space–time manifolds. Indeed, the four-dimensionality and $(+ - - -)$ signature of space–time, together with the desirable global properties of orientability, time-orientability, and existence of spin structure, may all, in a sense, be regarded as *derived* from two-component spinors, rather than just given. However, at this stage there is still only a limited sense in which these properties can be so regarded, because the manifold of space–time points itself has to be given beforehand, even though the nature of this manifold is somewhat restricted by its having to admit the appropriate kind of spinor structure. If we were to attempt to take totally seriously the philosophy that all the space–time concepts are to be derived from more primitive spinorial ones, then we would have to find some way in which the space–time points themselves can be regarded as derived objects.

Spinor algebra by itself is not rich enough to achieve this, but a certain extension of spinor algebra, namely twistor algebra, can indeed be taken as more primitive than space–time itself. Moreover, it is possible to use twistors to build up other physical concepts directly, without the need to pass through the intermediary of space–time points. The programme of twistor theory, in fact, is to reformulate the whole of basic physics in twistor terms. The concepts of space–time points and curvature, of energy-momentum, angular momentum, quantization, the structure of elementary particles with their various internal quantum numbers, wave functions, space–time fields (incorporating their possibly nonlinear interactions), can all be formulated – with varying degrees of speculativeness, completeness, and success – in a more or less direct way from primitive twistor concepts.

Twistor theory has, however, become rather mathematically elaborate. To cover in any thorough and comprehensible way all the above-mentioned aspects of the theory would in itself require a book considerably larger than the present volume. (Some of these topics are to be covered in a forthcoming book by Ward and Wells, 1986.) In any case, to appreciate twistors fully and to be able to calculate with them, one must first study spinor theory, somewhat along the lines that we have followed in Volume 1. Thus we do not attempt to be in any way complete in our description of twistor theory. The account given here will serve as an extended (perhaps somewhat lop-sided) introduction to the subject. We shall develop in detail mainly that part of twistor theory which relates to spinorial descriptions of twistors, and indicate some of their profound connections with energy-momentum/angular momentum and with massless fields. We do not enter into much discussion of how twistors may be regarded as more primitive than space–time points, nor do we discuss quantization at any length, or particle theory, or give much detail of the treatment of nonlinear fields. In this chapter, apart from giving a discussion of local twistors in general curved space–time (local twistors in fact lying somewhat outside the main development of twistor theory), we shall restrict our account of twistors almost entirely to Minkowski space–time $\mathbb{M}$, though many interrelations with curved space–time properties will be given. In §7.4 we indicate how certain twistor ideas can be applied in general space–time $\mathcal{M}$ (particularly in relation to a hypersurface $\mathcal{H} \subset \mathcal{M}$), and in §9.5 we show how twistors can be used in cosmological models, while in §9.9 we introduce the concept of 2-*surface twistors* and show how our flat-space discussion of §§6.3–6.5 can be adapted to a curved-space context and suggest a quasi-local (and asymptotic) definition of mass–momentum–angular momentum surrounded by a 2-surface (the asymptotic mass–momentum agreeing with the standard definition of Bondi–Sachs). However we have had to omit a good deal of the detailed theory of *general* curved-space twistor theory. The twistorial description of space–time curvature is one of the more elaborate and sophisticated (and, indeed, remarkable – though incomplete) parts of twistor theory and, regrettably, it must remain outside the scope of the present work. (See Penrose 1976*a*, 1979*a*, Hansen, Newman, Penrose and Tod 1978, Tod 1980, Tod and Ward 1979, Ward 1978, Penrose and Ward 1980, Porter 1982, Hitchin 1979, Atiyah, Hitchin and Singer 1978, Wells 1982 for details.) There is much that is striking and illuminating even in the application of twistor theory to the weak-field limit of general relativity, *cf.* §§6.4, 6.5. This will be of crucial significance also for the curved-space discussion of §9.9. (See also Huggett and Tod 1985).

The twistor equation

Our point of departure is the equation (*cf.* (4.12.46), (5.6.38)*)

$$\nabla_{A'}^{(A}\omega^{B)} = 0, \tag{6.1.1}$$

* Equations, propositions, etc. of Volume 1 (i.e. in Chapters 1–5) which are referred to in Volume 2 are all to be found in the preceding summary (except for a few parenthetic references distinguished by the explicit mention of 'Volume 1' when cited).

called the *twistor equation* (Penrose 1967*a*; *cf.* also Gårding 1945)*. We begin by investigating its formal properties, leaving its physical and geometrical significance to later sections. First, we easily prove it is conformally invariant. Choosing

$$\hat{\omega}^B = \omega^B, \tag{6.1.2}$$

we get, from (5.6.15), (5.6.14)

$$\hat{\nabla}_{AA'}\hat{\omega}^B = \nabla_{AA'}\omega^B + \varepsilon_A{}^B \Upsilon_{CA'}\omega^C, \tag{6.1.3}$$

whence

$$\hat{\nabla}_{A'}^{(A}\hat{\omega}^{B)} = \Omega^{-1}\nabla_{A'}^{(A}\omega^{B)}. \tag{6.1.4}$$

Thus, conformal invariance is established.

There is a severe consistency condition for (6.1.1) in curved space–time, analogous to (5.8.2). For we have (*cf.* (4.9.2), (4.9.11), (4.6.35), (5.1.44))

$$\nabla^{A'(C}\nabla_{A'}^{A}\omega^{B)} = -\square^{(CA}\omega^{B)} = -\Psi^{CA}{}_D{}^B\omega^D - ie\varphi^{(CA}\omega^{B)}, \tag{6.1.5}$$

allowing for the presence of an electromagnetic field, and the possibility that ω^B has charge e. Thus (6.1.1) implies

$$\Psi_{ABCD}\omega^D = -ie\varphi_{(AB}\omega_{C)}, \tag{6.1.6}$$

which is the analogue of (5.8.2). If $\omega^D \neq 0$ and $e = 0$, we see, by reference to Proposition (3.5.26), that ω^D is a four-fold principal spinor of Ψ_{ABCD}. Thus, non-zero uncharged solutions of the twistor equation can exist only at points where Ψ_{ABCD} is either zero or 'null' (i.e. possessing a four-fold principal spinor). If $e \neq 0$, the situation is no better. In view of these difficulties our discussion of (6.1.1) in this chapter will be restricted to conformally flat space–times (characterized by $\Psi_{ABCD} = 0$; see §§6.8,9), and most of our calculations will actually be done in Minkowski space $\mathbb{M}$. Their extension to conformally flat space then follows from conformal invariance. (For extensions to arbitrary curved space: local twistors, hypersurface twistors and 2-surface twistors, see §6.9, §7.4 and §9.9, respectively.) Even in flat space, (6.1.6) has no solutions other than zero if $e \neq 0$ and φ_{AB} is somewhere non-vanishing. *So we assume, from now on, that unless the contrary is stated all fields are uncharged.*

In Minkowski space, equation (6.1.1) indeed possesses non-trivial solutions. We shall now find these explicitly. We choose an arbitrary origin O in $\mathbb{M}$ and label points by their position vectors x^a relative to O. We regard x^a as a vector field on $\mathbb{M}$. At O it is zero, and everywhere it satisfies

$$\nabla_a x^b = g_a{}^b, \tag{6.1.7}$$

* A version of this equation, written in terms of γ-matrices, was found by Wess and Zumino (1974) in connection with supersymmetry theory. See Appendix to this volume: (B.94), (B.95).

since in standard Minkowski coordintes, $x^{\mathbf{a}}$, the components of x^a at a point, are the coordinates of that point. Now consider

$$\nabla^{A}_{A'}\nabla^{B}_{B'}\omega^{C}, \tag{6.1.8}$$

ω^C being a solution of (6.1.1). The expression is therefore skew in BC. But since $\mathbb{M}$ is flat, we can commute the derivatives and then the expression is seen to be skew in AC. It is therefore totally skew in ABC and so must vanish. This tells us that $\nabla^{B}_{B'}\omega^C$ is constant. Since it is skew, it must therefore be a constant multiple of ε^{BC}, say $-\mathrm{i}\pi_{B'}\varepsilon^{BC}$ for some constant spinor $\pi_{B'}$. (The factor $-\mathrm{i}$ is inserted for later convenience.) So we have

$$\nabla_{BA'}\omega^{C} = -\mathrm{i}\varepsilon_{B}{}^{C}\pi_{A'}. \tag{6.1.9}$$

Integrating this equation gives $\omega^C = x^{BA'}(-\mathrm{i}\varepsilon_B{}^C\pi_{A'}) + \text{constant}$, as can be seen by writing it in coordinate form, and so we find

$$\left.\begin{aligned} \omega^{A} &= \mathring{\omega}^{A} - \mathrm{i}x^{AA'}\mathring{\pi}_{A'} \\ \pi_{A'} &= \mathring{\pi}_{A'} \end{aligned}\right\} \tag{6.1.10}$$

where $\mathring{\omega}^A$ and $\mathring{\pi}_{A'}$ are to be understood as follows: since ω^A is a spinor *field*, the RHS of (6.1.10) (1) must be regarded as a spinor field also. This can be done by regarding $\mathring{\omega}^A$ and $\mathring{\pi}_{A'}$ as *constant spinor fields* whose values coincide with those of ω^A and $\pi_{A'}$, respectively, at the origin. A similar convention should be understood whenever we write a point symbol over a spinor kernel. (The symbol '∘' above $\pi_{A'}$ is, of course, redundant here, but it makes what follows more consistent.)

Twistor space

As in the case of all (complex) linear differential equations, the solutions of the twistor equation constitute a vector space over the complex numbers, with scalar multiplication and addition of solutions defined in the obvious way. In the case of a general linear equation this vector space is often infinite-dimensional. It is clear from (6.1.10), however, that the solutions ω^A of the twistor equation are fully determined by the four complex components at O of ω^A and $\pi_{A'}$ in a spin-frame at O. These solutions ω^A therefore constitute a *four-dimensional* vector space $\mathbb{T}^\alpha$ over the complex numbers called *twistor space* (and they thus have eight real degrees of freedom). The elements of twistor space are called $\left[\begin{smallmatrix}1\\0\end{smallmatrix}\right]$-twistors, and we shall usually denote them by sans-serif capital kernel symbols with small Greek (four-dimensional) abstract indices, e.g., Z^α. If we denote the particular solution ω^A of (6.1.1) by Z^α we express this as follows:

$$\mathsf{Z}^{\alpha} = [\omega^{A}]. \tag{6.1.11}$$

Multiplication by a complex number and addition of twistors are defined in the obvious way:

$$\lambda[\omega^A] = [\lambda\omega^A] \quad (\lambda \in \mathbb{C}), \quad [\omega^A] + [\xi^A] = [\omega^A + \xi^A]. \qquad (6.1.12)$$

From these $[{}^1_0]$-twistors we can build up twistors of arbitrary valence $[{}^p_q]$ according to the standard rules of constructing tensor systems such as given in Chapter 2.

Thus we have abstract-index copies $\mathbb{T}^\beta, \mathbb{T}^\gamma, \ldots$, of $\mathbb{T}^\alpha$, and other spaces $\mathbb{T}_\alpha$, $\mathbb{T}_\beta^{\gamma\delta}, \ldots$ It turns out, however, that higher-valence twistors cannot in general be represented by single fields of spinors. In order to make the algebra of higher-valence twistors more systematic and manageable in terms of their spinor-field descriptions, it is much more convenient to use the *pair* of spinor fields ω^A, $\pi_{A'}$ to represent Z^α rather than to use ω^A alone. When concerned with such descriptions we shall, as an alternative to (6.1.11), write*

$$Z^\alpha = (\omega^A, \pi_{A'}), \qquad (6.1.13)$$

where ω^A and $\pi_{A'}$ are related by (6.1.9) (or, equivalently, by (6.1.10)). But, unlike (6.1.11), the description (6.1.13) is not conformally invariant. (However, see §6.9.)

Since knowing ω^A is fully equivalent to knowing the constant spinors $\mathring{\omega}^A$ and $\mathring{\pi}_{A'}$ (*cf.* (6.1.9), (6.1.10)), we can also represent the field ω^A and hence Z^α by the *values* $\omega^A(O)$ and $\pi_{A'}(O)$ of the spinor fields ω^A and $\pi_{A'}$ at O. We then write

$$Z^\alpha \overset{O}{\leftrightarrow} (\omega^A(O), \pi_{A'}(O)), \qquad (6.1.14)$$

the symbol $\overset{O}{\leftrightarrow}$ reminding us that the correspondence (6.1.14) is not Poincaré-invariant, but depends on the choice of a particular space–time origin O. Occasionally we shall use the notation

$$Z^A = \omega^A, \quad Z_{A'} = \pi_{A'}, \qquad (6.1.15)$$

where ω^A, $\pi_{A'}$ could be either spinor fields (description (6.1.13)) or point-spinors at O (description (6.1.14)). They are, in either case, called the 'spinor parts' of Z^α ('at O'). By (6.1.14) and (6.1.10) we have

$$\lambda(\omega^A, \pi_{A'}) = (\lambda\omega^A, \lambda\pi_{A'}), \quad (\omega^A, \pi_{A'}) + (\xi^A, \eta_{A'}) = (\omega^A + \xi^A, \pi_{A'} + \eta_{A'}). \qquad (6.1.16)$$

* Any temptation to identify the twistor (6.1.13) with a Dirac spinor (*cf.* (5.10.15) and the Appendix) should be resisted. Though there is a certain formal resemblance *at one point*, the vital twistor dependence (6.1.10) on position has no place in the Dirac formalism.

One might choose to regard $\mathbb{T}^\alpha$ (non-Poincaré-invariantly) as the direct sum $\mathfrak{S}^A[O] \oplus \mathfrak{S}_{A'}[O]$ of the spaces of spinors of type ω^A and $\pi_{A'}$ at O. (We recall that $\mathfrak{S}^{\mathscr{A}}[P]$ is the complex vector space of spinors of index type $\mathscr{A}$ at the point P.) The index α of Z^α would then be viewed as a kind of direct sum of the abstract index A and the abstract index A' in reverse position. (Note that this is quite different from the 'clumping' of §2.2.) In practice we can often treat A and A' as the two 'values' taken by α. When viewed in this way, α is somewhat intermediate between being fully abstract and being numerical. Note that the 'components' (6.1.15) corresponding to these 'values' A and A' of α get transformed when we change the origin (*cf.* (6.1.10)), so we prefer not to adopt this view formally.

If we choose an arbitrary spin-frame (o^A, ι^A) at O, we may construct a twistor basis as follows:

$$\delta_0^\alpha \overset{O}{\leftrightarrow} (o^A, 0) \qquad \delta_1^\alpha \overset{O}{\leftrightarrow} (\iota^A, 0)$$

$$\delta_2^\alpha \overset{O}{\leftrightarrow} (0, -\iota_{A'}) \qquad \delta_3^\alpha \overset{O}{\leftrightarrow} (0, o_{A'}). \tag{6.1.17}$$

The linear independence of these twistors is manifest. Now, since

$$(\omega^A(O), \pi_{A'}(O)) = (\omega^0(O) o^A + \omega^1(O)\iota^A, \pi_{1'}(O) o_{A'} - \pi_{0'}(O)\iota_{A'}), \tag{6.1.18}$$

it is evident from (6.1.12), (6.1.16), and (6.1.17) that

$$Z^\alpha = Z^{\boldsymbol{\alpha}} \delta_{\boldsymbol{\alpha}}^\alpha, \tag{6.1.19}$$

with

$$Z^{\boldsymbol{\alpha}} = (\omega^0(O), \omega^1(O), \pi_{0'}(O), \pi_{1'}(O)), \quad \boldsymbol{\alpha} = 0, 1, 2, 3. \tag{6.1.20}$$

From this and (6.1.15) we have the following explicit equations:

$$Z^0 = \omega^0(O), \quad Z^1 = \omega^1(O), \quad Z^2 = \pi_{0'}(O) = Z_{0'}, \quad Z^3 = \pi_{1'}(O) = Z_{1'}. \tag{6.1.21}$$

So Z^0 and Z^1 can be consistently interpreted *either* as the 0, 1 components of Z^α *or* as the components of the spinor part Z^A at O. We shall make the convention that components of spinor parts of any twistor (unless otherwise stated) are always to be evaluated at the origin.

Dual twistors

Since $[^1_0]$-twistor space is effectively the direct sum of the spaces of spinors of type ω^A and $\pi_{A'}$ at O, the dual, $[^0_1]$-twistor space, must effectively be the direct sum of spaces of spinors of type λ_A, $\mu^{A'}$ at O. Typically, we may write

$$W_\alpha \overset{O}{\leftrightarrow} (\lambda_A(O), \mu^{A'}(O)) \tag{6.1.22}$$

and the scalar product must then be defined as

$$\mathrm{W}_\alpha \mathrm{Z}^\alpha = \lambda_A(O)\omega^A(O) + \mu^{A'}(O)\pi_{A'}(O). \tag{6.1.23}$$

Analogously to (6.1.13) we wish to represent $[{}^0_1]$-twistors by two spinor *fields* λ_A and $\mu^{A'}$, so that the dependence on the origin O is removed. We write

$$\mathrm{W}_\alpha = (\lambda_A, \mu^{A'}), \tag{6.1.24}$$

(6.1.22) giving the values at O. We require (6.1.23) to hold not just at O, but at every point of $\mathbb{M}$:

$$\begin{aligned}\lambda_A\omega^A + \mu^{A'}\pi_{A'} = \mathrm{W}_\alpha \mathrm{Z}^\alpha &= \lambda_A(O)\omega^A(O) + \mu^{A'}(O)\pi_{A'}(O)\\ &= \mathring{\lambda}_A\mathring{\omega}^A + \mathring{\mu}^A\mathring{\pi}_{A'},\end{aligned} \tag{6.1.25}$$

where, as before, $\mathring{\lambda}_A$ and $\mathring{\mu}^{A'}$ are constant spinors whose values at O are $\lambda_A(O)$ and $\mu^{A'}(O)$, respectively. Substituting (6.1.10) into this equation yields

$$\lambda_A(\mathring{\omega}^A - \mathrm{i}x^{AA'}\mathring{\pi}_{A'}) + \mu^{A'}\mathring{\pi}_{A'} = \mathring{\lambda}_A\mathring{\omega}^A + \mathring{\mu}^{A'}\mathring{\pi}_{A'}.$$

And since this relation must hold for arbitrary constant $\mathring{\omega}^A$, $\mathring{\pi}_{A'}$, the 'coefficients' of these spinors must be equal; this gives the following form for the fields λ_A, $\mu^{A'}$:

$$\begin{aligned}\lambda_A &= \mathring{\lambda}_A,\\ \mu^{A'} &= \mathring{\mu}^{A'} + \mathrm{i}x^{AA'}\mathring{\lambda}_A.\end{aligned} \tag{6.1.26}$$

We can verify at once from (6.1.26) that the field $\mu^{A'}$ satisfies (and is, in fact, the general solution of) the conjugate twistor equation

$$\nabla_A^{(A'}\mu^{B')} = 0, \tag{6.1.27}$$

and that, analogously to (6.1.9), λ_A can be obtained from $\mu^{A'}$ by

$$\nabla_{AA'}\mu^{B'} = \mathrm{i}\varepsilon_{A'}{}^{B'}\lambda_A. \tag{6.1.28}$$

Thus, the λ_A in (6.1.24) is redundant and W_α is fully determined by $\mu^{A'}$. In fact, we can, alternatively to (6.1.24), identify W_α with $\mu^{A'}$ (*cf.* (6.1.11)) and write

$$\mathrm{W}_\alpha = [\mu^{A'}], \tag{6.1.29}$$

this being conformally invariant, like (6.1.11), though less convenient than (6.1.24) for building up twistors of higher valences.

It is worth noting the form which the inner product $\mathrm{W}_\alpha \mathrm{Z}^\alpha$ takes directly in terms of the spinor fields ω^A and μ^A. We need only substitute from (6.1.9) and (6.1.28) into (6.1.25) to find

$$[\mu^{A'}]\cdot[\omega^A] := \mathrm{W}_\alpha \mathrm{Z}^\alpha = \tfrac{1}{2}\mathrm{i}(\mu^{A'}\nabla_{BA'}\omega^B - \omega^A\nabla_{AB'}\mu^{B'}). \tag{6.1.30}$$

Each solution $\mu^{A'}$ of (6.1.27) can be obtained by simply complex-conjugating a solution of (6.1.1): $\omega^A \mapsto \mu^{A'} = \bar{\omega}^{A'}$. This is evident from a simple inspection of the two differential equations, or, alternatively, from an inspection of their respective general solutions (6.1.10), (6.1.26). This suggests that we identify the $[^0_1]$-twistors W_α with the complex conjugates of the $[^1_0]$-twistors Z^α, and vice versa. Consequently we define

$$\overline{\mathrm{Z}^\alpha} = \bar{\mathrm{Z}}_\alpha := (\bar{\pi}_A, \bar{\omega}^{A'}),$$
$$\overline{\mathrm{W}_\alpha} = \bar{\mathrm{W}}^\alpha := (\bar{\mu}^A, \bar{\lambda}_{A'}). \tag{6.1.31}$$

Note that the choice of the arbitrary factor in (6.1.9) as $-\mathrm{i}$ enables us to pass via complex conjugation from (6.1.9) to (6.1.28), and from (6.1.10) to (6.1.26). The complex conjugation of an *arbitrary* twistor will be discussed presently.

Analogously to (6.1.15), we sometimes write

$$\mathrm{W}_A = \lambda_A, \quad \mathrm{W}^{A'} = \mu^{A'}, \tag{6.1.32}$$

(either as spinor fields or as point-spinors at O) for the spinor parts of W_α.

We shall want to define a dual basis $\delta^{\mathbf{a}}_\alpha$ to (6.1.17). It must satisfy

$$\delta^{\mathbf{a}}_\alpha \delta^\alpha_{\mathbf{b}} = \delta^{\mathbf{a}}_{\mathbf{b}}, \tag{6.1.33}$$

and this is easily verified for

$$\delta^0_\alpha \overset{O}{\leftrightarrow} (-\iota_A, 0) \qquad \delta^1_\alpha \overset{O}{\leftrightarrow} (o_A, 0)$$

$$\delta^2_\alpha \overset{O}{\leftrightarrow} (0, o^{A'}) \qquad \delta^3_\alpha \overset{O}{\leftrightarrow} (0, \iota^{A'}). \tag{6.1.34}$$

We then find that

$$\mathrm{W}_\alpha = \mathrm{W}_{\mathbf{a}} \delta^{\mathbf{a}}_\alpha, \tag{6.1.35}$$

with

$$\mathrm{W}_{\mathbf{a}} = (\lambda_0(O), \lambda_1(O), \mu^{0'}(O), \mu^{1'}(O)), \quad \mathbf{a} = 0, 1, 2, 3. \tag{6.1.36}$$

Explicitly, from this and (6.1.32), we have

$$\mathrm{W}_0 = \lambda_0(O), \quad \mathrm{W}_1 = \lambda_1(O), \quad \mathrm{W}_2 = \mu^{0'}(O) = \mathrm{W}^{0'}, \quad \mathrm{W}_3 = \mu^{1'}(O) = \mathrm{W}^{1'}. \tag{6.1.37}$$

(Compare the remarks after (6.1.21).)

Higher-valence twistors

Now consider the outer product $\mathrm{X}^\alpha \mathrm{Z}^\beta$ of two twistors such as

$$\mathrm{X}^\alpha = (\xi^A, \eta_{A'}), \quad \mathrm{Z}^\alpha = (\omega^A, \pi_{A'}), \tag{6.1.38}$$

represented at O by

$$\mathrm{X}^\alpha \overset{O}{\leftrightarrow} (\xi^A(O), \eta_{A'}(O)), \quad \mathrm{Z}^\alpha \overset{O}{\leftrightarrow} (\omega^A(O), \pi_{A'}(O)). \tag{6.1.39}$$

Reference to (6.1.20) shows that its components $X^\alpha Z^\beta$ will consist of all the components at O of the following four spinors:

$$\xi^A\omega^B, \quad \xi^A\pi_{B'}, \quad \eta_{A'}\omega^B, \quad \eta_{A'}\pi_{B'}. \tag{6.1.40}$$

The spinor fields (6.1.40) are the spinor parts of $X^\alpha Z^\beta$ and they have a position dependence determined by (6.1.10) (as applied to (6.1.38)). The general $[\begin{smallmatrix}2\\0\end{smallmatrix}]$-twistor $S^{\alpha\beta}$, being a sum of products of type $X^\alpha Z^\beta$, will be fully characterized by four independent spinors at O, namely the values at O of the four fields

$$S^{AB}, \quad S^A{}_{B'}, \quad S_{A'}{}^B, \quad S_{A'B'}. \tag{6.1.41}$$

These are said to constitute the spinor parts of $S^{\alpha\beta}$. They are of a very special interrelated type, being sums of expressions as in (6.1.40), whose constituents (6.1.38) have the position dependence (6.1.10).

In using the notation (6.1.41) it is vital to keep the *order* of the spinor indices unchanged, e.g., never to write $S^A{}_{B'} = S_{B'}{}^A$ (contrary to our usual conventions, *cf.* (2.5.33), since this order is our only notational indication of which spinor part is meant. But the notation may become confusing when indices are raised and lowered with εs. For this reason, if the spinor parts of some twistor are to be used extensively it is often convenient to introduce separate symbols for the various spinor parts. But for the general discussion of twistors the present notation is very economical. We often write

$$S^{\alpha\beta} = \begin{pmatrix} S^{AB} & S^A{}_{B'} \\ S_{A'}{}^B & S_{A'B'} \end{pmatrix}. \tag{6.1.42}$$

In the same way we find the typical patterns* of $[\begin{smallmatrix}1\\1\end{smallmatrix}]$- and $[\begin{smallmatrix}0\\2\end{smallmatrix}]$-twistors:

$$E^\alpha{}_\beta = \begin{pmatrix} E^A{}_B & E^{AB'} \\ E_{A'B} & E_{A'}{}^{B'} \end{pmatrix}, \quad R_{\alpha\beta} = \begin{pmatrix} R_{AB} & R_A{}^{B'} \\ R^{A'}{}_B & R^{A'B'} \end{pmatrix}. \tag{6.1.43}$$

It is clear that the general $[\begin{smallmatrix}p\\q\end{smallmatrix}]$-twistor will have 2^{p+q} spinor parts, which, however, cannot be exhibited as conveniently as the above. For example, a $[\begin{smallmatrix}2\\1\end{smallmatrix}]$-twistor $T^{\alpha\beta}{}_\gamma$ has eight independent spinor parts of the following form:

$$T^{AB}{}_C, \quad T_{A'}{}^B{}_C, \quad T^A{}_{B'C}, \quad T^{ABC'}, \quad T^A{}_{B'}{}^{C'}, \quad T_{A'}{}^{BC'}, \quad T_{A'B'C}, \quad T_{A'B'}{}^{C'}. \tag{6.1.44}$$

The particular spinor part which has all its indices at the upper level (in the

* The use of staggered indices on twistors, such as for $E^\alpha{}_\beta$ and $T^{\alpha\beta}{}_\gamma$ here, serves no 'purely twistorial' purpose, there being no 'metric' to raise or lower twistor indices, but it is helpful for keeping track of the various spinor parts. Accordingly we shall tend to adopt such staggering only when we are concerned with the taking of spinor parts.

case above, $T^{ABC'}$) is called the *primary spinor part* of the twistor. That part with all lower indices is called the *projection part*.

The definition (6.1.23) of $W_\alpha Z^\alpha$ leads to definitions for contractions of arbitrary twistors. In practice this amounts to contracting the 'relevant' spinor parts over A and A' for each contracted twistor index α. For example,

$$T^{\alpha\beta}{}_\alpha = (T^{AB}{}_A + T_{A'}{}^{BA'},\quad T^{A}{}_{B'A} + T_{A'B'}{}^{A'}). \tag{6.1.45}$$

It is now easy to see how we can take components of general twistors relative to the basis (6.1.17), analogously to (6.1.21) and (6.1.37). In the general case, exemplified by

$$T^{\boldsymbol{\alpha\beta}}{}_{\boldsymbol{\gamma}} = T^{\alpha\beta}{}_\gamma \delta^{\boldsymbol{\alpha}}_\alpha \delta^{\boldsymbol{\beta}}_\beta \delta^\gamma_{\boldsymbol{\gamma}}, \tag{6.1.46}$$

we find (spinor field components being taken at O, in accordance with our remark after (6.1.21))

$$\begin{aligned} &T^{00}{}_0 = T^{00}{}_0,\quad T^{01}{}_1 = T^{01}{}_1, \text{ etc.},\\ &T^{20}{}_1 = T_{0'}{}^{0}{}_1,\quad T^{20}{}_3 = T_{0'}{}^{01'},\quad T^{31}{}_2 = T_{1'}{}^{10'}, \text{ etc.}, \end{aligned} \tag{6.1.47}$$

where the *left* member of each equation is the *twistor* component, and the *right* member is a spinor-part component. Since twistor indices are not primed and spinor indices are never 2 or 3, there is no ambiguity of meaning in the second line of (6.1.47). In the first line, where there might be ambiguity, there is in fact none. Indeed, we have the following rule:

$$\begin{aligned} &0(\text{up/down}) \leftrightarrow 0(\text{up/down}),\quad 1(\text{up/down}) \leftrightarrow 1(\text{up/down})\\ &2(\text{up/down}) \leftrightarrow 0'(\text{down/up}),\quad 3(\text{up/down}) \leftrightarrow 1'(\text{down/up}) \end{aligned} \tag{6.1.48}$$

We next examine the position dependence of the spinor parts (6.1.41) of the $[{}^2_0]$-twistor $S^{\alpha\beta}$. The mutual and position dependences of the (field-)spinor parts of 1-valent twistors determine the form of the (field-)spinor parts of all twistors. For $S^{\alpha\beta}$, these can be found from the requirement that, for two arbitrary $[{}^0_1]$-twistors U_α, W_α, the (scalar) field represented by

$$S^{\alpha\beta} U_\alpha W_\beta \tag{6.1.49}$$

is constant and is therefore equal to its value at the origin. Proceeding exactly as in (6.1.25), we easily find the desired relations. In order to exhibit these conveniently we prefer to introduce a more specific notation for the field-spinor parts of $S^{\alpha\beta}$, namely

$$S^{\alpha\beta} = \begin{pmatrix} \sigma^{AB} & \rho^{A}{}_{B'} \\ \tau_{A'}{}^{B} & \kappa_{A'B'} \end{pmatrix}, \tag{6.1.50}$$

(although S^{AB}, $S^{A}{}_{B'}$, etc. would be perfectly legitimate). Then the relations between these are found to be as follows:

$$\begin{aligned}
\kappa_{A'B'} &= \mathring{\kappa}_{A'B'},\\
\tau_{A'}^{B} &= \mathring{\tau}_{A'}^{B} - \mathrm{i}x^{BB'}\mathring{\kappa}_{A'B'},\\
\rho_{B'}^{A} &= \mathring{\rho}_{B'}^{A} - \mathrm{i}x^{AA'}\mathring{\kappa}_{A'B'},\\
\sigma^{AB} &= \mathring{\sigma}^{AB} - \mathrm{i}x^{AA'}\mathring{\tau}_{A'}^{B} - \mathrm{i}x^{BB'}\mathring{\rho}_{B'}^{A} - x^{AA'}x^{BB'}\mathring{\kappa}_{A'B'}.
\end{aligned} \tag{6.1.51}$$

The content of (6.1.51) can also be expressed in terms of the following differential equations (*cf.* (6.1.9), (6.1.28)):

$$\begin{aligned}
\nabla_{CC'}\sigma^{AB} &= -\mathrm{i}\varepsilon_C{}^A\tau_{C'}^{B} - \mathrm{i}\varepsilon_C{}^B\rho_{C'}^{A},\\
\nabla_{CC'}\rho_{B'}^{A} &= -\mathrm{i}\varepsilon_C{}^A\kappa_{C'B'},\\
\nabla_{CC'}\tau_{A'}^{B} &= -\mathrm{i}\varepsilon_C{}^B\kappa_{A'C'},\\
\nabla_{CC'}\kappa_{A'B'} &= 0.
\end{aligned} \tag{6.1.52}$$

One way to establish these is to differentiate (6.1.51) at the origin, and then to recall that the origin is arbitrary, so that (6.1.52) holds generally. In this special case the twistor $\mathsf{S}^{\alpha\beta}$ is still fully determined by a single field, namely by its primary spinor part σ^{AB}. For (6.1.52)(1) yields

$$\begin{aligned}
\nabla_{CC'}\sigma^{CB} &= -2\mathrm{i}\tau_{C'}^{B} - \mathrm{i}\rho_{C'}^{B},\\
\nabla_{CC'}\sigma^{BC} &= -\mathrm{i}\tau_{C'}^{B} - 2\mathrm{i}\rho_{C'}^{B},
\end{aligned} \tag{6.1.53}$$

whence we get $\tau_{C'}^{B}$ and $\rho_{C'}^{B}$. And these, via (6.1.52)(2) or (3), yield $\kappa_{A'B'}$.

Relations analogous to (6.1.51) and (6.1.52) evidently hold for twistors of *any* valence, and can be obtained in an analogous way. Let us consider one more special case, namely the $[^1_1]$-twistor $\mathsf{E}^\alpha{}_\beta$ of (6.1.43) for whose field-spinor parts we now again introduce a more specific notation:

$$\mathsf{E}^\alpha{}_\beta = \begin{pmatrix} \theta^A{}_B & \xi^{AB'} \\ \eta_{A'B} & \zeta_{A'}{}^{B'} \end{pmatrix}. \tag{6.1.54}$$

For these fields we find

$$\begin{aligned}
\eta_{A'B} &= \mathring{\eta}_{A'B},\\
\zeta_{A'}{}^{B'} &= \mathring{\zeta}_{A'}{}^{B'} + \mathrm{i}x^{BB'}\mathring{\eta}_{A'B},\\
\theta^A{}_B &= \mathring{\theta}^A{}_B - \mathrm{i}x^{AA'}\mathring{\eta}_{A'B},\\
\xi^{AB'} &= \mathring{\xi}^{AB'} - \mathrm{i}x^{AA'}\mathring{\zeta}_{A'}{}^{B'} + \mathrm{i}x^{BB'}\mathring{\theta}^A{}_B + x^{AA'}x^{BB'}\mathring{\eta}_{A'B},
\end{aligned} \tag{6.1.55}$$

which (analogously to (6.1.51), (6.1.52)) are equivalent to

$$\begin{aligned}
\nabla_{CC'}\xi^{AB'} &= \mathrm{i}\varepsilon_{C'}{}^{B'}\theta^A{}_C - \mathrm{i}\varepsilon_C{}^A\zeta_{C'}{}^{B'},\\
\nabla_{CC'}\theta^A{}_B &= -\mathrm{i}\varepsilon_C{}^A\eta_{C'B},\\
\nabla_{CC'}\zeta_{A'}{}^{B'} &= \mathrm{i}\varepsilon_{C'}{}^{B'}\eta_{A'C},\\
\nabla_{CC'}\eta_{A'B} &= 0.
\end{aligned} \tag{6.1.56}$$

We note that equations (6.1.55) remain valid if we subject $\theta^A{}_C$ and $\zeta_{C'}{}^{B'}$ to the changes

$$\theta^A{}_B \mapsto \theta^A{}_C + \lambda\varepsilon_C{}^A, \qquad \zeta_{C'}{}^{B'} \mapsto \zeta_{C'}{}^{B'} + \lambda\varepsilon_{C'}{}^{B'} \quad (\lambda = \text{constant}), \tag{6.1.57}$$

and so we see that in this case the primary spinor part $\xi^{AB'}$ of $\mathsf{E}^\alpha{}_\beta$ does *not* uniquely determine $\mathsf{E}^\alpha{}_\beta$. The transformation (6.1.57) in fact changes the trace

$$\mathsf{E}^\alpha{}_\alpha = \theta^A{}_A + \zeta_{A'}{}^{A'} \tag{6.1.58}$$

of the twistor:

$$\mathsf{E}^\alpha{}_\alpha \mapsto \mathsf{E}^\alpha{}_\alpha + 4\lambda. \tag{6.1.59}$$

Only if the trace is independently known (e.g., known to be zero), is $\mathsf{E}^\alpha{}_\beta$ in fact determined by $\xi^{AB'}$.

Equations analogous to (6.1.51), (6.1.55) hold for all twistors. For some, such as for $\mathsf{S}^{\alpha\beta}$, all the information of the twistor is contained in the primary part. But the example of $\mathsf{E}^\alpha{}_\beta$ shows that this need not be the case. A class of twistors for which the primary part *does* carry all the information is that of the trace-free symmetric twistors:

$$\mathsf{T}^{\alpha\ldots\delta}{}_{\rho\ldots\tau} = \mathsf{T}^{(\alpha\ldots\delta)}{}_{(\rho\ldots\tau)}, \tag{6.1.60}$$

$$\mathsf{T}^{\alpha\beta\ldots\delta}{}_{\alpha\sigma\ldots\tau} = 0. \tag{6.1.61}$$

The equation satisfied by the primary part $\mathsf{T}^{A\ldots DR'\ldots T'} = \lambda^{A\ldots DR'\ldots T'}$ is

$$\nabla^{(E}_{(U'}\lambda^{A\ldots D)}_{R'\ldots T')} = 0. \tag{6.1.62}$$

At the other extreme, the alternating twistors $\varepsilon_{\alpha\beta\gamma\delta}$, $\varepsilon^{\alpha\beta\gamma\delta}$, satisfying

$$\varepsilon_{\alpha\beta\gamma\delta} = \varepsilon_{[\alpha\beta\gamma\delta]}, \quad \varepsilon^{\alpha\beta\gamma\delta} = \varepsilon^{[\alpha\beta\gamma\delta]}, \quad \varepsilon_{0123} = \varepsilon^{0123} = 1, \tag{6.1.63}$$

each have only six non-zero spinor parts, out of a total of sixteen, namely the parts with two unprimed and two primed indices. These are

$$\varepsilon^{A'B'}{}_{CD} = \varepsilon^{A'B'}\varepsilon_{CD}, \quad \varepsilon^{A'}{}_B{}^{C'}{}_D = -\varepsilon^{A'C'}\varepsilon_{BD}, \text{ etc.}, \tag{6.1.64}$$

for $\varepsilon_{\alpha\beta\gamma\delta}$, and

$$\varepsilon_{A'B'}{}^{CD} = \varepsilon_{A'B'}\varepsilon^{CD}, \quad \varepsilon_{A'}{}^B{}_{C'}{}^D = -\varepsilon_{A'C'}\varepsilon^{BD}, \text{ etc.}, \tag{6.1.65}$$

for $\varepsilon^{\alpha\beta\gamma\delta}$. In these cases the primary part *vanishes*. (It also vanishes in the case of a skew twistor $\mathsf{X}^{\alpha\beta\gamma} = \mathsf{X}^{[\alpha\beta\gamma]}$.) In fact, *only* the trace-free symmetric twistors can be represented in this way by a *single* spinor field (namely by their primary spinor part) subject to a single first-order differential equation. In certain other cases (e.g., that of a skew $[{}^2_0]$-twistor) this primary spinor part does determine the twistor completely, but is not characterized by a first-order differential equation. In most other cases the primary spinor part is insufficient to define the twistor. A symmetric twistor (6.1.60) which is *not* trace-free also has a primary part which satisfies (6.1.62), but the various trace parts, e.g., expressions of the form $\delta^{(\alpha}_{(\rho}\mathrm{U}^{\beta\ldots\delta)}_{\sigma\ldots\tau)}$, are not determined by the primary part (i.e., such expressions can be added to $\mathrm{T}^{\alpha\ldots\delta}{}_{\rho\ldots\tau}$ without changing the primary part).

We shall be particularly interested in *symmetric, skew-symmetric*, and *Hermitian* 2-valent twistors. The twistor $\mathrm{S}^{\alpha\beta}$ of (6.1.50) is symmetric if $\mathrm{S}^{\alpha\beta} = \mathrm{S}^{\beta\alpha}$, i.e., if

$$\sigma^{AB} = \sigma^{BA}, \quad \rho^{A}_{B'} = \tau^{A}_{B'}, \quad \kappa_{A'B'} = \kappa_{B'A'}. \tag{6.1.66}$$

It can be seen from (6.1.52) that the mere symmetry of σ^{AB} *at all points* forces the second and third of equations (6.1.66), and is therefore sufficient for the symmetry of $\mathrm{S}^{\alpha\beta}$. For a skew-symmetric twistor ($\mathrm{S}^{\alpha\beta} = -\mathrm{S}^{\beta\alpha}$) we have a minus sign on the right of all equations (6.1.66), and again the mere skew-symmetry of σ^{AB} at *all* points forces the rest.

We say that the twistor $\mathrm{E}^{\alpha}{}_{\beta}$ of (6.1.54) is *Hermitian* if

$$\bar{\mathrm{E}}_{\beta}{}^{\alpha} = \mathrm{E}^{\alpha}{}_{\beta}, \tag{6.1.67}$$

i.e., if

$$\xi^{AB'} = \bar{\xi}^{AB'}, \quad \eta_{AB'} = \bar{\eta}_{AB'}, \quad \theta^{A}{}_{B} = \bar{\zeta}_{B}{}^{A}. \tag{6.1.68}$$

Analogously to the remark we made about symmetric spinors, it can now be seen from (6.1.55) that the Hermiticity of $\xi^{AB'}$ *at all points* forces the second of equations (6.1.68), but permits the members of (6.1.68)(3) to differ by an imaginary multiple of $\varepsilon_{B}{}^{A}$. Consequently, however, it suffices to ensure the Hermiticity of $\mathrm{E}^{\alpha}{}_{\beta}$ if $\mathrm{E}^{\alpha}{}_{\beta}$ is known to be trace-free.

An important observation is the following: the primary spinor part σ^{AB} of *any* twistor $\mathrm{S}^{\alpha\beta}$ automatically satisfies the differential equation

$$\nabla^{(C}_{C'}\sigma^{AB)} = 0, \tag{6.1.69}$$

as is clear from (6.1.52)(1). Also the primary spinor part $\xi^{AB'}$ of any twistor $\mathrm{E}^{\alpha}{}_{\beta}$ automatically satisfies the differential equation

$$\nabla^{(B}_{(B'}\xi^{A)}_{A')} = 0 \tag{6.1.70}$$

(which is actually the conformal Killing equation, see §6.5 below), as follows from (6.1.56)(1). Just as with (6.1.1), which is satisfied by the primary spinor part of Z^{α}, these two differential equations are conformally invariant, with

$$\hat{\sigma}^{AB} = \sigma^{AB}, \quad \hat{\xi}^{AB'} = \xi^{AB'} \tag{6.1.71}$$

(as is implicit, in the case of conformally flat space, in their twistor origins). This can be established independently of space–time flatness as for (6.1.1). Furthermore, by an argument analogous to that leading to (6.1.10), it can be shown that the general solutions in $\mathbb{M}$ of equations (6.1.69) and (6.1.70) are, with σ^{AB} symmetric, given by (6.1.51) (4), (6.1.55) (4), respectively. Note that (6.1.69) and (6.1.70) are both special cases of (6.1.62), which is again conformally invariant (*cf.* (5.6.15)), with

$$\hat{\chi}^{A\dots CR'\dots T'} = \chi^{A\dots CR'\dots T'}. \tag{6.1.72}$$

Equations (6.1.69), (6.1.70) and (6.1.62) will be further discussed below (see §6.7; also (6.4.1)). Arguments from §6.7 show that *all symmetric solutions of* (6.1.62) *are primary parts of trace-free symmetric twistors.*

We now return to the question of defining complex conjugation of the general twistor. The rule is, in fact, determined by the definitions (6.1.31) for 1-valent twistors, together with the requirement that complex conjugation commute with product and sum, e.g.,

$$\overline{\mathsf{V}^\alpha \mathsf{W}_\beta + \mathsf{X}^\alpha \mathsf{Y}_\beta} = \overline{\mathsf{V}^\alpha \mathsf{W}_\beta} + \overline{\mathsf{X}^\alpha \mathsf{Y}_\beta}$$

Consider, for example, the twistor

$$\mathsf{P}^\alpha{}_\beta = \mathsf{Z}^\alpha \mathsf{W}_\beta = \begin{pmatrix} \mathsf{Z}^A \mathsf{W}_B & \mathsf{Z}^A \mathsf{W}^{B'} \\ \mathsf{Z}_{A'} \mathsf{W}_B & \mathsf{Z}_{A'} \mathsf{W}^{B'} \end{pmatrix}.$$

We must have

$$\overline{\mathsf{P}^\alpha{}_\beta} = \bar{\mathsf{P}}_\alpha{}^\beta = \bar{\mathsf{Z}}_\alpha \bar{\mathsf{W}}^\beta = \begin{pmatrix} \bar{\mathsf{Z}}_A \bar{\mathsf{W}}^B & \bar{\mathsf{Z}}_A \bar{\mathsf{W}}_{B'} \\ \bar{\mathsf{Z}}^{A'} \bar{\mathsf{W}}^B & \bar{\mathsf{Z}}^{A'} \bar{\mathsf{W}}_{B'} \end{pmatrix}. \tag{6.1.73}$$

Since the general $[{}^1_1]$-twistor is a sum of twistors of the type of $\mathsf{P}^\alpha{}_\beta$, and since the most general $[{}^p_q]$-twistor can be dealt with analogously, we recognize the following general rule: *To conjugate a twistor, we conjugate all its spinor parts, and then place each conjugated part into the correct position, namely that appropriate for a twistor with all original twistor indices at reverse level.*

Conformal invariance of helicity and scalar product

We define the *helicity* of a twistor Z^α by

$$s := \tfrac{1}{2}\mathsf{Z}^\alpha \bar{\mathsf{Z}}_\alpha = \tfrac{1}{2}(\omega^A \bar{\pi}_A + \pi_{A'} \bar{\omega}^{A'}) \tag{6.1.74}$$

(likewise, for W_α it is $s = \tfrac{1}{2}\bar{\mathsf{W}}^\alpha \mathsf{W}_\alpha$), and this is evidently real, although it can be positive or negative. We say Z^α (or W_α) is *null* if its helicity vanishes, *right handed* if $s > 0$, and *left handed* if $s < 0$. Twistor space $\mathbb{T}^\bullet (= \mathbb{T} = \mathbb{T}^\alpha)$ is thus composed of three pieces $\mathbb{T}^0 = \mathbb{N}$, $\mathbb{T}^+$ and $\mathbb{T}^-$, consisting of null, right-handed and left-handed twistors, respectively. Similarly, dual twistor space $\mathbb{T}_\bullet (= \mathbb{T}^* = \mathbb{T}_\alpha)$ is composed of $\mathbb{T}_0$, $\mathbb{T}_+$ and $\mathbb{T}_-$. For the *projective* versions of these spaces, i.e. the systems of one-dimensional linear subspaces contained in them (together with the origin), the prefix $\mathbb{P}$ is adjoined (see Fig. 6-1 and §9.3).

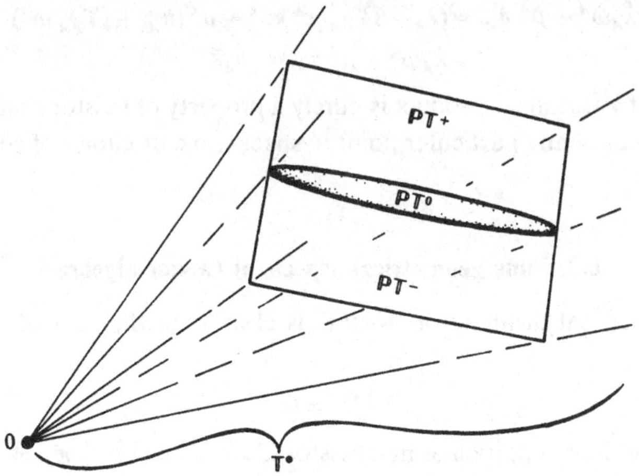

Fig. 6-1. Projective twistor space $\mathbb{PT}^\bullet$ is the space of one-dimensional linear subspaces of twistor space $\mathbb{T}^\bullet$. It consists of three regions $\mathbb{PT}^+$, $\mathbb{PT}^-$ and $\mathbb{PT}^0$ ($=\mathbb{PN}^\bullet$).

From (6.1.9) and (6.1.3) we see that, under conformal rescaling, the second of the following equations holds; the first is simply (6.1.2) again:

$$\hat{\omega}^A = \omega^A$$
$$\hat{\pi}_{A'} = \pi_{A'} + \mathrm{i}\Upsilon_{AA'}\omega^A. \tag{6.1.75}$$

Thus $\pi_{A'}$ is *not* a conformally weighted spinor field. (Note the formal similarity of the above equation with (6.1.10).) Equation (6.1.75) describes the effects of changes of the conformal scaling in the given manifold $\mathbb{M}$. The twistor Z^α itself is regarded as unaffected by this change. But its representation by spinor parts changes, unless we adopt an appropriate view of the spinor $\pi_{A'}$. In fact, $\pi_{A'}$ can be regarded not as a conformally weighted spinor field, but as something that behaves in the more complicated way (6.1.75) under conformal rescaling of the metric. (Viewed in this way, $\pi_{A'}$ cannot be considered independently of ω^A.) With this interpretation, it is still legitimate to write $\mathrm{Z}^\alpha = (\hat{\omega}^A, \hat{\pi}_{A'})$. This viewpoint is the one we shall adopt when we consider local twistors in §6.9.

The analogue of (6.1.75) for the spinor parts of a $[^0_1]$-twistor W_α is

$$\hat{\lambda}_A = \lambda_A - \mathrm{i}\Upsilon_{AA'}\mu^{A'},$$
$$\hat{\mu}^{A'} = \mu^{A'}, \tag{6.1.76}$$

(*cf.* (6.1.26)). We can now immediately verify the conformal invariance of the twistor inner product (6.1.23) and hence of the helicity given by (6.1.74):

$$\hat{\lambda}_A\hat{\omega}^A + \hat{\mu}^{A'}\hat{\pi}_{A'} = (\lambda_A - \mathrm{i}\Upsilon_{AA'}\mu^{A'})\omega^A + \mu^{A'}(\pi_{A'} + \mathrm{i}\Upsilon_{AA'}\omega^A)$$
$$= \lambda_A\omega^A + \mu^{A'}\pi_{A'} = \mathsf{W}_\alpha \mathsf{Z}^\alpha. \tag{6.1.77}$$

Thus the twistor inner product is purely a property of twistor space and is independent of any particular point in space–time or choice of conformal scale.

6.2 Some geometrical aspects of twistor algebra

The geometrical meaning of twistors is clearest in the case of *null* $\left[\begin{smallmatrix}1\\0\end{smallmatrix}\right]$-twistors:

$$\mathsf{Z}^\alpha \bar{\mathsf{Z}}_\alpha = 0. \tag{6.2.1}$$

Suppose we have a particular null twistor $\mathsf{Z}^\alpha = (\omega^A, \pi_{A'})$ with $\pi_{A'} \neq 0$. Let us first determine the locus Z of points in $\mathbb{M}$ at which $\omega^A = 0$: the geometry of the field ω^A is best described in terms of this locus. On Z the position vector must satisfy (*cf.* (6.1.10))

$$\mathrm{i}x^{AA'}\mathring{\pi}_{A'} = \mathring{\omega}^A. \tag{6.2.2}$$

We shall wish to assume that ω^A and $\bar{\pi}^A$ are not proportional at O. If by chance they *are*, we use the freedom we had in solving (6.1.9) for (6.1.10), and choose a *different* point as origin, where ω^A and $\bar{\pi}^A$ are *not* proportional. (This is always possible, since, by (6.1.10), $\bar{\pi}_A\omega^A = \bar{\pi}_A\mathring{\omega}^A - \mathrm{i}x^{AA'}\bar{\pi}_A\pi_{A'}$; so if $\bar{\pi}_A\mathring{\omega}^A = 0$ we need merely go to a point at which $x^{AA'}\bar{\pi}_A\pi_{A'} \neq 0$ to achieve $\bar{\pi}_A\omega^A \neq 0$.) Assume this done. Then a particular solution of (6.2.2) is given by

$$x^a = (\mathrm{i}\mathring{\bar{\omega}}^{B'}\pi_{B'})^{-1}\mathring{\omega}^A\mathring{\bar{\omega}}^{A'}. \tag{6.2.3}$$

This vector is real, since the parenthesis is real, by (6.2.1) and (6.1.74). The remaining solutions of (6.2.2) must be such that their differences from (6.2.3) annihilate $\pi_{A'}$. So since x^a is real, these differences must be real multiples of $\bar{\pi}^A\pi^{A'}$. Consequently the general solution of (6.2.2) has the form

$$x^a = (\mathrm{i}\mathring{\bar{\omega}}^{B'}\pi_{B'})^{-1}\mathring{\omega}^A\mathring{\bar{\omega}}^{A'} + h\bar{\pi}^A\pi^{A'}, \quad h \in \mathbb{R}. \tag{6.2.4}$$

This describes a *null straight line* Z, hereafter called a *ray*, in the direction of the flagpole of $\bar{\pi}^A$; it passes through a point Q, given by $h = 0$ in (6.2.4) whose displacement from O is along the flagpole of $\mathring{\omega}^A$ so Q lies on the light cone of O (see Fig. 6-2).

Note that the ray Z is independent of the scaling of Z^α: if we replace Z^α by $\lambda\mathsf{Z}^\alpha$ $(\lambda \neq 0)$ then Z is unchanged. Conversely we easily see that the ray Z *determines* Z^α up to proportionality, since (6.2.2) is homogeneous in $\mathring{\omega}^A$, $\mathring{\pi}_{A'}$.

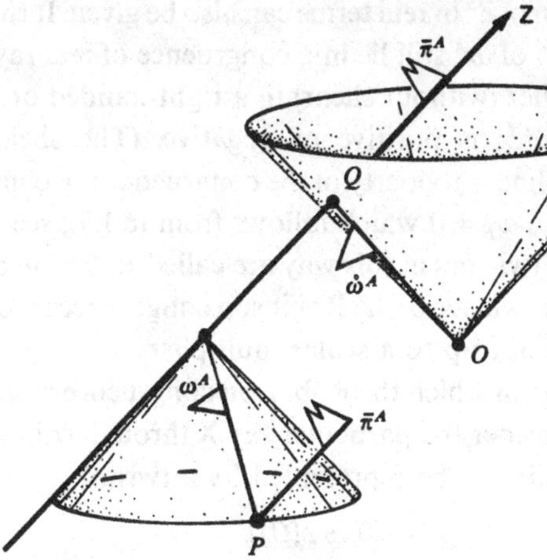

Fig. 6-2. The ray Z, determined by a null twistor Z^α, points in the direction of the flagpole of $\bar{\pi}^A$ and passes through a point Q whose displacement from the origin O is in the direction of the flagpole of $\mathring{\omega}^A$. At a general point P, the flagpole of ω^A lies in a ray meeting Z.

If $\pi_{A'} = 0$ there is *no* finite locus Z (*cf.* (6.2.2)). Instead (provided $\mathring{\omega}^A$ does not also vanish, in which case $\omega^A \equiv 0$ and so $Z^\alpha = 0$) one can then interpret the locus Z as a generator of the 'light cone at infinity'. This will be discussed in Chapter 9.

Flagpole field of ω^A: Robinson congruence

Having found the locus Z, it is easy to describe the general geometric pattern of the flagpoles of the field ω^A. Recalling the freedom we have in the choice of origin O, we see that a construction similar to the above can be carried out at any point P at which ω^A is not proportional to $\bar{\pi}^A$. Where it *is* proportional, the flagpole of ω^A is in a direction parallel to Z. For a general point P however, the flagpole direction of ω^A will lie along the (unique) ray joining P to the point at which Z cuts its light cone. Thus the field of flagpole directions of ω^A consists simply of all the null directions in all the light cones whose vertices lie on Z, together with those in the limiting light cone where the vertex goes to infinity on Z – the (unique) null hyperplane containing Z. This hyperplane is the locus of points at which the flagpole direction of ω^A is parallel to Z (ω^A proportional to $\bar{\pi}^A$).

If $Z^\alpha = (\omega^A, \pi_{A'})$ is not null, it is still possible to regard it as representing a locus in 'complexified' space–time. This viewpoint will be pursued in §9.3.

However, a realization of Z^α in real terms can also be given. It turns out that the flagpole directions of ω^A still lie in a congruence of real rays. The rays twist about one another (without shear) in a right-handed or left-handed sense according as $Z^\alpha\bar{Z}_\alpha$ is positive or negative. (The shear-free – and geodetic, i.e., straight-line – property of the congruence is a consequence of the equation $\omega^A\omega^B\nabla_{AA'}\omega_B=0$ which follows from (6.1.9); see §7.1.) Congruences arising from twistors in this way are called *Robinson congruences* (*cf* Penrose 1967*a*). Knowledge of the Robinson congruence associated with a twistor fixes the twistor up to a scalar multiplier.

There is another way in which the Robinson congruence associated with a twistor Z arises. Consider the particular ray X through O in the flagpole direction of $\omega^A(O)$. This can be represented by a twistor

$$X^\alpha \overset{O}{\leftrightarrow} (0, \bar{\omega}_{A'}(O)),$$

or by any non-zero multiple of this. Evidently $X^\alpha\bar{Z}_\alpha=0$, which condition characterizes, at O, the flagpole direction of ω^A. But the origin is arbitrary, so at *any* point the flagpole direction of ω^A is the direction of a ray X through that point described by a twistor X^α subject to

$$X^\alpha\bar{Z}_\alpha=0=X^\alpha\bar{X}_\alpha, \tag{6.2.5}$$

X^α being, of course, necessarily null. Thus the flagpoles of the ω-field all point along the rays given by (6.2.5) (for fixed $\bar{Z}_\alpha$), whence (6.2.5) describes the Robinson congruence.

To get a picture of a Robinson congruence we choose a particular $Z^\alpha=(\omega^A,\pi_{A'})$ with $Z^\alpha\bar{Z}_\alpha=2s$, given in the standard coordinate system and spin-frame, *cf.* (3.1.31) (and Chapter 1), by

$$Z^\alpha=(0,s,0,1)\quad (s\in\mathbb{R}). \tag{6.2.6}$$

The equation of the ω-field is given (*cf.* (6.1.10)) by

$$(\omega^0,\omega^1)=(0,s)-\frac{\mathrm{i}}{\sqrt{2}}\begin{pmatrix} t+z & x+\mathrm{i}y\\ x-\mathrm{i}y & t-z\end{pmatrix}\begin{pmatrix}0\\1\end{pmatrix}, \tag{6.2.7}$$

so

$$\omega^0:\omega^1=x+\mathrm{i}y:t-z+\mathrm{i}s\sqrt{2}. \tag{6.2.8}$$

We can find a differential equation for the rays of the congruence, the tangent direction defined by $\mathrm{d}t:\mathrm{d}x:\mathrm{d}y:\mathrm{d}z$ being that of the flagpole of ω^A, i.e.

$$\begin{aligned} \mathrm{d}t+\mathrm{d}z:\mathrm{d}x+\mathrm{i}\mathrm{d}y &= \mathrm{d}x-\mathrm{i}\mathrm{d}y:\mathrm{d}t-\mathrm{d}z\\ &=\bar{\omega}^{0'}:\bar{\omega}^{1'}\\ &=x-\mathrm{i}y:t-z-\mathrm{i}s\sqrt{2}. \end{aligned} \tag{6.2.9}$$

The general solution of these equations can be written down directly. But it is in any case known from (6.2.5): let us put

$$\mathsf{X}^{\alpha} = (\lambda, -s, \mu, 1)$$

so that (6.2.5) is satisfied when

$$\mathrm{Re}\,(\lambda\bar{\mu}) = s. \tag{6.2.10}$$

Then (6.2.2), with X^{α} for Z^{α}, is the equation of the ray X of the congruence (solution of (6.2.9)) determined by $\lambda, \mu \in \mathbb{C}$ (subject to (6.2.10)). Explicitly,

$$\begin{pmatrix} \lambda \\ -s \end{pmatrix} = \frac{\mathrm{i}}{\sqrt{2}} \begin{pmatrix} t+z & x+\mathrm{i}y \\ x-\mathrm{i}y & t-z \end{pmatrix} \begin{pmatrix} \mu \\ 1 \end{pmatrix}. \tag{6.2.11}$$

But for visualization purposes we are more interested in a three-dimensional description. Consider the intersection of the congruence with a spacelike hyperplane $t = \tau$ (= constant). Each ray meets this in one point (τ, x, y, z), and we can describe the direction of the ray in terms of its orthogonal projection into $t = \tau$ at this intersection point. The projected directions are then tangent to a series of curves in $t = \tau$ which give a picture in Euclidean 3-space of the structure of the Robinson congruence. The differential equation of these curves is obtained by replacing t by τ in (6.2.9), and $\mathrm{d}t$ by $\mathrm{d}l = (\mathrm{d}x^2 + \mathrm{d}y^2 + \mathrm{d}z^2)^{\frac{1}{2}}$. This yields

$$(x - \mathrm{i}y)(\mathrm{d}l - \mathrm{d}z) = (\tau - z - \mathrm{i}s\sqrt{2})(\mathrm{d}x - \mathrm{i}\,\mathrm{d}y),$$

whose solutions are given by

$$\begin{aligned} &x^2 + y^2 + (z-\tau)^2 - 2^{\frac{3}{2}}s(x\sin\phi + y\cos\phi)\tan\theta = 2s^2 \\ &z - \tau = (x\cos\phi - y\sin\phi)\tan\theta, \end{aligned} \tag{6.2.12}$$

where θ and ϕ are constants defining the different curves. These curves are evidently circles, being intersections of spheres with planes. They twist around one another (hence the term twistor!) in such a way that every pair of circles is linked (see Fig. 6-3). The twisting has a positive screw sense if $s > 0$, i.e., if Z^{α} is right-handed. They lie on the set of coaxial tori obtained by eliminating ϕ between the two equations. These tori are the rotations about the z-axis of a system of coaxial circles in the (x, z)-plane.

From the point of view of the compactified space–time that we shall discuss in §9.2 we should regard the hyperplane $t = \tau$ as being compactified (conformally) by a point at infinity. It then becomes topologically a three-dimensional sphere S^3 (of which the hyperplane $t = \tau$ may be regarded as the stereographic projection). The vector field on S^3 is everywhere non-singular and nowhere vanishing. The circles constitute what is known as a *Hopf fibring* of S^3. With a suitable scaling they become *Clifford parallels* on S^3 (Clifford 1882, *cf.* Hopf 1935; also Veblen and Young 1918).

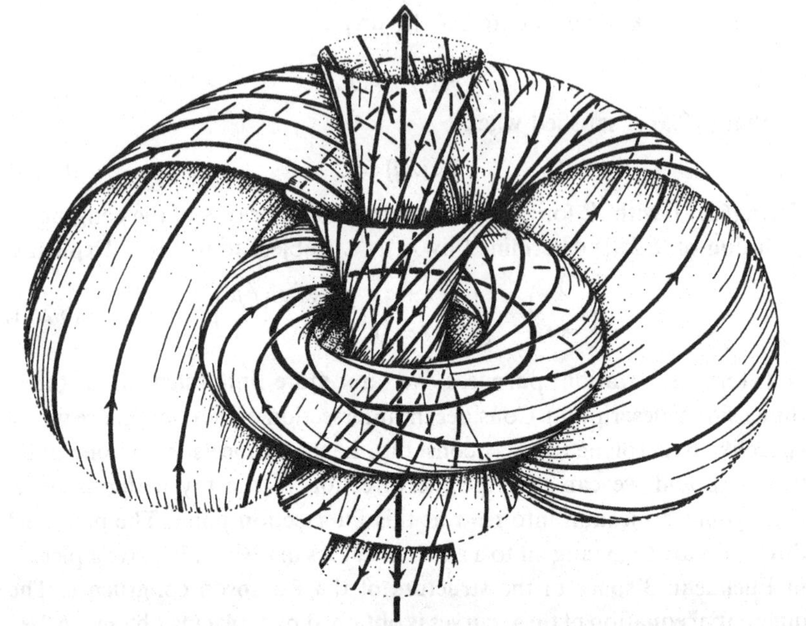

Fig. 6-3. For a general twistor $\mathbf{Z}^\alpha$, the flagpole directions of ω^A point, at any one time, in directions whose spatial projections are tangent to a family of circles, and one straight line, constituting stereographically projected Clifford parallels on S^3. As time progresses, the whole configuration moves with the speed of light in the negative direction of the straight line (i.e. in the projected direction of $\bar{\pi}^A\pi^{A'}$).

Note that all the circles in the hyperplane thread through the particular (smallest) circle of radius* $|s|\sqrt{2}$ and centre $z=\tau$, $x=y=0$ given when $\theta=0$. If s is small, this circle describes, as τ increases, a path approximating that of a ray. If s is zero, the path is exactly the ray $z=t, x=y=0$. For small s, we may think of the lines of the Robinson congruence as defining an 'approximate' ray, but the lines twist around one another and never quite meet. The twisting has a positive or negative screw sense according as $\mathbf{Z}^\alpha$ is right- or left-handed. In the limit $s\to 0$, the circles (6.2.12) all touch the z-axis at $z=\tau$. The tangents to these circles are orthogonal to the spheres touching the (x, y)-plane at $z=\tau$, these spheres being the intersections of $t=\tau$ with the null cones with vertices on $z=t$, $x=y=0$, which is now the ray $\mathbf{Z}$. The lines of the congruence are then just the generators of these null cones,

* It should be remarked that whereas the helicity s is an invariant of the description of the twistor, the radius of this smallest circle is not. This radius is given, generally, by $|s|(t^{AA'}\bar{\pi}_A\pi_{A'})^{-1}$ where t^a is the unit time-axis (i.e., as we shall see in §6.3, by the spin divided by the energy).

which is the case we considered earlier (Z null). Such a congruence, i.e., the system of rays meeting a given ray, is called a *special* Robinson congruence. (This includes any system of *parallel* rays, that being the limiting case when Z recedes to infinity – see §9.2.)

Incidence of twistors

From what we have just proved in the case of a special Robinson congruence, it is clear that two *null* twistors X^α, Z^α represent rays X, Z that *intersect* (possibly at infinity, which is the case when X and Z lie in a common null hyperplane) if and only if

$$X^\alpha \bar{Z}_\alpha = 0 \quad \text{i.e.,} \quad \bar{X}_\alpha Z^\alpha = 0. \tag{6.2.13}$$

For, requiring intersection is equivalent to requiring that X belong to the congruence defined by Z^α, and vice versa. We also call (6.2.13) the twistor 'orthogonality condition', or, more usually, *incidence* between Z^α and $\bar{X}_\alpha$.

As an alternative argument for the intersection/orthogonality condition (6.2.13), we observe that the position vector $x^a = r^a$ of any common point *R* of the rays X, Z must be given by (*cf.* (6.2.2)):

$$\begin{aligned} \mathrm{i}r^{AA'}\pi_{A'} &= \overset{\circ}{\omega}{}^A, \\ \mathrm{i}r^{AA'}\eta_{A'} &= \overset{\circ}{\xi}{}^A, \end{aligned} \tag{6.2.14}$$

where $X^\alpha = (\xi^A, \eta_{A'})$, $Z^\alpha = (\omega^A, \pi_{A'})$. These equations can be solved by translating them into component form, using an arbitrary dyad. The condition for a unique (complex) solution is that the 2×2 matrix $(\pi_{A'}, \eta_{A'})$ be non-singular, i.e., that $\pi_{A'}$ be not proportional to $\eta_{A'}$. (In case of proportionality, X and Z are parallel and so, if they meet at all, they do so at infinity, in the sense of lying in a common null hyperplane.) As may be easily verified by direct substitution, the solution of (6.2.14) is then given by

$$r^a = -\frac{\mathrm{i}}{\pi_{B'}\eta^{B'}}(\omega^A\eta^{A'} - \xi^A\pi^{A'}), \tag{6.2.15}$$

at the point *O*. But since the origin is arbitrary, we may omit the '∘' in (6.2.14) and consequently in (6.2.15), and regard r^a as the position vector of *R* relative to the field point at which ω^A and ξ^A are taken. Now, it is the reality of r^a that is equivalent to (6.2.13); more precisely, it is equivalent to the three conditions $X^\alpha\bar{X}_\alpha = 0$, $Z^\alpha\bar{Z}_\alpha = 0$, $Z^\alpha\bar{X}_\alpha = 0$, of which the first two are assumed. For these conditions state the reality of $r^{AA'}\bar{\eta}_A\eta_{A'}$, $r^{AA'}\bar{\pi}_A\pi_{A'}$, and the relation $\overline{r^{AA'}\bar{\pi}_A\eta_{A'}} = r^{AA'}\pi_{A'}\bar{\eta}_A$, respectively, from which it follows that $r^{AA'} - \bar{r}^{AA'}$ has vanishing components relative to the spinor basis $(\bar{\pi}_A, \bar{\eta}_A)$.

Complex geometry

Note that (6.2.15) defines a complex vector, and therefore a point in the *complexification** $\mathbb{CM}$ of Minkowski space $\mathbb{M}$, whether X^α or Z^α are null or not, and whether they are orthogonal or not, provided only that their projection parts (see remark after (6.1.44)) are not proportional to one another. In fact it is often fruitful to study twistors in relation to complex loci in $\mathbb{M}$ (i.e., loci in $\mathbb{CM}$). The equation (6.2.2), which in $\mathbb{M}$ defines a ray when Z^α is null (and $\pi_{A'} \neq 0$), can also be interpreted as the equation of a locus in $\mathbb{CM}$. In fact, this locus is not just the complexification, in the ordinary sense, of a ray** but is a two-complex-dimensional (and therefore four-real-dimensional) locus (whether Z^α is null or not), called an α-plane (see §9.3 below). In a similar way, a dual twistor W_α defines a β-plane (the complex conjugate of an α-plane) as the locus of solutions in $\mathbb{CM}$ of

$$-\mathrm{i}\mathring{\lambda}_A x^{AA'} = \mathring{\mu}^{A'}.$$

An α-plane [β-plane] is the locus, in $\mathbb{CM}$, along which the primary part of some Z^α [some W_α] vanishes. The point defined by (6.2.15) is then the *unique* intersection of the α-planes defined by X^α and Z^α. We say that Z^α[*or* W_α] is *incident* with a point $R \in \mathbb{M}$ if its α-plane [β-plane] contains R; moreover, Z^α and W_α are incident with one another iff their α- and β-planes meet.

The general twistor whose α-plane passes through the intersection point of the α-planes X^α, Z^α is

$$\mathrm{Y}^\alpha = \beta \mathrm{X}^\alpha + \gamma \mathrm{Z}^\alpha \tag{6.2.16}$$

where $\beta, \gamma \in \mathbb{C}$ and β, γ are not both zero. For (6.2.16) implies that the common (complex) zeros of the primary parts of X^α and Z^α are also zeros of the primary part of Y^α, and to establish the converse we need only interchange the role of Y^α with that of X^α or Z^α.

Note that, if X^α and Z^α are null and intersecting, Y^α is also null. In that case (6.2.16) represents the light cone with vertex R. We may regard the *family* of α-planes $\beta\mathrm{X}^\alpha + \gamma\mathrm{Z}^\alpha$ as representing the *point* R. In fact, *any* two-dimensional subspace of twistor space $\mathbb{T}^\alpha$ may be thought of as representing a point, of a

* Here $\mathbb{CM}$ is the eight-real-dimensional space obtained by allowing the coordinates t, x, y, z of $\mathbb{M}$ to become complex numbers. The metric is still $\mathrm{d}t^2 - \mathrm{d}x^2 - \mathrm{d}y^2 - \mathrm{d}z^2$, i.e., it is the holomorphic (complex-analytic) rather than the Hermitian extension of the metric of $\mathbb{M}$. Note that $\mathbb{CM}$ is the same as the complexification of four-dimensional Euclidean space, which itself is isometric with a subspace of $\mathbb{CM}$, namely that given by t = real, x, y, z = pure imaginary.

** *That* would be given by allowing h in (6.2.4) to become complex, so the complex x^a describes a two-real-dimensional locus, and provides the solution of the *simultaneous* equations $\mathring{\omega}^A = \mathrm{i}x^a\pi_{A'}$ and $\bar{\mathring{\omega}}^{A'} = -\mathrm{i}x^a\bar{\pi}_A$ (see (6.2.2) and (9.3.22)).

kind, in Minkowski space. But in general it will be a *complex* point (i.e., a point of $\mathbb{CM}$) since the r^a of (6.2.15) will not be real unless X^α and Z^α are null and orthogonal. Only if *every* member of the linear set is null will the set represent a point of $\mathbb{M}$, and then it will be a *finite* point only if the expression (6.2.15) is finite, i.e., when there are two twistors with non-proportional projection parts in the set. (For a more complete discussion of this geometry, see §9.3 below.)

One of the basic ideas of the twistor formalism is that it leads to an alternative description of physics in which the space–time points, or 'events', are no longer fundamental. The twistor space $\mathbb{T}^\alpha$ is itself regarded as more primitive than the space–time. Events are regarded as *derived* from the twistor space structure. We have seen above, in fact, how events may be interpreted in terms of certain linear sets in $\mathbb{T}^\alpha$. These ideas can be carried a great deal further, even to curved space–time. But we shall not pursue them here.

Simple skew twistors to represent points

Now consider the $[^2_0]$-twistor

$$\mathsf{R}^{\alpha\beta} = \mathsf{Z}^\alpha\mathsf{X}^\beta - \mathsf{X}^\alpha\mathsf{Z}^\beta, \tag{6.2.17}$$

where X^α, Z^α are the twistors occurring in (6.2.16). If we know the set (6.2.16) then we know $\mathsf{R}^{\alpha\beta}$ up to proportionality, and conversely. Hence we may represent the point R of intersection of the set (6.2.16) by $\mathsf{R}^{\alpha\beta}$ up to proportionality. The spinor parts of $\mathsf{R}^{\alpha\beta}$ are found to be as follows (*cf.* (6.1.40), (6.1.42)):

$$\mathsf{R}^{\alpha\beta} = \begin{pmatrix} \omega^A\xi^B - \xi^A\omega^B & \omega^A\eta_{B'} - \xi^A\pi_{B'} \\ \pi_{A'}\xi^B - \eta_{A'}\omega^B & \pi_{A'}\eta_{B'} - \eta_{A'}\pi_{B'} \end{pmatrix} = \pi_{D'}\eta^{D'}\begin{pmatrix} -\tfrac{1}{2}\varepsilon^{AB}r_c r^c & \mathrm{i}r^A{}_{B'} \\ -\mathrm{i}r_{A'}{}^B & \varepsilon_{A'B'} \end{pmatrix}. \tag{6.2.18}$$

Remember that r^c is the position vector of R with respect to a general field point.

It is useful to consider the *dual* $\mathsf{R}_{\alpha\beta}$ of $\mathsf{R}^{\alpha\beta}$. We write it without an asterisk, since there is no danger of confusion: there is no twistor metric with which to raise and lower indices. We have

$$\mathsf{R}_{\alpha\beta} = \tfrac{1}{2}\varepsilon_{\alpha\beta\gamma\delta}\mathsf{R}^{\gamma\delta}, \quad \mathsf{R}^{\alpha\beta} = \tfrac{1}{2}\varepsilon^{\alpha\beta\gamma\delta}\mathsf{R}_{\gamma\delta}, \tag{6.2.19}$$

where $\varepsilon_{\alpha\beta\gamma\delta}$, $\varepsilon^{\alpha\beta\gamma\delta}$ are the *alternating twistors* introduced in (6.1.63). We have, from (6.2.18) and (6.1.64),

$$\mathsf{R}_{\alpha\beta} = \pi_{D'}\eta^{D'}\begin{pmatrix} \varepsilon_{AB} & \mathrm{i}r_A{}^{B'} \\ -\mathrm{i}r^{A'}{}_B & -\tfrac{1}{2}\varepsilon^{A'B'}r_c r^c \end{pmatrix}. \tag{6.2.20}$$

If we form the twistor complex conjugate of (6.2.18) according to the rule enunciated after (6.1.73), we find that, apart from an overall factor, $\bar{\mathsf{R}}_{\alpha\beta}$ is just $\mathsf{R}_{\alpha\beta}$ with $\bar{r}^a$ in place of r^a (taking r^a to be a complex world vector). Thus the condition for r^a to be real is simply

$$\bar{\mathsf{R}}_{\alpha\beta} \propto \mathsf{R}_{\alpha\beta}. \tag{6.2.21}$$

Note that for $\mathsf{R}^{\alpha\beta}$ to represent a point at all, it must be *skew* and *simple* (meaning that the form (6.2.17) holds for some X^α, Z^α). This we can express as

$$\mathsf{R}^{\alpha\beta} = -\mathsf{R}^{\beta\alpha}, \quad \mathsf{R}^{\alpha\beta}\mathsf{R}_{\gamma\beta} = 0. \tag{6.2.22}$$

The second of these conditions, the condition for simplicity, can be expressed alternatively in any one of the following forms (*cf.* (3.5.30), (3.5.35)):

$$\mathsf{R}^{\alpha\beta}\mathsf{R}_{\alpha\beta} = 0, \quad \mathsf{R}^{\alpha[\beta}\mathsf{R}^{\gamma\delta]} = 0, \quad \mathsf{R}^{[\alpha\beta}\mathsf{R}^{\gamma\delta]} = 0, \quad \det(\mathsf{R}^{\alpha\beta}) = 0. \tag{6.2.23}$$

Finally, in addition to (6.2.21) and (6.2.22), if $\mathsf{R}^{\alpha\beta}$ is to represent a *finite* point of $\mathbb{M}$ we require

$$\mathsf{R}^{\alpha\beta}\mathsf{I}_{\alpha\beta} \neq 0, \tag{6.2.24}$$

where $\mathsf{I}_{\alpha\beta}$ is one of the *infinity twistors* $\mathsf{I}_{\alpha\beta}$, $\mathsf{I}^{\alpha\beta}$ – duals of each other and also twistor complex conjugates of each other – defined by

$$\mathsf{I}_{\alpha\beta} = \begin{pmatrix} 0 & 0 \\ 0 & \varepsilon^{A'B'} \end{pmatrix}, \quad \mathsf{I}^{\alpha\beta} = \begin{pmatrix} \varepsilon^{AB} & 0 \\ 0 & 0 \end{pmatrix}. \tag{6.2.25}$$

For it is clear that $\mathsf{R}^{\alpha\beta}\mathsf{I}_{\alpha\beta}$ in fact equals $2\pi_{D'}\eta^{D'}$; if this vanishes, $\pi_{D'}$ and $\eta_{D'}$ are proportional, and Z and X are parallel. That $\mathsf{I}^{\alpha\beta}$ is indeed a twistor (and so $\mathsf{I}_{\alpha\beta}$ also) can be seen by observing that it is of the form (6.2.18) with $\pi_{A'} = \eta_{A'} = 0$ and $\omega_A\xi^A = 1$. Alternatively we may observe that $\mathsf{I}^{\alpha\beta}$ satisfies (6.1.52).

The twistor equation is conformally invariant, and so twistor space is defined using only the conformal structure of $\mathbb{M}$. Duals and complex conjugates are also conformally invariant and so all the basic operations of twistor algebra possess this property. (We have checked this explicitly for the twistor inner product $\mathsf{W}_\alpha\mathsf{Z}^\alpha$.) Moreover, the geometric concepts used in our discussion are all conformally invariant: ray (i.e., null geodesic), null cone, point, intersection, etc. The metric scaling of $\mathbb{M}$ was nowhere used. In fact, twistors are the 'reduced spinors' for a certain pseudo-orthogonal group in *six* dimensions: $O(2,4)$ (the preserved quadratic form having two plus signs and four minus signs). This group is locally isomorphic with the fifteen-parameter (so-called) *conformal group* of Minkowski space (*cf.* Dirac 1936*b*, Penrose 1974), i.e., with the group of point mappings of $\mathbb{M}$ (compactified by the addition of suitable elements at infinity) onto itself which preserves the conformal structure of $\mathbb{M}$ (see §9.2). These mappings induce linear transformations on twistor space which preserve the form $\mathrm{Z}^\alpha\bar{\mathrm{Z}}_\alpha$ and also the alternating twistors. Since the signature of $\mathrm{Z}^\alpha\bar{\mathrm{Z}}_\alpha$ turns out to be $(++--)$, this means that the resulting twistor group may be identified with $SU(2,2)$ (the group of pseudo-unitary $(++--)$ unimodular 4×4 matrices *cf.* Cartan 1914). If we introduce $\mathrm{I}_{\alpha\beta}$ and $\mathrm{I}^{\alpha\beta}$ as basic elements into the twistor algebra (rather in the way that g_{ab} and g^{ab} are introduced into tensor algebra) and consider the reduced group which leaves $\mathrm{I}_{\alpha\beta}$ and $\mathrm{I}^{\alpha\beta}$ invariant also, then we get a group locally isomorphic with the *Poincaré group* on $\mathbb{M}$ (inhomogeneous Lorentz group). Thus $\mathrm{I}_{\alpha\beta}$ and $\mathrm{I}^{\alpha\beta}$ have the effect of enriching the twistor algebra so as to bring the metric structure of Minkowski space within its scope, rather than merely its conformal structure. Indeed, the basic concept of Minkowski *distance* between two points Q and R of $\mathbb{M}$ (*cf.* (1.1.22)) can be expressed in terms of the twistors $\mathrm{Q}^{\alpha\beta}$ and $\mathrm{R}^{\alpha\beta}$ which represent these points respectively, by the formula

$$\frac{4\mathrm{Q}^{\alpha\beta}\mathrm{R}_{\alpha\beta}}{\mathrm{Q}^{\gamma\delta}\mathrm{I}_{\gamma\delta}\mathrm{R}^{\rho\sigma}\mathrm{I}_{\rho\sigma}} = -(q^a - r^a)(q_a - r_a), \tag{6.2.26}$$

q^a and r^a being the position vectors of Q and R (with respect to the general field point). In view of this formula, and for other reasons, it is often convenient to normalize the twistor representing a point R of $\mathbb{M}$ according to

$$\mathrm{R}^{\alpha\beta}\mathrm{I}_{\alpha\beta} = 2, \tag{6.2.27}$$

which amounts to leaving out the factor $\pi_{D'}\eta^{D'}$ in (6.2.18), or, more accurately, to setting

$$\pi_{D'}\eta^{D'} = 1. \tag{6.2.28}$$

Then we have, dually,

$$\mathrm{R}_{\alpha\beta}\mathrm{I}^{\alpha\beta} = 2 \tag{6.2.29}$$

and (6.2.26) becomes

$$\mathrm{Q}^{\alpha\beta}\mathrm{R}_{\alpha\beta} = -(q^a - r^a)(q_a - r_a). \tag{6.2.30}$$

By (6.2.20) and (6.2.28), the proportionality constant in (6.2.21) is now fixed to be unity, so the condition for r^a to be real becomes

$$\bar{\mathrm{R}}_{\alpha\beta} = \mathrm{R}_{\alpha\beta}. \tag{6.2.31}$$

The basic properties of twistor algebra and its relation to Minkowski geometry have now been given. Much more can be said,* and some further discussion is given in §9.3. The interplay between the geometry of twistor space and that of $\mathbb{M}$ leads to many fruitful ideas. The interested reader is referred to the literature (Penrose 1967*a*, Penrose and MacCallum 1972, Ward and Wells 1986, Huggett and Tod 1985).

6.3 Twistors and angular momentum

We now turn to the physical interpretation of a twistor. In a sense the physical interpretation is a little more complete and a little more natural than the geometrical one given so far. It allows the twistor $\mathbf{Z}^\alpha$ to be interpreted up to phase, rather than merely the proportionality class $\{\lambda\mathbf{Z}^\alpha | 0 \neq \lambda \in \mathbb{C}\}$. But more importantly, the non-null twistors ($\mathbf{Z}^\alpha\bar{\mathbf{Z}}_\alpha \neq 0$) arise physically in a very natural way.

Let $\mathbf{Z}^\alpha$ be represented as usual:

$$\mathbf{Z}^\alpha = (\omega^A, \pi_{A'}), \tag{6.3.1}$$

and assume $\pi_{A'} \neq 0$. Define

$$p_a := \bar{\pi}_A \pi_{A'}, \quad M^{ab} := \mathrm{i}\omega^{(A}\bar{\pi}^{B)}\varepsilon^{A'B'} - \mathrm{i}\bar{\omega}^{(A'}\pi^{B')}\varepsilon^{AB}. \tag{6.3.2}$$

Then p_a is a future-null vector field and M^{ab} a real skew tensor field. Let $\mathring{p}_a$ and $\mathring{M}^{ab}$ be constant fields taking the same values as p_a and M^{ab}, respectively, at the origin O. The *position* dependence of the tensors (6.3.2) is given by (*cf.* (6.1.10)):

$$p_a = \mathring{p}_a, \quad M^{ab} = \mathring{M}^{ab} - x^a p^b + x^b p^a. \tag{6.3.3}$$

These are precisely the transformation formulae satisfied by the 4-momentum and the 6-angular momentum in special relativity (*cf.* Synge 1955), giving these quantities relative to the *general* point P in terms of the same quantities relative to the origin O. For later reference, we note from (6.3.3) that the dual ${}^*M_{ab} = \frac{1}{2}e_{abcd}M^{cd}$ satisfies

$$\nabla_{(a}{}^*M_{b)c} = 0, \quad \text{i.e. } \nabla_a{}^*M_{bc} = \nabla_{[a}{}^*M_{bc]}. \tag{6.3.4}$$

In fact, (6.3.4) conversely implies the position dependence (6.3.3)(2).

* For example, if X^α and Y^α are null twistors, then $|\mathrm{X}^\alpha\bar{\mathrm{Y}}_\alpha|/|\mathrm{X}^\beta\mathrm{Y}^\gamma I_{\beta\gamma}|$ has an interpretation in Euclidean 3-space terms, being equal to

$$r\{\tfrac{1}{2} + \cos\theta\cos\phi(1-\cos\psi)^{-1}\}^{\frac{1}{2}},$$

where r is the distance, at a given time, between the two 'photons' whose world-lines are X and Y. Here ψ is the angle between their 3-velocities, and θ and ϕ are the angles that these 3-velocities make with the line from one particle to the other. In the frame for which their velocities are equal and opposite, this expression reduces to $2^{-\frac{1}{2}}$ times their distance of closest approach. The implied Poincaré invariance of the general expression is by no means obvious!

Consider, then, a physical system with momentum and angular momentum as in (6.3.2). Since the 4-momentum is future-null, we may think of the system as (equivalent to) a massless particle (i.e. of zero rest-mass). Its intrinsic spin may be expressed in terms of the Pauli–Lubanski spin vector:

$$S_a = \tfrac{1}{2} e_{abcd} p^b M^{cd}, \tag{6.3.5}$$

which, in consequence of (6.3.3), is position independent:

$$S_a = \mathring{S}_a.$$

For massless particles occurring in nature, the Pauli–Lubanski vector must be a multiple of the 4-momentum:

$$S_a = s p_a. \tag{6.3.6}$$

The real number s is called the *helicity* of the particle and $|s|$, or $|s|\hbar^{-1}$, the *spin*. For quantum systems s is an integer multiple of $\tfrac{1}{2}\hbar$. If we substitute (6.3.2) into (6.3.5), we find (*cf.* (3.4.22)):

$$\begin{aligned} S_a &= {}^*M_{ab} p^b = (\omega_{(A} \bar{\pi}_{B)} \varepsilon_{A'B'} + \bar{\omega}_{(A'} \pi_{B')} \varepsilon_{AB}) \bar{\pi}^B \pi^{B'} \\ &= \tfrac{1}{2}(\omega^B \bar{\pi}_B + \bar{\omega}^{B'} \pi_{B'}) \bar{\pi}_A \pi_{A'}. \end{aligned} \tag{6.3.7}$$

Thus (6.3.6) is indeed satisfied and we have, in agreement with (6.1.74),

$$s = \tfrac{1}{2} Z^\alpha \bar{Z}_\alpha.$$

The requirements for p_a and M^{ab} to be the 4-momentum and 6-angular momentum of a massless particle are precisely that p_a should be future-null, and that M^{ab} should, with p_a, satisfy (6.3.3) and (6.3.6). As we have seen, these requirements are automatically satisfied by the definitions (6.3.2), given the twistor Z^α ($\pi_{A'} \neq 0$). Conversely, given these physical requirements, we can, in effect, reverse the above argument and construct the twistor Z^α from the pair (p_a, M^{ab}). The resulting Z^α is then determined up to phase: p_a and M^{ab} are evidently unchanged by the replacement

$$Z^\alpha \mapsto e^{i\theta} Z^\alpha, \quad (\theta \text{ real}) \tag{6.3.8}$$

(*cf.* (6.3.2)). Thus:

(6.3.9) PROPOSITION

The momentum and angular momentum of any massless particle is precisely described, according to (6.3.2), *by a twistor* Z^α (*for which* $\pi_{A'} \neq 0$) *up to the phase indeterminacy* (6.3.8).

Accordingly, the general twistor Z^α describes a classical massless particle of helicity $\tfrac{1}{2} Z^\alpha \bar{Z}_\alpha$.

When the helicity is zero, this description is closely related to the geometrical description given earlier. The ray $\mathbf{Z}$ is the locus of points of which $\omega^A = 0$, and by (6.3.2) this is the locus of points where $M^{ab} = 0$. Thus we may think of $\mathbf{Z}$ as the *world-line of the particle*. The information contained in Z^α which is not contained in the ray $\mathbf{Z}$ is the strength ('extent') of the 4-momentum (flagpole of $\pi^{A'}$) and the phase (flag plane of $\pi^{A'}$). We may think of the flag plane of $\pi^{A'}$ as defining a kind of 'polarization plane' for the particle – but it is not clear how significant such an interpretation is. One would expect also that the phase should have a relation to the phase of the quantum-mechanical state vectors. This is actually the case, and is connected with the descriptions of §6.4, but we shall not pursue these matters here.

When the helicity is non-zero, there are no real points where $\omega^A = 0$ (*cf.* (6.1.74)), i.e., where $M^{ab} = 0$. This is to be expected since the particle possesses an intrinsic spin which contributes to its angular momentum. One might expect, however, that there would be *some* ray singled out which could be regarded as the particle's world-line. But this turns out not to be the case. The best one can do is to determine a *null hyperplane* Π, given by $p_a M^{ab} = 0$. (For a *massive* particle, this equation *does* represent a unique timelike line, which is identified with the world-line of the particle (Synge 1955).) Every null generator of Π turns out to be on an equal footing with every other null generator. In fact, there is a Poincaré transformation sending Z^α to itself and sending any given point P of Π into any other given point Q of Π. For $p_a M^{ab} = 0$ is equivalent to $\mathrm{Im}(\omega^A \bar{\pi}_A) = 0$, by (6.3.2). So by (6.1.74) we have $\omega^A \bar{\pi}_A = s$ on Π, whence $\omega^A(P)\bar{\pi}_A(P) = \omega^A(Q)\bar{\pi}_A(Q)$. Therefore the required Poincaré transformation is achieved by first translating P to Q and then applying the Lorentz transformation (in fact, a null rotation) that takes the dyad $\omega^A(P)$, $\bar{\pi}^A(P)$ at Q to $\omega^A(Q)$, $\bar{\pi}^A(Q)$ at Q. Since Z^α is determined by its spinor parts at any one of its points, this Poincaré transformation sends Z^α to itself. So, in this sense, a classical massless particle with non-zero intrinsic spin is *not* localized.

It is of some interest that the Robinson congruence defined by Z^α also has direct relevance to the angular momentum structure of a spinning massless particle. The angular momentum about any point of $\mathbb{M}$ is determined by ω^A and $\bar{\pi}^A$, the PND of M^{ab} (*cf.* (3.5.18) *et seq.* and (3.4.20), (6.3.2)) being in fact the flagpole directions of these spinors. The flagpole direction of $\bar{\pi}^A$ is constant and is the direction of the 4-momentum, whereas the flagpole direction of ω^A is precisely the direction of the Robinson congruence at the point in question.

Angular momentum for general systems

We now briefly discuss the twistor description of momentum and angular momentum of *massive* particles or systems. Assume p_a to be real and M^{ab} to be real and skew, with position dependence given by (6.3.3) so (6.3.4) is satisfied. (For physical systems p_a will be future-timelike, but this does not affect the discussion.) Then, for some $\mu^{A'B'} \in \mathfrak{S}^{(A'B')}$ we have (*cf.* (3.4.20)):

$$M^{ab} = \bar{\mu}^{AB}\varepsilon^{A'B'} + \mu^{A'B'}\varepsilon^{AB}. \tag{6.3.10}$$

We now define (Penrose and MacCallum 1972) the *angular momentum twistor* (or *moment* twistor or *kinematic* twistor*) $\mathsf{A}_{\alpha\beta} \in \mathbb{T}_{(\alpha\beta)}$

$$\mathsf{A}_{\alpha\beta} := \begin{pmatrix} 0 & p_A{}^{B'} \\ p^{A'}{}_B & 2i\mu^{A'B'} \end{pmatrix}, \quad \bar{\mathsf{A}}^{\alpha\beta} = \begin{pmatrix} -2i\bar{\mu}^{AB} & p^A{}_{B'} \\ p_{A'}{}^B & 0 \end{pmatrix}. \tag{6.3.11}$$

(Without confusion, we could, if desired, define $\mathsf{A}^{\alpha\beta} := \bar{\mathsf{A}}^{\alpha\beta}$, as with $\mathsf{I}^{\alpha\beta}$ and $\varepsilon^{A'B'}$.) That it *is* a twistor can be verified easily by using (6.3.3) and (6.3.10) in checking the position dependence (6.1.51). It possesses two special properties. The first is its *symmetry* (*cf.* (6.1.66)). The second is that

$$\mathsf{I}_{\gamma\beta}\bar{\mathsf{A}}^{\alpha\beta} = \mathsf{I}^{\alpha\beta}\mathsf{A}_{\gamma\beta}, \tag{6.3.12}$$

which may be verified directly. It can easily be shown that, as a consequence of symmetry and (6.3.12), $\mathsf{A}_{\alpha\beta}$ is expressible in the form

$$\mathsf{A}_{\alpha\beta} = 2\mathsf{E}^{\gamma}{}_{(\alpha}\mathsf{I}_{\beta)\gamma} \tag{6.3.13}$$

where $\mathsf{E}^{\gamma}{}_{\alpha}$ is some Hermitian $[^1_1]$-twistor (*cf.* (6.1.68)) subject to a 'gauge freedom'

$$\mathsf{E}^{\alpha}{}_{\beta} \mapsto \mathsf{E}^{\alpha}{}_{\beta} + \mathsf{L}^{\alpha\gamma}\mathsf{I}_{\gamma\beta} + \bar{\mathsf{L}}_{\beta\gamma}\mathsf{I}^{\gamma\alpha} \tag{6.3.14}$$

$\mathsf{L}^{\alpha\beta}$ being an arbitrary element of $\mathbb{T}^{[\alpha\beta]}$. The decomposition (6.3.13) is, in fact, achieved by choosing the spinor parts $\theta^A{}_B = \mathsf{E}^A{}_B$ and $\eta_{BA'} = \mathsf{E}_{A'B}$ of $\mathsf{E}^{\alpha}{}_{\beta}$ (*cf.* (6.1.54)) to satisfy

$$\eta_{AB'}(O) = p_{AB'}(O), \quad i\theta^{(AB)}(O) = \bar{\mu}^{AB}(O),$$

$\theta^{[AB]}$, $\xi^{AB'}$ being arbitrary at O (corresponding to (6.3.14)), but of course related to these parts via twistor differential equations. The relation (6.3.13) between $\mathsf{A}_{\alpha\beta}$ and $\mathsf{E}^{\alpha}{}_{\beta}$ can, in effect, be expressed in the spinor form

$$\nabla^{(C}_{C'}\xi^{A)}_{A'} = \varepsilon_{C'A'}\bar{\mu}^{AC}, \tag{6.3.15}$$

where $\xi^{AB'} = \mathsf{E}^{AB'}$ is the primary part of $\mathsf{E}^{\alpha}{}_{\beta}$ (*cf.* (6.1.54)). For, by (6.1.56)(1),

* Corresponding twistors (of higher valence) for describing the higher multipole structure of a system have been given by Curtis (1978*b*).

the LHS is $i\varepsilon_{C'A'}\theta^{(AC)}$, which, by (6.3.13), equals the RHS. The equation $\eta_{AB'} = p_{AB'}$ is actually a consequence of $i\theta^{(AB)} = \bar{\mu}^{AB}$ holding everywhere: for if we know $\theta^{(AB)}$ everywhere, we know $\xi_{AB'}$ (*cf.* (6.1.55)(3), (4)), and $\bar{\mu}^{AB}$ similarly determines $p_{AB'}$.

Notice that from (6.3.15) and (6.3.10) we have

$$\begin{aligned} M^{ab} &= \nabla^{[A'|(A}\xi^{B)|B']} + \nabla^{(A'|[A}\xi^{B]|B')} \\ &= \nabla^{[a}\xi^{b]}. \end{aligned} \tag{6.3.16}$$

Recalling that (6.1.70) is necessarily satisfied by the primary part of $\mathrm{E}^{\alpha}{}_{\beta}$, and referring forward to §6.5 for the significance of this equation, called the *conformal Killing equation*, with solutions ξ^a called *conformal Killing vectors*, we can now state the following:

(6.3.17) PROPOSITION

The dependence on position of the angular momentum of a system is precisely such that M^{ab} should be the curl of a conformal Killing vector.

In the case of a massless system we may choose, by (6.3.24) (to be given presently),

$$\xi^a = \omega^A \bar{\omega}^{A'}. \tag{6.3.18}$$

We can verify directly (from (6.1.9)) that (6.3.16) holds for the angular momentum tensor (6.3.2). We have seen that the flagpole directions of ω^A are tangents to the lines of a Robinson congruence. We see from (6.3.18) that these directions are the directions of a future-null conformal Killing vector field. We shall prove the converse in a moment, so that we have:

(6.3.19) PROPOSITION

Any future-null conformal Killing vector field in Minkowski space is the flagpole field of the primary part ω^A of a twistor Z^{α}; and conversely.

Propositions (6.3.17) and (6.3.19) now yield:

(6.3.20) PROPOSITION

M^{ab} is the angular momentum of a massless particle (subject to (6.3.6)) if and only if M^{ab} is the curl of a future-null conformal Killing vector field.

The proof of the first half of (6.3.19) is not quite trivial. If $\nabla^{(A}_{(A'}\xi^{B)}_{B')} = 0$ and $\xi^a = \rho^A \bar{\rho}^{A'}$, then

$$\bar{\rho}^{(B'}\nabla^{A')(A}\rho^{B)} + \rho^{(B}\nabla^{A)(A'}\bar{\rho}^{B')} = 0. \tag{6.3.21}$$

We shall have established our result if we can find a real χ such that $\omega^A = e^{i\chi}\rho^A$ satisfies the twistor equation (6.1.1): $\nabla^{(A}_{A'}\omega^{B)} = 0$. So consider the equation $\nabla^{(A}_{A'}\{\rho^{B)}e^{i\chi}\} = 0$, i.e.,

$$\rho^{(B}\nabla^{A)}_{A'}\chi = i\nabla^{(A}_{A'}\rho^{B)}. \tag{6.3.22}$$

Now (6.3.21) implies

$$\rho_A\rho_B\rho^{(B'}\nabla^{A')A}\rho^B = 0,$$

so by reference to Proposition (3.5.15) we conclude

$$\rho_A\rho_B\nabla^{A'A}\rho^B = 0.$$

Then Proposition (3.5.27) shows that $\nabla^{A'(A}\rho^{B)}$ indeed has the form

$$i\nabla^{A'(A}\rho^{B)} = \rho^{(B}V^{A)A'} \tag{6.3.23}$$

needed for the satisfaction of (6.3.22), where V^a is uniquely determined, by (3.5.15). It remains to show that V^a is a real gradient. Equations (6.3.23) and (6.3.21) imply

$$\rho^{(B}V^{A)(A'}\bar{\rho}^{B')} = \rho^{(B}\bar{V}^{A)(A'}\bar{\rho}^{B')},$$

whence, by (3.5.15), $V^{AA'} = \bar{V}^{AA'}$, i.e., V^a is real. If we differentiate (6.3.23), contract and symmetrize the resulting equation (commuting derivatives – we are in $\mathbb{M}$), we get

$$\begin{aligned} 0 = i\nabla_{A'(C}\nabla^{A'}_{A}\rho_{B)} &= \nabla_{A'(C}\{\rho_B V^{A'}_{A)}\} \\ &= -iV^{A'}_{(A}\rho_B V_{C)A'} + \nabla_{A'(C}V^{A'}_{A}\rho_{B)}. \end{aligned}$$

The first term on the right vanishes, by symmetry; and so by (3.5.15) we can conclude

$$\nabla_{A'(C}V^{A'}_{A)} = 0.$$

But since V_a is real, this is equivalent (*cf.* (5.1.46)) to $\nabla_{[a}V_{c]} = 0$, whence there exists a χ such that $V_a = \nabla_a\chi$ and our result is established.

Decomposition of $\mathsf{A}_{\alpha\beta}$ in terms of 1-index twistors

Returning to the discussion of $\mathsf{A}_{\alpha\beta}$, if that is the angular momentum twistor of a massless particle, we may take (*cf.* (6.3.2))

$$\mathsf{E}^\alpha{}_\beta = \mathsf{Z}^\alpha\bar{\mathsf{Z}}_\beta = \begin{pmatrix} \omega^A\bar{\pi}_B & \omega^A\bar{\omega}^{B'} \\ \pi_{A'}\bar{\pi}_B & \pi_{A'}\bar{\omega}^{B'} \end{pmatrix}. \tag{6.3.24}$$

Thus, for a massless particle, $\mathsf{A}_{\alpha\beta}$ takes the form

$$\mathsf{A}_{\alpha\beta} = 2\bar{\mathsf{Z}}_{(\alpha}\mathsf{I}_{\beta)\gamma}\mathsf{Z}^\gamma. \tag{6.3.25}$$

When p_a is future-timelike, $\mathrm{E}^{\alpha}{}_{\beta}$ can be taken to be a sum of expressions of this form (and sums and differences of such expressions when p_a is general). In fact, it is not hard to show that any angular momentum twistor with future-timelike momentum can be expressed in the form

$$\mathrm{A}_{\alpha\beta} = 2\bar{\mathrm{X}}_{(\alpha}\mathrm{I}_{\beta)\gamma}\mathrm{X}^{\gamma} + 2\bar{\mathrm{Z}}_{(\alpha}\mathrm{I}_{\beta)\gamma}\mathrm{Z}^{\gamma}, \tag{6.3.26}$$

with considerable freedom in the choice of X^{α} and Z^{α}. This freedom is given by

$$\begin{aligned} \mathrm{X}^{\alpha} &\mapsto a\mathrm{X}^{\alpha} + b\mathrm{Z}^{\alpha} + \lambda\bar{c}\mathrm{I}^{\alpha\beta}\bar{\mathrm{X}}_{\beta} + \lambda\bar{d}\mathrm{I}^{\alpha\beta}\bar{\mathrm{Z}}_{\beta} \\ \mathrm{Z}^{\alpha} &\mapsto c\mathrm{X}^{\alpha} + d\mathrm{Z}^{\alpha} - \lambda\bar{a}\mathrm{I}^{\alpha\beta}\bar{\mathrm{X}}_{\beta} - \lambda\bar{b}\mathrm{I}^{\alpha\beta}\bar{\mathrm{Z}}_{\beta}, \end{aligned} \tag{6.3.27}$$

where $\left(\begin{smallmatrix} a & b \\ c & d \end{smallmatrix}\right)$ is a unitary matrix and $\lambda \in \mathbb{C}$. According to the *twistor particle programme**, a single massive particle is described in terms of two or more twistors, its angular momentum twistor being given in the case of two twistors by (6.3.26), and generally by

$$\mathrm{A}_{\alpha\beta} = 2\bar{\mathrm{X}}_{(\alpha}\mathrm{I}_{\beta)\gamma}\mathrm{X}^{\gamma} + \cdots + 2\bar{\mathrm{Z}}_{(\alpha}\mathrm{I}_{\beta)\gamma}\mathrm{Z}^{\gamma} \tag{6.3.28}$$

and, in particular, the rest-mass m of the system is given by a sum of $\frac{1}{2}n(n-1)$ terms:

$$m^2 = -\tfrac{1}{2}\mathrm{A}_{\alpha\beta}\bar{\mathrm{A}}^{\alpha\beta} = 2|\mathrm{X}^{\alpha}\mathrm{Y}^{\beta}\mathrm{I}_{\alpha\beta}|^2 + \cdots + 2|\mathrm{X}^{\alpha}\mathrm{Z}^{\beta}\mathrm{I}_{\alpha\beta}|^2 + \cdots + 2|\mathrm{Y}^{\alpha}\mathrm{Z}^{\beta}\mathrm{I}_{\alpha\beta}|^2 + \cdots$$

The freedom (6.3.27) generalizes, for n twistors, to

$$\begin{pmatrix} \mathrm{X}^{\alpha} \\ \vdots \\ \mathrm{Z}^{\alpha} \end{pmatrix} \mapsto \mathbf{U}\begin{pmatrix} \mathrm{X}^{\alpha} \\ \vdots \\ \mathrm{Z}^{\alpha} \end{pmatrix} + \mathbf{\Lambda}\bar{\mathbf{U}}\begin{pmatrix} \mathrm{I}^{\alpha\beta}\bar{\mathrm{X}}_{\beta} \\ \vdots \\ \mathrm{I}^{\alpha\beta}\bar{\mathrm{Z}}_{\beta} \end{pmatrix} \tag{6.3.29}$$

where $\mathbf{U}$ is an $(n \times n)$ unitary matrix and $\mathbf{\Lambda}$ an $(n \times n)$ complex skew-symmetric matrix. (Note that when $n = 1$ this gives the freedom (6.3.8) discussed before.) The transformations (6.3.29) constitute the *n-twistor internal symmetry group*,** or simply the *n-twistor group*. Its multiplication law is

$$(\mathbf{U}, \mathbf{\Lambda}) \times (\tilde{\mathbf{U}}, \tilde{\mathbf{\Lambda}}) = (\mathbf{U}\tilde{\mathbf{U}}, \mathbf{\Lambda} + \mathbf{U}\tilde{\mathbf{\Lambda}}\mathbf{U}^{\mathrm{T}}).$$

Finally it may be remarked that the future-timelike [future-causal†] condition on p_a can be expressed twistorially as the positive-definiteness

* See Penrose (1975*b*), Hughston (1979, 1980), Perjés (1975, 1977, 1982), Perjés and Sparling (1979), Sparling (1981); also Hughston and Ward (1979) and references cited therein.

** See, for example, Penrose and Sparling (1979) and references cited in the previous footnote.

† Recall (Volume 1) that the word 'causal' stands for 'timelike or null' and 'future-causal' asserts that the vector is future-pointing.

[semi-definiteness] (in Z^{α}) of the expression

$$\mathrm{Z}^{\alpha}\mathrm{A}_{\alpha\gamma}\mathrm{I}^{\beta\gamma}\mathrm{Z}_{\beta}. \tag{6.3.31}$$

It follows that the positive-definiteness of (6.3.31), together with symmetry and the relation (6.3.12), is a necessary and sufficient condition on the twistor $\mathrm{A}_{\alpha\beta}$ that it be expressible in the form (6.3.28) with more than one independent term.

6.4 Symmetric twistors and massless fields

We now return to a discussion of the differential equation (6.1.62) for the primary part of a trace-free symmetric twistor, and study some of its remarkable relations to the massless field equation (4.12.42). First consider the case of an r-unprimed-index symmetric spinor $\lambda^{AB\ldots D}$ which is the primary part of a symmetric $[{}^{r}_{0}]$-twistor, and therefore subject to (6.1.62):

$$\nabla_{M'}^{(M}\lambda^{AB\ldots D)}=0. \tag{6.4.1}$$

(Equations (6.1.1) and (6.1.69) are examples of this.) Now if $\phi_{A\ldots L}$ is symmetric with n indices ($n>r$) and satisfies (4.12.42), and $\lambda^{A\ldots D}$ is any symmetric solution of (6.4.1), then one sees that

$$\nabla^{EE'}(\phi_{A\ldots DE\ldots L}\lambda^{A\ldots D})=0. \tag{6.4.2}$$

(Refer to (6.4.31) for the case $n=r$.) The parenthesis represents a new massless free field of spin $\frac{1}{2}(n-r)$, so we have achieved a *spin lowering* by $\frac{1}{2}r$ units. This construction of new massless fields works formally in curved as well as flat space–time, but as it stands it is generally uninteresting in non-conformally flat curved space–time because consistency conditions like (6.1.6) arise for (6.4.1) which tend to prevent non-zero solutions. (However, see §9.9.)

Conserved integrals for linear gravity

The particular case with $n=4$ and $r=2$ is of interest in linearized Einstein theory. (The important generalization to the full theory will be given in (9.9.16).) We have seen in (5.7.4), (5.7.8) that a symmetric spinor ϕ_{ABCD} subject to (4.12.42) in $\mathbb{M}$ describes a free gravitational field in the weak-field limit. Thus, for each symmetric solution σ^{AB} of (6.1.69), the spinor

$$\chi_{AB}=\phi_{ABCD}\sigma^{CD} \tag{6.4.3}$$

satisfies Maxwell's free-space equations (5.1.57), even though it may have nothing to do with electromagnetism. Now suppose the sources of ϕ_{ABCD} lie in a limited region spatially surrounded by vacuum. The 'Maxwell' field χ_{AB} must also have its sources in that region, since it satisfies the free-field equations when ϕ_{ABCD} does. In Maxwell theory there is a well-known procedure for evaluating the *charge* of the source, both electric and magnetic, allowing the latter as a theoretical possibility. We wish to apply this procedure to χ_{AB} and thereby obtain information about the sources of ϕ_{ABCD}. This information will turn out to provide us with a definition of the *4-momentum* and *6-angular momentum* of these sources.

The procedure depends on the fundamental theorem of exterior calculus (4.3.25) and consists in expressing the total charge in terms of an integral over a closed 2-surface $\mathscr{S}$ surrounding the source region. Suppose J^a is a charge–current vector describing the source for a Maxwell field F_{ab}. Then the total charge in a (say spacelike) 3-volume $\mathscr{V}$ is

$$q = \tfrac{1}{6}\int_{\mathscr{V}} J^{\mathbf{a}} e_{\mathbf{apqr}}\, \mathrm{d}x^{\mathbf{p}} \wedge \mathrm{d}x^{\mathbf{q}} \wedge \mathrm{d}x^{\mathbf{r}} = -\tfrac{1}{6}\int_{\mathscr{V}} {}^{\dagger}\boldsymbol{J}$$

where the differential form notation of (4.3.20), (5.9.5) is being used. (Note that for $\mathscr{V}$ in T = constant, with standard Minkowski coordinates in $\mathbb{M}$, this would be $q = \int_{\mathscr{V}} J^0\, \mathrm{d}X \wedge \mathrm{d}Y \wedge \mathrm{d}Z$.) The Maxwell equation $\mathrm{d}\,{}^*\boldsymbol{F} = \tfrac{4}{3}\pi\,{}^{\dagger}\boldsymbol{J}$ (*cf.* (5.9.13)), together with (4.3.25), now gives

$$q = -\frac{1}{8\pi}\oint_{\mathscr{S}} {}^*\boldsymbol{F}$$

where $\mathscr{V}$ is taken to span $\mathscr{S}$ (i.e. to be compact with boundary $\partial\mathscr{V} = \mathscr{S}$). We can rewrite this as

$$q = \frac{1}{4\pi}\mathrm{Im}\oint_{\mathscr{S}} \varphi_{AB}\varepsilon_{A'B'}\, \mathrm{d}x^a \wedge \mathrm{d}x^b$$

where $F_{ab} = \varphi_{AB}\varepsilon_{A'B'} + \varepsilon_{AB}\bar{\varphi}_{A'B'}$, ${}^*F_{ab} = -\mathrm{i}\varphi_{AB}\varepsilon_{A'B'} + \mathrm{i}\varepsilon_{AB}\bar{\varphi}_{A'B'}$ as in (3.4.20), (3.4.22) and where $\mathrm{d}x^a$ stands for $g_{i_1}{}^a$, $\mathrm{d}x^a \wedge \mathrm{d}x^b$ for $g_{[i_1}{}^a g_{i_2]}{}^b$, etc. Similarly, were there to be a 'magnetic charge' μ this would be given by

$$\mu = \frac{1}{4\pi}\mathrm{Re}\oint_{\mathscr{S}} \varphi_{AB}\varepsilon_{A'B'}\, \mathrm{d}x^a \wedge \mathrm{d}x^b$$

whence

$$\mu + \mathrm{i}q = \frac{1}{4\pi}\oint_{\mathscr{S}} \varphi_{AB}\varepsilon_{A'B'}\, \mathrm{d}x^a \wedge \mathrm{d}x^b. \tag{6.4.4}$$

The integral (6.4.4), as a consequence of Maxwell's free-space equations, is insensitive to deformation of $\mathscr{S}$ through a source-free region. (In particular it is constant in time if 'charges' do not cross the region – an expression of 'charge' conservation). For if $\mathscr{S}$ moves to $\mathscr{S}'$, sweeping out the

3-volume $\mathscr{V}$, we have $\partial\mathscr{V} = \mathscr{S}' - \mathscr{S}$ and the result follows from the above since ${}^{\dagger}J = 0$ on $\mathscr{V}$.

We now apply this to χ_{AB}, where in place of F_{ab} we have the real bivector P_{ab} corresponding to χ_{AB} (*cf.* (3.4.38)):

$$P_{ab} = \chi_{AB}\varepsilon_{A'B'} + \varepsilon_{AB}\bar{\chi}_{A'B'}, \quad \chi_{AB}\varepsilon_{A'B'} = {}^{-}P_{ab} = \tfrac{1}{2}(P_{ab} + \mathrm{i}{}^{*}P_{ab}). \quad (6.4.5)$$

For each arbitrary symmetric solution of (6.1.69), in conjunction with the given field ϕ_{ABCD}, we obtain one complex 'charge integral', i.e., two real conserved quantities. We have seen that the symmetric solutions of (6.1.69) are given by (6.1.51)(4), subject to (6.1.66). Consequently they possess *ten complex* degrees of freedom: three for $\mathring{\sigma}^{AB}$, four for $\mathring{\rho}^{A}_{B'}$, and three for $\mathring{\kappa}_{A'B'}$. In this way we get ten independent complex charge integrals, i.e. twenty independent real charge integrals, for the field ϕ_{ABCD}.

However, for a ϕ_{ABCD} arising physically, we would expect only *ten real* conserved quantities, namely the 4-momentum and 6-angular momentum of the source. Indeed, it actually turns out that ten linearly independent real integrals of the above kind vanish identically provided ϕ_{ABCD} arises from a real h_{ab} according to (5.7.12). Moreover, as we shall see in §6.5, this conclusion also directly follows if a tensor K_{abcd} with Riemann tensor symmetries, subject to the Bianchi identity equation $\nabla_{[a}K_{bc]de} = 0$, exists throughout $\mathscr{V}$ where K_{abcd} reduces to the form (5.7.8), with the given ϕ_{ABCD}, at $\mathscr{S}$. The source is given by the E_{ac} of (5.7.6). The ten surviving independent integrals arise from the three complex components of $\mathring{\kappa}_{A'B'}$ and the four real components of $(\mathring{\rho}^{a} + \bar{\mathring{\rho}}^{a})$. The integrals which vanish come from $\mathring{\sigma}^{AB}$ and $\mathrm{i}(\bar{\mathring{\rho}}^{a} - \mathring{\rho}^{a})$. Compare Sachs and Bergmann (1958).

It may seem surprising that these *vanishing* integrals arise from just those spinor parts of $\mathrm{S}^{\alpha\beta}$ which *survive* when $\mathrm{S}^{\alpha\beta}$ is specialized to $\mathrm{i}\bar{\mathrm{A}}^{\alpha\beta}$, i.e., to the form i × (an angular momentum twistor), so they are the parts with a *direct* interpretation as 4-momentum and 6-angular momentum; whereas the 4-momentum and 6-angular momentum *integrals* actually come from the 'rest' of $\mathrm{S}^{\alpha\beta}$. This apparent paradox will be explained later (pp. 88, 93).

We shall later need the tensor version of (6.1.69) when σ^{AB} is symmetric:

$$\nabla^{(a}Q^{b)c} - \nabla^{(a}Q^{c)b} + g^{a[b}\nabla_{d}Q^{c]d} = 0, \quad (6.4.6)$$

where the skew tensor Q^{ab} is given by

$$Q^{ab} := \mathrm{i}\sigma^{AB}\varepsilon^{A'B'} - \mathrm{i}\bar{\sigma}^{A'B'}\varepsilon^{AB}. \quad (6.4.7)$$

Proof of the equivalence of (6.4.6) with (6.1.69) can be obtained directly by substituting (6.4.7) into (6.4.6). Moreover, the general solution of (6.4.6) is

$$Q^{ab} = \mathring{Q}^{ab} + 4U^{[a}x^{b]} - 2e^{abcd}V_{c}x_{d} - \tfrac{1}{2}K^{ab}x_{c}x^{c} + 2K^{c[b}x^{a]}x_{c}, \quad (6.4.8)$$

where the constant tensors U^a, V^a, K^{ab} are defined by

$$U^a + \mathrm{i}V^a = \overset{\circ}{\rho}{}^a, \quad K_{ab} = \mathrm{i}\kappa_{A'B'}\varepsilon_{AB} - \mathrm{i}\bar{\kappa}_{AB}\varepsilon_{A'B'}. \tag{6.4.9}$$

The solution (6.4.8) is obtained by translating the corresponding spinor solution, using (3.4.53) etc.

In the same way that contracting σ^{AB} with ϕ_{ABCD} yields a spin-1 zero rest-mass field, so also contracting Q^{ab} with the tensor version K_{abcd} of ϕ_{ABCD} (*cf.* (5.7.8)) yields a skew two-index tensor satisfying the Maxwell free-field equations. This is easily seen since, by (5.7.8) and (6.4.7),

$$K_{abcd}Q^{cd} = 2(\mathrm{i}\chi_{AB}\varepsilon_{A'B'} - \mathrm{i}\bar{\chi}_{A'B'}\varepsilon_{AB}) = -2\,{}^*P_{ab} \tag{6.4.10}$$

where χ_{AB} is given by (6.4.3). And since χ_{AB} satisfies the free-field Maxwell equations in spinor form, our assertion follows.

The general case; potentials

We next consider how the general equation (6.1.62),

$$\nabla^{(M}_{(M'}\lambda^{AB\ldots D)}_{P'Q'\ldots S')} = 0 \tag{6.4.11}$$

relates to the massless field equation (4.12.42). Equation (6.4.11) applies to a symmetric spinor $\lambda^{A\ldots DP'\ldots S'}$ of valence $[\begin{smallmatrix} p & q \\ 0 & 0 \end{smallmatrix}]$ which is the primary part of a trace-free symmetric $[\begin{smallmatrix} q \\ p \end{smallmatrix}]$-twistor. Let $\phi_{A\ldots L}$ (with $n > p$ indices) be a symmetric solution of (4.12.42), and define a new field

$$\chi^{P'\ldots S'}_{EF\ldots L} = \phi_{A\ldots DEF\ldots L}\lambda^{A\ldots DP'\ldots S'} \tag{6.4.12}$$

Then, by (6.4.11) and (4.12.42), this field satisfies

$$\nabla^{E(T'}\chi^{P'\ldots S')}_{EF\ldots L} = 0. \tag{6.4.13}$$

(In fact, equation (6.4.13) is conformally invariant, as we shall see in (6.7.31).) Suppose now that some symmetric field $\psi^{T'\ldots V'}_{A\ldots L}$ of valence $[\begin{smallmatrix} 0 & r \\ n-r & 0 \end{smallmatrix}]$ (with $n - r > p$) satisfies (6.4.13) – of which the massless field equation is a special case with $r = 0$ – and we use $\psi^{\cdots}_{\ldots}$ rather than $\phi_{\ldots}$ in the definition of $\chi^{\cdots}_{\ldots}$:

$$\chi^{P'\ldots S'T'\ldots V'}_{E\ldots L} = \lambda^{A\ldots D(P'\ldots S'}\psi^{T'\ldots V')}_{A\ldots DE\ldots L}, \tag{6.4.14}$$

then $\chi^{\cdots}_{\ldots}$ again satisfies (6.4.13).

A particular case of interest is given when $n - r = 1$. For it follows from results of twistor theory (*cf.* Eastwood, Penrose and Wells 1981, Penrose and Ward 1980), and can also be shown directly (using an argument due to G.A.J. Sparling), that if $\psi^{B'\ldots L'}_{A}$ is symmetric and of valence $[\begin{smallmatrix} 0 & n-1 \\ 1 & 0 \end{smallmatrix}]$, and subject to the equation

$$\nabla^{A(A'}\psi^{B'\ldots L')}_{A} = 0, \tag{6.4.15}$$

then (in flat space–time) the field

$$\phi^{AB\ldots L} = \nabla^{(B}_{B'}\cdots\nabla^{L}_{L'}\psi^{A)B'\ldots L'} \tag{6.4.16}$$

satisfies the massless field equation (4.12.42). Consequently, such a $\psi^{\cdots}_{\cdot}$, subject to (6.4.15), acts as a kind of potential for $\phi_{\ldots}$ (compare (5.7.12)). Furthermore,the general massless $\phi_{\ldots}$ can locally be expressed in this way. The gauge freedom turns out to be

$$\psi_A^{B'C'\ldots L'} \mapsto \psi_A^{B'C'\ldots L'} + \nabla_A^{(B'}\theta^{C'\ldots L')}, \tag{6.4.17}$$

for arbitrary (symmetric) $\theta^{C'\ldots L'}$ of valence $[^{0}_{0}\,{}^{n-2}_{0}]$, unless $n = 1$, when there is no gauge freedom, or $n = 0$, when no potential exists. In fact, (6.4.16) is conformally invariant in the *weak* sense that the relation goes over into itself for any conformal rescaling that sends the metric into another which is again flat, $\psi_A^{B'\ldots L'}$ being taken to have conformal weight -1.

Spin raising

Any symmetric $[^{0}_{q}]$-twistor has a symmetric primary part $\lambda^{P'\ldots S'}$ which satisfies

$$\nabla_M^{(M'}\lambda^{P'\ldots S')} = 0 \tag{6.4.18}$$

(this being a special case of (6.4.11) – also the conjugate of (6.4.1)), and which can therefore be used, as in (6.4.14), to convert a potential for an n-unprimed-index massless field into a potential for an $(n+q)$-unprimed-index massless field:

$$\chi_A^{B'\ldots E'F'\ldots L'} = \psi_A^{(B'\ldots E'}\lambda^{F'\ldots L')}. \tag{6.4.19}$$

Differentiating $\psi^{\cdots}_{\cdot}$ $(n-1)$ times to get the first field, and differentiating $\chi^{\cdots}_{\cdot}$ $(n+q-1)$ times to get the second field, we find an alternative ('dual') way to that of (6.4.2) to construct a new massless field from another of different spin. The general expression is complicated, but in the special case when $q = 1$, this process leads from the n-index massless field $\phi_{\ldots}$ to the $(n+1)$-index massless field

$$\gamma_{A\ldots LM} = \lambda^{M'}\nabla_{MM'}\phi_{A\ldots L} + \mathrm{i}(n+1)\rho_{(M}\phi_{A\ldots L)}, \tag{6.4.20}$$

where $\lambda^{A'}$ is the primary part of the $[^{0}_{1}]$-twistor $(\rho_A, \lambda^{A'})$, so $\rho_A\varepsilon_{B'}{}^{C'} = -\mathrm{i}\nabla_{AB'}\lambda^{C'}$ (Penrose 1965). Compare this with (6.4.2) in the case $r = 1$, where the $[^{1}_{0}]$-twistor $(\lambda^A, \rho_{A'})$ effects the 'lowering of spin'

$$\gamma_{A\ldots K} = \phi_{A\ldots KL}\lambda^{L}. \tag{6.4.21}$$

The spin-lowering and -raising formulae (6.4.2), (6.4.3), (6.4.12), (6.4.20), (6.4.21), (also (6.4.31) below) and certain instances of (6.4.12), (6.4.14) illustrate ways that a trace-free symmetric twistor may be used to shift the spin of a massless field. We could, of course, also have written down the complex conjugates of these expressions. There is an essential unity in all these results which will be illuminated in §6.10 (*cf.* (6.10.37), (6.10.38)). In terms of the concept of *helicity* of a (positive frequency) massless field (*cf.* after (5.7.3)) the spin shift occurs in a uniform way: for a trace-free symmetric twistor of valence $[{}^p_q]$, the helicity increase is precisely $\frac{1}{2}(p-q)\hbar$.

Other types of potential; relations to $\square$

It may be remarked that there is another form of (completely symmetric) potential $\psi_{\cdots}^{\cdots}$ for a massless $\phi_{\ldots}$ of arbitrary spin (in flat space–time) that is somewhat easier to handle, the relation between field and potential being

$$\phi_{A\ldots EF\ldots L} = \nabla_{AA'}\ldots\nabla_{EE'}\psi^{A'\ldots E'}_{F\ldots L}, \tag{6.4.22}$$

and $\psi_{\cdots}^{\cdots}$ now being subject to a somewhat stronger equation than (6.4.13) (namely, lacking the symmetrization):

$$\nabla^{EE'}\psi^{A'\ldots D'}_{EG\ldots L} = 0. \tag{6.4.23}$$

This equation does not have the conformal invariance properties of (6.4.13). On the other hand, for a given $\phi_{\ldots}$, we can find such potentials (locally) with *any* number of primed and unprimed indices totalling n (not just one unprimed). In fact, if $\psi_{\cdots}^{\cdots}$ has at least one unprimed index (the contrary case will be dealt with below), then (6.4.23) implies that

$$\beta^{A'\ldots C'}_{DE\ldots L} = \nabla_{DD'}\psi^{A'\ldots C'D'}_{E\ldots L} \tag{6.4.24}$$

is symmetric in DE, and therefore totally symmetric; and also that $\beta_{\cdots}^{\cdots}$ satisfies the corresponding equation (6.4.23). (In proving this, we must commute derivatives, which restricts us to flat space–time.) Continuing, we thus arrive at (6.4.22), the symmetry of $\phi_{\ldots}$ and its satisfaction of the massless field equation being the final result. We may remark that nowhere did we use the symmetry of $\psi_{\cdots}^{\cdots}$ in the primed indices, and this symmetry need not be assumed; but since all skew parts drop out of (6.4.22), symmetry can be imposed without loss of generality. We also note that (6.4.23) implies

$$\nabla^{E}_{A'}\psi^{A'B'\ldots D'}_{EG\ldots L} = 0, \tag{6.4.25}$$

which generalizes the *Lorenz gauge* condition (5.1.47) on the potential of electromagnetic theory. Furthermore, (6.4.23) implies that

$$\nabla^{M'}_{M}\nabla^{N'}_{N}\psi^{A'\ldots D'}_{E\ldots L}$$

is symmetric in N, E, and also in M, E (since derivatives commute in flat space) and therefore in M, N, so $\psi^{\cdots}_{\cdots}$ satisfies the *wave equation*

$$\Box \psi^{A'\ldots D'}_{E\ldots L} = 0, \tag{6.4.26}$$

where

$$\Box = \varepsilon^{MN}\varepsilon_{M'N'}\nabla^{M'}_{M}\nabla^{N'}_{N} = \nabla_a\nabla^a$$

is the D'Alembertian operator of (5.10.6) (*cf.* also (6.8.26) below).

We can even find a 'Hertz'-type potential $\chi^{A'\ldots L'}$ with *no* unprimed indices, assumed symmetric without loss of generality, and subject *only* to

$$\Box \chi^{A'\ldots L'} = 0. \tag{6.4.27}$$

As in (5.10.7), (6.4.27) implies $\nabla^{M'}_{[A}\nabla^{N'}_{B]}\chi^{A'\ldots L'} = 0$, whence

$$\nabla^{AM'}(\nabla_{AA'}\chi^{A'B'\ldots L'}) = 0.$$

The quantity in parentheses can now take the place of $\psi^{\cdots}_{\cdots}$ in (6.4.24), and the previous argument can be repeated, leading finally to

$$\phi_{A\ldots L} = \nabla_{AA'}\ldots\nabla_{LL'}\chi^{A'\ldots L'}. \tag{6.4.28}$$

Each of the expressions (6.4.22) for the various $\psi^{\cdots}_{\cdots}$ satisfying the required conditions (including (6.4.28)) represents (locally) the general massless field. Each has a gauge freedom, which can be shown to involve only massless fields of lower spin than $\phi_{\ldots}$, and the various derivatives of such fields (Penrose 1965). In fact this gauge freedom can be used to reduce the $\chi^{A'\ldots L'}$ of (6.4.27), (6.4.28) still further, to the form $\chi P^{A'\ldots L'}$, locally, where $P^{A'\ldots L'}$ is a constant spinor, independent of the field $\phi_{A\ldots L}$ and where $\Box\chi = 0$. This shows that the free field $\phi_{A\ldots L}$ has the same amount of 'freedom' for each spin, locally, as a complex *scalar* massless (D'Alembert) field χ, a fact of which we are already aware from the discussion of §5.11.

It is of some interest to note that the primary part of any symmetric $[^{r}_{0}]$-twistor satisfies the wave equation in flat space–time:

$$\nabla^{(M}_{M'}\lambda^{AB\ldots L)} = 0, \quad \lambda^{A\ldots L} = \lambda^{(A\ldots L)} \quad \textit{implies } \Box\lambda^{A\ldots L} = 0 \tag{6.4.29}$$

(Gårding 1945). To prove this we first note that, in $\mathbb{M}$,

$$\nabla^{M'N}\nabla^{M}_{M'} = -\tfrac{1}{2}\varepsilon^{NM'}\Box \tag{6.4.30}$$

(which follows from (2.5.24), the LHS being skew in NM because the ∇ operators commute; see also (6.8.33)). Applying (6.4.30) to the derivative of the LHS of (6.4.29), contracted over M', we obtain

$$\varepsilon^{N(M}\Box\lambda^{AB\ldots L)} = 0$$

whence by (3.5.15) we deduce $\Box\lambda^{AB\ldots L} = 0$ as required.

As a corollary of (6.4.29), we can easily derive an extension of (6.4.2) to the case when $\lambda^{A\ldots D}$ and $\phi_{A\ldots D}$ have the same number of (symmetric) indices:

$$\nabla^{(M}_{M'}\lambda^{A\ldots D)} = 0, \quad \nabla^{AA'}\phi_{AB\ldots D} = 0 \quad \textit{implies} \ \Box(\phi_{A\ldots D}\lambda^{A\ldots D}) = 0. \tag{6.4.31}$$

We have been working in Minkowski space $\mathbb{M}$, but by the conformal invariance of the two equations in the hypothesis of (6.4.31) ($\lambda^{A\ldots D}$ and $\phi_{A\ldots D}$ having respective conformal weights 0 and -1) it follows from (6.8.30) below that the conclusion in (6.4.31) still holds in *conformally* flat space–time provided that the operator $\Box$ is replaced by $\Box + \frac{1}{6}R$ $(= \Box + 4\Lambda)$.

6.5 Conformal Killing vectors, conserved quantities and exact sequences

In the last section we saw one way of obtaining the ten 'Poincaré' conserved quantities (energy–momentum, angular momentum) for the sources of a linearized gravitational field. In this section we shall examine another way of obtaining these quantities, and relate them to five more conserved quantities that exist when appropriate conformal invariance properties hold. The relations between the ten Poincaré and the fifteen conformal quantities are intriguing and somewhat intricate. To gain a more complete understanding of these relations – but also for later use in a different context – we shall briefly introduce the important concept of *exact sequences.*

Our overall objective is to relate two integrals for the Poincaré quantities. First, the 2-surface integral of §6.4 combines two ingredients, each linearly: the massless spin 2 field, ϕ_{ABCD} surrounding some source, and the primary part σ^{AB} of a symmetric twistor $\mathrm{S}^{\alpha\beta}$. The integral of ϕ_{ABCD} effectively gives the angular momentum twistor $\mathrm{A}_{\alpha\beta}$ of the surrounded sources. Second, a 3-surface integral combines another two ingredients linearly: the source E_{ab} for ϕ_{ABCD} (analogous to the source T_{ab} for Ψ_{ABCD}) and a Killing vector $\xi^{AB'}$ – the primary part of a Hermitian twistor $\mathrm{F}^{\alpha}{}_{\beta}$ (called $\mathrm{E}^{\alpha}{}_{\beta}$, to begin with). The integral of E_{ab} is effectively another Hermitian twistor $\mathrm{E}^{\alpha}{}_{\beta}$, a 'potential' according to (6.3.14) for the angular momentum twistor $\mathrm{A}_{\alpha\beta}$ of E_{ab}. Similarly $\mathrm{S}^{\alpha\beta}$ is a 'potential' (6.3.15) for $\mathrm{F}^{\alpha}{}_{\beta}$. Our final equation (6.5.53) succinctly sums up the relation between these twistors.

Conformal Killing vectors

First, let us return to a discussion of (6.1.70), $\nabla^{(C}_{(C'}\xi^{A)}_{A')} = 0$. By reference to (3.4.5) and (3.4.9) we can interpret this tensorially as the vanishing of the trace-free symmetric part of $\nabla_c\xi_a$:

$$\nabla_{(a}\xi_{b)} - \tfrac{1}{4}g_{ab}\nabla_c\xi^c = 0. \tag{6.5.1}$$

This is the *conformal Killing equation*. It is equivalent to

$$\nabla_{(a}\xi_{b)} \propto g_{ab}, \tag{6.5.2}$$

since the proportionality factor can be recovered from the latter equation by taking the trace. Another way of writing (6.5.2) is (*cf.* (4.3.3)):

$$\pounds_{\xi} g_{ab} \propto g_{ab}. \tag{6.5.3}$$

If ξ^a is taken to be a *real* vector field, this equation states that when the metric is 'dragged' along the vector field, it is altered only by a factor. Thus ξ^a expresses a local conformal (active) symmetry of the space-time. Such ξ^a is sometimes called a generator of infinitesimal conformal motions of the space-time, each point being displaced by $\varepsilon\xi^a$ where ε is infinitesimal: $x^{\mathbf{a}} \mapsto x^{\mathbf{a}} + \varepsilon\xi^{\mathbf{a}}$.

In a general curved space-time there will be no local conformal symmetries and consequently no (non-zero) solutions of (6.1.70). At the other extreme is flat (or conformally flat) space-time in which there is the maximum possible number (fifteen) of linearly independent solutions. In fact, we have the solutions explicitly in (6.1.55)(4). Since ξ^a is to be real, the conditions (6.1.68) will hold, i.e., the associated twistor $\mathsf{E}^\alpha{}_\beta$ given by (6.1.54) and (6.1.56) will be Hermitian. Counting the degrees of freedom, we see that there are four in $\mathring{\xi}^{AB'}$, eight in $\mathring{\zeta}_{A'}{}^{B'} = \bar{\mathring{\theta}}^{B'}{}_{A'}$, and four in $\mathring{\eta}_{A'B}$, making sixteen in all. However, the field $\xi^{AB'}$ does not quite determine all these coefficients. As we saw in §6.1, it is invariant precisely under

$$\mathring{\theta}^A{}_B \mapsto \mathring{\theta}^A{}_B + h\varepsilon_B{}^A \quad (h \in \mathbb{R}) \tag{6.5.4}$$

(the reality of h being needed to preserve the Hermiticity of $\mathsf{E}^\alpha{}_\beta$), which affects only the trace of $\mathsf{E}^\alpha{}_\beta$. Thus the *trace-free* part of $\mathsf{E}^\alpha{}_\beta$,

$$\mathsf{E}^\alpha{}_\beta - \tfrac{1}{4}\mathsf{E}^\gamma{}_\gamma \delta^\alpha{}_\beta, \tag{6.5.5}$$

is uniquely determined by the conformal Killing vector ξ^a. We now have fifteen degrees of freedom (the number of independent real components of (6.5.5)).

The solution (6.1.55)(4) (for Hermitian $\xi^{AB'}$) may be re-expressed tensorially as follows:

$$\xi_a = T_a + L_{ab}x^b + Rx_a + S_b(x^c x_c g_a{}^b - 2x_a x^b), \tag{6.5.6}$$

where

$$T_a = \mathring{\xi}_a, \quad R = -\operatorname{Im}(\mathring{\theta}^A{}_A), \quad L^{ab} = -\mathrm{i}\mathring{\theta}^{(AB)}\varepsilon^{A'B'} + \mathrm{i}\bar{\mathring{\theta}}^{(A'B')}\varepsilon^{AB}, \quad S_a = -\tfrac{1}{2}\mathring{\eta}_a. \tag{6.5.7}$$

We may interpret these four terms as giving, respectively, the infinitesimal generators of the translations (four parameters), the Lorentz rotations (six parameters), the dilations (one parameter), and the special conformal transformations (four parameters sometimes misleadingly called 'uniform acceleration transformations'). If R and S_a both vanish, then ξ^a is a Killing vector

$$\nabla_{(a}\xi_{b)}=0, \quad \text{i.e.,} \ \underset{\xi}{\pounds} g_{ab}=0, \tag{6.5.8}$$

and is a generator of the infinitesimal Poincaré group (ten parameters). In a curved space–time solutions of (6.5.8) will exist if the space–time permits local continuous metric-preserving motions into itself (not just conformal ones), as do stationary space–times, for example.

Conservation laws

There is an important relation between (conformal) Killing vectors and conservation laws in general relativity. Suppose we have a symmetric energy tensor T_{ab}, satisfying the usual divergence relation,

$$T_{ab}=T_{ba}, \quad \nabla^a T_{ab}=0. \tag{6.5.9}$$

If there is a Killing vector field ξ^a in the space–time, then the expression

$$C_a=T_{ab}\xi^b \tag{6.5.10}$$

satisfies

$$\nabla^a C_a=\xi^b\nabla^a T_{ab}+T_{ab}\nabla^a\xi^b=0, \tag{6.5.11}$$

by (6.5.9) and (6.5.8), since $T_{(ab)}\nabla^a\xi^b=T_{ab}\nabla^{(a}\xi^{b)}$. Thus C^a satisfies a vector divergence law like the $\nabla_a J^a=0$ for the charge–current vector J^a (*cf.* (5.1.54)).

Now the divergence theorem, which tells us that the 3-volume integral of such a J^a or C^a describes a *conserved quantity*, applies in curved space–time as well as in flat. For we have seen in (5.9.6) that $\nabla_a J^a=0$ has the differential-form expression $\mathrm{d}{}^{\dagger}\boldsymbol{J}=0$, where ${}^{\dagger}\boldsymbol{J}=e_{i_1i_2i_3a}J^a$ as in (5.9.5), our integral being $\int{}^{\dagger}\boldsymbol{J}$ over the 3-volume. The divergence theorem is a particular case of the fundamental theorem of exterior calculus (4.3.25): the integral of ${}^{\dagger}\boldsymbol{J}$ over the 3-volume boundary of any compact 4-volume necessarily vanishes – this boundary representing an initial and final location of our original 3-volume. Thus the same applies to ${}^{\dagger}\boldsymbol{C}=e_{i_1i_2i_3a}C^a$. But a tensor T_{ab} subject to (6.5.9) does not, by itself, generally provide any conserved quantity. The extra index on T^{ab}, i.e. on $e_{i_1i_2i_3a}T^{ab}$, prevents application of the divergence theorem in a general space–time. But when a Killing vector ξ^a is present, the above procedure is valid.

In $\mathbb{M}$ there are ten independent Killing vectors. Each of these may be used in (6.5.10) to provide a conserved quantity. The generators of the four independent translations give four independent such quantities, namely the energy and the three momenta. The generators of the six Lorentz rotations define the six components of relativistic angular momentum about the origin of rotation, three of these being the ordinary non-relativistic angular momentum and three describing the mass centre in its uniform motion.

Now suppose that ξ^a is merely a conformal Killing vector. Then, by (6.5.2), the calculation (6.5.11) is still valid, provided T_{ab} is trace-free:

$$T_a{}^a = 0. \tag{6.5.12}$$

We recall the trace-freeness is needed for the divergence equation $\nabla^a T_{ab} = 0$ to be conformally invariant (*cf.* (5.9.2) *et seq.*). Trace-freeness is also a property of the Maxwell, Dirac–Weyl neutrino, and 'new improved' massless scalar field energy tensors (*cf.* (5.2.4), (5.8.3), (6.8.36) below). There are general arguments based on Noether's theorem in the Lagrangian formalism which imply that conformally invariant fields such as these should lead to trace-free energy tensors (see e.g. Wald 1984, p. 448). Such fields will therefore give rise to one conserved quantity for each *conformal* Killing vector in the space–time.* In Minkowski space we have fifteen independent conformal Killing vectors and therefore fifteen independent conserved quantities. This gives five new conservation laws (beyond those for energy, momentum, and angular momentum) corresponding to the generators of the infinitesimal dilations (one) and special conformal transformations (four) (*cf.* (6.5.6), (6.5.7)).

Twistor description

The interrelations between the standard ten conserved quantities of Poincaré invariant theories and the fifteen conserved quantities of the conformally invariant theories in Minkowski space are very elegantly brought out in the twistor formalism. We have seen (*cf.* (6.5.5)) that each of the fifteen linearly independent solutions of the conformal Killing equation (6.5.1) is the primary part of a Hermitian $[^1_1]$-twistor $\mathrm{E}^\alpha{}_\beta$, whose trace-free part it uniquely determines. The twistors $\mathrm{E}^\alpha{}_\beta$ which arise in this way from Killing vectors constitute a ten-real-dimensional subspace of $\mathbb{T}^\alpha_\beta$. On the other hand, we also saw in (6.3.11) that the ten energy–momentum and angular-momentum components of a system can be collected together in the form of a symmetric twistor $\mathrm{A}_{\alpha\beta} \in \mathbb{T}_{(\alpha\beta)}$, which has the special form

* The Bel–Robinson tensor T_{abcd} is also trace-free symmetric and subject to vanishing divergence in vacuum. The equation $\nabla^a(T_{abcd}\xi^b\xi^c\xi^d) = 0$ holds for any conformal Killing vector ξ^a and therefore also gives rise to a conserved quantity (*cf.* Bel 1959).

(6.3.13) involving a Hermitian twistor $\mathsf{E}^{\alpha}{}_{\beta}$. (Furthermore, the trace part $\frac{1}{4}\mathsf{E}^{\gamma}{}_{\gamma}\delta^{\alpha}_{\beta}$ of $\mathsf{E}^{\alpha}{}_{\beta}$ does not contribute to $\mathsf{A}_{\alpha\beta}$.) The operation (6.3.13) thus extracts precisely the information of the ten energy–momentum and angular-momentum quantities out of a set of fifteen conformal quantities stored in $\mathsf{E}^{\alpha}{}_{\beta}$.

It would be natural to suppose, therefore, that there is a connection between these two procedures for reducing fifteen conformal quantities to ten Poincaré quantities. In fact there is. But the connection is not quite direct, and involves passing from a space to its dual space. Indeed, the passage from a general trace-free $\mathsf{E}^{\alpha}{}_{\beta}$ to one which represents a Killing vector, rather than merely a conformal Killing vector, is a passage from a space to a linear subspace. On the other hand, the map (6.3.13) taking $\mathsf{E}^{\alpha}{}_{\beta}$ to $\mathsf{A}_{\alpha\beta}$ is a projection taking the space of twistors $\mathsf{E}^{\alpha}{}_{\beta}$ to a factor space. Subspaces and factor spaces are dual concepts.

To see why duals must be involved, consider the physical meaning of a quantity like $\mathsf{A}_{\alpha\beta}$ and of a Killing vector. A particular twistor $\mathsf{A}_{\alpha\beta}$ represents the *entire* 4-momentum/angular-momentum structure of a *particular* physical system, say a billiard ball, while a particular Killing vector refers to a *single* component of this structure, say the energy, as applied to an *arbitrary* physical system. If we apply the one to the other – say, if we ask for the energy of the billiard ball – then we obtain a real number. This real number arises as a suitable scalar product, over the reals, between the twistor representing the Killing vector and that representing the 4-momentum/angular momentum of the physical system, since it is $\mathbb{R}$-linear in both. In the same way, we may consider a particular conformal physical system (such as a given free electromagnetic field) whose conformal conserved quantity structure is described by some trace-free Hermitian $\mathsf{E}^{\alpha}{}_{\beta}$. Then, given some conformal Killing vector ξ^{a}, which is now taken to be the primary part of a trace-free Hermitian twistor $\mathsf{F}^{\alpha}{}_{\beta}$, we should be able to extract the conserved quantity corresponding to the particular vector ξ^{a}, for this conformal system, as the result of some $\mathbb{R}$-bilinear 'scalar product' between $\mathsf{F}^{\alpha}{}_{\beta}$ and $\mathsf{E}^{\alpha}{}_{\beta}$. Conformal invariance demands that this be a twistor-scalar (not involving $\mathsf{I}^{\alpha\beta}$ or $\mathsf{I}_{\alpha\beta}$), and so it must be proportional to

$$\mathsf{F}^{\alpha}{}_{\beta}\mathsf{E}^{\beta}{}_{\alpha}. \tag{6.5.13}$$

This is clearly real because of the Hermiticity of $\mathsf{F}^{\alpha}{}_{\beta}$ and $\mathsf{E}^{\alpha}{}_{\beta}$. Also it is unaffected if a trace $(\lambda\delta^{\alpha}{}_{\beta})$ is added to $\mathsf{F}^{\alpha}{}_{\beta}$ or to $\mathsf{E}^{\alpha}{}_{\beta}$, but not to both.

In the particular case when the vectors ξ^{a} are restricted to be Killing vectors, the twistors $\mathsf{F}^{\alpha}{}_{\beta}$ will specialize to a particular form, so that only certain of the components of $\mathsf{E}^{\alpha}{}_{\beta}$ will be determined by scalar products of the type (6.5.13). These components will be precisely those conserved quantities possessed by a general Poincaré system (as opposed to a conformal system), namely the energy–momentum and angular momentum, i.e., the ten independent real components of $\mathsf{A}_{\alpha\beta}$, as defined by (6.3.13):

$$\mathsf{A}_{\alpha\beta} = 2\mathsf{E}^{\gamma}{}_{(\alpha}\mathsf{I}_{\beta)\gamma}. \tag{6.5.14}$$

The components of $\mathsf{E}^{\alpha}{}_{\beta}$ that are annihilated in the passage to $\mathsf{A}_{\alpha\beta}$ are the ones that are *not* well defined by a general Poincaré system – only conserved quantities can be well defined in this sense, that is, without reference to the value of some time coordinate, say. A Poincaré system defines an $\mathsf{A}_{\alpha\beta}$ but not an $\mathsf{E}^{\alpha}{}_{\beta}$.

Comparison of (6.5.14) with (6.5.13) suggests that for a Killing vector ξ^a, $\mathsf{F}^{\alpha}{}_{\beta}$ may have to specialize to the form $\mathsf{F}^{\alpha}{}_{\beta} = 2\mathsf{S}^{\alpha\gamma}\mathsf{I}_{\gamma\beta}$, where $\mathsf{S}^{\beta\gamma} \in \mathbb{T}^{(\beta\gamma)}$, or more correctly to

$$\mathsf{F}^{\alpha}{}_{\beta} = \mathsf{S}^{\alpha\gamma}\mathsf{I}_{\gamma\beta} + \bar{\mathsf{S}}_{\beta\gamma}\mathsf{I}^{\gamma\alpha}, \tag{6.5.15}$$

since $\mathsf{F}^{\alpha}{}_{\beta}$ must be Hermitian. The trace-free condition $\mathsf{F}^{\alpha}{}_{\alpha} = 0$ follows from the symmetry of $\mathsf{S}^{\alpha\beta}$ since $\mathsf{I}_{\alpha\beta}$ is skew. Substituting (6.5.15) into (6.5.13), and using (6.5.14), we find

$$\mathsf{F}^{\alpha}{}_{\beta}\mathsf{E}^{\beta}{}_{\alpha} = \mathrm{Re}\,\{\mathsf{S}^{\alpha\beta}\mathsf{A}_{\alpha\beta}\}. \tag{6.5.16}$$

Thus our scalar product can now be expressed, bilinearly over the reals, directly in terms of $\mathsf{S}^{\alpha\beta}$ and $\mathsf{A}_{\alpha\beta}$.

To see that $\mathsf{F}^{\alpha}{}_{\beta}$ indeed has the form (6.5.15) when ξ^a is a Killing vector, we note that the extra condition on $\mathsf{F}^{\alpha}{}_{\beta}$ implied by (6.5.8) is

$$\nabla_a \xi^a = 0. \tag{6.5.17}$$

From (6.1.56) we then get, using the notation (6.1.54),

$$\theta^A{}_A = \zeta_{A'}{}^{A'}.$$

We assume that the trace of $\mathsf{F}^{\alpha}{}_{\beta}$ is zero, so that it is determined by its primary part ξ^a:

$$\mathsf{F}^{\alpha}{}_{\alpha} = \theta^A{}_A + \zeta_{A'}{}^{A'} = 0$$

Then $\theta^A{}_A = \xi_{A'}{}^{A'} = 0$, and so, from (6.1.56)(2) and (6.1.68)

$$\xi^{AB'} = \bar{\xi}^{AB'}, \quad \theta^{AB} = \theta^{BA} = \bar{\zeta}^{AB} = \bar{\zeta}^{BA}, \quad \eta_{AB'} = 0. \tag{6.5.18}$$

These equations state that

$$\mathsf{F}^{[\alpha}{}_{\beta}\mathsf{I}^{\gamma]\beta} = 0. \tag{6.5.19}$$

By examining the meaning of (6.5.15), (6.5.19) in terms of spinor parts at some origin O, it is easy to show that, as a consequence of (6.5.19), $\mathsf{F}^{\alpha}{}_{\beta}$ is indeed necessarily of the form (6.5.15), with $\mathsf{S}^{\alpha\beta} \in \mathbb{T}^{(\alpha\beta)}$, as we asserted.

Thus, $S^{\alpha\beta}$ can be used, in place of $F^{\alpha}{}_{\beta}$, as a twistorial representation of a Killing vector. In some ways $S^{\alpha\beta}$ is more convenient. The difference is that whereas $F^{\alpha}{}_{\beta}$ is subject to a *restriction* (see (6.5.19)), $S^{\alpha\beta}$ is defined up to a 'gauge freedom'. In fact, one can readily verify that $S^{\alpha\beta}$ is defined up to

$$S^{\alpha\beta} \mapsto S^{\alpha\beta} + i2G^{(\alpha}{}_{\gamma}I^{\beta)\gamma}, \qquad (6.5.20)$$

$G^{\alpha}{}_{\beta}$ being arbitrary Hermitian:

$$G^{\alpha}{}_{\beta} = \bar{G}_{\beta}{}^{\alpha}. \qquad (6.5.21)$$

Relations to linear gravity source integrals

We are now in a position to illuminate the results we obtained in §6.4, in connection with the conserved quantities of a weak gravitational field ϕ_{ABCD}. The symmetric spinor field σ^{AB} of (6.4.3), arising from the $S^{\alpha\beta}$ of (6.1.50) and therefore satisfying (6.1.69), was used with ϕ_{ABCD} to construct ten complex integrals (i.e. twenty real integrals) of which only ten real values were independent. The ten *vanishing* real integrals in fact arise from the 'gauge term'

$$2G^{(\alpha}{}_{\gamma}I^{\beta)\gamma} \qquad (6.5.22)$$

in (6.5.20). Notice that this term has exactly the same form as $\bar{A}^{\alpha\beta}$, as given by (6.3.13). This is the cause of the *apparent paradox* we encountered on p. 77. In effect, it is just that 'part' of $S^{\alpha\beta}$ which has the form of $i\bar{A}^{\alpha\beta}$ that does *not* contribute to the energy–momentum or angular momentum of the source; it is when we pass to the *dual* space that we find a twistor with the structure of $A_{\alpha\beta}$ which *does* represent the energy–momentum and angular momentum of the system. We shall return to this curious matter shortly.

Let us see how the Killing vector ξ^a is, in fact, related to $S^{\alpha\beta}$. From (6.5.15) and the forms (6.1.50) and (6.1.54) for $S^{\alpha\beta}$ and $F^{\alpha}{}_{\beta}$, respectively, we obtain

$$\xi^{AB'} = -\rho^{AB'} - \bar{\rho}^{AB'}. \qquad (6.5.23)$$

Hence, by (6.1.52), with (6.1.66),

$$\nabla_{CC'}\sigma^{AB} = -2i\varepsilon_C{}^{(A}\rho^{B)}_{C'}. \qquad (6.5.24)$$

Combining the last two equations, we get

$$\xi^{AA'} = \tfrac{1}{3}(-i\nabla^{A'}_{B}\sigma^{AB} + i\nabla^{A}_{B'}\bar{\sigma}^{A'B'}). \qquad (6.5.25)$$

Thus σ^{AB}, subject to (6.1.69), acts as a kind of *potential* for ξ^a. With (6.1.69) assumed, the Killing equation for ξ^a is actually a consequence of (6.5.25). Furthermore, ρ^a is itself a special type of *complex* Killing vector, and so its imaginary part is also a Killing vector. This second Killing vector is associated with ξ^a but is not uniquely defined by it. Its existence and non-uniqueness is related to the 'gauge freedom' (6.5.20).

Exact sequences

To help understand questions of (linear) gauge freedom, it is useful to give a discussion in terms of *exact sequences* (*cf.* Strooker 1978). These are sequences of maps between vector spaces (or, more generally, between modules, Abelian groups, etc., – but here we need only consider vector spaces):

$$\cdots \to P \to Q \to R \to S \to \cdots, \tag{6.5.26}$$

which may extend infinitely in both directions or perhaps terminate at one or both ends with 0 (the zero vector space). The maps are all to be *linear* (with respect to the commutative ring concerned – here either $\mathbb{R}$ or $\mathbb{C}$) and to satisfy the following two properties:

(i) the composition of any two successive maps yields only the zero element,
(ii) any element mapping to zero in one of the maps must have a pre-image in the previous map.

We can combine (i) and (ii) in the form: the kernel of any map (i.e., the pre-image of zero – the set of elements mapping to zero) is exactly the image of the preceding map. Sometimes it is helpful in visualizing these sequences to have the diagrammatic description given in Fig. 6-4.

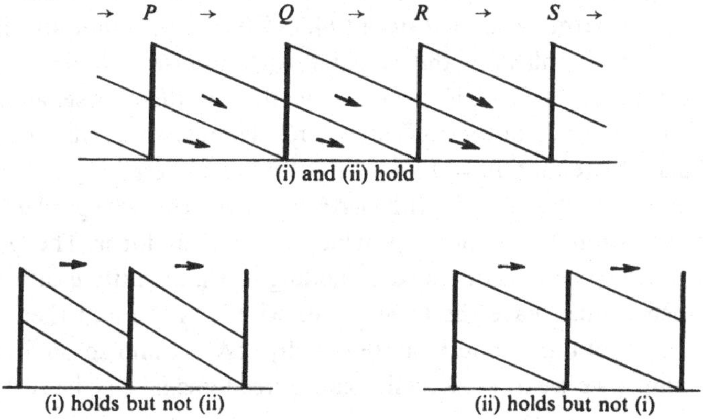

Fig. 6-4. Schematic description of an exact sequence (both (i) and (ii) holding) and of the two partial conditions separately. (If (i) holds, but not (ii), the sequence is called a *complex*.)

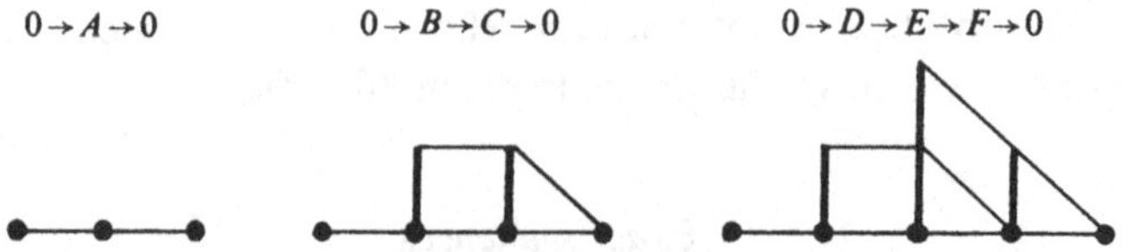

Fig. 6-5. Simple cases of exact sequences: $A = 0$, $B \cong C$, $F \cong E/D$.

Note that if

$$0 \to A \to 0, \quad 0 \to B \to C \to 0, \quad 0 \to D \to E \to F \to 0$$

are exact sequences, then $A = 0$, $B \cong C$, and F is the factor space of E by D, i.e., $F \cong E/D$ (see Fig. 6-5). Note also that if we have just *two* maps $D \to E \to F$ such that the kernel of the second is the image of the first, then we can always define arbitrarily many other spaces and maps to make a complete exact sequence of which $D \to E \to F$ forms a part. By linearity, if 0 is the beginning of an exact sequence, its image is necessarily the zero element; if it is the end, its pre-image is the entire space.

A familiar and important example of an exact sequence is the (*Poincaré*–) *de Rham sequence*

$$0 \to \mathbb{R} \xrightarrow{k} \Lambda^0 \xrightarrow{d} \Lambda^1 \xrightarrow{d} \Lambda^2 \xrightarrow{d} \cdots, \tag{6.5.27}$$

where Λ^r stands for the vector space (over $\mathbb{R}$) of C^∞ real r-forms, all defined on some n-dimensional C^∞ manifold $\mathcal{U}$ with Euclidean topology. Note that the sequence terminates with $\Lambda^{n+1} = 0$. The map k simply takes a real number f and maps it to the 0-form which has the constant value f on $\mathcal{U}$. The maps d are exterior differentiation (4.3.14). Since the constants are precisely those 0-forms that are annihilated by d, we have exactness up to Λ^1. Furthermore, $\mathrm{d}^2 = 0$ (recall (4.3.15)(viii)), so condition (i) for exactness is satisfied all along the sequence. And because we have assumed Euclidean topology* for $\mathcal{U}$, if, at any stage, $\mathrm{d}w = 0$, there exists a v such that $w = \mathrm{d}v$ – by (the converse of) Poincaré's lemma; so condition (ii) is satisfied also. Note that $\mathrm{d}w = 0$ for $w \in \Lambda^{p+1}$ is the integrability condition for solving the equation $w = \mathrm{d}v$ with $v \in \Lambda^p$. If, conversely, we regard $\mathrm{d}w = 0$ as a 'field equation' for w, then v provides a 'potential' for w. The 'gauge freedom' in v, $v \mapsto v + u$, is obtained by adding to v a quantity u satisfying $\mathrm{d}u = 0$. Such u must have the form $u = \mathrm{d}t$ with $t \in \Lambda^{p-1}$, and there is a 'gauge freedom of the second kind' $t \mapsto t + \mathrm{d}r$, $r \in \Lambda^{p-2}$, and so on. In this simple example we see how successive gauge freedoms take us back down

* Without Euclidean topology the exactness could fail at some stage. For such spaces this sequence is used at the *local* level only ('sheaf exact sequence'). Its globalization leads to cohomology (*cf.* §6.10, after (6.10.55)).

the exact sequence one step at a time; the reverse direction is associated with integrability conditions.

A simple but important exact sequence involving only *finite*-dimensional vector spaces is

$$0 \to \mathbb{S}^A \xrightarrow{1} \mathbb{T}^\alpha \xrightarrow{2} \mathbb{S}_{A'} \to 0, \tag{6.5.28}$$

where $\mathbb{S}^A$ and $\mathbb{S}_{A'}$ denote the spaces of *constant* spinor fields of the types indicated by the indices (*cf.* Penrose and MacCallum 1972). The map 1 takes any constant spinor field κ^A into the twistor $(\kappa^A, 0)$. Map 2 takes any twistor $(\omega^A, \pi_{A'})$ into the constant spinor field $\pi_{A'}$:

$$\kappa^A \overset{1}{\mapsto} (\kappa^A, 0), \quad (\omega^A, \pi_{A'}) \overset{2}{\mapsto} \pi_{A'}. \tag{6.5.29}$$

The exactness of (6.5.28) is obvious. So is its Poincaré invariance: each of the maps (6.5.29) is canonically defined in a Poincaré invariant (but not conformally invariant) way, so it is unaffected by any change of origin.

An important property of exact sequences is that their *duals* are also exact,* but with the maps running in the opposite direction. Thus, for the dual vector spaces P^*, Q^*,... of (6.5.26) we have

$$\cdots \leftarrow P^* \leftarrow Q^* \leftarrow R^* \leftarrow S^* \leftarrow \cdots \tag{6.5.30}$$

as another exact sequence.** The dual of (6.5.28) is

$$0 \leftarrow \mathbb{S}_A \leftarrow \mathbb{T}_\alpha \leftarrow \mathbb{S}^{A'} \leftarrow 0 \tag{6.5.31}$$

with $(\lambda_A, \mu^{A'}) \overset{1^*}{\mapsto} \lambda_A$, $\mu^{A'} \overset{2^*}{\mapsto} (0, \mu^{A'})$; and it is interesting to note that this is also the complex conjugate of (6.5.28) (read in the opposite direction).

The moment sequences

Let us now examine the exact sequence that arises from (6.5.15), (6.5.19), and (6.5.20). Consider the following partial sequence:

$$\mathbb{H}^\alpha_\beta \xrightarrow{p} \mathbb{T}^{(\alpha\beta)} \xrightarrow{q} \mathbb{H}^\alpha_\beta \xrightarrow{r} \mathbb{T}^{[\alpha\beta]}, \tag{6.5.32}$$

where $\mathbb{H}^\alpha_\beta$ is the vector space (over $\mathbb{R}$) of Hermitian $[^1_1]$-twistors, and $\mathbb{T}^{(\alpha\beta)}$ and $\mathbb{T}^{[\alpha\beta]}$ are the spaces of symmetric and antisymmetric $[^2_0]$-twistors, regarded for this purpose as vector spaces over $\mathbb{R}$. The map r is defined so

* In the general case, this property requires the Axiom of Choice.

** The reverse map from Q^* to P^*, say, associated with each element $q \in Q^*$ the element $p \in P^*$ such that $p \cdot x = q \cdot y$ for all $x \in P$, y being the image of x in the original map $P \to Q$.

that its kernel is (6.5.19),

$$r: \mathbb{H}^{\alpha}_{\beta} \to \mathbb{T}^{[\alpha\beta]} :: \mathrm{F}^{\alpha}{}_{\beta} \mapsto 2\mathrm{F}^{[\alpha}{}_{\gamma}\mathrm{I}^{\beta]\gamma}, \qquad (6.5.33)$$

where $\mathrm{F}^{\alpha}{}_{\beta}$ is here a general (not necessarily trace-free) element of $\mathbb{H}^{\alpha}_{\beta}$. To solve (6.5.19) we use (6.5.15), which provides the map q,

$$q: \mathbb{T}^{(\alpha\beta)} \to \mathbb{H}^{\alpha}_{\beta} :: \mathrm{S}^{\alpha\beta} \mapsto \mathrm{S}^{\alpha\gamma}\mathrm{I}_{\gamma\beta} + \bar{\mathrm{S}}_{\beta\gamma}\mathrm{I}^{\gamma\alpha}. \qquad (6.5.34)$$

And the gauge freedom in $\mathrm{S}^{\alpha\beta}$ is given by (6.5.20), which provides the map p,

$$p: \mathbb{H}^{\alpha}_{\beta} \to \mathbb{T}^{(\alpha\beta)} :: \mathrm{G}^{\alpha}{}_{\beta} \mapsto 2\mathrm{i}\mathrm{G}^{(\alpha}{}_{\gamma}\mathrm{I}^{\beta)\gamma}. \qquad (6.5.35)$$

With these maps, (6.5.32) is a possible portion of an exact sequence, since the image of q is the kernel of r, and the image of p is the kernel of q.

We can extend the sequence (6.5.32) in both directions, and obtain an infinite exact sequence which turns out to be periodic,

$$\cdots \to \mathbb{T}^{[\alpha\beta]} \xrightarrow{s} \mathbb{H}^{\alpha}_{\beta} \xrightarrow{p} \mathbb{T}^{(\alpha\beta)} \xrightarrow{q} \mathbb{H}^{\alpha}_{\beta} \xrightarrow{r} \mathbb{T}^{[\alpha\beta]} \xrightarrow{s} \mathbb{H}^{\alpha}_{\beta} \to \cdots, \qquad (6.5.36)$$

the one remaining map s being defined by

$$s: \mathbb{T}^{[\alpha\beta]} \to \mathbb{H}^{\alpha}_{\beta} :: \mathrm{K}^{\alpha\beta} \to \mathrm{i}\mathrm{K}^{\alpha\gamma}\mathrm{I}_{\gamma\beta} - \mathrm{i}\bar{\mathrm{K}}_{\beta\gamma}\mathrm{I}^{\gamma\alpha} \qquad (6.5.37)$$

(where $\mathrm{K}^{\alpha\beta}$ is skew). The exactness of this completed sequence is straightforward to verify (e.g., by examining the spinor parts at O).

The *dual* of the sequence (6.5.36) is

$$\cdots \leftarrow \mathbb{T}_{[\alpha\beta]} \xleftarrow{\tilde{r}} \mathbb{H}^{\beta}_{\alpha} \xleftarrow{\tilde{q}} \mathbb{T}_{(\alpha\beta)} \xleftarrow{\tilde{p}} \mathbb{H}^{\beta}_{\alpha} \xleftarrow{\tilde{s}} \mathbb{T}_{[\alpha\beta]} \xleftarrow{\tilde{r}} \mathbb{H}^{\beta}_{\alpha} \leftarrow \cdots \qquad (6.5.38)$$

where the map $\tilde{p}$ [*or* $\tilde{q}, \tilde{r}, \tilde{s}$ respectively], which is the dual of q [*or* p, s, r], is related to the map p [*or* q, r, s] as follows: first multiply by $-\mathrm{i}$, then apply p [*or* q, r, s], then take the complex conjugate. The resulting sequence (6.5.38) is exact, and indeed, as well as being the dual of (6.5.36), it is simply (when read from right to left) a trivial modification of (6.5.36) itself (with incorporation of factor $-\mathrm{i}$ and complex conjugation). It may be remarked that the scalar products between the spaces occurring in (6.5.36) and their duals are defined as the real part of the contracted product of corresponding elements from the two spaces. (Of course, when the two spaces are $\mathbb{H}^{\alpha}_{\beta}$ and $\mathbb{H}^{\beta}_{\alpha}$, the contracted product is real anyway.) It is this definition of scalar products that governs the definitions of the maps $\tilde{p}, \tilde{q}, \tilde{r}, \tilde{s}$.

Note that the relation between $\mathrm{E}^{\alpha}{}_{\beta}$ and the angular momentum twistor $\mathrm{A}_{\alpha\beta}$ that was given in (6.5.14) is precisely that defined by the map $\tilde{p}$, taking $\mathrm{E}^{\beta}{}_{\alpha} \in \mathbb{H}^{\beta}_{\alpha}$ to $\bar{\mathrm{A}}^{(\alpha\beta)} \in \mathbb{T}^{(\alpha\beta)}$ and the freedom (6.3.14) is just that defined by $\tilde{s}$. Again, the agreement (6.5.16) between the (real) scalar product of $\mathrm{S}^{\alpha\beta}$ with $\mathrm{A}_{\alpha\beta}$ and of $\mathrm{F}^{\alpha}{}_{\beta}$ with $\mathrm{E}^{\alpha}{}_{\beta}$ is a consequence of the fact (as applied to the two middle terms of the sequences as we have written them) that (6.5.36) and

(6.5.38) are dual sequences over $\mathbb{R}$. Recall the apparent paradox of pp. 77, 88, in which the expression (6.5.14) and the identical expression (6.5.22) seemed to be playing opposite roles, the first as the map giving the Poincaré conserved quantities from the conformal ones, and the second as a mere gauge freedom: this is now seen in a broader context. The relation (6.5.14) is the map $\tilde{p}$ in the sequence (6.5.38), while the gauge terms are the image of p in the sequence (6.5.36), i.e., the kernel of q, where q is the dual of the map $\tilde{p}$. The sequence (6.5.38) is the one referring to *particular physical objects* (e.g., the billiard ball of the discussion on p. 86, whose angular momentum twistor $\mathsf{A}_{\alpha\beta}$ would be a particular element of $\mathbb{T}_{(\alpha\beta)}$). The sequence (6.5.36) is the one referring to (conformal) Killing vectors, etc., in their roles as selecting *particular conserved quantities* (such as the Killing vector representing the energy component of a system).

So far we have treated the sequences (6.5.36), (6.5.38) in an algebraic way (that is, merely as examples of twistor algebra). But, as we saw in (6.5.25), we can also treat these twistor linear maps at the level of differential equations on spinor fields in $\mathbb{M}$. Moving back down the sequences then corresponds to finding potentials in the more-or-less ordinary sense (as σ^{AB} is a potential for ξ^a in (6.5.25)). In Fig. 6-6 the entire transformation scheme for the sequence (6.5.36) is exhibited. The equations are written in terms of the primary parts σ^{AB}, $\kappa\varepsilon^{AB}$ of twistors $\mathsf{S}^{\alpha\beta}\in\mathbb{T}^{(\alpha\beta)}$, $\mathsf{K}^{\alpha\beta}\in\mathbb{T}^{[\alpha\beta]}$, respectively (these being the cases where the primary part alone determines the twistor), and in terms of the pairs (γ^a,μ) and (ξ^a,λ), describing $\mathsf{G}^\alpha{}_\beta\in\mathbb{H}^\alpha_\beta$ and $\mathsf{F}^\alpha{}_\beta\in\mathbb{H}^\alpha_\beta$, respectively, where $\gamma^{AA'}$, $\xi^{AA'}$ are the respective primary parts and $\mu=\mathsf{G}^\alpha{}_\alpha$, $\lambda=\mathsf{F}^\alpha{}_\alpha$ are the (independent) traces which are needed to define these twistors completely.

Most of the relations in Fig. 6-6 are given in tensor form. The translation to tensor form of the remaining relations – those occurring in the central portion of the figure – can be achieved by setting

$$Q^{ab}=\mathrm{i}\sigma^{AB}\varepsilon^{A'B'}-\mathrm{i}\bar{\sigma}^{A'B'}\varepsilon^{AB}, \tag{6.5.39}$$

as in (6.4.7). The tensor form of the relation $\nabla^{(A}_{A'}\sigma^{BC)}=0$ was given in (6.4.6). The tensor form of the map q is

$$\xi^a=\tfrac{1}{3}\nabla_bQ^{ba} \tag{6.5.40}$$

(together with $\lambda=0$), and that of p,

$$Q^{ab}=-e^{abcd}\nabla_c\gamma_d, \tag{6.5.41}$$

the image of p (i.e., the kernel of q) being the set of Q^{ab} subject to (6.4.6) and

$$\nabla_bQ^{ab}=0. \tag{6.5.42}$$

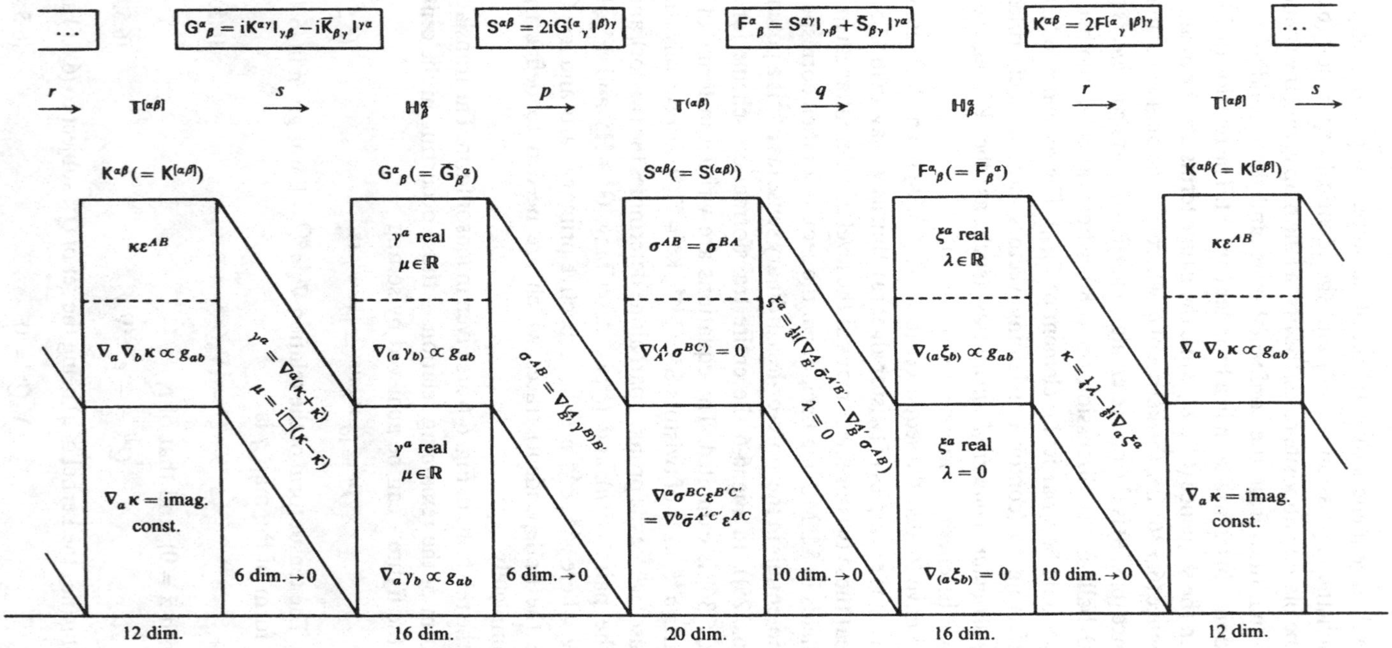

Fig. 6-6. The (dual) periodic moment sequence.

As we have seen, the dual sequence (read from right to left) is a trivial modification of the original one. In particular, the map $\tilde{p}$ which produced the angular-momentum twistor $\mathsf{A}_{\alpha\beta}$ from the conformal quantity $\mathsf{E}^{\alpha}{}_{\beta}$, takes essentially the form (6.5.41), in which, however, the quantity corresponding to Q^{ab} is replaced by its dual, because of the factor $-\mathrm{i}$ in the definition of $\tilde{p}$ (after (6.5.38)). The primary part of $\mathsf{A}_{\alpha\beta}$ is (*cf.* (6.3.11)) $2\mathrm{i}\mu^{A'B'}$, so replacing $\bar{\sigma}^{A'B'}$ by this in the definition (6.5.39) of Q^{ab}, we get twice the angular-momentum tensor (6.3.10): $2M^{ab}$. The conformal Killing vector $\xi^{AB'}$, now taken to be the primary part of $\mathsf{E}^{\alpha}{}_{\beta}$, is related to the dual of $2M^{ab}$ by (6.5.41). We see, therefore, that M^{ab} arises as the *curl* of the conformal Killing vector ξ^{a}, a fact already noted in (6.3.16). The relation corresponding to (the dual of) (6.5.42) simply states that the curl of M^{ab} vanishes.

The sequence (6.5.36) and its dual (6.5.38), which we refer to as the *periodic moment sequences* (or, strictly, to (6.5.38) as the periodic moment sequence and (6.5.36) as its dual), can be usefully split into overlapping successions of shorter exact sequences, each of nine terms. The nine-term sequence, which we call simply the *moment sequence*, is exhibited in (the dual of) Fig. 6-7. Here the space $\mathbb{H}^{\alpha}_{\beta}$ of Hermitian twistors is split into a direct sum

$$\mathbb{H}^{\alpha}_{\beta} = \hat{\mathbb{H}}^{\alpha}_{\beta} \oplus \mathbb{R}\delta^{\alpha}_{\beta},$$

where $\hat{\mathbb{H}}^{\alpha}_{\beta}$ consists of *trace-free* Hermitian twistors, and $\mathbb{R}$ represents the trace. The space $\mathbb{T}^{[\alpha\beta]}$ is represented as

$$\mathbb{T}^{[\alpha\beta]} = \mathbb{R}^{[\alpha\beta]} \oplus \mathrm{i}\mathbb{R}^{[\alpha\beta]},$$

where $\mathbb{R}^{[\alpha\beta]}$ is the six-dimensional real vector space of *twistor-real* skew twistors, i.e. of $\mathsf{R}^{\alpha\beta}$ subject to $\mathsf{R}^{\alpha\beta} = \bar{\mathsf{R}}^{\alpha\beta}$ where $\bar{\mathsf{R}}^{\alpha\beta} = \frac{1}{2}\varepsilon^{\alpha\beta\gamma\delta}\bar{\mathsf{R}}_{\gamma\delta}$, *cf.* (6.2.19), (6.2.31), so that its primary part $\kappa\varepsilon^{AB}$ has *real* κ.

The moment sequence has been shown by Hughston and Hurd (1983) to arise in a natural way as an example of a type of exact sequence known as a Koszul sequence (*cf.* Grothendieck and Dieudonné 1961):

$$\begin{array}{ccccccccccccc} 0 \rightarrow & \mathbb{V} & \rightarrow & \mathbb{V}^{\Phi} & \rightarrow & \mathbb{V}^{[\Phi_1\Phi_2]} & \rightarrow & \mathbb{V}^{[\Phi_1\Phi_2\Phi_3]} & \rightarrow \cdots \rightarrow & \mathbb{V}^{[\Phi_1\ldots\Phi_n]} & \rightarrow 0 \\ & \updownarrow & & \updownarrow & & \updownarrow & & \updownarrow & & \updownarrow & \\ 0 \leftarrow & \mathbb{V} & \leftarrow & \mathbb{V}_{\Phi} & \leftarrow & \mathbb{V}_{[\Phi_1\Phi_2]} & \leftarrow & \mathbb{V}_{[\Phi_1\Phi_2\Phi_3]} & \leftarrow \cdots \leftarrow & \mathbb{V}_{[\Phi_1\ldots\Phi_n]} & \leftarrow 0. \end{array} \tag{6.5.43}$$

Here $\mathbb{V}^{\Phi}$ is an n-dimensional vector space over a division ring $\mathbb{V}$, and the maps are defined in terms of one fixed element I^{Φ} of $\mathbb{V}^{\Phi}$. The upper maps are

$$A \mapsto AI^{\Phi}, \quad B^{\Phi} \mapsto B^{[\Phi_1}I^{\Phi_2]}, \quad C^{\Phi_1\Phi_2} \mapsto C^{[\Phi_1\Phi_2}I^{\Phi_3]}, \text{ etc.}$$

and the lower ones

$$Z_{\Phi} \mapsto Z_{\Phi}I^{\Phi}, \quad Y_{\Phi_1\Phi_2} \mapsto Y_{\Phi_1\Phi_2}I^{\Phi_2}, \quad X_{\Phi_1\Phi_2\Phi_3} \mapsto X_{\Phi_1\Phi_2\Phi_3}I^{\Phi_3}, \text{ etc.}$$

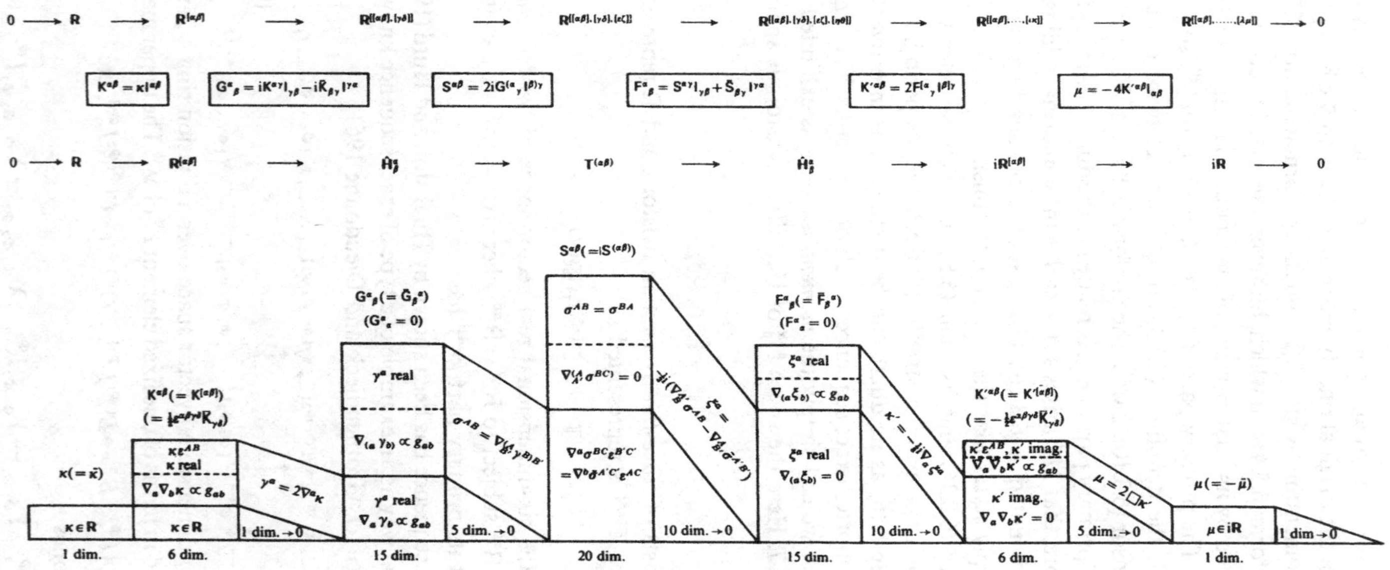

Fig. 6-7. The (dual) moment sequence.

Exactness is easy to verify, as is the fact that the sequences are indeed dual to one another, corresponding dual spaces being indicated by the vertical broken-lined double arrow. To obtain the moment sequence we take $\mathbb{V} = \mathbb{R}$ and $\mathbb{V}^{\Phi} = \mathbb{R}^{[\alpha\beta]}$, where Φ is the abstract index representing the clumped skew pair $[\alpha\beta]$. The particular element I^{Φ} is now $\mathrm{I}^{\alpha\beta}$ and, up to simple factors, the sequences in (6.5.43) turn out to reduce to that given in Fig. 6-7. The details of this are left to the reader.

Hughston and Hurd point out that this procedure can also be applied in the cosmological context of §9.5, where alternative selections of 'infinity twistor' may be made which need not be simple (or real). This leads to some modification of our twistor expressions.

Linear gravity integrals

We have given two methods of constructing integrals for the energy–momentum and angular momentum of a physical system in $\mathbb{M}$, and we wish to examine the relation between these methods. In the first, as described in §6.4, we considered the situation of a weak gravitational vacuum field spatially surrounding a region of source. Selecting a symmetric solution σ^{AB} of (6.1.69), we constructed from it and the (weak) gravitational spinor ϕ_{ABCD} an integral over a closed 2-surface $\mathscr{S}$ lying entirely in the vacuum surrounding the source (see (6.4.3), (6.4.4) *et seq.*). The value C' of this integral is unaltered when $\mathscr{S}$ is continuously moved, provided it does not cross any region of source. Moreover, C' does not depend upon any particular choice of 3-volume spanning $\mathscr{S}$. In these senses C' represents a conserved quantity.

In the second method (described in (6.5.9), (6.5.10) *et seq.*) we selected a Killing vector ξ^a and constructed from it and the energy tensor T_{ab} an integral over a compact 3-surface (volume) $\mathscr{V}$ with boundary $\partial\mathscr{V}$. The region of integration $\mathscr{V}$ is now allowed to intercept the source region. The value C of the integral now represents the total flux of $C^a = T^a{}_b\xi^b$ across $\mathscr{V}$. The quantity C is now conserved in the sense that it remains unchanged whenever $\mathscr{V}$ is continuously deformed without altering $\partial\mathscr{V}$. This is equivalent to saying that the integral of C^a-flux over any *closed* 3-surface (bounding a compact 4-volume) must vanish (see Fig. 6-8). If $\partial\mathscr{V}$ is entirely in the vacuum region, say $\partial\mathscr{V} = \mathscr{S}$, then it can be moved too, without affecting the result, provided $\mathscr{S}$ remains in the vacuum throughout the motion; this is because the region swept out by $\mathscr{S} = \partial\mathscr{V}$ contributes zero to the integral.

By analogy with electromagnetism, we would expect to be able to pass

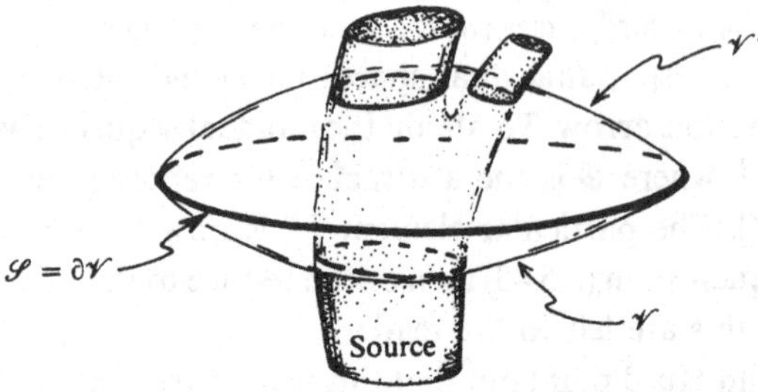

Fig. 6-8. Charge conservation: the total integrated flux across the compact 3-surface $\mathscr{V}$, with boundary $\mathscr{S}$, is equal to that across $\mathscr{V}'$. The union of $\mathscr{V}$ and $\mathscr{V}'$, the two being taken with opposite orientations, constitutes the boundary of a compact 4-volume.

from the first to the second method of generating conserved quantities by means of the fundamental theorem of exterior calculus (*cf.* 4.3.25). Indeed, it is by finding such a relationship that we may make the appropriate identifications between the results of the two methods. However, we here have the additional feature that the two methods involve *different* kinds of spinor, namely σ^{AB} and ξ^a, for the specification of the particular conserved quantity C with which we are concerned. But (6.5.25) supplies the connecting link.

The argument is most easily carried out in tensor form and the relevant formulae have been obtained by translation from the spinor formalism. We shall use the skew tensor Q^{ab} which was defined in terms of σ^{AB} by (6.4.7), and the required formulae are (6.4.6) and the tensor form (6.5.40) of (6.5.25): $\xi^a = \frac{1}{3}\nabla_b Q^{ba}$. In the fundamental formula of exterior calculus (4.3.25)

$$\oint_{\partial\mathscr{V}} \mathbf{\Theta} = \int_{\mathscr{V}} \mathrm{d}\mathbf{\Theta} \tag{6.5.44}$$

we substitute for the 2-form $\mathbf{\Theta}$ the expression

$$\mathbf{\Theta} = K^{*}_{i_1 i_2 ab} Q^{ab}, \tag{6.5.45}$$

where K_{abcd} is the weak-field curvature tensor defined in (5.7.4) and K^{*}_{abcd} is its right dual (*cf.* (4.6.11)). Then

$$\begin{aligned}\mathrm{d}\mathbf{\Theta} &= \nabla_{[i_1}\{K^{*}_{i_2 i_3]ab} Q^{ab}\} \\ &= \nabla_{[i_1} Q^{ab} K^{*}_{i_2 i_3]ab},\end{aligned} \tag{6.5.46}$$

by the linearized Bianchi identity (5.7.9), the dualization affecting only the last two indices a, b of $K\ldots$. Dualizing the form indices in (6.5.46) we get

$$^{\ddagger}\mathrm{d}\mathbf{\Theta} = \tfrac{1}{3}\nabla^c Q^{ab}\, {}^{*}K^{*}_{i_1 cab} \tag{6.5.47}$$

(*cf.* (3.4.30) and (3.4.25)). But by (6.4.6), (6.5.40) and (4.6.11), we have

$$2\,{}^*K^*_{dcab}\nabla^{(c}Q^{a)b} = -\,{}^*K^*_{dcab}g^{ca}\nabla_e Q^{be}$$
$$= -3(K_{deb}{}^e - \tfrac{1}{2}g_{db}K_{ef}{}^{ef})\xi^b = 24\pi G E_{db}\xi^b, \qquad (6.5.48)$$

where E_{ab} is the weak-field energy–momentum tensor (related to K_{abcd} as T_{ab} is to R_{abcd} *cf.* (5.7.6), G being the gravitational constant). Now ${}^*K^*_{abcd}$ satisfies the cyclic identity (*cf.* (4.6.9)), whence

$$2\,{}^*K^*_{dcab}\nabla^a Q^{cb} = {}^*K^*_{dcab}\nabla^c Q^{ab},$$

and so, by (6.5.48),

$$\tfrac{3}{2}\,{}^*K^*_{dcab}\nabla^c Q^{ab} = 24\pi G E_{db}\xi^b.$$

When we substitute in (6.5.47), we now have

$${}^{\ddagger}\mathrm{d}\Theta = \tfrac{16}{3}\pi G E_{i_1 a}\xi^a. \qquad (6.5.49)$$

On the other hand, translating (6.5.45) into spinors, using (5.7.8) and (4.6.11), we get, in the vacuum region

$$\Theta = (-\mathrm{i}\phi_{I_1I_2AB}\varepsilon_{I_1'I_2'}\varepsilon_{A'B'} + \mathrm{i}\bar{\phi}_{I_1'I_2'A'B'}\varepsilon_{I_1I_2}\varepsilon_{AB})Q^{ab}$$
$$= 2(\phi_{ABPQ}\sigma^{AB}\varepsilon_{P'Q'} + \bar{\phi}_{A'B'P'Q'}\bar{\sigma}^{A'B'}\varepsilon_{PQ})\mathrm{d}x^p \wedge \mathrm{d}x^q. \qquad (6.5.50)$$

Hence the LHS of (6.5.44) is proportional to the expression (6.4.4) (with the χ_{AB} of (6.4.3) in place of φ_{AB}) plus its complex conjugate – in other words, we obtain the integral for our first method of obtaining conserved quantities. The RHS of (6.5.44), when we substitute the dual of (6.5.49) and use (3.4.29) and (3.4.31)(1), is seen to be proportional to the flux integral for the relevant quantity (6.5.10) for our second method. Thus, choosing our overall multiplying factor so that when ξ^a is the standard time-translation $\partial/\partial t$ we get the total energy intercepted by $\mathscr{V}$, (6.5.44) is obtained in the form

$$\frac{-1}{32\pi G}\oint_{\partial\mathscr{V}} K^*_{abcd}Q^{cd}\mathrm{d}x^a \wedge \mathrm{d}x^b = -\frac{1}{6}\int_{\mathscr{V}} \mathrm{e}_{abc}{}^d E_{df}\xi^f \mathrm{d}x^a \wedge \mathrm{d}x^b \wedge \mathrm{d}x^c.$$
$$(6.5.51)$$

This is the desired relation between the two methods in the general case, where $\partial\mathscr{V}$ need not lie in a vacuum region. Note that the spinor translation of the LHS of (6.5.51) directly involves the source parts of the linearized curvature (corresponding to $\Phi_{ABC'D'}$ and Λ) unlike the analogous situation for the electromagnetic field. When these vanish at $\partial\mathscr{V} = \mathscr{S}$ ($\mathscr{S}$ in the vacuum region), the LHS may, by (6.5.50), be written as

$$-\frac{1}{8\pi G}\oint_{\mathscr{S}} P_{ab}\mathrm{d}x^a \wedge \mathrm{d}x^b \qquad (6.5.52)$$

with P_{ab} given by (6.4.5).

The ten vanishing integrals

Note that the RHS of (6.5.51) vanishes identically for *all* source distributions if $\xi^a = 0$, i.e. if $\nabla_b Q^{ab} = 0$. By (6.5.23), this occurs when the ρ^a of (6.1.50) (and therefore also the τ^a – *cf.* (6.1.66)) is purely imaginary everywhere. By (6.1.50) and (6.1.51) we deduce that $S^{\alpha\beta}$ has the form $i \times$ (an angular momentum twistor) (*cf.* (6.3.11)) – ignoring any future-causal restriction on the 4-momentum – and therefore arises from the 'gauge term' in (6.5.20), as expected. This verifies a claim made on p. 77 that these particular ten integrals (6.5.52) always vanish for a ϕ_{ABCD} subject to the massless free-field equation (4.12.42) in a neighbourhood $\mathscr{U}$ of $\mathscr{S}$, *provided that* ϕ_{ABCD} extends to a K_{abcd} with Riemann tensor symmetries and subject to the 'Bianchi identity' equation (5.7.9), where K_{abcd} is defined throughout a spanning 3-volume $\mathscr{V}$ of $\mathscr{S}$. Another way of stating this condition on ϕ_{ABCD} at $\mathscr{S}$ is the existence of a metric perturbation h_{ab} $(= h_{ba})$ defined in $\mathscr{U}$ such that ϕ_{ABCD} arises from it according to (5.7.12). For any such h_{ab} can be extended smoothly (in an arbitrary way) throughout $\mathscr{V}$, and the required K_{abcd} can be defined from it according to (5.7.4).

It is interesting to observe that whereas the massless free-field equation (4.12.42) is conformally invariant (*cf.* (5.7.17)), this further condition on ϕ_{ABCD} cannot be so. For if a conformal rescaling is applied to $\mathbb{M}$ (in the region $\mathscr{U}$), which sends the metric again to a flat one, this will correspond to a transformation in $\mathbb{T}^\alpha$ for which $I^{\alpha\beta}$ is not generally left invariant. (See §§9.3–9.5 for details.) The condition that $S^{\alpha\beta}$ take the form (6.5.22), from which these vanishing integrals arise, explicitly involves $I^{\alpha\beta}$ and is manifestly not invariant. In fact solutions of the massless equation can be found (for suitable $\mathscr{U}$) for which all *twenty* of the integrals (6.5.52) take independently arbitrary values.

Relations to duality in the moment sequences

Note that if we keep K_{abcd}, E_{ab} and the 3-surface $\mathscr{V}$ (with its boundary $\partial\mathscr{V}$) fixed, but allow Q^{ab} and ξ^a to vary, then the two sides of (6.5.51) provide us with *linear maps* (over the reals) to $\mathbb{R}$ from their respective twistor spaces, namely from $\mathbb{T}^{(\alpha\beta)}$ for Q^{ab} and from the image $\mathbb{K}^\alpha_\beta$, in $\mathbb{H}^\alpha_\beta$, of q in (6.5.34) for ξ^a. If ξ^a is not restricted to be a Killing vector but is allowed to be any *conformal* Killing vector, then the RHS of (6.5.51) provides a map from the *whole* of $\mathbb{H}^\alpha_\beta$ to $\mathbb{R}$. In this latter case the map depends on the choice of $\mathscr{V}$ (spanning a fixed boundary 2-surface $\partial\mathscr{V}$), in general – though in the particular case when $E_a{}^a = 0$ the map *would* actually be 3-surface independent in this sense.

These maps are achieved by particular elements of the *dual* spaces (over $\mathbb{R}$), namely by $\mathsf{A}_{\alpha\beta} \in \mathbb{T}_{(\alpha\beta)}$ and $\mathsf{E}^{\beta}{}_{\alpha} \in \mathbb{H}^{\beta}{}_{\alpha}$, respectively, and (6.5.51) now takes the form

$$\mathrm{Re}(\mathsf{A}_{\alpha\beta}\mathsf{S}^{\alpha\beta}) = \mathsf{E}^{\beta}{}_{\alpha}\mathsf{F}^{\alpha}{}_{\beta} \tag{6.5.53}$$

(identical to (6.5.16)), where σ^{AB} is the primary part of $\mathsf{S}^{\alpha\beta}$ and ξ^a the primary part of $\mathsf{F}^{\alpha}{}_{\beta}$ (with $\mathsf{F}^{\alpha}{}_{\alpha} = 0$). The LHS defines the angular momentum twistor $\mathsf{A}_{\alpha\beta}$ of the total matter intercepted by $\mathscr{V}$ (i.e. surrounded by $\partial\mathscr{V}$) as an integral over $\partial\mathscr{V}$. Allowing ξ^a to be a general *conformal* Killing vector in the RHS of (6.5.51) we obtain $\mathsf{E}^{\beta}{}_{\alpha}$ in terms of an integral over $\mathscr{V}$. In the general case (when $E_a{}^a \neq 0$) we get an $\mathsf{E}^{\beta}{}_{\alpha}$ which depends upon the choice of $\mathscr{V}$ (for fixed $\partial\mathscr{V}$). The various possible twistors $\mathsf{E}^{\beta}{}_{\alpha}$ arising in this way differ from one another by a 'gauge' transformation (6.3.14). The relationship between $\mathsf{A}_{\alpha\beta}$ and $\mathsf{E}^{\beta}{}_{\alpha}$ provided by (6.5.53) (where now $\mathsf{F}^{\alpha}{}_{\beta} \in \mathbb{K}^{\alpha}_{\beta}$, so ξ^a is a Killing vector – related to σ^{AB} by (6.5.25), whence $\mathsf{F}^{\alpha}{}_{\beta}$ and $\mathsf{S}^{\alpha\beta}$ are related by (6.5.15)), is just the standard one (6.3.13). Likewise, the scalar products in (6.5.53) are the ones occurring in the relationships between the moment sequences and their duals (*cf.* (6.5.38) *et seq.*).

It is instructive to write out the various integral expressions for all the components of $\mathsf{A}_{\alpha\beta}$ that putting (6.5.51) equal to (6.5.53) yields, but we do not spell this out here. The overall factor in (6.5.51) is chosen so as to give exact numerical agreement with (6.5.53). This is easily checked by putting ξ^a equal to the standard time-translation, as above.

6.6 Lie derivatives of spinors

The existence of a conformal Killing vector ξ^a in a space–time $\mathscr{M}$ expresses the fact that $\mathscr{M}$ possesses continuous motions over itself (ignoring global considerations) in which the null cone structure (i.e. the conformal structure) of $\mathscr{M}$ remains unaltered. In ordinary tensor analysis there is the concept of 'dragging' a vector or tensor with respect to an arbitrary vector field ξ^a: This means exactly that the Lie derivative of the vector or tensor with respect to ξ^a vanishes. One may extend this concept to spinors, but here one needs the condition that ξ^a be a conformal Killing vector field. For, otherwise, a null vector – and therefore a null flag – would be 'dragged off the null cone'. It is precisely the condition that ξ^a is a conformal Killing vector field which ensures that null cones are dragged into null cones, so that the dragging of a spinor may be expected to be meaningful. We now examine this formally. (We take ξ^a to be *real.*)

To find the effect of $\pounds = \underset{\xi}{\pounds}$ on a spin-vector κ^A, consider first its effect on the (complex) bivector $\kappa^A\kappa^B\varepsilon^{A'B'}$. We have, by (4.3.3)

$$£(\kappa^A\kappa^B\varepsilon^{A'B'}) = \xi^c\nabla_c(\kappa^A\kappa^B\varepsilon^{A'B'}) - \kappa^D\kappa^B\varepsilon^{D'B'}\nabla_d\xi^a - \kappa^A\kappa^D\varepsilon^{A'D'}\nabla_d\xi^b. \tag{6.6.1}$$

If £ is to behave formally as a derivative operator (Leibniz rule!), (6.6.1) must yield

$$2\varepsilon^{A'B'}\kappa^{(A}£\kappa^{B)} + \kappa^A\kappa^B£\varepsilon^{A'B'} = 2\varepsilon^{A'B'}\xi^c\kappa^{(A}\nabla_c\kappa^{B)} + \kappa^D\kappa^B\nabla_D^{B'}\xi^{AA'} - \kappa^D\kappa^A\nabla_D^{A'}\xi^{BB'}. \tag{6.6.2}$$

Skew-symmetrizing over AB, and changing dummies, we find

$$\kappa^A\kappa^B\nabla_A^{(A'}\xi_B^{B')} = 0.$$

Since this must hold for all κ^A, we must have (*cf.* (3.3.23))

$$\nabla_{(A}^{(A'}\xi_{B)}^{B')} = 0, \tag{6.6.3}$$

i.e., ξ^a must be a conformal Killing vector field, as expected.

Next, transvecting (6.6.2) with $\varepsilon_{A'B'}$ we find

$$\kappa^{(A}£\kappa^{B)} = \xi^c\kappa^{(A}\nabla_c\kappa^{B)} - \tfrac{1}{2}\kappa^D\kappa^{(A}\nabla_{DB'}\xi^{B)B'} - \tfrac{1}{2}\bar{\lambda}\kappa^A\kappa^B, \tag{6.6.4}$$

where λ is defined, $£\varepsilon^{AB}$ being skew in AB, by

$$£\varepsilon^{AB} = \lambda\varepsilon^{AB}. \tag{6.6.5}$$

By Proposition (3.5.15), (6.6.4) yields

$$£\kappa^A = \xi^c\nabla_c\kappa^A - \kappa^D h_D{}^A, \tag{6.6.6}$$

where

$$h_D{}^A = \tfrac{1}{2}\{\nabla_{DB'}\xi^{AB'} + \bar{\lambda}\varepsilon_D{}^A\}. \tag{6.6.7}$$

Applying the usual rules for generalizing the domain of a derivative to the whole spinor system (and *cf.* (4.3.3)) we obtain the following formula for the Lie derivative of any spinor:

$$£\chi_{D\ldots S'\ldots}^{A\ldots P'\ldots} = \xi^y\nabla_y\chi_{D\ldots S'\ldots}^{A\ldots P'\ldots} - h_{A_0}{}^A\chi_{D\ldots S'\ldots}^{A_0\ldots P'\ldots} - \cdots - \bar{h}_{P'_0}{}^{P'}\chi_{D\ldots S'\ldots}^{A\ldots P'_0\ldots} - \cdots + h_D{}^{D_0}\chi_{D_0\ldots S'\ldots}^{A\ldots P'\ldots} + \cdots + \bar{h}_{S'}{}^{S'_0}\chi_{D\ldots S'_0\ldots}^{A\ldots P'\ldots} + \cdots. \tag{6.6.8}$$

Applied to the special case when $\chi_{\ldots}^{\ldots} = \varepsilon^{AB}$ we get, using (6.6.5),

$$£\varepsilon^{AB} = \lambda\varepsilon^{AB} = -h_C{}^A\varepsilon^{CB} - h_C{}^B\varepsilon^{AC},$$

i.e., by (6.6.7),

$$\lambda + \bar{\lambda} = -\tfrac{1}{2}\nabla_c\xi^c. \tag{6.6.9}$$

This result can also be obtained directly from the tensor expression for $£g^{ab}$, using

$$£g^{ab} = £(\varepsilon^{AB}\varepsilon^{A'B'}) = \varepsilon^{AB}£\varepsilon^{A'B'} + \varepsilon^{A'B'}£\varepsilon^{AB} = (\lambda + \bar{\lambda})g^{ab}.$$

Note that (6.6.9) only fixes the *real* part of λ. In fact, one cannot, by such formal considerations as those given above, determine the imaginary part of λ uniquely from the conformal Killing vector ξ^a. However there is the *geometrically natural* choice that this imaginary part be zero. The reasons are basically the same as those given in §5.6 leading up to (5.6.2). There the factor Ω was chosen to be *real* in order that the geometric interpretation of a spin-vector should not be affected by a conformal rescaling*. The same argument can be used here. The choice of a real λ in (6.6.5) ensures that ε^{AB} is 'stretched' by a *real* factor (corresponding to a real Ω) as we drag it along ξ^a. It may be that for some purposes it could turn out to be convenient to drop this 'geometrically natural' restriction, but for the time being we shall adopt it; so by (6.6.9),

$$\lambda = -\tfrac{1}{4}\nabla_c\xi^c. \tag{6.6.10}$$

Substituting this into (6.6.7) we get

$$\begin{aligned} h_D{}^A &= \tfrac{1}{2}\nabla_{DB'}\xi^{AB'} - \tfrac{1}{8}\varepsilon_D{}^A\nabla_{CB'}\xi^{CB'}, \\ h^{DA} &= \tfrac{1}{2}\nabla^{(D}_{B'}\xi^{A)B'} + \tfrac{1}{4}\nabla^{[D}_{B'}\xi^{A]B'}, \\ h_D{}^A &= \tfrac{3}{8}\nabla_{DB'}\xi^{AB'} + \tfrac{1}{8}\nabla^A_{B'}\xi^{B'}_D, \end{aligned} \tag{6.6.11}$$

as various alternative ways of expressing the required relation giving $h_D{}^A$ in terms of ξ^a.

As a check, we can compare the effect of £ on a vector V^a according to (6.6.8) with the standard expression (4.3.2). For the two methods to give the same answer we require

$$h_B{}^A\varepsilon_{B'}{}^{A'} + \bar{h}_{B'}{}^{A'}\varepsilon_B{}^A = \nabla_b\xi^a. \tag{6.6.12}$$

Substituting (6.6.11)(3) (say) into this, we get a relation which is not immediately obviously an identity, but its validity may be checked quickly by raising the indices BB' and examining the parts $[AB][A'B']$, $(AB)[A'B']$, $[AB](A'B')$ and $(AB)(A'B')$ separately, the last giving zero on the right because of (6.6.3).

Relation to twistor theory

The formulae (6.6.11) are perhaps surprising in their complexity. It is therefore noteworthy that they considerably simplify if, for the case of conformal Killing vectors in Minkowski space, we adopt some of the twistor expressions we obtained earlier. With $\xi^{AB'}(=\mathsf{E}^{AB'})$ as the primary spinor part of a Hermitian twistor $\mathsf{E}^\alpha{}_\beta$ we had (with $\theta^A{}_B = \mathsf{E}^A{}_B$, $\zeta_{A'}{}^{B'} = \mathsf{E}_{A'}{}^{B'}$)

$$\nabla_{CC'}\xi^{AA'} = \mathrm{i}\varepsilon_{C'}{}^{A'}\theta^A{}_C - \mathrm{i}\varepsilon_C{}^A\zeta_{C'}{}^{A'}, \tag{6.6.13}$$

where the Hermiticity of $\mathsf{E}^\alpha{}_\beta$ (reality of ξ^a) gave

$$\zeta_{C'}{}^{A'} = \bar{\theta}^{A'}{}_{C'}. \tag{6.6.14}$$

* A possible interpretation of a complex Ω is considered by Penrose (1983*a*) (*cf.* also Volume 1, footnotes on pp. 353,.356, 360).

Imposing the trace-free condition $\mathsf{E}^{\alpha}{}_{\alpha}=0$ we also get

$$\theta^{A}{}_{A'}+\xi_{A'}{}^{A'}=0. \tag{6.6.15}$$

Substituting this into (6.6.13) once contracted we get

$$\nabla_{DB'}\xi^{AB'}=2\mathrm{i}\theta^{A}{}_{D}-\mathrm{i}\varepsilon_{D}{}^{A}\zeta_{B'}{}^{B'}=\mathrm{i}(2\theta^{A}{}_{D}+\varepsilon_{D}{}^{A}\theta^{B}{}_{B}). \tag{6.6.16}$$

Hence

$$\begin{aligned}\tfrac{3}{8}\nabla_{DB'}\xi^{AB'}+\tfrac{1}{8}\nabla^{A}_{B'}\xi^{B'}_{D}&=\mathrm{i}\{\tfrac{3}{4}\theta^{A}{}_{D}+\tfrac{1}{4}\theta_{D}{}^{A}+(\tfrac{3}{8}-\tfrac{1}{8})\varepsilon_{D}{}^{A}\theta^{B}{}_{B}\}\\&=\mathrm{i}\{\tfrac{3}{4}\theta^{A}{}_{D}+\tfrac{1}{4}\theta_{D}{}^{A}+\tfrac{1}{4}\theta^{A}{}_{D}-\tfrac{1}{4}\theta_{D}{}^{A}\}.\end{aligned}$$

Thus (6.6.11)(3) gives

$$h_{D}{}^{A}=\mathrm{i}\theta^{A}{}_{D}. \tag{6.6.17}$$

So, apart from the factor i, the spinor Lie derivative quantity $h_{A}{}^{B}$ is directly one of the spinor parts of the trace-free Hermitian twistor $\mathsf{E}^{\alpha}{}_{\beta}$ whose primary part is $\xi^{AB'}$; its complex conjugate $\bar{h}_{A'}{}^{B'}$ is another.

In this connection it is perhaps worth remarking that if we *drop* the condition (6.6.15) that $\mathsf{E}^{\alpha}{}_{\beta}$ be trace-free, the effect is simply to allow the trace of $\theta^{A}{}_{B}$ to have a real part. (It is (6.6.15), with (6.6.14), which states that $\theta^{A}{}_{A}$ is purely imaginary.) To pass from the case we considered before, when $\mathsf{E}^{\alpha}{}_{\alpha}=0$, to this more general situation, we substitute

$$\theta^{A}{}_{B}\mapsto\theta^{A}{}_{B}+k\varepsilon_{B}{}^{A} \tag{6.6.18}$$

with k real (and, in fact, constant). If we retain (6.6.17), we obtain

$$h_{B}{}^{A}\mapsto h_{B}{}^{A}+\mathrm{i}k\varepsilon_{B}{}^{A}, \tag{6.6.19}$$

which means that the λ of (6.6.5) has acquired an imaginary part:

$$\lambda\mapsto\lambda+2\mathrm{i}k. \tag{6.6.20}$$

This is the situation yielding a 'geometrically unnatural' Lie derivative of a spinor – which we had previously excluded in obtaining (6.6.11).

If ξ^{a} is a proper Killing vector, then, (*cf.* (6.5.1), (6.5.17)) $\nabla_{c}\xi^{c}=0$, whence, from (6.6.7), $h_{A}{}^{A}=0$, so that h_{AB} is symmetric:

$$h_{AB}=h_{BA}. \tag{6.6.21}$$

The tensor $\nabla_{b}\xi_{a}$ is then skew, so that (6.6.12) becomes a particular example of the representation (3.4.20) of a skew tensor in terms of a symmetric spinor.

6.7 Particle constants; conformally invariant operators

We saw in §6.5 how, given a conformal Killing vector ξ^{a} in a space-time $\mathcal{M}$, any *continuous* physical system with trace-free symmetric energy tensor T_{ab} subject to $\nabla^{a}T_{ab}=0$ gives rise to a conserved quantity. A similar property holds for a system of test particles in $\mathscr{M}$. We assume that the test particles describe geodesics in $\mathscr{M}$ between collisions and that 4-momentum is conserved in collisions. The geodesic condition is

$$p^a\nabla_a p^b = 0, \tag{6.7.1}$$

where p^a is a particle's 4-momentum; it is taken to be tangent to the particle's world-line and parallelly propagated along it. We have

$$p^a\nabla_a(\xi_b p^b) = \xi_b p^a\nabla_a p^b + p^a p^b\nabla_a\xi_b = 0, \tag{6.7.2}$$

where either we assume that ξ^a is a Killing vector, in which case the symmetry in $p^a p^b$ may be transferred to $\nabla_a\xi_b$ to make the last term zero; or we assume that ξ^a is a conformal Killing vector and p^a is null, so that the last term, being a multiple of $p^a p^b g_{ab}$, again vanishes. In each of these two cases we have the quantity

$$Q = \xi_a p^a \tag{6.7.3}$$

conserved along each particle's world-line. Also, owing to the linearity of (6.7.3) in p^a, if the sum of vectors p^a is conserved in collisions, so also will the corresponding sum of quantities Q be conserved. Thus we have a conservation law for the system: the total flux of Q across the boundary $\partial\mathscr{W}$ of a compact 4-volume $\mathscr{W}$ must vanish. This shows the similarity (and essential identity) with the continuous case.

Polynomial constants; Killing spinors

The quantity (6.7.3) admits a generalization to nonlinear expressions of the following kind:

$$Q = \xi_{ab\ldots d}p^a p^b \cdots p^d, \tag{6.7.4}$$

where, without loss of generality, $\xi_{a\ldots d}$ is taken to be symmetric:

$$\xi_{ab\ldots d} = \xi_{(ab\ldots d)}. \tag{6.7.5}$$

To generalize the first condition under which (6.7.2) holds, we assume

$$\nabla_{(e}\xi_{ab\ldots d)} = 0. \tag{6.7.6}$$

A tensor $\xi_{a\ldots d}$ subject to (6.7.5) and (6.7.6) is often called a *Killing tensor* and we shall follow this usage here (although there also exists a different concept for which the same name is sometimes used; see (6.7.19) and (6.7.20) below). It has no simple geometric interpretation like that of a Killing vector. However, if p^a has the same significance as before, we have, by (6.7.1) and (6.7.6),

$$p^a\nabla_e Q = p^e\nabla_e(\xi_{a\ldots d}p^a\cdots p^d) = p^e p^a\cdots p^d\nabla_e\xi_{a\ldots d} = 0. \tag{6.7.7}$$

Thus Q is constant along each particle's world-line between collisions. But, because of the nonlinearity of Q in p^a, the total Q is *not* conserved in collisions.

To generalize the second condition under which (6.7.2) holds, we assume

$$\nabla_{(e}\xi_{ab\ldots d)} = g_{(ea}\eta_{b\ldots d)}, \tag{6.7.8}$$

for some $\eta_{b\ldots d}$. Then for *null* p^a the calculation (6.7.7) is still valid. Equation (6.7.8) states that the trace-free symmetric part of $\nabla_e\xi_{a\ldots d}$ vanishes; hence, by the argument at the end of §3.3 (*cf.* (3.3.58)–(3.3.60)), we can write (6.7.8) in spinor form as

$$\nabla^{(E}_{(E'}\xi^{AB\ldots D)}_{A'B'\ldots D')} = 0. \tag{6.7.9}$$

Without loss of generality we can suppose $\xi^{A\ldots D}_{A'\ldots D'}$ to be symmetric in $A\ldots D$ and $A'\ldots D'$; this leaves (6.7.4) unaffected for null p^a (when p^a is of the form $\bar{\pi}^A\pi^{A'}$). Notice that (6.7.9), although now not restricted to flat space, is an example of (6.1.62) (see also (6.4.11)), which in flat space characterizes $\xi^{A\ldots D}_{A'\ldots D'}$ as the primary part of a trace-free symmetric twistor. The case of just one primed and one unprimed index (not necessarily in flat space) is the conformal Killing equation (6.1.70). In (6.7.9) the numbers of primed and unprimed indices are equal, but we can generalize further, to the case when they are unequal. Let $\xi^{A\ldots DK'\ldots N'}$ be symmetric and satisfy

$$\nabla^{(P}_{(Q'}\xi^{A\ldots D)}_{K'\ldots N')} = 0 \tag{6.7.10}$$

where $\xi^{A\ldots DK'\ldots N'}$ has r unprimed indices and s primed ones. Then this generalizes not only (6.1.70) but also (6.1.69) and its extension to (6.4.1). Such a $\xi^{A\ldots N'}$ is sometimes called a *Killing spinor*, a terminology we shall adopt here. When p^a is future-null and thus of the form $\bar{\pi}^A\pi^{A'}$, and

$$Q = \xi^{A\ldots DK'\ldots N'}\bar{\pi}_A\ldots\bar{\pi}_D\pi_{K'}\ldots\pi_{N'}, \tag{6.7.11}$$

then, by (6.7.10),

$$p^y\nabla_y Q = p^y\nabla_y(\xi^{A\ldots N'}\bar{\pi}_A\ldots\pi_{N'}) = \bar{\pi}_Y\bar{\pi}_A\ldots\bar{\pi}_D\pi_{Y'}\pi_{K'}\ldots\pi_{N'}\nabla^{YY'}\xi^{A\ldots N'} = 0, \tag{6.7.12}$$

provided we assume, in addition to p^a being parallelly propagated, that the flag plane of $\pi^{A'}$ is also parallelly propagated:

$$p^y\nabla_y\pi^{A'} = 0. \tag{6.7.13}$$

When $r = s$ and $\xi^{A\ldots DK'\ldots N'}$ is Hermitian, we are back at (6.7.7) with a real Q conserved along the geodetic world-lines. When $r \neq s$, we have an *essentially* complex Q; the information contained in Q concerns the flag plane direction of $\pi_{A'}$. Under

$$\pi_{A'} \mapsto e^{-i\theta}\pi_{A'}$$

we have

$$Q \mapsto e^{i(r-s)\theta} Q.$$

Thus Q is a *spin-weighted* quantity when referred to $\pi_{A'}$ as a basis spinor (*cf.* (4.12.9), *et seq.*).

Sometimes a $\xi_{\cdots}^{\cdots}$, subject to (6.7.10), is known explicitly in a space–time. If $r = s$, this gives us a 'first integral' of the equation for null geodesics and so is useful if we wish to integrate the null geodesics explicitly. In the same way, a quantity satisfying (6.7.6) gives us a 'first integral' of the geodesic equation, not necessarily assumed to be null. The metric g_{ab} is one quantity, satisfying (6.7.6) trivially, that is always available. So if we have three more such 'first integrals' we have enough information, in a four-dimensional space, to determine the geodesics explicitly up to quadratures. (Three such integrals exist, for example, for the Kerr solution, as we shall see shortly.) If we have a $\xi_{\cdots}^{\cdots}$ satisfying (6.7.10) with $r \neq s$, then we have additional information which enables us to determine the propagation of *polarization* planes along null geodesics. Thus, if we know $\xi_{\cdots}^{\cdots}$ at two points on a null geodesic γ, we can read off at once (without integration) the result of parallelly propagating the flag planes of $\pi_{A'}$, from the one point to the other along γ: $\arg(Q)$ must be the same at the two points for a parallelly propagated $\pi_{A'}$. Note that we also get a real 'first integral' from Q when $r \neq s$, namely $|Q|$, via the quantity $\tilde{Q} = Q\bar{Q}$ associated with the solution of (6.7.10) $\tilde{\xi}_{A'\ldots D'K'\ldots N'}^{A\ldots DK\ldots N} = (-1)^{r+s}\xi_{(K'\ldots N'}^{(A\ldots D}\bar{\xi}_{A'\ldots D')}^{K\ldots N)}$ (See Walker and Penrose 1970.)

Killing spinors for type {22} vacuums

In Chapter 8 we shall discuss the classification of the gravitational (Weyl) spinor Ψ_{ABCD} of a curved space–time in terms of the concidence pattern of its four PNDs (*cf.* (3.5.18) *et seq.*). The particular case of a vacuum space–time $\mathcal{M}$ (with or without cosmological constant) in which the PNDs coincide in *pairs*

$$\Psi_{ABCD} = 6\psi\alpha_{(A}\alpha_B\beta_C\beta_{D)} \quad (\alpha_A\beta^A = 1) \tag{6.7.14}$$

(referred to as type {22} or D) has special interest for us here, since the object defined by

$$\kappa_{AB} = \psi^{-\frac{1}{3}}\alpha_{(A}\beta_{B)} \tag{6.7.15}$$

turns out always to be a Killing spinor:

$$\nabla_{(A}^{A'}\kappa_{BC)} = 0. \tag{6.7.16}$$

(See Walker and Penrose 1970. The proof amounts, in effect, to rearranging the Bianchi identity equation for (6.7.14), *cf.* (4.10.9), so that it becomes (6.7.16); *cf.* also Hughston, Penrose, Sommers and Walker* 1972.) Thus by the foregoing discussion we have an explicit (complex) conserved quantity along any null geodesic in any such space–time. This fact is of particular value in the case of the Kerr solution, which is {22} and vacuum and which represents the gravitational field of a general (rotating) stationary black hole (*cf.* Hawking and Ellis 1973, Chandrasekhar 1983, Wald 1984). In particular, rotation effects on the polarization planes of photons can be read off directly in this solution by use of κ_{AB} (*cf.* Connors and Stark 1977). Other examples of {22} vacuums are Schwarzschild's solution (a special case of Kerr's, but here the {22} nature can be inferred without calculation, see p. 229), the NUT-space and the C-Metric (*cf.* Kramer, Stephani, MacCallum and Herlt 1980).

Moreover, we have (Hughston and Sommers 1973):

(6.7.17) PROPOSITION

For any vacuum space–time (cosmological constant allowed) possessing a Killing spinor $\chi^{AB}\in\mathfrak{S}^{(AB)}$, *the vector*

$$k^a=\nabla_B^{A'}\chi^{AB}$$

is a (complex) Killing vector.

Proof: We need to show that $\nabla_{(a}k_{b)}=0$ (*cf.* (6.5.8)). In spinor terms, this relation can be split into two equations, namely the vanishing of the part of $\nabla_{AA'}k_{BB'}$ *skew* in AB and in $A'B'$ and of the part *symmetric* in AB and in $A'B'$. Consider the first, which is equivalent to $\nabla_{AA'}k^{AA'}=0$. We have

$$\nabla_{AA'}k^{AA'}=\nabla_{AA'}\nabla_B^{A'}\chi^{AB}=\square_{AB}\chi^{AB}$$

(by (4.9.2) and the symmetry of χ^{AB}) and this clearly vanishes (*cf.* (4.9.7) *et seq.*) – even without calculation (and irrespective of the vacuum condition) since there are no scalars that can be constructed bilinearly from χ^{AC} and curvature spinors. For the symmetric–symmetric part, consider

$$\nabla_{A(A'}\nabla_{B')B}\chi_{CD}.$$

This is symmetric in AB since the curvature terms arising from the commutator of derivatives here involves only $\Phi_{...}$, which is zero. The part symmetric in BCD vanishes by the Killing spinor equation $\nabla_{A'(B}\chi_{CD)}=0$,

* In this reference it is also shown that if instead of the vacuum equations the Einstein–Maxwell equations hold with $\varphi_{AB}=2\phi\alpha_{(A}\beta_{B)}$, where α_A, β_A are SFR (*cf.* §7.3) and (6.7.14) holds, then $\phi^{-\frac{1}{2}}\alpha_{(A}\beta_{B)}$ is a Killing spinor.

and so also does the part symmetric in ACD (because of symmetry in AB). Thus Young tableau symmetry holds in AB, CD, corresponding to the partition (2, 2) (*cf.* the discussion in the extended 'footnote' in §3.3, some paragraphs before (3.3.62)). This involves anti-symmetrization in *two pairs* of unprimed indices, so all four unprimed indices can be 'split off' in terms of ε-spinors. It follows that

$$\nabla_{(A|(A'}\nabla_{B')[B}\chi_{C]|D)} = 0$$

which (with $[BC]$ replaced by a contraction) is the final condition required. (If preferred, this last equation can be obtained by a direct but more complicated calculational argument.)

Properties of the Kerr solution

In the case of the Kerr solution (and therefore also the Schwarzschild solution – but in no other case), the Killing vector k^a constructed as in (6.7.17), with κ^{AB} in place of χ^{AB}, is *real* (and actually generates the standard time-translation symmetry* of these solutions). This reality implies that the bivector (Floyd 1973, *cf.* Penrose 1973).

$$K_{ab} = \mathrm{i}\kappa_{AB}\varepsilon_{A'B'} - \mathrm{i}\varepsilon_{AB}\bar{\kappa}_{A'B'} \tag{6.7.18}$$

satisfies

$$\nabla_{(a}K_{b)c} = 0 \quad \text{i.e. } \nabla_a K_{bc} = \nabla_{[a}K_{bc]} \tag{6.7.19}$$

as is not hard to verify directly. (The parts of the LHS of (6.7.19)(1) symmetric in ABC or in $A'B'C'$ vanish by (6.7.16). The remaining part, skew in AB and $A'B'$, vanishes by the reality of k^a.) Note that (6.7.19) is formally identical with the condition (6.3.4) for a bivector in $\mathbb{M}$ to have the position-dependence of the dual of an angular-momentum tensor. This might lead us to expect that $^*K_{ab}$ describes, in some sense, the angular-momentum structure of the Kerr solution. In fact things are not quite so direct as this. In the weak-field limit of the Kerr solution the 'angular momentum' described by $^*K_{ab}$ has the opposite sign for spin-to-momentum ratio from that of the actual angular momentum (*cf.* Floyd 1973). The twistor whose primary part is κ^{AB} is actually proportional to the *inverse* of the angular-momentum twistor $\mathrm{A}_{\alpha\beta}$. This will, to some extent, be illuminated in §7.4, p. 205.

* In the general Kerr solution $H^{ab}k_b$ (*cf.* (6.7.20)) turns out to be a second independent Killing vector (though this vanishes in the Schwarzschild case) and, with k^a, spans the entire 2-space of Killing vectors (including that describing the axi-symmetry). (See Hughston and Sommers 1973.)

A skew K_{ab}, subject to (6.7.19), has been sometimes also referred to as a Killing tensor, since (6.7.19) is another generalization of the Killing equation, distinct from (6.7.6). More usually, a $K_{ab} \in \mathfrak{S}_{[ab]}$ subject to (6.7.19) is now called a *Killing–Yano tensor*, a terminology we shall follow here. The concept applies also to a $K_{\alpha\beta\ldots\delta}$ (in n dimensions) with any number of skew indices, and we have a 'conserved' object $p^\alpha K_{\alpha\beta\ldots\delta}$ defined along any geodesic with parallelly propagated tangent vector p^α. (Killing–Yano tensors in space–time have been studied by, for example, Dietz and Rudiger (1980, 1981), and Killing spinors, by Walker and Penrose (1970), Hughston, Penrose, Sommers and Walker (1972), Jeffryes (1984*a*); *cf.* also Sommers (1973), Hughston and Sommers (1973).)

As has been pointed out by Floyd 1973 (*cf.* Penrose 1973) the product

$$H_{ab} = K_{ac}K^c{}_b \tag{6.7.20}$$

satisfies

$$H_{ab} = H_{ba}, \quad \nabla_{(a}H_{bc)} = 0 \tag{6.7.21}$$

and thus constitutes a Killing tensor of our original kind (6.7.6). The conserved quantity $Q = H_{ab}p^a p^b = -(p^a K_{ac})(p^b K_b{}^c)$ is referred to as *Carter's constant* for the Kerr solution. There are also two independent Killing vectors in this solution and so, with g_{ab}, we have the required number, four, of constants enabling the geodesics to be obtained explicitly up to quadratures (Carter 1968*a*, *b*, *cf.* also Chandrasekhar 1983).

The physical meaning of Carter's constant is not altogether clear, but in some sense Q describes a multiple of the 'total angular momentum' of a test particle in orbit around the black-hole source. (This is 'total' in the sense of squared length of angular-momentum 3-vector.) Indeed, in the special case of the Schwarzschild solution this interpretation is certainly valid, there being an identity

$$(mG)^{\frac{4}{3}}H^{ab} = x^a x^b + y^a y^b + z^a z^b \tag{6.7.22}$$

relating H_{ab} to the three Killing vectors x^a, y^a, z^a which generate rotations about the standard axes. The conserved quantities $p_a x^a$, $p_a y^a$, $p_a z^a$ are the corresponding angular momenta of a test particle about these axes. The (constant) quantity m is the mass of the source (and G, the gravitational constant). We note in passing that in the Schwarzschild solution, with standard r-coordinate,

$$\psi = -\frac{mG}{r^3}. \tag{6.7.23}$$

The Kerr solution is also very remarkable in that the various standard field equations all separate in a Kerr background (Carter 1968*a*, Chandrasekhar

1983). These properties are related to the existence of the Killing–Yano tensor K_{ab} (Carter and McLenaghan 1979).

Conformally invariant first derivatives

As we have seen in (6.1.60)–(6.1.62), equation (6.7.10) is intimately connected with twistors. Any symmetric solution of (6.7.10) in $\mathbb{M}$ is precisely the primary spinor part of some uniquely defined trace-free symmetric twistor $\mathrm{X}^{\alpha\ldots\delta}_{\kappa\ldots\nu}$:

$$\mathrm{X}^{\alpha\ldots\delta}_{\kappa\ldots\nu} = \mathrm{X}^{(\alpha\ldots\delta)}_{(\kappa\ldots\nu)}, \quad \mathrm{X}^{\xi\beta\ldots\delta}_{\xi\lambda\ldots\nu} = 0. \tag{6.7.24}$$

Another way of putting this is to say that for each null Z^α the quantity

$$\mathrm{X}^{\alpha\ldots\delta}_{\kappa\ldots\nu}\mathrm{Z}^\kappa\ldots\mathrm{Z}^\nu\bar{\mathrm{Z}}_\alpha\ldots\bar{\mathrm{Z}}_\delta \tag{6.7.25}$$

is equal to (6.7.11) at every point P of the corresponding null line Z (since $\omega^A = 0$ on Z and the only spinor part of $\mathrm{X}^{\alpha\ldots\delta}_{\kappa\ldots\nu}$ contracted with *no* ω^A is $\xi^{A\ldots D}_{K'\ldots N'}$.) It follows that the general symmetric solution of (6.7.10) is given by the appropriate generalization of (6.1.10), (6.1.26), (6.1.51), (6.1.55). One example will suffice to illustrate this. Consider the case of a trace-free symmetric $[{}^2_1]$-twistor $\mathrm{X}^{\alpha\beta}{}_\gamma$. Its primary spinor part $\mathrm{X}^{ABC'}$ is given by

$$\begin{aligned}\mathrm{X}^{ABC'} = {} & \mathring{\mathrm{X}}^{(AB)C'} - 2\mathrm{i}x^{A'(A}\mathring{\mathrm{X}}_{A'}{}^{B)C'} + \mathrm{i}x^{CC'}\mathring{\mathrm{X}}^{(AB)}{}_C - x^{A'(A}x^{B)B'}\mathring{\mathrm{X}}_{A'B'}{}^{C'} \\ & + 2x^{CC'}x^{A'(A}\mathring{\mathrm{X}}_{A'}{}^{B)}{}_C - \mathrm{i}x^{A'(A}x^{B)B'}x^{CC'}\mathring{\mathrm{X}}_{A'B'C}\end{aligned} \tag{6.7.26}$$

for some constant spinors $\mathring{\mathrm{X}}^{ABC'}, \ldots, \mathring{\mathrm{X}}_{A'B'C}$. By differentiating this expression and symmetrizing we at once verify that it is in fact a solution of (6.7.10).

Since trace-free symmetric twistors *are* defined by symmetric solutions of (6.7.10), we expect that this equation is conformally invariant (*cf.* (6.1.72)). Indeed, provided

$$\hat{\xi}^{A\ldots DK'\ldots N'} = \xi^{A\ldots DK'\ldots N'}, \tag{6.7.27}$$

the symmetrized derivative occurring in (6.7.10) is itself a conformal spinor. This may be checked by directly applying (5.6.15). The result is

$$\hat{\nabla}^{(Q'|(P}\hat{\xi}^{A\ldots D)|K'\ldots N')} = \Omega^{-2}\nabla^{(Q'|P}\xi^{A\ldots D|K'\ldots N')}, \tag{6.7.28}$$

and it holds just as well in general curved space as in flat or conformally flat space.

The conformal invariance of this particular type of symmetrized derivative for *some* conformal weight for $\xi^{\cdots}$ is not surprising. The fact that the result of the symmetrization is an irreducible spinor at each point (*cf.* §3.3), as is $\xi^{\cdots}$ itself, implies that only one kind of term in Υ_a can appear. The choice of conformal weight zero in (6.7.27) is made to ensure that the term in Υ_a cancels out.

For the same reason, the other possibilities of constructing symmetric spinors by forming a contracted or symmetrized derivative can also be made to have conformal weight. Taking the most natural arrangements of upper and lower indices in each case, we obtain, for symmetric $\xi^{\cdots}_{\cdots}$,

$$\hat{\nabla}^{K'(P}\hat{\xi}^{A\dots D)}_{K'L'\dots N'} = \Omega^{-3}\nabla^{K'(P}\xi^{A\dots D)}_{K'L'\dots N'}, \qquad (6.7.29)$$

provided

$$\hat{\xi}^{A\dots D}_{K'\dots N'} = \Omega^{-1}\xi^{A\dots D}_{K'\dots N'}; \qquad (6.7.30)$$

whereas

$$\hat{\nabla}^{A(Q'}\hat{\xi}^{K'\dots N')}_{AB\dots D} = \Omega^{-3}\nabla^{A(Q'}\xi^{K'\dots N')}_{AB\dots D}, \qquad (6.7.31)$$

provided

$$\hat{\xi}^{K'\dots N'}_{A\dots D} = \Omega^{-1}\xi^{K'\dots N'}_{A\dots D} \qquad (6.7.32)$$

(which is not the same as (6.7.30), unless $r = s$); and, on the other hand,

$$\hat{\nabla}^{AK'}\hat{\xi}_{AB\dots DK'L'\dots N'} = \Omega^{-4}\nabla^{AK'}\xi_{AB\dots DK'L'\dots N'}, \qquad (6.7.33)$$

provided

$$\hat{\xi}_{A\dots DK'\dots N'} = \Omega^{-2}\xi_{A\dots DK'\dots N'}. \qquad (6.7.34)$$

All these properties hold in general curved space–time as well as in flat or conformally flat space–time. Several special cases have been noted earlier. For example, (6.7.31) with (6.7.32) establishes the conformal invariance of the massless equations (4.12.42), which correspond to $K'\dots N'$ being vacuous. Again, the conformal invariance of (6.4.13), (6.4.15) is now established. Also the conformal invariance of the divergence equation of a convector of weight -2 or of a symmetric trace-free tensor of valence $\left[\begin{smallmatrix}0\\2\end{smallmatrix}\right]$ and weight -2 were noted in §5.9, and this is seen to be implicit in (6.7.33).

Twistor solutions of the various equations

In Minkowski space $\mathbb{M}$ it turns out that not only (6.7.10) can be solved using twistors, but so also can the equations obtained by setting to zero (6.7.29), (6.7.31), and (6.7.33). We shall not enter into a fully detailed discussion of all these equations here since this would take us too far afield. Nevertheless it will be worth while merely to state formulae which give general classes of (analytic) solutions to these equations, and to explain why they give solutions. For completeness, we also include the solution to (6.7.10) in a slightly different form. Those particular cases of (6.7.29) and (6.7.31) whose vanishing corresponds to the massless field equations will be discussed in greater detail in §6.10, where the solutions in these cases will also be given in a different and more general form.

Let us take twistors $\mathrm{Z}^\alpha=(\omega^A,\pi_{A'})$ and $\mathrm{W}_\alpha=(\lambda_A,\mu^{A'})$ as in (6.1.13) and (6.1.24). We shall be concerned with holomorphic functions in Z^α and W_α, i.e. with holomorphic functions of the components of Z^α, W_α in some twistor basis (6.1.17), (6.1.34). We allow these functions to have singularities in some suitably arranged regions. Let

$$f(\mathrm{W}_\alpha,\mathrm{Z}^\alpha)$$

be holomorphic, and homogeneous, of the following respective degrees in W_α and Z^α:

$$\begin{aligned}&(p,q) \text{ for case } (6.7.28)\\&(p,-2-q) \text{ for case } (6.7.29)\\&(-2-p,q) \text{ for case } (6.7.31)\\&(-2-p,-2-q) \text{ for case } (6.7.33)\end{aligned} \qquad (6.7.35)$$

where $\xi_{\dots}^{\dots}$ has p unprimed and q primed indices, and where we further require that f be a *polynomial* in the relevant variables when the homogeneity degree is *non-negative*. We consider only the subset of $\mathbb{T}_\alpha\times\mathbb{T}^\alpha$ in which the incidence relation

$$\mathrm{W}_\alpha\mathrm{Z}^\alpha=0 \qquad (6.7.36)$$

holds (*cf.* (6.2.13), (9.3.17)), and we use only the restriction of f to this region. Referring W_α and Z^α to the origin $O\in\mathbb{M}$, then provided $\lambda_A\neq 0$ and $\pi_{A'}\neq 0$ (now both taken as point spinors at O, *cf.* (6.1.14) *et seq.*), we can find r^a at O (not necessarily real) such that

$$\mathrm{W}_\alpha\overset{O}{\leftrightarrow}(\lambda_A,-\mathrm{i}r^{AA'}\lambda_A) \qquad (6.7.37)$$
$$\mathrm{Z}^\alpha\overset{O}{\leftrightarrow}(\mathrm{i}r^{AA'}\pi_{A'},\pi_{A'}).$$

(This is easily seen by taking components.) However, r^a is not unique but has the freedom $r^a\mapsto r^a+k\lambda^A\pi^{A'}$, $k\in\mathbb{C}$. The points R of complexified Minkowski space $\mathbb{CM}$ with these position vectors r^a therefore constitute a complex null straight line. (The geometric significance of this fact is discussed in §9.3 below.) Note that, by the discussion of §§6.1, 6.2 these points R are just those *incident* with *both* W_α and Z^α in $\mathbb{CM}$.

The function f can now be re-expressed in terms of r^j, λ_A, and $\pi_{A'}$ and we write (in the region defined by (6.7.36))

$$f(\mathrm{W}_\alpha,\mathrm{Z}^\alpha)=F(r^j,\lambda_A,\pi_{A'}) \qquad (6.7.38)$$

for some suitable range of the variables r^j, λ_A, $\pi_{A'}$, and where F is now homogeneous, according to (6.7.35), in the respective spinors λ_A, $\pi_{A'}$.

Next, consider the expressions:

$$\xi^{A\ldots DK'\ldots N'}(r^j)\lambda_A\ldots\lambda_D\pi_{K'}\ldots\pi_{N'} = f(\mathrm{W}_\alpha, \mathrm{Z}^\alpha) \tag{6.7.39}$$

$$\xi^{A\ldots D}_{K'\ldots N'}(r^j)\lambda_A\ldots\lambda_D = \frac{1}{2\pi \mathrm{i}}\oint \pi_{K'}\ldots\pi_{N'}f(\mathrm{W}_\alpha, \mathrm{Z}^\alpha)\pi_{Q'}\mathrm{d}\pi^{Q'} \tag{6.7.40}$$

$$\xi^{K'\ldots N'}_{A\ldots D}(r^j)\pi_{K'}\ldots\pi_{N'} = \frac{1}{2\pi \mathrm{i}}\oint \lambda_A\ldots\lambda_D f(\mathrm{W}_\alpha, \mathrm{Z}^\alpha)\lambda_P\mathrm{d}\lambda^P \tag{6.7.41}$$

$$\xi_{A\ldots DK'\ldots N'}(r^j) = \frac{1}{(2\pi \mathrm{i})^2}\oint \lambda_A\ldots\lambda_D\pi_{K'}\ldots\pi_{N'}f(\mathrm{W}_\alpha, \mathrm{Z}^\alpha)\lambda_P\mathrm{d}\lambda^P \wedge \pi_{Q'}\mathrm{d}\pi^{Q'}. \tag{6.7.42}$$

The right-hand sides are evaluated by substituting F for f according to (6.7.38). The occurrence, in each case, of a $\xi^{\cdots}_{\cdots}$ on the left-hand side depending only on r^j and not on λ_A or $\pi_{A'}$ follows either from the assumed polynomial nature of f or from the fact that the dependence on the spinor variables has been integrated out.

In (6.7.40) we perform an integral over a closed one-dimensional contour in the $\pi_{A'}$-spin-space, for each fixed value of r^j and λ_A, the contour suitably surrounding singularities of F and moving continuously as r^j and λ_A are varied. In (6.7.41) the integral is over a closed one-dimensional contour in λ_A-spin-space for each fixed r^j and $\pi_{A'}$, which again moves continuously. In (6.7.42) the integral is over a two-dimensional contour in the product of the spin-spaces for $\pi_{A'}$ and λ_A, for each fixed r^j, and it moves continuously with r^j. In each case the homogeneity degrees (6.7.35) have been chosen to ensure that these integrals are in fact 'contour integrals' in the sense that the value of the integral is unchanged as the contour is continuously deformed within the relevant spin-space, provided it does not encounter singularities of F during the deformation. This is because the exterior derivative (*cf.* (4.3.14) and (4.3.25)) of the integrand, with r^j held constant, vanishes in each case when the homogeneities are as given (see the paragraph following (6.10.4)).

If we assume that each $\xi^{\cdots}_{\cdots}$ is chosen to be symmetric, then each will be uniquely defined as a function of r^j throughout any region of r^j in which suitable contours for the integrals exist; and in the case given by (6.7.39), for all r^j. We are to interpret r^j as the position vector at O of some point R in $\mathbb{CM}$. Also, we have seen that the relations (6.7.37) express the incidence of each of the twistors W_α and Z^α with the point R. Now we may regard f as depending on eight independent complex variables when expressed as F, i.e., in terms of r^a, λ_A, $\pi_{A'}$. But since the restriction (6.7.36) on W_α and Z^α is then implicit, the values of f with which we are concerned depend only on a seven-complex-dimensional manifold of arguments. Thus f is subject to a restriction when expressed in terms of r^j, λ_A and $\pi_{A'}$. This restriction is

$$\lambda_P \pi_{Q'} \nabla^{PQ'} F = 0, \tag{6.7.43}$$

where the operator ∇_a is the derivative with respect to r^a, treating $\lambda_A, \pi_{A'}$ as constant. For

$$\nabla^{PQ'} \mathrm{W}_\alpha \overset{o}{\leftrightarrow} (\nabla^{PQ'} \lambda_A, \nabla^{PQ'}(-\mathrm{i} r^{AA'} \lambda_A)) = (0, -\mathrm{i}\lambda^P \varepsilon^{Q'A'}),$$

whence $\lambda_P \nabla^{PQ'} \mathrm{W}_\alpha = 0$, and

$$\nabla^{PQ'} \mathrm{Z}^\alpha \overset{o}{\leftrightarrow} (\nabla^{PQ'}(\mathrm{i} r^{AA'} \pi_{A'}), \nabla^{PQ'} \pi_{A'}) = (\mathrm{i}\pi^{Q'} \varepsilon^{PA}, 0),$$

whence $\pi_{Q'} \nabla^{PQ'} \mathrm{Z}^\alpha = 0$. We can write (6.7.43) alternatively as

$$\lambda_P \nabla^{PK'} F \propto \pi^{K'} \tag{6.7.44}$$

or

$$\pi_{Q'} \nabla^{AQ'} F \propto \lambda^A. \tag{6.7.45}$$

Applying the operator in (6.7.43) to (6.7.39), that in (6.7.44) to (6.7.40), that in (6.7.45) to (6.7.41), and $\nabla^{AK'}$ to (6.7.42), we see that in each case the expression is annihilated. Since this holds for all $\lambda_A, \pi_{A'}$ in these expressions, the vanishing of the derivatives (6.7.28), (6.7.29), (6.7.31), (6.7.33) is automatically ensured by the respective constructions (6.7.39)–(6.7.42).

Note that in (6.7.39) we can write

$$f(\mathrm{W}_\alpha, \mathrm{Z}^\alpha) =: \mathrm{F}^{\alpha\ldots\delta}_{\kappa\ldots\nu} \mathrm{W}_\alpha \ldots \mathrm{W}_\delta \mathrm{Z}^\kappa \ldots \mathrm{Z}^\nu$$

because of the presumed polynomial nature of F. Since (6.7.36) holds, it is only the trace-free and symmetric part of $\mathrm{F}^{\alpha\ldots\delta}_{\kappa\ldots\nu}$ which is involved in defining the field $\xi^{A\ldots DK'\ldots N'}$. It follows that such fields $\xi^{\cdots}$ correspond to trace-free symmetric twistors, and so we are back at the description that we had earlier for solutions of (6.7.10). It is easy to see that we can also write

$$f(\mathrm{W}_\alpha, \mathrm{Z}^\alpha) =: \mathrm{F}^{\alpha\ldots\delta}(\mathrm{Z}^\nu) \mathrm{W}_\alpha \ldots \mathrm{W}_\delta$$

in (6.7.40), and

$$f(\mathrm{W}_\alpha, \mathrm{Z}^\alpha) =: \mathrm{F}_{\kappa\ldots\nu}(\mathrm{W}_\beta) \mathrm{Z}^\kappa \ldots \mathrm{Z}^\nu$$

in (6.7.41).

As we observed earlier, the vanishing of (6.7.31) with $q = 0$ [of (6.7.29) with $p = 0$] is the massless free-field equation for spin $\frac{1}{2}p$ [spin $\frac{1}{2}q$]. Thus (6.7.31) [*or* (6.7.29)] affords a method of expressing the solutions of these equations in $\mathbb{M}$ in a concise way in terms of twistor holomorphic functions (cf. also §6.10 below).

Repeated operations; resolutions of equations

Note that the only circumstance in which *more* than one of the operations (6.7.28), (6.7.29), (6.7.31), (6.7.33) can be applied to the same spinor (with indices suitably raised or lowered) is that (6.7.29) and (6.7.31) can both be applied if $\xi^{\cdots}_{\cdots}$ has an equal number of unprimed and primed indices ($p = q$). Observe also that, although the result of *any* of those four operations is a spinor with the correct symmetries so that we would envisage applying such an operation again, the resulting weights normally forbid this if we require conformal invariance for the result. Indeed, the only possibilities for a conformally invariant second application of such an operation are:

$$\left.\begin{array}{l}\text{(6.7.28) followed by (6.7.29) with } p=0\\ \text{(6.7.28) followed by (6.7.31) with } q=0\\ \text{(6.7.29) followed by (6.7.31) with } p=q>0\\ \text{(6.7.31) followed by (6.7.29) with } p=q>0\\ \text{(6.7.29) followed by (6.7.33) with } p=0,\ q>1\\ \text{(6.7.31) followed by (6.7.33) with } q=0,\ p>1\end{array}\right\} \qquad (6.7.46)$$

In these particular cases we therefore appear to get conformally invariant second-derivative operations – whether the space–time is conformally flat or general. Some other possibilities for conformally invariant second derivatives will be considered shortly.

However, cases (6.7.46)(1), (2), (5), and (6) are readily seen to give *zero* when the ∇ operators commute (i.e. in flat space without electromagnetic interaction). From the point of view of finding conformally invariant second-derivative operations, this may seem disappointing. But as we saw in §6.5, compositions of derivative operations may have another significance, namely one in terms of exact sequences. In fact, it can be shown (provided we restrict ourselves to a suitably 'convex' region of $\mathbb{M}$ – or of $\mathbb{CM}$, in which case all fields are taken to be holomorphic) that in each of these cases we not only have two maps whose composition is zero, but the kernel of one is precisely the image of the preceding one. Hence each pair satisfies the prerequisites for two successive maps of an exact sequence. Moreover, these partial exact sequences can be usefully extended. We shall display the entire sequences explicitly here in the cases (6.7.46)(1) and (6). The cases (2) and (5) are simply the complex conjugates of (1) and (6), respectively.

Consider the case (6) first. The operation (6.7.31) with $q = 0$ is the map on symmetric $\phi_{A\ldots E}$ whose kernel gives us the massless free fields:

$$\phi_{AB\ldots E} \mapsto \nabla^{AA'}\phi_{AB\ldots E}. \qquad (6.7.47)$$

This is to be followed by (6.7.33) (on symmetric $\theta^{A'}_{B\ldots E}$):

$$\theta^{A'}_{BC\ldots E} \mapsto \nabla^{B}_{A'}\theta^{A'}_{BC\ldots E}. \qquad (6.7.48)$$

This pair constitutes the third and fourth maps of the sequence

$$0 \rightarrow \mathfrak{F}_{ABC\ldots E} \xrightarrow{\iota} \mathfrak{D}_{(ABC\ldots E)} \xrightarrow{\nabla^{AA'}} \mathfrak{D}^{A'}_{(BC\ldots E)} \xrightarrow{\nabla^{B}_{A'}} \mathfrak{D}_{(C\ldots E)} \rightarrow 0 \qquad (6.7.49)$$

whose notation we shall now explain. If $\mathscr{U}$ is some open set in $\mathbb{M}$ (or in $\mathbb{CM}$), then $\mathfrak{D}^{A\ldots CP'\ldots R'}_{D\ldots FS'\ldots U'}$ may, for the time being, be thought of as the set $\mathfrak{S}^{A\ldots CP'\ldots R'}_{D\ldots FS'\ldots U'}[\mathscr{U}]$ of spinor fields on $\mathscr{U}$, regarded as a vector space over $\mathbb{C}$; a $\mathfrak{D}^{\cdots}_{\cdots}$ with symmetrization brackets on its indices may be thought of as that subject to $\mathfrak{S}^{\cdots}_{\cdots}[\mathscr{U}]$ having the indicated symmetries. (A more precise characterization of each $\mathfrak{D}$, which is actually a 'sheaf', will be indicated presently.) The space $\mathfrak{F}_{AB\ldots E}$ may (preliminarily) be thought of as the vector space over $\mathbb{C}$ of solutions of the massless field equations in $\mathscr{U}$, and thus as the kernel of (6.7.47). The map ι in (6.7.49) is simply 'injection', i.e., it takes any massless field into itself. With that, the first three map-junctions in the sequence (6.7.49) are exact. Exactness at the final stage demands that every symmetric $\psi_{C\ldots E}$ can be represented in the form $\nabla^{B}_{A'}\theta^{A'}_{BC\ldots E}$ for some symmetric $\theta^{\cdot}_{\cdots}$. The proof of this is straightforward in terms of components relative to an arbitrary basis (along the lines of (6.7.50), (6.7.51) below.)

The set $\mathscr{U}$ would have to be thought of as sufficiently small (or of suitable shape) that no *global* obstructions to the solutions of our differential equations arise. Recall that in (6.5.27) we restricted the region $\mathscr{U}$ to having Euclidean topology. For other differential equations it might be necessary to impose restrictions also on the *shape* of $\mathscr{U}$. Such a viewpoint would be adequate for a tentative interpretation of the meaning of (6.7.49). To be more accurate, that sequence should really be interpreted as an exact sequence of *sheaves*. It is beyond the scope of the present book to discuss sheaves, or even to give a precise definition of a sheaf. The essential point, however, is that the exactness of the sequence (6.7.49) (and of the other sheaf exact sequences that we shall consider later) is to be interpreted entirely *locally*. Rather than single out one *particular* open set $\mathscr{U}$ and consider spinor fields on this $\mathscr{U}$, we ask only that each point in $\mathbb{M}$ (or $\mathbb{CM}$) should possess *some* open neighbourhood in which our equations are soluble. Thus, in particular, for exactness of the pair of maps (6.7.47) and (6.7.48), we ask that, at each point $P \in \mathbb{M}$, if $\nabla^{B}_{A'}\theta^{A'}_{BC\ldots E} = 0$ in some neighbourhood $\mathscr{U}$ of P, then there exists some neighbourhood $\mathscr{U}' \subset \mathscr{U}$ of P in which we can find a $\phi_{AB\ldots E}$ with $\nabla^{AA'}\phi_{AB\ldots E} = \theta^{A'}_{BC\ldots E}$. (This is a localization of condition (ii) for exactness (*cf.* after (6.5.26). We have already seen that condition (i) holds – and no localization requirement is needed for that.) The proof that this is indeed the case is straightforward with the help of a basis. We set

$$\begin{aligned}&\phi_0=\phi_{00\ldots0},\quad \phi_1=\phi_{10\ldots0},\ldots,\phi_p=\phi_{11\ldots1};\\ &\theta_1=\theta^{0'}_{0\ldots0},\ldots,\theta_p=\theta^{0'}_{1\ldots1};\quad \chi_1=\theta^{1'}_{0\ldots0},\ldots,\\ &\chi_p=\theta^{1'}_{1\ldots1};\quad x^{00'}=u,\quad x^{11'}=v,\quad x^{01'}=\zeta,\quad x^{10'}=\bar{\zeta},\end{aligned}$$

and we have to show that we can solve

$$\left.\begin{aligned}&\frac{\partial}{\partial v}\phi_i-\frac{\partial}{\partial \xi}\phi_{i+1}=\theta_i\\ \text{and}\qquad &\\ &-\frac{\partial}{\partial \zeta}\phi_i+\frac{\partial}{\partial u}\phi_{i+1}=\chi_i\end{aligned}\right\}\tag{6.7.50}$$

locally, whenever

$$\frac{\partial\theta_i}{\partial\bar{\zeta}}-\frac{\partial\theta_{i+1}}{\partial u}-\frac{\partial\chi_{i+1}}{\partial\zeta}+\frac{\partial\chi_i}{\partial v}=0\tag{6.7.51}$$

($i=0,\ldots,p-1$). By use of (6.7.51) we can actually integrate the system (6.7.50) step by step, thus establishing our assertion. Accordingly, also, the equations (6.7.51) are seen as the integrability conditions for the system (6.7.50), giving us the usual situation in going 'up' an exact sequence.

The sequence (6.7.49) is what is called a *resolution* of the sheaf $\mathfrak{F}_{A\ldots E}$, or a resolution of the massless field equations. This means here that each of the sheaves $\mathfrak{D}^{\ldots}_{\ldots}$ is *free* in the sense that no differential equations are imposed on it. (Again, a more technical statement is beyond our present scope; resolutions are important in the theory of differential equations and in cohomology theory.)

The importance of the other case, (6.7.46)(1), is that it forms part of a resolution of the complex conjugate of (6.4.1), i.e. for the equation for the primary spinor part of a symmetric twistor of the type $\mathrm{T}_{\alpha\ldots\gamma}$. The subset (sheaf) of $\mathfrak{D}^{A'\ldots C'}$ consisting of solutions of (6.4.18) (Killing spinors) is denoted by $\mathfrak{R}^{A'\ldots C'}$. The operation (6.7.28) with $p=0$ is the map on symmetric $\lambda^{A'\ldots C'}$ whose kernel is the space of all such primary parts:

$$\lambda^{A'\ldots C'}\mapsto\nabla_E^{(D'}\lambda^{A'\ldots C')}.\tag{6.7.52}$$

This constitutes the third map in the following sequence, whose exactness up to that map is therefore established:

$$0\to\mathfrak{R}^{A'\ldots C'}\xrightarrow{\iota}\mathfrak{D}^{(A'\ldots C')}\xrightarrow{\nabla_E^{D'}}\mathfrak{D}_E^{(A'\ldots C'D')}\xrightarrow[\nabla^{EE'}]{\nabla_{AA'}\cdots\nabla_{DD'}\,\oplus}\begin{matrix}\mathfrak{D}_{(AB\ldots E)}\\ \oplus\\ \mathfrak{D}^{(A'B'\ldots E')}\end{matrix}$$

$$\xrightarrow[-\nabla_{BB'}\cdots\nabla_{EE'}]{\nabla^{AA'}}\mathfrak{D}^{A'}_{(B\ldots E)}\xrightarrow{\nabla^B_{A'}}\mathfrak{D}_{(C\ldots E)}\to 0.\tag{6.7.53}$$

(The symmetrizations to be performed after the various derivative operators are applied are implicit in the displayed symmetries of the image spaces.) Note that the final stage of the sequence (6.7.53) is the same as that of (6.7.49). The space $\mathfrak{D}_{(A\ldots E)}\oplus\mathfrak{D}^{(A'\ldots E')}$ (written vertically in (6.7.53)) is simply the space of pairs $(\phi_{A\ldots E},\eta^{A'\ldots E'})$ (each component being symmetric); and the maps on either side of it are

$$\xi_A^{A'\ldots D'}\mapsto(\nabla_{(A|A'|}\cdots\nabla_{D|D'|}\xi_{E)}^{A'\ldots D'},\ \nabla^{A(E'}\xi_A^{A'\ldots D')})\tag{6.7.54}$$

and

$$(\phi_{AB\ldots E},\eta^{A'B'\ldots E'})\mapsto\nabla^{AA'}\phi_{AB\ldots E}-\nabla_{BB'}\cdots\nabla_{EE'}\eta^{A'B'\ldots E'}.\tag{6.7.55}$$

One readily verifies that condition (i) for exactness is satisfied at each stage. But the proof that condition (ii) also holds (at the required local level) is much more involved than in the previous case, and we shall not go into it. (The sequence (6.7.53) was introduced by Eastwood 1985*a* and Buchdahl 1982.)

It is of interest that, if we assign conformal weights to the spinors in each (non-zero) sheaf in the sequence as follows

$$0,-1,(-1,-3),-3,-4,\tag{6.7.56}$$

then we obtain conformal invariance for each map in the sequence, but only for the class of conformal rescalings which send a flat g_{ab} (locally) into a flat $\hat{g}_{ab}$. As we shall see in (6.8.27), this class is characterized by

$$\nabla_a\Upsilon_b=\Upsilon_{AB'}\Upsilon_{BA'},\tag{6.7.57}$$

and this relation is needed in the verification of our assertion. For general rescalings, the derivatives of order higher than the first that occur in (6.7.54) and (6.7.55) would have to be modified by terms involving the Ricci curvature and derivatives of the Ricci curvature (*cf.* paragraph after (6.8.29)).

The exactness of the sequence (6.7.53) codes a considerable amount of information in a compact form. We note, as just one implication, the fact that was stated in (6.4.15), (6.4.16), (6.4.17), namely that *all* massless fields $\phi_{A\ldots L}$ can be locally obtained from (6.4.16), subject to (6.4.15), with the complete gauge freedom for $\psi_{\ldots}$ being given by (6.4.17). To see this, note that the part of the kernel of the map (6.7.55) having $\eta^{\cdots}=0$ is $(\phi_{\ldots},0)$ where $\phi_{\ldots}$ is a massless field; so this is precisely that part of the image of (6.7.54) for which $\nabla^{A(E'}\xi_A^{A'\ldots D')}=0$ (which is (6.4.15)); and the gauge freedom is the kernel of *that* map, i.e. the kernel of (6.7.52), giving precisely the gauge freedom in (6.4.17).

6.8 Curvature and conformal rescaling

As we have seen in the last section, the operations (6.7.46)(1), (2), (5), and (6) are not useful as second-derivative operations, since they give zero when the ∇ operators commute. We nevertheless found them to be important. Also in curved space–time (or in the presence of electromagnetic interaction) these operations are not really second-derivative operations. They simply yield contracted products of $\xi^{\cdots}_{\cdots}$ with curvature (or with the electromagnetic field) quantities. Their importance to us *here* is that their conformal invariance yields the conformal behaviour of the curvature. Note that the operation (6.7.46)(6) is precisely that considered in (5.8.1) for obtaining the consistency relation for massless fields, whence the result

$$\nabla^{B}_{A'}\nabla^{AA'}\xi_{ABC\ldots L} = -ie\varphi^{AB}\xi_{ABC\ldots L} - (p-2)\xi_{ABM(C\ldots K}\Psi_{L)}{}^{ABM} \quad (6.8.1)$$

($r>1$) is obtained. Similarly, (6.7.46)(1) is just the operation used for the consistency relation for the twistor equation (6.1.1) or its generalization (6.4.1). Taking for simplicity the case $r=1$ (the cases $r>1$ are essentially similar) we get from (6.1.5)

$$\nabla^{A'(C}\nabla^{A}_{A'}\xi^{B)} = -\Psi^{ABC}{}_{D}\xi^{D} + ie\varphi^{(AB}\xi^{C)}. \quad (6.8.2)$$

The operations (6.7.46)(5) and (2) arise simply as complex conjugates of the above.

Conformal invariance of Ψ_{ABCD}

Now the significance of the conformal invariance of the relations (6.7.46) for our present purposes lies in the fact that the conformal behaviour of Ψ_{ABCD} may be read off directly from them. The argument is most straightforward in the case of (6.8.2). We take $e=0$ (since the conformal behaviour $\hat{\varphi}_{AB} = \Omega^{-1}\varphi_{AB}$ has already been demonstrated – *cf.* (5.9.8)). Now, by (6.7.28) and (6.7.29), we have

$$\hat{\nabla}^{A'(C}\hat{\nabla}^{A}_{A'}\xi^{B)} = \Omega^{-3}\nabla^{A'(C}\nabla^{A}_{A'}\xi^{B)}, \quad (6.8.3)$$

whence

$$\hat{\Psi}^{ABC}{}_{D}\xi^{D} = \Omega^{-3}\Psi^{ABC}{}_{D}\xi^{D}.$$

But ξ^{D} is arbitrary, and so

$$\hat{\Psi}_{ABCD} = \Psi_{ABCD}. \quad (6.8.4)$$

This conformal invariance shows that Ψ_{ABCD} is a measure of that part of the curvature which remains invariant under conformal rescaling. Since $\Psi_{ABCD}=0$ in flat space–time, it vanishes also in conformally flat space–time. In fact, as we shall see later (in §6.9), the condition $\Psi_{ABCD}=0$

is also *sufficient* for conformal flatness (i.e. for each point of space–time to have a neighbourhood in which there exists a conformal factor Ω which makes the metric $\Omega^2 g_{ab}$ flat in that neighbourhood). It is for the above reasons that Ψ_{ABCD} is called the (Weyl) *conformal* spinor. Because of the relation (4.6.41) of Ψ_{ABCD} to the Weyl tensor C_{abcd}, we have

$$\hat{C}_{abcd} = \Omega^2 C_{abcd}, \quad \hat{C}^a{}_{bcd} = C^a{}_{bcd}, \text{ etc.} \tag{6.8.5}$$

Let us compare (6.8.4) with the equation (*cf.* (5.7.17))

$$\hat{\phi}_{ABCD} = \Omega^{-1}\phi_{ABCD} \tag{6.8.6}$$

which gives conformal invariance for a spin-2 massless field. The difference in the conformal behaviour of (6.8.4) and (6.8.6) is noteworthy. This is in contra-distinction to the case of electromagnetism, where the *same* conformal weight, namely -1, results from the requirement that φ_{AB} be obtainable from commutators of derivatives of charged fields, as results from the requirement that the massless free-field equation on φ_{AB} be preserved (*cf.* paragraph above (5.9.8)). The difference in the behaviours (6.8.4) and (6.8.6) is related to the fact that the equations of general relativity are not conformally invariant. One manifestation of this fact is that the RHS of the spinor Bianchi identity (4.10.7) is not a conformal density. For suppose we have a vacuum solution of Einstein's equations, with Weyl spinor Ψ_{ABCD} therefore satisfying

$$\nabla^{AA'}\Psi_{ABCD} = 0 \tag{6.8.7}$$

Putting $\phi_{ABCD} = \Psi_{ABCD}$ and transforming according to (6.8.6) we must have $\hat{\nabla}^{AA'}\hat{\phi}_{ABCD} = 0$. Therefore, by (6.8.4)

$$\hat{\nabla}^{AA'}(\Omega^{-1}\Psi_{ABCD}) = 0, \tag{6.8.8}$$

i.e. (*cf.* (5.6.14)),

$$\hat{\nabla}^{AA'}\Psi_{ABCD} = \Upsilon^{AA'}\Psi_{ABCD}, \tag{6.8.9}$$

and so the conformally rescaled Ricci spinor must satisfy

$$\hat{\nabla}^{B'}_{(B}\hat{\Phi}_{CD)A'B'} = \Upsilon^{A}_{A'}\Psi_{ABCD} \tag{6.8.10}$$

by (4.10.7), and, since it does not generally vanish whereas $\Phi_{ABC'D'}$ does, the Ricci spinor cannot have a conformal weight. A conformal rescaling applied to a solution (a space–time) of Einstein's equations will, therefore, generally destroy the satisfaction of the vacuum equations.

Conformal behaviour of $\Phi_{ABC'D'}$ and Λ

Let us find how $\Phi_{ABC'D'}$ and Λ in fact transform under conformal rescaling. For this purpose it is convenient to use an identity derived from an expression we had earlier, namely (5.8.4):

$$\nabla^A_{A'}\nabla_{B'(A}\xi_{B)} = \nabla_{B'B}\nabla^A_{A'}\xi_A - \tfrac{1}{2}\nabla_{A'B}\nabla^A_{B'}\xi_A + 3\Lambda\varepsilon_{A'B'}\xi_B - \Phi_{ABA'B'}\xi^A.$$

Adding $\frac{2}{3}$ of this equation to $\frac{1}{3}$ of this equation with A' and B' reversed, we get

$$\tfrac{2}{3}\nabla^A_{A'}\nabla_{B'(A}\xi_{B)} + \tfrac{1}{3}\nabla^A_{B'}\nabla_{A'(A}\xi_{B)} = \tfrac{1}{2}\nabla_{B'B}\nabla^A_{A'}\xi_A - \xi^A \mathrm{P}_{ABA'B'} \qquad (6.8.11)$$

where

$$\mathrm{P}_{ABA'B'} = \Phi_{ABA'B'} - \Lambda\varepsilon_{AB}\varepsilon_{A'B'} = \tfrac{1}{12}Rg_{ab} - \tfrac{1}{2}R_{ab}. \qquad (6.8.12)$$

Now consider the effect of a conformal rescaling on (6.8.11), where we take

$$\hat{\xi}^A = \xi^A, \quad \text{i.e.} \quad \hat{\xi}_A = \Omega\xi_A.$$

The two terms on the left of (6.8.11) involve derivatives of expressions like $\nabla_{B'(A}\xi_{B)}$ which, as we have seen, are conformal densities:

$$\hat{\nabla}_{B'(A}\hat{\xi}_{B)} = \Omega\nabla_{B'(A}\xi_{B)}. \qquad (6.8.13)$$

We now take the difference between (6.8.11) and its conformally rescaled version. Because of (6.8.13) the LHS of this difference can involve only terms which contain first derivatives of ξ_A. Such terms must necessarily cancel from the entire calculation. For, only undifferentiated and once-differentiated terms in ξ_A can occur after the difference is taken; but since ξ_A and $\nabla_{BB'}\xi_A$ are independent at each given point (ξ_A being arbitrary), these two types of term must *separately* give identical relations. Thus we need only consider the RHS of (6.8.11) in our calculation, and we can ignore the once-differentiated terms in ξ_A which result from the conformal rescaling.

We have

$$\hat{\nabla}^A_{A'}\hat{\xi}_A = \nabla^A_{A'}\xi_A + 2\Upsilon^A_{A'}\xi_A,$$

so taking the rescaled (6.8.11) minus the original (6.8.11) – and ignoring once-differentiated terms in ξ_A – we get

$$0 = (\nabla_{B'B}\Upsilon^A_{A'})\xi_A - \Upsilon_{A'B}(\Upsilon^A_{B'}\xi_A) - \xi^A\hat{\mathrm{P}}_{ABA'B'} + \xi^A \mathrm{P}_{ABA'B'}.$$

But ξ^A is arbitrary, and so we obtain

$$\hat{\mathrm{P}}_{ab} = \mathrm{P}_{ab} - \nabla_b\Upsilon_a + \Upsilon_{AB'}\Upsilon_{BA'} \qquad (6.8.14)$$

for the conformal rescaling behaviour of P_{ab}. Since

$$R_{ab} = -2\mathrm{P}_{ab} - g_{ab}\mathrm{P}_c{}^c, \qquad (6.8.15)$$

(6.8.14) also gives the rescaling behaviour of R_{ab}. However, it turns out that the particular combination of trace and trace-free parts of R_{ab} which occur in P_{ab} often arise in formulae which have to do with conformal rescalings.

In this connection we point out that the expression (4.8.2) giving $C_{ab}{}^{cd}$ in terms of the Riemann tensor, the Ricci tensor and the scalar curvature, becomes

$$C_{ab}{}^{cd} = R_{ab}{}^{cd} + 4\mathrm{P}_{[a}{}^{[c} g_{b]}{}^{d]}, \tag{6.8.16}$$

while for the Bianchi identity (4.10.1) we get

$$\nabla^a C_{abcd} = -2\nabla_{[c}\mathrm{P}_{d]b}, \tag{6.8.17}$$

with spinor equivalent (*cf.* (4.10.3)

$$\nabla^D_{A'}\Psi_{ABCD} = \nabla^{B'}_{(B}\mathrm{P}_{A)CA'B'}. \tag{6.8.18}$$

The symmetry

$$\mathrm{P}_{ab} = \mathrm{P}_{ba}, \quad \text{i.e. } \mathrm{P}_{ABA'B'} = \mathrm{P}_{BAB'A'} \tag{6.8.19}$$

is manifest. The symmetry of (6.8.14) in this respect arises from the fact that Υ_a is a gradient (*cf.* (5.6.14)) whence

$$\nabla_a \Upsilon_b = \nabla_b \Upsilon_a. \tag{6.8.20}$$

Alternative versions of (6.8.14) which are sometimes useful are

$$\mathrm{P}_{ab} = \hat{\mathrm{P}}_{ab} + \hat{\nabla}_a \Upsilon_b + \Upsilon_{AB'}\Upsilon_{BA'} \tag{6.8.21}$$

(obtained from (6.8.14) by interchanging the roles of g_{ab} and $\hat{g}_{ab}$ which entails $\Omega \to \Omega^{-1}$ and therefore $\Upsilon_a \mapsto -\Upsilon_a$), and

$$\begin{aligned}\hat{\mathrm{P}}_{ab} &= \mathrm{P}_{ab} - \Omega^{-1}\nabla_a\nabla_b\Omega + 2\Omega^{-2}\nabla_a\Omega\nabla_b\Omega - \tfrac{1}{2}g_{ab}\Omega^{-2}\nabla_c\Omega\nabla^c\Omega \\ &= \mathrm{P}_{ab} + \Omega\nabla_a\nabla_b\Omega^{-1} - \tfrac{1}{2}g_{ab}\Omega^{-2}\nabla_c\Omega\nabla^c\Omega\end{aligned} \tag{6.8.22}$$

(obtained by substituting (5.6.14)(1) into (6.8.14), and using (3.4.13)).

By taking the trace and trace-free parts of (6.8.14), (6.8.22), and using (*cf.* (6.8.12))

$$\Phi_{ABA'B'} = \mathrm{P}_{(AB)(A'B')}, \quad \Lambda = -\tfrac{1}{4}\mathrm{P}_a{}^a, \tag{6.8.23}$$

we get

$$\begin{aligned}\hat{\Phi}_{ABA'B'} - \Phi_{ABA'B'} &= -\nabla_{A(A'}\Upsilon_{B')B} + \Upsilon_{A(A'}\Upsilon_{B')B} \\ &= \Omega\nabla_{A(A'}\nabla_{B')B}\Omega^{-1} \\ &= -\Omega^{-1}\hat{\nabla}_{A(A'}\hat{\nabla}_{B')B}\Omega,\end{aligned} \tag{6.8.24}$$

and

$$\begin{aligned}\Omega^2\hat{\Lambda} &= \Lambda + \tfrac{1}{4}\nabla^a\Upsilon_a + \tfrac{1}{4}\Upsilon^a\Upsilon_a \\ &= \Lambda + \tfrac{1}{4}\Omega^{-1}\Box\Omega,\end{aligned} \tag{6.8.25}$$

where (*cf.* (5.10.6))

$$\Box = \nabla^a\nabla_a. \tag{6.8.26}$$

Conformal rescalings preserving flatness

One immediate consequence of (6.8.14) may be pointed out here, namely that the condition

$$\nabla_a \Upsilon_b = \Upsilon_{AB'} \Upsilon_{BA'}, \tag{6.8.27}$$

for the Ricci tensor to remain unchanged under conformal rescaling, is, in the case of flat space $\mathbb{M}$, the condition for the rescaling to yield another flat metric. The general solution of (6.8.27) is not hard to find. It is

$$\Omega = \frac{P}{(x^a - Q^a)(x_a - Q_a)}, \tag{6.8.28}$$

(and limiting cases) giving

$$\Upsilon_b = \frac{-2(x_b - Q_b)}{(x^a - Q^a)(x_a - Q_a)}, \tag{6.8.29}$$

since the right-hand side of (6.8.27) is $\Upsilon_a \Upsilon_b - \frac{1}{2}\Upsilon_c \Upsilon^c g_{ab}$, by (3.4.13)

Conformally invariant wave equation and $\Box$

Now that we have the transformation of all parts of the curvature under conformal rescaling, we are in a position to investigate the conformal invariance of equations involving second and higher derivatives. The reason that the curvature transformation formulae are needed for this is that curvature 'correction terms' are often required in order that such equations can become conformally invariant. The simplest example of this occurs with the wave equation, which in flat space–time is $\Box \phi = 0$. As it stands, this equation is not conformally invariant (unless we rescale our flat space–time so that $\hat{g}_{ab}$ is again flat, or rescale from one space with vanishing Ricci scalar to another). For conformal invariance we require the modification of this simple equation to

$$(\Box + \tfrac{1}{6}R)\phi = 0, \quad \text{i.e.} \ (\Box + 4\Lambda)\phi = 0, \tag{6.8.30}$$

where, as with the massless equations of positive spin,* we choose

$$\hat{\phi} = \Omega^{-1}\phi. \tag{6.8.31}$$

The invariance of (6.8.30) follows easily from (6.8.25). Indeed, if ϕ scales according to (6.8.31) but is otherwise arbitrary, then we have

$$(\hat{\Box} + 4\hat{\Lambda})\hat{\phi} = \Omega^{-3}(\Box + 4\Lambda)\phi. \tag{6.8.32}$$

Note that (6.8.25) can itself be regarded as a special case of (6.8.32), where

* A recent result due to Eastwood and Singer (1985) exhibits a modification of the square $\Box^2$ of the D'Alembertian which is conformally invariant when acting on scalars ϕ of conformal weight *zero* (i.e. $\hat{\phi} = \phi$), namely

$$\nabla_b(\nabla^b \nabla^a - 2R^{ab} + \tfrac{2}{3}Rg^{ab})\nabla_a.$$

(Note that $-2R^{ab} + \frac{2}{3}Rg^{ab} = 4\mathrm{P}^{AB'BA'}$.) A new theory by Eastwood and Rice provides many further examples, such as $(\nabla^{(A}_{(A'}\nabla^{B)}_{B')} + \Phi^{AB}_{A'B'})\,\phi$ for ϕ of conformal weight *one*.

$\phi = \Omega$ and $\hat{\phi} = \Omega^{-1}\Omega = 1$:

$$(\hat{\Box} + 4\hat{\Lambda})1 = \Omega^{-3}(\Box + 4\Lambda)\phi$$

Equation (6.8.30) is, in fact, the natural extension to zero spin of the positive-spin massless equation (4.12.42). Since ϕ has no indices, the exact analogue of (4.12.42) cannot be constructed. However, there are second order relations satisfied by $\phi_{AB\ldots L}$ in consequence of (4.12.42) and these can be extended to zero spin. In flat space–time we have the relation

$$\nabla_{AA'}\nabla_B^{A'} = \tfrac{1}{2}\varepsilon_{AB}\Box, \tag{6.8.33}$$

since the commutability of derivatives implies skewness in AB (*cf.* (2.5.24)). The analogous relation in curved space–time (or in the presence of electromagnetic interaction) is

$$\nabla_{AA'}\nabla_B^{A'} = \tfrac{1}{2}\varepsilon_{AB}\Box + \Box_{AB}, \tag{6.8.34}$$

with $\Box_{AB}$ as in (4.9.2). Applying this to (4.12.42). we get for a spin-$\tfrac{1}{2}n$ massless field of charge e,

$$\begin{aligned} 0 = \nabla_{AA'}\nabla^{MA'}\phi_{MB\ldots L} &= \tfrac{1}{2}\varepsilon_A{}^M\Box\phi_{MB\ldots L} + \Box_A{}^M\phi_{MB\ldots L} \\ &= \tfrac{1}{2}\Box\phi_{AB\ldots L} + \mathrm{X}_{AM}{}^{NM}\phi_{NB\ldots L} - \mathrm{X}_A{}^M{}_B{}^N\phi_{MNC\ldots L} \\ &\quad - \mathrm{X}_A{}^M{}_C{}^N\phi_{MBN\ldots L} - \cdots - ie\varphi_A{}^M\phi_{MB\ldots L}, \end{aligned}$$

where we have used (4.9.13) and (5.1.44). By reference to (4.6.19) and (4.6.34) this yields

$$\begin{aligned} (\Box + 2(n+2)\Lambda)\phi_{AB\ldots L} = 2(n-1)&\Psi_{(AB}{}^{MN}\phi_{C\ldots L)MN} \\ &+ 2ie\varphi_A{}^M\phi_{MB\ldots L}. \end{aligned} \tag{6.8.35}$$

This equation has a well-defined 'limit as $n \to 0$' only if Ψ_{ABCD} and $e\varphi_{AB}$ both vanish. But, even for positive (and sufficiently large) spin we have the consistency conditions (5.8.2) unless these quantities both vanish. So, in any case, (4.12.42) is of interest for general spin only in conformally flat space in the absence of electromagnetic interaction. And then, indeed, its specialization to $n = 0$, via (6.8.35), is (6.8.30). The 'correction term' 4Λ in the limit could of course have been anticipated from the fact that it is needed for conformal invariance.

It may be pointed out that there is a conserved energy tensor for solutions of (6.8.30):

$$\begin{aligned} T_{ab} &= \tfrac{1}{6}k\{2\nabla_{A(A'}\phi\nabla_{B')B}\phi - \phi\nabla_{A(A'}\nabla_{B')B}\phi + \phi^2\Phi_{ABA'B'}\} \\ &= \tfrac{1}{12}k\{4\nabla_a\phi\nabla_b\phi - g_{ab}\nabla_c\phi\nabla^c\phi - 2\phi\nabla_a\nabla_b\phi \\ &\quad + 4\Lambda\phi^2 g_{ab} - \phi^2 R_{ab}\}, \end{aligned} \tag{6.8.36}$$

(k = constant) which is not only symmetric, with vanishing divergence, but also trace-free (see: Newman and Penrose 1968, Callan, Coleman and Jackiw 1970). It has been referred to as 'the new improved energy tensor'. The more familiar expression

$$T'_{ab} = \tfrac{1}{2}k\nabla_{AB'}\phi\nabla_{BA'}\phi$$
$$= \tfrac{1}{4}k(2\nabla_a\phi\nabla_b\phi - g_{ab}\nabla_c\phi\nabla^c\phi) \tag{6.8.37}$$

is divergence-free when the (not conformally invariant) equation $\Box\phi = 0$ holds, but it is not trace-free. However, T'_{ab} has the definiteness property $T'_{ab}l^a n^b \geqslant 0$ (for future-null vectors l^a, n^a) not shared by T_{ab}. In flat space, the difference $T_{ab} - T'_{ab}$ is the 'divergence' $\nabla^c\{\tfrac{1}{3}k\phi g_{a[b}\nabla_{c]}\phi\}$, which contributes nothing to the *total* energy–momentum. For massless fields of spins $\tfrac{1}{2}$ and 1, the (trace-free) energy tensors were given in (5.8.3) and (5.2.4). For massless fields of higher spin, no (local) energy tensor exists.

We note, in passing, two special cases of (6.8.35). For a Maxwell field in curved charge-free space–time we have

$$\Box\varphi_{AB} = 2\Psi_{ABCD}\varphi^{CD} - 8\Lambda\varphi_{AB} \tag{6.8.38}$$

which may be compared with

$$\Box F_{ab} = R_{abcd}F^{cd} \tag{6.8.39}$$

(*cf.* Eddington 1924, §74. Eddington's factor 2 is in error).

In the case of gravitation we have

$$\Box\Psi_{ABCD} = 6\Psi_{(AB}{}^{EF}\Psi_{CD)EF} - 2\lambda\Psi_{ABCD} \tag{6.8.40}$$

in empty space with cosmological term λ (*cf.* (4.10.9) and (4.10.10)). When there exists a right-hand side in the Bianchi identities (*cf.* (4.10.12)) we get, instead,

$$\Box\Psi_{ABCD} = 6\Psi_{(AB}{}^{EF}\Psi_{CD)EF} - (2\lambda + 4\pi G T_q{}^q)\Psi_{ABCD}$$
$$- 8\pi G\nabla^{A'}_{(A}\nabla^{B'}_{B}T_{CD)A'B'}, \tag{6.8.41}$$

where we have used (4.6.32)(2). Regarding the last two terms on the right as 'source terms', the first term on the right is a 'correction term' to the gravitational wave equation. Its vanishing happens to be necessary and sufficient for Ψ_{ABCD} to be *null* (i.e. for all its principal null directions to coincide). This will be proved in §8.6. It is worth noting that there are exact null vacuum wavelike solutions that are very closely analogous to the corresponding linear fields but the analogy is less close for the non-null solutions (*cf.* Robinson and Trautman 1962, Kramer, Stephani, MacCallum and Herlt 1980, Sommers 1976, 1977).

Bach's conformally invariant tensor

Another interesting conformally invariant quantity is the *Bach tensor* (Bach 1921, *cf.* Schouten 1954, Kozameh, Newman and Tod 1985).

$$B_{ab} = 2(\nabla^C_{A'}\nabla^D_{B'} + \Phi^{CD}_{A'B'})\Psi_{ABCD} \tag{6.8.42}$$

(here given in Dighton's 1972, 1974 form). It satisfies

$$B_{ab} = B_{ba}, \quad B_{ab} = \bar{B}_{ab}, \quad \nabla^a B_{ab} = 0, \tag{6.8.43}$$

and, under conformal rescaling,

$$\hat{B}_{ab} = B_{ab}. \tag{6.8.44}$$

Tensor expressions for B_{ab} are provided by

$$\begin{aligned} B_{ab} &= \nabla^c\nabla^d C_{acbd} - \tfrac{1}{2}R^{cd}C_{acbd} \\ &= 2\nabla^c\nabla_{[a}\mathrm{P}_{c]b} + \mathrm{P}^{cd}C_{acbd}. \end{aligned} \tag{6.8.45}$$

6.9 Local twistors

Our discussion of twistors so far has been concerned mainly with flat space-time. This is partly due to the fact that the consistency relation (6.1.6) precludes the use of the twistor equation (6.1.1) unless $\Psi_{ABCD} = 0$, and partly to the fact that even in conformally flat space-time (i.e., when we *do* have $\Psi_{ABCD} = 0$) it is often more *convenient* to go over to flat space-time and describe twistors in terms of the position vector x^a of a point in $\mathbb{M}$ relative to an origin O. The formalism of *local* twistors which we describe in this section enables us to have twistors of a kind in an *arbitrary* curved space-time $\mathcal{M}$. In particular, when $\mathcal{M}$ is conformally flat, this new formalism enables us to deal with the original twistors (which in this section we shall call 'global' twistors) in a way which does not require us to transform first to a flat metric in order to do calculations. As an application of the local twistor formalism we shall show that the vanishing of Ψ_{ABCD} is sufficient, as well as necessary, for a space-time to be patchwise conformal to $\mathbb{M}$.

Complex space-time

Most of this section, as indeed most of this book, is concerned with ordinary real space-times. However, much of modern twistor theory is discussed against a background of *complex* space-times, and we shall therefore make a few remarks on that subject now. As we have noted before (see p. 64; footnotes), a *complexified* space-time is one which originates from an 'ordinary' real analytic space–time (which therefore has real analytic coordinates x^a and a real analytic metric of Lorentzian signature $+ - - -$), by allowing the coordinates to become complex and by extending the metric coefficients holomorphically into the complex domain. The original real space–time, however, retains a role as a privileged subspace. A *complex* space–time, on the other hand, is indistinguishable from a general four-complex-dimensional complex-Riemannian manifold, and in it no four-real-dimensional subspace has been singled out (and conventionally declared to be 'real') to give it a reality structure. (Indeed, no candidate for such a subspace will exist in general, *cf.* Woodhouse 1977.) In complexified space–times there exists the operation of complex conjugation, which is the map taking any point with (complexified) coordinates x^a into the point with coordinates $\overline{x^a}$ (and similarly for tensors). This map is invariant under real analytic transformations of the coordinates of the original real space. (But it is, of course, *not* invariant under the general holomorphic coordinate transformations in a *complex* space, and that is why the operation is undefined there.) Only the original real points (and real tensors at real points) are left invariant under this map (such invariance being the criterion for 'reality'). It should be noted that there exists no criterion for the reality of tensors – or the complex conjugateness of pairs of tensors – at *complex* points even in a complexified space, since the conjugate tensor is situated at the complex conjugate *point*. (Only in *flat* complexified spaces can we use – rather artificially – distant parallelism to bring the tensors to the same point.)

By retracing the arguments used in building up the spinor formalism, we easily see that there is the following rule for translating the spinor formulae of this book from a real to a complex space–time:* the algebraic and differential operations (except complex conjugation) formally go over unchanged; but whenever a real quantity appears it is replaced by a complex one, and whenever a complex quantity λ appears together with its conjugate $\bar{\lambda}$, it and its conjugate are replaced by two *independent* complex quantities λ and $\tilde{\lambda}$, say. For example, in (1.2.19) we had the complex coordinate $\zeta = (X + \mathrm{i}Y)/(T - Z)$ defined on the null cone. Similarly we had $\bar{\zeta} = (X - \mathrm{i}Y)/(T - Z)$. But when T, X, Y, Z become complex, these two expressions simply yield independent quantities ζ and $\tilde{\zeta}$. Non-holomorphic (but real-analytic) expressions in ζ, such as $(\zeta\bar{\zeta} + 1)^2$, become, when complexified, holomorphic expressions in the *two* complex variables ζ, $\tilde{\zeta}$ – in this case $(\zeta\tilde{\zeta} + 1)^2$. The operation of complex conju-

* For further discussion, see Flaherty (1976, 1980); also Lind and Newman (1974).

gation disappears, but, of course, when a doubly conjugated quantity $\tilde{\bar{\lambda}}$ appears, then that is identified with the original quantity λ.

The spaces $\mathfrak{S}^A$ and $\mathfrak{S}^{A'}$ become unrelated to one another, the pair of spinors ξ^A, $\bar{\xi}^{A'}$ – which previously determined one another – being replaced by a pair of *independent* spinors ξ^A, $\tilde{\xi}^{A'}$. The general (future-) null vector $\xi^A\bar{\xi}^{A'}$ becomes, upon complexification, the general complex null vector $\xi^A\tilde{\xi}^{A'}$ (*cf.* (3.2.6)). Quantities Y that are originally real, yield no *new* quantities $\tilde{Y}$, since $Y=\bar{Y}$ goes over into $Y=\tilde{Y}$. Thus, for example, the covariant derivative operator ∇_a, being originally real, yields no new operator $\tilde{\nabla}_a$. Instead, ∇_a becomes a complex holomorphic operator. The complex curvature quantities Ψ_{ABCD}, $\Phi_{ABC'D'}$, Λ, become accompanied by quantities $\tilde{\Psi}_{A'B'C'D'}$, $\tilde{\Phi}_{CDA'B'}$, $\tilde{\Lambda}$; but because of the original reality conditions (4.6.4) and (4.6.17), the latter give us nothing new, i.e.,

$$\tilde{\Phi}_{ABC'D'}=\Phi_{ABC'D'},\quad \tilde{\Lambda}=\Lambda. \tag{6.9.1}$$

On the other hand, $\tilde{\Psi}_{A'B'C'D'}$ is a new curvature quantity independent of Ψ_{ABCD}.

In fact, it is possible to have complex space–times in which, say, $\tilde{\Psi}_{A'B'C'D'}=0$, without Ψ_{ABCD} necessarily vanishing. These are called *right conformally flat* (or *conformally anti-self-dual*) spaces, whereas if only Ψ_{ABCD} necessarily vanishes, the space is called *left conformally flat* (or *conformally self-dual*). If, in addition, $\Phi_{ABC'D'}=0$, and $\Lambda=0$, then we call these spaces, respectively, *right-flat* (or *anti-self-dual*) and *left-flat* (or *self-dual*). Remarkably, such spaces arise naturally (Newman's $\mathscr{H}$-space, *cf.* Newman 1976, Ko, Ludvigsen, Newman and Tod 1981, Hansen, Newman, Penrose and Tod 1978) in the study of asymptotically flat *real* space–times. (We very briefly discuss these in Chapter 9, see p. 389 *et seq.*) Note that for a general complex space–time, the PNDs of Ψ_{ABCD} and of $\tilde{\Psi}_{A'B'C'D'}$ are quite independent of each other, and the coincidence scheme for the PNDs of Ψ_{ABCD} can be quite different from that of the PNDs of $\tilde{\Psi}_{A'B'C'D'}$. So at each point we have *two independent* classification schemes like that given in Chapter 8 below for a single Ψ_{ABCD}.*

* In a real (pseudo-)Riemannian 4-space with signature $(++++)$, $(++--)$, or $(----)$, we have a quite different type of complex conjugation operation on spinors, which takes each of the modules $\mathfrak{S}^A$, $\mathfrak{S}^{A'}$ to itself, while leaving them independent of each other. (In the – essentially equivalent – definite cases $(++++)$ and $(----)$, there exist no real non-zero elements of $\mathfrak{S}^A$, i.e., elements sent to themselves under complex conjugation, whereas in the case $(++--)$ there do exist such real elements. Thus the classifications of Ψ_{ABCD} and $\tilde{\Psi}_{A'B'C'D'}$ are independent of each other, but each is restricted by reality conditions. In particular, non-trivial right-flat and left-flat solutions of Einstein's equations exist with all these signatures, some of which are referred to as *gravitational instantons* (*cf.* Hawking 1977, Gibbons and Hawking 1979, Hitchin 1979, 1982, Atiyah, Hitchin and Singer 1978).

Consistency of the twistor equation

In our discussion of local twistors we shall phrase our arguments as though we were solely concerned with real Lorentzian space–times, but at each stage the generalization to complex space–times can be effected in a straightforward way, according to the rules given in the preceding paragraph. However, certain provisos must be borne in mind. For example, the condition '$\Psi_{ABCD}=0$' for conformal flatness must be interpreted, in the context of a complex space–time, as '$\Psi_{ABCD}=0$ *and* $\tilde{\Psi}_{A'B'C'D'}=0$'. In this connection a specific point should be made. We have seen in (6.1.6) that a consistency condition for the *twistor equation* (6.1.1) to have several (more than one) linearly independent solutions is $\Psi_{ABCD}=0$. One might be tempted to conclude that the disappearance of this consistency condition in a left conformally flat space–time would lead to a full family of solutions existing. Indeed, for the *massless field equations* (4.12.42), the Buchdahl–Plebanski conditions (5.8.2) are satisfied if $\Psi_{ABCD}=0$ but $\tilde{\Psi}_{A'B'C'D'}\neq 0$, and in this case it is true that, for fields with arbitrarily many *unprimed* indices, as wide a family of solutions exists locally in left conformally flat as in flat (complex) space–times. (A corresponding result holds, of course, for fields with primed indices in right conformally flat spaces.) But the twistor equation is a much more stringent restriction than the massless field equation (as evidenced just by the fact that even in flat space–time only a *finite*-dimensional solution space exists). Observe that, by (4.9.8)(2), the part skew in BC of

$$\nabla^A_{A'}\nabla^B_{(B'}\nabla^C_{C')}\omega^D \tag{6.9.2}$$

vanishes whenever $\Phi_{ABC'D'}=0$; thus if the twistor equation (6.1.1) holds, (6.9.2) is zero, being skew in CD. So if the twistor equation holds in a left-flat complex space–time, we have, by (4.9.14),

$$0=\nabla_{A(A'}\nabla^A_{B'}\nabla^C_{C')}\omega^D=\tilde{\Psi}_{A'B'C'D'}\nabla^{CD'}\omega^D. \tag{6.9.3}$$

It then follows, as in (6.1.6), that there can be only a restricted family of solutions of the twistor equation unless $\tilde{\Psi}_{A'B'C'D'}$ also vanishes. In fact, in any left-flat complex space–time there are always (locally) *two* linearly independent solutions of

$$\nabla_a\omega^B=0. \tag{6.9.4}$$

For, by (4.9.7), the commutators of derivatives vanish when acting on an unprimed spinor: $(\nabla_a\nabla_b-\nabla_b\nabla_a)\omega^C=0$. Thus ω^B can be chosen arbitrarily at one point and carried parallelly, in a consistent way, to all points in a neighbourhood. But in general these are the *only* solutions of (6.9.3) and therefore are the only solutions of the twistor equation in a general left-flat space. (There exists *one* other linearly independent solution if $\Psi_{A'B'C'D'}$ is null and constant.) Here we end our explicit digression on complex space–times and return to the discussion of local twistors.

Definition of a local twistor

The essential difference between local and global twistors is that, whereas a global twistor is defined over the whole space–time $\mathcal{M}$ (i.e. as a solution of the twistor equation) and does not require specific points of $\mathcal{M}$ to be called upon for its definition, a local twistor is defined essentially at *points* of $\mathcal{M}$. There is a four-complex-dimensional local twistor space at *each* point of $\mathcal{M}$, and the various local twistor spaces are independent of each other, apart from requirements of continuity and differentiability. Accordingly, the space of local twistors is a (four-complex-dimensional) *vector bundle* over $\mathcal{M}$ and not simply a four-complex-dimensional vector space. (One can also consider local twistor *fields*, as cross-sections of this bundle in the ordinary way.) Thus a local twistor cannot serve one purpose that strongly motivated the development of global twistors, namely to provide an alternative formalism in physics in which the concept of a space–time point is no longer regarded as primitive. (For such purposes yet a different concept of a twistor may be used in curved space–time, such as that of an asymptotic twistor – which we shall describe only briefly at the end of §9.8 – a concept which itself makes use of that of a local twistor. See also the discussion of 2-surface twistors given in §9.9.) On the other hand, the local twistor formalism provides us with a convenient calculus for the discussion of the conformal geometry of an arbitrary space–time manifold $\mathcal{M}$. (The necessary concepts were basically introduced by Cartan 1923, 1932; *cf.* also Veblen and Taub 1934, Penrose and MacCallum 1972.)

Let $P \in \mathcal{M}$ with metric g_{ab}. A *local twistor* Z^α at P is a quantity which can be represented by a pair of spinors $(\omega^A, \pi_{A'})$ *at* P (and we stress that these spinors are now *point*-spinors, not spinor fields). We require that under a conformal rescaling $g_{ab} \mapsto \hat{g}_{ab} = \Omega^2 g_{ab}$ the pair $(\omega^A, \pi_{A'})$ is replaced by a new pair $(\hat{\omega}^A, \hat{\pi}_{A'})$ according to the previous formula (6.1.75). We write

$$\mathrm{Z}^\alpha \overset{g}{=} (\omega^A, \pi_{A'})$$
$$\mathrm{Z}^\alpha \overset{\hat{g}}{=} (\hat{\omega}^A, \hat{\pi}_{A'}), \tag{6.9.5}$$

and require

$$\hat{\omega}^A = \omega^A, \quad \hat{\pi}_{A'} = \pi_{A'} + \mathrm{i}\Upsilon_{AA'}\omega^A. \tag{6.9.6}$$

Thus, while ω^A is a conformally weighted quantity (with weight 0), $\pi_{A'}$ is not, but undergoes a more complicated kind of conformal behaviour. If we take the point of view that, given Z^α, (ω^A and) $\pi_{A'}$ are not simply fixed point-spinors but quantities that have a *functional dependence on the metric g*, then it is legitimate to drop the 'g' or '$\hat{g}$' above the equality signs in (6.9.5) and simply write

$$\mathrm{Z}^\alpha = (\omega^A, \pi_{A'}) \tag{6.9.7}$$

as usual. The space of all local twistors is the *vector bundle* of all such Z^α at all the various points P. The *fibre* over P is a four-complex-dimensional vector space: the *space of local twistors* at P.

Local twistor transport

Next we construct a means of comparing local twistors at different points (i.e. a bundle connection *cf.* (5.4.17)–(5.4.19)). For this we are guided by two requirements. First we want a certain agreement with global twistors. This guides us in the construction of the parallel transport for local twistors – called *local twistor transport*. Given any smooth curve τ in $\mathcal{M}$, joining points P and Q in $\mathcal{M}$, and a local twistor Z^α at P, we define a corresponding local twistor at Q. Only in conformally flat space–time will this correspondence be path-independent. But in that case we shall require agreement with the global twistor concept, i.e., a local twistor which is in this sense 'parallel' everywhere must be a global twistor. Secondly, we want invariance under conformal rescalings. It may be noted that the simple expedient of parallelly transporting the point-spinor parts of Z^α would not fulfil these requirements.

Let us first examine the original concept of global twistor, as introduced in §6.1, in the present context of local twistors. A global twistor Z^α in conformally flat space–time is identified with a spinor field ω^A satisfying the twistor equation

$$\nabla_{A'}^{(A}\omega^{B)} = 0. \tag{6.9.8}$$

Let us define $\pi_{A'}$ by

$$\pi_{A'} = \tfrac{1}{2}\mathrm{i}\nabla_{AA'}\omega^A, \tag{6.9.9}$$

which, together with (6.9.8), is equivalent to (6.1.9). From (6.8.11) we find at once that

$$0 = \mathrm{i}\nabla_{BB'}\pi_{A'} - \omega^A \mathrm{P}_{ABA'B'}$$

(with P_{ab} as in (6.8.12)). Thus a global twistor is represented by a $(\omega^A, \pi_{A'})$

field subject to

$$\nabla_{AA'}\omega^B + \mathrm{i}\varepsilon_A{}^B\pi_{A'} = 0,$$
$$\nabla_{AA'}\pi_{B'} + \mathrm{iP}_{ABA'B'}\omega^B = 0. \qquad (6.9.10)$$

(Note that in the particular case of flat space–time these equations reduce to (6.1.9).)

We shall use (6.9.10) to motivate our concept of parallel transport of local twistors in an *arbitrary* space–time $\mathscr{M}$. We say that the *local* twistor $\mathrm{Z}^\alpha = (\omega^A, \pi_{A'})$ is *constant* throughout $\mathscr{M}$ (and that it is a *global* twistor) if (6.9.10) is satisfied. In a general space–time there will be no non-trivial local twistors satisfying (6.9.10) at all points of $\mathscr{M}$. On the other hand, transvecting (6.9.10) with a vector t^a we obtain the weaker relations

$$t^a\nabla_a\omega^B + \mathrm{i}t^{BA'}\pi_{A'} = 0,$$
$$t^a\nabla_a\pi_{B'} + \mathrm{i}t^a\mathrm{P}_{ab}\omega^B = 0, \qquad (6.9.11)$$

which state that Z^α is constant *in the direction* t^a. If we have a curve τ in $\mathscr{M}$ with tangent vector t^a, then Z^α is said to be *constant along* τ – or carried along τ by *local twistor transport* – if (6.9.11) holds at every point of τ. It is clear that the propagation equations (6.9.11) serve to define Z^α uniquely at every point of τ, given the value of $(\omega^A, \pi_{A'})$ at any one point of τ; and at that point ω^A and $\pi_{A'}$ may be chosen freely. (This is assuming that we have the 'normal' situation in which τ is not self-intersecting and t^a is nowhere zero.)

Having the concept of a local twistor which is constant along a curve, it is a small step to define *the rate of change* $\underset{t}{\nabla}\mathrm{Z}^\alpha$ of Z^α along τ by

$$\underset{t}{\nabla}\mathrm{Z}^\alpha = (t^b\nabla_b\omega^A + \mathrm{i}t^{AB'}\pi_{B'},\ t^b\nabla_b\pi_{A'} + \mathrm{i}t^{BB'}\mathrm{P}_{ABA'B'}\omega^A). \qquad (6.9.12)$$

Then it can be verified directly from (6.8.14) that the transformation behaviour (6.9.5), (6.9.6) holds for $\underset{t}{\nabla}\mathrm{Z}^\alpha$ if it holds for Z^α. Thus $\underset{t}{\nabla}\mathrm{Z}^\alpha$ is a *local twistor** at each point of τ. Indeed, this is an expression of the fact that the operator $\underset{t}{\nabla}$ is *conformally invariant* in this sense. As a consequence, the concept of local twistor transport is conformally invariant, since the above definitions (6.9.11) and (6.9.12) work in *any* space–time. A constant local twistor defined throughout $\mathscr{M}$ (i.e., a global twistor) must satisfy $\underset{t}{\nabla}\mathrm{Z}^\alpha = 0$ at every point of $\mathscr{M}$ and for every vector $\boldsymbol{t}$.

So far we have been concerned only with $[^1_0]$-local twistors. The generalization to $[^p_q]$-local twistors follows the normal pattern. We may consider local twistors at just one point of $\mathscr{M}$ or we may consider fields of local twistors on $\mathscr{M}$. In the first case, the $[^1_0]$-local twistors give a four-

* Local twistors, with this connection, can under certain circumstances be thought of as defining a kind of Yang–Mills theory, *cf.* Merkulov (1984) (Bach tensor current).

dimensional vector space over $\mathbb{C}$ and in the second, a (totally reflexive four-dimensional) module over the complex scalar fields $\mathfrak{S}$. We define $\lambda \mathsf{Z}^\alpha$ and $\mathsf{X}^\alpha + \mathsf{Z}^\alpha$ in the obvious way in each case. The abstract labelling system and general scheme for constructing quantities of arbitrary valence given in Chapter 2 applies here also; the notation and splitting of quantities into their spinor parts follows the same scheme as that for global twistors (*cf.* §6.1). As with global twistors, we have an operation of complex conjugation which takes a $[^p_q]$-twistor into a $[^q_p]$-twistor. Let us show this explicitly in the case $p=1, q=0$ and examine the implications with respect to local twistor transport. With Z^α as in (6.9.5) we have $\bar{\mathsf{Z}}_\alpha = (\bar{\pi}_A, \bar{\omega}^{A'})$ as in (6.1.31), so that when we replace g_{ab} by $\hat{g}_{ab} = \Omega^2 g_{ab}$, we get the new representation $\bar{\mathsf{Z}}_\alpha = (\bar{\pi}_A - \mathrm{i}\Upsilon_{AA'}\bar{\omega}^{A'}, \bar{\omega}^{A'})$. Thus, for $[^0_1]$-local twistors,

$$\mathsf{W}_\alpha = (\lambda_A, \mu^{A'})$$

with respect to g_{ab} implies

$$\mathsf{W}_\alpha = (\hat{\lambda}_A, \hat{\mu}^{A'})$$

with respect to the new metric $\hat{g}_{ab}$, where

$$\hat{\lambda}_A = \lambda_A - \mathrm{i}\Upsilon_{AA'}\mu^{A'}, \quad \hat{\mu}^{A'} = \mu^{A'}. \tag{6.9.13}$$

Hence, for example, the spinor parts of a $[^1_1]$-local twistor $\mathsf{Q}_\alpha{}^\beta$ transform as

$$\begin{pmatrix} \hat{\mathsf{Q}}_A{}^B & \hat{\mathsf{Q}}_{AB'} \\ \hat{\mathsf{Q}}^{A'B} & \hat{\mathsf{Q}}^{A'}{}_{B'} \end{pmatrix} =$$

$$\begin{pmatrix} \mathsf{Q}_A{}^B - \mathrm{i}\Upsilon_{AA'}\mathsf{Q}^{A'B} & \mathsf{Q}_{AB'} - \mathrm{i}\Upsilon_{AA'}\mathsf{Q}^{A'}{}_{B'} + \mathrm{i}\Upsilon_{BB'}\mathsf{Q}_A{}^B + \Upsilon_{AA'}\Upsilon_{BB'}\mathsf{Q}^{A'B} \\ \mathsf{Q}^{A'B} & \mathsf{Q}^{A'}{}_{B'} + \mathrm{i}\Upsilon_{BB'}\mathsf{Q}^{A'B} \end{pmatrix}$$

and correspondingly for local twistors with other index structures.

From the complex conjugate of (6.9.12) we have

$$\underset{t}{\nabla}\, \mathsf{W}_\alpha = (t^b\nabla_b\lambda_A - \mathrm{i}t^{BB'}\mathrm{P}_{ABA'B'}\mu^{A'}, t^b\nabla_b\mu^{A'} - \mathrm{i}t^{BA'}\lambda_B). \tag{6.9.14}$$

As with $[^1_0]$-twistors, $\underset{t}{\nabla}$ is (of necessity) conformally invariant in the sense that if W_α has the correct rescaling behaviour (6.9.13) for a local twistor, so has $\underset{t}{\nabla}\,\mathsf{W}_\alpha$. Furthermore, the scalar product $\mathsf{W}_\alpha \mathsf{Z}^\alpha := \lambda_A\omega^A + \mu^{A'}\pi_{A'}$ is invariant under conformal rescalings (as it must be by virtue of the agreement of (6.9.6) and (6.9.13) with the corresponding global twistor expressions (6.1.75) and (6.1.76)), and has the property

$$\underset{t}{\nabla}\,(\mathsf{W}_\alpha \mathsf{Z}^\alpha) = \mathsf{W}_\alpha \underset{t}{\nabla}\,\mathsf{Z}^\alpha + \mathsf{Z}^\alpha \underset{t}{\nabla}\,\mathsf{W}_\alpha, \tag{6.9.15}$$

as is easily verified from (6.9.12) and (6.9.14). The term on the left is defined simply as the ordinary directional derivative ($\underset{t}{\nabla} = t^a\nabla_a$) of a scalar (*cf.* (4.3.31)). One implication of (6.9.15) is that the helicity $\frac{1}{2}\mathsf{Z}^\alpha\bar{\mathsf{Z}}_\alpha$ of a local twistor is unchanged under parallel transport. In particular, a null twistor remains null. This property is analogous to the corresponding property of Riemannian geometry, to the effect that norms and scalar products of vectors are preserved under parallel transport. In twistor theory it is complex conjugation which plays an analogous role to the Riemannian metric: local twistor transport commutes with twistor complex conjugation.

The extension of the definition of $\underset{t}{\nabla}$ to local twistors of arbitrary valence is governed by the normal requirements of additivity and the Leibniz law. A particular example should suffice to illustrate the general pattern:

$$\underset{t}{\nabla}\mathrm{Q}_\alpha{}^\beta = \underset{t}{\nabla}\begin{pmatrix} \mathrm{Q}_A{}^B & \mathrm{Q}_{AB'} \\ \mathrm{Q}^{A'B} & \mathrm{Q}^{A'}{}_{B'} \end{pmatrix} \tag{6.9.16}$$

$$= t^c\begin{pmatrix} \nabla_c\mathrm{Q}_A{}^B + \mathrm{i}\varepsilon_C{}^B\mathrm{Q}_{AC'} - \mathrm{i}\mathrm{P}_{ac}\mathrm{Q}^{A'B} & \nabla_c\mathrm{Q}_{AB'} + \mathrm{i}\mathrm{P}_{bc}\mathrm{Q}_A{}^B - \mathrm{i}\mathrm{P}_{ac}\mathrm{Q}^{A'}{}_{B'} \\ \nabla_c\mathrm{Q}^{A'B} + \mathrm{i}\varepsilon_C{}^B\mathrm{Q}^{A'}{}_{C'} - \mathrm{i}\varepsilon_{C'}{}^{A'}\mathrm{Q}_C{}^B & \nabla_c\mathrm{Q}^{A'}{}_{B'} + \mathrm{i}\mathrm{P}_{bc}\mathrm{Q}^{BA'} - \mathrm{i}\varepsilon_{C'}{}^{A'}\mathrm{Q}_{CB'} \end{pmatrix}.$$

Local twistor curvature

We have mentioned earlier that we do not expect the concept of local twistor parallel transport to be 'integrable' (i.e., path-independent) in a general curved space–time. This non-integrability can be made quantitative by examining the effect of going around a small loop spanned by vectors $\boldsymbol{t}$, $\boldsymbol{u}$ (see Fig. 6-9). If $\boldsymbol{t}$ and $\boldsymbol{u}$ are two vector fields, the Lie bracket $[\boldsymbol{t},\boldsymbol{u}]$ expresses the 'gap' encountered when we attempt to construct a small quadrilateral out of neighbouring vectors of the vector fields. Unless this 'gap' closes* ($[\boldsymbol{t},\boldsymbol{u}]=0$) we get, in effect, an infinitesimal pentagon. The operator $\underset{t}{\nabla}\underset{u}{\nabla} - \underset{u}{\nabla}\underset{t}{\nabla} - \underset{[t,u]}{\nabla}$ expresses the increment in a quantity carried around this infinitesimal pentagon by parallel transport (*cf.* Dodson and

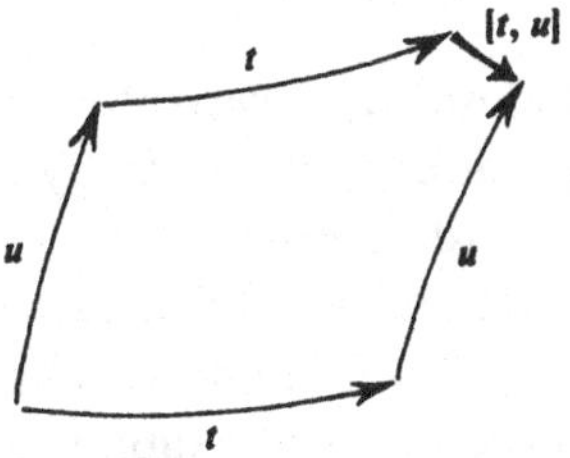

Fig. 6-9. The Lie bracket $[\boldsymbol{t},\boldsymbol{u}]$, between vector fields $\boldsymbol{t}$ and $\boldsymbol{u}$ measures the 'gap' in small quadrilaterals constructed from $\boldsymbol{t}$ and $\boldsymbol{u}$.

* If $\boldsymbol{u}$ and $\boldsymbol{t}$ in the diagram are $O(\varepsilon)$, then the 'gap' is $O(\varepsilon^2)$, generally, since $[\boldsymbol{t},\boldsymbol{u}]$ is bilinear in $\boldsymbol{t}$, $\boldsymbol{u}$. 'Closing' means that the gap is $O(\varepsilon^3)$.

Poston 1977) and serves to define the relevant curvature quantity. We already considered the corresponding expression for ordinary parallel transport in (4.3.33). Now, we define the *local twistor curvature* $\mathrm{K}_\alpha{}^\beta(\boldsymbol{t},\boldsymbol{u})$ by

$$\mathrm{i}(\underset{t}{\nabla}\underset{u}{\nabla} - \underset{u}{\nabla}\underset{t}{\nabla} - \underset{[t,u]}{\nabla})\mathrm{Z}^\beta = \mathrm{Z}^\alpha \mathrm{K}_\alpha{}^\beta(\boldsymbol{t},\boldsymbol{u}) \tag{6.9.17}$$

and find,* after some calculation,

$$\mathrm{K}_\alpha{}^\beta(\boldsymbol{t},\boldsymbol{u}) = \begin{pmatrix} \mathrm{i}t^P u^q \varepsilon_{P'Q'} \Psi_{PQA}{}^B & t^P u^q(\varepsilon_{PQ}\nabla_A^{A'}\bar{\Psi}_{B'P'Q'A'} + \varepsilon_{P'Q'}\nabla_{B'}^B \Psi_{APQB}) \\ 0 & -\mathrm{i}t^P u^q \varepsilon_{PQ} \bar{\Psi}_{P'Q'}{}^{A'}{}_{B'} \end{pmatrix}. \tag{6.9.18}$$

The factor i in (6.9.17) is chosen to ensure the Hermiticity of $\mathrm{K}_\alpha{}^\beta$ (like the $F_{\dots}$ of (5.5.30)).

It may be remarked that the dependence of $\mathrm{K}_\alpha{}^\beta$ on the two vector fields is, of course, bilinear; so a curvature quantity $\mathrm{K}_{pq\alpha}{}^\beta$ independent of $\boldsymbol{t}$ and $\boldsymbol{u}$ may be defined by the equation

$$\mathrm{K}_\alpha{}^\beta(\boldsymbol{t},\boldsymbol{u}) =: t^p u^q \mathrm{K}_{pq\alpha}{}^\beta. \tag{6.9.19}$$

Similarly, the dependence of $\underset{t}{\nabla}$ on $\boldsymbol{t}$ is linear; so a covariant derivative operator ∇_p acting on local twistors may be defined, where

$$t^p \nabla_p = \underset{t}{\nabla} \tag{6.9.20}$$

for any $\boldsymbol{t}$ (thus indeed providing us with a bundle connection as in (5.4.17)–(5.4.19)). Then (6.9.17) may be expressed in the form

$$(\mathrm{i}\nabla_p\nabla_q - \mathrm{i}\nabla_q\nabla_p)\mathrm{Z}^\beta = \mathrm{K}_{pq\alpha}{}^\beta \mathrm{Z}^\alpha \tag{6.9.21}$$

(which is an instance of (5.4.23)).

This is, perhaps, more in keeping with the development of §§4.2, 5.4, 5.5, and we may construct higher derivatives of curvature quantities:

$$\nabla_p \dots \nabla_r \mathrm{K}_{st\alpha}{}^\beta. \tag{6.9.22}$$

However these are not conformally invariant objects because ∇_a acting on a quantity with tensor indices is not generally conformally invariant. One may construct higher-order conformally invariant derivatives by adapting Dighton's (1974) procedure (*cf.* Penrose and MacCallum 1972) whereby tensor and spinor indices are eliminated by translating them into local twistor form, each (lower) primed [unprimed] index appearing as an upper [lower] twistor index. This is achieved via the (second) map

* Note that both $\Psi_{\dots}$ and $\bar{\Psi}_{\dots}$ appear in this expression. Thus in a complex space–time, both $\Psi_{\dots}$ and $\tilde{\Psi}_{\dots}$ occur in the local twistor curvature, as would be expected from (6.1.6), (6.9.3).

(e.g. $\pi_{A'} \mapsto (0, \pi_{A'})$) of each of the conformally invariant exact sequences

$$0 \to \mathfrak{S}_{A'}[P] \to \mathbb{T}^{\alpha}[P] \to \mathfrak{S}^{A}[P] \to 0$$
$$0 \to \mathfrak{S}_{A}[P] \to \mathbb{T}_{\alpha}[P] \to \mathfrak{S}^{A'}[P] \to 0$$

where '$[P]$' refers to the local space at $P \in \mathcal{M}$ (*cf.* p. 5; contrast p. 91). Thus $\nabla_{AB'}$ is translated to a conformally invariant operator $\nabla_{\alpha}{}^{\beta}$ – though a complication arises from the existence of a (constant) torsion. Conformally invariant spinors can be extracted (via the third map) by taking primary parts, or secondary parts when this is zero, or tertiary parts when all these are zero, etc. For this procedure to be useful, the twistor indices must be subjected to various symmetry operations; and in this way Dighton (1972, 1974) has been able to obtain the Bach tensor (6.8.42) and other conformally invariant objects (*cf.* also du Plessis 1970 for an alternative procedure).

Vanishing Weyl curvature implies conformal flatness

Observe that the local twistor curvature (6.9.18) involves the Weyl curvature and its (contracted) derivative. If $\mathcal{M}$ is conformally flat then these quantities must vanish. For $\Psi_{ABCD} = 0$ in flat space–time; and being conformally invariant (*cf.* (6.8.4)), Ψ_{ABCD} must also vanish in any conformally flat space–time. Thus (as was evident, in any case, from the conformal invariance of local twistor transport), the local twistor curvature must also vanish in any conformally flat space–time. We wish now to establish a converse result, namely:

(6.9.23) THEOREM

If $\Psi_{ABCD} = 0$ throughout the space–time $\mathcal{M}$, then each point of $\mathcal{M}$ has a neighbourhood $\mathcal{U}$ in which a conformal factor can be found that rescales the metric in $\mathcal{U}$ to that of a portion of Minkowski space. Such a space–time $\mathcal{M}$ will be said to be 'patchwise conformal' to Minkowski space.*

Proof: Suppose $\Psi_{ABCD} = 0$ throughout $\mathcal{M}$. Then the local twistor curvature vanishes at each point of $\mathcal{M}$. Let O be a point of $\mathcal{M}$ and choose a neighbourhood $\mathcal{V}$ of O in $\mathcal{M}$ which is simply-connected and has spinor structure (*cf.* §1.5). Then if P is any other point in $\mathcal{V}$, any two curves in $\mathcal{V}$ from O to P can be continuously deformed one into the other. Let Z^{α} be a

* It is not necessarily true that a conformal factor can be found for the whole of $\mathcal{M}$, which rescales the metric to a flat metric everywhere.

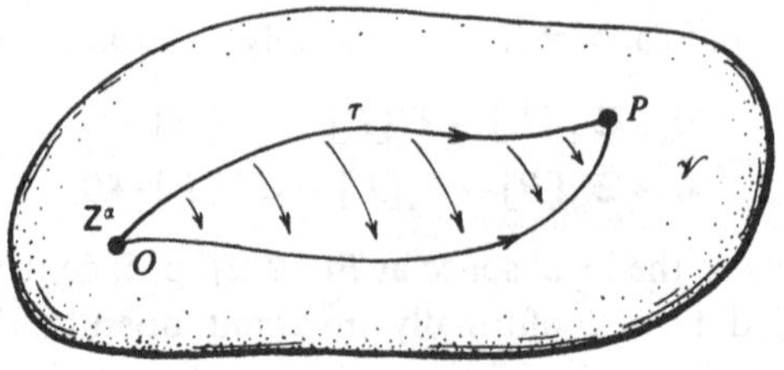

Fig. 6-10. The vanishing of local twistor curvature implies that the result of local twistor transport of Z^α along τ, from O to P, is unaffected when τ is varied continuously.

local twistor at O. For any smooth curve τ, in $\mathscr{V}$, from O to P, we can define a corresponding local twistor at P simply by carrying Z^α along τ from O to P by local twistor transport. Now let this curve be continuously deformed within $\mathscr{V}$, keeping the end-points O and P fixed. Because of the vanishing of the local twistor curvature and the interpretation of this curvature in terms of local twistor transport around infinitesimal loops, it follows that the result of transporting Z^α from O to P along τ will be unaffected as τ is thus varied continuously (see Fig. 6-10). (It is possible to make this argument considerably more rigorous, but the main point should now be clear.) Because of the simple connectedness of $\mathscr{V}$, the parallel transport of Z^α from O to P is completely path-independent within $\mathscr{V}$. We can vary P also, within $\mathscr{V}$, and obtain a field of local twistors Z^α in $\mathscr{V}$ which satisfies (6.9.11) at each point and for each vector t^a. Thus, (6.9.10) is satisfied and we have a *global* twistor Z^α.

The same applies *whatever* local twistor is chosen at O. Thus the local twistor space at O is in one-to-one correspondence with the space of global twistors in $\mathscr{V}$. Any $(\omega^A, \pi_{A'})$ pair at the point O defines a global twistor in $\mathscr{V}$ and vice versa. If the ω^A-part of a global twistor Z^α vanishes at a point P of $\mathscr{V}$, then we say that Z^α *is incident with* P. Since the local twistor description of Z^α is then *null* at P and since nullity is preserved under local twistor transport, it follows that Z^α must be null at O also. If X^α is another global twistor incident with P, then clearly $X^\alpha \bar{Z}_\alpha = 0$ at P. Again, since twistor orthogonality is preserved under local twistor transport (*cf.* (6.9.15)), we must also have $X^\alpha \bar{Z}_\alpha = 0$ at O. The system of all global twistors incident with P must therefore be represented, at O, by a (two-complex-dimensional) system of local twistors at O which are all *null*, and *orthogonal* to one another.

We may regard the tangent space at O as a Minkowski space $\mathbb{M}$ and the $(\omega^A, \pi_{A'})$ spinor pairs at O as defining ordinary Minkowski-space twistors for $\mathbb{M}$. The above two-complex-dimensional system of local twistors at O, representing a point P of $\mathscr{V}$, is then precisely of the kind which is characterized as the system of all twistors in $\mathring{\mathbb{M}}$ incident with some point $\mathring{P} \in \mathring{\mathbb{M}}$, although exceptionally $\mathring{P}$ could lie at infinity in $\mathring{\mathbb{M}}$. Let us avoid this latter possibility by choosing a neighbourhood $\mathscr{U} \subset \mathscr{V}$ of O, sufficiently small for each point P of $\mathscr{U}$ to have an image $\mathring{P}$ in $\mathring{\mathbb{M}}$ which is indeed a finite point of $\mathring{\mathbb{M}}$. Furthermore we choose $\mathscr{U}$ so small that no two points of $\mathscr{U}$ have the same image in $\mathring{\mathbb{M}}$. We then have a map from $\mathscr{U}$ to some portion of $\mathring{\mathbb{M}}$ which is one-to-one and evidently continuous. Furthermore it is a *conformal* map. To see this, consider a global twistor Z^α incident with P and having the spinor parts $(0, \pi_{A'})$ at P. Choose τ to be the null geodesic through P in the direction of the flagpole of $\pi_{A'}$ at P (take $t^a = \bar{\pi}^A \pi^{A'}$). Then it is evident from (6.9.11) that the same form $(0, \pi_{A'})$ will be retained all along τ. Thus Z^α will be incident with every point of τ. This establishes the fact that in our map from $\mathscr{U}$ to $\mathring{\mathbb{M}}$, null geodesics in $\mathscr{U}$ correspond to null straight lines (points of fixed null twistors) in $\mathring{\mathbb{M}}$. Thus null cones in $\mathscr{U}$ are certainly mapped to null cones in $\mathring{\mathbb{M}}$, establishing that the map is indeed conformal. This establishes theorem (6.9.23), since the point O was chosen arbitrarily in $\mathscr{M}$.

6.10 Massless fields and twistor cohomology

In §6.7 we briefly described a contour integral method for solving (among other equations) the massless field equations of each spin in $\mathbb{M}$. In this section we give a somewhat deeper discussion of that construction of massless fields, which has importance as one of the cornerstones of most of the subsequent developments of twistor theory. Our discussion will lead us to a sampling of twistor sheaf cohomology theory, though it is beyond the scope of the present book to enter into this in very much detail.

Contour integrals for massless fields

Let us return, first, to the formula (6.7.41) in the particular case $q = 0$, when there are no upper (i.e., primed) indices. It then represents the solution of the massless field equations* in $\mathbb{M}$ or $\mathbb{CM}$. The 'f' in the formula has then

* The particular form of this formula, for general spin, as given in (6.10.1), was first written down in Penrose and MacCallum (1972), this formula being based on earlier ones given by Penrose (1968*a*), (1969). However, in the case of spin-0 a virtually equivalent expression had been found much earlier by Bateman (1904), (1944), as developed from an expression due to Whittaker (1903) for solving the three-dimensional Laplace equation. The case of spin-1 (Maxwell theory) was also treated by Bateman (1944).

only a 'constant' dependence on the twistor Z^α, i.e., it is a function of W_α only; and it is homogeneous of degree $-p-2$, where p is the number of indices of the field $\phi_{A\ldots D}$ to be constructed (and also the number of $\lambda_A,\ldots,\lambda_D$ under the integral sign):

$$\phi_{A\ldots D}(R)=\frac{1}{2\pi\mathrm{i}}\oint\lambda_A\ldots\lambda_D f(\mathrm{W}_\alpha)\lambda_E\mathrm{d}\lambda^E. \tag{6.10.1}$$

The function f must be holomorphic in some suitable region of $\mathbb{T}_\alpha$ (which will be made more precise later). The integral is taken, for each space-time point R, over a suitable one-dimensional closed contour in the space of $[{}^0_1]$-twistors W_α incident with R. This space may be identified with $\mathfrak{S}_A[R]$, the vector space of spin-(co)vectors at R or, equivalently, with the spin-space $\mathbb{S}_A$ of *constant* spinor fields λ_A arising from $\mathrm{W}_\alpha=(\lambda_A,\mu^{A'})$. The first point of view is the appropriate one for curved conformally flat space-times, but the second is more convenient for explicit calculations in $\mathbb{M}$ or $\mathbb{CM}$. In the conformally flat case, we can use a constant local twistor description of W_α, and at any point R incident with it, W_α then takes the form $(\lambda_A,0)$, where now $\lambda_A\in\mathfrak{S}_A[R]$. In the case of $\mathbb{M}$ or $\mathbb{CM}$, on the other hand, it is simplest to use the description

$$\mathrm{W}_\alpha\overset{O}{\leftrightarrow}(\lambda_A,-\mathrm{i}r^{AA'}\lambda_A) \tag{6.10.2}$$

where r^a denotes the position vector of R relative to the point $O\in\mathbb{M}$, and where both r^a and λ_A are taken as point spinors at O. More invariantly, we may replace (6.10.2) by

$$\mathrm{W}_\alpha=(\lambda_A,-\mathrm{i}r^{AA'}\lambda_A)$$

with the understanding that λ_A is a constant spinor ($\in\mathbb{S}_A$) and r^a is (as in (6.2.15), (6.2.18)) the position vector of R relative to the general field point. So, in particular, *at* R, W_α takes the form $(\lambda_A,0)$.

Quite similarly, in the case of a massless field with q primed indices ($p=0$ in (6.7.40)), we have the formula

$$\phi_{A'\ldots D'}(R)=\frac{1}{2\pi\mathrm{i}}\oint\pi_{A'}\ldots\pi_{D'}f(\mathrm{Z}^\alpha)\pi_{E'}\mathrm{d}\pi^{E'}, \tag{6.10.3}$$

where f now is homogeneous of degree $-q-2$ and holomorphic in some suitable region of $\mathbb{T}^\alpha$ (to be made more precise shortly); and where the integral is taken over a one-dimensional closed contour in the space of $[{}^1_0]$-twistors Z^α incident with R. This space can be thought of as $\mathfrak{S}_{A'}[R]$ i.e. of point-spinors at R – especially in curved conformally flat space – with $\mathrm{Z}^\alpha=(\omega^A,\pi_{A'})(=(0,\pi_{A'})$ locally at R). Alternatively, in the case of $\mathbb{M}$ or $\mathbb{CM}$, it can be taken as the spin-space $\mathbb{S}_{A'}$ of constant spinor fields $\pi_{A'}$, with $Z^\alpha = (\mathrm{i}r^{AA'}\pi_{A'}, \pi_{A'})$ and r^a the position vector of R relative to the general field point. As in (6.10.2) it will be convenient to fix the description of the twistor at O and to represent Z^α by

$$Z^\alpha \overset{O}{\longleftrightarrow} (\mathrm{i}r^{AA'}\pi_{A'}, \pi_{A'}), \tag{6.10.4}$$

r^a and $\pi_{A'}$ being point spinors at O.

Holding r^a constant and using the fact that f is holomorphic with homogeneity degree as given, we can easily check that the exterior derivative of the integrand in (6.10.1) and (6.10.3) vanishes, so that the result of the integration is unaffected by continuous deformations of the contour through (non-singular) regions of the domain of the integrand. In each case the components of the integrand have the form $h(u, v)(u\mathrm{d}v - v\mathrm{d}u)$ with h holomorphic and homogeneous of degree -2, whence, by Euler's theorem,

$$u\frac{\partial h}{\partial u} + v\frac{\partial h}{\partial v} = -2h,$$

and so

$$d((u\mathrm{d}v - v\mathrm{d}u)h) = \left(u\frac{\partial h}{\partial u} + v\frac{\partial h}{\partial v} + h + h\right)\mathrm{d}u \wedge \mathrm{d}v = 0.$$

Also, the massless field equations follow at once from

$$\nabla_a f(W_\alpha) = -\mathrm{i}\lambda_A \frac{\partial f}{\partial \mu^{A'}} \tag{6.10.5}$$

$$\nabla_a f(Z^\alpha) = \mathrm{i}\pi_{A'} \frac{\partial f}{\partial \omega^{A}}, \tag{6.10.6}$$

where ∇_a stands for $\partial/\partial r^a$ and where the point spinors $\mu^{A'}$ and ω^A at O are defined, according to (6.10.2) and (6.10.4), by

$$\mu^{A'} = -\mathrm{i}r^{AA'}\lambda_A, \quad \omega^A = \mathrm{i}r^{AA'}\pi_{A'}. \tag{6.10.7}$$

(The concept of partial derivative with respect to a spinor with abstract indices in such a context should be self-explanatory; one can always take components in some basis and then convert back to abstract indices.)

The non-projective form

Sometimes it is convenient to use a form of integral that does not require f to be homogeneous (though still holomorphic), but which instead projects out a part of f that has the correct homogeneity:

$$\phi_{A\ldots D}(R) = \frac{1}{(2\pi\mathrm{i})^2}\oint \lambda_A \ldots \lambda_D f(W_\alpha)\mathrm{d}^2\lambda \tag{6.10.8}$$

or

$$\phi_{A'\ldots D'}(R) = \frac{1}{(2\pi i)^2} \oint \pi_{A'} \ldots \pi_{D'} f(Z^\alpha) d^2\pi, \tag{6.10.9}$$

where

$$d^2\lambda = d\lambda_0 \wedge d\lambda_1 = \tfrac{1}{2} d\lambda_A \wedge d\lambda^A$$
$$d^2\pi = d\pi_{0'} \wedge d\pi_{1'} = \tfrac{1}{2} d\pi_{A'} \wedge d\pi^{A'}, \tag{6.10.10}$$

and where the integrals are now taken over two-dimensional contours in the appropriate spin-space. Irrespective of homogeneity, the exterior derivatives of the integrands (with r^a held constant) clearly vanish, so if the contour is deformed continuously throughout the domain of f, the result of the integration is unaffected. If f has the *wrong* homogeneity, the integrals must vanish, since replacing W_α by kW_α or Z^α by kZ^α (k constant) would multiply the whole integrand (including the differential) by some non-trivial power of k, whereas the result cannot depend on k. For f of the correct homogeneity (and integrals over the appropriate contours), (6.10.8) agrees with (6.10.1), and (6.10.9) agrees with (6.10.3). For if $h(u, v)$ has homogeneity degree -2, then

$$h(u, v)\, du \wedge dv = h(1, v/u) \frac{du}{u} \wedge d(v/u),$$

and integrating out u by Cauchy's theorem, we get

$$2\pi i h(1, v/u) d(v/u) = 2\pi i h(u, v)(u dv - v du).$$

This assumes that the contour avoids $u = 0$, but even if it does not, the result is the same, since we can interchange u and v in the above argument, (and $u = v = 0$ is necessarily singular for h).

Twistor quantization

At this point it is illuminating to refer to a certain result of twistor quantization theory, since this clarifies the roles of the twistor functions f and sheds light on the reason for their homogeneity degrees. We recall the standard procedure of quantum theory ('first quantization') according to which the position x^a and the momentum p_a of a particle become operators satisfying the commutation law

$$p_a x^b - x^b p_a = i\hbar g_a{}^b. \tag{6.10.11}$$

In the x-space description a particle wave function is a complex function $\psi(x^a)$, dependent on x^a but not on p_a, the operator p_a being represented as differentiation,

$$p_a:\psi\mapsto i\hbar\frac{\partial\psi}{\partial x^a}, \qquad (6.10.12)$$

and x^a simply as multiplication,

$$x^a:\psi\mapsto x^a\psi. \qquad (6.10.13)$$

In the p-space description the wave function is a complex function $\tilde{\psi}(p_a)$ of p_a but not of x^a, and the operators p_a and x^a reverse their roles:

$$p_a:\tilde{\psi}\mapsto p_a\tilde{\psi} \qquad (6.10.14)$$

and

$$x^a:\tilde{\psi}\mapsto -i\hbar\frac{\partial\tilde{\psi}}{\partial p_a}. \qquad (6.10.15)$$

Whichever description is used, we get back to the mathematics of commuting variables, provided we use *either* x^a or p_a, but not both, as our variable, the other being replaced by a differential operator.

In the analogous quantization procedure for twistor descriptions ('twistor first quantization'), the twistors Z^α and $\bar{\mathrm{Z}}_\alpha$ (or, equivalently, $\bar{\mathrm{W}}^\alpha$ and W_α) become operators satisfying the commutation law

$$\mathrm{Z}^\alpha\bar{\mathrm{Z}}_\beta-\bar{\mathrm{Z}}_\beta\mathrm{Z}^\alpha=\hbar\delta^\alpha_\beta. \qquad (6.10.16)$$

In the $\mathbb{T}^\alpha$-description of a particle we have a wave function f which is a complex function of Z^α but is independent of $\bar{\mathrm{Z}}_\alpha$, where 'independence of $\bar{\mathrm{Z}}_\alpha$' is to be interpreted as

$$\frac{\partial f}{\partial\bar{\mathrm{Z}}_\alpha}=0, \qquad (6.10.17)$$

which means that the function f is to be *holomorphic* in Z^α. The operator $\bar{\mathrm{Z}}_\alpha$ is represented as differentiation,

$$\bar{\mathrm{Z}}_\alpha:f\mapsto -\hbar\frac{\partial f}{\partial\mathrm{Z}^\alpha}, \qquad (6.10.18)$$

and Z^α simply as multiplication,

$$\mathrm{Z}^\alpha:f\mapsto\mathrm{Z}^\alpha f. \qquad (6.10.19)$$

In the $\mathbb{T}_\alpha$-description, Z^α and $\bar{\mathrm{Z}}_\alpha$ change roles. The twistor wave function f is now holomorphic in $\bar{\mathrm{Z}}_\alpha$ (i.e., anti-holomorphic in Z^α); or relabelling $\bar{\mathrm{Z}}_\alpha$ as W_α and Z^α as $\bar{\mathrm{W}}^\alpha$, f is a holomorphic function of W_α with the operators W_α and $\bar{\mathrm{W}}^\alpha$, respectively, represented as

$$\mathrm{W}_\alpha:f\mapsto\mathrm{W}_\alpha f \qquad (6.10.20)$$

and

$$\bar{W}^\alpha : f \mapsto \hbar \frac{\partial f}{\partial W_\alpha}. \qquad (6.10.21)$$

Again, we are back with the mathematics of commuting variables provided we do not use *both* Z^α and $\bar{Z}_\alpha$ or *both* W_α and $\bar{W}^\alpha$, i.e., provided we work with *holomorphic* twistor functions and operations, the conjugated variable being always replaced by a differential operator.

The explicit relation between the twistor variables and the more familiar space–time variables is achieved via the momentum and angular-momentum relations given in §6.3. For a single-twistor system the expressions are as given in (6.3.2) and the rest-mass is necessarily zero. The *twistor wave functions* are, in essence, the functions f appearing in the integrals (6.10.8) and (6.10.9) (or (6.10.1) and (6.10.3)), the integrals themselves giving the translation to the normal x-space wave functions, these being spinor fields in space–time, in accordance with the usual descriptions.

Helicity

The quantum operator representing the helicity s, as defined by (6.3.5), (6.3.6), accordingly turns out to be, when the non-commutation of variables is taken into account,

$$s = \tfrac{1}{4}(Z^\alpha \bar{Z}_\alpha + \bar{Z}_\alpha Z^\alpha). \qquad (6.10.22)$$

Notice that when Z^α and $\bar{Z}_\alpha$ commute, we are back with (6.1.74). Bearing in mind that the commutation law (6.10.16) gives us $Z^\alpha \bar{Z}_\alpha - \bar{Z}_\alpha Z^\alpha = 4\hbar$, we see that s is represented in the $\mathbb{T}^\alpha$-description by

$$s : f \mapsto -\tfrac{1}{2}\hbar\left(Z^\alpha \frac{\partial}{\partial Z^\alpha} + 2\right) f \qquad (6.10.23)$$

and in the $\mathbb{T}_\alpha$-description by

$$s : f \mapsto \tfrac{1}{2}\hbar\left(W_\alpha \frac{\partial}{\partial W_\alpha} + 2\right) f. \qquad (6.10.24)$$

Thus, if we want a twistor wave function which is in an eigenstate of helicity, we require that $f(Z^\alpha)$ be in an eigenstate of (6.10.23), or that $f(W_\alpha)$ be in an eigenstate of (6.10.24). Since the operations appearing in these relations are trivial modifications of the Euler homogeneity operation, this means that, in either case, f must be *homogeneous.* Taking the homogeneity degree of $f(Z^\alpha)$ to be $-q-2$, or that of $f(W_\alpha)$ to be $-p-2$, we find that the *eigenvalue of helicity* is, respectively,

$$\tfrac{1}{2}\hbar q \quad \text{and} \quad -\tfrac{1}{2}\hbar p. \qquad (6.10.25)$$

It is gratifying to notice that when we pass to the space–time wave function $\phi_{\ldots}$, the values (6.10.25) are precisely those already assigned in §5.7 (*cf.* after (5.7.3)) to a massless positive-frequency field with p unprimed indices and with q primed indices respectively, the positive-frequency requirement being necessary in order that the wave function describe a physical (positive energy) particle.

Reversed helicity integrals

As things stand, an unsatisfactory aspect of the twistor function description of a wave function is that we have to switch from a $\mathbb{T}^\alpha$-description to a $\mathbb{T}_\alpha$-description when changing from positive to negative helicity. Clearly it would be better to have a uniform description in terms of just one space or the other. Indeed, the quantization rules (6.10.18) and (6.10.21) tell us how this is to be achieved. For a $\mathbb{T}^\alpha$-description of a positive-helicity wave function (p unprimed indices), we would be tempted to take the complex conjugate of (6.10.3) or (6.10.9) and to replace $\bar{f}(\bar{Z}_\alpha)$ by a holomorphic function of Z^α of homogeneity degree $p-2$, for consistency with (6.10.23). But the product of spinors

$$\bar{\pi}_A \ldots \bar{\pi}_D \tag{6.10.26}$$

that appears in the complex conjugated integrand, and which seems to be necessary in order to produce the unprimed indices of $\phi_{A\ldots D}$, would spoil the holomorphicity of the integrand. In addition, the use of the primary part ω_A of Z^α at O in place of $\bar{\pi}_A$ would be wrong for a variety of reasons: mainly, because the resulting $\phi_{\ldots}$ would not satisfy the massless field equations; but even the spirit would have been wrong, since the integrand would lack Poincaré invariance owing to the position dependence of ω_A.

The clue to resolving this difficulty (following a suggestion made by L.P. Hughston in 1973; *cf.* Hughston 1979, Penrose 1975*b*) lies in the quantization rule (6.10.18), the projection spinor part of which suggests the replacement

$$\bar{\pi}_A \mapsto -\hbar\frac{\partial}{\partial\omega^A}. \tag{6.10.27}$$

Applying this substitution to (6.10.26) (and ignoring signs and factors of $\hbar$), we arrive at the formula

$$\begin{aligned}\phi_{A\ldots D}(R) &= \frac{1}{2\pi i}\oint\frac{\partial}{\partial\omega^A}\cdots\frac{\partial}{\partial\omega^D}f(Z^\alpha)\pi_{A'}d\pi^{A'}\\ &= \frac{1}{(2\pi i)^2}\oint\frac{\partial}{\partial\omega^A}\cdots\frac{\partial}{\partial\omega^D}f(Z^\alpha)d^2\pi.\end{aligned} \tag{6.10.28}$$

It is a simple matter to verify that the massless field equations (4.12.42) are satisfied by the $\phi_{A\ldots D}$ so defined, just using the relation (6.10.6) again. Indeed, it is interesting to note that the relation (6.10.6) (and similarly (6.10.5)) already has an aspect of the quantization rules implicit in it. For the operator ∇_a occurring on the left is essentially $\partial/\partial x^a$, which by (6.10.12) is the quantum replacement of $-i\hbar^{-1}p_a$. On the RHS of (6.10.6) we have the operator $i\pi_{A'}\partial/\partial\omega^A$ which, by (6.10.27), is the quantum replacement of $-i\hbar^{-1}\pi_{A'}\bar{\pi}_A$, in remarkable agreement with (6.3.2).

There is clearly a corresponding expression to (6.10.28) for positive-helicity massless wave functions in the $\mathbb{T}_\alpha$-description, namely

$$\phi_{A'\ldots D'}(R) = \frac{1}{2\pi i}\oint \frac{\partial}{\partial\mu^{A'}}\cdots\frac{\partial}{\partial\mu^{D'}} f(W_\alpha)\lambda_A d\lambda^A$$

$$= \frac{1}{(2\pi i)^2}\oint \frac{\partial}{\partial\mu^{A'}}\cdots\frac{\partial}{\partial\mu^{D'}} f(W_\alpha)d^2\lambda, \tag{6.10.29}$$

$\mu^{A'}$ being the primary part of W_α (at O).

Note that $\partial/\partial\omega^A$ and $\partial/\partial\mu^{A'}$ are both origin-independent, whereas $\partial/\partial\pi_{A'}$ and $\partial/\partial\lambda_A$ are not. This can be seen from the fact that the first two operators are projection parts of the respective twistor operators $\partial/\partial Z^\alpha$ and $\partial/\partial W_\alpha$. It can also be seen directly by changing the variables $\omega^A, \pi_{A'}$ representing Z^α at O to the pair $\tilde{\omega}^A, \tilde{\pi}_{A'}$ representing Z^α at some fixed point Q ($Z^\alpha \overset{Q}{\leftrightarrow} (\tilde{\omega}^A, \tilde{\pi}_{A'})$), whose position vector relative to O is q^a, so

$$\tilde{\omega}^A = \omega^A - iq^{AA'}\pi_{A'}, \quad \tilde{\pi}_{A'} = \pi_{A'}. \tag{6.10.30}$$

One readily verifies that (if q^a is held constant)

$$\frac{\partial}{\partial\tilde{\omega}^A} = \frac{\partial}{\partial\omega^A}, \quad \frac{\partial}{\partial\tilde{\pi}_{A'}} = \frac{\partial}{\partial\pi_{A'}} + iq^{AA'}\frac{\partial}{\partial\omega^A}, \tag{6.10.31}$$

which is the same position dependence as that given in (6.1.26) for the spinor parts of W_α, verifying that the parts of $\partial/\partial Z^\alpha$ do indeed have the correct dependence on position required for a $[^0_1]$-twistor. Similarly, if $\tilde{\lambda}_A$, $\tilde{\mu}^{A'}$ represent W_α at Q ($W_\alpha \overset{Q}{\leftrightarrow} (\tilde{\lambda}_A, \tilde{\mu}^{A'})$), we have

$$\tilde{\lambda}_A = \lambda_A, \quad \tilde{\mu}^{A'} = \mu^{A'} + iq^{AA'}\lambda_A, \tag{6.10.32}$$

so

$$\frac{\partial}{\partial\tilde{\lambda}_A} = \frac{\partial}{\partial\lambda_A} - iq^{AA'}\frac{\partial}{\partial\mu^{A'}}, \quad \frac{\partial}{\partial\tilde{\mu}^{A'}} = \frac{\partial}{\partial\mu^{A'}}, \tag{6.10.33}$$

this being the correct position dependence for a $[^1_0]$-twistor.

Suppose now that we place *both* types of quantity $\pi_{A'}$ and $\partial/\partial\omega^A$ inside the integral and define

$$\phi_{A\ldots FK'\ldots P'}(R)=\frac{1}{(2\pi i)^2}\oint \pi_{K'}\ldots\pi_{P'}\frac{\partial}{\partial\omega^A}\cdots\frac{\partial}{\partial\omega^F}f(Z^\alpha)d^2\pi. \quad (6.10.34)$$

The equation satisfied by this symmetric spinor $\phi_{\ldots}$ is

$$\nabla^{Q'}_{G}\phi^{K'\ldots P'}_{A\ldots F}=\nabla^{(Q'}_{(G}\phi^{K'\ldots P')}_{A\ldots F)}, \quad (6.10.35)$$

or, equivalently,

$$\nabla^{AA'}\phi_{AB\ldots FK'\ldots P'}=0, \quad \nabla^{KK'}\phi_{A\ldots FK'L'\ldots P'}=0. \quad (6.10.36)$$

Equation (6.10.36) is just (5.10.9), so we can apply Proposition (5.10.10) and obtain the result that $\phi_{A\ldots FK'\ldots P'}$ is (locally) simply the uncontracted rth derivative of a massless free field. Here $r=\min(u,v)$, where $\phi_{\ldots}$ has valence $\left[\begin{smallmatrix}0&0\\u&v\end{smallmatrix}\right]$, and the massless field has $|u-v|$ indices, primed or unprimed according as $u<v$ or $u>v$ (and is a scalar wave field if $u=v$). Indeed, it is manifest that an expression like (6.10.34) arises by repeated application of uncontracted operations $\nabla_{\ldots}$ to (6.10.9) or (6.10.28)(2). What we have just established shows that (6.10.34) gives the *general* analytic solution of (6.10.35) (for symmetric $\phi_{\ldots}$), assuming that this is true for the special cases (6.10.9), (6.10.28)(2).

Helicity raising and lowering

These contour integral expressions also serve to illuminate some of the results of §6.4. In particular, it was remarked (after (6.4.21)) that a trace-free symmetric $\left[\begin{smallmatrix}p\\q\end{smallmatrix}\right]$-twistor $\mathrm{T}^{\alpha\ldots\delta}_{\rho\ldots\tau}$

$$\mathrm{T}^{\alpha\ldots\delta}_{\rho\ldots\tau}=\mathrm{T}^{(\alpha\ldots\delta)}_{(\rho\ldots\tau)}, \quad \mathrm{T}^{\alpha\beta\ldots\delta}_{\alpha\sigma\ldots\tau}=0, \quad (6.10.37)$$

having primary part $\lambda^{A\ldots DR'\ldots T'}\in\mathfrak{S}^{(A\ldots D)(R'\ldots T')}$, which determines it completely and satisfies

$$\nabla^{(M}_{(M'}\lambda^{A\ldots D)}_{R'\ldots T')}=0,$$

can (in various ways) be used to increase the helicity of a positive-frequency massless field by a (possibly negative) amount $\frac{1}{2}(p-q)\hbar$. In terms of a twistor function f, using the $\mathbb{T}^\alpha$-description, these amount to instances of

$$f\mapsto \mathrm{T}^{\alpha\ldots\delta}_{\rho\ldots\tau}Z^\rho\ldots Z^\tau\frac{\partial}{\partial Z^\alpha}\cdots\frac{\partial}{\partial Z^\delta}f \quad (6.10.38)$$

(and the $\mathbb{T}_\alpha$-description is similar). Note that the homogeneity in Z^α is increased by $q-p$, which is consistent with (6.10.23). Note also that the trace-free condition (6.10.37)(2) ensures that the factor ordering in (6.10.38) is unimportant.

There are also expressions, analogous to (6.10.34) for obtaining the potentials of §6.4 directly. This will not be entered into here (*cf.* Eastwood, Penrose and Wells 1981).

Massive twistor wave functions

The foregoing discussion indicates that twistors provide an elegant formalism for the description of massless fields – or of wave functions for massless particles. It is natural to ask whether *massive*-particle wave functions can also be described in terms of twistors. In fact they can. A clue is provided by the expressions (6.3.26), (6.3.28), which show how the general massive angular-momentum twistor can be built up from a family of 1-valent twistors. This suggests analogues of the following form for (6.10.8), (6.10.9), (6.10.28)(2) and (6.10.29)(2):

$$\psi_{A\ldots E'J'\ldots N}(R) = \frac{1}{(2\pi i)^{2n}} \oint \lambda_A \cdots \pi_{E'} \frac{\partial}{\partial \mu^{J'}} \cdots \frac{\partial}{\partial \omega^N} f(W_\alpha, \ldots, Z^\alpha) d^2\lambda \wedge \cdots \wedge d^2\pi, \qquad (6.10.39)$$

involving, say, n twistors $W_\alpha, \ldots, Z^\alpha$. There are many possible different combinations of terms in the integrand, and, in contrast to (6.10.34), the derivatives in (6.10.39) can yield many new fields which are not just derivatives of simpler ones.

The considerable freedom in the choices of these expressions reflects the freedom in the n-twistor internal symmetry group (6.3.29) in its quantized form. Polynomial expressions constructed from the generators of this group provide reasonably plausible models for quantum numbers for particles (Penrose 1975*b*, Perjés 1975, 1977, 1982, Perjés and Sparling 1979, Hughston 1979, 1980). However, there appears to be no very simple way, analogous to (6.10.1) etc. of automatically satisfying the Schrödinger–Klein–Gordon equation (5.10.20) or the Dirac equations (5.10.15), (5.10.35), (5.10.36). We shall not pursue these matters further here, but refer the interested reader to the above literature (*cf.* also Penrose 1975*a*, Hodges 1982, 1985*a*, *b*, Hodges and Huggett 1980).

Geometry of contour integrals

Instead, we return to examine the massless case in more detail. Our discussion so far has been purely formal. We now want to consider the explicit structure of twistor functions and their singularity sets. For integrals* of the type (6.10.3) and (6.10.28)(1), the geometric discussion will be considerably clarified if we pass to the *projective* twistor space $\mathbb{PT}^\alpha$ (or $\mathbb{PT}^\bullet$) in our descriptions, i.e., the space of proportionality classes of non-zero $[^1_0]$-twistors identified according to the equivalence relation

$$Z^\alpha \equiv \kappa Z^\alpha \quad (0 \neq \kappa \in \mathbb{C}). \tag{6.10.40}$$

As coordinates for $\mathbb{PT}^\alpha$ we take the three independent complex ratios

$$Z^0 : Z^1 : Z^2 : Z^3. \tag{6.10.41}$$

The geometry of this space and its relation to $\mathbb{CM}$ will be discussed at much greater length in §9.3 below, and the reader is referred to that discussion for further details. For our immediate purposes it will suffice to observe that any *point* R in $\mathbb{CM}$ is represented by a complex projective *line* $\mathbf{R}$ in $\mathbb{PT}^\bullet$. This follows from the discussion of §6.2 (*cf.* (6.2.16)). The points of the line $\mathbf{R}$ correspond to the family of twistors incident with the point R in $\mathbb{CM}$. Now recall that the integrals (6.10.3) and (6.10.28)(1) were taken over a one-real-dimensional contour in the space of these twistors. This is, in effect, a one-dimensional contour in the complex projective 1-space of ratios

$$\pi_{0'} : \pi_{1'}, \tag{6.10.42}$$

which may be identified with the projective line $\mathbf{R}$. Recall next that the space of ratios (6.10.42) is topologically a sphere S^2, the *Riemann sphere* of ratios (6.10.42) (*cf.* §1.2). As we have seen in Chapter 1, for a real point $R \in \mathbb{M}$ this sphere is precisely the celestial sphere of an observer at R since it represents the space of null rays through R. But also a *complex* point $R \in \mathbb{CM}$ will be represented in $\mathbb{PT}^\bullet$ by a locus $\mathbf{R}$ whose topological – and indeed complex-analytic – structure is again that of a Riemann sphere (see Fig. 6-11). To evaluate our integrals, we have to find a contour Γ in this Riemann sphere which links the singularities of f and so cannot be shrunk to a point continuously over the sphere without crossing a singularity (for otherwise the result of the integration is necessarily zero). It is therefore of particular importance to study the nature of the singularity sets of the twistor wave functions.

A simple example (to which Fig. 6-11 refers) will illuminate the situation.

* There are also other types of contour integrals occurring in twistor theory. Especially noteworthy are those which involve all the twistor variables and not just $\pi_{0'}$ and $\pi_{1'}$. For example, integrals of the form $\int f \mathrm{d}^4 Z$ can be used to express the electric charge, when f has homogeneity -4 in Z^α; and of the form $\int Z^\alpha Z^\beta f \mathrm{d}^4 Z$ to express $\bar{A}^{\alpha\beta}$ when f has homogeneity -6. (Here $\mathrm{d}^4 Z = \mathrm{d}Z^0 \wedge \mathrm{d}Z^1 \wedge \mathrm{d}Z^2 \wedge \mathrm{d}Z^3$.) The theory of these expressions involves *relative sheaf cohomology* (Bailey 1985). Their generalizations to several twistor variables leads to *twistor diagram theory* (Penrose and MacCallum 1972, Penrose 1975*a*, Sparling 1975, Hughston and Hurd 1981, Eastwood and Ginsberg 1981, Ginsberg 1983, Qadir 1978, Hodges and Huggett 1980, Hodges 1982, 1985, *a*, *b*)

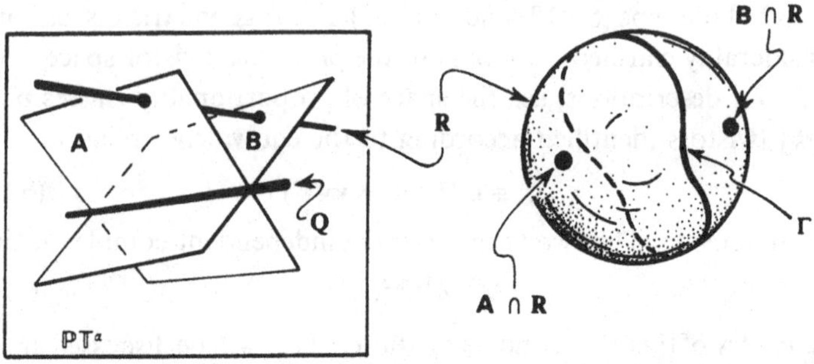

Fig. 6-11. The function to be integrated has poles along the planes **A** and **B** in $\mathbb{PT}^{\alpha}$. The point R in $\mathbb{CM}$ is represented by **R** in $\mathbb{PT}^{\alpha}$ whose topology is S^2. The poles on S^2 are points to be separated by the contour Γ.

Consider the case of spin 0, and let us try to generate a solution of the wave equation in $\mathbb{M}$ from a twistor function of homogeneity degree -2. Let us take

$$f(\mathsf{Z}^{\alpha}) = \frac{1}{\mathsf{A}_{\alpha}\mathsf{Z}^{\alpha}\mathsf{B}_{\beta}\mathsf{Z}^{\beta}}. \tag{6.10.43}$$

Then the singularities of f lie on two hyperplanes in the space $\mathbb{T}^{\alpha}$, namely those given by

$$\mathsf{A}_{\alpha}\mathsf{Z}^{\alpha} = 0 \tag{6.10.44}$$

and

$$\mathsf{B}_{\alpha}\mathsf{Z}^{\alpha} = 0. \tag{6.10.45}$$

These equations define two planes in $\mathbb{PT}^{\alpha}$ which we call **A** and **B**. We assume that these planes are distinct (i.e., that A_{α} and B_{α} are not proportional), and that the line **R** does not meet their intersection. Then the two points in which **R** meets **A** and **B** are distinct, being the points on the sphere **R** at which f becomes singular. Thus we choose Γ to be a topological circle S^1 such that just one of $\mathbf{A}\cap\mathbf{R}$ and $\mathbf{B}\cap\mathbf{R}$ lies on either side of Γ.

Let us now explicitly calculate the contour integral (6.10.3) for the present wave function f. We have:

$$\begin{aligned}2\pi i\phi(R) &= \oint f(\mathsf{Z}^{\alpha})\pi_{E'}\mathrm{d}\pi^{E'}\\ &= \oint \frac{\pi_{E'}\mathrm{d}\pi^{E'}}{\mathsf{A}_{\alpha}\mathsf{Z}^{\alpha}\mathsf{B}_{\beta}\mathsf{Z}^{\beta}}\\ &= \oint \frac{\pi_{E'}\mathrm{d}\pi^{E'}}{(\mathsf{A}_{C}\mathrm{i}r^{CC'}\pi_{C'} + \mathsf{A}^{C'}\pi_{C'})(\mathsf{B}_{D}\mathrm{i}r^{DD'}\pi_{D'} + \mathsf{B}^{D'}\pi_{D'})}.\end{aligned}$$

Now

$$\oint \frac{\xi \mathrm{d}\eta - \eta \mathrm{d}\xi}{(a\xi + c\eta)(b\xi + e\eta)} = \oint \frac{\mathrm{d}(\eta/\xi)}{(a + c\eta/\xi)(b + e\eta/\xi)}$$
$$= 2\pi\mathrm{i}\frac{1}{c(b + e(-a/c))} = \frac{2\pi\mathrm{i}}{cb - ae},$$

by Cauchy's theorem, using an appropriate orientation of the contour, and therefore

$$\phi(R) = \{\varepsilon_{D'C'}(\mathsf{A}_C \mathrm{i} r^{CC'} + \mathsf{A}^{C'})(\mathsf{B}_D \mathrm{i} r^{DD'} + \mathsf{B}^{D'})\}^{-1}$$
$$= 2\{\mathsf{A}_C\mathsf{B}^C(r^a - q^a)(r_a - q_a)\}^{-1}, \tag{6.10.46}$$

where

$$q^e = \frac{\mathrm{i}\mathsf{A}^{E'}\mathsf{B}^E - \mathrm{i}\mathsf{A}^E\mathsf{B}^{E'}}{\mathsf{A}_C\mathsf{B}^C}, \tag{6.10.47}$$

and where A_C, $\mathsf{A}^{C'}$, B_C, $\mathsf{B}^{C'}$ are the actual spinor parts, the other versions being obtained from these (unambiguously here) by raising and lowering indices. (We may compare (6.10.47) with (6.2.15); they are basically the same.)

The point Q having position vector q^a is in fact that whose representation in $\mathbb{PT}^\bullet$ is the line **Q** of intersection of **A** with **B**. The contour integral is not well-defined (and $\phi(R)$ becomes infinite) only when the line **R** meets **Q**, which is the condition (*cf.* §9.3) for R and Q to be null-separated points in $\mathbb{CM}$, i.e., for R to lie on Q's light cone.

Positive-frequency fields

By arranging that q^a be a complex vector with timelike imaginary part, we can ensure that (6.10.46) does not become singular for any *real* vector r^a. If the imaginary part of q^a is *future*-timelike, then by a result to be given later (Proposition (9.3.24)) **Q** lies entirely in that portion of $\mathbb{PT}^\bullet$, denoted by $\mathbb{PT}^-$, which arises from the space $\mathbb{T}^-$ of twistors Z^α restricted by $\mathsf{Z}^\alpha\bar{\mathsf{Z}}_\alpha < 0$. (We similarly define spaces $\mathbb{PT}^+$, $\mathbb{T}^+$, $\mathbb{PT}_-$, $\mathbb{T}_-$, $\mathbb{PT}_+$, $\mathbb{T}_+$, where, for example, $\mathsf{W}_\alpha \in \mathbb{T}_-$ iff $\mathsf{W}_\alpha\bar{\mathsf{W}}^\alpha < 0$, etc. All pairs $\mathbb{T}_+$, $\mathbb{T}^+$; $\mathbb{PT}_+$, $\mathbb{PT}^+$; etc., are complex conjugates.) The field ϕ is then well-defined, in particular, at all points R whose position vector r^a has an imaginary part that is *past*-timelike. (For then $r^a - q^a = u^a - \mathrm{i}t^a$, with u^a, t^a real and t^a future-timelike, while $(r^a - q^a)(r_a - q_a)$ can vanish only if $t^a u_a = 0$; but that implies that u^a is spacelike, so $\mathrm{Re}\{(r^a - q^a)(r_a - q_a)\} = u^a u_a - t^a t_a < 0$).

This region of $\mathbb{CM}$, whose points have position vectors with past-timelike imaginary parts, is called the *forward tube* in $\mathbb{CM}$. Positive-frequency fields can be characterized as fields which have non-singular holomorphic extensions into that tube. (In §5.7 a different characterization of positive-frequency fields was implied, in terms of their Fourier decomposition; but the two are essentially equivalent (*cf.* Bailey, Ehrenpreis and Wells 1982).) Thus, with $\mathbf{Q} \subset \mathbb{PT}^-$, our field ϕ turns out to be of positive frequency according to this criterion. The forward tube is represented in $\mathbb{PT}^\bullet$ by lines in the subspace $\mathbb{PT}^+$ (which we defined above), as follows once more by use of Proposition (9.3.24). From this point of view it is clear that ϕ is non-singular whenever $\mathbf{R} \subset \mathbb{PT}^+$ (with $\mathbf{Q} \subset \mathbb{PT}^-$), because then $\mathbf{R}$ does not meet $\mathbf{Q}$ and our integral is well-defined.

It is important for the twistor description that we should be able to state what positive frequency means in terms of twistor wave functions. We observe, from the example of (6.10.43)$\mapsto$(6.10.46) that while $\phi(R)$ is holomorphic in the forward tube, and $\mathbb{PT}^+$ is the region of $\mathbb{PT}^\bullet$ corresponding to the forward tube, the wave function f is *not* holomorphic in the whole of $\mathbb{T}^+$. (In fact, it follows from certain results of complex function theory (Griffiths and Harris 1978, Field 1982) that the only homogeneous functions which are holomorphic on the whole of $\mathbb{T}^+$ are polynomials in Z^α and therefore necessarily of non-negative homogeneity; and they are useless in our integrals since they give zero.) What appears to be required, from the example considered, is that there should be two separated regions of singularity of f which are intersected by all lines of $\mathbb{PT}^+$, so that, as in Fig. 6-11, the Riemann sphere representing $\mathbf{R}$ meets these singularities also in two separated regions, between which the required contour can be drawn. Similar remarks apply if we are interested in fields defined on general open regions $\mathscr{U}$ of $\mathbb{CM}$. The points of $\mathscr{U}$ are represented by a family of lines in $\mathbb{PT}^\bullet$ which fill some open region $\boldsymbol{\mathscr{U}}$ in $\mathbb{PT}^\bullet$. We would not ask for the corresponding twistor function to be holomorphic throughout $\boldsymbol{\mathscr{U}}$; this would be too strong a restriction and lead to no useful integrals.

Many examples of suitable twistor functions exist with suitable singularity regions for generating positive-frequency fields. For instance, there are certain direct generalizations of (6.10.43):

$$f(\mathsf{Z}^\alpha) = \frac{(\mathsf{C}_\alpha \mathsf{Z}^\alpha)^c (\mathsf{D}_\alpha \mathsf{Z}^\alpha)^d}{(\mathsf{A}_\alpha \mathsf{Z}^\alpha)^{a+1} (\mathsf{B}_\alpha \mathsf{Z}^\alpha)^{b+1}} \tag{6.10.48}$$

which, provided A_α, B_α, C_α, D_α are linearly independent and $\mathbf{A} \cap \mathbf{B} \subset \mathbb{PT}^-$, generate positive-frequency fields referred to as *elementary states*. Their helicity is

$$\tfrac{1}{2}\hbar(a + b - c - d), \tag{6.10.49}$$

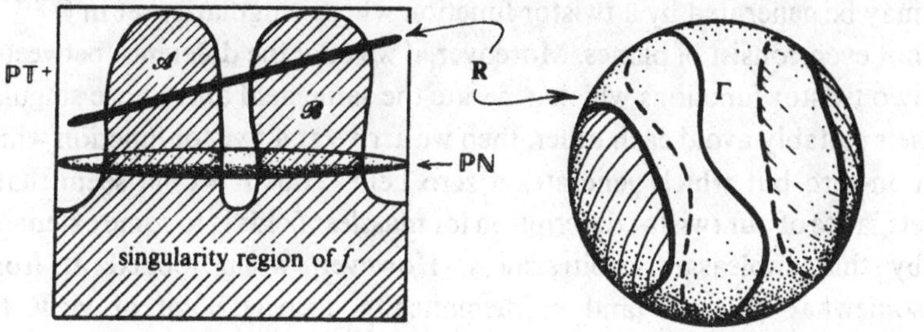

Fig. 6-12. The singularity region for a twistor function generating a positive-frequency field. The singularities in $\mathbb{PT}^+$ are contained in two disconnected closed sets.

where each of a, b, c, d is taken to be a non-negative integer. The resulting fields are similar to (6.10.46) but more complicated.

We could also add together several twistor functions of the form (6.10.48) with different choices of A_α and B_α, provided we can separate all the $\mathbf{A}$ singularities into one set and all the $\mathbf{B}$ singularities into another, to give the required two separated singularity regions in $\mathbb{PT}^+$. More generally, we could 'add' together (i.e. integrate over some parameter space) a whole continuous infinity of expressions of the form (6.10.48), provided that the (closed) regions $\mathscr{A}$, $\mathscr{B}$, filled by $\mathbf{A}$ and $\mathbf{B}$, respectively, do not intersect one another in $\mathbb{PT}^+$. In this case, the singularities that arise on the Riemann sphere need not simply be poles, but may well be extended (closed) sets on the sphere with non-vacuous interiors (see Fig. 6-12).

Motivation for cohomology

The above process could give us a wide class of twistor functions generating positive-frequency fields, but there is clearly something unsatisfactory about such a description since the actual location of the singularities of any particular twistor function does not uniquely reflect the location of the singularities of the resulting field in $\mathbb{CM}$. This is particularly manifest in the case of the spin-0 example given above (in (6.10.43)), since if we replace A_α and B_α by any other pair

$$\rho\mathsf{A}_\alpha + \sigma\mathsf{B}_\alpha, \quad \tau\mathsf{A}_\alpha + \kappa\mathsf{B}_\alpha, \tag{6.10.50}$$

with $\rho, \sigma, \tau, \kappa \in \mathbb{C}$, $\rho\kappa - \sigma\tau = 1$, then the field ϕ as given in (6.10.46) remains unaltered. The singularity sets in this case consist of two *arbitrary* distinct planes through $\mathbf{Q}$. In fact, if we do not restrict ourselves to the specific form (6.10.43) for our twistor function, then this same field ϕ (in the forward tube) may be generated by a twistor function whose singularity set in $\mathbb{PT}^+$ does not even consist of planes. Moreover, if we take the difference between any two twistor functions which generate the same field and whose singularity sets suitably avoid each other, then we arrive at a twistor function which is non-zero but which generates a zero field. Thus it would seem that the elegance of our twistor description for massless fields is to some extent offset by this unpleasant arbitrariness. However, when looked at from a somewhat different (and mathematically deeper) point of view, these apparently unpleasant features are seen, instead, as a necessary part of a very elegant mathematical formalism, called 'sheaf cohomology theory'. From this point of view, a twistor wave function should not be thought of as a function in the ordinary sense, but as a kind of 'second-order' function in a sense to be explained shortly. These second-order functions are, technically, elements of 'first sheaf cohomology groups' – and we shall refer to them here as 1-functions – while functions in the ordinary sense are elements of 'zeroth sheaf cohomology groups', and in that connotation we shall refer to them as 0-functions.

To elaborate these matters, let us consider once more the situation depicted in Fig. 6-12 (or in Fig. 6-11). We are concerned here with the region $\mathbb{PT}^+$ since it is this which corresponds to the forward tube in $\mathbb{CM}$, the region in which ordinary space–time single-particle (positive-frequency) wave functions are holomorphically defined. We recall that the singularities of f were required to lie within two disconnected closed sets in $\mathbb{PT}^+$, which we shall now refer to generally as $\mathscr{A}$ and $\mathscr{B}$. It is more appropriate, however, to fix attention not on the singularity regions, but on the region in which f is holomorphic. This can be conveniently described as

$$\mathscr{U} \cap \mathscr{V}, \tag{6.10.51}$$

where

$$\mathscr{U} = \mathbb{PT}^+ - \mathscr{A}, \quad \mathscr{V} = \mathbb{PT}^+ - \mathscr{B}. \tag{6.10.52}$$

Note that

$$\mathbb{PT}^+ = \mathscr{U} \cup \mathscr{V}, \tag{6.10.53}$$

so that the sets $\mathscr{U}$ and $\mathscr{V}$ together provide an open covering of $\mathbb{PT}^+$.

If we were to consider merely functions defined on the *fixed* set $\mathscr{U} \cap \mathscr{V}$ as our twistor wave functions (where in the passage to the corresponding space–time fields we could envisage adopting a contour-integral description using, for each **R**, a *definite* contour, independent of the function), then we come up against various difficulties. For example, certain types of function of this kind always yield zero when the integration is performed, namely those which are holomorphically extendible to the entire region $\mathscr{U}$

or to the entire region $\mathscr{V}$ (since in these cases the contour would collapse to a point on one side of the Riemann sphere or on the other), or any constant linear combinations of such functions. A simple example, where the regions $\mathscr{A}$ and $\mathscr{B}$ are the planes indicated in Fig. 6-11, is this:

$$f = \frac{1}{(\mathsf{A}_\alpha \mathsf{Z}^\alpha)^2} - \frac{1}{(\mathsf{B}_\alpha \mathsf{Z}^\alpha)^2}. \tag{6.10.54}$$

Each term has only one singularity on the Riemann sphere, so integrating them separately lets the contour collapse on the side opposite to the singularity in each case, and we get zero. Another example is the difference between (6.10.43) and the twistor function which is obtained from (6.10.43) by replacing A_α and B_α by the expressions (6.10.50). For when $\rho\kappa - \sigma\tau = 1$, we have the identity

$$\frac{1}{AB} - \frac{1}{(\rho A + \sigma B)(\tau A + \kappa B)} = \frac{\sigma}{A(\rho A + \sigma B)} + \frac{\tau}{B(\tau A + \kappa B)}, \tag{6.10.55}$$

which is to be applied with $A = \mathsf{A}_\alpha \mathsf{Z}^\alpha$, $B = \mathsf{B}_\alpha \mathsf{Z}^\alpha$. We define the region $\mathscr{A}$ as the union of the two planes $\mathsf{A}_\alpha \mathsf{Z}^\alpha = 0$ and $(\rho \mathsf{A}_\alpha + \sigma \mathsf{B}_\alpha)\mathsf{Z}^\alpha = 0$ and $\mathscr{B}$ correspondingly. (We must take $\rho \neq 0 \neq \kappa$ so that $\mathscr{U} \cup \mathscr{V}$ in fact covers $\mathbb{PT}^+$.) The first term on the RHS of (6.10.55) is holomorphic on $\mathscr{U}$ and the second term is holomorphic on $\mathscr{V}$, and so again the integration yields zero. Thus functions on $\mathscr{U} \cap \mathscr{V}$ cannot *uniquely** correspond to wave functions.

The concept of a 1-function

This suggests that we ought to consider not simply functions on $\mathscr{U} \cap \mathscr{V}$, but functions on $\mathscr{U} \cap \mathscr{V}$ *modulo* the restriction to $\mathscr{U} \cap \mathscr{V}$ of functions defined holomorphically on $\mathscr{U}$, or holomorphically on $\mathscr{V}$. In other words, two functions on $\mathscr{U} \cap \mathscr{V}$ will be deemed to be *equivalent* if they differ by sums of functions which are either extendible holomorphically to $\mathscr{U}$ or to $\mathscr{V}$. (The point is, that the integral of the difference vanishes *manifestly*, by collapse of the contour.) As an example, the function defined in (6.10.54) is equivalent to zero. Indeed, the system of these equivalence classes of functions constitutes the first holomorphic sheaf cohomology group with respect to the covering $\{\mathscr{U}, \mathscr{V}\}$ of $\mathbb{PT}^+$. We shall refer to these classes as 1-functions

* In fact there are other difficulties with considering functions on such a *fixed* set. For no matter how this set is chosen there will always be wave functions *not* obtainable from that particular set. Moreover the class of wave functions that *is* obtainable will not be conformally (or Poincaré) invariant. All these difficulties are removed by the sheaf cohomology viewpoint.

with respect to the covering $\{\mathcal{U}, \mathcal{V}\}$. By contrast, 'ordinary' functions will be referred to as 0-functions.

However, we are now quickly forced to extend the definition of 1-functions to more complicated coverings of $\mathbb{PT}^+$, and ultimately to remove their dependence on any specific covering altogether. For even the simple task of adding together a number of 1-functions defined on *different* pairs of coverings $\{\mathcal{U}, \mathcal{V}\}$ of $\mathbb{PT}^+$ entails consideration of 1-functions defined on the common refinement (*cf.* later: (6.10.62)) of these different coverings. Let us therefore consider *general* open coverings of $\mathbb{PT}^+$ by open sets, though we shall restrict these coverings to be *locally finite*, which means that every point is contained in a finite number of sets of the covering. (However, a non-compact set like $\mathbb{PT}^+$ admits coverings that are locally finite yet contain an infinite number of sets.) In Fig. 6-13 a covering of $\mathbb{PT}^+$ is shown which consists of many (possibly infinitely many) open sets $\mathcal{U}_i$. For such a covering we need a whole family $\{f_{ij}\}$ of holomorphic functions, defined on the intersections of *pairs* of these open sets:

$$f_{ij} \textit{ defined on } \mathcal{U}_i \cap \mathcal{U}_j. \tag{6.10.56}$$

Now **R**, being compact, intersects this locally finite system in only a *finite* number of the sets $\mathcal{U}$, so only a finite number of functions f_{ij} have non-vacuous restrictions to **R**. We must now generalize our integral formulae to apply to such a situation. The intuitive picture of how this might be done is supplied by Fig. 6-13: we perform a *branched* contour integral, where the contour is now a network of segments. Each segment lies in an overlap of two of the sets of the covering, and the corresponding function (6.10.56) is integrated over it. Each segment is terminated at both ends (unless it is itself a closed loop) by vertices lying in a triple overlap of sets $\mathcal{U}$, and each vertex is to belong to precisely three segments, one segment associated with each *pair* from the three sets $\mathcal{U}$. By the properties of contour integrals, the result of such an integration will be unaffected if any segment is varied

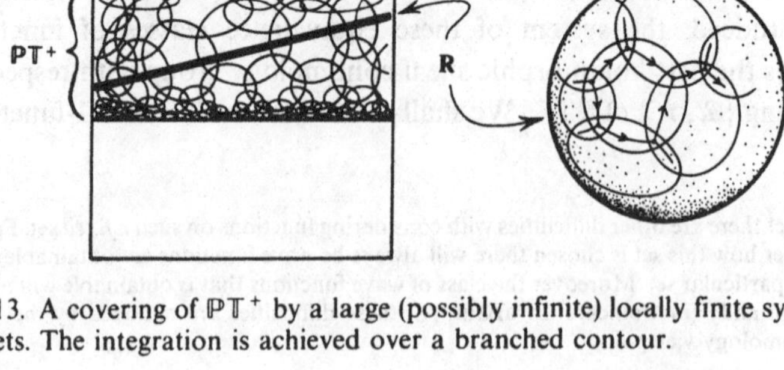

Fig. 6-13. A covering of $\mathbb{PT}^+$ by a large (possibly infinite) locally finite system of open sets. The integration is achieved over a branched contour.

continuously within its given intersection domain, provided the end-points are not moved (and if all intersections have Euclidean topology, the word 'continuously' can be removed).

There are two further requirements on the family of functions f_{ij} if such a branched integral is to be unambiguously defined. In the first place, since the integration along a given segment can be performed in either direction, the sign of the contribution would seem to depend on the direction of integration. This difficulty can be avoided by reversing the sign of the function to be integrated whenever we reverse the direction of the contour. This is best done by accompanying each function f_{ij} by its negative, which we denote by f_{ji}:

$$f_{ij} = -f_{ji}. \tag{6.10.57}$$

The rule which tells us which of the functions (6.10.57) is to be associated with which direction of integration is indicated in Fig. 6-14 for a simple type of covering.

Our second requirement on the f_{ij} stems from not wanting our integrals to depend on the exact location of the end points within the triple intersection regions. As should be clear from Fig. 6-15, such dependence can be eliminated by requiring the functions within any triple region $\mathcal{U}_i \cap \mathcal{U}_j \cap \mathcal{U}_k$ to satisfy

$$f_{ij} - f_{ik} + f_{jk} = 0. \tag{6.10.58}$$

Any family of functions f_{ij} defined on intersections $\mathcal{U}_i \cap \mathcal{U}_j$ of pairs of open sets of an open covering $\{\mathcal{U}_i\}$, and satisfying conditions (6.10.57) and (6.10.58), is called a 1-*cocycle* with respect to this covering. We have motivated this concept in terms of contour integrals, but in fact it is much

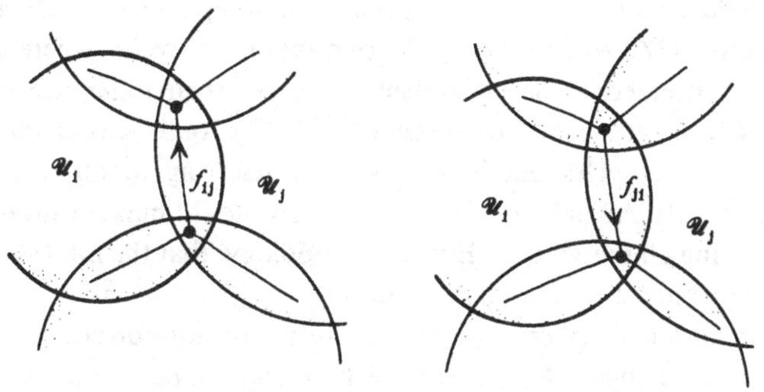

Fig. 6-14. To integrate f_{ij}, keep $\mathcal{U}_i$ to the left and $\mathcal{U}_j$ to the right. For $f_{ji} = -f_{ij}$, the contour direction is reversed, so the result is the same.

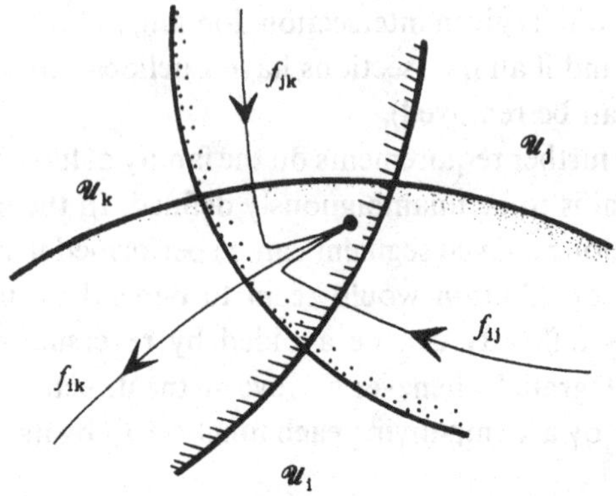

Fig. 6-15. The point in $\mathscr{U}_i \cap \mathscr{U}_j \cap \mathscr{U}_k$ at which three contour segments are joined may be moved continuously without affecting the result. This is because of the cocycle condition (6.10.58).

broader than this and applies in many mathematical situations where integration is not being envisaged. We shall also need the concept of 1-*coboundary* and a method of factoring out 1-cocycles by 1-coboundaries. (This will generalize our earlier definition of 1-functions associated with coverings consisting of just two sets.) A 1-coboundary is a 1-cocycle which can be represented as

$$f_{ij} = h_i - h_j \tag{6.10.59}$$

for some family of functions (in our context holomorphic) with

$$h_i \textit{ defined on } \mathscr{U}_i, \tag{6.10.60}$$

where (6.10.59) refers to the restriction of h_i and h_j to the overlap region $\mathscr{U}_i \cap \mathscr{U}_j$. Clearly every 1-coboundary satisfies the 1-cocycle conditions (6.10.57) and (6.10.58).

We now define a *1-function (or 1-cohomology group element) with respect to the covering* $\{\mathscr{U}_i\}$ to be an equivalence class of 1-cocycles, where two 1-cocycles are deemed to be equivalent whenever their difference is a 1-coboundary (all with respect to the covering $\{\mathscr{U}_i\}$). Again we can motivate factoring the 1-cocycles out in this way by referring to our branched contour integrals. An examination of Fig. 6-16 should make it intuitively clear (assuming all the $\mathscr{U}_i$ have Euclidean topology) that the integration of any (holomorphic) 1-coboundary gives zero.

As we mentioned earlier, part of the motivation for considering more complicated coverings of $\mathbb{PT}^+$ (than just $\mathscr{U}, \mathscr{V}$) was to be able to construct sums of 1-functions defined with respect to different coverings. We are now in a position to deal with this question. For convenience, let us introduce

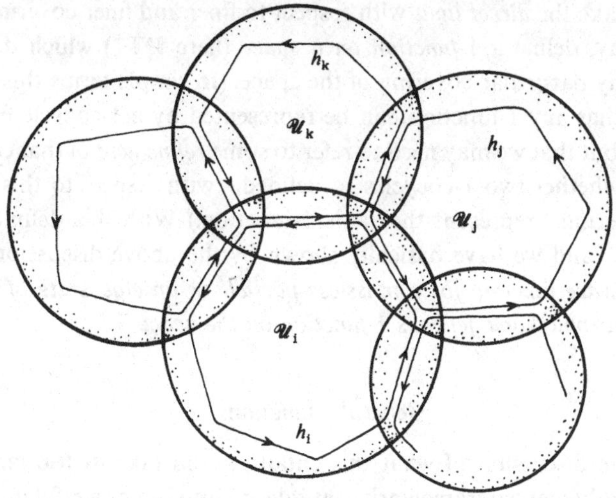

Fig. 6-16. When the coboundary condition holds, the integral reduces to one performed over a family of closed contours which can be shrunk to nothing. The result of the integral is therefore zero.

the notation

$$F_{i\ldots k}|_{l\ldots n} \quad (=F_{i\ldots k}|_{(l\ldots n)}) \tag{6.10.61}$$

for the restriction of a function $F_{i\ldots k}$ defined on $\mathcal{U}_i\cap\cdots\cap\mathcal{U}_k$ to $\mathcal{U}_i\cap\cdots\cap\mathcal{U}_k\cap\mathcal{U}_l\cap\cdots\cap\mathcal{U}_n$. Now let f_{ij} and $g_{\hat{i}\hat{j}}$ be two cocycles defined on coverings $\{\mathcal{U}_i\}$ and $\{\mathcal{V}_{\hat{i}}\}$, respectively, of $\mathbb{PT}^+$. Define the covering $\{\mathcal{W}_I\}$ as the *common refinement* of $\{\mathcal{U}_i\}$ and $\{\mathcal{V}_{\hat{i}}\}$:

$$\mathcal{W}_I = \mathcal{W}_{i\hat{i}} = \mathcal{U}_i \cap \mathcal{V}_{\hat{i}} \tag{6.10.62}$$

where 'I' is a composite index standing for the pair of independent indices i, î. (We shall similarly use J for j, ĵ, etc.) Also define

$$f_{IJ} = f_{ij}|_{\hat{i}\hat{j}}, \quad g_{IJ} = g_{\hat{i}\hat{j}}|_{ij}. \tag{6.10.63}$$

The cocycle conditions (6.10.57) and (6.10.58) are easily seen to remain true for these restrictions (composite indices taking the place of simple indices), and a coboundary remains a coboundary when referred to the refinement. But it may happen that new coboundaries appear in the refinement which widen the equivalence classes, so that a non-zero 1-function with respect to some covering may become zero when referred to a refinement. By referring 1-functions to refinements, we now have a way of adding them: $f_{IJ} + g_{IJ}$ can represent the sum of f_{ij} and $g_{\hat{i}\hat{j}}$. But since some 1-functions may be sent to zero when a refinement is taken, we need really to take the *direct limit* with respect to finer and finer coverings and, in this way, define a 1-*function on a space* (here $\mathbb{PT}^+$) which does not refer to any particular covering of the space. (In simple terms this means, in effect, that any 1-function can be represented by a 1-cocycle for *some* covering, but that we may need to refer to some *refinement* of that covering to know whether two 1-cocycles so defined – with respect to this or any other covering – represent the same 1-function.) With this definition* it turns out – and we have basically shown by the above discussion – that *a twistor wave function for a massless particle in an eigenstate of helicity is a holomorphic homogeneous 1-function on the space* $\mathbb{T}^+$.

General r-functions

A complete discussion of what this entails would take us too far afield. But the mathematical framework that it leads into is a powerful one: there are many deep and multifaceted mathematical results that can be called into play for the study of this kind of (sheaf) cohomology. A calculus of *r*-functions (or *r*th cohomology group elements) exists (*cf.* Griffiths and Harris 1978, Chern 1979) according to which they may be added together, integrated, and multiplied in various ways. Twistor theory indicates that this calculus should have importance in understanding the nature of quantum particles. Accordingly, we shall provide at least a brief general definition of an *r*-function.

These functions can be discussed in the context of a general complex manifold $\mathcal{Q}$, or indeed in any Hausdorff paracompact topological space. But for our purposes the former is relevant, since we are interested in holomorphic functions.** Let $\{\mathcal{U}_i\}$ be a locally finite covering of $\mathcal{Q}$ and define, for any given non-negative integer *r*, a (Čech) *r-cochain* with respect to $\{\mathcal{U}_i\}$ to be a collection of (holomorphic) functions $\{f_{i_0\ldots i_r}\}$ such that we have

$$f_{i_0\ldots i_r} \textit{ defined on } \mathcal{U}_{i_0}\cap\cdots\cap\mathcal{U}_{i_r}, \qquad (6.10.64)$$

* There are also alternative procedures to defining 1-functions, such as that of Dolbeault cohomology (*cf.* Morrow and Kodaira 1971, Wells 1980). For a development of twistor theory along these lines, see Woodhouse (1985).

** More generally, one can discuss cohomology with respect to any sheaf, not just, as here, the sheaf of holomorphic functions. It is worth while just to indicate what the term 'sheaf' really entails. Basically, one has some concept of 'function' (which, however, need not be a point function in any ordinary sense, but which is nevertheless *locally* defined) assigned to any open set in the space $\mathcal{Q}$. For this, there must be a concept of *restriction*, to any smaller open set. For any covering $\{\mathcal{U}_i\}$ of $\mathcal{Q}$, the following two properties must hold: $f_{k[i}|_{j]} = 0$ implies $f_{ki} = g_k|_i$, for some g_k; and $f_k|_i = g_k|_i$, for all i implies $f_k = g_k$.

and

$$f_{i\ldots k} = f_{[i\ldots k]}, \tag{6.10.65}$$

ths r-cochain being an *r-cocycle* if, in addition,

$$f_{[i\ldots k}|_{l]} = 0. \tag{6.10.66}$$

Here the vertical bar denotes *restriction*, as in (6.10.61), and the square brackets skew-symmetrization as usual. Note that (6.10.58) is a special case of (6.10.66) since the former should, strictly, be written $f_{ij}|_k - f_{ik}|_j + f_{jk}|_i = 0$; and clearly (6.10.57) is a special case of (6.10.65). An *r-coboundary* with respect to $\{\mathscr{U}_i\}$ is an r-cocycle expressible as

$$f_{i\ldots jk} = h_{[i\ldots j}|_{k]} \tag{6.10.67}$$

for some holomorphic $(r-1)$-cochain $\{h_{i\ldots j}\}$. Our earlier equation (6.10.59) is an example of this (with a factor 2 omitted). An argument identical to that used in (4.3.17) to establish $\mathrm{d}^2 = 0$ (d = exterior derivative) shows that every r-coboundary is indeed an r-cocycle. For if the *coboundary operator* δ is defined* as taking an $(r-1)$-cochain h to an r-cochain f according to (6.10.67), then

$$\delta^2 = 0 \quad \text{i.e. } h_{[i\ldots j}|_{kl]} = 0, \tag{6.10.68}$$

since the order in which a restriction is taken is clearly immaterial (*cf.* 6.10.61)).

We can now define an r-function with respect to the covering $\{\mathscr{U}_i\}$ to be an r-cocycle modulo r-coboundaries, and then take the *direct limit*, exactly as we did in the case of a 1-function, to define a (holomorphic) r-function on the space $\mathscr{Q}$. With this definition, a 0-function turns out indeed to be a holomorphic function in the ordinary sense.

Of course it is a nuisance to have to take a direct limit, and there are theorems that can be invoked (Griffiths and Harris 1978, Godement 1964, Hirzebruch 1962) which imply that in the (present) case of *holomorphic* sheaf cohomology we need not actually take this limit, provided our sets $\mathscr{U}_i$ are, in an appropriate sense, 'holomorphically convex' (or, more strictly, 'Stein manifolds', *cf.* Gunning and Rossi 1965).

The viewpoint that the twistor wave functions of single massless particles should be holomorphic 1-functions on $\mathbb{T}^+$ (or $\mathbb{T}_+$) leads to the associated view that a quantum state consisting of n massless particles should be described twistorially by an n-function defined on a suitable region in the product of n twistor spaces, i.e., by an n-function of n twistor variables. This

* Our definition differs by a simple factor from the conventional one. The notational devices employed here are based on a suggestion due to L. P. Hughston.

arises from the fact that when an r-function and an s-function are multiplied together in the appropriate way ('cup product'), the result is an $(r+s)$-function (Griffiths and Harris 1978). Multiplying together the various 1-functions describing the separate particles (and forming linear combinations), we arrive at the above n-function description. It is yet unclear how the twistor functions for *massive* particles are to be regarded. They also appear to be r-functions, of some sort, of several twistor variables, and it is tempting to suppose that a single massive particle should be described twistorially by a 1-function. This would give a clear way of distinguishing such descriptions cohomologically from the descriptions of systems of massless particles and would suggest that perhaps, generally, an n-particle twistor wave function should be an n-function.

Nonlinear 1-functions

But there is another reason for regarding a 1-function as playing a basic role analogous to that of a single particle. This is related to the way that interactions enter the theory. Recall that a 1-function is defined in terms of a family of ordinary functions themselves defined on the overlaps between open sets, with a certain compatibility condition (the cocycle relation (6.10.58)) holding on triple overlaps. This is somewhat reminiscent of the way that a manifold can be constructed by patching together coordinate neighbourhoods: on each overlap between a pair of such neighbourhoods one must define a *transition function* which gives the translation from one system of coordinates to the next, and one also has a compatibility relation between the three transition functions that appear at the triple overlaps of neighbourhoods. Furthermore, if we are interested in this manifold only intrinsically, there is the freedom which allows us to relabel the coordinates on each entire coordinate patch separately. This is analogous to the coboundary freedom (6.10.59), the h_i of (6.10.60) corresponding to the relabelling functions. The main difference is that the relations (6.10.58) and (6.10.59) are *linear*, whereas the relations connecting transition functions are generally nonlinear.

This analogy is not just a superficial one. Consider, for a moment, instead of building a manifold from scratch, merely making an infinitesimal *deformation* of an existing (complex) manifold $\mathcal{Q}$. We can take $\mathcal{Q}$ to be pieced together from a system of open sets $\{\mathcal{U}_i\}$ which we then imagine being separated from one another and put back together in a slightly different way. If the shift is infinitesimal, then the displacement is described by a vector field on each overlap – and the compatibility relation becomes

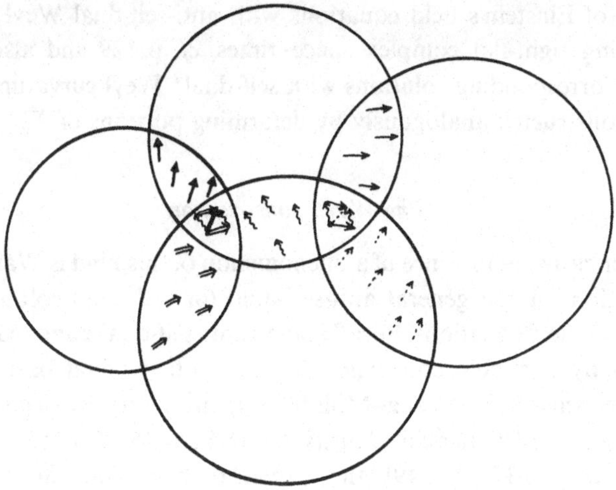

Fig. 6-17. Deforming a complex manifold.

precisely (6.10.58) (for the vector fields) on the triple overlaps (see Fig. 6-17). Furthermore, the condition for such a deformation to be trivial (i.e., for the deformed manifold to be equivalent to the original) – in the sense which is appropriate here – is precisely (6.10.59) (for the vector fields). So we see that linear deformations are described by holomorphic 1-functions – or rather, by 1-function vector fields.

Thus, in suitable situations, it may be possible to regard 1-functions as corresponding to linearized deformations, perhaps arising as *approximations* to a more correct theory involving finite nonlinear deformations. A view has been emerging in twistor theory that interactions should always be expressed in terms of such deformations, the free-particle 1-functions arising as, essentially, the linearized limit. Since individual particles rather than systems of particles are the objects which interact directly, it is tempting to suppose that a 1-function description of single particles would be related to this.

In fact, such a view receives some tantalizing support from two instances. The $\mathbb{T}^\bullet$-space twistor 1-functions of homogeneity $+2$ describe $-2\hbar$ helicity particles (gravitons), the contour integral giving space–time wave functions ϕ_{ABCD} satisfying the equations of linearized Einstein theory in the anti-self-dual case. It turns out that such 1-functions can also be used to generate linearized deformations of portions of $\mathbb{T}^\bullet$, and that these exponentiate to *finite* deformations which, when re-interpreted in space–time terms, yield (though somewhat implicitly) the general *nonlinear* (local) solution of Einstein's field equations with anti-self-dual Weyl curvature (thus being right-flat complex space–times, *cf.* p. 129 and also Penrose 1976*a*). Corresponding solutions with self-dual* Weyl curvature (left-flat) can be constructed analogously by deforming portions of $\mathbb{T}_\bullet$.

The Ward construction

The other clear-cut instance of a phenomenon of this kind is Ward's (1977) construction for the general anti-self-dual (or self-dual) solution of the Yang–Mills field equations. Recall that an anti-self-dual Yang–Mills field is provided by a bundle connection ∇_a, the anti-self-dual part of whose curvature vanishes. The Yang–Mills field equations are then a consequence of the Yang–Mills Bianchi identity (*cf.* (5.5.35)–(5.5.50) *et seq.*). In the notation of (5.5.37), (5.5.49), the curvature condition can be written $\chi_{A'B'\Theta}{}^{\Psi}=0$, i.e., by (5.5.40)(2) (with $\square_{A'B'}=\nabla_{A(A'}\nabla^{A}_{B')}$, as usual),

$$\square_{A'B'}\mu^{\Psi}=0, \qquad (6.10.69)$$

for any Yang–Mills-charged field μ^{Ψ}. We take the vector space $\mathscr{V}^{\Psi}$ to which μ^{Ψ} belongs at each point, to be a finite-dimensional complex vector space, and the Yang–Mills vector bundle $\mathscr{B}$ to be holomorphic over some suitable region of $\mathbb{CM}$. The condition (6.10.69) can be re-expressed (in Lie bracket notation, *cf.* (4.3.26)) as

$$[\pi^{A'}\nabla_{A'0},\pi^{B'}\nabla_{B'1}]\mu^{\Psi}=0 \qquad (6.10.70)$$

(using a constant spin-frame), for each choice of $\pi_{A'}\in\mathbb{S}_{A'}$. Now in $\mathbb{CM}$, an α-plane is a null complex 2-surface whose tangent plane at each point is spanned by $o^A\pi^{A'}$ and $\iota^A\pi^{A'}$ for some fixed $\pi_{A'}$ (*cf.* p. 64 and §9.3). Equation (6.10.70) therefore states that the Yang–Mills connection is integrable on any α-plane. Hence there exists on each α-plane a global parallelism for Yang–Mills-charged fields – assuming the field to be defined on a non-empty, connected and simply-connected portion of the α-plane. Since each α-plane Z corresponds to a unique point $\mathbf{Z}$ of $\mathbb{PT}^\bullet$ (*cf.* p. 64, §9.3 and

* In accordance with the general twistor programme, one would hope also to find a nonlinear version of the 1-functions of homogeneity -6 in $\mathbb{T}^\bullet$ (rather than merely $+2$ in $\mathbb{T}_\bullet$). This still-elusive putative construction is referred to as the 'googly graviton', in accordance with an appropriate cricketing analogy (*cf.* Penrose 1979*a*, 1980*a*, Law 1985). One would hope also eventually to find a combined construction in which general solutions (i.e. not constrained to be either self-dual or anti-self-dual) could be generated. An alternative approach is that based on the concept of 'ambitwistors', where standard *flat* (projective) ambitwistor space consists of pairs of (projective) twistors Z^α, W_α subject to $\mathsf{Z}^\alpha\mathsf{W}_\alpha=0$ (*cf.* Penrose 1975*a*, LeBrun 1983, 1985 for the gravitational case and Isenberg, Yasskin and Green 1978, Witten 1978, Eastwood 1985*b* for Yang–Mills fields; *cf.* also Manin 1982, Gindikin 1982).

Table (9.3.22)), the various Yang–Mills-charged fields μ^{Ψ} which are constant on $\mathbf{Z}$ constitute a vector space which we may think of as a *fibre* over the point $\mathbf{Z} \in \mathbb{PT}^{\bullet}$. This fibre is isomorphic with the original vector space V^{Ψ}, the fibre of the bundle $\mathscr{B}$. The α-planes (or portions of them) on which this global parallelism is defined provide us with a subset $\mathscr{Y}$ of $\mathbb{PT}^{\bullet}$, and corresponding to each point of $\mathscr{Y}$ we have a copy of the original Yang–Mills vector space $\mathscr{V}^{\Psi}$. Thus we have a $\mathscr{V}^{\bullet}$-bundle $\mathscr{C}$ over $\mathscr{Y} \subset \mathbb{PT}^{\bullet}$. In fact, this bundle is a holomorphic vector bundle, as follows easily from our construction.

What is less obvious, however, is that the Yang–Mills connection ∇_a (and with it the Yang–Mills curvature $F_{ab\Phi}{}^{\Psi}$) is uniquely defined merely from the holomorphic nature of the bundle $\mathscr{C}$ over $\mathscr{Y}$. No connection need be defined on $\mathscr{C}$, but the information of the original bundle $\mathscr{B}$, *with* its Yang–Mills connection ∇_a, is already contained in the structure of $\mathscr{C}$. To see, roughly, how this comes about, we first indicate how from $\mathscr{C}$ we can reconstruct $\mathscr{B}$ (over some appropriate region of $\mathbb{CM}$, which may be somewhat different from the original base space of $\mathscr{B}$.) Each point $R \in \mathbb{CM}$ corresponds to a projective line $\mathbf{R}$ in $\mathbb{PT}^{\bullet}$, and provided $\mathbf{R} \subset \mathscr{Y}$, there is over $\mathbf{R}$ a portion $\mathscr{C}_{\mathbf{R}}$ of the holomorphic bundle $\mathscr{C}$. The nature of $\mathscr{C}_{\mathbf{R}}$ is such that it *only admits constant holomorphic cross-sections.* (This follows from general results of holomorphic vector bundle theory (Chern 1979, Gunning 1967) if a certain 'stability' condition – true in the generic case, and also always true when $\mathscr{C}$ is constructed from a *given* $\mathscr{B}$ as above – is taken to hold for $\mathscr{C}_{\mathbf{R}}$.) These constant cross-sections define the vector space providing the required fibre of the bundle $\mathscr{C}$ over $R \in \mathbb{CM}$. To see that the connection on $\mathscr{B}$ is already implicit in this construction, consider the family of lines $\mathbf{R} \subset \mathscr{Y} \subset \mathbb{PT}^{\bullet}$ through a fixed point $\mathbf{Z} \in \mathscr{Y}$ (see Fig. 6-18). Because the fibre over $\mathbf{Z}$ is common to all the portions $\mathscr{C}_{\mathbf{R}}$ of $\mathscr{C}$ above the various lines $\mathbf{R}$, we can relate their constant cross-sections to one another via the fibre over $\mathbf{Z}$. Thus we have a natural isomorphism between all the fibres over the α-plane, i.e., we have a global parallelism for the μ^{Ψ} over the α-plane. This defines ∇_a restricted to the α-plane, and by allowing the α-plane to vary, we get ∇_a over the entire space (though, of course, not necessarily globally integrable). This establishes our assertion that the Yang–Mills connection ∇_a is fully determined by the holomorphic nature of the bundle $\mathscr{C}$.

Since no connection is needed for the bundle $\mathscr{C}$, it is possible to piece $\mathscr{C}$ together simply by using free holomorphic functions in suitable domains (that is, if the Yang–Mills group is $GL(n, \mathbb{C})$ – otherwise we need to solve whatever *algebraic*, but not differential, equations are needed for defining generic elements of the Yang–Mills group). This piecing together is done (as

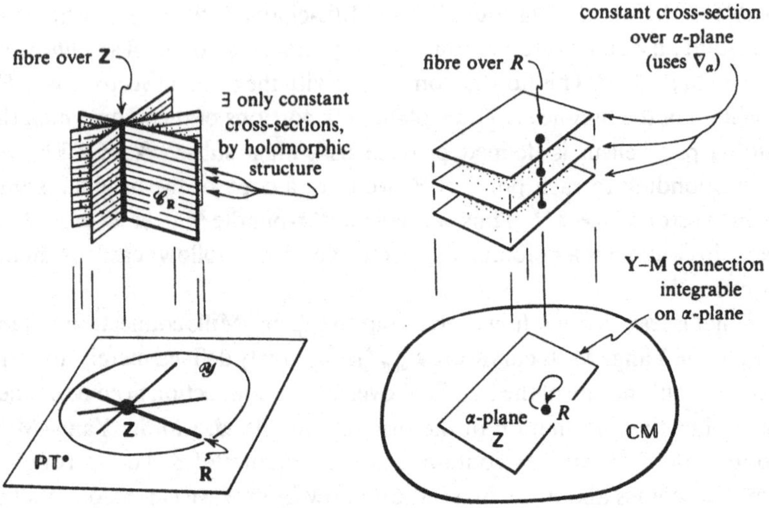

Fig. 6-18. The Ward construction for (anti-) self-dual Yang–Mills fields.

in §5.4) by defining transition functions for the various patches of the bundle lying over the sets $\mathscr{U}_i$ of some (locally finite) open covering of $\mathscr{Y}$. If the group is Abelian (as it is in the case of electromagnetism), then this patching can be achieved in an essentially linear way and therefore by a suitable 1-function. But with a non-Abelian gauge group we have an example of the nonlinear generalization of 1-functions involved in piecing together a manifold that we referred to earlier.

Explicit procedure

Let us end by being a little more explicit about the Ward construction. Let the Yang–Mills gauge group be $\mathscr{G}$, which we take to be a complex matrix group ($\mathscr{G} \subset GL(n, \mathbb{C})$). To build $\mathscr{C}$, choose some suitable covering $\{\mathscr{U}_i\}$ of $\mathscr{Y}$ and choose, for each overlap $\mathscr{U}_i \cap \mathscr{U}_j$, an $n \times n$ matrix $\mathbf{F}_{ij}$ of twistor functions, all homogeneous of degree zero, such that

$$\mathbf{F}_{ij}(Z^\alpha) \subset \mathscr{G}, \tag{6.10.71}$$

$$\mathbf{F}_{ij} = \mathbf{F}_{ji}{}^{-1}, \tag{6.10.72}$$

and, on each triple overlap,

$$\mathbf{F}_{ij}\mathbf{F}_{jk} = \mathbf{F}_{ik} \quad \text{(no sum)}. \tag{6.10.73}$$

(Equation (6.10.73) should, strictly speaking, be written $(\mathbf{F}_{ij}|_k)(\mathbf{F}_{jk}|_i) = \mathbf{F}_{ik}|_j$, and similarly for some subsequent formulae. Also, in the notation of §5.4, we should write $\mathbf{F}_{ij}$ as $\mathbf{F}_{ij\Theta}{}^{\Psi}$, but for simplicity we use a different (matrix) script for the kernel and omit the Yang–Mills indices – which are all numerical here, since we are concerned with an explicit construction.) (Equations (6.10.72) and (6.10.73) are the nonlinear versions of (6.10.57) and (6.10.58).)

Now write

$$\mathbf{G}_{jk}(r^a, \pi_{A'}) = \mathbf{F}_{jk}(\mathrm{i}r^{AA'}\pi_{A'}, \pi_{A'}). \tag{6.10.74}$$

It follows from general theory (which requires the 'stability' condition referred to earlier) that we can always 'split' $\mathbf{G}_{jk}$ as follows:

$$\mathbf{G}_{jk}(r^a, \pi_{A'}) = \mathbf{H}_j(r^a, \pi_{A'})[\mathbf{H}_k(r^a, \pi_{A'})]^{-1}, \tag{6.10.75}$$

where $\mathbf{H}_j \in \mathscr{G}$ is homogeneous of degree 0 in $\pi_{A'}$ and, given $\mathbf{R}$, is defined on $\mathscr{U}_j \cap \mathbf{R}$. (This is the nonlinear analogue of a condition like (6.10.59), stating that a cocycle is a coboundary.) In practice, to achieve the actual splitting (6.10.75) may be very difficult: in this sense, the Ward construction is not really 'explicit'. The splitting is not unique and we always have the 'gauge freedom'

$$\mathbf{H}_j(r^a, \pi_{A'}) \mapsto \mathbf{H}_j(r^a, \pi_{A'})\Lambda(r^a), \tag{6.10.76}$$

where $\Lambda \in \mathscr{G}$ is holomorphic in r^a.

By (6.10.74) (*cf.* (6.10.6)), $\pi^{A'}\nabla_{AA'}$ annihilates $\mathbf{G}_{jk}$. So applying this operation to (6.10.75), we get, on each overlap,

$$\mathbf{H}_j^{-1}\pi^{A'}\nabla_{AA'}\mathbf{H}_j = \mathbf{H}_k^{-1}\pi^{A'}\nabla_{AA'}\mathbf{H}_k \quad \text{(no sum of j, k)}. \tag{6.10.77}$$

On each overlap the quantities (6.10.77) agree, so we can piece them all together to obtain a ($\mathfrak{S}_A$-valued) matrix $\mathbf{\Theta}_A$ which is holomorphic and homogeneous of degree unity in $\pi_{A'}$ for each r^a. This piecing together of the $\mathscr{U}_i \cap \mathbf{R}$ provides a quantity which is defined globally in $\pi_{A'}$ (since r^a is to be the position vector of a point R whose entire image $\mathbf{R}$ in $\mathbb{PT}$ lies in the required region $\mathscr{Y}$). It follows that

$$\mathbf{\Theta}_A(r^a, \pi_{A'}) = \mathrm{i}\pi^{A'}\mathbf{\Phi}_{AA'}(r^a) \tag{6.10.78}$$

for some space–time field $\mathbf{\Phi}_a(r^b)$. Operating on (6.10.77) with $\pi^{B'}\nabla^A_{B'}$, we obtain

$$\nabla^A_{(B'}\mathbf{\Phi}_{A')A} + \mathrm{i}\mathbf{\Phi}^A_{(B'}\mathbf{\Phi}_{A')A} = 0. \tag{6.10.79}$$

We can now put back the Yang–Mills indices, so $\mathbf{\Theta}_A$ becomes $\Theta_{A\Theta}{}^{\Psi}$ and $\mathbf{\Phi}_a$ becomes $\Phi_{a\Theta}{}^{\Psi}$ satisfying

$$-\chi_{A'B'\Theta}{}^{\Psi} = \nabla^A_{(B'}\Phi_{A')A\Theta}{}^{\Psi} + \mathrm{i}\Phi^{A\Omega}_{\Theta(B'}\Phi_{A')A\Omega}{}^{\Psi} = 0, \tag{6.10.80}$$

which is the required anti-self-dual condition, by (5.5.41)(2). Under (6.10.76) we have

$$\Phi_a \mapsto \Lambda^{-1}\Phi_a\Lambda - i\Lambda^{-1}\nabla_a\Lambda, \tag{6.10.81}$$

i.e., the usual gauge freedom for Yang–Mills fields, (5.5.25). Clearly, self-dual (as opposed to anti-self-dual) fields can be constructed by an entirely analogous method, using $\mathbb{PT}_{\bullet}$ instead of $\mathbb{PT}^{\bullet}$. (General Yang–Mills fields – neither self-dual nor anti-self-dual – have been treated by a twistor-type method by Isenberg, Yasskin, and Green 1978 and by Witten 1978. Their procedure is not, however, so explicit or useful as the Ward construction.)

Note that in the Ward construction the local 'field' information in the space time description is coded in the *global* structure of the twistor description, whereas there is no local (differential) information in the twistor description. Thus whatever specific Yang–Mills field is coded into the bundle $\mathscr{C}$, the portion over a small enough, but finite, region of $\mathbb{PT}^{\bullet}$ is always the same, assuming the group $\mathscr{G}$ remains unchanged. This way in which local space–time field equations tend to 'evaporate' into global holomorphic structure is a characteristic (and somewhat remarkable) feature of twistor descriptions.

This is even more striking with the twistor construction for (anti-)self-dual solutions of Einstein's equations (the 'non-linear graviton') which we briefly outline next (Penrose 1976*a*; *cf.* Ward 1980*b* for non-zero cosmological constant). Let $\mathscr{M}$ be a complex space–time (p. 127) with anti-self-dual Weyl curvature ($\tilde{\Psi}_{A'B'C'D'} = 0$), ensuring local existence of a complex 3-parameter family of α-surfaces – complex 2-surfaces with tangent vectors $\lambda^A\pi^{A'}$ for fixed $\pi^{A'}$ and λ^A. The α-surfaces give the points of curved projective twistor space $\mathbb{P}\mathscr{T}$; and when scaled by covariantly constant $\pi_{A'}(\pi^{B'}\nabla_{BB'}\pi_{A'} = 0)$, the points of $\mathscr{T}$. Einstein's $R_{ab} = 0$ and self-duality yield integrability for parallelism of primed spinors ($[\nabla_a, \nabla_b]\pi_{A'} = 0$), providing a projection $\Pi: \mathscr{T} \to \mathbb{S}_{A'} - \{0\}$ ($\mathbb{S}_{A'}$ being the space of constant $\pi_{A'}$s). Π is locally determined by the simple (p. 14) 2-form $\tau(=\frac{1}{2}\text{`}\varepsilon^{A'B'}\pi_{A'} \wedge \pi_{B'}\text{'})$. There is also a volume 4-form $\boldsymbol{\varepsilon}(=\frac{1}{24}\text{`}\varepsilon_{\alpha\beta\gamma\delta}\mathrm{d}Z^\alpha \wedge \mathrm{d}Z^\beta \wedge \mathrm{d}Z^\gamma \wedge \mathrm{d}Z^\delta\text{'})$ and an Euler vector field $\Upsilon(=\text{`}Z^\alpha\partial/\partial Z^\alpha\text{'})$, the integral curves of which give the projection $\mathscr{T} \to \mathbb{P}\mathscr{T}$. There are relations $\pounds_\Upsilon\tau = 2\tau$, $\pounds_\Upsilon\varepsilon = 4\varepsilon$ defining respective homogeneity degrees 2, 4 for $\tau, \boldsymbol{\varepsilon}$. To reconstruct $\mathscr{M}$ from $\mathscr{T}$, we identify $\mathscr{M}$'s points as *cross-sections of the fibration* defined by Π, two points in $\mathscr{M}$ being null separated iff their cross-sections meet. The forms $\tau, \boldsymbol{\varepsilon}$ serve to fix the metric of $\mathscr{M}$. (Even without $\tau, \boldsymbol{\varepsilon}, \Upsilon$ or Π the construction, applied to $\mathbb{P}\mathscr{T}$, yields the general half *conformally* flat $\mathscr{M}$.) Infinitesimally, $\mathscr{T}$ is constructed from regions $\mathscr{U}_i \subset \mathbb{T}$ according to $Z^\alpha_{[i\,j]} = \varepsilon I^{\alpha\beta}\partial f_{ij}/\partial Z^\beta_j$, where ε is infinitesimal, $Z^\alpha_i \in \mathscr{U}_j$ and f_{ij} defines a 1-function of homogeneity 2.

7

Null congruences

7.1 Null congruences and spin-coefficients

Congruences of null curves in space–time – referred to here as *null congruences* – and especially of *rays* (geodetic* null curves), play an important part in electromagnetic and gravitational radiation theory and in the construction of exact solutions of Einstein's equations. We recall that a congruence is a family of curves, surfaces, etc., with the property that precisely *one* member of the family passes through each point of a given domain of the space under consideration. (The tangent-space elements to a congruence constitute what is known as a *foliation cf.* Hermann 1968.) In fact, all calculations in this chapter are *local* in space–time, so it will not matter if certain congruences globally violate this one-point one-member condition. The null congruences one encounters are frequently many-sheeted globally, in the sense that as one moves continuously from a point of the space–time and returns to that point, one may find that the associated line of the congruence has shifted; but such behaviour will not affect our local considerations. Moreover, there are likely to be specific points (such as branch loci of the congruence, or 'source' world-lines from each of whose points many rays diverge) which have to be regarded as singularities of the congruence and must lie outside the domain of interest. We shall here study some of the geometric properties of null congruences, many of which have significant relations to physical properties. Considerable use will be made of the spin-coefficient formalism, in its original (and also compacted) form.

A line congruence $\mathscr{C}$ may be specified by giving a vector field l^a on a space–time $\mathscr{M}$ (or on an open subset of $\mathscr{M}$), the integral curves μ of l^a defining $\mathscr{C}$. In other words, the vectors l^a are tangents to the curves μ. Once a suitable parameter u is chosen for each curve of $\mathscr{C}$, the scaling of the vectors l^a may be defined by the relation

$$l^a \nabla_a u = 1. \qquad (7.1.1)$$

* The word 'geodetic' is used here consistently as the adjectival form of the noun 'geodesic'.

Then the derivative with respect to u of a smooth scalar function f defined along the curves $\mu\in\mathscr{C}$ is

$$\frac{\partial f}{\partial u} = l^a\nabla_a f = l(f). \tag{7.1.2}$$

In terms of a coordinate system $(x^{\mathbf{a}})$ on $\mathscr{M}$, for which the curves μ are given by

$$x^{\mathbf{a}} = x^{\mathbf{a}}(u, y^1, y^2, y^3), \tag{7.1.3}$$

where the $y^{\mathbf{i}}(\mathbf{i} = 1, 2, 3)$ distinguish the various curves and u distinguishes the points on a given curve, we have

$$l^{\mathbf{a}} = \frac{\partial x^{\mathbf{a}}}{\partial u}, \quad \frac{\partial f}{\partial u} = \frac{\partial f}{\partial x^{\mathbf{a}}}\frac{\partial x^{\mathbf{a}}}{\partial u} = l^{\mathbf{a}}\nabla_{\mathbf{a}} f. \tag{7.1.4}$$

We shall here investigate only *null* congruences, i.e. those in which

$$l_a l^a = 0. \tag{7.1.5}$$

We assume u to be chosen so that l^a points into the future and never vanishes. With the vector l^a we associate a spin-vector o^A in the standard way,

$$l^a = o^A o^{A'},$$

where the choice of the flag plane of o^A is for the moment left arbitrary.

Null geodesic; interpreting κ and ε

The condition for the congruence to be *geodetic* (i.e., for each of its members to be a geodesic) is that the directions of l^a are propagated parallelly:

$$l^a\nabla_a l^b \propto l^b. \tag{7.1.6}$$

The parameter u is then called *affine* if, in fact,

$$l^a\nabla_a l^b = 0. \tag{7.1.7}$$

Such a parametrization is always possible if (7.1.6) holds. In the first case we have, with $D = l^a\nabla_a$ (*cf.* (4.5.23))

$$Do^A \propto o^A, \tag{7.1.8}$$

while in the second case, and only then, o^A *can* be chosen so that

$$Do^A = 0. \tag{7.1.9}$$

Geometrically this is achieved by arranging the flag planes (in addition to the flagpoles) to be parallel along $\mathscr{C}$ (*cf.* discussion in §§3.2, 4.4 on the geometric interpretation of spin-vectors and of parallel transport of spin-vectors).

Note that the geodetic condition (7.1.8) is equivalent to

$$o^A D o_A = 0, \tag{7.1.10}$$

i.e. (*cf.* (4.5.21))

$$\kappa = 0. \tag{7.1.11}$$

Thus we may regard κ as a measure of the curvature of each curve μ of $\mathscr{C}$. But owing to the spin- and boost-weight behaviour

$$o^A \mapsto \lambda o^A \Rightarrow \kappa \mapsto \lambda^3 \bar{\lambda} \kappa, \tag{7.1.12}$$

the actual value of κ (except $\kappa = 0$) does not have a geometrical meaning for the unscaled curve μ itself. On the other hand, for the *scaled* curve μ (i.e. μ together with a smooth choice of tangent vectors, one at each point; or, equivalently, μ together with a choice of parameter u up to $u \mapsto u + k$ with k constant) the modulus $|\kappa|$ of κ does acquire an invariant meaning; for under (7.1.12) we have, for an arbitrary smooth positive function r on μ where we can take $r = \lambda\bar{\lambda}$,

$$l^a \mapsto r l^a \text{ (or } u \mapsto r^{-1} u\text{) } \textit{implies } |\kappa| \mapsto r^2 |\kappa|. \tag{7.1.13}$$

Scaling μ with l^a or with u prevents rescaling o^A except in a way such that $r = 1$. The need for *choice* of scaling for μ (rather than being given a canonical one) arises, of course, from the null character of μ: there is no natural parameter analogous to proper length or time with respect to which l^a can be normalized. To assign a meaning to $\arg \kappa$ we not only need the scaling for μ but also a choice of flag plane at each point of μ. We shall return to the question of the geometrical meaning of κ and other spin-coefficients shortly.

Suppose for the moment that $\mathscr{C}$ is geodetic. The condition (7.1.8) for this is equivalent, as we have seen, to (7.1.11):

$$o^A o^B o^{A'} \nabla_{AA'} o_B = 0, \tag{7.1.14}$$

and hence to each of the following:

$$o^A o^{A'} \nabla_{AA'} o_B = \varepsilon o_B, \tag{7.1.15}$$

$$o^B o^{A'} \nabla_{AA'} o_B = \rho o_A, \tag{7.1.16}$$

$$o^A o^B \nabla_{AA'} o_B = \sigma o_{A'}, \tag{7.1.17}$$

for some ε, ρ, σ. In fact, these proportionality coefficients are precisely the corresponding spin-coefficients in (4.5.21), (though in the case of σ this is true only if $\chi = o_A \iota^A$ is real). This follows by transvection with ι^B, ι^A, $\iota^{A'}$, respectively. To simplify matters, we shall henceforth assume throughout this discussion, that o^A and ι^A constitute a spin-frame, i.e., that

$$\chi = 1$$

so that the specializations (4.5.29) hold.

Notice that ε, ρ and σ are now defined *without* reference to ι^A, by virtue of the fact that $\kappa = 0$. They refer, therefore to the geometry of the o^A-field alone.

Comparing (7.1.15) with (7.1.9), we see that the necessary and sufficient condition for the geodetic congruence $\mathscr{C}$ to have both affine parametrization and parallelly propagated flag planes is

$$\varepsilon = 0. \tag{7.1.18}$$

The condition for affine parametrization alone is

$$\varepsilon + \bar{\varepsilon} = 0 \tag{7.1.19}$$

(*cf.* (7.1.7), (4.5.21)). The condition for parallelly propagated flag planes alone is

$$\varepsilon - \bar{\varepsilon} = 0; \tag{7.1.20}$$

for this states that by some rescaling of o^A,

$$o^A \mapsto \lambda o^A, \tag{7.1.21}$$

with λ *real*, we can achieve $\varepsilon = o$, while affecting only the extent of the flagpole.

For future reference we collect these results and one other in the following table:

$$\begin{aligned} &\mathscr{C} \textit{ geodetic} \Leftrightarrow \kappa = 0 \\ &\mathscr{C} \textit{ geodetic, u affine} \Leftrightarrow \kappa = 0,\ \varepsilon + \bar{\varepsilon} = 0 \\ &\mathscr{C} \textit{ geodetic, flag planes parallel} \Leftrightarrow \kappa = 0,\ \varepsilon - \bar{\varepsilon} = 0 \\ &Do^A = 0 \Leftrightarrow \kappa = 0, \varepsilon = 0 \\ &D\iota^A = 0 \Leftrightarrow \tau' = 0, \varepsilon = 0. \end{aligned} \tag{7.1.22}$$

The last is obtained by taking dyad components of its LHS and referring to (4.5.21). Also, by the first and (5.6.28): *rays are conformally invariant.*

We remark that in the general case of a geodetic null congruence, equation (7.1.14), together with the discussion of §3.2, tells us that $2\,\mathrm{Re}(\varepsilon)$ measures the rate of proportional increase of the flagpoles of o^A along the congruence while $2\,\mathrm{Im}(\varepsilon)$ measures the rate of rotation of its flag planes in a right-handed sense along the congruence.

The geometrical interpretation of κ for a not necessarily geodetic $\mathscr{C}$ can now be obtained as follows. Owing to the nullity of l^a, the derivative Dl^a is orthogonal to l^a. Thus we have

$$Dl^a = (\varepsilon + \bar{\varepsilon})l^a - \kappa\bar{m}^a - \bar{\kappa}m^a, \tag{7.1.23}$$

since by (4.5.22),

$$m_a Dl^a = \kappa, \quad \bar{m}_a Dl^a = \bar{\kappa}, \quad n_a Dl^a = \varepsilon + \bar{\varepsilon}. \tag{7.1.24}$$

From (7.1.24), incidentally, we find the formula

$$Dl^a Dl_a = -2\kappa\bar{\kappa}. \tag{7.1.25}$$

It can easily be seen that a suitable scaling ($o^A \mapsto \lambda o^A$, $\iota^A \mapsto \lambda^{-1}\iota^A$) of μ will reduce ε to zero. (Alternatively this can be achieved geometrically by choosing ι^A parallel along μ.) So the first term on the right in (7.1.23) has no geometric significance for $\mathscr{C}$. The turning of the curve is therefore measured by $-\kappa\bar{m}^a - \bar{\kappa}m^a$. In fact, we shall find it useful to describe the geometry of $\mathscr{C}$ with reference to the plane Π spanned by the real and imaginary parts of m^a. We can regard this as the Argand plane of a complex variable $2^{\frac{1}{2}}\zeta = X - \mathrm{i}Y$ where X and Y are coordinates in the Minkowski tetrad associated with o^A, ι^A (*cf.* (3.1.20)). The general vector in Π is given by

$$v^a = Xx^a + Yy^a = \zeta\bar{m}^a + \bar{\zeta}m^a. \tag{7.1.26}$$

The vector v^a for which $\zeta = -\kappa$ measures the rate of turning of l^a, with $\varepsilon = 0$ (*cf.* (7.1.23)). Thus the magnitude $2^{\frac{1}{2}}|\zeta| = 2^{\frac{1}{2}}|\kappa|$ of v^a is a measure of the amount of curvature, while $\arg\kappa$ is a measure of the direction of curvature relative to, say x^a, that is, to the flag plane of o^A (*cf.* Fig. 1-17).

Interpreting ρ, σ and τ; abreastness of rays

Unlike ε, the ρ and σ have spin- and boost-weights, such that

$$\rho \mapsto \lambda\bar{\lambda}\rho, \quad \sigma \mapsto \lambda^3\bar{\lambda}^{-1}\sigma$$

under (7.1.21) i.e. (*cf.* (4.12.13) with $\chi = 1$) ρ and σ have respective types $\{1, 1\}$ and $\{3, -1\}$. Thus they cannot be made to vanish by a rescaling. For a geodetic $\mathscr{C}$ they are independent of the choice of ι^A, as we have seen. In fact, we can prove from the definitions that, if $Dl^a = 0$, then

$$\rho = \tfrac{1}{2}(-\nabla_a l^a \pm \mathrm{i}\sqrt{2\nabla_{[a}l_{b]}\nabla^{[a}l^{b]}}), \tag{7.1.27}$$

$$\sigma\bar{\sigma} = \tfrac{1}{2}(\nabla_{(a}l_{b)}\nabla^{(a}l^{b)} - \tfrac{1}{2}(\nabla_a l^a)^2) \tag{7.1.28}$$

(*cf.*, for example, Robinson 1961). Thus we expect (and shall find) that the vanishing of ρ or of σ corresponds to a geometric property of a geodetic $\mathscr{C}$. For a non-geodetic $\mathscr{C}$, however, any one of ρ, σ, ε can be made to vanish without rescaling, simply by a choice of ι^A. For we then have $o^A o^B o^{A'}\nabla_{AA'}o_B \neq 0$, hence, for example, $o^B o^{A'}\nabla_{AA'}o^B$ is not proportional to o_A, so we need merely choose $\iota_A \propto o^B o^{A'} \nabla_{AA'} o_B$ to ensure $\rho = \iota^A o^B o^{A'} \nabla_{AA'} o_B = 0$. Similarly we can achieve $\sigma = 0$ or $\varepsilon = 0$.

Let us now examine how the lines of $\mathscr{C}$ relate to their neighbours. For this we introduce the concept of a *connecting vector* q^a, which is a vector field defined along a particular ray μ of $\mathscr{C}$; at $P \in \mu$, q^a is the displacement from P to a point P' on a neighbouring ray μ' of $\mathscr{C}$, where P and P' have the same parameter value u. Mathematically this is expressed as the vanishing of the Lie derivative of q^a with respect to l^a, since q^a is 'dragged' along by the vector field l^a (*cf.* (4.3.3)):

$$\pounds_{l} q^a = 0, \quad \text{that is} \quad Dq^a = q^b \nabla_b l^a \tag{7.1.29}$$

on μ. (This does not require q^a to be defined anywhere except on μ. It also does not require l^a to be null, although *we* shall continue to do so.) We can give a simple proof of this formula by using the representation (7.1.3), (7.1.4) of $\mathscr{C}$, in which

$$q^{\mathbf{a}} = \frac{\partial x^{\mathbf{a}}}{\partial y^{\mathbf{i}}} \delta y^{\mathbf{i}} \quad (\delta y^{\mathbf{i}} = \text{constant}). \tag{7.1.30}$$

We have, trivially,

$$\frac{\partial}{\partial u}\left(\frac{\partial x^{\mathbf{a}}}{\partial y^{\mathbf{i}}}\right) = \frac{\partial}{\partial y^{\mathbf{i}}}\left(\frac{\partial x^{\mathbf{a}}}{\partial u}\right).$$

Multiplied by the (constant) $\delta y^{\mathbf{i}}$ and contracted over $\mathbf{i}$, this is equivalent to

$$\frac{\partial q^{\mathbf{a}}}{\partial u} = \frac{\partial l^{\mathbf{a}}}{\partial y^{\mathbf{i}}} \delta y^{\mathbf{i}},$$

which, in turn, is equivalent to (7.1.29)(2), since by the invariance of (4.3.2) under change of (torsion-free) ∇, a Lie bracket is unchanged when passing from coordinate derivative to covariant derivative.

We are primarily concerned with properties of the congruence $\mathscr{C}$ which are 'not too' dependent on the scaling; in fact, we are interested in quantities which simply scale by a factor as $l^a \mapsto r l^a$ or $o^A \mapsto \lambda o^A$. The tensor $\nabla_b l^a$ which governs the propagation of q^a according to (7.1.29), involves quantities that do scale in this way and others that do not. We shall see that we can pick out just those quantities (or their complex conjugates) that do so scale, and therefore belong to the compacted spin-coefficient formalism, by forming the vector

$$S_b = o^A \nabla_b o_A, \tag{7.1.31}$$

which scales according to

$$S_b \mapsto \lambda^2 S_b. \tag{7.1.32}$$

We can also obtain S_b from $\nabla_b l^a$ by

$$o^A \nabla_b l_a = o^A \nabla_b (o_A o_{A'}) = S_b o_{A'}, \qquad (7.1.33)$$

and, in a form only apparently dependent on ι^A, by

$$S_b = o^A \iota^{A'} \nabla_b l_a = m^a \nabla_b l_a. \qquad (7.1.34)$$

The components $o^A o^{A'} \nabla_b l_a$, $o^A \iota^{A'} \nabla_b l_a$, $\iota^A o^{A'} \nabla_b l_a$ of $\nabla_b l_a$ all scale (the first, in fact, vanishes) whereas $\iota^A \iota^{A'} \nabla_b l_a$ does not. Those which scale are expressible in terms of S_b and $\bar{S}_b$. The quantity $S_b v^b$ therefore expresses the scaling part of $\nabla_b l_a$ as we move in the direction v^b. Since the dyad components of S_b are just the spin-coefficients κ, ρ, σ, τ:

$$\kappa = S_{00'}, \quad \rho = S_{10'}, \quad \sigma = S_{01'}, \quad \tau = S_{11'}, \qquad (7.1.35)$$

we conclude that κ, τ, $2^{-\frac{1}{2}}(\sigma + \rho)$, $2^{-\frac{1}{2}}(\sigma - \rho)$ measure, roughly speaking, how the direction of l^a changes as we move along l^a, n^a, x^a, and y^a, respectively. However, S_a is a more 'invariant' indicator of this information than is the collection of spin-coefficients, since S_a does not depend on the arbitrary choice of an ι^A.

The role of S_a in the propagation equation (7.1.29) is obtained by transvecting that equation with o^A:

$$o^A D q_a = o_{A'} S_b q^b. \qquad (7.1.36)$$

A more precise interpretation of κ, ρ, σ and τ is now implicit in (7.1.36). To make this explicit, it is convenient to choose ι^A parallel along μ, i.e. $D\iota^A = 0$, which is equivalent to $\varepsilon = \tau' = 0$ (*cf.* (7.1.22)). Under this assumption, the components of (7.1.36) give

$$Dh = \bar{\kappa}\zeta + \kappa\bar{\zeta} \qquad (7.1.37)$$

$$D\zeta = -\rho\zeta - \sigma\bar{\zeta} - \tau h, \qquad (7.1.38)$$

where h and ζ are defined by the following form of q^a.

$$q^a = g l^a + \zeta \bar{m}^a + \bar{\zeta} m^a + h n^a. \qquad (7.1.39)$$

The propagation of g is of less geometric interest, being essentially the non-scaling part of (7.1.29). It is obtained by transvecting that equation with ι^A; for the sake of completeness we state the result:

$$Dg = (\alpha + \bar{\beta})\zeta + (\bar{\alpha} + \beta)\bar{\zeta} + (\gamma + \bar{\gamma})h$$

(*cf.* (4.5.21), (4.5.29)). We note also the transformation of g, ζ, h under a general change of ι^A, namely $\iota^A \mapsto \lambda^{-1}\iota^A + \omega o^A$ and $o^A \mapsto \lambda o^A$:

$$g \mapsto \lambda^{-1}\bar{\lambda}^{-1} g - \bar{\lambda}^{-1}\omega\zeta - \lambda^{-1}\bar{\omega}\bar{\zeta} + \omega\bar{\omega}h, \quad \zeta \mapsto \lambda\bar{\lambda}^{-1}\zeta - \bar{\omega}\lambda h, \quad h \mapsto \lambda\bar{\lambda}h. \qquad (7.1.40)$$

Equations (7.1.37) and (7.1.38) can be expressed in the following occasionally useful matrix form:

$$D\begin{bmatrix} h \\ \zeta \\ \bar{\zeta} \end{bmatrix} = -\begin{bmatrix} 0 & -\bar{\kappa} & -\kappa \\ \tau & \rho & \sigma \\ \bar{\tau} & \bar{\sigma} & \bar{\rho} \end{bmatrix}\begin{bmatrix} h \\ \zeta \\ \bar{\zeta} \end{bmatrix} \tag{7.1.41}$$

We shall see that (7.1.38) leads to a precise interpretation of ρ and σ whenever $h=0$. However, by (7.1.37), if $\kappa \neq 0$ (and $\zeta \neq 0$) we cannot expect h to remain zero along μ. In any case, we have seen that ρ and σ are independent of ι^A only if $\kappa = 0$, so only in this case can we expect a clear-cut interpretation of ρ and σ. But this *is* the case of greatest interest.

Let us consider, then, a *geodetic* null congruence $\mathscr{C}$, and assume that the dyad is parallelly transported* along each member curve (*cf.* (7.1.22)(4), (5)):

$$\kappa = \varepsilon = \tau' = 0. \tag{7.1.42}$$

Now equation (7.1.37) becomes

$$Dh = 0, \tag{7.1.43}$$

which is equivalent to

$$D(q_a l^a) = 0. \tag{7.1.44}$$

Neighbouring pairs of rays satisfying

$$q_a l^a = 0, \tag{7.1.45}$$

i.e. $h=0$, will be called *abreast*. That this property is independent of the parametrization (as well as being obviously independent of ι^A) can be seen by making the substitution

$$q^a \mapsto q^a + \psi l^a, \tag{7.1.46}$$

which leaves (7.1.45) invariant.

The reason for the term 'abreast' in the above connection is this: if we realize the congruence physically by a cloud of photons, then two abreast rays correspond to the world-lines of two neighbouring photons which in some (and therefore *any*) observer's local 3-space lie in a 2-plane element perpendicular to their paths (i.e. which move 'abreast').

Moreover, *any* two local observers will judge the photons to be at the same distance from each other if and only if the rays are abreast. For if we regard only the *direction* of l^a as given, and the other three tetrad vectors n^a, m^a, $\bar{m}^a$ as free up to the required normalization, then these can be chosen to

* A slight generalization is achieved if we assume only that the flagpole *directions* of o^A and ι^A are parallelly transported, i.e. merely $\kappa = \tau' = 0$. The 'D' equations of this section remain valid if D is replaced by þ throughout (*cf.* (4.12.15)).

define a Minkowski tetrad for an observer with arbitrary velocity. His local 2-space is spanned by m^a, $\bar{m}^a$, $l^a - n^a = 2^{\frac{1}{2}}z^a$ (*cf.* (3.1.20) and Fig. 1-17). The tracks of the rays in this 3-space are in the z^a direction, and the orthogonal 2-plane is spanned by m^a and $\bar{m}^a$. A reparametrization (7.1.46) will achieve $g = 0$ and so, by (7.1.39), the connecting vector lies in this 2-plane and its (invariant) magnitude $(2\zeta\bar{\zeta})^{\frac{1}{2}}$ represents the distance between the photons. Changing the observer amounts to changing ι^A, and we see from (7.1.40) that $\zeta\bar{\zeta}$ is invariant if and only if $h = 0$. This establishes our assertion.

It may be noted that the invariant aspect of 'abreast' photons discussed here is closely related to the 'invisibility of the Lorentz contraction': photographs taken by parallel light at a given event, by differently moving observers, are all identical, (see Terrell 1959, Rindler 1982 and Volume 1, p. 26).

Recall that τ measures the change of l^a in the direction of n^a. However it is not such a pleasant quantity to work with as ρ and σ for a ray congruence. To a large extent this is because of its dependence on ι^A. Only when ρ and σ both vanish does τ acquire a meaning independent of ι^A. This is illustrated by its transformation behaviour (taking $\kappa = 0$) under $\iota^A \mapsto \iota^A + \omega o^A$:

$$\tau \mapsto \tau + \omega\sigma + \bar{\omega}\rho.$$

As we see from (7.1.38), if $\rho = \sigma = 0$ and we consider non-abreast rays, τ (multiplied by the constant 'degree of non-abreastness' h) specifies the propagation of ζ.

We are mainly interested in the relation of a ray μ to those of its neighbours which are abreast with it. We shall study the changing pattern of the intersection of a 'bundle' of rays – all abreast with μ – with the spatial 2-plane Π of m^a and $\bar{m}^a$ which is orthogonal to μ and carried parallelly along μ. Since the transverse distances from μ are the same to all observers. the divergence and shear of this pattern have direct physical meaning; so also has its twist.

Since $h = 0$ is the condition for abreastness (7.1.38) reduces to

$$D\zeta = -\rho\zeta - \sigma\bar{\zeta}. \tag{7.1.47}$$

This equation describes the motion of the intersection point in Π of a neighbouring curve μ' relative to μ (which itself corresponds to $\zeta = 0$) (see Fig. 7-1). To interpret ρ and σ in detail it is convenient to consider the total effect as composed of parts due to $\mathrm{Re}(\rho)$, $\mathrm{Im}(\rho)$, and σ separately. We write

$$\rho = k + it, \quad \sigma = se^{2i\theta}, \tag{7.1.48}$$

where k, t, s, θ are real and $s \geqslant 0$. Setting $t = s = 0$ in (7.1.47), we obtain

$$D\zeta = -k\zeta,$$

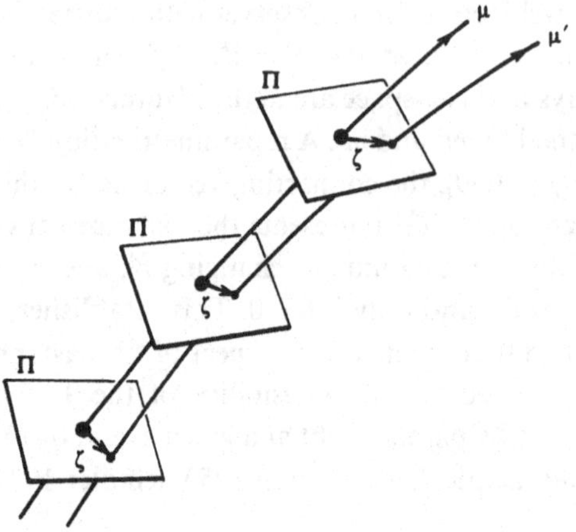

Fig. 7-1. Two abreast neighbouring rays. Their separation, as time progresses, is measured by ζ (an Argand plane coordinate when viewed from *behind*).

showing that $k=\mathrm{Re}(\rho)$ measures the rate of contraction of the simultaneous bundle of rays, i.e. the *convergence* of $\mathscr{C}$. Setting $k=s=0$ we obtain

$$D\zeta = -\mathrm{i}t\zeta,$$

showing that $t=\mathrm{Im}(\rho)$ measures the *twist* (or rotation). Similarly, setting $\rho=0$ gives

$$D\zeta = -se^{2\mathrm{i}\theta}\bar{\zeta},$$

which shows that $D\zeta$ is a real multiple of ζ when $\arg\zeta=\theta, \theta+\pi$, or $\theta\pm\frac{1}{2}\pi$; in the first two cases we get contraction towards the origin, in the other two cases dilation. Thus $|\sigma|$ is a measure of the degree of shearing while $\frac{1}{2}\arg\sigma$ defines the angle that the maximum inward shear makes with the flag-plane direction of o^A (see Fig. 7-2)

By considering propagation along $\mathscr{C}$ of the area

$$\tfrac{1}{2}\mathrm{i}(\zeta_1\bar{\zeta}_2-\zeta_2\bar{\zeta}_1) \tag{7.1.49}$$

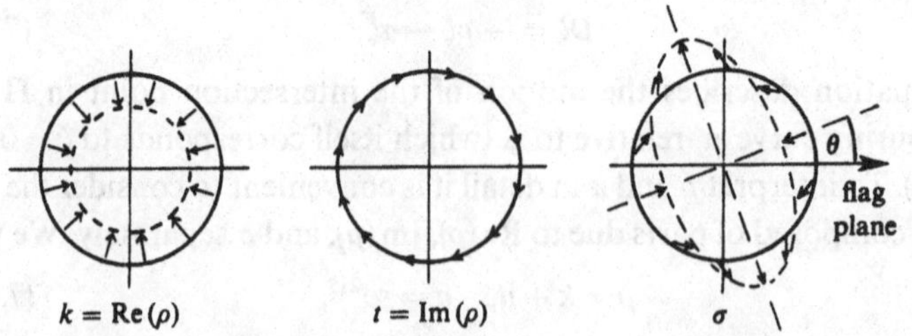

Fig. 7-2. The geometrical interpretation of ρ and σ in terms of behaviour in the ζ-plane. In the pictures, the photons are moving *away* from the viewer.

of a small triangle formed by the points 0, ζ_1, ζ_2 in Π, we find by use of (7.1.47) that, for *any* small area δA,

$$D(\delta A) = -2k\delta A = -(\rho + \bar{\rho})\delta A. \tag{7.1.50}$$

This makes clear the role of k as a measure of convergence and also shows that the local effects of t and σ are area-preserving.

It is occasionally useful to define a 'luminosity parameter' (Bondi, van der Burg and Metzner 1962, Sachs 1962*a*) L along a ray congruence by requiring $L^2 \propto \delta A$. As we see at once from (7.1.50), this implies

$$DL = -kL, \quad \text{i.e. } D\log L = -k. \tag{7.1.51}$$

Comparison with (7.1.47) shows that, in the absence of shear, and only then,

$$L \propto |\zeta|, \tag{7.1.52}$$

so in this case L varies as the length of the connecting vectors between abreast rays.

We also note an alternative formula for ρ which holds if $\varepsilon = 0$:

$$\rho = -o^{A'}\nabla_{AA'}o^A \tag{7.1.53}$$

since by (2.5.54), (4.5.21),

$$o^{A'}\nabla_{AA'}o^A = o^{A'}(o_A\iota^B - \iota_A o^B)\nabla_{BA'}o^A = \varepsilon - \rho.$$

Similarly we find, when $\varepsilon + \bar{\varepsilon} = 0$, that

$$\nabla_a l^a = -\rho - \bar{\rho} = -2k. \tag{7.1.54}$$

Hypersurface orthogonality; null hypersurfaces

The vanishing of twist turns out to have another geometrical significance, being the condition for $\mathscr{C}$ to be *hypersurface-orthogonal*, i.e. proportional to a gradient field:

$$l_a \propto \nabla_a f \quad \textit{for some } f \in \mathfrak{T}. \tag{7.1.55}$$

(Recall that $\mathfrak{T}$ denotes the ring of real scalar fields on $\mathscr{M}$.) To see this, consider first the condition for $\mathscr{C}$ to be actually a gradient:

$$l_a = \nabla_a f \quad \textit{for some } f \in \mathfrak{T}, \tag{7.1.56}$$

namely

$$l_a \textit{ is (locally) a gradient} \Leftrightarrow \nabla_{[a}l_{b]} = 0 \Leftrightarrow \begin{cases} \kappa = 0 \\ \rho = \bar{\rho} \\ \varepsilon = -\bar{\varepsilon} \\ \tau = \bar{\alpha} + \beta \end{cases} \tag{7.1.57}$$

where the last set of equations is obtained – the nullity of l_a being understood – by taking components of $\nabla_{[a}l_{b]}$ with respect to the null tetrad and using (4.5.22). If l_a is merely proportional to a gradient field the last two of equations (7.1.57) need not hold (since they are not invariant under rescaling* of l^a). In fact, we have (Newman and Penrose 1962; and *cf.* (4.14.2))

$$l_a \text{ is (locally) hypersurface-orthogonal} \Leftrightarrow l_{[a}\nabla_b l_{c]} = 0 \Leftrightarrow \begin{cases} \kappa = 0 \\ \rho = \bar{\rho} \end{cases} \tag{7.1.58}$$

The last two equations can be obtained by taking components of $l_{[a}\nabla_b l_{c]}$ directly, but it is simpler to observe (*cf.* (3.4.26) and (3.4.23))

$$\begin{aligned} l_{[a}\nabla_b l_{c]} = 0 &\Leftrightarrow {}^*\{\nabla_{[a} l_{b]}\} l^a = 0 \\ &\Leftrightarrow l^a\{\nabla_{AB'} l_{BA'} - \nabla_{BA'} l_{AB'}\} = 0 \\ &\Leftrightarrow o^A o_B o^{A'} \nabla_{AB'} o_{A'} = o^{A'} o_{B'} o^A \nabla_{BA'} o_{A'}, \end{aligned} \tag{7.1.59}$$

and to take dyad components of the final relation. We can state (7.1.58) as

(7.1.60) PROPOSITION

A null congruence is hypersurface-orthogonal iff it is geodetic and twist-free.

For an alternative characterization, we have

(7.1.61) PROPOSITION

A null congruence is hypersurface-orthogonal iff it is null-hypersurface forming (i.e., iff there exists a one-parameter family of *null hypersurfaces* (hypersurfaces whose normals are null) to which l^a is tangent at each point).

For, if a null congruence l^a is hypersurface-orthogonal, then, by definition, the hypersurfaces $\mathcal{N}$ to which it is orthogonal must be null, and l^a is also tangent to them. Moreover, since the normal direction to a particular $\mathcal{N}$ is unique at each point, and since l^a is the only null direction orthogonal to l^a, l^a is the unique future-pointing null tangent direction at each point of $\mathcal{N}$. These directions in $\mathcal{N}$ have a two-parameter family of integral curves called *generators*: they 'form' $\mathcal{N}$. So, conversely, the generators of a one-

* Thus the statement (7.1.57) does not find natural expression in terms of the compacted spin-coefficient formalism whereas, on the other hand, (7.1.58) does. (Recall the roles of the equations $\rho = \bar{\rho}$ and $\kappa = 0$ encountered in §§4.14 and 5.12 of Volume 1, *cf.* especially (4.14.75), (4.14.76), (5.12.11), (5.12.13).) We note also that *apart from* the final spin-coefficient formulation, (7.1.57) and (7.1.58) hold perfectly well when l_a is not restricted to be null.

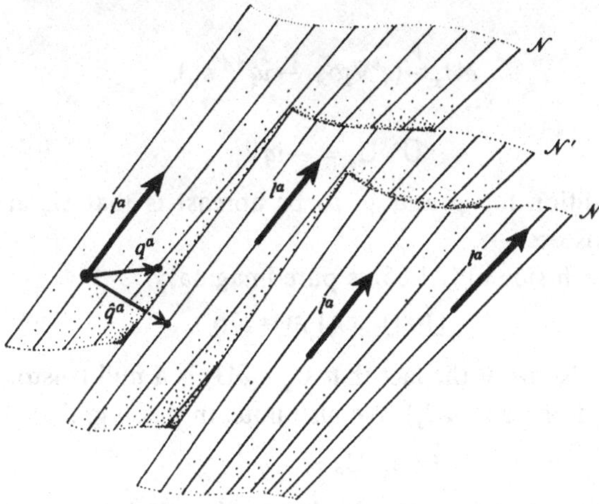

Fig. 7-3. A one-parameter family of null hypersurfaces. These are hypersurfaces whose normals l^a are *null* vectors and which are therefore also tangent vectors. Neighbouring rays within one hypersurface are abreast ($p_a q^a = 0$), and the rotation vanishes.

parameter family of null hypersurfaces $\mathcal{N}$ constitute a three-parameter family of null lines which are hypersurface-orthogonal. Thus the equivalence is established (see Fig. 7-3). Note that the generators of the hypersurfaces $\mathcal{N}$, being null and hypersurface-orthogonal, must be *geodetic* (by (7.1.58)). Since any vector q^a connecting points of neighbouring generators of a given $\mathcal{N}$ lies in $\mathcal{N}$, it must be orthogonal to the l^a, and so any two neighbouring generators of $\mathcal{N}$ are abreast. For this reason, the quantities ρ and σ refer as well to the geometry of $\mathcal{N}$ (with $\rho = \bar{\rho}$, by (7.1.58)) as to the entire ray congruence. We can therefore speak of the *convergence and shear of a single null hypersurface.*

Abreastness in twistor terms

We end this section by pointing out the relevance to twistor theory of the concept of null rays being abreast. Let U_α and $\mathsf{U}_\alpha + \delta\mathsf{U}_\alpha$ be null twistors describing a ray μ and a neighbouring ray μ' belonging to a null congruence in Minkowski space $\mathbb{M}$. Choose $P \in \mu$ and a connecting vector q^a at P which labels a neighbouring point P' on μ', as before; then we have (*cf.* (6.1.22))

$$\mathsf{U}_\alpha \overset{P}{\leftrightarrow} (o_A, 0) \tag{7.1.62}$$

and

$$\mathsf{U}_\alpha + \delta\mathsf{U}_\alpha \overset{P}{\leftrightarrow} (o_A + q^b\nabla_b o_A, \, -\mathrm{i}q^{AA'}o_A), \tag{7.1.63}$$

so that

$$\delta \mathsf{U}_\alpha \overset{P}{\leftrightarrow} (q^b \nabla_b o_{A'}, -\mathrm{i} q^{AA'} o_A), \tag{7.1.64}$$

whence

$$\bar{\mathsf{U}}^\alpha \delta \mathsf{U}_\alpha = -\mathrm{i} q^a l_a. \tag{7.1.65}$$

Thus the condition for μ and μ' to be abreast is that U_α and $\delta\mathsf{U}_\alpha$ be orthogonal twistors.

Note that each side of (7.1.65) is pure imaginary:

$$\bar{\mathsf{U}}^\alpha \delta \mathsf{U}_\alpha + \mathsf{U}_\alpha \delta \bar{\mathsf{U}}^\alpha = 0, \tag{7.1.66}$$

which follows also from the fact that $\mathsf{U}_\alpha + \delta\mathsf{U}_\alpha$ is a null twistor (ignoring terms of second order in $\delta\mathsf{U}_\alpha$). We also note, in passing, that

$$\mathsf{I}^{\alpha\beta} \mathsf{U}_\alpha \delta \mathsf{U}_\beta = -q^b S_b, \tag{7.1.67}$$

with S_b as in (7.1.31) (*cf.* (6.2.25)). Further results relating null congruences and twistor theory will be given in §7.4.

7.2 Null congruences and space–time curvature

We now turn to the discussion of the effect of space–time curvature on the geometry of a null congruence $\mathscr{C}$. For this we examine the second propagation derivative of the connecting vector q^a, obtained by operating with D on (7.1.29). This yields

$$\begin{aligned} D^2 q^d &= D\{q^b \nabla_b l^d\} \\ &= \underset{l}{\nabla}\underset{q}{\nabla} l^d = \{\underset{q}{\nabla}\underset{l}{\nabla} + \underset{[l,q]}{\nabla}\} l^d + R_{abc}{}^d l^a q^b l^c \\ &= q^a \nabla_a (D l^d) + R_{abc}{}^d l^a q^b l^c, \end{aligned} \tag{7.2.1}$$

by (4.3.31), (4.3.33). We shall consider only the case when $\mathscr{C}$ is geodetic, i.e. $Dl^d = 0$, giving the geodesic-deviation (or Jacobi) equation (for rays)

$$D^2 q_d = R_{abcd} l^a q^b l^c. \tag{7.2.2}$$

A field of vectors q^a, defined along a geodesic (not necessarily null) with tangent vector l^a, and satisfying (7.2.2) is referred to as a *Jacobi field* (*cf.* Hawking and Ellis 1973; Hicks 1965). Referring to (7.1.39) and taking components with respect to a parallelly propagated* dyad o^A, ι^A (with

* Again (*cf.* footnote on p. 176) some slight generalization may be achieved if we demand merely that the flagpole *directions* of the dyad be propagated, and use þ in place of D throughout. There is, however, a certain complication: for þ can be applied only to weighted quantities, whereas the affine parameter u, as used here, is not such a quantity. This problem can be circumvented by 'decoupling' u from l^a, so that the equation $Du = 1$ is replaced by $þu = U$, where U is a $\{-1, -1\}$-scalar. The affine nature of the parameter u can be stated as $þ\,U = 0$, i.e. $þ^2 u = 0$. This is the type of approach that we shall adopt in §§9.8–9.10 (and adopted previously in §4.14, Volume 1).

$l^a = o^A o^{A'}$) we get

$$D^2 h = 0 \tag{7.2.3}$$

$$D^2 \zeta = -\Phi_{00}\zeta - \Psi_0\bar{\zeta} - (\Psi_1 + \Phi_{01})h \tag{7.2.4}$$

$$D^2 g = (\Psi_1 + \Phi_{10})\zeta + (\bar{\Psi}_1 + \Phi_{01})\bar{\zeta} + (\Psi_2 + \bar{\Psi}_2 - 2\Lambda + 2\Phi_{11})h, \tag{7.2.5}$$

where we have used (4.11.9) and (4.11.10).

Abreast rays; the Sachs equations

For abreast rays we set $h = 0$ in (7.1.38) and obtain from (7.1.41) the useful (2 × 2) matrix form

$$D\mathbf{z} = -\mathbf{P}\mathbf{z} \tag{7.2.6}$$

where

$$\mathbf{z} = \begin{pmatrix} \zeta \\ \bar{\zeta} \end{pmatrix}, \quad \mathbf{P} = \begin{pmatrix} \rho & \sigma \\ \bar{\sigma} & \bar{\rho} \end{pmatrix}. \tag{7.2.7}$$

Ignoring (7.2.5), since the value of g is of no consequence if we are concerned only with the intersection of the rays with Π (*cf.* before (7.1.47)), we obtain, for the significant content of (7.2.2), equation (7.2.4) in the form

$$D^2\mathbf{z} = -\mathbf{Q}\mathbf{z}, \tag{7.2.8}$$

where

$$\mathbf{Q} = \begin{pmatrix} \Phi_{00} & \Psi_0 \\ \bar{\Psi}_0 & \Phi_{00} \end{pmatrix}. \tag{7.2.9}$$

Differentiating (7.2.6) once more, and using (7.2.8), we get

$$D^2\mathbf{z} = -\mathbf{Q}\mathbf{z} = -(D\mathbf{P})\mathbf{z} - \mathbf{P}D\mathbf{z} = (-D\mathbf{P} + \mathbf{P}^2)\mathbf{z}. \tag{7.2.10}$$

Since this holds for arbitrary ζ and arbitrary complex linear combinations of vectors $\mathbf{z}$ we must have

$$D\mathbf{P} = \mathbf{P}^2 + \mathbf{Q}, \tag{7.2.11}$$

which, when written out in full, becomes the set of *Sachs equations* (Sachs 1961, 1962*a*, *c*):

$$D\rho = \rho^2 + \sigma\bar{\sigma} + \Phi_{00}$$
$$D\sigma = (\rho + \bar{\rho})\sigma + \Psi_0. \tag{7.2.12}$$

(These equations can also be obtained directly from (4.12.32)(*a*), (*b*), with (4.12.15), when the conditions $\kappa = \varepsilon = \tau' = 0$ for a parallel dyad (*cf.* (7.1.22)) are incorporated. We have $\gamma' = -\varepsilon$ since $\chi = 1$, *cf.* (4.5.29).)

The Sachs equations tell us, in effect, that it is Φ_{00} which controls the propagation of convergence, and Ψ_0 which controls the propagation of shear. Two simple consequences of (7.2.12) are the following: first,

$$\Phi_{00}=0 \text{ (i.e. } R_{ab}l^a l^b=0), \rho\equiv 0 \textit{ on } \mu \Rightarrow \sigma\equiv 0 \textit{ on } \mu, \qquad (7.2.13)$$

which states that if $\Phi_{00}=0$ (e.g. in flat or Einstein space) a null congruence cannot shear without somewhere contracting (or expanding) or twisting (in fact $\Phi_{00}\geqslant 0$ also implies this); and second,

$$l^a \textit{ shear-free } (\sigma=0) \Rightarrow l^a \textit{ principal null direction of } \Psi_{ABCD} \qquad (7.2.14)$$

since $\sigma=0$ implies $\Psi_0=0$ which, by reference to (4.11.6) and (3.5.22), is equivalent to the RHS of (7.2.14).

The twist propagation, on the other hand, does not depend on the curvature,* since the difference between the first of equations (7.2.12) and its complex conjugate reads (*cf.* (7.1.48))

$$Dt=2kt, \qquad (7.2.15)$$

Φ_{00} being real.

If L is a luminosity parameter satisfying (7.1.51), we see that

$$D(L^2t)=0, \qquad (7.2.16)$$

and so L^2t gives a measure of twist which is constant along the rays. Hence we note:

$$t=0 \textit{ at some point on } \mu \Rightarrow t\equiv 0 \textit{ on } \mu \qquad (7.2.17)$$

It is worth mentioning here that, in addition to the geometric quantities h and L^2t which are constant along rays (*cf.* (7.1.43)), there is also a 'symplectic invariant' Σ of *two* rays near μ, whose connecting vectors q^a and $\tilde{q}^a$ independently satisfy (7.2.2), namely

$$2\Sigma=\tilde{q}_a Dq^a - q_a D\tilde{q}^a, \qquad (7.2.18)$$

the constancy of which along μ (Lagrange identity) is a consequence of the interchange symmetry of R_{abcd} (*cf.* Hicks 1965). Considering abreast rays only, we have what is effectively a restatement of (7.2.16):

$$\begin{aligned} 2\Sigma &= -(D\mathbf{z}^*)\tilde{\mathbf{z}}+\mathbf{z}^*D\tilde{\mathbf{z}}=\mathbf{z}^*(\mathbf{P}^*-\mathbf{P})\tilde{\mathbf{z}} \\ &= (\rho-\bar{\rho})(\bar{\tilde{\zeta}}\zeta-\tilde{\zeta}\bar{\zeta})=\text{constant} \end{aligned} \qquad (7.2.19)$$

(*cf.* (7.1.49)), where $*$ denotes Hermitian transpose.

This fact has a number of significant consequences, some of which have

* It is worth remarking, however, that if *torsion* is present (and the concept of 'ray' is defined in terms of the torsion connection; *cf.* §4.2 and the discussion of the E.C.S.K. theory, §4.7, Volume 1) then Φ_{00} need not be real and (7.2.15), (7.2.16) need not hold in general. Thus torsion can induce a rotation effect along rays (and *cf.* Penrose 1983*a* for a possible physical significance of this fact).

importance for twistor theory as will be indicated at the end of §7.4. A somewhat different formulation of the constancy of Σ along μ is provided in Penrose (1966), where a matrix $\mathbf{V}_{12}$ is examined, which relates two bundles of abreast neighbouring rays to μ, this matrix being constant along μ. One consequence is the (non-obvious) fact that the 'luminosity distance' between two points on a (scaled) ray μ is symmetrical between the two points (see Volume 1, p. 399; and Bondi, van der Burg and Metzner 1962, Kristian and Sachs 1966).

The matrix formulation (7.2.11) of the Sachs equations leads to a simple derivation of the behaviour of ρ and σ for ray congruences in flat space–time; the same would also hold in a curved space–time in which $\mathbf{Q} = 0$ for the particular rays under consideration. We wish to solve

$$D\mathbf{P} = \mathbf{P}^2. \tag{7.2.20}$$

This equation is equivalent to

$$\mathbf{P}^{-1}(D\mathbf{P})\mathbf{P}^{-1} = \mathbf{I} \tag{7.2.21}$$

where $\mathbf{I}$ is the (2×2) unit matrix; but

$$0 = D(\mathbf{P}^{-1}\mathbf{P}) = (D\mathbf{P}^{-1})\mathbf{P} + \mathbf{P}^{-1}D\mathbf{P},$$

hence, by (7.2.21),

$$D\mathbf{P}^{-1} = -\mathbf{I},$$

i.e.

$$\mathbf{P}^{-1} = \mathbf{A} - u\mathbf{I}$$

where $\mathbf{A} =$ constant and u is affine along μ with $Du = 1$. Thus

$$\mathbf{P} = (\mathbf{A} - u\mathbf{I})^{-1}. \tag{7.2.22}$$

Writing

$$\mathbf{A} = \begin{pmatrix} \rho_0 & \sigma_0 \\ \bar{\sigma}_0 & \bar{\rho}_0 \end{pmatrix}^{-1}$$

where ρ_0 and σ_0 are the values of ρ and σ at $u = 0$, we get, explicitly,

$$\rho = \frac{\rho_0 - u(\rho_0\bar{\rho}_0 - \sigma_0\bar{\sigma}_0)}{1 - u(\rho_0 + \bar{\rho}_0) + u^2(\rho_0\bar{\rho}_0 - \sigma_0\bar{\sigma}_0)},$$

$$\sigma = \frac{\sigma_0}{1 - u(\rho_0 + \bar{\rho}_0) + u^2(\rho_0\bar{\rho}_0 - \sigma_0\bar{\sigma}_0)}. \tag{7.2.23}$$

Non-abreast rays

The matrices we have used above can be generalized, as in (7.1.41), to include the case of non-abreast rays near μ. ($\mathscr{C}$ is still geodetic, with

$Do^A = D\iota^A = 0$.) For this, we redefine the matrices as follows

$$\mathbf{z} = \begin{bmatrix} h \\ \zeta \\ \bar{\zeta} \end{bmatrix}, \quad \mathbf{P} = \begin{bmatrix} 0 & 0 & 0 \\ \tau & \rho & \sigma \\ \bar{\tau} & \bar{\sigma} & \bar{\rho} \end{bmatrix},$$

$$\mathbf{Q} = \begin{bmatrix} 0 & 0 & 0 \\ \Psi_1 + \Phi_{01} & \Phi_{00} & \Psi_0 \\ \bar{\Psi}_1 + \Phi_{10} & \bar{\Psi}_0 & \Phi_{00} \end{bmatrix}, \tag{7.2.24}$$

and, by (7.1.41), (7.2.4) and (7.2.5), we again obtain (7.2.6) and (7.2.8), as before. The same calculation (7.2.10) leads us back to the matrix form (7.2.11) of the Sachs equations, giving us, in addition to the original Sachs formulae (7.2.12), also the τ equation

$$D\tau = \tau\rho + \bar{\tau}\sigma + \Psi_1 + \Phi_{01}. \tag{7.2.25}$$

This is, in fact, (4.12.32)(*c*) with (4.12.15) and the specialization $\kappa = \varepsilon(= -\gamma') = \tau' = 0$ (*cf.* (7.1.22), (4.5.29)).

The flat-space solution (7.2.22) of the equation (7.2.20) will not now work, because the $\mathbf{P}$ of (7.2.24) is a singular matrix. However, an alternative procedure, valid in $\mathbb{M}$ (or in any curved space–time $\mathcal{M}$ for which Ψ_0, Ψ_1, Φ_{00}, Φ_0, all vanish along the ray μ under consideration), is as follows. Consider any three independent solutions of the equation (7.2.6), but where the matrices appearing are the *three*-rowed ones of (7.2.24), and consider the (3×3) matrix whose columns are these solutions:

$$\mathbf{Z} = \begin{bmatrix} h_1 & h_2 & h_3 \\ \zeta_1 & \zeta_2 & \zeta_3 \\ \bar{\zeta}_1 & \bar{\zeta}_2 & \bar{\zeta}_3 \end{bmatrix}. \tag{7.2.26}$$

We have, by (7.2.8),

$$D^2\mathbf{Z} = 0 \tag{7.2.27}$$

so $\mathbf{Z}$ is linear in u:

$$\mathbf{Z} = \mathbf{Z}_1 u + \mathbf{Z}_2 \tag{7.2.28}$$

where $\mathbf{Z}_1$ and $\mathbf{Z}_2$ are constant (3×3) matrices. Hence, applying (7.2.6) we get

$$\mathbf{Z}_1 = -\mathbf{P}(\mathbf{Z}_1 u + \mathbf{Z}_2) \tag{7.2.29}$$

so

$$\mathbf{P} = -\mathbf{Z}_1(\mathbf{Z}_1 u + \mathbf{Z}_2)^{-1}. \tag{7.2.30}$$

We could obtain our formula (7.2.22) for the previous case by putting $\mathbf{Z}_1 = \mathbf{I}$, $\mathbf{Z}_2 = -\mathbf{A}$. Note that there are also many other ways of obtaining that same formula since (7.2.30) is invariant under $\mathbf{Z}_1 \mapsto \mathbf{Z}_1\mathbf{T}$, $\mathbf{Z}_2 \mapsto \mathbf{Z}_2\mathbf{T}$, with any non-singular (3×3) matrix $\mathbf{T}$. In our present case, $\mathbf{Z}_1$ must have zeros for its first row – for consistency between (7.2.24)(2) and (7.2.29) – but it is not restricted in any other way. We choose $\mathbf{Z}_2 = -\mathbf{I}$ instead, and take

$$\mathbf{Z}_1 = \begin{bmatrix} 0 & 0 & 0 \\ \tau_0 & \rho_0 & \sigma_0 \\ \bar{\tau}_0 & \bar{\sigma}_0 & \bar{\rho}_0 \end{bmatrix}, \qquad (7.2.31)$$

giving the general solution of (7.2.20), with (7.2.24)(2), in the form

$$\mathbf{P} = \mathbf{Z}_1(\mathbf{I} - u\mathbf{Z}_1)^{-1}. \qquad (7.2.32)$$

Description in terms of Minkowski 3-space

We now return to (7.2.11) in a general curved space–time $\mathcal{M}$ where, for simplicity, we consider abreast rays only. The matrices are again (2×2), with $\mathbf{Q}$ arbitrary (2×2) Hermitian. Let us further simplify to the case of non-twisting (i.e. null-hypersurface forming) rays so that $\rho = \bar{\rho}$, whence $\mathbf{P}$ is also Hermitian. Following a suggestion of N.F. Ross, we consider the three-dimensional 'Minkowski' metric

$$\mathrm{d}S^2 = \mathrm{d}\rho^2 - \mathrm{d}\sigma \mathrm{d}\bar{\sigma} \qquad (7.2.33)$$

on the space $\mathscr{P}$ whose points are labelled by such matrices $\mathbf{P}$, where we are really interested only in the *conformal*, i.e. null-cone, structure of $\mathscr{P}$ that is provided by (7.2.33). Two points of $\mathscr{P}$ are null separated (i.e. joined by a 'light ray' in $\mathscr{P}$ – a null geodesic with respect to the metric (7.2.33)) if and only if the matrices $\mathbf{P}_1$ and $\mathbf{P}_2$ satisfy

$$(\rho_1 - \rho_2)^2 - (\sigma_1 - \sigma_2)(\bar{\sigma}_1 - \bar{\sigma}_2) = 0 \qquad (7.2.34)$$

i.e.

$$\det(\mathbf{P}_1 - \mathbf{P}_2) = 0 \qquad (7.2.35)$$

i.e.

$$\mathbf{P}_1\mathbf{z} = \mathbf{P}_2\mathbf{z} \qquad (7.2.36)$$

for some non-zero (2×1) complex matrix $\mathbf{z}$. This means that the two bundles of abreast rays neighbouring μ, described by $\mathbf{P}_1$ and $\mathbf{P}_2$ respectively, have a ray (other than μ itself) in common. Thus the conformal metric (equivalent to the light-ray structure) defined by (7.2.33) has an invariant geometrical meaning independent of the particular choice of point P on μ at which the values of ρ and σ are being determined. As P is moved along μ the quantities ρ and σ defined at P will change, but we are to regard the space $\mathscr{P}$ as being fixed, the different elements of $\mathscr{P}$ representing the different possible bundles of rays in the neighbourhood of μ, non-twisting and abreast with μ. The various choices of P provide various choices of 'coördinates' ρ and σ for the space $\mathscr{P}$. For each such choice we get a metric (7.2.33). These will differ as P is moved but, by the preceding discussion, the metrics must be conformally related to one another.

Indeed, this conformal property may be verified directly by considering the Sachs equations (7.2.12) applied to dS^2:

$$\begin{aligned} D(dS^2) &= D(d\rho^2 - d\sigma d\bar{\sigma}) \\ &= 2d\rho D d\rho - d\sigma D d\bar{\sigma} - d\bar{\sigma} D d\sigma \\ &= 2d\rho d(\rho^2 + \sigma\bar{\sigma} + \Phi_{00}) - d\sigma d(2\rho\bar{\sigma} + \bar{\Psi}_0) - d\bar{\sigma} d(2\rho\sigma + \Psi_0) \\ &= 4\rho(d\rho^2 - d\sigma d\bar{\sigma}) = 4\rho dS^2. \end{aligned} \tag{7.2.37}$$

Here 'd' refers to changes in the choice of ray lying abreast with μ and not to changes in position of the point P (the latter variation being the role of D). Thus we set $d\Psi_0 = 0 = d\Phi_{00}$ in the calculation (7.2.37) to obtain our result. (It may be remarked that the calculation also works if the condition $\rho = \bar{\rho}$, for absence of twist, is dropped and the metric (7.2.33) is replaced by $dS^2 = d\rho d\bar{\rho} - d\sigma d\bar{\sigma}$, the result now being $D(dS^2) = 2(\rho + \bar{\rho})dS^2$. Thus, for abreast rays with twist allowed, we have a four-dimensional space in place of $\mathscr{P}$ with a conformal metric of signature $(++--)$ in place of the 'Minkowskian' $(+--)$ one for $\mathscr{P}$.)

An advantage of our present point of view (pointed out by Ross) is that we can allow those positions of P for which ρ and (perhaps) σ become infinite (caustic points), such as the rays neighbouring μ at the vertex of a light cone containing μ. For this, we must consider the *conformal compactification* $\mathscr{P}^\#$ of $\mathscr{P}$, following the procedure that we shall describe in §9.2 (*cf.* footnote on p. 299) for obtaining $\mathbb{M}^\#$ from $\mathbb{M}$. Indeed, the conformal rescalings that apply to $\mathscr{P}$, as P moves along μ really refer to $\mathscr{P}^\#$ rather than to $\mathscr{P}$. This is regarding the transformations on $\mathscr{P}^\#$ as *passive* i.e. changing from one ρ, σ-coordinate system to another on $\mathscr{P}^\#$, neither of which would quite cover the whole of $\mathscr{P}^\#$.

The point P itself will now have an *interpretation within* $\mathscr{P}$ as the point of $\mathscr{P}$ which represents the rays, neighbouring μ, generating the *light cone of* P in $\mathscr{M}$. As P moves along μ, its image in $\mathscr{P}$ moves smoothly along a curve which is timelike with respect to the metric dS^2.

Many interesting results can be derived in this connection, but the matter will not be pursued further here. It is, however, worth pointing out that $\mathscr{P}^\#$ can also be obtained in another way as the space of *Lagrangian 2-planes* in the four-dimensional symplectic vector space representing the possible neighbouring rays to μ which lie abreast with μ. The symplectic form on this space is precisely that given by (7.2.18). (See Woodhouse 1980, for details concerning these concepts – and particularly p. 307 of that work, for a remark which implies the compactified Minkowski 3-space structure of $\mathscr{P}^{\#}$ when it is viewed in this way.)

7.3 Shear-free ray congruences

A null congruence or null vector field (or its related spin-vector field) which is both geodetic and shear-free will here be called *SFR* (both adjective and noun!). The conditions for this are

$$l^a = o^A o^{A'} \text{SFR} \Leftrightarrow \kappa = \sigma = 0 \Leftrightarrow o^A o^B \nabla_{AA'} o_B = 0. \tag{7.3.1}$$

SFRs play an important role in relativity theory. They occur in relation to solutions of the massless field equations and they have a special significance for the theory of twistors. We shall be concerned here with their relation to massless fields. In §7.4 we shall illustrate the role of SFRs in twistor theory, showing, among other things, how twistors can be used to generate the general SFRs in Minkowski space (Kerr's theorem). But we may note at once that every spinor field ω^B that satisfies the twistor equation (6.1.1) automatically satisfies the SFR condition (7.3.1); in other words, Robinson congruences are SFR (*cf.* *p.* 60).

Relations to PNDs; totally null complex 2-surfaces

We now note two preliminary results about SFRs. As a restatement of (7.2.14) we have

$$l^a \text{ SFR} \Rightarrow l^a \textit{ principal null vector of } \Psi_{ABCD}, \tag{7.3.2}$$

i.e. $\Psi_0 = 0$, which is, of course, evident from (7.2.12). If, in addition, $\Phi_{00} = 0$ (i.e. $R_{ab} l^a l^b = 0$), as, for example, in flat space or in an Einstein space, the calculation leading to (7.2.23) applies. The result, with $\sigma = 0$ and $\rho = k + it$, can be written

$$k = \frac{k_0 - u(k_0^2 + t_0^2)}{1 - 2k_0 u + u^2(k_0^2 + t_0^2)}, \quad t = \frac{t_0}{1 - 2k_0 u + u^2(k_0^2 + t_0^2)}. \tag{7.3.3}$$

In particular, we observe that, if l^a is SFR,

$$\rho = 0 \textit{ at some point on } \mu \Rightarrow \rho \equiv 0 \textit{ on } \mu. \tag{7.3.4}$$

Recall that a massless field of spin $\frac{1}{2}n$ is described by a symmetric spinor $\phi_{AB\ldots L}$ of valence $\left[\begin{smallmatrix} 0 & 0 \\ n & 0 \end{smallmatrix}\right]$, so that the canonical decomposition (3.5.18) applies:

$$\phi_{AB\ldots L} = \alpha_{(A}\beta_B \ldots \lambda_{L)}. \tag{7.3.5}$$

If two or more of the principal null directions (PNDs) of $\phi_{\ldots}$ (i.e. flagpole directions of $\alpha_A, \ldots, \lambda_L$) coincide, then $\phi_{\ldots}$ is called *algebraically special* ('along' the repeated PNDs). If all n PNDs coincide (and $n > 1$) the field is called *null* ('along' the coincident PNDs). Recall, from (3.5.29), that in the case of an electromagnetic field, nullity is equivalent to the vanishing of its complex scalar invariant:

$$\varphi_{AB}\varphi^{AB} = 0 \tag{7.3.6}$$

(i.e. of its two real scalar invariants). The corresponding condition in the case of spin-2 will be given in (8.6.3). PNDs of Ψ_{ABCD} are called *gravitational principal null directions*, or 'GPNDs'.

From (3.5.26) we find the following conditions:

$$\phi_{AB\ldots L} \text{ is algebraically special along } \xi^A\ (\neq 0)$$
$$\Leftrightarrow \phi_{AB\ldots L}\xi^B \ldots \xi^L = 0 \tag{7.3.7}$$
$$\phi_{AB\ldots L} (n > 1) \text{ is null along } \xi^A \Leftrightarrow \phi_{AB\ldots L}\xi^A = 0. \tag{7.3.8}$$

Algebraically special and null fields are closely related to SFRs as the following two propositions show. We assume that $\phi_{A\ldots L}$ satisfies the massless free-field equation (4.14.42).

(7.3.9) PROPOSITION

If $\nabla^{AA'}\phi_{AB\ldots L} = 0$ *and* $\phi_{AB\ldots L}$ *is algebraically special along* l^a *then (in regions where* $\phi_{A\ldots L}$ *does not vanish)* l^a *is SFR.*

Proof: Consider an open region where $l^a = \xi^A \bar{\xi}^{A'}$ is a k-fold PND of $\phi_{\ldots}$. (We shall ignore the boundaries between such regions; the SFR condition must apply there also, by continuity.) By (3.5.26) we have

$$\phi_{A\ldots FGH\ldots L}\xi^G\xi^H \ldots \xi^L = 0, \tag{7.3.10}$$

where there are $(n-k+1)$ occurrences of ξ, whereas (*cf.* (3.5.24))

$$\phi_{A\ldots FGH\ldots L}\xi^H \ldots \xi^L = \nu\xi_A \ldots \xi_F\xi_G \tag{7.3.11}$$

with $\nu \neq 0$, there now being $(n-k)$ occurrences of ξ on the left and k on the right. Differentiating (7.3.10) and contracting over F (which is possible since $k \geqslant 2$) we get

$$\begin{aligned} 0 &= \phi_{A\ldots FGH\ldots L}\nabla^{FF'}(\xi^G\xi^H \ldots \xi^L) \\ &= (n-k+1)\phi_{A\ldots FGH\ldots L}(\nabla^{FF'}\xi^G)\xi^H \ldots \xi^L \\ &= \nu(n-k+1)\xi_A \ldots \xi_F\xi_G\nabla^{FF'}\xi^G \end{aligned}$$

by (7.3.11). Since $\nu \neq 0$, $n \geqslant k$, and $\xi_A \neq 0$, we get

$$\xi_F\xi_G\nabla^{FF'}\xi^G = 0, \tag{7.3.12}$$

which, on comparison with (7.3.1), yields the required result (*cf.* also Mariot 1954, Lichnerowicz 1958, Jordan, Ehlers and Sachs 1961).

We shall see that the converse to (7.3.9) is true for the case $n=2$ (*Robinson's theorem, cf.* Robinson 1961: l^a SFR $\Rightarrow$ such ϕ_{AB} exists); but when $n>2$ it is true only under a certain servere restriction on the curvature. The latter is a result of the Buchdahl–Plebanski relations (5.8.2) and is expressed by the following:

(7.3.13) PROPOSITION

If a massless free field $\phi_{A\ldots L}$ of spin $\frac{1}{2}n>1$ is null along l^a, then the Weyl spinor is algebraically special along l^a (in regions where $\phi_{A\ldots L}\neq 0$).

Proof: By (5.8.2), with $e=0$ and with

$$\phi_{AB\ldots L}=\chi\xi_A\xi_B\cdots\xi_L,$$

we get

$$\xi_A\xi_B\xi_M\xi_{(C}\cdots\xi_K\Psi_{L)}{}^{ABM}=0,$$

whence, by Proposition (3.5.15),

$$\Psi_{ABCD}\xi^B\xi^C\xi^D=0.$$

The result follows from (7.3.7), with $l^a=\xi^A\bar{\xi}^{A'}$.

We can now state the joint converse of (7.3.9) and (7.3.13):

(7.3.14) THEOREM (Robinson 1961, Sommers 1976)

If l^a is SFR and analytic then there is a non-zero solution of Maxwell's equations $\nabla^{AA'}\varphi_{AB}=0$ which is null along l^a; furthermore, if the Weyl spinor is algebraically special along l^a then for each spin $\frac{1}{2}n>1$ there is a non-zero symmetric solution of $\nabla^{AA'}\phi_{A\ldots L}=0$ which is null along l^a. In each case there is a freedom in the solution which can be described by a single holomorphic function of two complex variables.*

Before proceeding to the proof (which follows lines suggested by P. Sommers, after (7.3.22)) it is helpful to establish some lemmas concerning

* For real functions of real variables 'analytic' means Taylor-expandable. For complex functions the term 'analytic' will be used here to indicate that the real and imaginary parts are Taylor-expandable in the real and imaginary parts of the arguments. The term 'holomorphic' is reserved for the usual 'complex analytic'. Whenever we speak of any object being analytic on the manifold $\mathcal{M}$, we shall presume that $\mathcal{M}$ together with its metric g_{ab} are also analytic.

SFRs. They are useful in other contexts and have relevance to the problem of finding exact solutions of Einstein's equations and to the theory of twistors. Our first lemma is in effect due to Robinson (1961) (whereby if $\phi_{A\ldots L}$ satisfies (7.3.14), so does $f(\omega_1, \omega_2)\phi_{A\ldots L}$):

(7.3.15) LEMMA

Let ξ^A be an analytic SFR. Then there exist functionally independent complex scalars ω_1, ω_2 such that

$$\xi^A \nabla_{AA'} \omega_1 = 0 = \xi^A \nabla_{AA'} \omega_2$$

and then every solution ω of $\xi^A \nabla_{AA'} \omega = 0$ is a holomorphic function of ω_1, ω_2.

(Note that, as a useful implication of this result, three convenient coordinates for $\mathcal{M}$ 'tailored' to the SFR can be chosen from among the real and imaginary parts of ω_1, ω_2. These are all constant along the rays of the SFR, and so a fourth coordinate would have to be chosen in another way. Note that the gradient of each of ω_1 and ω_2 has the form $\xi_A \eta_{A'}$, so at each point some linear combination of the gradients equals $\xi_A \bar{\xi}_{A'}$. At the end of the present section we shall show how a specially natural choice of ω_1 can be made when $\mathcal{M}$ is vacuum.)

Proof: Let us first complexify $\mathcal{M}$ by allowing the coordinates in any analytic coordinate neighbourhood to take on complex values. Since all relevant quantities are analytic, this will yield (for small enough coodinate imaginary parts) a complex manifold $\mathbb{C}\mathcal{M}$ with holomorphic metric and holomorphic SFR ξ^A. (Quantities which are already complex are 'complexified' simply by regarding their complex conjugates as independent quantities, *cf.* the discussion before (6.9.1).)

Next we introduce an analytic spinor basis $o^{A'}$, $\iota^{A'}$ on $\mathcal{M}$ and allow it to be complexified also, so that it becomes holomorphic on $\mathbb{C}\mathcal{M}$. Now consider the holomorphic vector fields

$$U^a = \xi^A o^{A'}, \quad V^a = \xi^A \iota^{A'}. \tag{7.3.16}$$

We wish to show that there exists a two-complex-parameter system of complex 2-surfaces Σ to which the vector fields $\boldsymbol{U}$, $\boldsymbol{V}$ are tangent. For then, taking ω_1 and ω_2 to parametrize the system, and to be constant on each Σ, we have $\boldsymbol{U}(\omega_i) = 0 = \boldsymbol{V}(\omega_i)$, i = 1, 2, whence (7.3.15) follows. The necessary and sufficient condition that $\boldsymbol{U}$ and $\boldsymbol{V}$ be indeed 2-surface forming is that which the following lemma asserts to be true of $\boldsymbol{U}$, $\boldsymbol{V}$ (Kobayashi and Nomizu 1963):

(7.3.17) LEMMA

$[U, V] = \lambda U + \mu V$ for some holomorphic scalar fields λ, μ on $\mathbb{C}\mathscr{M}$.

Proof: We have $[U, V]^{BB'} = \xi^A \nabla_{A0'}(\xi^B \iota^{B'}) - \xi^A \nabla_{A1'}(\xi^B o^{B'})$, so $\xi_B [U, V]^{BB'} = 0$ by the SFR condition. This tells us that $[U, V]^{BB'}$ is of the form $\xi^B v^{B'}$, for some holomorphic $v^{B'}$, and so must be a holomorphic linear combination of U and V, establishing (7.3.17).

Incidentally, we have now established:

(7.3.18) PROPOSITION

The necessary and sufficient condition that an analytic ξ^A be SFR is that all vectors of the form $\xi^A \lambda^{A'}$ (with ξ^A analytically extended to $\mathbb{C}\mathscr{M}$) are tangent to a two-complex-parameter system of complex 2-surfaces in $\mathbb{C}\mathscr{M}$.

This property would seem to be the essence of the SFR condition on ξ^A.
Two more lemmas will be needed in this connection:

(7.3.19) LEMMA

The integrability condition for solution of the simultaneous differential equations $U(x) = a$, $V(x) = b$ (cf. (4.1.40)), a and b being fields on $\mathscr{M}$, is $U(b) - V(a) = \lambda a + \mu b$, where λ, μ are as in (7.3.17)).

Proof: This is a standard result (Kobayashi and Nomizu 1963). The argument is essentially as follows: integrate $U(x) = a$ on some initial curve in Σ; then propagate away from this curve in the direction of V, solving $V(x) = b$. When $U(b) - V(a) = \lambda a + \mu b$, we have

$$\begin{aligned} V(U(x) - a) &= U(V(x)) - \lambda U(x) - \mu V(x) - V(a) \\ &= U(b) - \lambda U(x) - \mu b - V(a) \\ &= -\lambda(U(x) - a), \end{aligned}$$

whence $U(x) = a$ on Σ. (The argument is only local and may not apply to the whole of Σ.)

(7.3.20) LEMMA (Sommers 1976)

The integrability condition for the equation $\xi^A \nabla_{AA'} x = \alpha_{A'}$ to be soluble in (complex) x (for ξ^A analytic SFR) is $\xi^A \xi^B \nabla_A^{A'} \alpha_{A'} = \alpha_{A'} \xi^A \nabla_A^{A'} \xi^B$. The *general solution, given a particular solution x_0, is then $x = x_0 + f(\omega_1, \omega_2)$, where f is holomorphic and ω_1, ω_2 are as in* (7.3.15).

Proof: The equation to be solved, in component form relative to the basis o^A, ι^A, is the same as that of (7.3.19), with $a = \alpha_{0'}$, $b = \alpha_{1'}$. Thus, by (7.3.19), the integrability condition in (7.3.19) is

$$\xi^A\nabla_{A0'}\alpha_{1'} - \xi^A\nabla_{A1'}\alpha_0 = \lambda\alpha_{0'} + \mu\alpha_{1'} \tag{7.3.21}$$

where λ and μ are given by

$$\xi^A\nabla_{A0'}(\xi^B\iota^{B'}) - \xi^A\nabla_{A1'}(\xi^B o^{B'}) = \lambda\xi^B o^{B'} + \mu\xi^B\iota^{B'}. \tag{7.3.22}$$

Transvecting (7.3.22) with $\alpha_{B'}$ and equating to (7.3.21), we find

$$\xi^A\xi^B\{\nabla_{A0'}(\alpha_{B'}\iota^{B'}) - \nabla_{A1'}(\alpha_{B'}o^{B'})\} = \alpha_{B'}\xi^A\{\nabla_{A0'}(\xi^B\iota^{B'}) - \nabla_{A0'}(\xi^B o^{B'})\},$$

which, after some simplification, gives the required integrability condition. For the general solution, we refer to (7.3.15).

We are now in a position to give a proof of (7.3.14). Let $l^a = \xi^A\bar{\xi}^{A'}$ be the given SFR congruence. We require a non-zero solution of the massless equation with $\phi_{A\ldots L} = \chi\xi_A\cdots\xi_L$, i.e.,

$$\begin{aligned} 0 &= \nabla^{AA'}(\chi\xi_A\xi_B\cdots\xi_L) \\ &= (\xi_A\nabla^{AA'}\chi)\xi_B\cdots\xi_L + \chi\{\nabla^{AA'}\xi_A - (n-1)\eta^{A'}\}\xi_B\cdots\xi_L, \end{aligned} \tag{7.3.23}$$

where $\eta^{A'}$ is defined as a proportionality factor in the equation

$$\xi^A\nabla_{AA'}\xi_B = \xi_B\eta_{A'}, \tag{7.3.24}$$

which follows from (7.3.1). Thus the equation to be solved is

$$\xi^A\nabla_{AA'}\log\chi = -\nabla_{AA'}\xi^A - (n-1)\eta_{A'}. \tag{7.3.25}$$

We now substitute the RHS of this equation for $\alpha_{A'}$ in the integrability conditions of Lemma (7.3.20), to check that they are indeed satisfied. This substitution yields, after some suffix manipulation and use of (7.3.24),

$$\xi^A\nabla_{AA'}\{\nabla_D^{A'}\xi^D + (n-1)\eta^{A'}\} = \eta_{A'}\nabla_D^{A'}\xi^D. \tag{7.3.26}$$

For the various terms of this condition we can find the following expressions (P. Sommers):

$$\lambda_B := \xi_B\xi^A\nabla_{AA'}\eta^{A'} = \Psi_{ABCD}\xi^A\xi^C\xi^D, \tag{7.3.27}$$

$$\mu_B := \xi_B\xi^A\nabla_{AA'}\nabla_D^{A'}\xi^D - \xi_B\eta_{A'}\nabla_D^{A'}\xi^D = -\Psi_{ABCD}\xi^A\xi^C\xi^D. \tag{7.3.28}$$

To establish (7.3.27) we have, from (7.3.24) and finally from (4.9.15),

$$\begin{aligned} \lambda_B &= \lambda_B + \xi^A\eta^{A'}\nabla_{AA'}\xi_B = \xi^A\nabla_{AA'}(\xi^C\nabla_C^{A'}\xi_B) \\ &= (\nabla_C^{A'}\xi_B)\xi^C\eta_{A'} + \xi^A\xi^C\nabla_{AA'}\nabla_C^{A'}\xi_B \\ &= 0 + \xi^A\xi^C\Box_{AC}\xi_B = \Psi_{ABCD}\xi^A\xi^C\xi^D. \end{aligned}$$

To establish (7.3.28), we begin by taking a derivative of (7.3.1):

$$0 = \nabla_{BA'}(\xi^A\xi^D\nabla^{A'}_A\xi_D)$$
$$= \xi^A\xi^D\nabla_{BA'}\nabla^{A'}_A\xi_D + 2(\nabla_{BA'}\xi^D)\xi^A\nabla^{A'}_{(A}\xi_{D)}. \qquad (7.3.29)$$

The first term of the last line, by use of (2.5.23) and (4.9.15), can be written

$$\xi^A\xi^D\nabla_{DA'}\nabla^{A'}_A\xi_B - \xi^A\xi_B\nabla_{DA'}\nabla^{A'}_A\xi^D = \Psi_{ABCD}\xi^A\xi^C\xi^D + \xi^A\xi_B\nabla_{AA'}\nabla^{A'}_D\xi^D. \qquad (7.3.30)$$

Let us introduce, analogously to (7.3.24), the quantity $\zeta_{A'}$ by the first of the following equations:

$$\xi_A\zeta_{A'} := \xi^B\nabla_{AA'}\xi_B = \tfrac{1}{2}\xi^B(\nabla_{AA'}\xi_B + \nabla_{BA'}\xi_A + \varepsilon_{AB}\nabla_{DA'}\xi^D)$$
$$= \tfrac{1}{2}\xi_A(\zeta_{A'} + \eta_{A'} - \nabla_{DA'}\xi^D). \qquad (7.3.31)$$

Hence

$$\zeta_{A'} = \eta_{A'} - \nabla_{DA'}\xi^D. \qquad (7.3.32)$$

Then the last term in (7.3.29) becomes

$$(\nabla_{BA'}\xi^D)\xi_D(\eta^{A'} + \zeta^{A'}) = -\xi_B\zeta_{A'}\eta^{A'} = -\xi_B\eta_{A'}\nabla^{A'}_D\xi^D. \qquad (7.3.33)$$

By (7.3.29), we can set the sum of the right sides of (7.3.30) and (7.3.33) equal to zero: this establishes (7.3.28). Now if we use (7.3.27) and (7.3.28), the integrability condition (7.3.26) immediately reduces to

$$(n-2)\Psi_{ABCD}\xi^A\xi^C\xi^D = 0, \qquad (7.3.34)$$

which establishes the main part of Proposition (7.3.14). The final part, namely the freedom in the solution, follows from Lemma (7.3.20): $\log\chi$ (*cf.* (7.3.25)) can have *added* to it any holomorphic function of ω_1 and ω_2, i.e. ψ itself can be *multiplied* by such a function.

The (generalized) Goldberg–Sachs theorem

We next turn to another important result on SFRs, namely the Goldberg–Sachs (1962) theorem. We present it in a generalized form (due to Kundt and Thompson 1962, and Robinson and Schild 1963; our proof is adapted from one given by Sommers):

(7.3.35) PROPOSITION

Of the following three conditions:

(i) ξ^A *is a p-fold GPND* $(2 \leqslant p \leqslant 4)$
(ii) ξ^A *is SFR*
(iii) $\xi^A \ldots \xi^C\nabla^{DD'}\Psi_{ABCD} = 0$ ($(5-p)$ *occurrences of* ξ);

for $p = 2$, 3 *or* 4, (i) & (ii)$\Rightarrow$(iii) *and* (i) & (iii)$\Rightarrow$(ii)
for $p = 2$, (ii) & (iii)$\Rightarrow$(i)$' :=$ (i) *with* $p = 2$ *or* 3 *or* 4.

Note: the original Goldberg–Sachs theorem stated that a nowhere flat vacuum metric has a multiple GPND l^a if and only if l^a is SFR. Evidently this is a particular case of (7.3.35), since (iii) with $p = 2$ is a consequence of the vacuum Bianchi identity $\nabla^{DD'}\Psi_{ABCD} = 0$.

Proof: For specificity, assume $p = 2$. Rewriting (i) as $\xi^A\xi^B\xi^C\Psi_{ABCD} = 0$, taking a derivative and finally using the form $\xi_{(A}\xi_B\alpha_C\beta_{D)}$ (with neither α_A nor β_A proportional to ξ_A) for Ψ_{ABCD}, we have

$$\begin{aligned} 0 &= \nabla^{DD'}(\xi^A\xi^B\xi^C\Psi_{ABCD}) \\ &= \xi^A\xi^B\xi^C\nabla^{DD'}\Psi_{ABCD} + 3\Psi_{ABCD}\xi^A\xi^B\nabla^{DD'}\xi^C \\ &= \xi^A\xi^B\xi^C\nabla^{DD'}\Psi_{ABCD} + \Psi\xi_C\xi_D\nabla^{DD'}\xi^C \end{aligned}$$

where Ψ is a non-zero scalar. By reference to (7.3.1), it is therefore evident that (i) & (ii)$\Rightarrow$(iii) and (i) & (iii)$\Rightarrow$(ii). The proof is entirely similar for the cases $p = 3$ and $p = 4$.

To show that (ii) & (iii) ($p = 2$)$\Rightarrow$(i)$'$, we first note that, by (7.3.2), ξ^A is a GPND, which implies $\xi^A\xi^B\xi^C\Psi_{ABCD} = x\xi_D$, for some x. Our objective is to show $x = 0$. Taking a derivative gives

$$\xi^A\xi^B\xi^C\nabla^{DD'}\Psi_{ABCD} + 3\Psi_{ABCD}\xi^A\xi^B\nabla^{DD'}\xi^C = \xi_D\nabla^{DD'}x + x\nabla^{DD'}\xi_D. \quad (7.3.36)$$

Assuming (iii) ($p = 2$), the first term on the left vanishes. To transform the second, we assume $\Psi_{ABCD} = \xi_{(A}\alpha_B\beta_C\gamma_{D)}$. That term then is

$$\tfrac{3}{4}[\xi_C\alpha_{(A}\beta_B\gamma_{D)} + \xi_D\alpha_{(A}\beta_B\gamma_{C)}]\xi^A\xi^B\nabla^{DD'}\xi^C,$$

which, by reference to (7.3.24) and (7.3.31), is

$$-\tfrac{3}{4}[\xi^A\xi^B\xi^D\zeta^{D'}\alpha_{(A}\beta_B\gamma_{D)} + \xi^A\xi^B\xi^C\eta^{D'}\alpha_{(A}\beta_B\gamma_{C)}] = -3x(\zeta^{D'} + \eta^{D'}).$$

For the last term in (7.3.36), we refer to (7.3.32). Thus (7.3.36) now reads, provided $x \neq 0$,

$$\xi^A\nabla_{AA'}(\log x) = 2\eta_{A'} + 4\zeta_{A'} = 6\eta_{A'} - 4\nabla_{DA'}\xi^D. \quad (7.3.37)$$

To check if a solution $\log x$ of this equation exists we substitute its RHS for $\alpha_{A'}$ into the integrability condition of Lemma (7.3.20). This yields

$$-\xi^A\xi_B\nabla_{AA'}(6\eta^{A'} - 4\nabla^{A'}_D\xi^D) = \xi_B\alpha_{A'}\eta^{A'} = 4\xi_B\eta_{A'}\nabla^{A'}_D\xi^D \quad (7.3.38)$$

and the identities (7.3.27) and (7.3.28) now immediately convert (7.3.38) into $\Psi_{ABCD}\xi^A\xi^B\xi^D = 0$. This implies $x = 0$, contrary to our assumption, and so establishes the desired result.

It is perhaps instructive to give an alternative and direct proof of the

Goldberg–Sachs theorem which uses the compacted spin-coefficient formalism of §4.12. (The proof is essentially that of Newman and Penrose 1962.) For simplicity we deal with the original theorem only, though the generalized theorem can be treated similarly. Thus we prove that, under the assumption of the vacuum Bianchi identity, (i)′⇔(ii). First assume (i) with $p=2$, i.e., (*cf.* (4.11.6)) $\Psi_0=\Psi_1=0$, $\Psi_2\neq 0$. Recalling that in vacuum $\Phi_{ij}=0$ and $\Lambda=0$, the Bianchi identity equation (4.12.36) then shows $\kappa=0$ while* (4.12.39)′, shows $\sigma=0$, i.e., (ii). Similarly, if $p=3$, i.e. $\Psi_0=\Psi_1=\Psi_2=0$, $\Psi_3\neq 0$, (4.12.37) and (4.12.38)′ respectively, yield $\kappa=0$, $\sigma=0$. And when $p=4$, i.e. $\Psi_0=\Psi_1=\Psi_2=\Psi_3=0$, $\Psi_4\neq 0$, the same service is performed by (4.12.38) and (4.12.37)′, respectively.

Now, conversely, assume (ii), i.e. $\kappa=\sigma=0$. By (7.3.2) we know that Ψ_0 will then vanish. We must therefore prove $\Psi_1=0$. Let us first dispense with the case where $\rho\equiv 0$ in an open region: then (4.12.32*d*) immediately yields the desired $\Psi_1=0$. If throughout an open region $\rho\neq 0$ (boundaries between such regions are dealt with by continuity) we can apply a transformation $\iota^A\mapsto\iota^A+\omega o^A$. This leaves κ, σ, unchanged but effects $\tau\mapsto\tau+\bar{\omega}\rho$ and can thus be used to achieve $\tau=0$. Suppose now that $\Psi_1\neq 0$. Equation (4.12.36) then yields

$$Þ\Psi_1=4\rho\Psi_1 \tag{7.3.39}$$

while (4.12.39)′ yields

$$ð\Psi_1=0 \tag{7.3.40}$$

Taking mixed derivations and subtracting the resulting equations, we get

$$(Þð-ðÞ)\Psi_1=0-ð(4\rho\Psi_1)=4(\Psi_1)^2 \tag{7.3.41}$$

after using (7.3.40), (4.12.32*d*). But also (4.12.34) gives an expression for this mixed derivative, namely

$$(Þð-ðÞ)\Psi_1=(\bar{\rho}ð-\bar{\tau}'Þ-2\Psi_1)\Psi_1=(-4\bar{\tau}'\rho-2\Psi_1)\Psi_1$$

by (7.3.39) and (7.3.40), where we use the fact that Ψ_1 has type $\{2,0\}$ (*cf.* (4.12.25)). Equating these two expressions we get $\Psi_1=-\frac{2}{3}\bar{\tau}'\rho$ or $\Psi_1=0$. On the other hand, according to (4.12.32*c*), $\Psi_1=\bar{\tau}'\rho$ $(=-\frac{3}{2}\Psi_1)$. Hence $\Psi_1=0$, which completes our proof.

We may note that the original Goldberg–Sachs theorem applies not only to a vacuum metric but also to any metric conformal to a vacuum metric. For, under a conformal rescaling, $\kappa=\sigma=0$ is preserved, and so is nullity; hence SFRs correspond. Since, moreoever, the Weyl spinor is

* Here, as elsewhere, a *primed* equation number, such as (4.12.39)′, denotes the result of an application of the *priming operation* (4.5.17) to the equation in question.

unchanged, so are its PNDs, and thus the assertion is established.

Under conformal rescaling, (i) and (ii) of the Goldberg–Sachs theorem are invariant. The conformal behaviour of (iii) can be found from the relation

$$\hat{\nabla}^{AA'}\Psi_{ABCD} = \Omega^{-2}\nabla^{AA'}\Psi_{ABCD} + \Omega^{-2}\Upsilon^{AA'}\Psi_{ABCD}, \qquad (7.3.42)$$

which follows from (5.6.15). It shows that (iii) is conformally invariant if and only if (i) holds.

An interesting corollary of the generalized Goldberg–Sachs theorem is the following result due to Robinson and Schild (1963):

(7.3.43) PROPOSITION

The gravitational field due to any Maxwell field with shear-free rays is algebraically special.

One says a Maxwell field has shear-free rays if at least one of the PNDs of φ_{AB} (*cf.* (5.1.39)) constitutes an SFR. Note that Proposition (7.3.9) tells us that all electromagnetic *null* fields have shear-free rays; but non-null fields can also possess this property – e.g. the Reissner–Nordström ('charged Schwarzschild') field. Note also that Proposition (7.3.13) cannot be regarded as covering the present proposition in the null case, since (7.3.13) refers to *test* fields, i.e. to fields which are *not* sources of the Einstein field equations. For proof of the present proposition we simply observe that condition (iii) in (7.3.35) with $p = 2$ is satisfied as a consequence of the Einstein–Maxwell Bianchi identity (5.2.7), the assumption $\varphi_{AB} = \xi_{(A}\alpha_{B)}$, and the condition (7.3.1) on ξ^A.

Kerr coordinates

We end this section by giving a result due to R.P. Kerr (*cf.* Kerr 1963, Kerr and Schild 1965) which is particularly useful for the study of algebraically special exact solutions of Einstein's vacuum equations.

(7.3.44) LEMMA (Kerr)

Let ξ^A be an analytic SFR and suppose $\Psi_{ABCD}\xi^A\xi^B\xi^C = 0$, $\Phi_{ABC'D'}\xi^A\xi^B = 0$. Then there exist complex scalars χ, ω such that $\xi^A\nabla_b\xi_A = \chi\nabla_b\omega$.

Proof: Complexify $\mathcal{M}$ to $\mathbb{C}\mathcal{M}$. We must show that

$$S_b := \xi^A\nabla_b\xi_A \qquad (7.3.45)$$

is (in the complex sense) proportional to a gradient on $\mathbb{C}\mathcal{M}$, the condition for which (*cf.* 7.1.59)) $*(\nabla_{[a}S_{b]})S^b = 0$, i.e., by (3.4.23) (*cf.* (7.1.60)),

$$S^{AB'}\nabla_{[a}S_{b]} = 0 \tag{7.3.46}$$

Now

$$\begin{aligned}\nabla_{[a}S_{b]} &= \xi^C\nabla_{[a}\nabla_{b]}\xi_C + (\nabla_{[a}\xi^C)(\nabla_{b]}\xi_C)\\ &= \tfrac{1}{2}\varepsilon_{A'B'}\mathrm{X}_{ABEC}\xi^E\xi^C + (\nabla_a\xi^C)(\nabla_b\xi_C),\end{aligned} \tag{7.3.47}$$

by (4.9.7) and the given condition on $\Phi_{ABC'D'}$. The SFR condition on ξ^A yields

$$S^{AB'} = \xi^A\zeta^{B'} \tag{7.3.48}$$

(*cf.* (7.3.31)), which, when transvected with (7.3.47), gives

$$S^{AB'}\nabla_{[a}S_{b]} = (\xi^A\nabla_{AA'}\xi^C)(\zeta^{B'}\nabla_{BB'}\xi_C), \tag{7.3.49}$$

the X term going out because of (4.6.35) and the given condition on Ψ_{ABCD}. The RHS (*cf.* (7.3.24)) is

$$\eta_{A'}\xi^C\zeta^{B'}\nabla_{BB'}\xi_C = \eta_{A'}\zeta^{B'}\zeta_{B'}\xi_B = 0$$

and thus (7.3.46) is established, and with it the theorem.

It may be recalled that S_b is the quantity defined in (7.1.31) which measures how the direction of the flagpole of ξ^A varies. Its scaling behaviour under $\xi^A \mapsto \lambda\xi^A$ is $S_b \mapsto \lambda^2 S_b$, so this freedom can be used to reduce χ to unity: $\xi^A\nabla_b\xi_A = \nabla_b\omega$. Note that if the Einstein vacuum equations or the Einstein–Maxwell equations hold (with or without cosmological term), and if in the latter case ξ^A is a PND of the electromagnetic spinor φ_{AB}, then the second of the curvature conditions of the Kerr lemma is satisfied (*cf.* (5.2.6)), as is (iii) in the generalized Goldberg–Sachs theorem (7.3.35), in consequence of the vacuum or the Einstein–Maxwell Bianchi identity (5.2.7) so that each of the remaining two conditions of the Kerr lemma implies the other. The real and imaginary parts of ω give coordinates which are related in a natural way to the structure of such space–times and have proved useful in the generation of exact solutions.* Similarly ξ^A and ζ^B can be used to define a natural spin-frame, in the case when the rays are diverging ($\rho \neq 0$) so that $\bar{\xi}^{B'}\xi^A\nabla_{BB'}\xi_A = \bar{\xi}^{B'}\zeta_{B'}\xi_B \neq 0$.

7.4 SFRs, twistors and ray geometry

There is an intimate relation between SFRs in Minkowski space and homogeneous holomorphic functions of twistors. We shall illustrate this fact

* For example, see Kerr and Schild 1965, 1967, Robinson and Schild 1963, Plebański and Schild 1976, Kramer, Stephani, MacCallum and Herlt 1980.

by establishing various results, the first of which was found by Kerr before the emergence of twistor theory. We give this result in Kerr's original form and then in the twistor version. Then, for the rest of the section we show how twistor ideas can be used generally in curved space–time. As a final result we obtain a form of Kerr's theorem that applies (relative to a hypersurface) in arbitrary space–times.

Original Kerr theorem

Consider Minkowski space $\mathbb{M}$ and introduce spinor coordinates

$$u = x^{00'}, \quad \zeta = x^{01'}, \quad \tilde{\zeta} = x^{10'}, \quad v = x^{11'}, \tag{7.4.1}$$

where the position vector x^a of a point is referred to a *constant* spin-frame o^A, ι^A. The metric therefore has the form

$$ds^2 = 2\,du\,dv - 2\,d\zeta\,d\tilde{\zeta}, \tag{7.4.2}$$

$x^{\mathbf{AA'}}$ being related to a standard Minkowski system $x^{\mathbf{a}}$ by (3.1.31). The notation $\tilde{\zeta}$, rather than $\bar{\zeta}$, is adopted here since we shall presently allow the coordinates to become complexified. Let ξ_A be SFR:

$$\xi^A \xi^B \nabla_{AA'} \xi_B = 0, \tag{7.4.3}$$

and choose the scaling of ξ_A so that

$$\xi_0 = -\xi^1 = 1, \quad \xi_1 = \xi^0 = \lambda. \tag{7.4.4}$$

The complex number λ thus defines the null direction of the SFR at each point, i.e. the flagpole direction of ξ^A. In terms of the spinor coordinates at the general point, this direction is given by the relations*

$$du + \lambda d\tilde{\zeta} = 0 = d\zeta + \lambda dv, \tag{7.4.5}$$

which state $\xi_{\mathbf{A}} dx^{\mathbf{AA'}} = 0$.

We now regard (7.4.3) as an equation on λ, where λ is a function of u, ζ, $\tilde{\zeta}$, v. Substituting (7.4.4) into (7.4.3) we get, since the spin-frame is constant,

$$\lambda \frac{\partial \lambda}{\partial u} = \frac{\partial \lambda}{\partial \tilde{\zeta}}, \quad \lambda \frac{\partial \lambda}{\partial \zeta} = \frac{\partial \lambda}{\partial v} \tag{7.4.6}$$

* In these (and other) equations the symbol 'd' may be interpreted either in 'old fashioned' terms, in which case dx^a (like δx^a) is thought of as a *vector* (or infinitesimal displacement) at the point whose position vector is x^a, or in more 'modern' terms as a (vector-valued) *differential form*. In the former case, an equation $A_a dx^a = 0$ is interpreted as the fact that the vector dx^a lies in the hyperplane element determined by A_a; in the latter, it is the vectors *annihilated* by the 1-form $A_a dx^a$ ($= A_{i_1}$) which have this required property.

Thus, our equation $\xi_A dx^{AA'} = 0$ asserts that the relevant vectors point in the direction of the flagpole of ξ^A.

as the equations to be solved. We can put

$$P = \frac{\partial \lambda}{\partial u}, \quad Q = \frac{\partial \lambda}{\partial \zeta}$$

so that

$$\lambda P = \frac{\partial \lambda}{\partial \bar{\zeta}}, \quad \lambda Q = \frac{\partial \lambda}{\partial v}.$$

then

$$d\lambda = \frac{\partial \lambda}{\partial u} du + \cdots = P(du + \lambda d\bar{\zeta}) + Q(d\zeta + \lambda dv),$$

from which it follows that $d\lambda = 0$ whenever $du + \lambda d\bar{\zeta} = 0 = d\zeta + \lambda dv$. Hence λ is constant whenever $u + \lambda\bar{\zeta}$ and $\zeta + \lambda v$ are both constant. In real variable theory it could then be deduced that a functional relation

$$\psi(\lambda, u + \lambda\bar{\zeta}, \zeta + \lambda v) = 0 \qquad (7.4.7)$$

must exist, where $\psi(\alpha, \beta, \gamma)$ is an arbitrary function of α, β, γ (*cf.* also Cox and Flaherty 1976). However, in the present context this procedure is not strictly applicable since in real Minkowski space only u and v are real variables while ζ and $\bar{\zeta}$ are complex conjugates, and λ is complex also. We can get around this difficulty provided we assume that λ is analytic in u, v and the real and imaginary parts of ζ. For then we can complexify $\mathbb{M}$ to $\mathbb{CM}$ in a procedure corresponding to that of §7.3, allowing u and v to be complex and ζ, $\bar{\zeta}$ to be independent complex variables. If λ is extended to being a holomorphic function of u, ζ, $\bar{\zeta}$, v, the formal method of solution remains valid provided $\psi(\alpha, \beta, \gamma)$ is now taken to be holomorphic. Furthermore, the converse result that (7.4.7) represents the *general* solution of (7.4.6) is also true with these restrictions on λ. Without such restrictions the result would not be quite true. (There are in fact non-analytic SFRs in $\mathbb{M}$, e.g. the non-twisting system of null rays meeting a non-analytic curve, and these cannot be generated in this way except in a certain sense as a limiting case of the construction. Twisting SFRs can also be non-analytic, though in this case there is a sense in which (7.4.8) can still be applied; *cf.* the remarks at the end of this section.) Thus we have:

(7.4.8) THEOREM (Kerr)

The general analytic SFR in $\mathbb{M}$ is obtainable locally by choosing an arbitrary holomorphic function $\psi(\alpha, \beta, \gamma)$ of three complex variables α, β, γ and setting $\psi(\lambda, u + \lambda\bar{\zeta}, \zeta + \lambda v) = 0$. Solving this equation for λ in terms of u, ζ, $\bar{\zeta}$, v gives the direction of the SFR at each point $(u, \zeta, \bar{\zeta}, v)$ of $\mathbb{M}$ via (7.4.5) (u, v real, ζ complex).

Twistor form of the Kerr theorem

We shall now express this result in twistor terms. We define a (dual) twistor W_α by*

$$\mathsf{W}_\alpha \overset{O}{\leftrightarrow} (\xi_A, -\mathrm{i}\xi_A x^{AA'}) \tag{7.4.9}$$

(*cf.* (6.1.22)), so that using (7.4.1) and (7.4.4) we obtain the components of W_α (with respect to the origin $x^a = 0$ and the given spin-frame) in the form

$$\mathsf{W}_\alpha = (1, \lambda, -\mathrm{i}(u + \lambda\bar{\zeta}), -\mathrm{i}(\zeta + \lambda v)). \tag{7.4.10}$$

Note that, by (6.1.26), the expression (7.4.9) ensures that the point P with position vector x^a lies on the β-plane $\mathbb{C}\mathsf{W} \subset \mathbb{CM}$, the locus, in $\mathbb{CM}$ along which the primary part of W_α vanishes (*cf.* §6.2, p. 64 and §9.3 below). This complex 2-plane defines W_α uniquely up to proportionality. When x^a is real, W_α defines a real null line W through P in the direction of the flagpole of ξ_A.

We observe that the arguments of ψ in (7.4.7) are, in effect, the components W_α. Equation (7.4.7) can indeed be re-expressed as

$$\psi(\mathsf{W}_1, \mathrm{i}\mathsf{W}_2, \mathrm{i}\mathsf{W}_3) = 0. \tag{7.4.11}$$

We can remove the normalization $\mathsf{W}_0 = 1$ on W_α (which is implied by (7.4.4): $\xi_0 = 1$) by replacing ψ by a homogeneous holomorphic function of W_α:

$$\chi(\mathsf{W}_\alpha) = (\mathsf{W}_0)^k \psi\left(\frac{\mathsf{W}_1}{\mathsf{W}_0}, \frac{\mathrm{i}\mathsf{W}_2}{\mathsf{W}_0}, \frac{\mathrm{i}\mathsf{W}_3}{\mathsf{W}_0}\right), \tag{7.4.12}$$

where the value of the integer k is at our disposal and defines the homogeneity degree of χ. The Kerr condition (7.4.11) now becomes

$$\chi(\mathsf{W}_\alpha) = 0 \tag{7.4.13}$$

and we have seen that this states that ξ_A is analytic and SFR. But the flagpole direction of ξ_A is the direction of the null ray W through P. Note also that, since an SFR is geodetic, the flagpole of ξ_A remains tangent to W at other points of W. The null twistors which satisfy $\chi(\mathsf{W}_\alpha) = 0$ define a congruence of rays W which coincides with our original SFR. The twistor form of Kerr's theorem is therefore

* A twistor of valence $\left[\begin{smallmatrix}0\\1\end{smallmatrix}\right]$ is dictated by our use of the unprimed spin-vector ξ_A to specify the congruence. With $\xi_{A'}$ instead, we would have defined a twistor W^α having $\xi_{A'}$ for its projection part, and the analysis would have been the 'conjugate' of the one to be given here.

(7.4.14) THEOREM

The general analytic SFR in $\mathbb{M}$ *is a system of rays* W *in* $\mathbb{M}$ *defined locally by the null twistors* W_α *satisfying* $\chi(\mathsf{W}_\alpha)=0$, *where* χ *is an unrestricted holomorphic and homogeneous function of* W_α.

There are also more geometrical (coordinate independent) ways of deriving (7.4.14). One of these is to appeal to (7.3.18). Each β-plane $\mathbb{C}\mathsf{W}$ consists of points whose position vectors have the form $x^{AA'}+\xi^A\zeta^{A'}$, where $x^{AA'}$ and ξ^A are fixed and are as in (7.4.9), whereas $\zeta^{A'}$ varies. The tangent vectors have the form $\xi^A\zeta^{A'}$ for fixed ξ^A, so $\mathbb{C}\mathsf{W}$ is a complex 2-surface of the type considered in (7.3.18). The condition for an analytic SFR, as stated in (7.3.18), is that there should be a complex 2-parameter holomorphic family of such 2-surfaces, to which the vectors $\xi^A\zeta^{A'}$ are tangent, for each $\zeta^{A'}$. Since each $\mathbb{C}\mathsf{W}$ can be specified by giving the twistor W_α up to proportionality, we can specify a holomorphic 2-parameter family of β-planes $\mathbb{C}\mathsf{W}$ by giving a holomorphic 3-parameter family of (dual) twistors W_α which is invariant under $\mathsf{W}_\alpha\mapsto\lambda\mathsf{W}_\alpha$ $(\lambda\neq0)$. Such a 3-parameter family is clearly defined by the vanishing of a homogeneous holomorphic function $\chi(\mathsf{W}_\alpha)$.

Linear and quadratic twistor functions; angular momentum

There are two classes of functions $\chi(\mathsf{W}_\alpha)$ which yield SFRs of special interest. In the first place, if χ is linear in W_α then (7.4.14) defines, in the general case, a type of congruence that we have encountered before in twistor theory, namely a *Robinson congruence.* This was defined, for a given (non-null) twistor Z^α, by the SFR spinor field ω^A associated with Z^α, where $\mathsf{Z}^\alpha\leftrightarrow(\omega^A,\pi_{A'})$ (*cf.* §6.2), and it was shown in (6.2.5) to consist of the null lines W associated with the twistors W_α satisfying $\bar{\mathsf{W}}^\alpha\mathsf{W}_\alpha=0=\mathsf{Z}^\alpha\mathsf{W}_\alpha$. Thus, theorem (7.4.14) gives us the Robinson congruence of Z^α when χ is chosen to be

$$\chi(\mathsf{W}_\alpha)=\mathsf{Z}^\alpha\mathsf{W}_\alpha. \tag{7.4.15}$$

So Robinson congruences arise from the choice of χ in theorem (7.4.14) – or, equivalently, of ψ in (7.4.8) – as arbitrary *linear* functions.

As a second class of interest, consider a general (conjugate) angular momentum twistor $\bar{\mathsf{A}}^{\alpha\beta}$, defined as in (6.3.11), and set

$$\chi(\mathsf{W}_\alpha)=\bar{\mathsf{A}}^{\alpha\beta}\mathsf{W}_\alpha\mathsf{W}_\beta. \tag{7.4.16}$$

With respect to the point P we have W_α given by

$$\mathsf{W}_\alpha\overset{P}{\leftrightarrow}(\lambda_A,0), \tag{7.4.17}$$

so by (6.3.11) we have

$$\bar{\mu}^{AB}\lambda_A\lambda_B = 0, \qquad (7.4.18)$$

at the point P, whenever the condition $\chi(\mathrm{W}_\alpha) = 0$ is satisfied. Here, $\bar{\mu}^{AB}$ $(= \bar{\mu}^{BA})$ is related to the angular momentum M^{ab} about P by (6.3.10),

$$M^{ab} = \bar{\mu}^{AB}\varepsilon^{A'B'} + \mu^{A'B'}\varepsilon^{AB}, \qquad (7.4.19)$$

and so (7.4.18) tells us that *the SFR generated by* (7.4.16) *consists of rays which are everywhere PNDs of the angular momentum tensor.* In the particular case when there is no intrinsic spin (but the mass is non-zero) the congruence is the system of null rays which intersect the centre-of-mass timelike (straight) world-line γ. The congruence is then hypersurface-orthogonal (and null-hypersurface forming), the hypersurfaces being the light cones of the points of γ. When intrinsic spin is present, the congruence is more general and twists in a sense associated with the spin. Every such congruence in fact arises as the linearized limit of the GPND congruence of Kerr's solution of the Einstein vacuum equations representing a rotating black hole or naked singularity. In the case when the rest-mass is zero, the congruence splits into two parts, since by (6.3.2) we have $\bar{\mu}^{AB} = \mathrm{i}\omega^{(A}\bar{\pi}^{B)}$; thus one part is the Robinson congruence defined by ω^A, and the other is the system of parallel rays defined by $\bar{\pi}^B$, in the direction of the 4-momentum.

In the general (timelike) momentum case there are locally two congruences (i.e two rays through each point), which are, however, globally connected in the sense that we can pass continuously from one local system of rays to the other. In the case when $\chi(\mathrm{W}_\alpha)$ is a polynomial of degree n rather than 2 as in (7.4.16), we get n such globally connected congruences.

Relations to twistor functions for massless fields

Let us now turn to applications to massless fields. We have seen in (7.3.9) and (7.3.14) that there is a close relation between SFRs and algebraically special massless fields. This relation shows up again in the two main applications of holomorphic functions of a twistor W_α that we have encountered, namely (7.4.14) and the use of (6.10.1) to generate solutions of the massless field equations. To see this relation in the latter case, suppose that the function f in the expression

$$\phi_{AB\ldots L} = \oint \lambda_A\lambda_B\ldots\lambda_L f(\lambda_{R'}, -\mathrm{i}\lambda_R x^{RR'})\lambda_P \mathrm{d}\lambda^P \qquad (7.4.20)$$

has a k-fold pole at some given x^a, about which the integral is to be performed. If this pole occurs where λ_P is proportional to ξ_P, we have

$$\phi_{A\ldots CD\ldots L}\xi^D\ldots\xi^L = \oint \lambda_A\ldots\lambda_C(\lambda_D\xi^D)^k f(\lambda_R, -\mathrm{i}\lambda_R x^{RR'})\lambda_P \mathrm{d}\lambda^P = 0 \quad (7.4.21)$$

(there being k occurrences of ξ on the left), since the pole is cancelled by the factor $(\lambda_D\xi^D)^k$. Thus, by Proposition (3.5.26), the field $\phi_{A\ldots L}$ has at least an $(n-k+1)$-fold PND along the flagpole of ξ^D at x^a. So if this k-fold pole persists as x^a is varied, the function f must be of the form

$$f(\mathrm{W}_\alpha) = \theta(\mathrm{W}_\alpha)\{\chi(\mathrm{W}_\alpha)\}^{-k}, \quad (7.4.22)$$

where θ is homogeneous and holomorphic and is regular at the poles we are concerned with, these being given by

$$\chi(\xi_R, -\mathrm{i}\xi_R x^{RR'}) = 0 \quad (7.4.23)$$

as x^r varies. The relevant zeros of χ are assumed to be *simple* zeros, χ being also homogeneous and holomorphic. Comparing this result with (7.4.9) and (7.4.13), we see that ξ_R is SFR as expected.

A special case of particular interest is given when f has the form

$$f(\mathrm{W}_\alpha) = (\mathrm{B}^{\alpha\beta}\mathrm{W}_\alpha\mathrm{W}_\beta)^{-3}.$$

For each $x^{RR'}$ we have two triple poles, and separating these with our contour we obtain a linear gravitational field which has two pairs of double PNDs at each point (type $\{22\}$). The linear limits of the Schwarzschild and Kerr solutions have this form (and the Coulomb electromagnetic field is analogous with the exponent -3 replaced by -2). If one applies the procedures of §§6.4, 6.10 for obtaining the angular momentum twistor $\mathrm{A}_{\alpha\beta}$ one finds that it is proportional to the matrix inverse of $\mathrm{B}^{\alpha\beta}$ divided by the square root of its determinant. This change from $\mathrm{B}^{\alpha\beta}$ to $\mathrm{A}_{\alpha\beta}$ explains the apparent anomaly concerning the sign of the spin associated with the Killing spinor for the Kerr solution mentioned on p. 109.

It may be remarked that if $k = n$ we still get ξ_R which is SFR whereas the PND is simple. This is not to say that simple PNDs of massless fields are always SFR, which is certainly not the case. However, the fields arising from (7.4.20) are of a very special type when f has the form (7.4.22). When $k = n$ this fact expresses itself not through the algebraic specialness of the field but through its having SFR rays (PNDs). To generate a general field $\phi_{A\ldots L}$ we would require a function f with singularities more complicated than poles.

It is of interest to compare (7.4.22) with Robinson's theorem on null fields (*cf.* (7.3.14)). The field (7.4.20) is null whenever $k = 1$ in (7.4.22) (when the contour surrounds a simple pole). The values of $\theta(\mathrm{W}_\alpha)$ which contribute to the integral (7.4.20) consequently are those giving the residues at the poles of χ. Thus we are concerned with θ only on $\chi = 0$. Now the locus $\chi = 0$ is three-complex-dimensional in $\mathbb{T}_\alpha$ and is swept out by lines through the origin, this family of lines being therefore two-complex-dimensional. Since θ is homogeneous, this means in effect that the relevant values of θ are defined by *two complex parameters*. So (7.4.20) enables us to generate null massless fields for each spin $\frac{1}{2}n \geqslant 1$ in $\mathbb{M}$ based on an analytic SFR. The SFR arises from the set of zeros of χ (Kerr's theorem) and the freedom in the resulting null field (Robinson's theorem) arises from the arbitrariness of θ at the zeros of χ. All analytic massless null fields in $\mathbb{M}$ are locally obtainable in this way.*

The generalization of this result to algebraically special fields of lower degeneracy (not *all* PNDs coincident) is also of some interest. Since we now have $k > 1$ in (7.4.22), it is not just the values of θ on $\chi = 0$ which concern us, but the values of the first $k - 1$ derivatives of θ in directions away from $\chi = 0$ at points of $\chi = 0$. This means, in effect, that we are concerned now with k holomorphic functions of two complex variables. Thus, for spin $\frac{1}{2}n$ (with $n \geqslant k$) the integral (7.4.20) can be used to generate massless fields for which a given analytic SFR is an $(n - k + 1)$-fold PND (at least), the freedom in the fields generated being that of k holomorphic functions of two complex variables.

It is of some interest to show how the twistor integral (7.4.20) can be re-expressed using the notation (7.4.1)–(7.4.4). Taking components of (7.4.20) with respect to the constant spin-frame, and writing

$$\phi_r = \phi_{0\ldots01\ldots1}, \tag{7.4.24}$$

where there are $n - r$ zeros and r ones on the right, we get

$$\phi_r = \oint \lambda^r F(\lambda, u + \lambda\bar{\zeta}, \zeta + \lambda v)\, d\lambda \quad (r = 0, \ldots, n), \tag{7.4.25}$$

where

$$F(\alpha, \beta, \gamma) = f(\mathrm{W}_\alpha), \quad \text{with } \mathrm{W}_\alpha = (1, \alpha, -i\beta, -i\gamma). \tag{7.4.26}$$

The function F is arbitrary holomorphic on some domain which excludes certain specified regions of singularity around which the integration is

* By way of a historical remark it may be pointed out that the transcription into twistor language of this combination of the theorems of Robinson and Kerr was highly instrumental in motivating the discovery, in the context of twistor theory, of the original contour integral expressions of §6.10 (*cf.* Penrose 1968*a*, 1986). In effect, these theorems state that a null field is described by a holomorphic function on a holomorphic surface in (projective) twistor space. The key realization was that this surface should be treated as a pole and the holomorphic function as its residue.

performed. One verifies immediately that the equations

$$\frac{\partial\phi_{r+1}}{\partial u}=\frac{\partial\phi_r}{\partial\bar{\zeta}},\quad \frac{\partial\phi_{r+1}}{\partial\zeta}=\frac{\partial\phi_r}{\partial v}\quad (r=0,\ldots,n-1)$$

are satisfied, these being the component form of the massless-field equation (4.12.42) for $n>0$. Furthermore we have

$$\frac{\partial^2\phi_r}{\partial u\partial v}=\frac{\partial^2\phi_r}{\partial\zeta\partial\bar{\zeta}},$$

which is the wave equation for the metric (7.4.2), showing that spin 0 is also incorporated by (7.4.25) (with $n=0=r$).

$\mathbb{P}\mathcal{N}$ and $\mathcal{N}$; invariant contact structure

Let us now turn to the case of a general space–time $\mathcal{M}$. We shall see that twistor ideas have a role to play in this case also.

In §7.1 we showed that the property of a neighbouring ray to μ being *abreast* with μ, though definable at any one point of μ, is actually a property of the rays as a *whole*: if the property holds at any one point of μ, then it also holds at all the others. In §7.2 we showed that the same is true for the property of being *twist-free*, for bundles of rays neighbouring μ and abreast with μ. Thus these two properties describe invariant information, or 'structure' for the space that we shall denote by $\mathbb{P}\mathcal{N}$, each point of which represents* a single ray in $\mathcal{M}$.

The notation is analogous to that of §§6.10, 9.3, where (*cf.* p. 313 below) $\mathbb{PN}$ stands for the space of projective null twistors for $\mathbb{M}$. An element of $\mathbb{PN}$ is a null twistor up to proportionality so (apart from the fact that $\mathbb{PN}$ also includes certain 'rays at infinity', *cf.* §9.3) we may regard $\mathbb{PN}$ *as* the space of rays in $\mathbb{M}$. But each element of the *non*-projective space $\mathbb{N}$ of actual null twistors requires the further specification of a spinor $\pi_{A'}$ at each point of the ray, with flagpole tangent to the ray and the entire spinor parallelly propagated along it. Setting $o^A=\bar{\pi}^A$, we are then provided with the kind of (affine) scaling for μ, and reference flag plane, that we have found to be useful earlier in this chapter. Strictly, $\pi_{A'}$ refers to the space $\mathbb{N}^\bullet$ (or $\mathbb{N}^\alpha$) of null $[^1_0]$-twistors. The unprimed spinor o_A correspondingly refers to $\mathbb{N}_\bullet$ (or $\mathbb{N}_\alpha$), the space of null $[^0_1]$-twistors, the two being related through complex

* In certain circumstances, the space $\mathbb{P}\mathcal{N}$, so defined, may exhibit non-Hausdorff properties (*cf.* Penrose 1972*b*; also p. 183, Volume 1). However, $\mathbb{P}\mathcal{N}$ will in fact be a Hausdorff manifold if $\mathcal{M}$ is globally hyperbolic (*cf.* Hawking and Ellis 1973, Penrose 1972*b*) or if our considerations are applied suitably locally.

conjugation. (The projective versions $\mathbb{PN}^\bullet$ and $\mathbb{PN}_\bullet$ of these spaces are essentially identical with each other.) As with the points of $\mathbb{N}_\bullet$, we now define a *point* of the space $\mathscr{N}_\bullet$ to be a *ray in* $\mathscr{M}$ *together with a spinor* o_A *whose flagpole is tangent to the ray and which is parallelly propagated along it.* The space $\mathscr{N}^\bullet$ is defined similarly, but with a primed spinor $\pi_{A'}$ in place of o_A. We shall be working almost exclusively with $\mathscr{N}_\bullet$ and $\mathbb{P}\mathscr{N}_\bullet$ here, and to avoid notational awkwardness we denote these spaces simply by $\mathscr{N}$ and $\mathbb{P}\mathscr{N}$ respectively, except when it becomes important that the distinctions from $\mathscr{N}^\bullet$ and $\mathbb{P}\mathscr{N}^\bullet$ be maintained.

One advantage that the space $\mathscr{N}$ has over $\mathbb{P}\mathscr{N}$ is that the actual measure of non-abreastness

$$h = q^a l_a \tag{7.4.27}$$

(*cf.* (7.1.39), (7.1.43)) itself has an interpretation, not just its vanishing. By (7.1.65), (7.1.66), (7.4.27) we have

$$h = \mathrm{i}\bar{\mathsf{U}}^\alpha \delta \mathsf{U}_\alpha = -\mathrm{i}\mathsf{U}_\alpha \delta \bar{\mathsf{U}}^\alpha \tag{7.4.28}$$

for $\mathsf{U}_\alpha \in \mathbb{N}_\alpha$, and a similar interpretation can be given for $\mathscr{N}$, where a *local twistor* description is used (*cf.* §6.9) at some point on the ray in question, as we shall indicate shortly. Likewise, the actual *measure* of twist – either $L^2 t$ (with L a luminosity parameter and t the twist, *cf.* (7.1.51), (7.1.48), (7.2.16)) or, slightly more generally, the symplectic invariant Σ of (7.2.18) – also has a direct interpretation in $\mathscr{N}$.

To establish the form of the latter, consider first a ray μ in $\mathbb{M}$, together with two rays ν, $\tilde{\nu}$ neighbouring to μ. Represent μ, ν, $\tilde{\nu}$ by U_α, $\mathsf{U}_\alpha + \delta\mathsf{U}_\alpha$, $\mathsf{U}_\alpha + \tilde{\delta}\mathsf{U}_\alpha \in \mathbb{N}_\alpha$, respectively, where taking the origin at $P \in \mu$ we have (with $q^a = \delta x^a$, $\tilde{q}^a = \tilde{\delta} x^a$, *cf.* (7.1.62)–(7.1.64))

$$\begin{aligned} \mathsf{U}_\alpha &\overset{P}{\leftrightarrow} (o_A, 0) \\ \delta\mathsf{U}_\alpha &\overset{P}{\leftrightarrow} (q^b \nabla_b o_A, -\mathrm{i} q^{AA'} o_A) + O(q^2) \\ \tilde{\delta}\mathsf{U}_\alpha &\overset{P}{\leftrightarrow} (\tilde{q}^b \nabla_b o_A, -\mathrm{i}\tilde{q}^{AA'} o_A) + O(\tilde{q}^2) \end{aligned} \tag{7.4.29}$$

Then, ignoring third-order terms,

$$\begin{aligned} \tilde{\delta}\bar{\mathsf{U}}^\alpha \delta\mathsf{U}_\alpha - \delta\bar{\mathsf{U}}^\alpha \tilde{\delta}\mathsf{U}_\alpha &\overset{P}{\leftrightarrow} (\mathrm{i}\tilde{q}^a o_{A'}, \tilde{q}^b \nabla_b o_{A'}) \cdot (q^b \nabla_b o_A, -\mathrm{i} q^a o_A) \\ &\quad - (\mathrm{i} q^a o_{A'}, q^b \nabla_b o_{A'}) \cdot (\tilde{q}^b \nabla_b o_A, -\mathrm{i}\tilde{q}^a o_A) \\ &= \mathrm{i}\tilde{q}^a q^b o_{A'} \nabla_b o_A - \mathrm{i}\tilde{q}^b q^a o_A \nabla_b o_{A'} - \mathrm{i} q^a \tilde{q}^b o_{A'} \nabla_b o_A + \mathrm{i} q^b \tilde{q}^a o_A \nabla_b o_{A'} \\ &= \mathrm{i}\tilde{q}^a q^b \nabla_b l_a - \mathrm{i} q^a \tilde{q}^b \nabla_b l_a \\ &= \mathrm{i}\tilde{q}^a \mathrm{D} q_a - \mathrm{i} q^a \mathrm{D}\tilde{q}_a = 2\mathrm{i}\Sigma \end{aligned} \tag{7.4.30}$$

(using the conditions $\underset{l}{\pounds} q^a = 0$, $\underset{l}{\pounds} \tilde{q}^a = 0$, see (7.1.29)). The penultimate line of (7.4.30) can be rewritten as

$$2iq^{[a}\tilde{q}^{b]}\nabla_{[a}l_{b]} \qquad (7.4.31)$$

which, by the results of §7.1, is an 'imaginary mutiple of squared luminosity parameter times twist' in the case when ν, $\tilde{\nu}$ are abreast with μ. For non-abreast rays, the further information contained in (7.4.31) is concerned only with how the scalings provided by l_a change as we pass from ray to ray. In particular, (7.4.31) vanishes for *all* q^a, $\tilde{q}^a$ if and only if l_a is a *gradient* field ($\rho = \bar{\rho}$, $\tau = \bar{\alpha} + \beta$; *cf.* (7.1.57)) – as opposed to being merely *proportional* to one ($\rho = \bar{\rho}$; *cf.* (7.1.58)), which is what is asserted by the vanishing of (7.4.31) just for *abreast* rays ($q^a l_a = 0 = \tilde{q}^a l_a$). (Note: $\kappa = 0 = \varepsilon$ automatically here.)

Now suppose that the rays lie in a curved space-time $\mathcal{M}$. Consider a local twistor description at P, and represent ν, $\tilde{\nu}$ by the local twistors which would be reduced to the form $(o_A, 0)$ when carried from P to the points displaced from P by q^a and $\tilde{q}^a$, respectively. Neglecting second-order terms in q^a, $\tilde{q}^a$, we find that the respective expressions $\mathrm{U}_\alpha + \delta\mathrm{U}_\alpha$ and $\mathrm{U}_\alpha + \tilde{\delta}\mathrm{U}_\alpha$, as given by (7.4.29), do indeed have the required property, as is seen immediately from (6.9.14). Hence the expression

$$\Sigma = \tfrac{1}{2}i(\tilde{\delta}\mathrm{U}_\alpha\delta\bar{\mathrm{U}}^\alpha - \delta\mathrm{U}_\alpha\tilde{\delta}\bar{\mathrm{U}}^\alpha) \qquad (7.4.32)$$

is valid in $\mathcal{M}$, in the sense of local twistors at P. Moreover, this holds also at any other point of μ, in view of the constancy (*cf.* after (7.2.18)) of Σ along μ, even though the local twistor description of each of ν, $\tilde{\nu}$ will generally *not* be constant (in the local twistor sense) along μ. Similar remarks apply to (7.4.28).

The structure that h and Σ provide for $\mathcal{N}$ is most satisfactorily described in terms of *differential forms* (*cf.* §4.3). Thus we have a 1-form

$$\boldsymbol{h} = i\bar{\mathrm{U}}^\alpha d\mathrm{U}_\alpha = -i\mathrm{U}_\alpha d\bar{\mathrm{U}}^\alpha \qquad (7.4.33)$$

and a 2-form

$$\boldsymbol{\Sigma} = i d\bar{\mathrm{U}}^\alpha \wedge d\mathrm{U}_\alpha \qquad (7.4.34)$$

canonically defined on $\mathcal{N}$, where in each case the RHS can be interpreted in terms of a local twistor description, as in (7.4.28) and (7.4.32), respectively. The relation between the forms $\boldsymbol{h}, \boldsymbol{\Sigma}$ and the expressions h, Σ is that the latter are simply the former applied to q^a (and $\tilde{q}^a$). This is easily checked (at any one point P of μ). Strictly q^a and $\tilde{q}^a$ should be regarded as *Jacobi fields* along the entire ray μ (*cf.* after (7.2.2)) since it is such a Jacobi field which specifies a tangent vector at a point of $\mathcal{N}$ (i.e. the displacement from one entire ray to a neighbouring one).

We note also the important relation:

$$\boldsymbol{\Sigma} = d\boldsymbol{h}. \qquad (7.4.35)$$

It is more or less evident from (7.4.33), (7.4.34) that (7.4.35) should hold,* but some care is needed in formally applying the operation 'd' to (7.4.33) because U_α and $\bar{\mathsf{U}}^\alpha$ should not be treated as independent variables, there being the constraint

$$\mathsf{U}_\alpha \bar{\mathsf{U}}^\alpha = 0 \tag{7.4.36}$$

on $\mathcal{N}$. However there is no difficulty in relaxing this constraint (locally) and regarding $\mathcal{N}$ as part of a larger (eight-real-dimensional) manifold $\mathcal{T}$ for which (7.4.36) need not hold. The equation (7.4.36) then serves to define $\mathcal{N}$ locally within $\mathcal{T}$. The extensions of $\boldsymbol{h}$ and Σ to $\mathcal{T}$ are in no way canonical, but this uncertainty does not affect the relation (7.4.35). The uncertainty can be expressed as the addition of a term which is $\mathsf{U}_\alpha \bar{\mathsf{U}}^\alpha$ multiplied by a smooth form. This does not show up on $\mathcal{N}$ after exterior differentiation because

$$\mathrm{d}(\mathsf{U}_\alpha \bar{\mathsf{U}}^\alpha) = \mathsf{U}_\alpha \mathrm{d}\bar{\mathsf{U}}^\alpha + \bar{\mathsf{U}}^\alpha \mathrm{d}\mathsf{U}_\alpha \tag{7.4.37}$$

and this vanishes along $\mathcal{N}$ by (7.4.33).

We have been somewhat cavalier, in the above, in treating the abstract-indexed local twistor quantities U_α, $\bar{\mathsf{U}}^\alpha$ as though they were coordinate functions on $\mathcal{N}$ and $\mathcal{T}$. In fact our procedures are not hard to justify. One way of doing this is to envisage that the region $\mathcal{U} \subset \mathcal{M}$ under consideration (assumed to be suitably small) is extended smoothly to a different** space–time manifold $\mathcal{M}'$ ($\supset \mathcal{U}$), where $\mathcal{M}'$ is taken to be *flat* in some open set $\mathcal{V}$ containing a portion of the extension of μ (the ray under consideration) into $\mathcal{M}'$ (see Fig. 7-4). We do not require any field equations to hold for $\mathcal{M}'$, so such extensions can be achieved in many ways. We choose one such way and set up a standard twistor coordinate frame $\delta_{\boldsymbol{\alpha}}^{\alpha}$ in $\mathcal{V}$ (*cf.* (6.1.17), (6.1.34)). The rays in the immediate neighbourhood of μ all extend into $\mathcal{V}$ and so can be assigned standard (null) twistor components with respect to $\delta_{\boldsymbol{\alpha}}^{\alpha}$. The forms $\boldsymbol{h}$ and Σ, being constant along μ, may be evaluated in this system and we obtain simply the coordinate versions of (7.4.33), (7.4.34) and therefore

* There is also a direct geometrical significance in (7.4.35). The generators of a null hypersurface in $\mathcal{M}$ constitute a system of rays abreast with their neighbours. Hence they are described in $\mathcal{N}$ by a region $\mathcal{Q}$ along which $\boldsymbol{h}$ vanishes (that is, $\boldsymbol{h}$ annihilates any tangent vector to $\mathcal{Q}$). By (7.4.35), Σ must also vanish along $\mathcal{Q}$, which tells us that the twist-free property of generators of a null hypersurface is a *consequence* of their abreastness, together with the integrability properties of the tangent elements to $\mathcal{Q}$ which ensure that the rays actually sweep out a hypersurface in $\mathcal{M}$. (It should perhaps be remarked that many of the ideas that we have been describing here can, in one form or another, be traced back to Lie; *cf.* Lie and Scheffers 1896).

** In our terminology, the phrase 'space–time manifold', and the corresponding notation '$\mathcal{M}$' or '$\mathcal{M}'$' is taken to include the metric as well as the space of points. Thus, it is sufficient simply to change the *metric* of $\mathcal{M}$, outside $\mathcal{U}$, in order to obtain the required 'different' space–time manifold $\mathcal{M}'$.

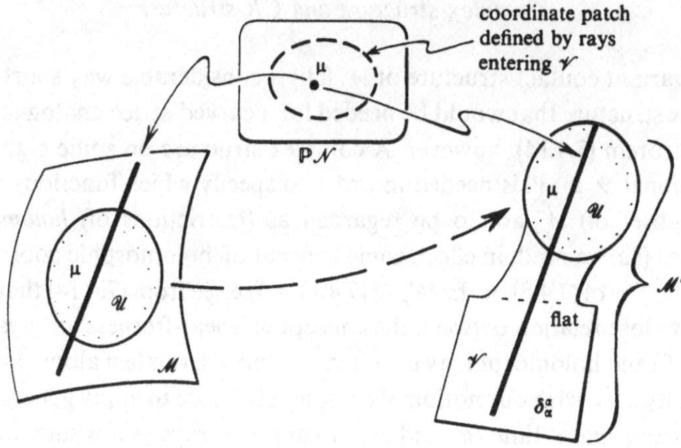

Fig. 7-4. Local coordinates for the space $\mathbb{P}\mathscr{N}$ may be assigned by attaching the curved region $\mathscr{U}$ of $\mathscr{M}$ to a flat space $\mathscr{V}$ (with an intermediate connecting region).

(7.4.35). Note that this procedure provides a *coordinate patch* for the manifold $\mathscr{N}$ in which the forms $\boldsymbol{h}$, $\boldsymbol{\Sigma}$ are represented as these standard expressions. There is clearly much freedom in this procedure (which we have adapted from Penrose 1972*c*). The procedure incidentally shows that *the forms* $\boldsymbol{h}$, $\boldsymbol{\Sigma}$ *on* $\mathscr{N}$ *contain no local information about the curvature of* $\mathscr{M}$, since they are equivalent to those for the flat region $\mathscr{V}$.

The structure that the forms $\boldsymbol{h}$ and $\boldsymbol{\Sigma}$ assign to $\mathscr{N}$ is of the type known as a *contact* structure (Arnol'd 1978). We shall refer to it as the *invariant contact structure** of $\mathscr{N}$ (*cf.* Penrose 1968*a*). The form $\boldsymbol{\Sigma}$ assigns a *symplectic* structure (*cf.* Arnol'd 1978, Woodhouse 1980) to the (non-canonically defined) space $\mathscr{T}$, being non-degenerate and closed:

$$\mathrm{d}\boldsymbol{\Sigma} = 0. \tag{7.4.38}$$

Moreover, we can regard $\boldsymbol{\Sigma}$ as assigning a symplectic structure to the six-real-dimensional space $\hat{\mathscr{N}}$ obtained from $\mathscr{N}$ by factoring out by the phase circles

$$o_A \mapsto e^{i\theta} o_A \quad (\theta \in \mathbb{R}). \tag{7.4.39}$$

(Symplectic structures exist only on even-dimensional manifolds.) The symplectic manifold $\hat{\mathscr{N}}$ is thus the space of affinely scaled rays in $\mathscr{M}$. The 1-form $\boldsymbol{h}$ is also well defined on $\hat{\mathscr{N}}$ and is related to $\boldsymbol{\Sigma}$ by (7.4.35) (see Penrose 1972*c*, Crampin and Pirani 1971 for details).

* The vector field defined by the Euler homogeneity operator $\mathrm{U}_\alpha \partial/\partial \mathrm{U}_\alpha + \bar{\mathrm{U}}^\alpha \partial/\partial \bar{\mathrm{U}}^\alpha$ is also part of the invariant contact structure of $\mathscr{N}$, but it does not represent additional information being, in effect, $\boldsymbol{h} \wedge \boldsymbol{\Sigma} \wedge \boldsymbol{\Sigma} \wedge \boldsymbol{\Sigma}$ 'divided by' $\boldsymbol{\Sigma} \wedge \boldsymbol{\Sigma} \wedge \boldsymbol{\Sigma} \wedge \boldsymbol{\Sigma}$.

Complex structure and CR-structure

The invariant contact structure of $\mathcal{N}$ falls a considerable way short of the *complex* structure that would be needed for a curved-space analogue of the Kerr theorem (7.4.14), however. A complex structure on some eight-real-dimensional $\mathcal{T} \supset \mathcal{N}$ is needed in order to specify which functions on (an open subset of) $\mathcal{N}$ are to be regarded as (restrictions of) *holomorphic* functions (i.e. we need, in effect, some concept of 'holomorphic coordinate' like the α, β, γ of (7.4.8) or the W_α of (7.4.11)). We see from (7.4.14) that there is a very close relation between the concept of 'shear-freeness' of rays in $\mathcal{M}$ (i.e. $\sigma = 0$) and holomorphicity in $\mathcal{N}$ ($\subset \mathcal{T}$). Indeed, this fact alone points to a difficulty if we wish our notion of complex structure to apply generally for an arbitrary space–time $\mathcal{M}$, and also to apply to rays as a whole, without reference to a selection of some arbitrary point on each ray. The Sachs equation (7.2.12)(2) tells us that unless $\Psi_0 = 0$ along a ray μ, then the 'shear-free' condition $\sigma = 0$ will *not* propagate along the ray. Only if $\mathcal{M}$ is conformally flat will such propagation occur generally for all rays.

As a way of circumventing this difficulty we shall ask only for a complex structure which *depends on a choice of hypersurface $\mathcal{H}$ in $\mathcal{M}$*. Then our concept of 'shear-freeness' need refer only to the intersections of the rays with $\mathcal{H}$. Our construction is clearest when $\mathcal{H}$ is chosen to be *spacelike*, and we shall phrase most of our discussion to suit this case. But a timelike or null $\mathcal{H}$ can also be used (with some caution). The null case has special significance when $\mathcal{H}$ is taken to infinity, to become one of the hypersurfaces $\mathscr{I}^\pm$ of §9.6, for an asymptotically flat $\mathcal{M}$. Our construction then yields the space of *asymptotic* twistors for $\mathcal{M}$ that we shall discuss briefly towards the end of §9.8.

Before entering into the details of our construction, we should be clearer about the type of 'complex structure' that we are really concerned with. Recall that for the Kerr theorem (7.4.14) we were interested in the *restriction to* $\mathbb{N}_\alpha$ of functions holomorphic in (some region of) $\mathbb{T}_\alpha$ or, in effect (since our functions were also homogeneous), with restrictions to $\mathbb{PN}_\alpha$ of functions holomorphic in $\mathbb{PT}_\alpha$. It is the odd-real-dimensional spaces $\mathbb{N}_\alpha$ and $\mathbb{PN}_\alpha$, rather than the even-real-dimensional complex spaces $\mathbb{T}_\alpha$ and $\mathbb{PT}_\alpha$, that have immediate interpretations in terms of rays in $\mathbb{M}$, and it is these odd-real-dimensional spaces that find ready generalizations, to become our $\mathcal{N}$ and $\mathbb{P}\mathcal{N}$ for $\mathcal{M}$. Thus we are more concerned with the *restriction*, to an (odd-real-dimensional) hypersurface, of the complex structure of its ambient (necessarily even-real-dimensional) complex manifold, than we are with that complex structure itself. The restriction to a real hypersurface of a complex structure is referred to as a (realizable) CR-structure* (*cf.* Folland and Kohn 1972, Nirenberg 1973, Penrose 1983*b*). We shall need to understand the geometrical properties of such a structure in some detail.

Let us first consider what, in real terms, characterizes a $2n$-real-dimensional manifold $\mathscr{W}$ as an n-dimensional complex manifold. The essential geometric property of $\mathscr{W}$ which is required is that we should be able to recognize the complex tangent vectors to $\mathscr{W}$. In real terms, such a complex tangent vector z will consist of two real tangent vectors, its 'real part' x and its 'imaginary part' y, and we can write

$$z = x + \mathrm{i}y. \tag{7.4.40}$$

When z undergoes the replacement

$$z \mapsto e^{\mathrm{i}\theta} z \quad (\theta \in \mathbb{R}) \tag{7.4.41}$$

the real vectors x and y must undergo

$$\begin{aligned} x &\mapsto x\cos\theta - y\sin\theta \\ y &\mapsto x\sin\theta + y\cos\theta \end{aligned} \tag{7.4.42}$$

and, in particular, when

$$z \mapsto \mathrm{i}z \tag{7.4.43}$$

we have

$$\begin{aligned} x &\mapsto -y \\ y &\mapsto x. \end{aligned} \tag{7.4.44}$$

The replacement (7.4.44) is usually denoted by the letter J:

$$J(x) = -y, \quad J(y) = x \tag{7.4.45}$$

and referred to as the *complex structure* of $\mathscr{W}$. The operator J acts in the (real) $2n$-dimensional tangent spaces of $\mathscr{W}$ and is *real-linear* satisfying

$$J^2 = -1, \tag{7.4.46}$$

as is evident from (7.4.45). The *complex* tangent vectors to $\mathscr{W}$ are then precisely the quantities of the form

$$t - \mathrm{i}J(t) \tag{7.4.47}$$

where t is a real tangent vector to $\mathscr{W}$ (and so, *given* J, the complex tangent vectors to $\mathscr{W}$ are in one-to-one correspondence with the real ones).

The property (7.4.46) is not actually sufficient to characterize $\mathscr{W}$ as a complex manifold, however, and defines $\mathscr{W}$ merely as an *almost complex*

* The more general situation of restriction to a real submanifold of dimension smaller than that of a hypersurface (i.e. dimension smaller than $2n-1$, where the ambient manifold is n-complex-dimensional) is also considered by some authors under the heading of 'CR-manifold'. We do not follow this terminology here, however.

manifold. For a *complex manifold* we need, in addition, an integrability condition* on J. One way of stating this extra condition is:

The Lie bracket of any two smooth fields of complex tangent vectors is again a field of complex tangent vectors. (7.4.48)

Here the term 'smooth field' means smooth only in the sense of real functions, with no concept of holomorphicity implied. For a general almost complex manifold, these Lie brackets would bring in the *complex conjugate tangent vectors*, which have the form

$$\bar{z} = x - iy = x + iJ(x) \tag{7.4.49}$$

in contrast with (7.4.47).

The theorem of Newlander and Nirenberg (1957) states that in any complex manifold $\mathscr{W}$, so defined, there always exist local *complex* coordinates $\zeta_1, \ldots, \zeta_n$ such that the complex tangent vectors at any point arise as the complex linear combinations of

$$\frac{\partial}{\partial \zeta_1}, \ldots, \frac{\partial}{\partial \zeta_n} \tag{7.4.50}$$

at that point. Such coordinates are referred to as *holomorphic coordinates* for $\mathscr{W}$, and they enable the concept of a *holomorphic function* to be defined, for regions in $\mathscr{W}$, as holomorphic (i.e. complex-analytic) functions of such coordinates. The holomorphic functions on $\mathscr{W}$ are then just those complex functions f satisfying $\bar{z}(f) = 0$ for every complex *conjugate* tangent vector $\bar{z}$ (*cf.* also Proposition (4.14.25) Volume 1).

This is clearly a concept that would be needed for an appropriate curved-space version of the holomorphic functions arising in the Kerr theorem. But, for the interpretation in terms of ray geometry, we need also to understand the geometry involved in the restriction of complex structure to some real hypersurface $\mathscr{X}$ in $\mathscr{W}$.

Now the tangent space $T[Q]$, to any such $\mathscr{X}$, at any point Q of $\mathscr{X}$, is $(2n-1)$-dimensional. It contains a $(2n-2)$-real-dimensional subspace $H[Q]$, referred to as the *holomorphic tangent space*, which is invariant under the action of J and so has a structure as an $(n-1)$-complex-dimensional vector space. $H[Q]$ is spanned by (the real and imaginary parts of) $n-1$ complex vectors $z_1, \ldots, z_{n-1}$. To span the whole of $T[Q]$ we need one further real vector $\boldsymbol{u}$ at Q (but the action of J in $\mathscr{W}$ does *not* take $\boldsymbol{u}$ to a vector tangent to $\mathscr{X}$).

* In fact, when written out explicitly in terms of J, this condition becomes the vanishing of the (Nijenhuis) expression given in (4.3.39) on p. 207, Volume 1 (*cf.* Schouten 1954).

Thus, $\mathscr{X}$ by itself has an *intrinsic* structure defined by an operator J, subject to (7.4.46) as before, but which now acts (real-linearly) *only* on each $H[Q]$. The integrability condition is the same as before, namely (7.4.48), but now the concept of complex tangent vector refers only to $H[Q]$. Such an 'integrable' structure J defines $\mathscr{X}$ intrinsically as a *CR-manifold.*

If $\mathscr{X}$ and J are *real-analytic* then it follows that an embedding manifold $\mathscr{W}$ ($\supset \mathscr{X}$) can be constructed, locally, $\mathscr{W}$ being a complex n-manifold whose complex structure reduces to the given J on $\mathscr{X}$. Holomorphic functions on $\mathscr{W}$ then restrict to what are known as *CR-functions* on $\mathscr{X}$ (annihilated by $\bar{z}$, for each complex conjugate tangent vector $\bar{z}$)*. However, without this analyticity restriction it is a delicate question as to whether such an embedding manifold $\mathscr{W}$ (or CR-functions) will exist for a CR-manifold $\mathscr{X}$. When such $\mathscr{W}$ exists we say that the CR-manifold $\mathscr{X}$ is *realizable* (or *embeddable*). Non-realizable CR-manifolds do occur in certain circumstances (*cf.* Nirenberg 1974, Jacobowitz and Trèves 1982, Penrose 1983*b*) and, indeed, can arise in the very cases that we are concerned with here (LeBrun 1984), when analyticity is not assumed.

Hypersurface twistors

Let us now consider the space $\mathbb{P}\mathscr{N}$ and exhibit the CR-structure that it naturally acquires in relation to a spacelike hypersurface $\mathscr{H}$ in $\mathscr{M}$. (We assume that $\mathscr{H}$ is suitable, in relation to the region of $\mathbb{P}\mathscr{N}$ that we are concerned with, namely that it should intersect each ray of that region exactly once.) The *tangent space* $T[\mu]$ to $\mathbb{P}\mathscr{N}$ at a point $\mu \in \mathbb{P}\mathscr{N}$ consists of the rays 'neighbouring' μ, in $\mathscr{M}$. The *holomorphic* tangent space $H[\mu]$, at μ, (as a real four-dimensional vector space) turns out to consist of the neighbouring rays to μ which are *abreast* with μ. Indeed, this much structure is already determined by the invariant contact structure of $\mathscr{N}$, and needs no reference to the hypersurface $\mathscr{H}$.

The role of $\mathscr{H}$ is to define the action of J on $H[\mu]$ and thus to provide $H[\mu]$ with the structure of a complex two-dimensional vector space. Let P be the intersection of μ with $\mathscr{H}$ and define Π to be the 2-plane element at P which is tangent to $\mathscr{H}$ and orthogonal to μ. Choose a neighbouring ray ν, abreast with μ, and take the connecting vector q^a within Π. The ray ν will be determined once we know not only q^a but also Dq^a. Now $l_a Dq^a = D(l_a q^a) = 0$, by (7.1.44), so Dq^a is necessarily orthogonal to l^a. By a change in the scaling

* We note here that not only in the Kerr theorem (7.4.8), (7.4.14) is one effectively concerned with CR-functions, but this is true also of Lemma (7.3.15), etc.

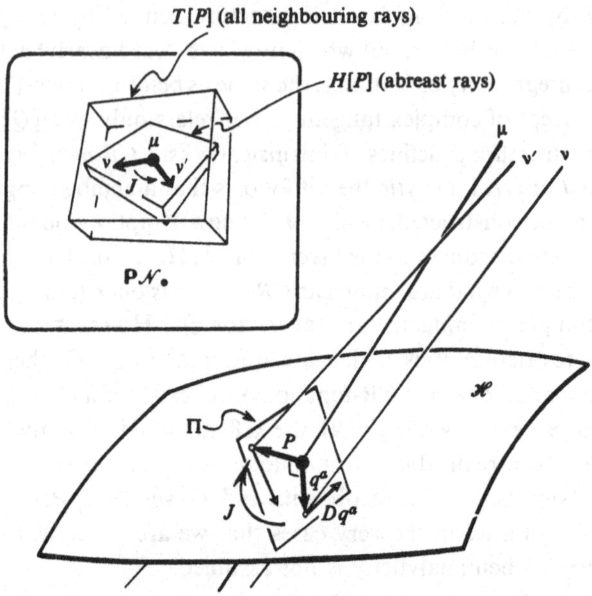

Fig. 7-5. For any choice of (spacelike) hypersurface $\mathscr{H}$ in $\mathscr{M}$, a CR-structure can be assigned to $\mathbb{P}\mathscr{N}_{\bullet}$. This CR-structure varies as $\mathscr{H}$ is moved unless $\mathscr{M}$ is conformally flat.

of ν we can add an arbitrary multiple of l^a to Dq^a so as to bring it into the plane Π. Thus, the different possible choices of neighbouring abreast rays ν to μ, i.e. the different points of $H[\mu]$, are labelled by the different choices of pairs of vectors q^a, Dq^a lying in Π. The action of J (Fig. 7-5) now turns out to be simply a *rotation through a right angle* (in a left-handed sense* about the spatial projection of l^a) of both vectors q^a, Dq^a within Π (see Penrose 1983*b*).

It is evident from Fig. 7-2 that a bundle of abreast neighbouring rays to μ will be shear-free at P if and only if it is invariant under the action of J, so defined. The first two diagrams of Fig. 7-2, representing the presence of convergence or twist, are invariant under a right-angle rotation whereas the third, representing the presence of shear, is not. It would in fact now follow that *if* such a definition of J assigns to $\mathbb{P}\mathscr{N}$ a realizable CR-structure, then a Kerr-type theorem would arise, describing the ray congruences which are shear-free at their intersections with $\mathscr{H}$.

It actually turns out that this definition of J *does* satisfy the integrability conditions of a CR-structure (realizable if $\mathscr{H}$ is embedded real-analytically

* This is for the space $\mathbb{P}\mathscr{N}_{\bullet}$ that we are considering here. For $\mathbb{P}\mathscr{N}^{\bullet}$, the sense would be right-handed.

in an analytic space–time $\mathcal{M}$). Direct arguments have been given by LeBrun (1984) and R.L. Bryant*. We outline a different argument here which, for the analytic case, gives the required complex manifold $\mathcal{T}(\mathcal{H})$ explicitly, in which $\mathcal{N}$ is embedded as a real hypersurface and from which it inherits its required CR-structure.

The manifold $\mathcal{T}(\mathcal{H})$ will be called the *hypersurface twistor space* for $\mathcal{H}$, and its definition, assuming appropriate analyticity for $\mathcal{H}$ in $\mathcal{M}$ is as follows. (Further details may be found in Penrose 1975*a*, 1983*b*.) First complexify $\mathcal{H}$ and $\mathcal{M}$, in the neighbourhood of $\mathcal{H}$ (i.e. choose local analytic coordinates and then allow them to acquire small imaginary parts, *cf.* §6.9, p. 127), to produce the slightly 'thickened' complex manifolds $\mathbb{C}\mathcal{H}$, $\mathbb{C}\mathcal{M}$. Next, consider complex curves in $\mathbb{C}\mathcal{H}$, referred to as *β-curves*, which have (complex null) tangent vectors

$$o^A o_B N^{A'B}, \qquad (7.4.51)$$

N^a being normal to $\mathcal{H}$, where o_C is parallelly propagated along the curve:

$$o^A o_B N^{A'B} \nabla_{AA'} o_C = 0. \qquad (7.4.52)$$

Equation (7.4.52) provides an ordinary differential equation defining the β-curves.

The significance of (7.4.51) is that it is automatically orthogonal to N^a, and therefore tangent to $\mathbb{C}\mathcal{H}$, and also is of the form $o^A \zeta^{A'}$. In flat or conformally flat (complex) space–time, the *β-planes* (*cf.* just after (7.4.10) and §9.3) are the totally null complex 2-surfaces with tangent vectors of this form where, at each point, holding o^A fixed and varying $\zeta^{A'}$ we get the entire tangent space. It we assign a *specific* o_A at each point of the β-plane, where o_A is parallelly propagated over the surface, then it defines a $[{}^0_1]$-twistor. Equations (7.4.51) and (7.4.52) are just the restrictions of these properties to $\mathcal{H}$, so that for $\mathcal{M}$ conformally flat the solutions of (7.4.51) are the intersections of $\mathcal{H}$ with β-planes and do, indeed, define the $[{}^0_1]$-twistors for $\mathcal{M}$. In the general case, because everything in the definition is holomorphic, the solutions of (7.4.52) constitute a *complex* 4-manifold, and this is the required space $\mathcal{T}(\mathcal{H})$. The β-curves themselves are the points of the complex 3-manifold $\mathbb{P}\mathcal{T}(\mathcal{H})$. The overall scaling for the parallelly propagated o_A defines the extra complex dimension in $\mathcal{T}(\mathcal{H})$. The elements of $\mathcal{T}(\mathcal{H})$ [*or* $\mathbb{P}\mathcal{T}(\mathcal{H})$] are our sought for [*projective*] *hypersurface twistors.*

* An interesting approach due to Sparling (1985) makes use of the concept of *Fefferman metric* (Fefferman 1976), which is defined on a certain circle bundle over the CR-manifold and from which the CR-structure can be derived. Sparling's Fefferman metric is $\mathrm{dU}_\alpha \mathrm{d\bar{U}}^\alpha$ (expressed locally) and it is defined on the factor space of $\mathcal{N}(\mathcal{H})$ by the *real* scalings for the twistors. See also Ko, Newman and Penrose 1977, Lugo 1982.

More correctly we should refer to this space as $\mathscr{T}_\bullet(\mathscr{H})$ to distinguish it from $\mathscr{T}^\bullet(\mathscr{H})$, which is defined similarly but now the tangent vectors to the complex curves in $\mathbb{C}\mathscr{H}$, called *α-curves*, have the form

$$\pi^{A'}\pi_{B'}N^{AB'} \tag{7.4.53}$$

and it is the *primed* spinor $\pi_{C'}$ which is parallelly propagated along the curves:

$$\pi^{A'}\pi_{B'}N^{AB'}\nabla_{AA'}\pi_{C'} = 0. \tag{7.4.54}$$

Complex conjugation interchanges α-curves with β-curves and takes $\mathscr{H}$ to itself.

A β-curve which contains a *real* point (point of $\mathscr{H}$) must *meet its complex conjugate α-curve at that point* – since the point is invariant under complex conjugation. (If $\mathbb{C}\mathscr{H}$ is 'thin' enough, $\mathscr{H}$ being spacelike, these will actually be the *only* points at which a β-curve can meet its conjugate α-curve.) The elements of $\mathscr{T}_\bullet(\mathscr{H})$ [*or* $\mathbb{P}\mathscr{T}_\bullet(\mathscr{H})$] which represent such β-curves (the β-curves meeting $\mathscr{H}$) are called *null* [projective] hypersurface twistors; as are, correspondingly, the elements of $\mathscr{T}^\bullet(\mathscr{H})$ [*or* $\mathbb{P}\mathscr{T}^\bullet(\mathscr{H})$] for which the α-curve meets $\mathscr{H}$. The spaces of these null twistors are denoted by $\mathscr{N}_\bullet(\mathscr{H})$ [*or* $\mathbb{P}\mathscr{N}_\bullet(\mathscr{H})$] and $\mathscr{N}^\bullet(\mathscr{H})$ [*or* $\mathbb{P}\mathscr{N}^\bullet(\mathscr{H})$], respectively.

To tie this in with our earlier discussion, we note that at the intersection P of a β-curve with its conjugate α-curve (the case of a null hypersurface twistor), we can choose

$$l_a = o_A\pi_{A'} = o_A o_{A'}$$

(since under complex conjugation o_A becomes $\pi_{A'}$ and $o_{A'} := \bar{o}_{A'}$). We then define μ to be the (o_A-scaled) ray whose direction is that of l^a at P (see Fig. 7-6). Thus $\mathscr{N}_\bullet(\mathscr{H})$ [*or* $\mathbb{P}\mathscr{N}_\bullet(\mathscr{H})$] can be identified (locally) with $\mathscr{N}$ [*or* $\mathbb{P}\mathscr{N}$] and provides the CR-structure that we required. We note, however, that this CR-structure is dependent upon the choice of $\mathscr{H}$, and (except in the case of conformally flat $\mathscr{M}$) will generally *vary* as $\mathscr{H}$ is moved within $\mathscr{M}$. Compatibility with the invariant contact structure of $\mathscr{N}$ (in the sense that the holomorphic tangent spaces remain the spaces of abreast rays, though their complex structure changes) is maintained as $\mathscr{H}$ moves.

An important fact of these constructions is their *conformal invariance*:

(7.4.55) PROPOSITION

*The spaces $\mathscr{N}(\mathscr{H})$ and $\mathbb{P}\mathscr{N}(\mathscr{H})$, and also the forms **h** and **Σ** of the invariant contact structure are unaffected by a conformal rescaling $g_{ab} \mapsto \Omega^2 g_{ab}$ of the metric of $\mathscr{M}$, where we choose $o_A \mapsto o_A$.*

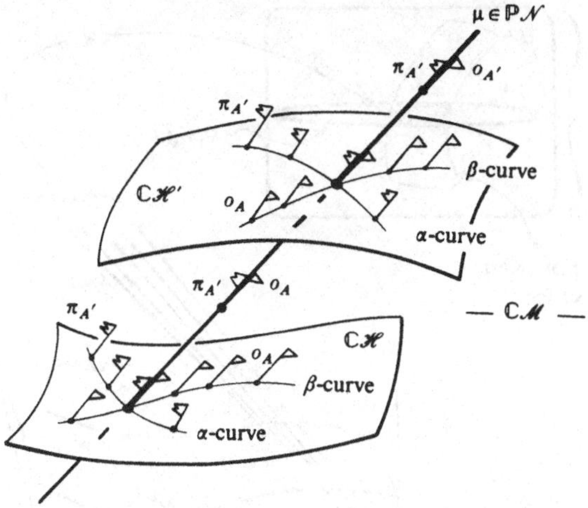

Fig. 7-6. The connection between hypersurface twistors, dual hypersurface twistors and null geodesics. There is no simple relation between the β-curves in different hypersurfaces in general. This corresponds to the fact that the CR-structure of Fig. 7-5 generally depends upon $\mathscr{H}$.

Proof: The conformal invariance of $\mathscr{N}(\mathscr{H})$ and $\mathbb{P}\mathscr{N}(\mathscr{H})$ follow from the fact that

$$o^A\nabla_{AA'}o_C \mapsto \Omega^{-1}o^A(\nabla_{AA'}o_C - \Upsilon_{CA'}o_A) = \Omega^{-1}o^A\nabla_{AA'}o_C \qquad (7.4.56)$$

under the given rescaling, by (5.6.2), (5.6.14), (5.6.15), so that their defining equation (7.4.52) is invariant. The invariance of $\boldsymbol{h}$ and Σ under conformal rescalings follow from their invariance under propagation along the ray μ, once it is observed (e.g. from (7.4.56)) that parallel propagation of o_A along μ is preserved under the rescaling. For if the rescaling is applied near one point of μ it obviously cannot affect the values of the forms at some distant point of μ. The invariance of $\boldsymbol{h}$ under conformal rescalings is also evident from (7.4.27) since $q^a \mapsto q_a$, $l_a \mapsto l_a$, and the invariance of Σ then follows from (7.4.35).

The hypersurface form of the Kerr theorem can be stated as follows:

Assuming appropriate analyticity of all quantities involved, a ray congruence is shear-free at its intersection with a hypersurface $\mathscr{H}$ iff it is defined by the vanishing of a [*homogeneous*] *holomorphic function in $\mathbb{P}\mathscr{T}_\bullet(\mathscr{H})$* [or $\mathscr{T}_\bullet(\mathscr{H})$]. (7.4.57)

Thus the ray congruences shear-free at $\mathscr{H}$ are given by the intersections of a complex (holomorphic) hypersurface in $\mathbb{P}\mathscr{T}_\bullet(\mathscr{H})$ with $\mathbb{P}\mathscr{N}_\bullet(\mathscr{H})$ (see Fig. 7-7). The essential geometric content of this fact, in terms of the

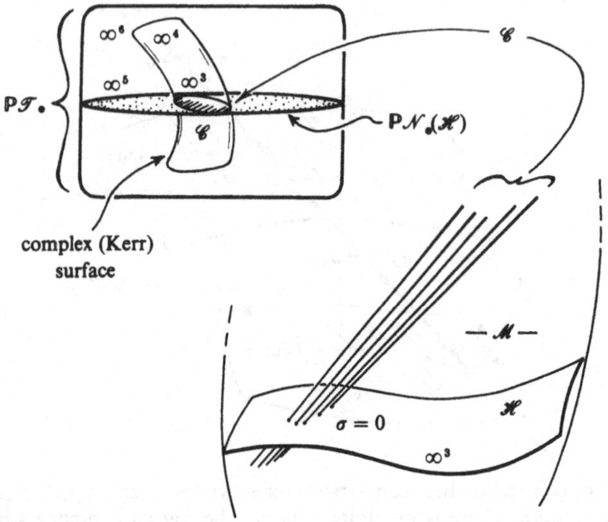

Fig. 7-7. The hypersurface version of the Kerr theorem for curved space–time. A ray congruence with vanishing shear at $\mathscr{H}$ corresponds (in the analytic case) to the intersection with $\mathbb{P}\mathscr{N}_\bullet(\mathscr{H})$ of a complex analytic surface in $\mathbb{P}\mathscr{T}_\bullet(\mathscr{H})$.

geometry of the CR-structure on $\mathbb{P}\mathscr{N}_\bullet(\mathscr{H})$ (*cf.* Fig. 7-5) is that such an intersection (a CR-hypersurface) must have tangent spaces which, where they intersect each $H[P]$, are invariant under J, this fact corresponding to the congruence being shear-free at $\mathscr{H}$. The detailed verification of these matters is straightforward and follows the lines of reasoning that we have been following earlier.

The role of analyticity

A remark concerning the analyticity requirements at $\mathscr{H}$ is appropriate here. We needed such analyticity in order to define the 'complex thickenings' $\mathbb{C}\mathscr{H}$ and $\mathbb{C}\mathscr{M}$ that were needed for $\mathscr{T}(\mathscr{H})$. However, our original geometrical definition of J for $\mathscr{N}(\mathscr{H})$ did not require such procedures, so the CR-structure for $\mathscr{N}(\mathscr{H})$ (perhaps non-realizable) could be directly defined, provided that the required integrability conditions could be established. Moreover, the fact that these integrability conditions must actually *be* satisfied can be established without further calculation once one has verified that in the *analytic* case our geometrical definition of J agrees with the one provided by the hypersurface twistor construction. (Indeed, this is easy enough to check by examining the local twistor descriptions at P.) The integrability conditions are just a system of differential relations which are *automatically* satisfied in the analytic case. Since these relations are precisely the same in the non-analytic case as in the analytic case, their satisfaction in the former case follows from that in the latter.

This establishes the fact that $\mathscr{N}(\mathscr{H})$ is a CR-manifold *whether or not* the analyticity conditions hold at $\mathscr{H}$. However, as remarked earlier, these 'integrability conditions' for a CR-manifold are not sufficient to ensure that it can be realized as a real hypersurface in a complex manifold, and without analyticity the spaces $\mathscr{T}(\mathscr{H})$ and $\mathbb{P}\mathscr{T}(\mathscr{H})$ will generally not exist. In such cases, $\mathscr{N}(\mathscr{H})$ will be lacking in CR-functions and, as a consequence, it may turn out (*cf.* Lewy 1957) that there are *no* ray congruences whatever that are shear-free at $\mathscr{H}$!

Even when the analyticity conditions *do* hold at $\mathscr{H}$ – or else in the case of (conformally) flat space–time, when $\mathscr{H}$ is irrelevant – considerations of analyticity play a role for the congruence itself. In our statements of the Kerr theorem we always needed to assume that the congruence was analytic, it having been pointed out (*cf.* the remark just before Theorem (7.4.8)) that non-analytic SFRs do in fact exist in $\mathbb{M}$. The example mentioned (the system of rays meeting a non-analytic curve) is *twist-free*, and it is of some interest to note that when twist *is* present, a 'one-sided' Kerr function always (locally) exists in the sense depicted in Fig. 7-8. The sense of the twist determines the side of $\mathbb{PN}$ (or $\mathbb{P}\mathscr{N}$) into which the Kerr surface locally extends, namely $\mathbb{PT}_+$ (or $\mathbb{PT}^+$) in the case of a right-handed twist, or $\mathbb{PT}_-$ (or $\mathbb{PT}^-$) in the case of a left-handed one. (The same would

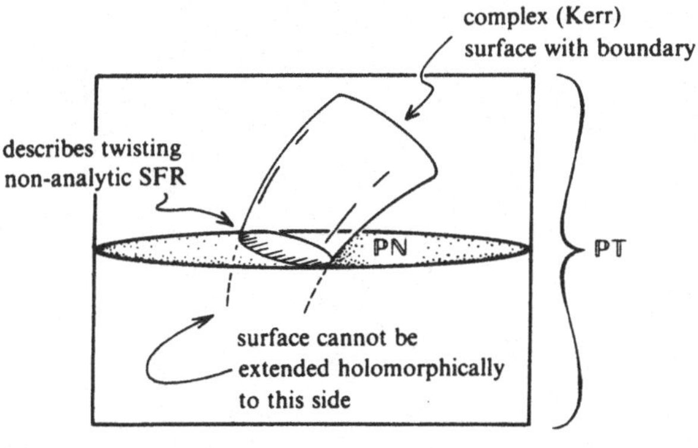

Fig. 7-8. Non-analytic rotating SFRs in $\mathbb{M}$ correspond to 3-surfaces in $\mathbb{PN}$ which are (locally) the boundaries of complex manifolds in $\mathbb{PT}$ on *one* side of $\mathbb{PN}$ only (the side depending upon the sign of the rotation) and which cannot be extended as complex manifolds in $\mathbb{PT}$ on the other side of $\mathbb{PN}$.

apply for $\mathbb{P}\mathscr{T}_{\pm}$ or $\mathbb{P}\mathscr{T}^{\pm}$, suitably defined.) These results follow from certain extension properties which hold for CR-functions in circumstances when the holomorphic 'convexity' properties are appropriate (Lewy 1956, *cf.* Hörmander 1966). These convexity properties turn out to be determined by the twist of the congruence.

This has relevance also to Robinson's theorem (*cf.* (7.3.14)). In (7.3.15) one is, in effect, looking for CR-functions on a three-real-dimensional CR-submanifold of, say, $\mathbb{P}\mathbb{N}$ (as restrictions of holomorphic functions on the Kerr surface). But without the assumption of analyticity, Robinson's procedure leads to a differential equation of the Hans Lewy type (Lewy 1957), which in general has no solutions (*cf.* Tafel 1985). A full discussion of all these matters would carry us too far from our purposes here.

8
Classification of curvature tensors

8.1 The null structure of the Weyl spinor

One of the most immediate and striking examples of the utility of spinor techniques in general relativity theory is the spinor classification of the Weyl tensor (Cartan 1922*a*, p. 194, Penrose 1960, 1962). This greatly simplifies the (originally more familiar) Petrov classification of the Weyl curvature tensor (Petrov 1954, 1966; *cf.* Pirani 1957, also Ludwig 1969, Kramer, Stephani, MacCallum and Herlt 1980 and references contained therein). As we have seen in (4.6.41) the Weyl tensor (or empty-space curvature tensor) is represented by a totally symmetric spinor Ψ_{ABCD}. In Proposition (3.5.18) we showed how any (non-zero) totally symmetric spinor of valence $\left[\begin{smallmatrix}0 & 0\\ n & 0\end{smallmatrix}\right]$ can be expressed, uniquely apart from factors and reorderings, as a symmetrized product of n spin-vectors. Their flagpole directions are the n principal null directions (PNDs) of the symmetric spinor. These define the spinor uniquely, apart from an overall complex scale factor. The pattern of coincidences among the PNDs provides a classification scheme for the spinor. In this chapter we shall investigate in some detail how this classification scheme applies to Ψ_{ABCD} and how it relates to the geometry and algebra of the gravitational field. In the final two sections we show how the scheme can be generalized so that it applies also to symmetric spinors generally and, in particular, to the Ricci spinor (or trace-free Ricci tensor).

The GPNDs and their multiplicities

We shall be concerned with descriptions *at a single point P of space–time only*. The canonical decomposition for Ψ_{ABCD} at P is

$$\Psi_{ABCD} = \alpha_{(A}\beta_B\gamma_C\delta_{D)}. \tag{8.1.1}$$

The GPNDs (gravitational principal null directions), being the flagpole directions of $\alpha_A, \beta_A, \gamma_A$ and δ_A, can be located according to (3.5.22) by finding the zeros of the polynomial

$$\Psi_{ABCD}\xi^A\xi^B\xi^C\xi^D = \Psi_0 + 4\Psi_1 z + 6\Psi_2 z^2 + 4\Psi_3 z^3 + \Psi_4 z^4$$
$$= (\alpha_0 + z\alpha_1)(\beta_0 + z\beta_1)(\gamma_0 + z\gamma_1)(\delta_0 + z\delta_1), \qquad (8.1.2)$$

an arbitrary spin-frame having been chosen (so $\chi := o_A\iota^A = 1$) and ξ^A taken as $(1, z)$. The multiplicities in the factorization (8.1.2) give the corresponding multiplicities in the GPNDs. Thus knowledge of the five quantities $\Psi_0, \ldots, \Psi_4$ (*cf.* (4.11.6)) will lead directly to the location and coincidence scheme of the GPNDs. Recall that $\Psi_0, \ldots, \Psi_4$ can be obtained directly from the Riemann tensor and null tetrad by means of (4.11.9). This affords a method of obtaining the GPNDs directly from the tensor expressions for the curvature.

Alternatively we can use the tensor expressions in Table (8.1.4), which gives equivalent spinor and tensor conditions for the null vector

$$v^a = \pm \xi^A \bar{\xi}^{A'} \qquad (8.1.3)$$

to be a simple, double, etc., GPND of a non-zero Ψ_{ABCD}.

At least simple	$\Psi_{ABCD}\xi^A\xi^B\xi^C\xi^D = 0$	$v_{[f}C_{a]bc[d}v_{e]}v^b v^c = 0$
At least double	$\Psi_{ABCD}\xi^A\xi^B\xi^C = 0$	$C_{abc[d}v_{e]}v^b v^c = 0$
At least triple	$\Psi_{ABCD}\xi^A\xi^B = 0$	$C_{abc[d}v_{e]}v^c = 0$
Quadruple	$\Psi_{ABCD}\xi^A = 0$	$C_{abcd}v^c = 0$

(8.1.4)

The equivalence between the first and second columns is a consequence of Proposition (3.5.26) (and was already used in Chapter 7). To establish the equivalence between the second and third columns we work out, successively, the following identities, using (4.6.41), (3.4.55), and (2.5.23):

$$C_{abcd}v^c = \Psi_{ABCD}\xi^C\varepsilon_{A'B'}\bar{\xi}_{D'} + \text{c.c.}$$
$$C_{abc[d}v_{e]}v^c = \tfrac{1}{2}\Psi_{ABCX}\xi^C\xi^X\varepsilon_{A'B'}\varepsilon_{DE}\bar{\xi}_{D'}\bar{\xi}_{E'} + \text{c.c.}$$
$$C_{abc[d}v_{e]}v^c v^b = -\tfrac{1}{2}\Psi_{ABCX}\xi^C\xi^X\xi^B\varepsilon_{DE}\bar{\xi}_{D'}\bar{\xi}_{E'}\bar{\xi}_{A'} + \text{c.c.}$$
$$v_{[f}C_{a]bc[d}v_{e]}v^c v^b = -\tfrac{1}{4}\Psi_{YBCX}\xi^C\xi^X\xi^B\xi^Y\varepsilon_{DE}\varepsilon_{AF}\bar{\xi}_{D'}\bar{\xi}_{E'}\bar{\xi}_{A'}\bar{\xi}_{F'} + \text{c.c.} \qquad (8.1.5)$$

where c.c. in each case stands for the complex conjugate of the *signed* preceding term. From these identities it is at once apparent that the conditions in column 2 of Table (8.1.4) imply the corresponding conditions in column 3. For the converse we observe that the vanishing of any of the expressions on the left in (8.1.5) implies the vanishing of *both* corresponding terms on the right, as is seen on successively transvecting with $\varepsilon^{A'B'}$ and ε^{AB} in the first two cases, and with $\varepsilon^{D'E'}$ and ε^{DE} in the last two cases.

We incidentally extract from this discussion the fact that each of the tensor conditions in Table (8.1.4) is equivalent to the same condition with ${}^{-}C_{abcd} = \frac{1}{2}(C_{abcd} + \mathrm{i}{}^{*}C_{abcd})$ (*cf.* (4.6.42)) taking the place of C_{abcd}, and thus, separating real and imaginary parts, is equivalent to the same condition with ${}^{*}C_{abcd}$ in place of C_{abcd}.

The possible coincidence schemes for the GPNDs at any one point are given by the five different partitions of the number 4. These, together with the remaining possibility that Ψ_{ABCD} vanishes, define the different *types* of Ψ_{ABCD}, specified as follows:

$$\begin{aligned}
\{1111\}&: \Psi_{ABCD} = \alpha_{(A}\beta_B\gamma_C\delta_{D)}\\
\{211\}&: \Psi_{ABCD} = \alpha_{(A}\alpha_B\gamma_C\delta_{D)}\\
\{22\}&: \Psi_{ABCD} = \alpha_{(A}\alpha_B\beta_C\beta_{D)}\\
\{31\}&: \Psi_{ABCD} = \alpha_{(A}\alpha_B\alpha_C\beta_{D)}\\
\{4\}&: \Psi_{ABCD} = \alpha_A\alpha_B\alpha_C\alpha_D\\
\{-\}&: \Psi_{ABCD} = 0
\end{aligned} \tag{8.1.6}$$

where it is assumed that $\alpha_A, \beta_A, \gamma_A$ *and* δ_A *are all non-proportional and non-vanishing.* The following notation is also used in the literature, and frequently referred to as the 'Petrov types':

$$\mathrm{I} = \{1111\},\quad \mathrm{II} = \{211\},\quad \mathrm{D} = \{22\},\quad \mathrm{III} = \{31\},\quad \mathrm{N} = \{4\}\quad \mathrm{O} = \{-\}, \tag{8.1.7}$$

where O denotes the zero Weyl spinor, D stands for 'double' (or, originally, 'degenerate'), and N for 'null', by analogy with the electromagnetic null field which is characterized by the coincidence of all (two) PNDs. The symbols I, ..., O are occasionally used as generic letters to denote Weyl spinors of the corresponding type, e.g. D_{abcd}. All types but $\{1111\}$ are algebraically special (*cf.* after (7.3.5)). Petrov originally classified I, D, and O as his type 1; II and N as his type 2; and III as his type 3.

Occasionally one needs to add together several spinors (or tensors) each of which has Weyl spinor (or tensor) symmetry and whose canonical decomposition is known, or partially known. Then the following proposition may be useful:

(8.1.8) PROPOSITION ('Addition Theorem')
If two or more Weyl-type summands have one or more PNDs in common, then the multiplicity of any such PND in the sum is not less than its smallest multiplicity in the summands.

For example,

$$\begin{aligned} &\alpha_{(A}\alpha_B\beta_C\gamma_{D)} + \alpha_{(A}\alpha_B\alpha_C\delta_{D)} + \alpha_{(A}\alpha_B\varepsilon_C\varepsilon_{D)} \\ &= \alpha_{(A}\alpha_B\{\beta_C\gamma_{D)} + \alpha_C\delta_{D)} + \varepsilon_C\varepsilon_{D)}\} \\ &= \alpha_{(A}\alpha_B\rho_C\sigma_{D)} \end{aligned}$$

(applying the canonical decomposition to the expression in curly brackets), showing that the sum is $\{211\}$ or more special. The general case is similar.

Specialization scheme

It is instructive to indicate by means of a diagram how the different types can arise as specializations of one another:

$$\begin{array}{ccccc} \{1111\} & \searrow & & & \\ & & \{211\} & & \\ \{22\} & \swarrow & & \searrow & \{31\} \\ & \searrow & \{4\} & \swarrow & \\ \{-\} & \swarrow & & & \end{array} \qquad (8.1.9)$$

The arrows point in the direction of further specialization (greater degeneracy), it being evident from (8.1.6) that each such specialization can indeed be achieved algebraically (i.e. at one point, satisfaction of field equations not being involved here). Specializations can also be achieved which are compositions of the primitive specializations depicted in (8.1.9). For example, we can achieve $\{1111\} \to \{4\}$ via the routes $\{1111\} \to \{211\} \to \{22\} \to \{4\}$ or $\{1111\} \to \{211\} \to \{31\} \to \{4\}$. Moreover such specializations can also always be achieved *directly*. For example, $\alpha_{(A}(\alpha_B + \varepsilon\beta_B)(\alpha_C + \varepsilon\gamma_C) \times (\alpha_{D)} + \varepsilon\delta_{D)})$ under $\varepsilon \to 0$ directly achieves $\{1111\} \to \{4\}$ without passing through any intermediate stages; $\varepsilon\alpha_{(A}\alpha_B\beta_C\gamma_{D)}$ under $\varepsilon \to 0$ directly achieves $\{211\} \to \{-\}$, etc. It should be emphasized that each specialization in (8.1.9) gives a *limiting* case rather than a *particular* case of its antecedent. By definition, the types are mutually exclusive.

8.2 Representation of the Weyl spinor on S^+

In order to visualize the GPNDs it is sometimes helpful to represent them as points on a sphere S^+, the Riemann sphere of the complex number

z. We recall from §1.2 how this sphere – a cross-section of the future null cone of P – conveniently represents the different future-null directions at a space–time point P. Each GPND at P corresponds to a single point on S^+, and so the entire Weyl spinor is specified – up to a complex multiplier – by the unordered set of four points A, B, C, D on S^+ which represent the four GPNDs at P. To achieve specializations of type (other than to $\{-\}$) one simply moves some of A, B, C, D into coincidence with one another on S^+.

It is instructive to consider the effect of a Lorentz transformation on the GPNDs. The reader is reminded of the discussion in §1.2: any active Lorentz transformation effects a conformal mapping of S^+ to itself. Such transformations (and mappings) are generated by the rotations of S^+ and the pure boosts, the latter corresponding to a pair of antipodal points Q^-, Q^+ on S^+ remaining fixed while all other points are transported along meridians with Q^-, Q^+ as poles – away from Q^- and towards Q^+. (See Fig. 1–7, volume 1, for a graphic illustration of this.)

These transformations can also be regarded as *passive* Lorentz transformations, i.e. as leaving the space–time unchanged but describing its changed appearance relative to an observer who changes his velocity and orientation. This is perhaps the simpler view here since it is a little tricky to define an active Lorentz transformation at a point of curved space–time. Let us in particular examine the effect of high-velocity passive Lorentz transformations on the representations of the GPNDs on S^+, and of *limiting* transformations as this velocity tends to unity. Evidently no *finite* Lorentz transformation can affect the type of the Weyl spinor (i.e. the coincidence scheme for A, B, C, D) but *in the limit* the type *can* change. For example, in the limit of a boost all points on S^+ except Q^- are carried into coincidence with Q^+, while Q^- stays put. Hence, in general the Weyl spinor is carried into one of type $\{4\}$ (*cf.* Penrose 1960, 1976*b*; and also Pirani 1959). In the particular cases when one, two, three or four of A, B, C, D coincide with Q^-, the limiting type will be $\{31\}, \{22\}, \{31\}$ or $\{4\}$, respectively.

So far we have discussed the principal null *directions* only. But the behaviour of the actual components of the Weyl spinor (and thereby of the Weyl tensor) under such boosts is also of interest. Again the spinor formalism greatly simplifies the discussion. Only if the limiting type is $\{22\}$ will the Weyl spinor components remain finite (i.e. neither zero nor infinite) in the limit. In the general case, or when only one of A, B, C, D is at Q^-, the Weyl spinor components become infinite, while in the cases when three or four of A, B, C, D are at Q^- they become zero. This is easily seen by choosing a spin-frame o^A, ι^A, where the flagpole of o^A corresponds to Q^- and that of ι^A to Q^+. Then the passive Lorentz transformations under consideration are defined by (*cf.* (1.2.37))

$$o^A \mapsto \varepsilon^{-1} o^A, \quad \iota^A \mapsto \varepsilon \iota^A, \tag{8.2.1}$$

the limit being approached as $\varepsilon \to 0$. Thus we have

$$\alpha_0 = \alpha_A o^A \mapsto \varepsilon^{-1} \alpha_0, \quad \alpha_1 = \alpha_A \iota^A \mapsto \varepsilon \alpha_1,$$

Whence α_A goes to infinity as ε^{-1}, unless $\alpha_0 = 0$ (i.e. unless the point A is at Q^-) in which case it goes to zero as ε. Similarly for β_A, γ_A, and δ_A, and so the above assertions follow at once. Evidently, in order to give rise to a finite limit in these situations, the Weyl spinor components would have to be continuously scaled down or up (as the case may be) during the passage to the limit. In physical terms, a limiting boost characterized by Q^- and Q^+ on S^+ corresponds to the world-view of an observer whose world-line approaches the null direction characterized by Q^-. Thus, if his velocity is in a general direction (relative to the GPNDs) he will 'observe' all the future-pointing GPNDs to come into coincidence behind* him, and most of the components of the Weyl spinor (and tensor) in his frame will tend to infinity. If his velocity approaches the (future) direction of a simple GPND, then that remains fixed and simple in the limit, the type becoming {31} and some of the Weyl spinor components becoming infinite. He must approach a double GPND if his limiting measurements of the Weyl spinor are to remain finite, in which case the limiting type is {22}. If he approaches a triple or quadruple GPND, his limiting measurements of the Weyl spinor become zero (he is 'following the wave') and the type stays the same.

The GPNDs in highly symmetrical space–times

The effect of ordinary finite Lorentz transformations on the Weyl spinor and on the GPNDs will be important in much of what follows. In particular, each type has its own characteristic symmetry under the Lorentz group. Even some very simple symmetry considerations can give useful information about the type. For example, in the well-known Friedmann–Robertson–Walker cosmological models (*cf.* §9.5), there exists at each point a timelike vector (corresponding to a galaxy 'at rest in the universe') relative to which the space–time is spherically symmetric. The Weyl tensor must share this symmetry. But no pattern of 1, 2, 3 or 4 points on S^+

* This corresponds to the fact that the *past* null directions – the directions the observer physically *sees* – come into coincidence in *front* of him.

can be spherically symmetric, whence the type must be $\{-\}$ at each point, i.e. $\Psi_{ABCD} = 0$. From Theorem (6.9.23) we thus have at once:

(8.2.2) PROPOSITION

Every Friedmann–Robertson–Walker model is conformally flat.

As a second example, consider the Schwarzschild metric. At each point the space–time is axially symmetric about the spatial direction of the source, and also time-symmetric. Since the pattern of GPNDs must share this symmetry, each one of them must point towards or away from the source, and thus the type must be $\{4\}$, $\{31\}$, or $\{22\}$. (The metric being vacuum and the curvature nowhere zero, $\{-\}$ is excluded.) Now under time-reflection an outward future-null direction becomes an inward future-null direction and vice versa (counting $\pm v^a$ as the *same* null direction).

Time-symmetry thus excludes $\{4\}$ and $\{31\}$. (We shall see shortly – after Fig. 8-3 – that axial symmetry suffices for this.) Hence the type is $\{22\}$ (*cf.* also p. 108). Corresponding considerations apply, for example, to the plane-wave space–times. In this case the type is $\{4\}$.

The fingerprint of the Weyl tensor

Before examining the various types in detail, there is one further general consideration which has, among other things, a bearing on the symmetries of Ψ_{ABCD}. In addition to the location of the GPNDs there is also the 'phase' and the 'magnitude' of Ψ_{ABCD} needed for a complete description. We shall first examine how this phase can be interpreted in relation to S^+. Consider the quantity

$$\Psi = \Psi_{ABCD}\xi^A\xi^B\xi^C\xi^D. \tag{8.2.3}$$

Provided the flagpole of ξ^A is *not* in a GPND (and $\xi^A \neq 0$), then we can arrange that

$$\Psi > 0 \tag{8.2.4}$$

by a suitable choice of flag plane of ξ^A, keeping the flagpole fixed. There will be four different such choices of ξ achieving the same effect, namely

$$\pm\xi^A, \quad \pm i\xi^A. \tag{8.2.5}$$

The first two of these have identical flag planes, as do the second two, but the direction of the flag plane of the first pair is opposite to that of the second pair. In terms of S^+, therefore, we have two special tangent directions (arising from those flag planes giving $\Psi > 0$) at each point of S^+ which do not correspond to a GPND, these two tangent directions being opposite to one another. The unoriented tangent direction (or line-element) is therefore unambiguously defined at those points. We thus have a field of such directions on S^+ which characterizes not only the GPNDs (as we shall see) but also the phase of Ψ_{ABCD}, because

$$\Psi_{ABCD} \mapsto e^{i\theta}\Psi_{ABCD} \quad \textit{then} \quad \xi^A \mapsto e^{-\frac{1}{4}i(\theta+2k\pi)}\xi^A \tag{8.2.6}$$

(k integral) in order to preserve (8.2.4). The tangent directions therefore rotate through an angle $-\frac{1}{2}\theta$ under the phase change (8.2.6).

We call this pattern of directions on S^+ the *fingerprint* of the Weyl tensor. The fingerprint determines Ψ_{ABCD} up to a positive factor. Hence it determines the Weyl tensor C_{abcd} up to a positive factor ($-C_{abcd}$ corresponding to the orthogonal fingerprint pattern).

The fingerprint tangent directions in fact have a straightforward physical interpretation. We saw from the Sachs equation (7.2.12) that the change in the shear of the rays of a null-geodetic congruence is effectively governed by the value of Ψ_0, while the change in convergence is governed by Φ_{00}. We can now consider a null-geodetic o^A-congruence $\mathscr{C}$ containing a particular ray which passes through a point P in the flagpole direction of ξ^A, and take $o^A = \xi^A$ at P. The curvature quantities Φ_{00} and Ψ_0 have the effect of a lens on these rays (*cf.* Penrose 1966), Φ_{00} effecting a positive focusing and Ψ_0 a purely astigmatic focusing. We refer to Fig. 7-2 (after 7.1.49)) describing the interpretations of $\mathrm{Re}(\rho)$ and σ in order to see this. However, whereas $\mathrm{Re}(\rho)$ and σ refer to the relative 'velocities', Φ_{00} and Ψ_0 refer, in effect, to the relative 'accelerations' of neighbouring rays. Only if a ray bundle passes P without shear and convergence ($\rho = \sigma = 0$ at P) will Φ_{00} and Ψ_0 give the increments in ρ and σ directly. The direction of maximum focusing is given by the minor axis of the ellipse in Fig. 7-2, this making an angle $\frac{1}{2}\arg\Psi_0$ with the o^A flag plane. However, we have arranged the o^A flag plane to be that of ξ^A, where Ψ_0 is the $\Psi > 0$ of (8.2.3) and (8.2.4), so the maximum focusing plane here coincides with the flag plane and thus with the fingerprint direction on S^+. Hence we have:

(8.2.7) PROPOSITION

The fingerprint gives the directions of astigmatism or maximum positive focusing directions (the maximum defocusing directions being orthogonal to these) in the lens effect of space–time curvature at P (see Fig. 8-1).

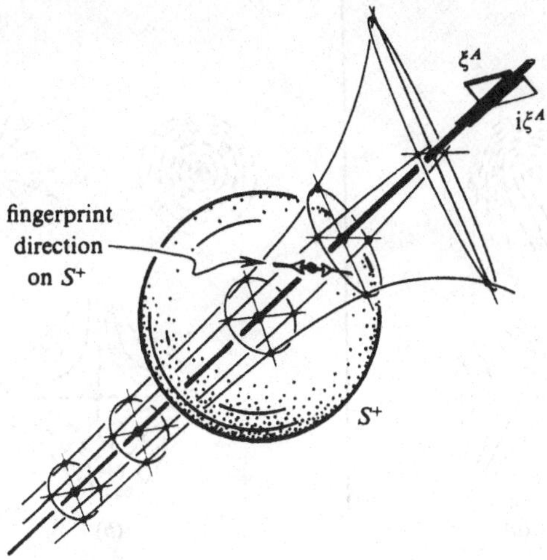

Fig. 8-1. The fingerprint direction on S^+ corresponds to the direction of maximum (inward) focusing by Weyl curvature.

The fact that there are four GPNDs at each point is closely related to a well-known topological property of vector fields on a sphere (S^+). This is usually stated (*cf.* Maunder 1980; also Chern 1979, Thorpe 1979) in the form that there is a net total of *two* points (correctly counted) at which a vector field* on S^+ has to vanish (some of these points possibly counting negatively). In the fingerprint we have a field of line elements which, being non-oriented, allow patterns like that indicated in Fig. 8-2*a*, surrounding the points (GPNDs) at which the line elements are not defined (*cf.* Penrose 1979*b*), in addition to those indicated in Fig. 8-2*b*. Arrows could not be consistently assigned to the directions in Fig. 8-2*a*. Since two of the non-orientable patterns could always be fused into one orientable pattern, this has the effect that there is now a net total of *four* vanishing points. In fact, each GPND counts positively, so the existence of four GPNDs conforms to the topological result. In the case of a multiple GPND the count has to be that of the multiplicity. The pattern of line elements is, indeed, different for each multiplicity, and is illustrated in Fig. 8-3. We observe that rotational symmetry about a GPND occurs only in the case $n = 2$. Hence rotational symmetry suffices in the Schwarzschild case to

* In fact, had the foregoing discussion been concerned with the electromagnetic PNDs we *should* have had a vector field on S^+, the *orientation* of the line element arising from the direction of the electric vector. Then we have *two* PNDs in agreement with the topological result. A corresponding topological theorem exists for each (integral or half-integral) spin.

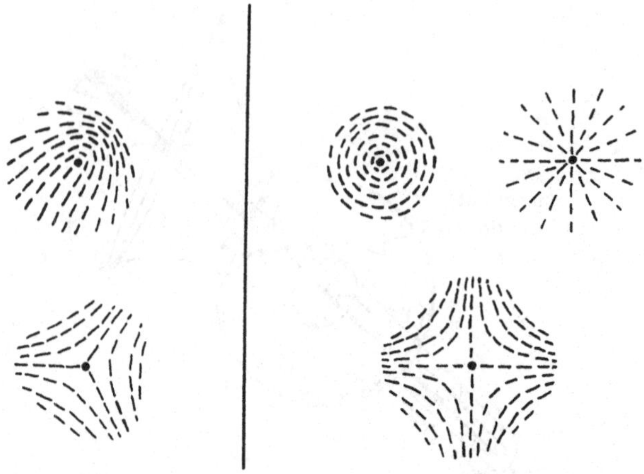

(a) (b)

Fig. 8-2. Some types of singularity that can occur in a 'fingerprint' pattern: (a) consistent orientation not possible (applies to spin two); (b) consistent orientation possible (applies to spins one or two). The bottom figure in each case counts negatively and does not feature in the present discussion.

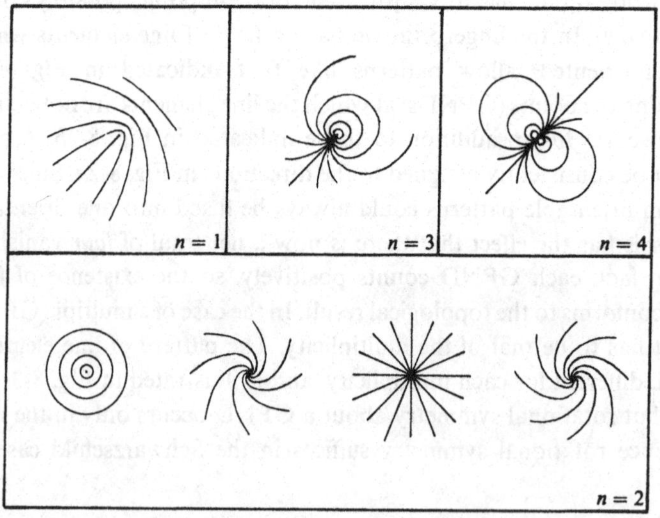

Fig. 8-3. The types of fingerprint singularity that can occur for the Weyl curvature, corresponding to PNDs with n-fold multiplicity.

exclude the types {31} and {4}, as we asserted at the time. The fact that the case $n = 2$ allows several essentially different patterns is connected with the fact that rotational symmetry loses us 'one parameter's worth' of information in the orientation of the pattern, this being regained in the variety of pattern.

It is worth remarking that the symmetry properties of the fingerprint configurations near a GPND point P are related to the spin-weight of the various Ψ_i. Taking $o^A = \xi^A$ we have $\Psi_0 = \cdots = \Psi_{n-1} = 0$ for an n-fold GPND. The form of the fingerprint pattern near P is governed by the leading non-zero term in the expansion of (8.2.3) for which ξ^A is replaced by $o^A + \varepsilon\iota^A$, which is Ψ_n. Thus there is local complete rotational symmetry in the case $n = 2$ (spin-weight zero); there is no rotational symmetry in the cases $n = 1, 3$ (spin-weight ± 1), the local pattern being characterized by a unique tangent direction to S^+; and there is simply central symmetry in the case $n = 4$ (spin-weight -2).

8.3 Eigenspinors of the Weyl spinor

Consider the complex three-dimensional space $\mathfrak{S}_{(AB)}$ of symmetric spinors* ϕ_{AB} of valence $[\begin{smallmatrix}0 & 0\\ 2 & 0\end{smallmatrix}]$ *at a given point* P of $\mathcal{M}$. When it is written in the form $\Psi_{AB}{}^{CD}$, the Weyl spinor may be regarded as effecting a linear transformation on $\mathfrak{S}_{(AB)}$:

$$\phi_{AB} \mapsto \Psi_{AB}{}^{CD}\phi_{CD}. \tag{8.3.1}$$

The space $\mathfrak{S}_{AB}$ has a complex metric canonically defined so that the scalar product between two elements $\phi_{AB}, \chi_{AB} \in \mathfrak{S}_{AB}$ is

$$\phi_{AB}\chi^{AB} = \{-\varepsilon^{A(A_0}\varepsilon^{B_0)B}\}\phi_{AB}\chi_{A_0B_0}, \tag{8.3.2}$$

the expression in { } playing the role of a 'metric tensor'. Under active spin transformations they undergo proper complex orthogonal transformations** for which the above metric is clearly invariant. Given any spin-frame o^A, ι^A, we can construct a corresponding orthonormal basis

* This space would be denoted by $\mathfrak{S}_{(AB)}[P]$ in accordance with our general notational scheme, but we drop the '$[P]$' consistently in this chapter since all considerations refer just to a single point in $\mathcal{M}$.

** This is the group $SO(3, \mathbb{C})$, which is isomorphic with the restricted Lorentz group and 1-2 homomorphic with the spin group $SL(2, \mathbb{C})$. This 1-2 homomorphism is a consequence of the 1-2 correspondence between standard bases in $\mathfrak{S}_{(AB)}$ and $\mathfrak{S}_A$ that we establish by means of (8.3.3).

$\overset{1}{\delta}_{AB}, \overset{2}{\delta}_{AB}, \overset{3}{\delta}_{AB}$ ($\overset{\alpha}{\delta}_{AB}\overset{\beta}{\delta}{}^{AB} = \delta_{\alpha\beta}, \alpha, \beta = 1, 2, 3$) for $\mathfrak{S}_{(AB)}$ thus:

$$\overset{1}{\delta}_{AB} = -\frac{\mathrm{i}}{\sqrt{2}}(o_A o_B - \iota_A \iota_B), \quad \overset{2}{\delta}_{AB} = \frac{1}{\sqrt{2}}(o_A o_B + \iota_A \iota_B), \quad \overset{3}{\delta}_{AB} = \mathrm{i}\sqrt{2}\, o_{(A}\iota_{B)}. \tag{8.3.3}$$

Of several different possibilities, we choose the above for later notational compatibility with the standard spin-frame relation to Minkowski space. We also note that because of the orthonormality of the basis we need make no distinction between upper and lower positions for the indices $\alpha, \beta, \ldots$

Conversely, given *any* orthonormal triad $\overset{1}{\delta}_{AB}$, $\overset{2}{\delta}_{AB}, \overset{3}{\delta}_{AB} \in \mathfrak{S}_{(AB)}$, we have a corresponding spin-frame o_A, ι_A, which is unique up to sign. For, orthonormality implies $\overset{2}{\delta}_{AB} - \mathrm{i}\overset{1}{\delta}_{AB}$ is null (i.e. $(\overset{2}{\delta}_{AB} - \mathrm{i}\overset{1}{\delta}_{AB}) \times (\overset{2}{\delta}{}^{AB} - \mathrm{i}\overset{1}{\delta}{}^{AB}) = 0$), which therefore defines a spinor ι_A uniquely up to sign, by

$$\iota_A \iota_B = \frac{1}{\sqrt{2}}(\overset{2}{\delta}_{AB} - \mathrm{i}\overset{1}{\delta}_{AB}).$$

This is orthogonal to $\overset{3}{\delta}_{AB}$ (but not proportional to it), so the canonical decomposition (8.3.3) (3) of $\overset{3}{\delta}_{AB}$ yields a unique spinor o_A, given the choice of sign for ι_A. Since $\overset{2}{\delta}_{AB} + \mathrm{i}\overset{1}{\delta}_{AB}$ is also null and orthogonal to $\overset{3}{\delta}_{AB}$ it must be proportional to $o_A o_B$. The normalization of $\overset{1}{\delta}_{AB}, \overset{2}{\delta}_{AB}$, and $\overset{3}{\delta}_{AB}$ is all that remains to be satisfied, and this is clearly achieved by the spin-frame condition $o_A \iota^A = 1$.

With respect to the basis (8.3.3) we can define 'Cartesian' components for any $\phi_{AB} \in \mathfrak{S}_{(AB)}$:

$$\overset{1}{\phi} = \frac{-\mathrm{i}}{\sqrt{2}}(\phi_{00} - \phi_{11}), \quad \overset{2}{\phi} = \frac{1}{\sqrt{2}}(\phi_{00} + \phi_{11}), \quad \overset{3}{\phi} = \mathrm{i}\sqrt{2}\,\phi_{01}. \tag{8.3.4}$$

We remark that if ϕ_{AB} defines an electromagnetic field according to (5.1.39), then these components are simply those of the complex 3-vector $-2^{-\frac{1}{2}}\mathrm{i}(\boldsymbol{E} - \mathrm{i}\boldsymbol{B})$ in the standard frame given by (3.1.20) (*cf.* (5.1.59), (5.1.60)).

Similarly we have the following matrix $\boldsymbol{\Psi}$ of components of $\Psi_{AB}{}^{CD}$ with respect to this basis:

$$\boldsymbol{\Psi} = \begin{bmatrix} \frac{1}{2}(-\Psi_0 + 2\Psi_2 - \Psi_4) & \frac{-\mathrm{i}}{2}(\Psi_0 - \Psi_4) & (\Psi_1 - \Psi_3) \\ \frac{-\mathrm{i}}{2}(\Psi_0 - \Psi_4) & \frac{1}{2}(\Psi_0 + 2\Psi_2 + \Psi_4) & \mathrm{i}(\Psi_1 + \Psi_3) \\ (\Psi_1 - \Psi_3) & \mathrm{i}(\Psi_1 + \Psi_3) & -2\Psi_2 \end{bmatrix} \tag{8.3.5}$$

where $\Psi_0, \ldots, \Psi_4$ are the standard Weyl spinor components considered earlier (*cf.* (4.11.6)). Notice that the matrix $\boldsymbol{\Psi}$ is trace-free. This follows, of course, from the symmetry of Ψ_{ABCD} which implies

$$\Psi_{AB}{}^{AB} = \overset{\alpha\alpha}{\Psi} = 0. \tag{8.3.6}$$

In fact, this is the only restriction on the complex matrix $\boldsymbol{\Psi}$ apart from symmetry, since this leaves us with five independent matrix elements linearly related to $\Psi_0, \ldots, \Psi_4$.

An *eigenspinor* of the Weyl spinor is a non-zero element $\phi_{AB} \in \mathfrak{S}_{(AB)}$ for which

$$\Psi_{AB}{}^{CD}\phi_{CD} = \lambda\phi_{AB} \tag{8.3.7}$$

for some complex λ called the corresponding *eigenvalue*. Expressing (8.3.7) in components according to the basis (8.3.3), we see that λ is also an eigenvalue in the ordinary sense of the matrix $\boldsymbol{\Psi}$. If $\lambda_1, \lambda_2, \lambda_3$ are the three eigenvalues of $\boldsymbol{\Psi}$ we thus have

$$\begin{aligned} \lambda_1 + \lambda_2 + \lambda_3 &= \Psi_{AB}{}^{AB} = 0 \\ \lambda_1^2 + \lambda_2^2 + \lambda_3^2 &= \Psi_{AB}{}^{CD}\Psi_{CD}{}^{AB} =: I \\ \lambda_1^3 + \lambda_2^3 + \lambda_3^3 &= \Psi_{AB}{}^{CD}\Psi_{CD}{}^{EF}\Psi_{EF}{}^{AB} =: J, \end{aligned} \tag{8.3.8}$$

and we note that $\lambda_1, \lambda_2, \lambda_3$ are the roots of

$$6\lambda^3 - 3I\lambda - 2J = 0.$$

Using the expression

$$\begin{aligned} \Psi_{ABCD} = {} & \Psi_0 \iota_A \iota_B \iota_C \iota_D - 4\Psi_1 o_{(A}\iota_B\iota_C\iota_{D)} + 6\Psi_2 o_{(A}o_B\iota_C\iota_{D)} \\ & - 4\Psi_3 o_{(A}o_B o_C \iota_{D)} + \Psi_4 o_A o_B o_C o_D, \end{aligned} \tag{8.3.9}$$

(whose validity follows at once by taking components of both sides), we can readily obtain the first of the following expressions for the invariant scalars I, J (the second will follow from (8.3.11) below):

$$\begin{aligned} I &= 2\Psi_0\Psi_4 - 8\Psi_1\Psi_3 + 6\Psi_2{}^2 \\ J &= 6\begin{vmatrix} \Psi_0 & \Psi_1 & \Psi_2 \\ \Psi_1 & \Psi_2 & \Psi_3 \\ \Psi_2 & \Psi_3 & \Psi_4 \end{vmatrix}. \end{aligned} \tag{8.3.10}$$

Furthermore, from (8.3.8) we get (by cubing the first line and subtracting three times the product of the first two)

$$J = 3\lambda_1\lambda_2\lambda_3, \tag{8.3.11}$$

which shows that the determinant of $\boldsymbol{\Psi}$ is $\frac{1}{3}J$ and this leads to the second of (8.3.10); and also (by cubing the second line)

$$I^3 - 6J^2 = 2(\lambda_1 - \lambda_2)^2(\lambda_2 - \lambda_3)^2(\lambda_3 - \lambda_1)^2, \tag{8.3.12}$$

which establishes that

$$\textit{two or more } \lambda\textit{s equal} \Leftrightarrow I^3 = 6J^2. \tag{8.3.13}$$

The relation between the eigenvalues and cross-ratios

To investigate the geometrical significance of $\lambda_1, \lambda_2, \lambda_3$, I and J, we recall from (1.3.9) that any ordered set of four null directions – in which no three coincide – has a uniquely defined cross-ratio (an element of $\mathbb{C} \cup \{\infty\}$) which is invariant under restricted Lorentz transformations. Translating (1.3.10) into spinor notation, we have, for the cross-ratio of the GPNDs

$$\{A, B, C, D\} = \frac{\alpha_P \beta^P \gamma_Q \delta^Q}{\alpha_R \delta^R \gamma_S \beta^S}, \tag{8.3.14}$$

the notation on the left agreeing with that of (1.3.9) (in which the points A, B, C, D of §8.2 would be regarded as elements of $\mathbb{C} \cup \{\infty\}$) if S^+ is taken to be the Riemann sphere for $\mathbb{C} \cup \{\infty\}$ according to the prescriptions of §1.2. We have

$$\{A, B, C, D\} = \{B, A, D, C\} = \{C, D, A, B\} = \{D, C, B, A\}. \tag{8.3.15}$$

Let us now assume that B, C, D are distinct; then we can send them into any other specified ordered set of distinct null directions by a unique restricted Lorentz transformation. The cross-ratio then uniquely defines the image of A under the transformation in relation to the images of B, C, D.

Set

$$\chi = \{A, B, C, D\}. \tag{8.3.16}$$

Then, by the above remarks, we can choose our spin-frame o^A, ι^A and the scaling of α etc. so that

$$\beta_A = o_A, \quad \gamma_A = o_A + \iota_A, \quad \delta_A = \iota_A \tag{8.3.17}$$

(giving $\beta_0 = 0$, $\beta_1 = 1$, $\gamma_0 = -1$, $\gamma_1 = 1$, $\delta_0 = -1$, $\delta_1 = 0$), whence, for some $\eta \neq 0$ (a factor 6 being inserted for convenience),

$$\alpha_A = 6\eta(o_A + \chi \iota_A) \tag{8.3.18}$$

by (8.3.14) and (8.3.16) (giving $\alpha_0 = -6\eta\chi$, $\alpha_1 = 6\eta$). The form (8.1.2) then becomes

$$\begin{aligned} \Psi_{ABCD}\xi^A\xi^B\xi^C\xi^D &= 6\eta(z - \chi)z(z-1)(-1) \\ &= -6\eta\{\chi z - (\chi + 1)z^2 + z^3\}, \end{aligned} \tag{8.3.19}$$

so that, in this particular frame,

$$\Psi_0 = 0,\quad \Psi_1 = -\tfrac{3}{2}\eta\chi,\quad \Psi_2 = \eta(\chi+1),\quad \Psi_3 = -\tfrac{3}{2}\eta,\quad \Psi_4 = 0. \tag{8.3.20}$$

Thus, by (8.3.10),

$$I = 6\eta^2(\chi^2 - \chi + 1) = 6\eta^2(\chi+\omega)(\chi+\omega^2), \tag{8.3.21}$$

(where $\omega = e^{2\pi i/3}$) and

$$J = -6\eta^3(\chi+1)(\chi-2)(\chi-\tfrac{1}{2}). \tag{8.3.22}$$

We observe at once from these expressions for I and J that:

$$\text{GPND } \textit{equianharmonic } (\chi = -\omega, -\omega^2) \Leftrightarrow J \neq 0 = I \tag{8.3.23}$$

and

$$\text{GPND } \textit{harmonic } (\chi = -1, 2, \tfrac{1}{2}) \Leftrightarrow J = 0 \neq I \tag{8.3.24}$$

Observe that in (8.3.8) we have three equations for the three λs in terms of I and J (the last two being replaceable by (8.3.11) and (8.3.12)), while in (8.3.21) and (8.3.22) we have I and J expressed in terms of χ and η. We can thus solve for the λs in terms of χ and η. By inspection, we find, in arbitrary order

$$\lambda_1 = \eta(1-2\chi),\quad \lambda_2 = \eta(1+\chi),\quad \lambda_3 = \eta(\chi-2). \tag{8.3.25}$$

We recall (*cf.* (1.3.12)) that a re-ordering of A, B, C, D, entails a replacement of χ by one of the values

$$1-\chi,\quad \chi^{-1},\quad (1-\chi)^{-1},\quad 1-\chi^{-1},\quad \chi(\chi-1)^{-1}. \tag{8.3.26}$$

One easily confirms from (8.3.25) that such a replacement, when accompanied by the corresponding replacement of η by

$$-\eta,\quad \eta\chi,\quad \eta(\chi-1),\quad -\eta\chi,\quad \eta(1-\chi) \tag{8.3.27}$$

respectively, merely permutes the λs (the corresponding permutations of the λs being $\lambda_1, \lambda_2, \lambda_3 \mapsto \lambda_1, \lambda_3, \lambda_2$; $\lambda_3, \lambda_2, \lambda_1$; $\lambda_2, \lambda_3, \lambda_1$; $\lambda_3, \lambda_1, \lambda_2$; $\lambda_2, \lambda_1, \lambda_3$, respectively.

Special cases

The solutions (8.3.25) were obtained on the assumption of distinct B, C, D (cf. (8.3.17)). Various special and limiting cases are of interest. If $\chi = 0$ or 1, A coincides with B or C, respectively (*cf.* (8.3.18)); and if $\chi = \infty$, $\eta = 0$ (keeping $\chi\eta$ finite), A coincides with D. In each of these three cases, two of the λs in (8.3.25) are equal, while the third is distinct. If after such a coincidence of A with one of B, C, D the remaining two directions are also brought into coincidence, distinct from A, χ remains unchanged (as is clear from (8.3.14)) and equations (8.3.25) remain valid. If, however, three of A, B, C, D are brought into coincidence, it is clear from (8.3.14) that, depending on the approach to coincidence, χ can take *any* limiting value; we shall write this as $\chi = 0/0$. (For example, if we put $\alpha_A = a\gamma_A + b\delta_A$, $\beta_A = c\gamma_A + d\delta_A$, we find $\chi = 1 - bc/ad$; now letting $b = \mu d \to 0$ for arbitrary μ achieves coincidence of A, B, C, with $\chi \to 1 - \mu c/a$.) For consistency, it is evident that all λs in (8.3.25) must now vanish in the limit. We have thus established:

$$\begin{aligned} &\textit{Two or more GPNDs coincident} \Leftrightarrow \chi = 0,\ 1,\ \infty, \text{ or } 0/0 \\ &\qquad \Leftrightarrow \textit{two or more } \lambda\textit{s coincident} \Leftrightarrow I^3 = 6J^2 \end{aligned} \tag{8.3.28}$$

(*cf.* (8.3.13)), and, in particular,

$$\begin{aligned} &\textit{Three or more GPNDs coincident} \Leftrightarrow \chi = 0/0 \\ &\qquad \Leftrightarrow \lambda_1 = \lambda_2 = \lambda_3 = 0 = I = J. \end{aligned} \tag{8.3.29}$$

If we know $\Psi_0, \ldots, \Psi_4$, the eigenspinors may be obtained by simply finding the eigenvectors (a_1, a_2, a_3) of the matrix (8.3.5), whereupon the corresponding eigenspinor is given by

$$\begin{aligned} \phi_{AB} &= a_1 \overset{1}{\delta}_{AB} + a_2 \overset{2}{\delta}_{AB} + a_3 \overset{3}{\delta}_{AB} \\ &= \frac{1}{\sqrt{2}}\{(a_2 - \mathrm{i}a_1)o_A o_B + 2\mathrm{i}a_3 o_{(A}\iota_{B)} + (a_2 + \mathrm{i}a_1)\iota_A \iota_B\}. \end{aligned} \tag{8.3.30}$$

(This is easily seen by multiplying the first line by $\Psi_{CD}{}^{AB}$.)

We note that the eigenspinors corresponding to distinct eigenvalues are orthogonal. This is a well-known result of matrix theory whose proof, in the present context, is provided by

$$\lambda_1 \phi_{(1)AB}\phi_{(2)}{}^{AB} = \phi_{(1)CD}\Psi_{AB}{}^{CD}\phi_{(2)}{}^{AB} = \phi_{(1)CD}\phi_{(2)}{}^{CD}\lambda_2,$$

where $\phi_{(1)AB}, \phi_{(2)AB}$ are eigenspinors corresponding to respective eigenvalues λ_1, λ_2; if $\lambda_1 \neq \lambda_2$ then $\phi_{(1)CD}\phi_{(2)}{}^{CD} = 0$.

Thus, in the situation where all eigenvalues $\lambda_1, \lambda_2, \lambda_3$ are distinct (i.e. in the case $\{1111\}$) we have a mutually orthogonal triad of eigenspinors in $\mathfrak{S}_{(AB)}$ (this is not a foregone conclusion since our matrix is not *real* symmetric), which can be chosen to be orthonormal. Labelling these $\overset{1}{\delta}_{AB}, \overset{2}{\delta}_{AB}, \overset{3}{\delta}_{AB}$, we obtain a spin-frame o^A, ι^A (*cf.* after (8.3.3)) according to which the matrix $\boldsymbol{\Psi}$ of (8.3.5) is diagonal, say diag $(\lambda_1, \lambda_2, \lambda_3)$. Thus we have, in this frame,

$$\Psi_0 = \Psi_4, \quad \Psi_1 = \Psi_3 = 0 \tag{8.3.31}$$

(from the off-diagonal zeros) and

$$\lambda_1 = -\Psi_0 + \Psi_2, \quad \lambda_2 = \Psi_0 + \Psi_2, \quad \lambda_3 = -2\Psi_2, \tag{8.3.32}$$

giving

$$\Psi_0 = \tfrac{3}{2}\eta\chi, \quad \Psi_1 = 0, \quad \Psi_2 = \tfrac{1}{2}\eta(2-\chi), \quad \Psi_3 = 0, \quad \Psi_4 = \tfrac{3}{2}\eta\chi. \tag{8.3.33}$$

Then, from (8.3.9),

$$\Psi_{ABCD} = \tfrac{3}{2}\eta\chi\iota_A\iota_B\iota_C\iota_D + 3\eta(2-\chi)o_{(A}o_B\iota_C\iota_{D)} + \tfrac{3}{2}\eta\chi o_A o_B o_C o_D, \tag{8.3.34}$$

and this gives a canonical form for any type $\{1111\}$ Weyl spinor. We may compare this with the alternative canonical form given by (8.3.20) and (8.3.9), for which the o_A and ι_A flagpoles are GPNDs (*cf.* (8.3.17)). Here the o_A and ι_A flagpoles are the PNDs of one of the eigenspinors (namely $\overset{3}{\delta}_{AB}$ in (8.3.3)). There is obviously a degree of arbitrariness in the choice of a canonical form, but the present choice (8.3.34) allows us to discuss the geometric symmetry properties of type $\{1111\}$ (see §8.5 below) rather more transparently than would other choices.

Note that if $\chi = 2$, which gives harmonic GPNDs, we have Ψ_{ABCD} expressed as a sum of two type $\{4\}$ symmetric spinors. Conversely we can see directly that the sum (or difference) of (non-proportional) type $\{4\}$ spinors must be harmonic, since

$$\begin{aligned}\alpha_A\alpha_B\alpha_C\alpha_D - \beta_A\beta_B\beta_C\beta_D \\ = \{\alpha_{(A} + \beta_{(A}\}\{\alpha_B + \mathrm{i}\beta_B\}\{\alpha_C - \beta_C\}\{\alpha_{D)} - \mathrm{i}\beta_{D)}\},\end{aligned} \tag{8.3.35}$$

the cross-ratio of the four factors on the right being harmonic (*cf.* (8.3.14)).

The canonical form (8.3.34) does not necessarily require that λ_1, λ_2 and λ_3 be all distinct, but merely that a set of three linearly independent eigenspinors (which can then be chosen orthonormal) exist for $\Psi_{AB}{}^{CD}$. In the cases $\{211\}$ and $\{22\}$ we may take $\chi = 0$ and substitute into (8.3.34) to see whether such a canonical form is possible. The result is

$$\Psi_{ABCD} = 6\eta o_{(A}o_B\iota_C\iota_{D)}, \tag{8.3.36}$$

which is $\{22\}$, showing that the eigenspinors of a $\{211\}$ Weyl spinor cannot span $\mathfrak{S}_{(AB)}$, while those of a $\{22\}$ Weyl spinor can. Indeed *clearly* any $\{22\}$ Weyl spinor has a canonical form (8.3.36), with $\eta = \Psi_2$. Equally clearly, $\{-\}$ can take the form (8.3.34) with $\eta = 0$, and the eigenspinors span $\mathfrak{S}_{(AB)}$. However in the cases $\{31\}$ and $\{4\}$ the eigenspinors cannot span $\mathfrak{S}_{(AB)}$ but all have the form $\alpha_{(A}\xi_{B)}$ with α_A the multiple principal spinor of Ψ_{ABCD} (and ξ_A proportional to α_A in case $\{31\}$, *cf.* Table (8.3.41) below). For we have $\alpha^A\alpha^B\Psi_{ABCD}=0$ and so $\Psi_{AB}{}^{CD}\phi_{CD}=\lambda\phi_{AB}$ implies $\lambda\alpha^A\alpha^B\phi_{AB}=0$; thus α^A is a principal spinor of ϕ_{AB} as asserted, or $\lambda=0$. But in the latter case it is still true that $\alpha^A\alpha^B\phi_{AB}=0$, as follows from $\Psi_{AB}{}^{CD}\phi_{CD}=0$ quite obviously in case {4} and only slightly less obviously in case {31} (transvect once with the other principal spinor of Ψ_{ABCD}).

Canonical forms for the cases {4} and {31} are obtained trivially by taking, respectively,

$$\Psi_{ABCD}=o_Ao_Bo_Co_D \tag{8.3.37}$$

and

$$\Psi_{ABCD}=-4o_{(A}o_Bo_C\iota_{D)}. \tag{8.3.38}$$

In either case the freedom $(o^A,\iota^A)\mapsto(\lambda o^A,\lambda^{-1}\iota^A)$ is available to absorb any overall factor. (Note that for {22}, on the other hand, this would not allow the factor 6η in (8.3.36) to be eliminated.) In the case {211} we have just three distinct GPNDs so we can choose the spin-frame such that the GPNDs lie in any three preassigned directions relative to the frame, say those of the flagpoles of $o^A, o^A\pm i\iota^A$, giving

$$\psi_{ABCD}=6\eta(o_Ao_Bo_Co_D+o_{(A}o_B\iota_C\iota_{D)}), \tag{8.3.39}$$

the overall factors being chosen to agree with (8.3.25) with $\chi=0$ (the eigenvalues $\lambda_1=\lambda_2=\eta$, $\lambda_3=-2\eta$ being those of the matrix (8.3.5) with $\Psi_0=\Psi_1=\Psi_3=0$, $\Psi_2=\eta$, $\Psi_4=6\eta$). Many other alternative canonical forms are clearly possible

Summary of canonical forms; Petrov types and Jordan forms

The canonical forms (8.3.34), (8.3.36), (8.3.37), (8.3.38), (8.3.39) for the various types can be summarized by the following table:

Table (8.3.40)

	Ψ_0	Ψ_1	Ψ_2	Ψ_3	Ψ_4
{1111}	$\frac{3}{2}\eta\chi$	0	$\frac{1}{2}\eta(2-\chi)$	0	$\frac{3}{2}\eta\chi$
{211}	0	0	η	0	6η
{31}	0	0	0	1	0
{22}	0	0	η	0	0
{4}	0	0	0	0	1
{–}	0	0	0	0	0

The eigenspinors and corresponding eigenvectors are readily found in each case by referring to the matrix (8.3.5) and translating the appropriate eigenvectors into spinor form by means of (8.3.30). The results are collected in Table (8.3.41) below, where each eigenvalue is placed immediately below its corresponding eigenspinor. In the cases where more than one eigenspinor corresponds to the same eigenvalue, alternative linear combinations of the eigenspinors could have been used.

Table (8.3.41)

$\{1111\}$	$o_Ao_B - \iota_A\iota_B$ $\eta(1-2\chi)$	$o_Ao_B + \iota_A\iota_B$ $\eta(1+\chi)$	$o_{(A}\iota_{B)}$ $\eta(\chi-2)$
$\{211\}$	o_Ao_B η, η	$o_{(A}\iota_{B)}$ -2η	
$\{31\}$	o_Ao_B $0,0,0$		
$\{22\}$	o_Ao_B η	$\iota_A\iota_B$ η	$o_{(A}\iota_{B)}$ -2η
$\{4\}$	o_Ao_B $0,0$	$o_{(A}\iota_{B)}$ 0	
$\{-\}$	o_Ao_B 0	$\iota_A\iota_B$ 0	$o_{(A}\iota_{B)}$ 0

Note that in the cases $\{1111\}$, $\{22\}$ and $\{-\}$ the eigenspinors span a three-complex-dimensional space; in the cases $\{211\}$ and $\{4\}$ they span a two-complex-dimensional space; and in case $\{31\}$ they span a one-complex-dimensional space. In the original terminology of Petrov (although he used tensor rather than spinor language), the three cases $\{1111\}$, $\{22\}$ and $\{-\}$ were all referred to as 'type I' essentially for this reason, the two cases $\{211\}$ and $\{4\}$ as 'type II', and the case $\{31\}$ as 'type III'. The scheme (8.1.9) is drawn in such a way that these different Petrov types are arranged in columns.

These Jordan normal forms, into which the matrix $\boldsymbol{\Psi}$ of (8.3.5) may be transformed by a similarity transformation, are readily obtained from (8.3.41). The results are shown in (8.3.42), which also includes the specialization scheme (8.1.9).

$$\{1111\}: \begin{bmatrix} \eta(1+\chi) & & \\ & \eta(1-2\chi) & \\ & & \eta(\chi-2) \end{bmatrix}$$

$$\{211\}: \begin{bmatrix} \eta & 1 & \\ & \eta & \\ & & -2\eta \end{bmatrix}$$

$$\{22\}: \begin{bmatrix} \eta & & \\ & \eta & \\ & & -2\eta \end{bmatrix} \qquad \{31\}: \begin{bmatrix} 0 & 1 & \\ & 0 & 1 \\ & & 0 \end{bmatrix}$$

$$\{4\}: \begin{bmatrix} 0 & 1 & \\ & 0 & \\ & & 0 \end{bmatrix}$$

$$\{-\}: \begin{bmatrix} 0 & & \\ & 0 & \\ & & 0 \end{bmatrix} \tag{8.3.42}$$

8.4 The eigenbivectors of the Weyl tensor and its Petrov classification

An eigenvalue equation similar to (8.3.7), but for the Weyl *tensor* C_{abcd} can be set up:

$$C_{ab}{}^{cd}X_{cd} = \mu X_{ab}, \tag{8.4.1}$$

where $X_{ab} \neq 0$ belongs to $\mathfrak{S}_{[ab]}$, the six-complex-dimensional space comprising the antisymmetric elements of $\mathfrak{S}_{ab}$ (our interest being essentially confined to *one* point in space–time, as it was all along in this chapter). The quantity X_{ab} is called an *eigenbivector* of $C_{ab}{}^{cd}$ corresponding to the eigenvalue μ. Writing C_{abcd} and X_{ab} out in terms of their irreducible parts (4.6.41), (3.4.17),

$$C_{abcd} = \Psi_{ABCD}\varepsilon_{A'B'}\varepsilon_{C'D'} + \varepsilon_{AB}\varepsilon_{CD}\bar{\Psi}_{A'B'C'D'} \tag{8.4.2}$$

$$X_{ab} = \phi_{AB}\varepsilon_{A'B'} + \varepsilon_{AB}\xi_{A'B'}, \tag{8.4.3}$$

equation (8.4.1) becomes

$$\Psi_{AB}{}^{CD}\phi_{CD} = \tfrac{1}{2}\mu\phi_{AB} \text{ and } \bar{\Psi}_{A'B'}{}^{C'D'}\xi_{C'D'} = \tfrac{1}{2}\bar{\mu}\xi_{A'B'}. \tag{8.4.4}$$

Comparing with (8.3.7), we see that the eigenvalues μ of $C_{ab}{}^{cd}$ are

$$\mu = 2\lambda_1, 2\lambda_2, 2\lambda_3, 2\bar{\lambda}_1, 2\bar{\lambda}_2, 2\bar{\lambda}_3, \tag{8.4.5}$$

where λ_1, λ_2, λ_3 are the eigenvalues of $\Psi_{AB}{}^{CD}$. The corresponding eigenbivectors are of the form

$$X_{ab} = \phi_{AB}\varepsilon_{A'B'} \tag{8.4.6}$$

in the first three cases, and

$$X_{ab} = \varepsilon_{AB}\bar{\phi}_{A'B'} \tag{8.4.7}$$

in the remaining three, ϕ_{AB} being the appropriate eigenspinor of $\Psi_{AB}{}^{CD}$.

Note that the eigenbivectors (8.4.6) and (8.4.7) are all *complex* and anti-self-dual or self-dual, respectively (*cf.* (3.4.41), (3.4.35)). We obtain real eigenbivectors only if they arise as linear combinations of these complex ones, an eigenbivector (8.4.6) being added to its complex conjugate (8.4.7). But this yields an eigenbivector only if the corresponding eigenvalues are equal, and thus real (e.g. $\lambda_1 = \bar{\lambda}_1$). For each real λ there will then be a whole two-dimensional array of real eigenbivectors of $C_{ab}{}^{cd}$ (or four- or six-dimensional, if λ corresponds to two or three linearly independent eigenspinors of $\Psi_{AB}{}^{CD}$, respectively). A particular situation of some interest which gives rise to a real λ may be pointed out, namely when the two other λs are complex conjugates of one another, say $\lambda_1 = \bar{\lambda}_2$, since $\lambda_1 + \lambda_2 + \lambda_3 = 0$ then implies that λ_3 is real. In this case the six eigenvalues of $C_{ab}{}^{cd}$ coincide in pairs (at least) to give extra possibilities for eigenbivectors (e.g., complex linear combinations of (8.4.6) for λ_1 with (8.4.7) for λ_2). One easily finds from (8.3.25) – by setting $\lambda_1\bar{\lambda}_1 = \lambda_2\bar{\lambda}_2$ etc. – that this situation can only arise if $|1-\chi| = 1$ (case $\lambda_1 = \bar{\lambda}_2$), $|\chi| = 1$ (case $\lambda_1 = \bar{\lambda}_3$) or $\mathrm{Re}(\chi) = \frac{1}{2}$ (case $\lambda_3 = \bar{\lambda}_2$).

Components of C_{abcd} in a bivector basis

We shall relate the canonical forms that we obtained for Ψ_{ABCD} in the different cases to corresponding canonical forms for C_{abcd}. But to do this, we first need to obtain a general translation scheme between bivector components in a standard Minkowski frame and those arising from the spinor basis $\overset{\alpha}{\delta}_{AB}$ for $\mathfrak{S}_{(AB)}$ (*cf.* (8.3.3)). Defining six real bivectors $\overset{\alpha}{V}_{ab}$, ${}^*\overset{\alpha}{V}_{ab}$ ($\alpha = 1, 2, 3$) by

$$\overset{\alpha}{V}_{ab} + \mathrm{i}\ {}^*\overset{\alpha}{V}_{ab} = \overset{\alpha}{\delta}_{AB}\varepsilon_{A'B'} \tag{8.4.8}$$

(the star indicating dualization, *cf.* (3.4.21), (3.4.38)), we obtain – by the orthonormality of the δs – a basis for the set $\mathfrak{T}_{[ab]}$ of real bivectors, which is pseudo-orthonormal in the sense

$$\overset{\alpha}{V}_{ab}\overset{\beta}{V}{}^{ab} = \delta_{\alpha\beta}, \quad {}^*\overset{\alpha}{V}_{ab}{}^*\overset{\beta}{V}{}^{ab} = -\delta_{\alpha\beta}, \quad \overset{\alpha}{V}_{ab}{}^*\overset{\beta}{V}{}^{ab} = 0. \tag{8.4.9}$$

The relation between the components of a bivector X_{ab} in this basis and the components (8.3.4) of the corresponding ϕ_{AB}, $\xi_{A'B'}$ (*cf.* (8.4.3)) with respect to the bases $\overset{\alpha}{\delta}_{AB}$ and $\overset{\alpha}{\delta}_{A'B'}$ for $\mathfrak{S}_{(AB)}$ and $\mathfrak{S}_{(A'B')}$, respectively, is given by

$$\overset{\alpha}{X} := X_{ab}\overset{\alpha}{V}{}^{ab} = \overset{\alpha}{\phi} + \overset{\alpha}{\xi} \tag{8.4.10}$$

$${}^*\overset{\alpha}{X} := X_{ab}{}^*\overset{\alpha}{V}{}^{ab} = -\mathrm{i}\overset{\alpha}{\phi} + \mathrm{i}\overset{\alpha}{\xi}, \tag{8.4.11}$$

$$X_{ab}(\overset{\alpha}{V}{}^{ab} + \mathrm{i}{}^*\overset{\alpha}{V}{}^{ab}) = 2\phi_{AB}\overset{\alpha}{\delta}{}^{AB}$$

and

$$X_{ab}(\overset{\alpha}{V}{}^{ab} - \mathrm{i}{}^*\overset{\alpha}{V}{}^{ab}) = 2\xi_{A'B'}\overset{\alpha}{\delta}{}^{A'B'}$$

Translating the $\overset{\alpha}{\phi}$ and $\overset{\alpha}{\xi}$ into spinor dyad form according to (8.3.4) and its complex conjugate, and comparing the resulting equations (8.4.10), (8.4.11) with the Minkowski components $X_{\mathbf{ab}}$ of X_{ab} in the standard frame (3.1.20) (*cf.* (3.1.49)), we obtain, after some calculation,

$$X_{23} = \frac{1}{\sqrt{2}}\overset{1}{X}, \quad X_{31} = \frac{1}{\sqrt{2}}\overset{2}{X}, \quad X_{12} = \frac{1}{\sqrt{2}}\overset{3}{X} \tag{8.4.12}$$

$$X_{10} = \frac{1}{\sqrt{2}}{}^*\overset{1}{X}, \quad X_{20} = \frac{1}{\sqrt{2}}{}^*\overset{2}{X}, \quad X_{30} = \frac{1}{\sqrt{2}}{}^*\overset{3}{X}. \tag{8.4.13}$$

Now the components of C_{abcd} in the basis $\overset{\alpha}{V}_{ab}$, ${}^*\overset{\alpha}{V}_{ab}$ are

$$2\overset{\alpha\beta}{A} := \overset{\alpha}{V}{}^{ab}C_{abcd}\overset{\beta}{V}{}^{cd} = -{}^*\overset{\alpha}{V}{}^{ab}C_{abcd}{}^*\overset{\beta}{V}{}^{cd} \tag{8.4.14}$$

and

$$2\overset{\alpha\beta}{B} := \overset{\alpha}{V}{}^{ab}C_{abcd}{}^*\overset{\beta}{V}{}^{cd} = {}^*\overset{\alpha}{V}{}^{ab}C_{abcd}\overset{\beta}{V}{}^{cd} \tag{8.4.15}$$

because ${}^*C^*_{abcd} = -C_{abcd}$ and $C^*_{abcd} = {}^*C_{abcd}$ (*cf.* (4.6.11) etc.). We shall presently see that they satisfy the trace-free and symmetric properties

$$\overset{\alpha\beta}{A} = \overset{\beta\alpha}{A}, \quad \overset{\alpha\alpha}{A} = 0; \quad \overset{\alpha\beta}{B} = \overset{\beta\alpha}{B}, \quad \overset{\alpha\alpha}{B} = 0 \tag{8.4.16}$$

(*cf.* (8.4.21) below, in conjunction with (8.3.6)). By (8.4.12) and (8.4.13) we obtain the standard Minkowski components in the form

$$\begin{aligned}
-C_{0101} &= C_{2323} = \overset{11}{A}, & -C_{0102} &= C_{2331} = \overset{12}{A},\\
-C_{0103} &= C_{2312} = \overset{13}{A}, & -C_{0202} &= C_{3131} = \overset{22}{A},\\
-C_{0203} &= C_{3112} = \overset{23}{A}, & -C_{0303} &= C_{1212} = \overset{33}{A},\\
C_{0123} &= \overset{11}{B}, & C_{0131} &= C_{2302} = \overset{12}{B},\\
C_{0112} &= C_{2303} = \overset{13}{B}, & C_{0231} &= \overset{22}{B},\\
C_{0212} &= C_{3103} = \overset{23}{B}, & C_{0312} &= \overset{33}{B}.
\end{aligned} \tag{8.4.17}$$

Denoting the matrices of components $\overset{\alpha\beta}{A}$ and $\overset{\alpha\beta}{B}$ simply by $\mathbf{A}$ and $\mathbf{B}$ respectively, we can express (8.4.17) in the form

$$C_{\mathbf{abcd}} = \begin{array}{c|ccc|ccc|} \mathbf{ab} \backslash \mathbf{cd} & 01 & 02 & 03 & 23 & 31 & 12 \\ \hline 01 & & & & & & \\ 02 & & -\mathbf{A} & & & \mathbf{B} & \\ 03 & & & & & & \\ \hline 23 & & & & & & \\ 31 & & \mathbf{B} & & & \mathbf{A} & \\ 12 & & & & & & \\ \hline \end{array} \tag{8.4.18}$$

The components of $C_{\mathbf{ab}}{}^{\mathbf{cd}}$ and $C^{\mathbf{abcd}}$ are obtained similarly:

$$C_{\mathbf{ab}}{}^{\mathbf{cd}} = \begin{bmatrix} \mathbf{A} & \mathbf{B} \\ -\mathbf{B} & \mathbf{A} \end{bmatrix}, \quad C^{\mathbf{abcd}} = \begin{bmatrix} -\mathbf{A} & -\mathbf{B} \\ -\mathbf{B} & \mathbf{A} \end{bmatrix}. \tag{8.4.19}$$

These matrices can be related very simply to the matrix $\mathbf{\Psi}$ of (8.3.5), whose components we now denote by $\overset{\alpha\beta}{\Psi}$. For,

$$\begin{aligned} \overset{\alpha\beta}{\Psi} &= \overset{\alpha}{\delta}{}^{AB}\Psi_{ABCD}\overset{\beta}{\delta}{}^{CD} \\ &= \tfrac{1}{4}(\overset{\alpha}{V}{}^{ab} + \mathrm{i}\,{}^*\overset{\alpha}{V}{}^{ab})C_{abcd}\overset{\beta}{V}{}^{cd} \\ &= \tfrac{1}{2}\overset{\alpha\beta}{A} + \tfrac{1}{2}\mathrm{i}\overset{\alpha\beta}{B}. \end{aligned} \tag{8.4.20}$$

i.e., $\mathbf{\Psi} = \frac{1}{2}(\mathbf{A} + \mathrm{i}\mathbf{B})$. Thus

$$\mathbf{A} = \mathbf{\Psi} + \bar{\mathbf{\Psi}}, \quad \mathbf{B} = -\mathrm{i}\mathbf{\Psi} + \mathrm{i}\bar{\mathbf{\Psi}}. \tag{8.4.21}$$

Petrov's canonical forms

The canonical forms for the various Weyl spinor types that we have obtained in the previous section can now be directly translated to give corresponding canonical forms for the matrix (8.4.18) of components $C_{\mathbf{abcd}}$. Referring to Table (8.3.40) and to (8.3.5), we obtain, for the first three rows of this matrix in the various cases (also writing $\mathrm{Re}\,\eta =: \eta_1$, $\mathrm{Im}\,\eta =: \eta_2$ in cases $\{211\}$ and $\{22\}$):

$$\{1111\}: \quad \begin{array}{|cccccc|} \hline 2\mathrm{Re}[\eta(2\chi-1)] & \cdot & \cdot & -2\mathrm{Im}[\eta(2\chi-1)] & \cdot & \cdot \\ \cdot & -2\mathrm{Re}[\eta(\chi+1)] & \cdot & \cdot & 2\mathrm{Im}[\eta(\chi+1)] & \cdot \\ \cdot & \cdot & 2\mathrm{Re}[\eta(2-\chi)] & \cdot & \cdot & -2\mathrm{Im}[\eta(2-\chi)] \\ \hline \end{array}$$

$$\{211\}: \quad \begin{array}{|cccccc|} \hline 4\eta_1 & 6\eta_2 & \cdot & -4\eta_2 & 6\eta_1 & \cdot \\ 6\eta_2 & -8\eta_1 & \cdot & 6\eta_1 & 8\eta_2 & \cdot \\ \cdot & \cdot & 4\eta_1 & \cdot & \cdot & -4\eta_2 \\ \hline \end{array}$$

$$\{22\}: \quad \begin{array}{|cccccc|} -2\eta_1 & \cdot & \cdot & 2\eta_2 & \cdot & \cdot \\ \cdot & -2\eta_1 & \cdot & \cdot & 2\eta_2 & \cdot \\ \cdot & \cdot & 4\eta_1 & \cdot & \cdot & -4\eta_2 \end{array}$$

$$\{31\}: \quad \begin{array}{|cccccc|} \cdot & \cdot & 2 & \cdot & \cdot & \cdot \\ \cdot & \cdot & \cdot & \cdot & \cdot & 2 \\ 2 & \cdot & \cdot & \cdot & 2 & \cdot \end{array}$$

$$\{4\}: \quad \begin{array}{|cccccc|} 1 & \cdot & \cdot & \cdot & 1 & \cdot \\ \cdot & -1 & \cdot & 1 & \cdot & \cdot \\ \cdot & \cdot & \cdot & \cdot & \cdot & \cdot \end{array} \tag{8.4.22}$$

These forms, except in the case {211}, are essentially identical with those obtained by Petrov (1954) using direct tensor methods. In the case {211}, Petrov's form differs slightly from ours and corresponds to replacing the row $(0\,0\,\eta\,0\,6\eta)$ in (8.3.40) by $(0\,0\,\eta\,0\,1)$. The change of spin-frame achieving this is easily found.

8.5 Geometry and symmetry of the Weyl curvature

In the present section we study the geometry of the eigenbivector structure of C_{abcd} (eigenspinor structure of Ψ_{ABCD}) in relation to the GPNDs and use this to discuss the discrete symmetries of the Weyl curvature in case {1111}. We also obtain the symmetries in the other cases, using more direct methods. For comparison, we give, in addition, a corresponding discussion for the classification of Maxwell field tensors.

Special planes and directions for type {1111}

As we have seen, the basic eigenbivectors (8.4.6) and (8.4.7) of the Weyl tensor occur in anti-self-dual self-dual pairs. Each such pair at a point P defines a pair of totally orthogonal* planes through the origin in the tangent space at P. These planes are determined by the two real simple bivectors which are linear combinations of the eigenbivector pair in question. One of these planes is timelike and may alternatively (and more

* The term 'totally orthogonal' means, here, that they are orthogonal complements, i.e. that *every* direction in one plane is orthogonal to *every* direction in the other (e.g. the (x, y)- and (z, t)-planes in Minkowski space). 'Orthogonal' means merely that *some* direction in one plane is orthogonal to every direction in the other.

simply) be described as that spanned by the two PNDs of the corresponding eigenspinor of $\Psi_{AB}{}^{CD}$. The second plane is spacelike and is just the orthogonal complement of the first. To see all this, let $\phi_{AB} = \alpha_{(A}\beta_{B)}$ be the eigenspinor and $\phi_{AB}\varepsilon_{A'B'}$ the corresponding self-dual eigenbivector. The real bivector

$$F_{ab} = \gamma\phi_{AB}\varepsilon_{A'B'} + \bar{\gamma}\varepsilon_{AB}\bar{\phi}_{A'B'}, \tag{8.5.1}$$

(where γ is some complex coefficient) is *simple* (*cf.* around (3.5.30)) if and only if $\gamma^2\phi_{AB}\phi^{AB}$ is real, as follows directly from (8.5.1) and (3.4.22) and the criterion (3.5.35)(ii). It is easy to see that when this condition is satisfied, $\gamma = \bar{\alpha}_{A'}\bar{\beta}^{A'}$ or $\gamma = \mathrm{i}\bar{\alpha}_{A'}\bar{\beta}^{A'}$ (up to real non-zero multiples, which can here be ignored). In either case F_{ab} then becomes expressible in the form

$$F_{ab} = p_a q_b - q_a p_b \tag{8.5.2}$$

where, for $\gamma = \bar{\alpha}_{A'}\bar{\beta}^{A'}$,

$$p_a = \alpha_A\bar{\alpha}_{A'}, \quad q_a = \beta_A\bar{\beta}_{A'} \tag{8.5.3)(a}$$

and, for $\gamma = \mathrm{i}\bar{\alpha}_{A'}\bar{\beta}^{A'}$,

$$p_a = \sqrt{2}\,\mathrm{Re}\,(\beta_A\bar{\alpha}_{A'}), \quad q_a = \sqrt{2}\,\mathrm{Im}\,(\beta_A\bar{\alpha}_{A'}). \tag{8.5.3)(b}$$

The (simple) bivector (8.5.2) is considered to represent the plane of p_a, q_a. In case (8.5.3)(*a*), it represents the timelike plane spanned by the PND of ϕ_{AB}; in case (8.5.3)(*b*) it represents the spacelike plane spanned by two vectors orthogonal to both the PND of ϕ_{AB} (as one easily verifies). Note also that the two simple bivectors above are $(\pm)$ duals of each other, as follows from the fact that the corresponding coefficients γ in (8.5.1) differ by a purely imaginary factor (*cf.* (3.4.22)).

Applying these results now to a discussion of Weyl spinors (or tensors) of type $\{1111\}$ we see that the three eigenspinors of $\Psi_{AB}{}^{CD}$ give us three pairs of totally orthogonal planes. In fact, every one of these six 'eigenplanes' is orthogonal to each of the others – as follows from the orthogonality of eigenspinors corresponding to different eigenvalues, and the consequent orthogonality of the corresponding simple bivectors, say $F_{ab} = p_a q_b - q_a p_b$ and $G_{ab} = r_a s_b - s_a r_b$, in the sense $F_{ab}G^{ab} = 0$. For it implies the existence of a non-zero pair of real coefficients a, b, such that $ap_a + bq_a$ is orthogonal to both r_a and s_a (that is $ap_a r^a + bq_a r^a = 0$, $ap_a s^a + bq_a s^a = 0$), for which the condition is $p_a r^a q_b s^b - p_a s^a q_b r^b = 0$, i.e. $F_{ab}G^{ab} = 0$.

We now show that the six eigenplanes intersect in four real lines having the characteristics of a Minkowski tetrad. The four directions so defined are called the *Riemann principal directions* of C_{abcd}. As might be expected, they stand in a special relation to the principal null directions of C_{abcd}. To prove our assertion, we first show that each eigenplane U intersects any other eigenplane V, except its own orthogonal complement U', in a line. For U contains a vector p orthogonal to V', the complement of V. But V contains *all* vectors orthogonal to V', hence also p. Since both U and V pass through the origin, they intersect along p. Consider next two totally orthogonal pairs U, U' and V, V' of eigenplanes. They intersect in four lines $U \cap V$, $U \cap V'$, $U' \cap V$, $U' \cap V'$. Any two of these lines are orthogonal, containing an eigenplane and its complement respectively. The remaining two eigenplanes W, W' intersect each of the four others, and so must intersect them along the lines $U \cap V$, $U' \cap V'$ and $U \cap V'$, $U' \cap V$ respectively (as one can easily see by adopting these directions as a basis and looking at the vectors spanning W or W'). Hence, as we asserted, there are just four intersection lines of the six eigenplanes, and they are mutually orthogonal. Being in Minkowski space, three must therefore be spacelike and one timelike. The latter is evidently the intersection line of the three timelike eigenplanes.

An alternative way of looking at the same situation is to represent the eigenplanes as lines in the projective 3-space of directions through the origin. The configuration of planes U, V, U', V' is represented by a skew quadrilateral, since opposite edges, representing complementary planes, can have no point in common (the planes having no line in common). W and W' are represented by two new lines, each of which intersects all the old four. Evidently W and W' must correspond to the diagonals of the quadrilateral, and no new intersection points (lines) are created. In all, the six eigenplanes correspond to the edges of a tetrahedron (see Fig. 8-4), opposite edges being orthogonally complementary and the vertices giving the directions of the axes.

It is now straightforward to verify that the Riemann principal directions as defined above coincide with the Minkowski tetrad relative to which the Petrov form (8.4.22)(1) of C_{abcd} was (implicitly) calculated.

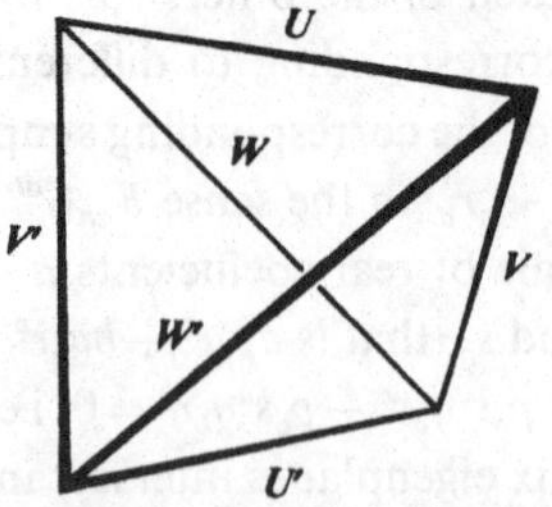

Fig. 8-4. In the projective 3-space, $\mathbb{PV}$, of directions through the origin, the eigenplanes of C_{abcd} are represented by the edges of a tetrahedron, the vertices of which correspond to the Riemann principal directions. (This tetrahedron is not to be confused with that of Fig. 8-5. The vertices here correspond to the three axes EF, GH and KL of that figure, together with the time-direction giving the rest-frame.)

From that form there follows certain discrete symmetries of $C_{\mathbf{abcd}}$ in the case $\{1111\}$, namely those (orientation preserving) reflections which reverse just two of the spatial axes. For the form (8.4.22) (1) gives components with the property

$$C_{\mathbf{abcd}} = 0 \quad \textit{whenever} \quad \mathbf{a} = \mathbf{c} \quad \textit{and} \quad \mathbf{b} \neq \mathbf{d}, \tag{8.5.4}$$

from which it follows that if precisely two axes are reversed and the others left unchanged all components $C_{\mathbf{abcd}}$ are unaltered.

The disphenoid for type $\{1111\}$

In fact it is possible to derive these symmetries directly without prior reference to the Petrov canonical form. Consider the sphere S^+ and the four points A, B, C, D on S^+ representing the GPND (*cf.* §8.2). There is a unique restricted Lorentz transformation $\mathscr{L}_1$ which sends A, B, C into B, A, D, and since the cross-ratios $\{A, B, C, D\}$ and $\{B, A, D, C\}$ are equal (*cf.* (8.3.15)) it follows that D is also sent into C by $\mathscr{L}_1$. The square of $\mathscr{L}_1$ is clearly the identity since is sends each of A, B, C, (and D) into itself. Therefore $\mathscr{L}_1$ cannot be a null rotation (*cf.* (3.6.47)) and so leaves precisely two points E, F invariant. Indeed, choosing a Lorentz frame in which E and F are antipodal on S^+, $\mathscr{L}_1$ is represented simply as a rotation through an angle π about the axis EF. Similarly there is a corresponding restricted Lorentz transformation $\mathscr{L}_2$ which sends A, B, C, D into C, D, A, B with invariant points G and H. Now $\mathscr{L}_2$ is invariant under $\mathscr{L}_1$ (i.e. $\mathscr{L}_1^{-1}\mathscr{L}_2\mathscr{L}_1 = \mathscr{L}_2$, which holds because the left-hand transformation effects $ABCD \mapsto BADC \mapsto DCBA \mapsto CDAB$). Thus G rotates into H under $\mathscr{L}_1$ and so the line GH is perpendicular to the axis EF. We can now apply a boost along the axis EF, moving the line GH so that it intersects the axis EF at its mid-point (the centre of S^+). Now both pairs E, F and G, H are antipodal (and together they form the vertices of a square). Similarly there is a transformation $\mathscr{L}_3$, which has invariant points K, L and effects $ABCD \mapsto DCBA$. The line KL intersects each of EF and GH at right angles, so we have a triad of mutually perpendicular lines through the centre of S^+ (the six points E, F, G, H, K, L forming the vertices of a regular octahedron).

The rotations through π about the axes EF, GH, KL each send the unordered set (A, B, C, D) into itself. In fact, given A, we can locate B, C and D simply by applying each rotation in turn. Thus, B is obtained by reflecting A through EF; also C, by reflecting A through GH; and D by reflecting A through KL. The final configuration of four points $ABCD$

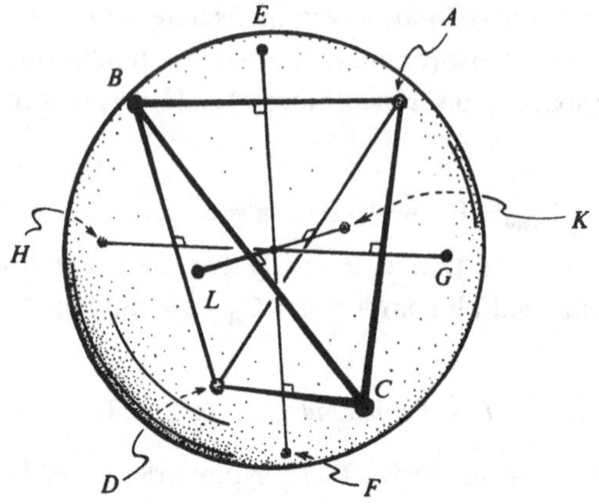

Fig. 8-5. The disphenoid, whose vertices correspond to the PND as represented on S^+ in a special frame (type $\{1111\}$).

constitutes the set of vertices of a special kind of tetrahedron known as a *disphenoid*, characterized by the fact that it has opposite edges equal in pairs, which property follows here from the existence of rotations sending each edge into its opposite. Consequently the joins of mid-points of opposite edges give three mutually orthogonal axes of two-fold symmetry – here the lines EF, GH and KL (see Fig. 8-5).

When the cross-ratio χ is real, the disphenoid $ABCD$ is flattened into a rectangle. When $\chi = -1, 2$, or $\frac{1}{2}$ (harmonic case) it becomes a square. When $\chi = e^{2\pi i/3}, e^{-2\pi i/3}$ (equianharmonic case) it becomes a regular tetrahedron. (See after (1.3.12).)

The rotations $\mathscr{L}_1, \mathscr{L}_2, \mathscr{L}_3$, each send the disphenoid $ABCD$ into itself. They each must therefore send the Weyl spinor into some multiple of itself. But since each transformation squares to the identity, this multiple must be ± 1, and, by symmetry, the same multiple for each of $\mathscr{L}_1, \mathscr{L}_2$ and $\mathscr{L}_3$. But $\mathscr{L}_1\mathscr{L}_2 = \mathscr{L}_3$, so each multiple must in fact be unity. It follows that Ψ_{ABCD} (and hence C_{abcd}) must itself be invariant under these three rotations.

The significance of the three pairs of points (E, F), (G, H) and (K, L) in relation to the earlier discussion in this chapter is that they are the representations on S^+ of the PNDs of the three eigenspinors of $\Psi_{AB}{}^{CD}$. This may be seen from the invariance of $\Psi_{AB}{}^{CD}$ under the three rotations $\mathscr{L}_1, \mathscr{L}_2, \mathscr{L}_3$. For the eigenspinors must also be invariant under *each* rotation (the eigenvalues being distinct). The only pairs of points which are so invariant are, in fact, (E, F), (G, H) and (K, L).

The spin transformations giving these rotations may be explicitly exhibited, namely

$$\xi_A \mapsto \sqrt{2}\phi_A{}^B\xi_B \tag{8.5.5}$$

where ϕ_{AB} is a normalized eigenspinor, $\phi_{AB}\phi^{AB} = 1$ (the factor $\sqrt{2}$ being inserted to obtain the correct normalization (3.6.30) for a spin transformation. From the discussion of §3.6 it follows that the PNDs of ϕ_{AB} are indeed fixed directions of the transformation (8.5.5) and that the square of (8.5.5) gives the negative identity spin transformation $((\sqrt{2}\phi_A{}^B)(\sqrt{2}\phi_B{}^C) = -\varepsilon_A{}^C)$ – which corresponds to the identity Lorentz transformation, as required. Choosing a spin-frame o^A, ι^A for which the canonical form (8.3.34) holds

$$\Psi_{ABCD} = \tfrac{3}{2}\eta\chi\iota_A\iota_B\iota_C\iota_D + 3\eta(2-\chi)o_{(A}o_B\iota_C\iota_{D)} + \tfrac{3}{2}\eta\chi o_A o_B o_C o_D,$$

we see that these transformations are indeed given when ϕ_{AB} takes the values (8.3.3) in turn, so that

$$o_A \mapsto i\iota_A, \quad \iota_A \mapsto io_A,$$

or

$$o_A \mapsto \iota_A, \quad \iota_A \mapsto -o_A,$$

or

$$o_A \mapsto io_A, \quad \iota_A \mapsto -io_A,$$

in the three cases, respectively.

The geometrical picture we have now set up gives us an independent way of establishing the Petrov canonical form for type $\{1111\}$. With the (unique) choice of time-axis as given above (in terms of the six eigenplanes), the four points A, B, C, D take on the symmetrical disphenoid configuration we have described. The resulting discrete rotational symmetries (which arise from the timelike planes spanned by the PNDs of the eigenspinors, i.e. the rotation axes EF, GH, KL) imply that the characteristic property (8.5.4) must hold for the components of C_{abcd} in the frame defined by the rotation axes.

Symmetries in special $\{1111\}$ *cases*

When, in special cases, the disphenoid possesses symmetries beyond these, it does not necessarily follow that C_{abcd} is invariant under them. For example, in the equianharmonic case, the regular tetrahedron $ABCD$ can be rotated into itself through $2\pi/3$ about an axis of three-fold symmetry – keeping one vertex, say A, fixed and permuting the remaining three. That such a rotation cannot leave C_{abcd} invariant becomes evident when we examine its fingerprint (*cf.* §8.2). Since A corresponds to a simple GPND, we have a configuration resembling the first diagram of Fig. 8-3 at A. This does not exhibit the three-fold symmetry which would be required for the rotation to send C_{abcd} to itself. In fact, such rotations effect duality rotations of C_{abcd} through angles $2\pi/3$, $4\pi/3$ (*cf.* (4.8.15), (4.8.16)). Similarly, in the harmonic case, the square $ABCD$ admits a rotation through $\pi/2$ about an axis through its centre and perpendicular to its plane. But this axis intersects S^+ in two points at which the fingerprint pattern cannot exhibit four-fold symmetry. Thus, again, this rotation effects a duality rotation of C_{abcd}, this time through an angle π, which sends C_{abcd} to its negative (since a single repetition of this rotation gives an allowable symmetry of C_{abcd} of the kind considered above).

The disphenoid possesses reflectional symmetries in the following two cases: (i) when it is flattened to a rectangle (χ real) and (ii) when four of its edges are equal: $|\chi|=1$, $|1-\chi|=1$, or $\mathrm{Re}(\chi)=\frac{1}{2}$ (*cf.* between (8.4.7) and (8.4.8)). In these cases, the reflectional symmetries may or may not apply to C_{abcd}, depending upon the particular value of η which accompanies the cross-ratio value χ, so that the fingerprint pattern possesses the appropriate symmeties. We have reflectional symmetries in case (i) if η is real and in case (ii) when η is such that one of the eigenvalues of Ψ_{ABCD} is the complex conjugate of another.* The necessity of these conditions follows from the fact that under a spatial reflection $\Psi_{AB}{}^{CD}$ becomes $\Psi_{A'B'}{}^{C'D'}$, so the unordered set $(\lambda_1, \lambda_2, \lambda_3)$ of eigenvalues of the former is sent into the unordered set $(\bar{\lambda}_1, \bar{\lambda}_2, \bar{\lambda}_3)$ of eigenvalues of the latter. The sufficiency follows also: if the λ-eigenvalues are sent into the $\bar{\lambda}$-eigenvalues in some order by the reflective transformation, then no duality rotation of Ψ_{ABCD} can be involved, since if Ψ_{ABCD} picks up a phase factor, so also would each λ.

Type {211}

Let us now consider the symmetries of C_{abcd} in the algebraically special cases. Consider {211} first. Any restricted Lorentz symmetry must send the three GPNDs to themselves, and since the repeated GPND is singled out, the only possibility is the discrete symmetry which interchanges the other two GPNDs. This again is a rotational symmetry of period 2 (rotation through π) in a suitable Lorentz frame and it actually sends C_{abcd} to itself rather than to its negative. This is easily seen, in many ways, e.g.

* In each case there is an extra reflectional symmetry when χ is harmonic.

from the fact that

$$o_A \mapsto \mathrm{i} o_A, \quad \iota_A \mapsto -\mathrm{i}\iota_A \tag{8.5.6}$$

(or its negative) leaves the canonical form (8.3.39)

$$\Psi_{ABCD} = 6\eta(o_A o_B o_C o_D + o_{(A} o_B \iota_C \iota_{D)}) \tag{8.5.7}$$

invariant while interchanging the flagpole directions of the principal spinors $o^A + \mathrm{i}\iota^A, o^A - \mathrm{i}\iota^A$. Alternatively, we need merely examine the fingerprint pattern of Fig. 8-3 for the repeated PND ($n = 2$) to see that a rotational symmetry which leaves the repeated point fixed cannot transform the pattern to the orthogonal pattern (which would have to be the case if C_{abcd} were sent to its negative – the only alternative possibility for a symmetry of period 2).

There are two orthochronous improper Lorentz transformations to consider, each of period 2. One of these leaves all three GPNDs invariant while the other interchanges the pair of simple ones. They can be represented spinorially by linear maps from $\mathfrak{S}_A$ to $\mathfrak{S}_{A'}$ generated by

$$o_A \mapsto \mathrm{i} o_{A'}, \quad \iota_A \mapsto -\mathrm{i}\iota_{A'} \tag{8.5.8}$$

and

$$o_A \mapsto o_{A'}, \quad \iota_A \mapsto \iota_{A'} \tag{8.5.9}$$

(and their negatives), respectively. The Weyl tensor is sent to itself [to its negative] if and only if Ψ_{ABCD} is sent to $\bar{\Psi}_{A'B'C'D'}$ [to $-\bar{\Psi}_{A'B'C'D'}$] and this occurs in both cases if and only if the eigenvalues are real [pure imaginary] (i.e. η is real [pure imaginary] in (8.5.7)).

Type {31}

Consider next the case {31}. The canonical form (8.3.38)

$$\Psi_{ABCD} = -4 o_{(A} o_B o_C \iota_{D)}$$

is invariant only under restricted Lorentz transformations for which o_A and ι_A are sent to multiples of themselves. Preserving the dyad normalization $o_A \iota^A = 1$, we leave ourselves only with $(o_A, \iota_A) \mapsto \pm(o_A, \iota_A)$, which both correspond to the identity Lorentz transformation. The only non-trivial orthochronous Lorentz transformation preserving C_{abcd} is, in fact, the reflection defined by (8.5.9). Examination of the fingerprint patterns for $n = 1, 3$ in Fig. 8-3 will also make it clear that no proper rotation can leave a {31} Weyl tensor invariant. Rotations leaving both PND invariant give duality rotations of C_{abcd}.

Type $\{22\}$

We now examine the case $\{22\}$. The canonical form (8.3.36):

$$\Psi_{ABCD} = 6\eta o_{(A}o_B\iota_C\iota_{D)} \tag{8.5.10}$$

is clearly invariant under the spin transformations

$$o_A \mapsto \lambda o_A, \quad \iota_A \mapsto \lambda^{-1}\iota_A, \tag{8.5.11}$$

and

$$o_A \mapsto \lambda\iota_A, \quad \iota_A \to -\lambda^{-1}o_A \tag{8.5.12}$$

(with $0 \neq \lambda \in \mathbb{C}$), and under no others. The transformations (8.5.11) form a connected group of two real dimensions, namely the *multiplicative group of non-zero complex numbers.* (In terms of Lorentz transformations, this is the multiplicative group of λ^2.) The transformations (8.5.12) form a two-real-parameter system disconnected from these. There are also improper orthochronous Lorentz transformations to consider, namely those given by (8.5.9) and its composition with (8.5.11) and (8.5.12). These are symmetries of C_{abcd} if and only if η is real (and then the fingerprint at the double GPND points has the form of the first or third $n = 2$ drawings in Fig. 8-3, so that reflectional symmetry is possible).

Much of the discussion given for the case $\{1111\}$ above will, of course also apply to the case $\{22\}$. But the disphenoid degenerates to a pair of repeated antipodal points on S^+ so that continuous rotations (and boosts) now become possible.

Type $\{4\}$

Next we consider the case $\{4\}$. The restricted Lorentz transformations defined by

$$o_A \mapsto o_A, \quad \iota_A \mapsto \iota_A + \lambda o_A \tag{8.5.13}$$

or by

$$o_A \mapsto \mathrm{i}o_A, \quad \iota_A \mapsto -\mathrm{i}\iota_A - \mathrm{i}\lambda o_A \tag{8.5.14}$$

(and their negatives) clearly preserve the canonical form (8.3.37)

$$\Psi_{ABCD} = o_A o_B o_C o_D \tag{8.5.15}$$

and there are no others. The transformations (8.5.13) are *null rotations* (*cf.* (3.6.47)) and form a connected group of two real dimensions, namely the *additive group of complex numbers.* The (non-null) transformations (8.5.14) form a two-real-parameter system disconnected from these. The improper orthochronous Lorentz transformations (8.5.8) and (8.5.9) and their compositions with (8.5.13) are clearly all symmetries of C_{abcd} (since they send Ψ_{ABCD} into $\bar{\Psi}_{A'B'C'D'}$), giving two additional two-real-parameter families. These four different possibilities can be understood in terms of their different behaviours in the neighbourhood of the multiple GPND. We refer to the fingerprint pattern of Fig. 8-3 for $n = 4$. The pattern has a discrete symmetry group of order 4, two of the symmetries being reflections.

Finally there is the case $\{-\}$. Its symmetry group is just the six-real-parameter Lorentz group (with or without reflections).

Dimensions of symmetry groups generally

The dimension of the symmetry group is related, in each case, to the number of independent scalars which can be formed and to the number of dimensions of the space of Weyl tensors of that particular type, as follows:

$$\text{dimension of space of Weyl tensors} + \text{dimension of symmetry group} \\ = \text{number of independent scalars} + 6. \tag{8.5.16}$$

To determine the (real) dimension of the space of Weyl tensors of a given type, we consider the freedom in choosing Ψ_{ABCD} with given multiplicity in its GPNDs. Let r be the number of distinct GPNDs. There is a $2r$-dimensional freedom in choosing these directions. Finally there is a two-dimensional freedom in choosing the complex overall scale factor for Ψ_{ABCD}. To calculate the number of independent scalars we merely observe what relations (if any) necessarily hold between the two complex scalars I and J by virtue of the particular choice of type. The remaining freedom in I and J gives the required answer. Alternatively, the number of independent eigenvalues may be counted, or, equivalently, the freedom allowed for η and χ. The significance of the '6' in (8.5.16) is that it is the dimension of the Lorentz group.

In Table (8.5.17) various pieces of information concerning the different Weyl curvature types are collected together. Relation (8.5.16), in particular, is illustrated. Note that the horizontal, vertical, and both sloping directions of (8.1.9) all have significance.

Classification of the Maxwell tensor

It is instructive to consider the corresponding facts and to exhibit a corresponding table for the electromagnetic case. Suppose the PNDs of the electromagnetic spinor ('EPNDs') are proportional to α_A, β_A, so that

$$\varphi_{AB} = \alpha_{(A}\beta_{B)}.$$

Table (8.5.17)

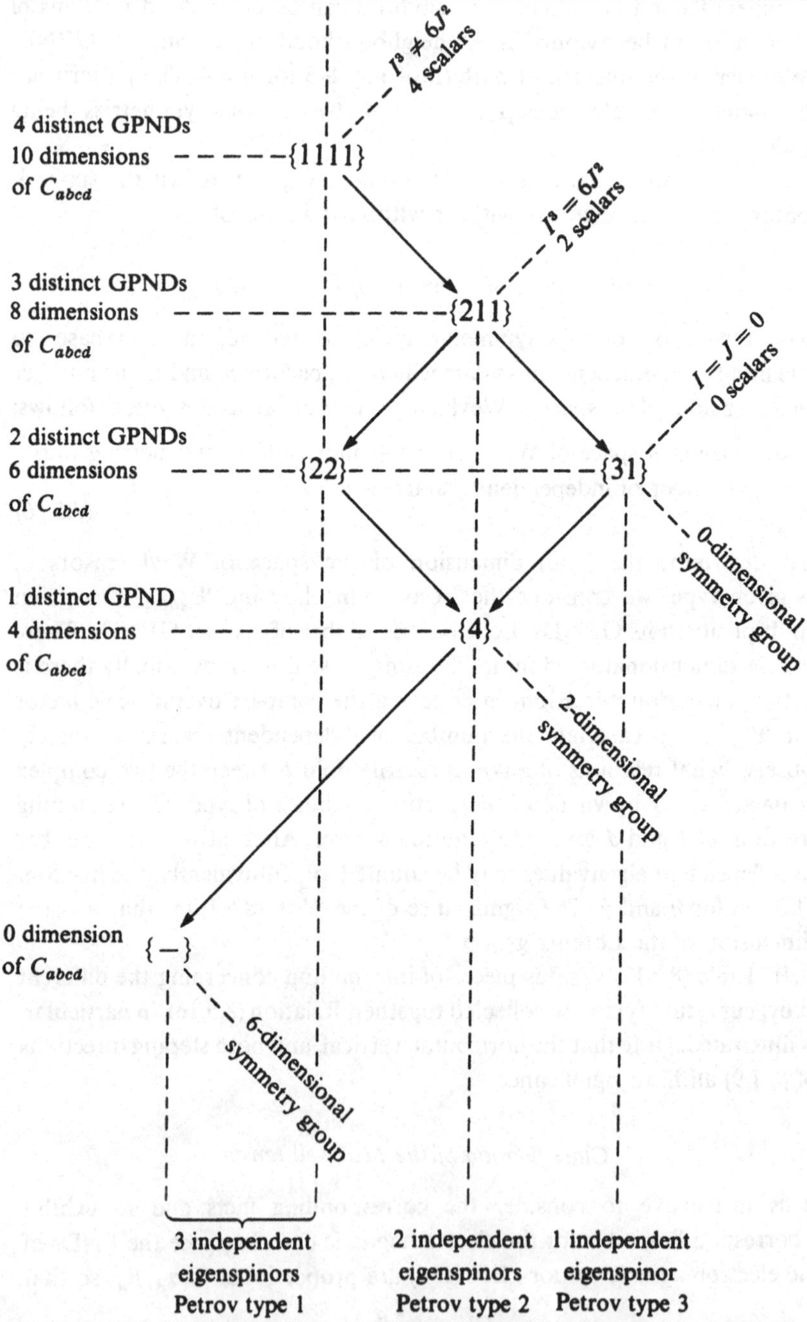
4 distinct GPNDs
10 dimensions
of C_{abcd}
{1111}
$I^3 \neq 6J^2$
4 scalars
3 distinct GPNDs
8 dimensions
of C_{abcd}
{211}
$I^3 = 6J^2$
2 scalars
2 distinct GPNDs
6 dimensions
of C_{abcd}
{22}
{31}
$I = J = 0$
0 scalars
0-dimensional
symmetry group
1 distinct GPND
4 dimensions
of C_{abcd}
{4}
2-dimensional
symmetry group
0 dimension
of C_{abcd}
{−}
6-dimensional
symmetry group
3 independent
eigenspinors
Petrov type 1
2 independent
eigenspinors
Petrov type 2
1 independent
eigenspinor
Petrov type 3

Then

$$\varphi_A{}^B\alpha_B = \tfrac{1}{2}\alpha_A\beta^B\alpha_B, \tag{8.5.18}$$

which shows that α_A is an eigenvector of $\varphi_A{}^B$ with eigenvalue

$$\lambda = \tfrac{1}{2}\alpha_A\beta^A. \tag{8.5.19}$$

Similarly β_A is an eigenvector, with eigenvalue $-\lambda$. (That the sum of the eigenvalues is zero is, of course, to be expected since $\varphi_A{}^A = 0$.) In the limiting case of a *null* field ($\alpha_A \propto \beta_A$) these eigenvectors coincide, and so do the eigenvalues: $\lambda = 0$. The necessary and sufficient condition for nullity (*cf.* (5.1.68) and just after (3.5.29)) is the vanishing of the invariant

$$K = P + \mathrm{i}Q = \varphi_{AB}\varphi^{AB} = -2\lambda^2. \tag{8.5.20}$$

The (real) electromagnetic tensor F_{ab} is given in terms of the electromagnetic spinor by (5.1.39):

$$F_{ab} = \varphi_{AB}\varepsilon_{A'B'} + \bar{\varphi}_{A'B'}\varepsilon_{AB}. \tag{8.5.21}$$

Its eigenvectors X_a and corresponding eigenvalues f are determined by the equation

$$F_a{}^bX_b = fX_a, \tag{8.5.22}$$

i.e.,

$$\varphi_A{}^BX_{BA'} + \varphi_{A'}{}^{B'}X_{AB'} = fX_{AA'}. \tag{8.5.23}$$

Now one easily checks by direct substitution that these eigenvectors and corresponding eigenvalues are

$$\begin{aligned}\alpha_A\bar{\alpha}_{A'}&:\lambda+\bar{\lambda}; \quad \beta_A\bar{\beta}_{A'}:-(\lambda+\bar{\lambda});\\ \alpha_A\bar{\beta}_{A'}&:\lambda-\bar{\lambda}; \quad \beta_A\bar{\alpha}_{A'}:-(\lambda-\bar{\lambda}).\end{aligned} \tag{8.5.24}$$

If α_A and β_A are distinct, let o_A, ι_A be a spin-frame proportional to them. Then the vectors in (8.5.24) are seen to be proportional to the null tetrad (3.1.21) associated with o_A, ι_A:

$$X_a = l_a, n_a, m_a, \bar{m}_a, \tag{8.5.25}$$

which well characterizes the mutual configuration of the four eigenvectors of F_{ab} in the generic case.

A special case of (8.5.24) arises when λ is either real or purely imaginary, which, by (8.5.20), is the same as saying $K =$ real. But that is the condition for F_{ab} to be simple (*cf.* after (5.1.70)), i.e., of the form

$$F_{ab} = p_{[a}q_{b]}. \tag{8.5.26}$$

Suppose first that $K > 0$, i.e., λ is purely imaginary. (As we have seen after (5.1.70), the field is then 'purely magnetic'.) In that case the eigenvalues in the first row of (8.5.24) are both zero. Nevertheless the corresponding eigenvectors do not become coincident: rather, they 'spread'. *Any* vector in the plane of l_a and n_a is now an eigenvector with eigenvalue zero. In fact, in this case we have

$$F_{ab} = -4\lambda m_{[a}\bar{m}_{b]}, \tag{8.5.27}$$

which, by (3.1.15)–(3.1.18), implies the following equations (which agree with (8.5.24) and so validate (8.5.27)):

$$F_{ab}l^b = 0, \quad F_{ab}n^b = 0, \quad F_{ab}m^b = 2\lambda m_a, \quad F_{ab}\bar{m}^b = -2\lambda\bar{m}_a. \tag{8.5.28}$$

Similarly, if $K < 0$, i.e., λ is real (and the field is 'purely electric'), we have

$$F_{ab} = 4\lambda l_{[a}n_{b]}, \tag{8.5.29}$$

and

$$F_{ab}l^b = 2\lambda l_a, \quad F_{ab}n^b = -2\lambda n_a, \quad F_{ab}m^b = 0, \quad F_{ab}\bar{m}^b = 0. \tag{8.5.30}$$

Note that in the purely magnetic case there is an entire *timelike* plane of eigenvectors totally orthogonal to the *spacelike* plane which contains the remaining two *distinct* eigenvectors and which is also the plane (8.5.27) of F_{ab}; whereas in the purely electric case it is just the other way around.

Only when the field is null ($\alpha_A \propto \beta_A$, $\lambda = 0$, $K = 0$) does the form (8.5.25) not apply. It is then seen from (8.5.24) that all the eigenvalues are zero, and all the eigenvectors of F_{ab} coincide.

We can now draw up a table (Table (8.5.31)) for the electromagnetic field analogous to Table (8.5.17) for the gravitational field. The continuous symmetry groups arising here are exactly the same as in the gravitational case, but there are now no discrete symmetries (except for $\{-\}$, trivially). In fact, Table (8.5.31) is virtually the same as the lower left-hand corner of Table (8.5.17). This is not really surprising, since we can subsume the classification of φ_{AB} – up to sign – under that of Ψ_{ABCD} if we formally put $\varphi_{(AB}\varphi_{CD)} =: \Psi_{ABCD}$.

8.6 Curvature covariants

The quantities I and J of (8.3.8) constitute what is known, in the classical theory of invariants (Grace and Young 1903), as a 'complete set of invariants' of the quartic form $P = \Psi_{ABCD}\xi^A\xi^B\xi^C\xi^D$. In effect, this means that every *scalar* expression formed from a general Ψ_{ABCD} by means of the four tensor-type operations of sum, outer product, contraction and

Table (8.5.31)

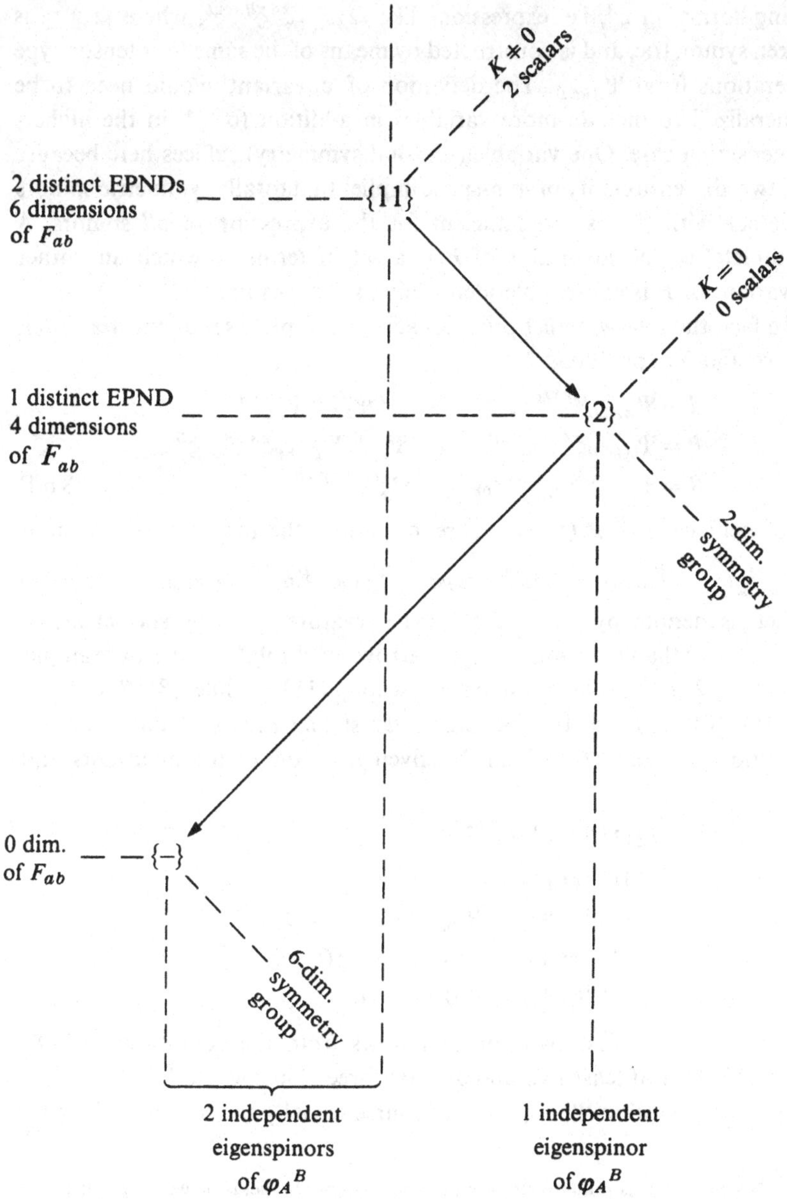

K ≠ 0
2 scalars
2 distinct EPNDs
6 dimensions
of F_{ab}
{11}
K = 0
0 scalars
1 distinct EPND
4 dimensions
of F_{ab}
{2}
2-dim.
symmetry
group
0 dim.
of F_{ab}
{–}
6-dim.
symmetry
group
2 independent
eigenspinors
of $\varphi_A{}^B$
1 independent
eigenspinor
of $\varphi_A{}^B$

index permutation* is expressible identically as a polynomial in I and J. The classical theory of invariants is also concerned with 'covariants', these being 'forms' in ξ^A, i.e. expressions like $\Omega_{AB\ldots L}\xi^A\xi^B\cdots\xi^L$, where $\Omega_{AB\ldots L}$ is taken symmetric, and is constructed by means of the same four tensor-type operations from Ψ_{ABCD}. The definition of 'covariant' would need to be generalized to include more variables in addition to 'ξ^A' in the higher-dimensional case. One variable (i.e. total symmetry) suffices here because the two-dimensionality of spin-space implies that totally symmetric spinors together with the εs are sufficient for the expression of *all* spinors. A 'complete' set of covariants of P is a set in terms of which any other covariant of P is expressible identically as a polynomial.

In fact, the following list provides such a complete set of covariants (*cf.* Grace and Young 1903):

$$\begin{aligned} I &= \Psi_{ABCD}\Psi^{ABCD}, \quad J = \Psi_{AB}{}^{CD}\Psi_{CD}{}^{EF}\Psi_{EF}{}^{AB} \\ P &= \Psi_{ABCD}\xi^A\xi^B\xi^C\xi^D, \quad Q = \Psi_{AB}{}^{EF}\Psi_{CDEF}\xi^A\xi^B\xi^C\xi^D \\ R &= \Psi_{ABC}{}^{K}\Psi_{DE}{}^{LM}\Psi_{FKLM}\xi^A\xi^B\xi^C\xi^D\xi^E\xi^F \end{aligned} \tag{8.6.1}$$

The coefficients of P, Q and R are, of course, the respective expressions

$$\Psi_{ABCD}, \quad \Psi_{(AB}{}^{EF}\Psi_{CD)EF}, \quad \Psi_{(ABC}{}^{K}\Psi_{DE}{}^{LM}\Psi_{F)KLM}. \tag{8.6.2}$$

Let us denote by $\{\ \}^{\text{cl}}$ the Weyl curvature types as special as, or more special than $\{\ \}$, following the arrows in (8.1.9).** Thus, for example, the case $\{211\}^{\text{cl}}$ includes all types except $\{1111\}$, while $\{31\}^{\text{cl}}$ consists of $\{31\}$, $\{4\}$, and $\{-\}$. It is of some interest that necessary and sufficient conditions for each $\{\ \}^{\text{cl}}$ can be given in terms of the invariants and covariants (8.6.1):

$$\begin{aligned} &\{211\}^{\text{cl}} \Leftrightarrow I^3 = 6J^2 \\ &\{31\}^{\text{cl}} \Leftrightarrow I = J = 0 \\ &\{22\}^{\text{cl}} \Leftrightarrow \Psi_{(ABC}{}^{K}\Psi_{DE}{}^{LM}\Psi_{F)KLM} = 0 \quad (R \equiv 0) \\ &\{4\}^{\text{cl}} \Leftrightarrow \Psi_{(AB}{}^{EF}\Psi_{CD)EF} = 0 \quad (Q \equiv 0) \\ &\{-\}^{\text{cl}} \Leftrightarrow \Psi_{ABCD} = 0 \quad (P \equiv 0). \end{aligned} \tag{8.6.3}$$

The truth of the first two entries follows from the conditions (8.3.28) and (8.3.29) for at least two and at least three coincident GPNDs respectively. The condition $\Psi_{ABCD} = 0$ is of course trivially the condition for $\{-\}$.

* Note that this is rather restrictive in various ways. Quotients between proportional indexed expressions can sometimes be used to give more general types of invariant (although not in the case of Ψ_{ABCD}). Also non-algebraic expressions and complex conjugates are not being considered here.

** 'cl' stands for topological *closure* in the space of Weyl spinors.

Consider next, $\Psi_{(AB}{}^{EF}\Psi_{CD)EF}=0$. This is equivalent to

$$0=Q=\eta^{EF}\eta_{EF}\quad where\quad \eta_{AB}=\Psi_{ABEF}\xi^E\xi^F \tag{8.6.4}$$

and so, by (3.5.29), to

$$\eta_{AB}=\eta_A\eta_B. \tag{8.6.5}$$

In particular, this holds when ξ^A is a principal spinor of Ψ_{ABCD}, in which case $\eta_{AB}\xi^A\xi^B=0$, so by (8.6.5) $\eta_A\xi^A=0$ and we deduce $\eta_{AB}\xi^B=0$. From this, (8.6.4) (2) gives $\Psi_{ABEF}\xi^A\xi^E\xi^F=0$, which is the condition for ξ^A to be a *repeated* principal spinor (*cf.* Proposition (3.5.26)). Hence *every* principal spinor is multiple, and this singles out $\{22\},\{4\},\{-\}$. However, $\Psi_{(AB}{}^{EF}\Psi_{CD)EF}$ cannot vanish in case $\{22\}$ (and in fact equals $-\eta\Psi_{ABCD}$, as one readily verifies from the canonical form (8.3.36)), whereas it obviously does vanish in the cases $\{4\}$ and $\{-\}$. Thus (8.6.3)(4) is established.

Lastly, consider the implications of $R=0$, when ξ^A is a principal spinor of Ψ_{ABCD}, so that

$$\Psi_{ABCD}\xi^B\xi^C\xi^D=\mu\xi_A \tag{8.6.6}$$

for some μ:

$$0=R=\mu\xi^K\Psi_{DE}{}^{LM}\Psi_{FKLM}\xi^D\xi^E\xi^F=\mu\eta^{LM}\eta_{LM}, \tag{8.6.7}$$

with η_{AB} as before. If $\mu\neq 0$, the same argument as before establishes ξ^A as a repeated principal spinor; if $\mu=0$, then (8.6.6) establishes this same fact. Thus, as before, the type must be $\{22\}$, $\{4\}$ or $\{-\}$. Now in this case, setting $\Psi_{ABCD}=\alpha_{(A}\alpha_B\beta_C\beta_{D)}$ we have

$$\Psi_{(ABC}{}^{K}\Psi_{DE}{}^{LM}\Psi_{F)KLM}=\nu\alpha_{(A}\alpha_B\alpha_C\beta_D\beta_E\beta_{F)}(\alpha^K\beta_K)^3 \tag{8.6.8}$$

for some ν (since no other type of term survives). But interchanging α_A and β_A leaves Ψ_{ABCD} unaltered while it reverses the sign of the RHS (8.6.8). Hence $\nu=0$ and we have established the last remaining condition in (8.6.3).

The expression (8.6.2)(3) is of interest for another reason: Provided Ψ_{ABCD} is not $\{22\}^{\text{cl}}$,

$$\Psi_{(ABC}{}^{K}\Psi_{DE}{}^{LM}\Psi_{F)KLM}=\zeta\phi_{(AB}\theta_{CD}\psi_{EF)} \tag{8.6.9}$$

for some $\zeta\neq 0$, where ϕ_{AB}, θ_{AB} and ψ_{AB} are the eigenspinors of $\Psi_{AB}{}^{CD}$ (with appropriate multiplicities in the cases $\{211\}$ and $\{31\}$ – *cf.* (8.3.41)). To see this, suppose

$$\phi_{AB}=\rho_{(A}\sigma_{B)}. \tag{8.6.10}$$

Then the eigenspinor condition gives

$$\Psi_{AB}{}^{CD}\rho_C\sigma_D=\lambda\rho_{(A}\sigma_{B)} \tag{8.6.11}$$

for some λ. Consequently

$$\Psi_{ABCD}\rho^A\rho^B\rho^C\sigma^D = 0, \tag{8.6.12}$$

whence

$$\Psi_{ABCD}\rho^A\rho^B\rho^C = \gamma\sigma_D \tag{8.6.13}$$

for some γ. By (8.6.13), (8.6.11) and (8.6.12) we have

$$\begin{aligned}&\Psi_{ABC}{}^K\Psi_{DE}{}^{LM}\Psi_{FKLM}\rho^A\rho^B\rho^C\rho^D\rho^E\rho^F\\ &= \gamma\Psi_{DE}{}^{LM}(\Psi_{FKLM}\rho^F\sigma^K)\rho^D\rho^E\\ &= \lambda\gamma\Psi_{DE}{}^{LM}\rho_L\sigma_M\rho^D\rho^E = 0,\end{aligned} \tag{8.6.14}$$

showing that every principal spinor of an eigenspinor of $\Psi_{AB}{}^{CD}$ is also a principal spinor of the LHS of (8.6.9). This establishes (8.6.9) in the case $\{1111\}$. The two remaining cases $\{211\}$ and $\{31\}$ then follow by going to the limits $\{1111\}\to\{211\}\to\{31\}$, since we are assured that the LHS of (8.6.9) will not go to zero, by our hypothesis (*cf.* (8.6.3)).

It may also be remarked that the vanishing of (8.6.2) (3) in the case $\{22\}$ is a consequence of (8.6.9). For in this case the eigenspinors are indeterminate up to a continuous symmetry group and no particular choice could be represented in the RHS of (8.6.9). It is also possible to infer the relation (8.6.9) in the general case without calculation (provided the expression is assumed not to vanish identically) because the six PNDs of the three eigenspinors must define an unordered set of six points on the sphere S^+ (possibly with some coincidences) which is defined by the unordered set (A, B, C, D) and hence invariant under all three rotations $\mathscr{L}_1, \mathscr{L}_2$ and $\mathscr{L}_3$ (*cf.* §8.5). This can only be (E, F, G, H, K, L).

Invariants for the full curvature

We end this section with a few brief remarks concerning invariants of the full curvature, and, in more detail, invariants of the combined gravitational and electromagnetic field. These latter lead to invariants of the full curvature under the special assumption of the Einstein–Maxwell equations. Our discussion will indicate some of the difficulties involved in finding complete sets of invariants in general.

If we do not assume that the vacuum field equations hold, we must, in general, consider all the curvature quantities

$$\Psi_{ABCD},\quad \Phi_{ABC'D'},\quad \Lambda \tag{8.6.15}$$

together. A discussion of the general classification problem for $\Phi_{ABC'D'}$ will be given in the next two sections (and this will be complicated enough). But in itself it would not be sufficient for a classification of the whole

Riemann curvature, since the interrelation between the structures of $\Phi_{ABC'D'}$ and of Ψ_{ABCD} must also be considered.

The number of invariants of R_{abcd} may be ascertained from (8.5.16). The space of tensors R_{abcd} is 20-dimensional (since there are 20 independent components) and there is no continuous symmetry group (clearly, since there is none for a general Weyl tensor). Hence there are $20-6=14$ invariants of some kind – although this gives no indication as to *how* these invariants might be constructed. In fact, it is not hard to string together curvature tensors or curvature spinors in many different ways, and to obtain 14 invariants which are in fact independent (*cf.* Witten 1959). However the problem of obtaining a 'complete set' is much more difficult (and, as far as we are aware, as yet unsolved). For it must be expected that a redundant set would be required, the members of this set being connected by a number of relations known as *syzygies*. We shall make no attempt here to consider the general problem, but merely point out that even in the case of the much simpler problem of simultaneous classification of gravitational and electromagnetic fields, some of these difficulties are already present.

Einstein–Maxwell case

We thus consider the combined system of spinors

$$\Psi_{ABCD}, \quad \varphi_{AB}, \tag{8.6.16}$$

each being symmetric. We expect just three complex invariants in addition to I and J, since Ψ_{ABCD} fixes a Minkowski frame (up to discrete symmetries) in terms of which the three complex components of φ_{AB} are now invariants. Indeed, we have

$$K=\varphi^{AB}\varphi_{AB}, \quad L=\varphi^{AB}\Psi_{AB}{}^{CD}\varphi_{CD}, \quad M=\varphi^{AB}\Psi_{AB}{}^{CD}\Psi_{CD}{}^{EF}\varphi_{EF} \tag{8.6.17}$$

as a possible set of invariants, these being in fact independent, as is not hard to show using matrix arguments. (One expands φ_{AB} in terms of eigenspinors of $\Psi_{AB}{}^{CD}$, the coefficients being arbitrary; then K, L and M are independent linear functions of the squares of these coefficients.) One may, indeed, verify that expressions such as

$$\varphi^{AB}\Psi_{AB}{}^{CD}\Psi_{CD}{}^{EF}\Psi_{EF}{}^{GH}\varphi_{GH} \tag{8.6.18}$$

are expressible as polynomials in terms of I, J, K, L and M. However, these five scalars do not form a complete set in the sense of classical invariant theory. For the quantity

$$N=\varphi^{AB}\Psi_{AB}{}^{CD}\Psi_{CD}{}^{EF}\varphi_{E}{}^{G}\Psi_{FG}{}^{PQ}\varphi_{PQ} \tag{8.6.19}$$

is clearly not so expressible in this way since it is of odd order in φ_{AB} whereas each of $I,\ldots,M$ is of even order. But N is dependent on $I,\ldots,M$ by virtue of the syzygy

$$N^2=\tfrac{1}{2}JKLM-\tfrac{1}{6}JL^3-\tfrac{1}{2}M^3-\tfrac{1}{8}I^2KL^2-\tfrac{1}{6}IJK^2L \\ -\tfrac{1}{18}J^2K^3+\tfrac{1}{4}IKM^2+\tfrac{1}{4}IL^2M. \tag{8.6.20}$$

(*cf.* Penrose 1960, Grace and Young 1903). In fact the whole system I,J,K,L,M,N does form a complete set of invariants for Ψ_{ABCD} and φ_{AB}.

We have seen (in the penultimate paragraph of §8.5) that $K=0$ is the condition for the two electromagnetic PNDs (EPNDs) to coincide. The condition for an EPND to coincide with a GPND is the vanishing of the *resultant* of the corresponding quartic and quadratic forms:

$$\begin{vmatrix} \Psi_0 & 4\Psi_1 & 6\Psi_2 & 4\Psi_3 & \Psi_4 & \\ & \Psi_0 & 4\Psi_1 & 6\Psi_2 & 4\Psi_3 & \Psi_4 \\ \varphi_0 & 2\varphi_1 & \varphi_2 & & & \\ & \varphi_0 & 2\varphi_1 & \varphi_2 & & \\ & & \varphi_0 & 2\varphi_1 & \varphi_2 & \\ & & & \varphi_0 & 2\varphi_1 & \varphi_2 \end{vmatrix}=0. \tag{8.6.21}$$

In terms of invariants this is (*cf.* Penrose 1960)

$$2K^2I-4KM+L^2=0. \tag{8.6.22}$$

For both EPNDs to lie along a GPND we therefore have the condition

$$K=0=L. \tag{8.6.23}$$

When the Einstein–Maxwell equations hold (say, with $\lambda=0$) then, by (5.2.6), we have

$$\Phi_{ABA'B'}=2G\varphi_{AB}\bar{\varphi}_{A'B'}$$

and it follows that the 4 real quantities $K\bar{K}, L\bar{L}, M\bar{M}, N\bar{N}$ and the three complex quantities $K\bar{L}, L\bar{M}, M\bar{N}$ are expressible as invariants of

$$\Phi_{ABA'B'},\ \Psi_{ABCD},\ \bar{\Psi}_{A'B'C'D'}. \tag{8.6.24}$$

There are obvious identities connecting these quantities (e.g. $(K\bar{K})(L\bar{L})=(K\bar{L})(\overline{K\bar{L}})$), making some of them redundant. There can be, in fact, only nine independent real invariants of the spinors (8.6.24), since there were ten from the real and imaginary parts of the independent invariants of φ_{AB} and Ψ_{ABCD}, but now we lose one real invariant because of the duality rotation freedom $\varphi_{AB}\mapsto e^{i\theta}\varphi_{AB}$ (θ real).

8.7 A classification scheme for general spinors

In the previous sections we have considered the structure of the Weyl tensor (spinor) in some detail. In this and the following section we turn our attention to the remaining portion of the space–time curvature, namely the Ricci tensor (without, however, considering its relation to the Weyl tensor structure). At first sight it might seem that the classification of the Ricci tensor should be less of a problem than that of the Weyl tensor. For R_{ab} has only two tensor indices as opposed to the four of C_{abcd}. It is therefore open to being treated directly as a matrix and classified in terms of the coincidence scheme of its eigenvalues and eigenvectors in a relatively straightforward way. However, particularly because of the indefinite signature of the space–time metric, this does not yield as simple or transparent a scheme as one might wish for (*cf.* Churchill 1932).

We present an alternative method, based on spinor techniques. However, because the spinor form of R_{ab} possesses both primed and unprimed indices, the spinor treatment yields a classification which is not nearly so simple as was that of the Weyl spinor (tensor). But it does not seem that this is really a drawback of the spinor technique as applied to symmetric two-index tensors. Rather, the spinor formalism reveals an essential complication present in such tensors, which is perhaps surprisingly absent in the case of the more complicated looking four-index Weyl tensor.

The method we describe in this section (in outline) is applicable to tensors or spinors with any number of indices. In the following section we apply it to the particular case of the (trace-free) Ricci spinor $\Phi_{ABC'D'}$, though not in full detail (see Penrose 1972*a* for a more complete discussion). The relation of this method to the eigenvectors and eigenvalues of $\Phi_a{}^b$ will be indicated. Also, we show how our method relates to an alternative approach due to Plebański (1964 *cf.* also Ludwig and Scanlan 1971 for a comparison with the matrix and spinor arguments). There the spinor

$$\Pi_{ABCD} = \Phi_{P'Q'(AB}\Phi^{P'Q'}_{CD)} \tag{8.7.1}$$

is classified according to the scheme that we have already given for the Weyl spinor, which allows one to give a characterization of $\Phi_{ABC'D'}$ itself.

The complex function Ω and locus ω

The general idea of our method follows naturally from that of the canonical decomposition (3.5.18) of a symmetric spinor $\phi_{AB\ldots L}$. A corresponding procedure, for a general symmetric (that is, symmetric in each of $A \ldots L$ and $P' \ldots V'$ separately: i.e. pointwise irreducible) spinor $\Omega_{A \ldots LP' \ldots V'}$, would be to consider the expression

$$\Omega(\xi, \bar{\xi}) := \Omega_{A \ldots LP' \ldots V'} \xi^A \ldots \xi^L \bar{\xi}^{P'} \ldots \bar{\xi}^{V'}. \qquad (8.7.2)$$

If $\Omega_{A \ldots V'}$ were not symmetric we should first have to express it in terms of its symmetric parts and classify each part separately. (However, this would not by itself give a full classification of a non-symmetric $\Omega_{A \ldots V'}$, since the interrelationships between the different parts would also have to be considered.) One thing we shall be concerned with particularly, is the set of spinors ξ for which

$$\Omega(\xi, \bar{\xi}) = 0. \qquad (8.7.3)$$

This will tell us something about the structure of $\Omega_{\ldots}$ up to proportionality. Defining a null vector

$$x^a := \xi^A \bar{\xi}^{A'}, \qquad (8.7.4)$$

we see that (8.7.3) defines a locus $\mathbb{R}\omega$ (a complex locus ω will be defined presently) on the sphere S^+ whose points represent the null vectors x^a up to proportionality. However, this real locus may not tell us much about the spinor $\Omega_{\ldots}$. (Consider, for example, the case when $\Omega_{AA'}$ is a timelike covector. Then $\mathbb{R}\omega$ is vacuous.) Instead we are led to consider *complex* vectors (*cf.* (3.2.6))

$$z^a := \xi^A \eta^{A'}, \qquad (8.7.5)$$

counted as equivalent if differing by a non-zero complex factor, which we may regard as the points Z of the complexification $\mathbb{C}S^+$ of S^+:

$$\mathbb{C}S^+ = \Sigma \times \Sigma', \qquad (8.7.6)$$

where Σ is the sphere of spinors ξ^A (up to a complex factor of proportionality) and Σ' the sphere of spinors $\eta^{A'}$ (up to proportionality). We thus examine the locus ω of *complex* points of S^+ (i.e. points of $\mathbb{C}S^+$) given by the vanishing of

$$\Omega(\xi, \eta) := \Omega_{A \ldots LP' \ldots V'} \xi^A \ldots \xi^L \eta^{P'} \ldots \eta^{V'}. \qquad (8.7.7)$$

Since ξ^A and $\eta^{A'}$ are now independent, it follows from two applications of (3.3.23) that ω does in fact define $\Omega_{\ldots}$ completely up to proportionality.

Reducibility of ω

The question of reducibility of this locus ω is of some interest. Suppose

$$\Omega_{A \ldots CD \ldots L}^{P' \ldots S'T' \ldots V'} = \Lambda_{(A \ldots C}^{(P' \ldots S'} \Gamma_{D \ldots L)}^{T' \ldots V')} \qquad (8.7.8)$$

then we have

$$\Omega(\xi,\eta)=(\Lambda_{A\ldots CP'\ldots S'}\xi^A\ldots\xi^C\eta^{P'}\ldots\eta^{S'})\times(\Gamma_{D\ldots LT'\ldots V'}\xi^D\ldots\xi^L\eta^{T'}\ldots\eta^{V'}) \tag{8.7.9}$$

so that

$$\omega=\lambda\cup\gamma \tag{8.7.10}$$

where λ is the locus on $\mathbb{C}S^+$ defined by $\Lambda_{\ldots}$ and γ is defined similarly by $\Gamma_{\ldots}$. A corresponding result obviously applies if ω is reducible into more than just two factors.

The crucial feature of symmetric spinors possessing just one type of index, upon which the canonical decomposition (3.5.18) was based, is that all such spinors are completely reducible, in the above sense, to linear factors. Thus if the present method is applied to the Weyl spinor Ψ_{ABCD}, the locus ω on $\mathbb{C}S^+$ is a set of four complex lines (in fact generators of the quadric surface $\mathbb{C}S^+$, as we shall presently see) which intersect the real section S^+ of $\mathbb{C}S^+$ in the real points A, B, C, D representing the GPNDs on S^+.

When both types of index are present, this decomposition is in general not possible (e.g. $\Omega_{AA'}$ decomposes only if null: $\Omega_{AA'}\Omega^{AA'}=0$), and the classification scheme is correspondingly more complicated. The structure of each irreducible factor must then be considered on its own merits.

Complex (p,q)-curves on $\mathbb{C}S^+$

The complex manifold $\mathbb{C}S^+=\Sigma\times\Sigma'$ is a quadric surface, being defined in complex projective 3-space $\mathbb{CP}^3$ by the quadratic equation $g_{ab}z^az^b=0$. It is generated by two systems of complex lines: those given by fixed $\eta^{A'}$ and varying ξ^A – called *Σ-generators*, being the different copies of Σ in $\Sigma\times\Sigma'$ – and those given by fixed ξ^A and varying $\eta^{A'}$, called *Σ'-generators*. Now each algebraic curve ω on $\mathbb{C}S^+$ can be classified, in the first instance, by a pair (p,q) of non-negative integers, as follows: p is the number of points (with multiplicities correctly counted) in which ω is met by a Σ-generator, and q the number of points in which ω is met by a Σ'-generator. In the case of ω arising from an $\Omega_{\ldots}$ as described above, p and q are simply the numbers of unprimed and primed indices, respectively. This is easily seen because, for a Σ-generator, we hold $\eta^{A'}$ fixed in (8.7.7) and ask how many solutions (correctly counted and up to proportionality) there are for ξ^A in $\Omega(\xi,\eta)=0$. This number is just the number of unprimed indices of $\Omega_{\ldots}$. Note that these solutions are given by

$$\xi^A=\alpha^A(\eta),\ldots,\xi^A=\lambda^A(\eta), \tag{8.7.11}$$

where $\alpha^A(\eta), \ldots, \lambda^A(\eta)$ are the p principal spinors of

$$\Omega_{A\ldots LP'\ldots V'}\eta^{P'}\ldots\eta^{V'}. \tag{8.7.12}$$

The corresponding result holds for the intersection of ω with a Σ'-generator giving, now, q points.

We observe that if a (p,q)-curve on $\mathbb{C}S^+$ is reducible to the union of an (r,s)-curve and a (t,u)-curve then we must have $r+t=p$, $s+u=q$. It is important always to keep track of multiplicities in such unions. For example, if $p=2r$, $q=2s$, it might be that ω consists of a double (r,s)-curve (i.e. one (r,s)-curve 'squared'). Or ω may have a number of components with different multiplicities. The canonical decomposition (3.5.18) expresses the fact that every $(p,0)$-curve is reducible to a set of p Σ'-generators, while every $(0,q)$ curve consists of q Σ-generators (with possible multiplicities).

Procedure for classification of $\Omega_{A\ldots LP'\ldots V'}$

The first stage in the classification of $\Omega_{\ldots}$ is the reducibility structure of ω. Thus (assuming $pq \neq 0$), in the general case ω is irreducible, while in certain special cases ω will split into various numbers of distinct components. In more special cases still, some of these components may be multiple. Consider, for example, the case $p=2$, $q=1$. Then we have a crude classification of $\Omega_{ABC'}$ into five types, the different bracket terms on the left defining the different types of irreducible factors:

$$\begin{aligned}
&(2,1)\colon \Omega_{ABC'} \ \textit{irreducible}\\
&(1,0)(1,1)\colon \Omega_{ABC'} = \Lambda_{(A}\Gamma_{B)C'} \neq 0, \Gamma_{BC'} \ \textit{irreducible}\\
&(1,0)(1,0)(0,1)\colon \Omega_{ABC'} = \Lambda_{(A}\Theta_{B)}\Upsilon_{C'} \neq 0\\
&(1,0)^2(0,1)\colon \Omega_{ABC'} = \Lambda_A\Lambda_B\Upsilon_{C'} \neq 0\\
&(-)\colon \Omega_{ABC'} = 0.
\end{aligned} \tag{8.7.13}$$

Note that the type $(2,0)(0,1)$ does not appear because of the canonical decomposition $(1,0)(1,0)$ of $(2,0)$. We also remark that if $p=q=1$ the corresponding classification of $\Omega_{AA'}$ gives three types, $(1,1)$, $(1,0)(0,1)$ and $(-)$, these being the cases when the complex vector Ω^a is non-null, null and zero, respectively.

When $p=q$, we may be concerned with the Hermiticity structure of $\Omega_{\ldots}$. For example, it might be specified that $\Omega_{a\ldots l} = \Omega_{A\ldots LA'\ldots L'}$ is a *real* tensor. In this case only symmetrical reductions can occur (e.g. $(2,1)(1,2)(3,3)$ but not $(3,1)(1,2)(2,3)$ or $(2,1)^2(1,2)(1,2)$, in the case $p=q=6$). This, of course, has relevance to the classification of $\phi_{ABC'D'}$.

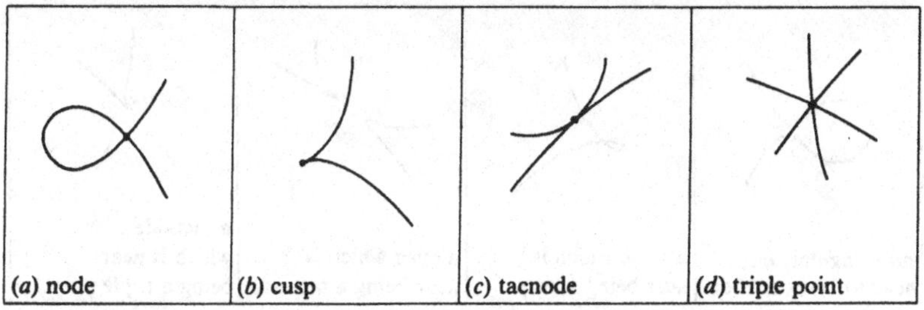

Fig. 8-6. Some types of singularity that can occur for a (complex) curve.

Multiple-point structure

The next point to consider is whether or not to subdivide the class of irreducible (r, s)-curves into subclasses; and if so, what should be the criteria for characterizing the different subclasses? One clear possibility is to use the multiple-point structure of the curve ω (*cf.* Walker 1950). This has the advantage that it is closely related to the question of reducibility. For the intersection point between two different components λ, γ of ω is always a multiple point of the combined curve ω. When λ and γ intersect with distinct tangents, that intersection point would be called a *node*, but nodes can occur more generally in a single irreducible curve ω – where the curve just crosses itself, the different branches of ω having distinct tangents at the crossing points (see Fig. 8-6*a*). In more special cases λ and γ may *touch* at an intersection point and, accordingly, an irreducible curve ω may have two branches which touch. Such a contact point is called a *tacnode* (see Fig. 8-6*c*). A degenerate form of node – called a *cusp* – can also occur where the tangents to the two branches coincide, but where, in contrast to the case of a tacnode, there is only one branch to the curve at the point, not two (see Fig. 8-6*b*). Nodes, tacnodes and cusps are all examples of *double points* of a curve ω, which means that a general curve through that point meets ω twice there. Triple points and points of higher multiplicity can also occur (see Fig. 8-6*d*). In fact, the case of a triple point may be regarded as a degenerate case of a tacnode, which in turn may be regarded as a degenerate case of a cusp; and a cusp, as we have just remarked, is a degenerate case of a node (see Fig. 8-7).

The precise definitions of these concepts for curves presented in ordinary (complex) Cartesian coordinates x, y – from which the above degeneration properties can readily be inferred – are as follows. Consider ω defined by the equation

$$\Omega(x, y) = 0,$$

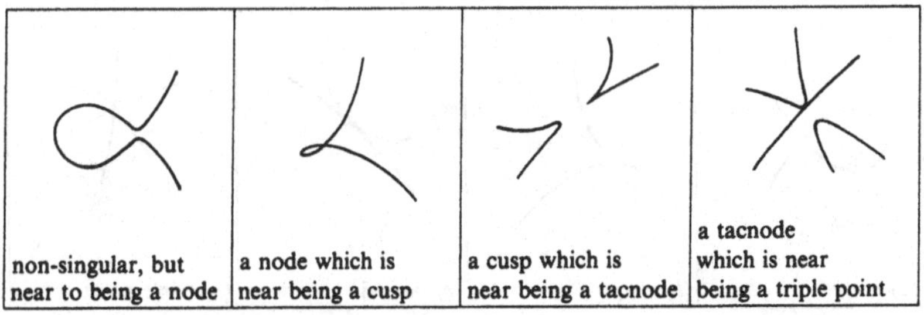

Fig. 8-7. How the singularities of Fig. 8-6 can arise as specializations

where Ω is analytic (holomorphic) in x, y at the origin O. Then we have, in some neighbourhood of O,

$$\Omega(x,y) \equiv A_0 + (B_0x + B_1y) + (C_0x^2 + 2C_1xy + C_2y^2) + (D_0x^3 + 3D_1x^2y + 3D_2xy^2 + D_3y^3) + \cdots \tag{8.7.14}$$

The point O lies on ω if $\Omega(0,0) = 0$, i.e., if

$$A_0 = 0,$$

in which case the equation of the tangent at O is

$$B_0x + B_1y = 0,$$

unless B_0 and B_1 both vanish. Its slope is then defined by the ratio of the coefficients

$$B_0{:}B_1, \quad i.e., \ \frac{\partial\Omega}{\partial x}{:}\frac{\partial\Omega}{\partial y} \quad at\ O. \tag{8.7.15}$$

In particular, ω touches the x-axis at O if

$$\frac{\partial\Omega}{\partial x} = 0 \quad at\ O, \tag{8.7.16}$$

and it touches the y-axis at O if

$$\frac{\partial\Omega}{\partial y} = 0 \quad at\ O. \tag{8.7.17}$$

The point O is at least double (i.e., a node) if, in addition to $A_0 = 0$,

$$B_0 = B_1 = 0, \quad i.e., \ \frac{\partial\Omega}{\partial x} = \frac{\partial\Omega}{\partial y} = 0 \quad at\ O,$$

in which case the equation of the *pair* of tangents to the two branches at O is

$$C_0x^2 + 2C_1xy + C_2y^2 = 0, \tag{8.7.18}$$

which is determined by the ratios

$$C_0{:}C_1{:}C_2, \quad \text{i.e.,} \frac{\partial^2\Omega}{\partial x^2}{:}\frac{\partial^2\Omega}{\partial x \partial y}{:}\frac{\partial^2\Omega}{\partial y^2} \quad \text{at } O.$$

The condition for ω to have a cusp (or more special point) at O is that these two tangents coincide, which happens if

$$C_0 C_2 = C_1^2, \quad \text{i.e.,} \frac{\partial^2\Omega}{\partial x^2}\frac{\partial^2\Omega}{\partial y^2} = \left(\frac{\partial^2\Omega}{\partial x \partial y}\right)^2 \quad \text{at } O. \tag{8.7.19}$$

A tacnode (or more special point) arises as a degenerate case of a cusp in which the repeated tangent at O intersects ω with a multiplicity of at least four instead of the multiplicity three for a generic cusp. Thus if, in accordance with (8.7.19), we put

$$C_0 = a^2, \quad C_1 = ab, \quad C_2 = b^2, \tag{8.7.20}$$

so that (8.7.18) becomes $(ax + by)^2 = 0$, we shall have, since $b(\partial/\partial x) - a(\partial/\partial y)$ now represents differentiation in the direction of the (repeated) tangent,

$$\left(b\frac{\partial}{\partial x} - a\frac{\partial}{\partial y}\right)^3 \Omega = 0 \quad \text{at } O \tag{8.7.21}$$

as the condition (with (8.7.20)) for a tacnode (or more special point), i.e.,

$$D_0 b^3 - 3D_1 b^2 a + 3D_2 b a^2 - D_3 a^3 = 0. \tag{8.7.22}$$

The condition for O to be at least a triple point is

$$C_0 = C_1 = C_2 = 0, \quad \text{i.e.,} \frac{\partial^2\Omega}{\partial x^2} = \frac{\partial^2\Omega}{\partial x \partial y} = \frac{\partial^2\Omega}{\partial y^2} = 0 \quad \text{at } O.$$

Multiple points of ω

Of course the curve ω of present interest to us is given not by an inhomogeneous equation like the vanishing of (8.7.14), but rather by the vanishing of the doubly homogeneous function (8.7.7). So the point X of $\mathbb{C}S^+$ represented by $\xi^A \eta^{A'}$ lies on ω if

$$\Omega(\xi^A, \eta^{A'}) = 0, \tag{8.7.23}$$

in which case the tangent direction at that point is defined by the ratio

$$\frac{\partial\Omega}{\partial\xi^A}{:}\frac{\partial\Omega}{\partial\eta^{A'}}. \tag{8.7.24}$$

The meaning of these abstract-index expressions is the obvious one in view of the definition (8.7.7):

$$\frac{\partial\Omega}{\partial\xi^A} = p\Omega_{AB\ldots LP'\ldots V'}\xi^B\ldots\xi^L\eta^{P'}\ldots\eta^{V'}$$

$$\frac{\partial\Omega}{\partial\eta^{P'}} = q\Omega_{A\ldots LP'Q'\ldots V'}\xi^A\ldots\xi^L\eta^{Q'}\ldots\eta^{V'},$$

and the definition of a ratio $\Psi_{\mathscr{A}}:\Phi_{\mathscr{B}}$ between abstract-index quantities is simply the equivalence class of pairs $(\Psi_{\mathscr{A}}, \Phi_{\mathscr{B}})$ under the relation $(\Psi_{\mathscr{A}}, \Phi_{\mathscr{B}}) \sim (\lambda\Psi_{\mathscr{A}}, \lambda\Phi_{\mathscr{B}})$ $(\lambda \neq 0)$.

Now we have, also by Euler's theorem on homogeneous functions,

$$\xi^A\frac{\partial\Omega}{\partial\xi^A} \equiv p\Omega, \quad \eta^{A'}\frac{\partial\Omega}{\partial\eta^{A'}} \equiv q\Omega. \tag{8.7.25}$$

When $\Omega = 0$, these relations imply, by (2.5.56),

$$\frac{\partial\Omega}{\partial\xi^A} = \theta\xi_A, \quad \frac{\partial\Omega}{\partial\eta^{A'}} = \zeta\eta_{A'}$$

for some θ, ζ. The ratio

$$\theta:\zeta \tag{8.7.26}$$

is thus the entire residual information from (8.7.24), and therefore defines the slope of ω at the point X. Note that the operator

$$\Delta_{AA'} = b\eta_{A'}\frac{\partial}{\partial\xi^A} - a\xi_A\frac{\partial}{\partial\eta^{A'}} \tag{8.7.27}$$

annihilates Ω at X if and only if

$$a:b = \theta:\zeta,$$

and so it corresponds to differentiation in the direction given by the ratio $a:b$.

As a first application of the above discussion, consider the condition for the Σ-generator defined by $\eta^{A'}$ to *touch* ω (i.e., to have two coincident intersections with ω) at X. Corresponding to (8.7.16), this condition is

$$\theta = 0, \quad \textit{i.e.}, \quad \frac{\partial\Omega}{\partial\xi^A} = 0 \quad \textit{at } X, \tag{8.7.28}$$

which is also the condition for the canonical decomposition of (8.7.12) to have ξ^A as a repeated principal spinor. Similarly, the Σ'-generator defined by ξ^A touches ω at X if, corresponding to (8.7.17),

$$\zeta = 0, \quad \textit{i.e.}, \quad \frac{\partial\Omega}{\partial\eta^{A'}} = 0 \quad \textit{at } X. \tag{8.7.29}$$

The condition for X to be a *double* (or more special) *point* of ω is that both these conditions hold simultaneously:

$$\frac{\partial \Omega}{\partial \xi^A} = 0, \quad \frac{\partial \Omega}{\partial \eta^{A'}} = 0 \quad \text{at } X. \tag{8.7.30}$$

Note that because of (8.7.25) the condition $\Omega = 0$ for X to lie on ω is implicit in *either* of the conditions in (8.7.30).

As in (8.7.25) we have, from Euler's theorem,

$$\xi^B \frac{\partial}{\partial \xi^B}\left(\frac{\partial \Omega}{\partial \xi^A}\right) = (p-1)\frac{\partial \Omega}{\partial \xi^A}.$$

For a double point, therefore,

$$\xi^B \frac{\partial}{\partial \xi^B}\left(\frac{\partial \Omega}{\partial \xi^A}\right) = 0, \quad \text{i.e.,} \quad \frac{\partial^2 \Omega}{\partial \xi^A \partial \xi^B} = \theta_A \xi_B \quad \text{at } X$$

but, by the symmetry of the derivative, we must have $\theta_A \propto \xi_A$, so that we get the first of the following relations, while the others are obtained similarly:

$$\frac{\partial^2 \Omega}{\partial \xi^A \partial \xi^B} = \rho \xi_A \xi_B, \quad \frac{\partial^2 \Omega}{\partial \xi^A \partial \eta^{A'}} = \sigma \xi_A \eta_{A'}, \quad \frac{\partial^2 \Omega}{\partial \eta^{A'} \partial \eta^{B'}} = \tau \eta_{A'} \eta_{B'} \quad \text{at } X, \tag{8.7.31}$$

for some ρ, σ, τ. The ratios $\rho:\sigma:\tau$ define the pair of slopes of the branches of ω at X. A direction defined by $a:b = \theta:\zeta$ (as in (8.7.26)) corresponds to one of these slopes if

$$\rho b^2 - 2\sigma ab + \tau a^2 = 0, \tag{8.7.32}$$

for this is the condition $\Delta_{AA'}\Delta_{BB'}\Omega = 0$, with $\Delta_{AA'}$ given by (8.7.27). These slopes *coincide* if (8.7.32) has repeated roots, i.e.,

$$\rho\tau = \sigma^2, \tag{8.7.33}$$

which is therefore the condition for X to be a *cusp* or more special point (*cf.* (8.7.19)).

By (8.7.32), the direction given by $a:b$ is that of the above repeated slope if (as in (8.7.20))

$$\rho = a^2, \quad \sigma = ab, \quad \tau = b^2. \tag{8.7.34}$$

So, as in (8.7.21), a cusp at X specializes to a *tacnode* (or more special point) if

$$\Delta_{AA'}\Delta_{BB'}\Delta_{CC'}\Omega = 0 \quad \text{at } X, \tag{8.7.35}$$

which yields a relation analogous to (8.7.22). (Note that in (8.7.35) a and b are taken as *fixed* numbers which satisfy (8.7.34), so they do *not* get differentiated.)

The condition for (at least) a *triple point* is, in addition to (8.7.30), and with the definitions (8.7.31),

$$\rho = \sigma = \tau = 0. \tag{8.7.36}$$

We may note that all the above considerations of multiple points also apply to curves on $\mathbb{C}S^+$ which are merely holomorphic and not necessarily algebraic.

The locus $\mathbb{R}\omega$ *on* S^+

Although we rejected (*cf.* after (8.7.4)) $\mathbb{R}\omega$ as the main route to classification of $\Omega_{\dots}$, that locus on S^+ may still be of interest. Its points are given by (8.7.3). In the general case, when $\Omega_{\dots}$ is not Hermitian (e.g., whenever $p \neq q$), $\mathbb{R}\omega$ will consist normally of discrete points on S^+, since (8.7.3) will be a complex equation whose real and imaginary parts give two relations to be satisfied for the points of S^+. When $\Omega_{\dots}$ is Hermitian, on the other hand, the locus $\mathbb{R}\omega$ will normally be a curve on S^+, although exceptionally $\mathbb{R}\omega$ may consist of, or contain, isolated (double) points. The classification scheme for $\Omega_{\dots}$ may involve examining the number of disconnected pieces (possibly zero) into which the locus $\mathbb{R}\omega$ falls. Note that $\mathbb{R}\omega$ has a quite direct significance, when $p = q$, since it then describes the set of real null solutions of

$$\Omega_{a\dots l}x^a\dots x^l = 0. \tag{8.7.37}$$

Thus, for example, if $p = q = 1$ and Ω^a is a real vector, the general case $(1,1)$ is naturally subdivided into two subclasses, namely that for which $\mathbb{R}\omega$ is vacuous, obtaining when the vector Ω^a is timelike (so $\Omega_a x^a = 0$ has no non-zero null solutions), and that for which $\mathbb{R}\omega$ is a circle, obtaining when Ω^a is spacelike (so the null solutions of $\Omega_a x^a = 0$ lie on the intersection with the null cone of the timelike hyperplane orthogonal to Ω^a). Since the timelike and spacelike vectors form systems disconnected from one another, where to pass continuously from one to the other it is necessary to pass through an algebraically distinct type (namely $(1,0)(0,1)$ – the null vectors), it seems reasonable to classify the timelike and spacelike vectors as two distinct types. Note that in order for the circle $\mathbb{R}\omega$ (Ω^a spacelike) to move continuously until it disappears (Ω^a timelike) it must pass through the situation in which it becomes a 'point circle' (Ω^a null). This is an isolated double point, which is actually a node on ω, with complex conjugate tangents. In fact this node is the intersection of the two generators of $\mathbb{C}S^+$ which make up ω (ω is $(1,0)(0,1)$ when Ω^a is null) and is the point of S^+ which represents the null direction of Ω^a.

The sign of Ω; black and white colouring of S^+

Finally, it may be felt that a classification which takes no account of the sign of $\Omega_{\dots}$ is still too crude. For example, the future-pointing and past-pointing vectors Ω^a have just been classified together. It is possible to take into account this sign, for Hermitian $\Omega_{\dots}$, by examining the sign of the expression in (8.7.37) for the various future-pointing null vectors x^a rather than just its zeros. This amounts to imagining that the two regions of S^+ into which it is divided by the curve $\mathbb{R}\omega$ are coloured differently, say black when the expression in (8.7.37) is negative and white when it is positive. We observe that whenever the expression in (8.7.37) is positive semi-definite (i.e., non-negative for all future-pointing causal vectors x^a), then the whole of S^+ is coloured white except possibly for some points of $\mathbb{R}\omega$ itself, which must be double. We can now distinguish the case when Ω^a is a future-pointing timelike vector from the case when it is past-pointing. For in the first case S^+ is white, while in the second case it is black. When Ω^a is spacelike, S^+ is black on one side of $\mathbb{R}\omega$ and white on the other.

8.8 Classification of the Ricci spinor

We shall now apply the methods developed in the last section to the case of the (traceless) Ricci spinor $\Omega_{ABC'D'} = \Phi_{ABC'D'}$. We shall not consider the trace of R_{ab}. To consider Λ simultaneously with $\Phi_{ABA'B'}$ would add still further complication to the classification. As it is, were we to follow through the above scheme completely in all detail, we should end up with a classification of $\Phi_{ABA'B'}$ into 41 different types (see Penrose 1972*b*). This may seem excessive, but, in fact, if certain rather natural criteria are adhered to for a classification scheme for $\Phi_{ABA'B'}$ (specifying how specialization generates new types) then it seems that we cannot reasonably make do with fewer. The existing alternative classification schemes all seem to behave anomalously in certain of these respects. The criteria require, in essence, a scheme which is Lorentz invariant, with the property that the elements belonging to each specific type form a connected manifold, and with the further property that elements which are the common specializations of two different types, neither of which consists solely of specializations of the other, shall not be considered to belong to either of these two types. These criteria, together with some broad requirements that certain $\Phi_{ABA'B'}$ should 'obviously' belong to separate types, seem to imply at least as fine a classification as that suggested here. Of course, it is always possible, whenever variable scalar invariants are present within one type, to continue to subdivide that type indefinitely. For example, in the case of the $\{1111\}$ Weyl tensors, we could, if desired, have regarded the harmonic ($J = 0$) and equianharmonic ($I = 0$) cases as types distinct from the general type. Or, less reasonably, we could have regarded those whose cross-ratio is rational, or real, or algebraic, or which satisfies any of an infinity of other possible conditions, as belonging to distinct types. Clearly the alternatives are endless and the criteria for judging which classification schemes are most reasonable must inevitably be somewhat subjective. For this reason we do not wish to be dogmatic or definitive about a classification scheme for $\Phi_{ABA'B'}$ but merely indicate how our method can be applied and how it relates to certain alternative methods which have been put forward by others.

Reducibility of $\Phi_{ABA'B'}$

First, let us examine the possible ways that $\Phi_{ABA'B'}$ might be reducible. Since it is Hermitian, we have just the following seven possibilities:

$$\begin{array}{ll}
(2,2) & \Phi_{ABA'B'} \textit{ irreducible} \\
(1,1)(1,1) & \Phi_{AB}^{A'B'} = \Lambda_{(A}^{(A'}\Upsilon_{B)}^{B')} \textit{ or } \Phi_{AB}^{A'B'} = \pm\Gamma_{(A}^{(A'}\bar{\Gamma}_{B)}^{B')} \\
(1,1)^2 & \Phi_{AB}^{A'B'} = \pm\Lambda_{(A}^{(A'}\Lambda_{B)}^{B')} \\
(1,1)(1,0)(0,1) & \Phi_{ABA'B'} = \rho_{(A}\Lambda_{B)(C'}\bar{\rho}_{D')} \\
(1,0)(1,0)(0,1)(0,1) & \Phi_{ABA'B'} = \pm\rho_{(A}\sigma_{B)}\bar{\rho}_{(A'}\bar{\sigma}_{B')} \\
(1,0)^2(0,1)^2 & \Phi_{ABA'B'} = \pm\rho_A\rho_B\bar{\rho}_{A'}\bar{\rho}_{B'} \\
(-) & \Phi_{ABA'B'} = 0,
\end{array} \tag{8.8.1}$$

where $\Lambda_{AA'}$ and $\Upsilon_{AA'}$ are Hermitian and where all of $\Lambda_{..}, \Upsilon_{..}, \Gamma_{..}, \rho_., \sigma_.$, are non-proportional and non-zero. To distinguish the two possibilities in the case $(1,1)(1,1)$ we shall henceforth denote only the first by that symbol, the second being written $|(1,1)|^2$ in order to indicate that it is of the form $(1,1)\overline{(1,1)}$, where the second curve is the complex conjugate of the first. For notational consistency, we shall also correspondingly re-designate the final three non-zero types:

$$\begin{array}{l}
(1,1)(1,0)(0,1) \mapsto (1,1)|(1,0)|^2 \\
(1,0)(1,0)(0,1)(0,1) \mapsto |(1,0)(1,0)|^2 \\
(1,0)^2(0,1)^2 \mapsto |(1,0)^2|^2.
\end{array} \tag{8.8.2}$$

The specialization diagram for these eight types is given in Table (8.8.3). The number on the left indicates the real dimension (or number of degrees

Table (8.8.3)

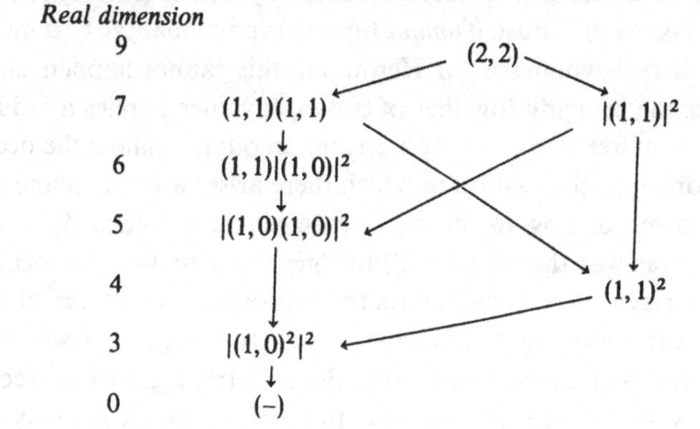
Real dimension
9
7
6
5
4
3
0
(2, 2)
(1, 1)(1, 1)
|(1, 1)|²
(1, 1)|(1, 0)|²
|(1, 0)(1, 0)|²
(1, 1)²
|(1, 0)²|²
(–)

of freedom) for each system of curves in the diagram; it is easily computed from the number of real and imaginary parts of the independent spinor components of each factor.

Eigenvalues and eigenvectors of $\Phi_a{}^b$; double points of ω

So far this classification is not refined enough to encompass other schemes, which have been suggested, such as one based on the coincidence of eigenvalues and on the dimension of the space spanned by the eigenvectors of $\Phi_a{}^b$ (e.g. Churchill 1932, Ludwig and Scanlan 1971; *cf.* Kramer, Stephani, MacCallum and Herlt 1980 for further references). The condition that $\Phi_a{}^b$ have equal eigenvalues actually reduces the dimension of the space of $\Phi_a{}^b$ by just one (rather than two, which would have been the case with a positive definite metric – as we shall explain shortly). Thus, we need to find a sub-case of the general type (2, 2) which is restricted by just *one* degree of freedom, so the curve ω must remain irreducible. Such a sub-case is obtained by making ω acquire a node.

To see that there is this relation between the presence of a node in ω and a repeated eigenvalue for $\Phi_a{}^b$, we consider first the general situation which occurs when a matrix $\mathbf{A}$ is specialized so that two of its eigenvalues λ_1 and λ_2 are brought into coincidence. If $\mathbf{A}$ is not assumed symmetric, then generically, the directions of the corresponding eigenvectors $\mathbf{x}_1$ and $\mathbf{x}_2$ will also come into coincidence. Only when $\mathbf{A}$ is made more special still can a pair of independent eigenvectors be found corresponding to the repeated eigenvalue (see (8.3.42) for an illustration of this fact). If $\mathbf{A}$ is symmetric, then when $\lambda_1 \neq \lambda_2$ it follows directly that $\mathbf{x}_1$ and $\mathbf{x}_2$ are orthogonal. Thus the 'generic' situation above, when $\mathbf{x}_1$ and $\mathbf{x}_2$ are brought into coincidence, occurs when each of $\mathbf{x}_1$, $\mathbf{x}_2$ approaches a vector orthogonal to itself, i.e. a *null* vector. Thus, any vector corresponding to a repeated eigenvalue must, *if unique* (up to proportionality), be a null vector. When **A** is real-symmetric (or Hermitian) this cannot happen, so in such situations the bringing together of two eigenvalues implies a reduction in the degrees of freedom for **A** by a greater amount, to allow the occurrence of the more special situation in which there arises a whole plane of eigenvectors corresponding to the repeated eigenvalue. When **A** is complex-symmetric (as was the case for $\boldsymbol{\Psi}$) the 'generic' situation *can* occur; coincidence of eigenvalues thus implies the loss of just one degree of freedom, and the corresponding eigenvector is null. In the present situation, $\Phi_a{}^b$ is symmetric with respect to the *indefinite* metric g_{ab}, so null vectors are again possible and the 'generic' situation can occur. (It is easily checked by example that this can happen for $\Phi_{\mathbf{a}}{}^{\mathbf{b}}$.)

From the above discussion we see that the first (i.e. most general) degenerate case must occur when $\Phi_a{}^b$ possesses a null eigenvector

$$\Phi_a{}^b z_b = \lambda z_a, \quad z_a z^a = 0, \quad 0 \neq z_a \in \mathfrak{S}_a. \tag{8.8.4}$$

(It follows from general matrix theory that, conversely, an eigenvector can be null *only* when λ is a repeated eigenvalue. A proof may be given in terms of a limiting argument, considering the non-degenerate case first.) Expressing z^a as in (8.7.5), we have $\Phi_{AA'}{}^{BB'}\xi_B\eta_{B'} = \lambda\xi_A\eta_{A'}$, which can be written as

$$\Phi_{ABC'D'}\xi^A\xi^B\eta^{D'} = 0, \quad \Phi_{ABC'D'}\xi^B\eta^{C'}\eta^{D'} = 0. \tag{8.8.5}$$

Comparison with (8.7.28), (8.7.29), (8.7.30) tells us that the point Z of $\mathbb{C}S^+$ represented by $\xi^A\eta^{A'}$ must be multiple for ω, i.e. in the most general case it must be a *node*. In fact, most generally, there will be only one such node, so the point Z must be *real* (i.e. $Z \in S^+$) and we can take $\eta^{A'} = \bar{\xi}^{A'}$ in (8.8.5).

Two distinct possibilities for real nodes can occur, namely those for which the two branches of ω at Z have *real* tangents and those for which the two branches have complex conjugate tangents. In the latter case, Z is an *isolated double point* of ω. An example of an isolated double point is the point circle considered earlier in connection with the discussion of a real null vector Ω^a. Defining, as in (8.7.31), a complex number α and a *real* number β by

$$\Phi_{ABC'D'}\bar{\xi}^{C'}\bar{\xi}^{D'} = \alpha\xi_A\xi_B, \quad \Phi_{ABC'D'}\xi^A\bar{\xi}^{C'} = \beta\xi_B\bar{\xi}_{D'}, \tag{8.8.6}$$

so that comparison with (8.7.31) gives

$$\beta = \tfrac{1}{4}\sigma = \tfrac{1}{4}\bar{\sigma}, \quad \alpha = \tfrac{1}{2}\rho = \tfrac{1}{2}\bar{\tau},$$

one sees from (8.7.32) that the condition for the node to have real branches (i.e., solutions of (8.7.32) with $|a|=|b|$) is

$$\alpha\bar{\alpha} > 4\beta^2, \tag{8.8.7}$$

while the condition for it to be isolated is

$$\alpha\bar{\alpha} < 4\beta^2. \tag{8.8.8}$$

Recall from (8.7.33) that the node degenerates to a cusp (or tacnode or more special point) if

$$\alpha\bar{\alpha} = 4\beta^2. \tag{8.8.9}$$

To examine the meaning of these conditions in terms of the eigenvalue structure of $\Phi_a{}^b$, suppose that $x_a = \xi_A \bar{\xi}_{A'}$ is an eigenvector of $\Phi_a{}^b$ corresponding to the double eigenvalue λ for which the only eigenvectors are multiples of x_a, and that the remaining two eigenvectors lie in a real spacelike plane Π, orthogonal to x_a. The plane Π coincides with a pencil of vectors

$$y_a = \frac{u}{\sqrt{2}}(\xi^A\bar{\theta}^{A'} + \theta^A\bar{\xi}^{A'}) - \frac{v\mathrm{i}}{\sqrt{2}}(\xi^A\bar{\theta}^{A'} - \theta^A\bar{\xi}^{A'}), \tag{8.8.10}$$

where u, v are real, and where

$$\xi_A\theta^A = 1 \tag{8.8.11}$$

fixes θ^A uniquely in terms of the scaling of ξ^A. (The two vectors in parentheses in (8.8.10) are orthogonal spacelike unit vectors, orthogonal also to x_a.) If $\overset{1}{y}_a$ and $\overset{2}{y}_a$ are two vectors of the pencil (8.8.10), corresponding to $(\overset{1}{u},\overset{1}{v})$ and $(\overset{2}{u},\overset{2}{v})$ respectively, then from (8.8.6) we get

$$\overset{1}{y}{}^a\Phi_a{}^b\overset{2}{y}_b = -(\overset{1}{u}\quad \overset{1}{v})\begin{pmatrix} -\beta-\frac{1}{2}(\alpha+\bar{\alpha}) & \frac{1}{2}\mathrm{i}(\bar{\alpha}-\alpha) \\ \frac{1}{2}\mathrm{i}(\bar{\alpha}-\alpha) & -\beta+\frac{1}{2}(\alpha+\bar{\alpha}) \end{pmatrix}\begin{pmatrix} \overset{2}{u} \\ \overset{2}{v} \end{pmatrix}. \tag{8.8.12}$$

The (2×2) matrix in (8.8.12) defines, in terms of u and v, a linear transformation in Π which is that induced by $\Phi_a{}^b$. For if $\overset{3}{y}_a = \Phi_a{}^b\overset{2}{y}_b$, its $(\overset{3}{u},\overset{3}{v})$ representation must, by comparison with the scalar product $\overset{1}{y}{}^a\overset{3}{y}_a = -\overset{1}{u}\overset{3}{u} - \overset{1}{v}\overset{3}{v}$ be represented by the final matrix $\times$ column-vector product in (8.8.12). The eigenvalues of this matrix, which are easily seen to be $-\beta \pm |\alpha|$, must equal those two eigenvalues λ_2, λ_3 of $\Phi_a{}^b$ which correspond to eigenvectors in Π. Furthermore, from (8.8.6) it follows that β is the (repeated) eigenvalue $(=\lambda_0=\lambda_1)$ which corresponds to the repeated eigenvector x_a, so we have

$$\lambda_0 = \beta, \quad \lambda_1 = \beta, \quad \lambda_2 = |\alpha| - \beta, \quad \lambda_3 = -|\alpha| - \beta. \tag{8.8.13}$$

The cusp

From this we see that the condition (8.8.9) for a cusp (or more special point) states that a triple coincidence of eigenvalues ($\lambda_0 = \lambda_1 = \lambda_2$ or $\lambda_0 = \lambda_1 = \lambda_3$) takes place. However, in the most general case, namely that of a cusp, it turns out that the eigenvectors corresponding to this triple eigenvalue ($= \beta$) are just the multiples of x^a and no others. The only remaining eigenvectors are those corresponding to the simple eigenvalue ($= -3\beta$). So in this case the spacelike plane Π of our previous discussion does not exist. (It has become null, containing x_a.) Instead, we have a null eigenvector x_a and merely a spacelike eigenvector y_a orthogonal to it. The general form for Φ_{ab} when ω possesses a *cusp* corresponding to the null vector x^a is

$$\Phi_{ab} = \beta g_{ab} + 4\beta y_a y_b + x_{(a} w_{b)}, \tag{8.8.14}$$

where $y_a y^a = -1$, and where w_a ($\neq 0$) is also spacelike and is orthogonal to both x^a and y^a ($\beta \neq 0$).

The tacnode

To obtain the condition for the cusp to specialize further, we can employ (8.7.35) (with (8.7.27)), choosing

$$a{:}b = \alpha{:}2\beta = 2\beta{:}\bar{\alpha}, \tag{8.8.15}$$

cf. (8.8.9). In fact, it is only the totally symmetric part of (8.7.35) which yields anything new. After some calculation, and use of Proposition (3.5.15) we get

$$2\beta \Phi_{ABC'(A'} \bar{\xi}^{C'} \bar{\xi}_{B')} = \alpha \xi_{(A} \xi^{C} \Phi_{B)CA'B'}, \tag{8.8.16}$$

as the required *condition for (at least) a tacnode*. As we see at once by transvecting (8.8.16) with ξ^A, it implies the condition (8.8.9). In fact, all we need to know about α and β in (8.8.16) is that they are not both zero, since their ratio is determined by transvection with $\bar{\xi}^{A'}$. Now, with a tacnode, ω must degenerate to a pair of circles; it therefore follows from (8.8.1) (type (1, 1)(1, 1) or $|(1, 1)|^2$) that $\Phi_{ABA'B'}$ has the form

$$\Phi_{AB}^{A'B'} = \Lambda_{(A}^{(A'} \Upsilon_{B)}^{B')} \text{ or } \pm \Gamma_{(A}^{(A'} \bar{\Gamma}_{B)}^{B')}, \tag{8.8.17}$$

where Λ_a and Υ_a are real (and, in fact, spacelike, by (8.8.22) below). Contact between these circles is expressed (for real circles) as

$$\Lambda_{AC'} \bar{\xi}^{C'} = \nu \xi_A, \quad \Upsilon_{AC'} \bar{\xi}^{C'} = \pm \nu \xi_A \tag{8.8.18}$$

or (for complex conjugate circles) as

$$\Gamma_{AC'} \bar{\xi}^{C'} = \nu \xi_A, \quad \bar{\Gamma}_{AC'} \bar{\xi}^{C'} = \pm \nu \xi_A \tag{8.8.19}$$

(using a suitable scaling between the two factors in (8.8.17)), where

$$\alpha = v^2, \quad 2\beta = \pm v\bar{v} \tag{8.8.20}$$

Suppose $\beta > 0$, so that the upper signs apply. (In the contrary case we make the same argument for $-\Phi_{ab}$.) The eigenvectors of $\Phi_a{}^b$ corresponding to the triple eigenvalue β consist of all the vectors orthogonal to both Λ_a and Υ_a [*or* Γ_a and $\bar{\Gamma}_a$] (a two-dimensional space), while those corresponding to the simple eigenvalue -3β are the multiples of $\Lambda_a + \Upsilon_a$ [*or* $\Gamma_a + \bar{\Gamma}_a$].

Instead of (8.8.17), we may prefer a tensor form resembling (8.8.14), namely

$$\Phi_{ab} = \beta g_{ab} + 4\beta y_a y_b \mp 4\beta x_a x_b, \tag{8.8.21}$$

where

$$y_a = \gamma(\Lambda_a + \Upsilon_a) \quad \textit{or}\ \gamma(\Gamma_a + \bar{\Gamma}_a),$$

with

$$16\beta\gamma^2 = 1$$

and ξ^A scaled so that

$$x_a = \xi_A \bar{\xi}_{A'} = \gamma(\Lambda_a - \Upsilon_a) \quad \textit{or}\ \mathrm{i}\gamma(\Gamma_a - \bar{\Gamma}_a).$$

(That x_a is *proportional* to these expressions follows from (8.8.18), (8.8.19).) The upper sign in (8.8.21) now refers to the first of these expressions for x_a and y_a, while the lower sign refers to the second. The required properties

$$y_a y^a = -1, \quad y_a x^a = 0$$

follow from the relations

$$\Lambda_a \Lambda^a = \Upsilon_a \Upsilon^a = \Lambda_a \Upsilon^a = -4\beta \tag{8.8.22}$$

or

$$\Gamma_a \Gamma^a = \Gamma_a \bar{\Gamma}^a = -4\beta, \tag{8.8.23}$$

which are consequences of (8.8.18), (8.8.19), (8.8.20). Using these relations, (8.8.21) can be readily derived from (8.8.17).

Since for the present case of a tacnode we again have a spacelike 2-plane of eigenvectors, in addition to x_a, the discussion of (8.8.10)–(8.8.12) applies, which it does not in the case of a cusp. Thus the specialization (8.8.9) in (8.8.12), yielding a triple eigenvalue, gives us the more special case of a tacnode rather than that of a cusp.

Types of node

Returning now to the general case of a node, we remark that the condition (8.8.8) for the node to be *isolated* corresponds, in terms of the eigenvalues (8.8.12), to the condition that the repeated eigenvalue should lie *outside* the interval bounded by the two simple eigenvalues, i.e.,

$$\lambda_0 = \lambda_1 > \max(\lambda_2, \lambda_3) \quad or \quad \lambda_0 = \lambda_1 < \min(\lambda_2, \lambda_3);$$

the condition (8.8.7) for the node to have *two real branches* corresponds to the condition that the repeated eigenvalue should lie *between* the two simple eigenvalues, i.e.,

$$\lambda_2 < \lambda_0 = \lambda_1 < \lambda_3 \quad or \quad \lambda_3 < \lambda_0 = \lambda_1 < \lambda_2.$$

Note that a different, additional, coincidence between eigenvalues occurs when $\alpha = 0$, *namely* $\lambda_2 = \lambda_3$. Referring back to (8.8.6), we see that this means

$$\Phi_{ABC'D'}\bar{\xi}^{C'}\bar{\xi}^{D'} = 0,$$

from which it follows that $\Phi_{ABC'D'}$ has the form

$$\Phi_{ABC'D'} = \xi_{(A}\Lambda_{B)(C'}\bar{\xi}_{D')}$$

for some Hermitian $\Lambda_{BC'}$ so ω is $(1,1)|(1,0)|^2$, or more special.

We observe that again our specialization has jumped some cases, this time the node-pairs $(1,1)(1,1)$ and $|(1,1)|^2$. The reason here is that these cases are characterized by the occurrence of *additional* eigenvectors, rather than by further *coincidences* of eigenvectors. In these cases (8.8.17) holds, as with the tacnode, but not (8.8.18) or (8.8.19). There is a 2-space of eigenvectors orthogonal to Λ_a and Υ_a [*or* Γ_a and $\bar{\Gamma}_a$], this being the space spanned by the two nodes, and there are two separate eigenvectors which are linear combinations of Λ_a and Υ_a [*or* Γ_a and $\bar{\Gamma}_a$]. Many distinct possibilities arise with regard to the eigenvectors being spacelike or timelike, real or complex.

The different 'categories' within one type

It is not our intention here to examine all the individual special cases in detail. But we remark that even the general situation of $(2,2)$, where ω has no multiple points, needs to be subdivided further if we are to adhere to our criterion that each type is to form a connected manifold. Thus we subdivide types into connected 'categories' having the same dimensionality. (Specialization of types, on the other hand, always leads to types of lower dimensions).

In the general case, when the eigenvalues are all distinct, if the eigenvalues are all real there will be one, say λ_0, which corresponds to a timelike eigenvector and three others, λ_1, λ_2, λ_3, which correspond to spacelike eigenvectors. (This is because the eigenvectors, which are real in the cases now under consideration, are all mutually orthogonal.) The special cases that we have been considering above will arise as limiting situations of these general cases when λ_0 comes into coincidence with one of the other eigenvalues. (For, with real eigenvectors, it is only when a timelike and a spacelike eigenvector come into coincidence that a null eigenvector can arise as a limit.)

Now there are four essentially different possible arrangements of distinct real eigenvalues, namely that λ_0 may be the largest, second largest, third largest, or smallest among the eigenvalues. Each of these four arrangements gives a different category; for in order to pass continuously from one to another it is necessary to pass through a degenerate situation corresponding to one of the more special types, say when ω acquires a node as discussed above – although it turns out, in fact, that it can only do so in this way if, at some stage, ω aquires *two* nodes at the same time,* which implies that ω is $(1,1)(1,1)$ or $|(1,1)|^2$, or more special.

These four cases of $(2,2)$ can be described in terms of ω on S^+ as follows. When λ_0 is the largest eigenvalue, the following positive-definiteness property** holds:

$$\Phi_{ab}x^a x^b > 0 \quad \textit{for all non-zero real null vectors } x^a$$

(*cf.* (8.8.33), (8.8.34) below). This is the situation when $\mathbb{R}\omega$ is vacuous and the whole of S^+ is coloured *white* (*cf.* end of §8.7).

When λ_0 is the second largest eigenvalue, it can be shown that $\mathbb{R}\omega$ consists of two separate loops, the interiors of which are coloured black while the annular region between them is white. When λ_0 is the second smallest, $\mathbb{R}\omega$ consists again of two loops, but the colours are reversed. When λ_0 is the smallest, $\mathbb{R}\omega$ is vacuous and the whole of S^+ is black.

There is yet another possibility for the curve $\mathbb{R}\omega$ in the general case: it can consist of just a single loop. Then one side must be black and the other white. In this situation there does *not* exist a real orthonormal eigentetrad for Φ_{ab}. Two of the eigenvalues are real and correspond to two spacelike eigenvectors; but the remaining two eigenvalues are

* Roughly speaking, when two eigenvalues 'collide' along the real axis, then, in the general case, they 'bounce' off into complex conjugate eigenvalues. The case of λ_1 'passing through' λ_2, say, along the real axis corresponds to the double 'accident' of a pair of nodes.

** With the strict inequality '$>$' replaced by '$\geqslant$', this is the *weak energy condition* in the terminology of Penrose (1972*b*: p. 63) (or the *null convergence condition* in the terminology of Hawking and Ellis 1973, who use '*weak energy condition*' for the stronger condition which obtains when x^a is allowed also to be timelike and the full energy tensor replaces Φ_{ab}; *cf.* also Volume 1, p. 327).

*conjugate complex** and correspond to a pair of eigenvectors which are complex conjugates of one another. This is the only remaining possibility for the non-singular (2, 2) case, giving *five* disconnected categories in all.

The *singular* cases of (2, 2) are also subdivided, giving eight more categories. Two of these occur when $\mathbb{R}\omega$ possesses a cusp, with alternative arrangements of black and white in relation to the cusp. For the remaining six, $\mathbb{R}\omega$ has a node, two categories arising when the node has real branches and four others when it is isolated – distinguished by the colouring and by the presence or absence of a loop of $\mathbb{R}\omega$ in addition to the isolated double point. All these different cases can be distinguished by their eigenvalue and eigenvector properties but we shall not enter into this discussion here.

Many of the cases when ω is reducible will also separate into subclasses and different categories. For example, in case (1, 1)(1, 1), $\mathbb{R}\omega$ consists of two separate circles either of which may have real or imaginary radius – according as, in the expression

$$\Phi_{AB}^{C'D'} = \Lambda_{(A}^{(C'}\Upsilon_{B)}^{D')}, \quad i.e., \; \Phi_{ab} = \Lambda_{(a}\Upsilon_{b)} - \tfrac{1}{4}g_{ab}\Lambda_c\Upsilon^c$$

the respective real vectors Λ^a, Υ^a are spacelike or timelike. These circles intersect in a pair of distinct real or conjugate complex points (according as Λ^a, Υ^a span a timelike or spacelike plane), or, as we have seen, in a more *special* circumstance (dimensionality reduced) the circles may *touch* (when Λ^a, Υ^a span a null plane, ω having a tacnode at the contact point). The *generic* case (1, 1)(1, 1) thus subdivides into a number of different categories, six in all (taking into account the colour) these all being cases when ω possesses two distinct nodes (the intersection points of the two circles). In the *tacnode* case, the circles have real radii and there are two colourings.

The generic case $|(1,1)|^2$ also divides into (four) categories (characterized by whether the two nodes are real and isolated, or conjugate complex, and also with respect to colour). The *special* case of $|(1,1)|^2$ is again distinguished by the presence of a tacnode (two categories).

Finally, the case $(1,1)|(1,0)|^2$ should be especially remarked upon. Again, the generic case subdivides into (four) categories. But there is also a special case arising when the intersection point of the pair of lines $|(1,0)|^2$ (which is generically an isolated double point) lies *on* the circle (1, 1) (in which case it becomes a *triple point* with one real and two conjugate complex

* The adjectival phrase 'conjugate complex' is a useful one referring to a pair of objects (numbers, curves, points, etc.) which are interchanged under the operation of complex conjugation.

branches). From looking at the real point set $\mathbb{R}\omega$ alone, one would not even recognise the existence of a triple point here.

Complete classification and specialization scheme

The complete scheme of types and their specializations is given in Penrose (1972*a*). It is somewhat complicated in its totality, but simple enough to reconstruct, if one bears in mind such features as the fact that the multiplicities of multiple points cannot decrease under specialization and that the only bits of $\mathbb{R}\omega$ which can appear discontinuously under specialization are multiple portions (with even multiplicity).

Ignoring the splitting of classes into different categories of the same generality (which would show up in the different eigenvalue orderings and reality structure), the various possibilities for the curve ω, together with the corresponding Segre characteristic* for $\Phi_a{}^b$ are listed in Table (8.8.25), and their scheme of specialization is given in Table (8.8.26). (The original notation of (8.8.1) has been reverted to since the Hermiticity structure of the factors is now being incorporated into the subdivisions into categories.)

Plebański spinor type

The only feature of Table (8.8.24) that has not yet been discussed here at all is the Plebański (1964) spinor type. In (8.7.1) we defined the symmetric spinor Π_{ABCD} such that

$$\Pi_{ABCD}\xi^A\xi^B\xi^C\xi^D = \Phi_{P'Q'AB}\xi^A\xi^B\Phi^{P'Q'}_{CD}\xi^C\xi^D. \qquad (8.8.24)$$

Plebański's method is to apply the standard Weyl spinor classification to Π_{ABCD} and to use this as a classification scheme for $\Phi_{ABC'D'}$. We first discuss the relation between the representations of $\Pi_{...}$ and $\Phi_{...}$ in terms of corresponding curves π and ω on $\mathbb{C}S^+$. Now equation (8.8.24) tells

* The Segre characteristic [ ] specifies the Jordan normal form of a matrix (*cf.* (8.3.42); Turnbull and Aitken 1948). Each set of equal eigenvalues down the diagonal of the matrix gives rise to a separate entry in [ ]. However, such a set may have units just above the diagonal in certain square blocks; the size of these blocks is indicated in the entries. For example, [(3 2 1)(4 4)] indicates a block of three λs down the diagonal (with 2 units above the diagonal), followed by a block of two λs (with one unit above the diagonal), followed by a single λ; then a block of four μs (with three units) and another of four μs (with three more units). Parentheses round single numbers are omitted. The dimension of the space spanned by all the eigenvectors is the total number of numerals appearing in [ ] (here five). For each eigenvalue, the dimension of the corresponding eigenvector space is the number of numerals within that particular bracket (here 3, 2).

Table (8.8.25)

Degree of freedom of $\Phi_{ABA'B'}$	Number of distinct categories	Structure of ω: Reducibility	Structure of ω: Singularities	Segre characteristic of $\Phi_a{}^b$	Plebański spinor type
9	5	(2, 2)	none	[1111]	{1111}
8	6	(2, 2)	node	[211]	{211}
7	2	(2, 2)	cusp	[31]	{31}
7	10	(1, 1)(1, 1)	two nodes	[(11)11]	{22}
6	4	(1, 1)(1, 1)	tacnode	[(21)1]	{4}
6	4	(1, 1)(1, 0)(0, 1)	three nodes	[2(11)]	{22}
5	1	(1, 1)(1, 0)(0, 1)	triple point	[(31)]	{4}
5	2	(1, 0)(1, 0)(0, 1)(0, 1)	four nodes	[(11)(11)]	{22}
4	4	$(1, 1)^2$	double curve	[(111)1]	{−}
3	2	$(1, 0)^2 (0, 1)^2$	double curve / quadruple point	[(211)]	{−}
0	1	(−)	$\mathbb{C}S^+$	[(1111)]	{−}

Table (8.8.26)

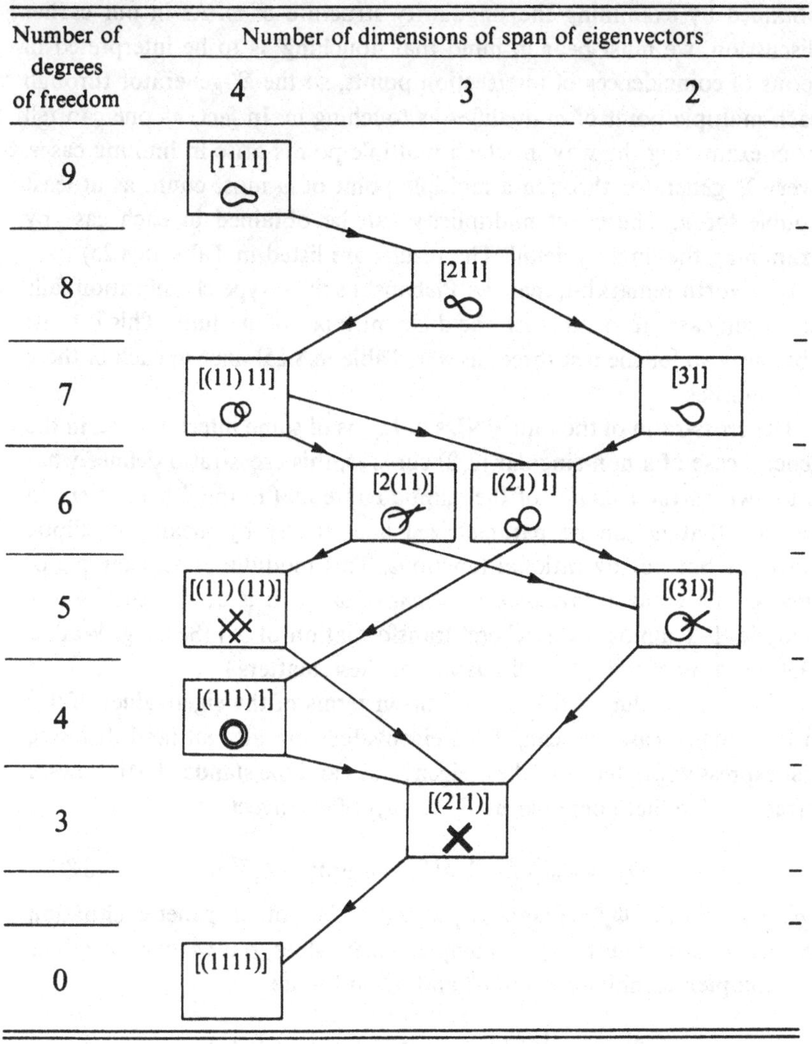

us that if ξ^A is a principal spinor of Π_{ABCD}, then, (*cf.* just after (3.5.29)),

$$\Phi_{ABC'D'}\xi^A\xi^B = \eta_{C'}\eta_{D'}, \tag{8.8.27}$$

for some $\eta_{C'}$. Thus

$$\Phi_{ABC'D'}\xi^A\xi^B\eta^{D'} = 0, \tag{8.8.28}$$

which, by (8.7.29), is the condition for the Σ'-generator defined by ξ^A to touch ω at $\xi^A\eta^{A'}$. Since this applies to each principal spinor of Π_{ABCD} we obtain the result that π consists of the set of four Σ'-generators which *touch* ω. The coincidence scheme for these generators is then readily obtained by examining the singularity structure of ω. As in our earlier discussion, we must bear in mind that 'touching' is to be interpreted in terms of coincidences of intersection points, so the Σ'-generator through each multiple point of ω qualifies as touching ω. In fact, as one can tell from examining the way in which multiple points arise in limiting cases, every Σ'-generator through a multiple point of ω must count as at least double for π. The exact multiplicity can be obtained in each case by examining the limit in detail. The results are listed in Table (8.8.25)

It is worth remarking that the Plebański spinor type classification fails in certain cases to distinguish the different types of $\Phi_{...}$ fully. This is most obviously so for the last three cases in Table (8.8.25) since in each of these $\Pi_{...}$ vanishes.

The cross-ratio of the four PNDs of $\Pi_{...}$ is of some interest since, in the general case of a non-singular (2, 2) curve ω, this cross-ratio defines what is known as the *modulus* of the elliptic curve ω. ('Elliptic' here refers to the fact that ω can be parametrized analytically by means of elliptic functions but not by rational functions. This modulus is an example of what is known as a *birational invariant* of ω, i.e. it is unchanged by any analytic '1–1 almost everywhere' transformation of ω.) (See, e.g. Walker 1950 for a more complete discussion of these matters.)

To find the value of this cross-ratio in terms of the eigenvalues of $\Phi_a{}^b$ in the general case, we can, if the eigenvalues are all real (and distinct), first express Φ_{ab} in terms of these eigenvalues and the standard Minkowski tetrad (3.1.20) (here denoted by $\delta_a^0, \ldots, \delta_a^3$) of eigenvectors:

$$\Phi_{ab} = \lambda_0 \delta_a^0 \delta_b^0 - \lambda_1 \delta_a^1 \delta_b^1 - \lambda_2 \delta_a^2 \delta_b^2 - \lambda_3 \delta_a^3 \delta_b^3 \tag{8.8.29}$$

(i.e., the matrix $\Phi_{\mathbf{a}}{}^{\mathbf{b}} = \mathrm{diag}(\lambda_0, \lambda_1, \lambda_2, \lambda_3)$). The other generic situation requires complex vectors and complex eigenvalues, but we can introduce two complex combinations of δ_a^0 and δ_a^3 and write

$$\begin{aligned}\Phi_{ab} = &-\tfrac{1}{2}\mathrm{i}\lambda_0(o_A o_{A'} + \mathrm{i}\iota_A \iota_{A'})(o_B o_{B'} + \mathrm{i}\iota_B \iota_{B'}) \\ &+\tfrac{1}{2}\mathrm{i}\lambda_3(o_A o_{A'} - \mathrm{i}\iota_A \iota_{A'})(o_A o_{A'} - \mathrm{i}\iota_A \iota_{A'}) \\ &-\lambda_1 \delta_a^1 \delta_b^1 - \lambda_2 \delta_a^2 \delta_b^2,\end{aligned} \tag{8.8.30}$$

where λ_1, λ_2 are real and $\lambda_3 = \bar{\lambda}_0$, and where the standard relation (3.1.20) between a spinor dyad and Minkowski tetrad is being employed. Using (3.1.20) also in the case (8.8.29), we can compute Π_{ABCD} directly to obtain, from (8.8.29), its expression in standard canonical form (8.3.34), where

$$\eta = \tfrac{1}{3}(\lambda_0 - \lambda_1)(\lambda_2 - \lambda_3) \quad \textit{and} \quad \chi = \frac{(\lambda_0 - \lambda_3)(\lambda_2 - \lambda_1)}{(\lambda_0 - \lambda_1)(\lambda_2 - \lambda_3)}. \qquad (8.8.31)$$

In the case (8.8.30), we obtain an expression for Π_{ABCD} which differs only inessentially from the standard canonical form (and which reverts to it under $o^A \mapsto e^{\pi i/8} o^A, \iota^A \mapsto e^{-\pi i/8} \iota^A$, and again we arrive at precisely (8.8.31) for the values of η and χ for Π_{ABCD}.

Note the striking fact that the cross-ratio of the PNDs for Π_{ABCD} is also the cross-ratio of the four eigenvalues of $\Phi_a{}^b$. This cross-ratio must therefore be real in the case when $\Phi_a{}^b$ has a real tetrad of eigenvectors. In the other case we have, by virtue of the reality properties of the λs, that χ satisfies $|\chi| = 1$, $|1 - \chi| = 1$, or $\mathrm{Re}(\chi) = \frac{1}{2}$. We recall, from the discussion of §8.6, that these four possibilities are just the cases in which the PNDs (of Π_{ABCD}) have reflective symmetries. Finally, using $\lambda_0 + \lambda_1 + \lambda_2 + \lambda_3 = 0$, we observe that the three eigenvalues of $\Pi_{AB}{}^{CD}$ are just the three expressions:

$$\lambda_0\lambda_2 + \lambda_1\lambda_3 + S, \quad \lambda_0\lambda_1 + \lambda_2\lambda_3 + S, \quad \lambda_0\lambda_3 + \lambda_2\lambda_1 + S,$$

where

$$S = \tfrac{1}{6}(\lambda_0^2 + \lambda_1^2 + \lambda_2^2 + \lambda_3^2) \qquad (8.8.32)$$

Physical energy tensors

It should now be clear that the Ricci tensor is really a much more complicated object for general classification than the Weyl tensor. However, for space–times of physical interest one may reasonably argue that the great majority of the types are to be ruled out. The weak energy condition* (in its usual non-strict form) asserts, of the energy–momentum tensor T_{ab}, that for each null vector x^a,

$$T_{ab}x^a x^b \geqslant 0. \qquad (8.8.33)$$

Thus if Einstein's equations (4.6.32) are adopted, we then have

$$\Phi_{ab}x^a x^b \geqslant 0. \qquad (8.8.34)$$

This is just the condition that none of S^+ be 'coloured black' i.e. that every point of S^+ be either 'white' or a point of $\mathbb{R}\omega$. And since the colour changes from one side of $\mathbb{R}\omega$ to the other wherever $\mathbb{R}\omega$ is a simple curve, this implies that $\mathbb{R}\omega$ consists entirely of *double* (or

* See second footnote on p. 283

quadruple) points. The only categories that can satisfy this criterion are the 'white' versions of the following:

$(2,2)$ *non-singular (and $\mathbb{R}\omega$ empty)*
$(2,2)$ *node (isolated; and $\mathbb{R}\omega$ otherwise empty)*
$(1,1)(1,1)$ *two nodes (complex and $\mathbb{R}\omega$ empty)*
$|(1,1)|^2$ *two nodes (both isolated)*
$|(1,1)|^2$ *two nodes (both complex)*
$|(1,1)|^2$ *tacnode (isolated)*
$(1,1)|(1,0)|^2$ *three nodes (one isolated; $\mathbb{R}\omega$ otherwise empty)*
$|(1,0)(1,0)|^2$ *four nodes (two isolated, two complex)*
$(1,1)^2$ *double curve (real)*
$(1,1)^2$ *double curve (complex)*
$|(1,0)^2|^2$ *double curve and quadruple point (isolated)*
$(-)$ $(\mathbb{R}\omega = S^+)$ (8.8.35)

The remaining categories may, if desired, be ruled out as 'non-physical'. On the other hand, one is often interested in symmetric 2-index tensors in any case, whether or not they are the Ricci tensors of 'physically reasonable' space-times. For example, one might be interested in the various types that can occur as terms in an asymptotic expansion of the Ricci tensor (compare §9.7). The individual terms need not satisfy an energy condition.

Of course, *generically*, in a physical space-time, the type will be $(2,2)$ non-singular, as above, almost everywhere. But when the Einstein–Maxwell equations hold, the generic type is $|(1,0)(1,0)|^2$ (as immediately follows from $\Phi_{ABA'B'} = 2G\varphi_{AB}\bar{\varphi}_{A'B'}$, *cf.* eq. (5.2.6)) or, in the case of a null field, $|(1,0)^2|^2$. For an isotropic medium, it follows from the spherical symmetry (three equal eigenvalues *cf.* also (8.8.1)) that the type is always $(1,1)^2$ with complex double curve.

9

Conformal infinity

9.1 Infinity for Minkowski space

One of the most useful areas of application of 2-spinor methods has turned out to be the study of *asymptotic* questions in relativity, important examples of which are the definition of the total energy–momentum contained in an asymptotically flat space–time and of gravitational radiation. For this, the spinor methods are particularly powerful when combined with a technique (Penrose 1963, 1964*b*, 1965) which employs conformal metric rescalings in order to 'make infinity finite'. According to this technique we rescale the metric of the space–time $\mathcal{M}$, replacing the original physical metric $\mathrm{d}s$ by a new 'unphysical' metric $\mathrm{d}\hat{s}$, which is conformally related to it,

$$\mathrm{d}\hat{s} = \Omega\,\mathrm{d}s, \tag{9.1.1}$$

Ω being a suitably smooth, everywhere positive function defined on $\mathcal{M}$. The metric tensor g_{ab} and its inverse g^{ab} are accordingly rescaled by

$$g_{ab} \mapsto \hat{g}_{ab} = \Omega^2 g_{ab}, \quad g^{ab} \mapsto \hat{g}^{ab} = \Omega^{-2} g^{ab}. \tag{9.1.2}$$

Provided that the asymptotic structure of $\mathcal{M}$ is suitable, and that Ω is chosen appropriately, it is possible to adjoin to $\mathcal{M}$ a certain boundary surface, denoted by $\mathscr{I}$ (and pronounced 'scrī' – a contraction of 'script I'), in such a way that the 'unphysical' metric $\hat{g}_{ab}$ extends non-degenerately and with some degree of smoothness to these new points. The function Ω can also be extended appropriately smoothly but becomes zero on $\mathscr{I}$. This implies that the physical metric would have to be infinite on $\mathscr{I}$, so *it* cannot be so extended. Thus, from the point of view of the physical metric, the new points (namely those on $\mathscr{I}$) are infinitely distant from their neighbours. Physically, they represent 'points at infinity'.

The adjoining of $\mathscr{I}$ to such a space–time $\mathcal{M}$ provides us with a smooth manifold-with-boundary,* denoted by $\bar{\mathcal{M}}$, with

* We do not go into the details of the precise definition of a manifold-with-boundary here but refer the reader to Lang (1972). Essentially, it is a space whose points have neighbourhoods which are either Euclidean spaces or Euclidean half-spaces (e.g. $\{(x^1, \ldots, x^n) \in \mathbb{R}^n \mid x^1 \geqslant 0\}$).

$$\mathscr{I} = \partial \bar{\mathscr{M}}, \quad \mathscr{M} = \text{int}\, \bar{\mathscr{M}}$$

(∂ = boundary, int = interior). The advantage is that the powerful *local* techniques of differential geometry and spinor algebra can now be employed on $\bar{\mathscr{M}}$ with implications for the asymptotics of $\mathscr{M}$. Thus we need not resort to complicated limiting arguments when studying the all-important detailed rates of fall-off of physical or geometrical quantities in, for example, radiation questions in asymptotically flat space-times. Indeed, the very definition of asymptotic flatness in general relativity can now be given in a convenient and 'coordinate-free' way. Conformal methods are particularly appropriate in relativity because many of the important concepts are conformally invariant. Among these are the massless free-field equations, the Weyl conformal tensor, null geodesics, null hypersurfaces, relativistic causality, and, most particularly for the case of Minkowski space, twistor theory. The technique here outlined is similar to that used in complex analysis, where the 'point at infinity' is adjoined to the Argand plane to obtain the Riemann sphere (*cf.* §1.2), and in projective geometry.

Explicit coordinate description; $\mathscr{I}^{\pm}$, $\mathrm{i}^{\pm}$ *and* i^0

Let us begin by examining the construction of conformal infinity for Minkowski space $\mathbb{M}$. The physical metric, in spherical polar coordinates, is

$$\mathrm{d}s^2 = \mathrm{d}t^2 - \mathrm{d}r^2 - r^2(\mathrm{d}\theta^2 + \sin^2\theta\, \mathrm{d}\phi^2). \tag{9.1.3}$$

For convenience, we introduce a retarded time parameter $u = t - r$ and an advanced time parameter $v = t + r$ to obtain

$$\mathrm{d}s^2 = \mathrm{d}u\, \mathrm{d}v - \tfrac{1}{4}(v-u)^2(\mathrm{d}\theta^2 + \sin^2\theta\, \mathrm{d}\phi^2). \tag{9.1.4}$$

There is much freedom in the choice of a conformal factor Ω. However, for the types of space-times that we shall consider here ('asymptotically simple' space-times), it turns out from general considerations (*cf.* just after (9.7.22)) that our choice of Ω must be such that along any ray it approaches zero (both in the past and in the future) like the reciprocal of an affine parameter λ on the ray (i.e. $\Omega\lambda \to$ constant as $\lambda \to \infty$ and $\Omega\lambda \to$ constant as $\lambda \to -\infty$, along the ray). Each u = constant hypersurface is a future light cone, generated by the rays (null straight lines) for which θ and ϕ are also constant. The coordinate v serves as an affine parameter into the future on each of these radial rays. Similarly, the coordinate u serves as an affine parameter into the past on these rays. Thus we shall require

$\Omega v \to$ constant as $v \to \infty$ on $u, \theta, \phi =$ constant, and $\Omega u \to$ constant as $u \to -\infty$ on $v, \theta, \phi =$ constant. If we wish also to keep Ω smooth over the finite parts of space–time, then the choice

$$\Omega = 2(1+u^2)^{-\frac{1}{2}}(1+v^2)^{-\frac{1}{2}}$$

suggests itself (the factor 2 being for later convenience), and then

$$d\hat{s}^2 = \Omega^2\, ds^2 = \frac{4\,du\,dv}{(1+u^2)(1+v^2)} - \frac{(v-u)^2}{(1+u^2)(1+v^2)}(d\theta^2 + \sin^2\theta\, d\phi^2).$$

Many other choices of Ω are equally possible, but this one is especially convenient, as we shall see shortly.

In order that our 'points at infinity' may be assigned finite coordinates, we can replace u and v by p and q, where

$$u = \tan p, \quad v = \tan q.$$

Then

$$d\hat{s}^2 = 4\,dp\,dq - \sin^2(q-p)(d\theta^2 + \sin^2\theta\, d\phi^2). \qquad (9.1.5)$$

The range of the variables p, q is as indicated in Fig. 9-1, in which each point represents a 2-sphere of radius $\sin(q-p)$. The vertical line

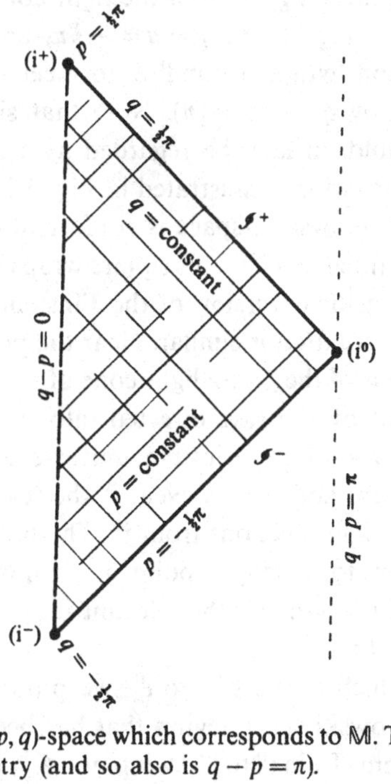

Fig. 9-1. The region of (p, q)-space which corresponds to $\mathbb{M}$. The line $q - p = 0$ is an axis of spherical symmetry (and so also is $q - p = \pi$).

$q-p=0$ represents the spatial origin ($r=0$) and is just a *coordinate* singularity. The space–time is, of course, non-singular on this line (as everywhere else). The sloping lines $p=-\frac{1}{2}\pi$ $(-\frac{1}{2}\pi<q<\frac{1}{2}\pi)$ and $q=\frac{1}{2}\pi$ $(-\frac{1}{2}\pi<p<\frac{1}{2}\pi)$ represent (null) infinity (denoted by $\mathscr{I}^-$ and $\mathscr{I}^+$, respectively) for Minkowski space (since they correspond to $u=-\infty$ and to $v=\infty$). However, the metric (9.1.5) is evidently perfectly regular on these regions. Indeed, the space–time and its metric $d\hat{s}$ can clearly be extended *beyond* these regions in a non-singular fashion. The vertical line $q-p=\pi$ is again a coordinate singularity – of precisely the same type as that at $q-p=0$. The entire vertical strip $0\leqslant q-p\leqslant\pi$ may be used to define a space–time $\mathscr{E}$ whose global structure is that of the product of a spacelike 3-sphere with an infinite timelike line (an 'Einstein static universe'). To see this, we choose new coordinates

$$\tau=p+q, \quad \rho=q-p$$

and obtain

$$d\hat{s}^2=d\tau^2-\{d\rho^2+\sin^2\rho(d\theta^2+\sin^2\theta\, d\phi^2)\}. \tag{9.1.6}$$

The part in curly brackets represents the metric of a unit 3-sphere.

The portion of $\mathscr{E}$ which is conformal to the original Minkowski space may be described as that lying between the light cones of two points i^- and i^+. The point i^- is given by $p=q=-\frac{1}{2}\pi$, and the point i^+ by $p=q=\frac{1}{2}\pi$. This portion 'wraps around' $\mathscr{E}$ to meet at the 'back' in the single point i^0 (given by $q=-p=\frac{1}{2}\pi$). Note that $\sin^2(q-p)=0$ at i^0, indicating that i^0 should, in fact, be regarded as a single point, rather than a 2-sphere. The situation is illustrated in Fig. 9-2, under suppression of two dimensions. Minkowski 2-space is conformal to the interior of a square (represented as tilted at 45°). This square wraps around the cylinder which is the two-dimensional version of the Einstein static universe. In higher dimensions the situation is similar. Near the point i^-, the relevant region lies in the *interior* of the *future* light cone of i^-. This light cone (i.e. the point-set swept out by the rays directed into the future from i^-) is focused around the back of the Einstein universe to a single point i^0 (which is spatially the antipode of i^-). Near i^0 the relevant ('Minkowski') region lies in the *spacelike* directions from i^0. The future light cone of i^0 is focused back again to a single point i^+, whose spatial location corresponds to that of i^-. Near i^+, the relevant region lies in the *interior* of the *past* light cone of i^+.

The ray segments which connect i^- to i^0 sweep out the portion of the boundary of the Minkowski space region that has been denoted by $\mathscr{I}^-$. Similarly the ray segments from i^0 to i^+ sweep out $\mathscr{I}^+$. The points i^-, i^0, i^+

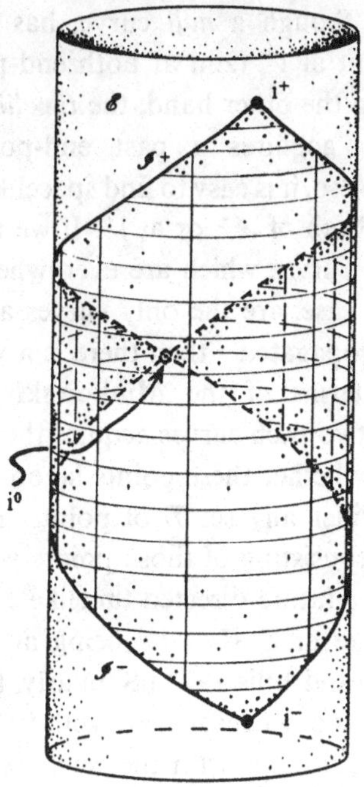

Fig. 9-2. The region on the Einstein cylinder $\mathscr{E}$ which corresponds to $\mathbb{M}$.

themselves are considered not to belong to $\mathscr{I}^-$ or to $\mathscr{I}^+$. Physically, we interpret i^- as representing past temporal infinity, $\mathscr{I}^-$ as past null infinity, i^0 as spatial infinity, $\mathscr{I}^+$ as future null infinity, and i^+ as future temporal infinity. The reason for this terminology becomes clear when we examine the behaviour of straight lines in Minkowski space (straight, that is, with respect to the Minkowski metric ds). A timelike straight line acquires a past end-point i^- and a future end-point i^+. A null straight line acquires a past end-point on $\mathscr{I}^-$ and a future end-point on $\mathscr{I}^+$. A spacelike straight line becomes a closed curve through i^0. (The detailed verification of these facts is straightforward.) Since rays remain rays after conformal rescalings, the *null* straight lines become rays according to the d$\hat{s}$ metric; but the timelike or spacelike straight lines are not, in general, geodesics with respect to d$\hat{s}$.

TIPs and TIFs

When we consider *curved* lines in Minkowski space, the question of which end-points they acquire is more complicated. For example, the helix given by $t = \phi$, $r = 1$, $\theta = \frac{1}{2}\pi$, though a *null* curve, has a past end-point at i^- and a future end-point at i^+ (and at both end-points the completed curve is not smooth). On the other hand, the *timelike* curve $r = (1 + t^2)^{\frac{1}{2}}$, $\theta = \frac{1}{2}\pi$, $\phi = 0$ smoothly acquires a past end-point on $\mathscr{I}^-$ and a future end-point on $\mathscr{I}^+$. Also, it is easy to find spacelike curves that acquire end-points on either or both of $\mathscr{I}^\pm$ or at $i^\pm$. If we restrict our attention to *causal* curves (that is, curves which are everywhere either timelike or null) – and these, in any case, are the only curves along which particles or information can be propagated – then there is a very simple criterion, expressible entirely in terms of the Minkowski space $\mathbb{M}$ itself, for deciding whether or not two such curves acquire the same past end-point or future end-point and whether these points lie on $\mathscr{I}^\pm$ or $i^\pm$.

To this purpose, consider any set Σ of points in $\mathbb{M}$, and let $I^+[\Sigma]$ denote the subset of $\mathbb{M}$ consisting of those points which can be reached from some point of Σ by a future-directed timelike curve; in other words, $I^+[\Sigma]$ is the (open) future of Σ. (For its being an open set in $\mathbb{M}$, see Penrose 1972*b*, Hawking and Ellis 1973.) Similarly, let $I^-[\Sigma]$ denote the (open) past of Σ. Now if α is any causal curve with a (finite) past end-point P in $\mathbb{M}$, the future of P coincides with the set $I^+[\alpha]$, a property which holds for no other point in $\mathbb{M}$ (as one may easily convince oneself – full proofs can be found in Geroch, Kronheimer and Penrose 1972). Thus, any other causal curve β also has P as its past end-point if and only if $I^+[\beta] = I^+[\alpha]$. The advantage of this criterion is that it applies also if α and β are *past-endless* (i.e. extend indefinitely into the past and do not attain finite past end-points in $\mathbb{M}$): such curves α and β acquire identical

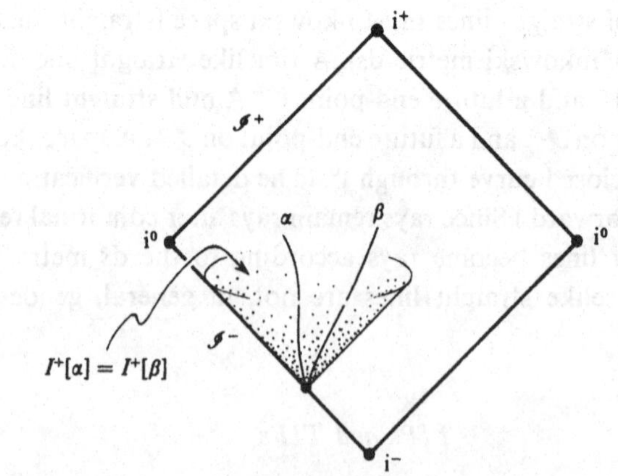

Fig. 9-3. Two causal curves α, β have the same past end-point on $\mathscr{I}^-$ if and only if they share the same future.

past end-points, either at i^- or on $\mathscr{I}^-$, if and only if $I^+[\alpha] = I^+[\beta]$ (see Fig. 9-3). (They clearly cannot have past end-points at i^0, i^+, or on $\mathscr{I}^+$.) From this we see that a past-endless causal curve α acquires a past end-point at i^- or on $\mathscr{I}^-$ according as $I^+[\alpha]$ is or is not the whole space $\mathbb{M}$ (since, for example, the future of the time-axis $t = 0$ is the whole of $\mathbb{M}$). In exactly the same way, a future-endless causal curve α reaches i^+ or a point of $\mathscr{I}^+$ according as $I^-[\alpha]$ is or is not the whole of $\mathbb{M}$; furthermore, a future-endless curve β reaches the same point at infinity as α if and only if $I^-[\alpha] = I^-[\beta]$.

These criteria are especially useful since they also apply in curved space. Sets of the form $I^+[\alpha]$, where α is a past-endless causal curve are called TIFs (terminal indecomposable future-sets); those of the form $I^-[\alpha]$ with α a future-endless causal curve are called TIPs (terminal indecomposable past-sets). They can be used to provide *definitions* of past/future boundaries in very general space–times (see Seifert 1971, Geroch, Kronheimer and Penrose 1972). The boundaries of TIFs and TIPs are also of interest. They are generated by rays which are, respectively, future-endless or past-endless (*cf.* Penrose 1972*b*, Hawking and Ellis 1973). In the case of Minkowski space $\mathbb{M}$, these boundaries (when non-vacuous) are the null hyperplanes in $\mathbb{M}$, i.e. sets with equations of the form $x^a A_a = B$, where A_a, B are constant, and A_a is null. This fact will have significance for the next section.

The required conformal manifold-with-boundary $\bar{\mathbb{M}}$ now consists of the original Minkowski space $\mathbb{M}$, with its conformal metric, together with the two boundary 3-surfaces $\mathscr{I}^+$ and $\mathscr{I}^-$. But the points i^+, i^0, and i^- have to be excluded from $\bar{\mathbb{M}}$ because the boundary would not be smooth at these points. We observe that the topology of each of $\mathscr{I}^+$ and $\mathscr{I}^-$ is $S^2 \times \mathbb{R}$, where S^2 is coordinatized by the spherical polar angles θ, ϕ and $\mathbb{R}$ by retarded time u, in the case of $\mathscr{I}^+$, and by advanced time v, in the case of $\mathscr{I}^-$.

9.2 Compactified Minkowski space

When we come to consider asymptotically flat space–times shortly we shall see that much of the above discussion still applies. However, one property which is very specific to the Minkowski space model is the fact that *every* null geodesic which originates at some point A^- on $\mathscr{I}^-$ will pass through the *same* point A^+ on $\mathscr{I}^+$ (see Fig. 9-4). This property may seem surprising at first, but it becomes clear when we recall that the future light cone of a point of $\mathscr{I}^-$ is simply a null hyperplane in the Minkowski space. (It is the limit of a light cone when the vertex recedes into the past

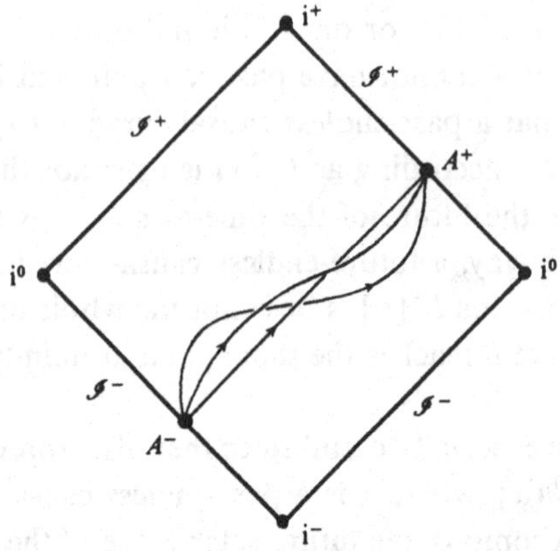

Fig. 9-4. The rays in $\mathbb{M}$ from a point $A^- \in \mathscr{I}^-$ all share the same future end-point $A^+ \in \mathscr{I}^+$. (This is a special property of $\mathbb{M}$ not possessed by general space-times.)

along a null straight line.) Similarly the past light cone of any point on $\mathscr{I}^+$ is also a null hyperplane. So a null hyperplane will acquire a past 'vertex' on $\mathscr{I}^-$ (say A^-) and a corresponding future 'vertex' (say A^+) on $\mathscr{I}^+$. We can also see this in terms of the Einstein-universe model $\mathscr{E}$. The future light cone of A^- is focused at a point A^+ which is spatially antipodal to A^-.

Having this natural association between points of $\mathscr{I}^-$ and $\mathscr{I}^+$, for Minkowski space, it is in some respects natural to make an identification between $\mathscr{I}^-$ and $\mathscr{I}^+$, the point A^- being identified with A^+ and $\mathscr{I}^-$ and $\mathscr{I}^+$ then being written as $\mathscr{I}$. If we do this, then for the sake of continuity we should also identify i^- with i^0, and i^0 with i^+. The three points $i^\pm$, i^0 thus become one, which we label I. We see from Fig. 9-5 that the Minkowski regions fit neatly together at the point I, so that I becomes simply a normal interior point of the identified manifold. The *compact* conformal manifold that we construct by means of these identifications is referred to as *compactified Minkowski space* $\mathbb{M}^\#$ (see Bôcher 1914, Coxeter 1936, Kuiper 1949, Penrose 1964*a*). For reasons that we shall see in more detail later, such identifications cannot be satisfactorily carried out in *curved* asymptotically flat spaces. (Not only is there apparently no *canonical* way of performing such identifications in general, but, when the total mass is non-zero *any* identification would lead to failure of the required regularity conditions along the identification hypersurface.) For many purposes, the identification of $\mathscr{I}^-$ with $\mathscr{I}^+$ may, even in Minkowski space, seem unphysical (and, of course, it need not be made). However, for various

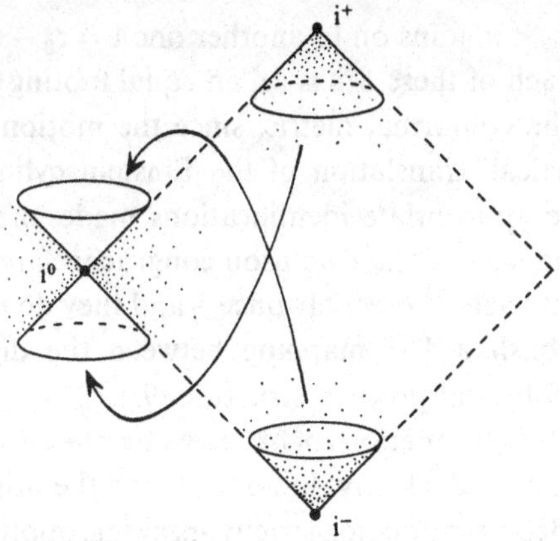

Fig. 9-5. In the identification of $\mathscr{I}^-$ with $\mathscr{I}^+$ which produces $\mathbb{M}^\#$, the points i^-, i^0 and i^+ all get identified. The resulting non-singular point I has a neighbourhood made from three separate pieces of $\mathbb{M}$.

mathematical purposes the identification is very useful,* and so it will be worth our while in this section to develop the geometry of compactified Minkowski space in a little detail, particularly in view of its relevance to twistor theory, and because it is the space on which the fifteen-parameter conformal group acts.

With $\mathscr{I}^-$ and $\mathscr{I}^+$ identified as explained above, each null geodesic in $\mathbb{M}^\#$ has the topology of a circle S^1, being a null straight line in $\mathbb{M}$ joined into a loop with a single point at infinity. The topology of the whole space $\mathbb{M}^\#$ turns out to be

$$\mathbb{M}^\# \cong S^3 \times S^1. \tag{9.2.1}$$

This is not immediately obvious but it can be seen if we select an arbitrary Robinson congruence in $\mathbb{M}^\#$ (*cf.* §6.2, p. 59 *et seq.* and Fig. 6-3), and consider the family of spacelike hypersurfaces given by $\tau = \text{constant}$ in (9.1.6). The identification of $\mathscr{I}^-$ with $\mathscr{I}^+$ entails that each hypersurface

* The analogous construction in *three* space–time dimensions (metric signature $(+--)$) also has importance in relativity theory. As was mentioned at the end of §7.2, the space $\mathscr{P}^\#$ of possible bundles of rays which neighbour a given ray μ, and which can lie on some null hypersurface through μ, is such a compactified Minkowski 3-space. The construction of $\mathscr{P}^\#$ from $\mathscr{P}$ (coordinatized, at some point P of μ, by real ρ and complex σ) is exactly analogous to that of $\mathbb{M}^\#$ from $\mathbb{M}$ (the only essential difference being that the resulting topology is not $S^2 \times S^1$ – analogously to (9.2.1) – but a non-orientable 3-space similar to a Klein bottle). The analogue of $\mathscr{I}$ now represents bundles of rays for which ρ and σ both diverge at P (the general case of a caustic point at P) whereas for the analogue of I, ρ diverges but σ remains finite (as at the vertex of a light cone).

$\tau = \tau_0$, with $0 < \tau_0 < \pi$, joins on to another one $\tau = \tau_0 - \pi$ to give a spacelike S^3 section. Each of these S^3s is on an equal footing with every other, with respect to the conformal metric, since the motion $\tau \mapsto \tau +$ constant (mod π) is a 'vertical' translation of the Einstein cylinder to itself (see Fig. 9-2), with the appropriate identifications made, preserving the conformal metric. The lines of the Robinson congruence are topologically S^1 and each intersects each S^3 precisely once – and they do not intersect each other. They establish a 1–1 mapping between the different S^3s, thus providing the topological product structure (9.2.1).

The conformal transformations of $\mathbb{M}^{\#}$ given by $\tau \to \tau +$ constant (mod π), that were just considered, clearly do not preserve the original Minkowski metric g of $\mathbb{M}$. (Indeed, they do not, strictly speaking, apply to $\mathbb{M}$ at all, since some points of $\mathbb{M}$ are mapped to $\mathscr{I}$ and vice versa.) The spacelike hyperplane $t = 0$ of the original Minkowski coordinates, which is also given by $\tau = 0$, is mapped to one of the hypersurfaces $\tau = \tau_0$ (mod π), which, on referral back to the coordinates (t, r, θ, ϕ), is seen to be a pair of branches of the spacelike 3-hyperboloid

$$B(t^2 - r^2) + 2t - B = 0, \tag{9.2.2}$$

with

$$B := \tan \tau_0 = \tan(p + q) = \frac{u + v}{1 - uv} = \frac{2t}{1 - t^2 + r^2}. \tag{9.2.3}$$

By combining such conformal transformations with the motions of the Poincaré group, we can, in fact, generate the entire identity-connected component $\mathrm{C}_+^\uparrow(1,3)$ of the fifteen-parameter group $C(1,3)$ of conformal motions of $\mathbb{M}^{\#}$, under which the five-parameter family of hypersurfaces that are space–time translations of (9.2.2), together with the spacelike hyperplanes in $\mathbb{M}$, are transformed into one another.

Description in terms of $\mathbb{P}^5$

There is another way of constructing the space $\mathbb{M}^{\#}$ and its conformal group of motions. Consider the six-dimensional pseudo-Euclidean space $\mathbb{E}^6$ with coordinates T, V, W, X, Y, Z and metric

$$ds^2 = dT^2 + dV^2 - dW^2 - dX^2 - dY^2 - dZ^2. \tag{9.2.4}$$

The null cone $\mathscr{K}$ of the origin has equation

$$T^2 + V^2 - W^2 - X^2 - Y^2 - Z^2 = 0. \tag{9.2.5}$$

As we see by substituting

$$V - W = 1 \tag{9.2.6}$$

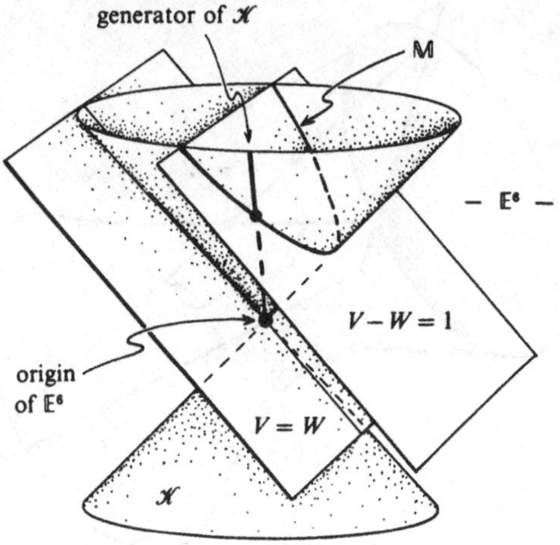

Fig. 9-6. The $(++----)$-cone $\mathscr{K}$ in $\mathbb{E}^6$. Minkowski space $\mathbb{M}$ is represented as a parabolic section of $\mathscr{K}$, and the compactified space $\mathbb{M}^\#$ as the space of generators of $\mathscr{K}$.

into (9.2.4), the intersection of $\mathscr{K}$ with the null 5-plane (9.2.6) has an induced metric of the Minkowskian form

$$ds^2 = dT^2 - dX^2 - dY^2 - dZ^2. \tag{9.2.7}$$

For this intersection the coordinates T, X, Y, Z suffice and are unrestricted in range; the intersection is therefore intrinsically identical with Minkowski space $\mathbb{M}$ and we shall so label it. As a subspace of $\mathbb{E}^6$, however, $\mathbb{M}$ has the form of a 'paraboloid' (see Fig. 9-6),* the remaining coordinates being defined in terms of T, X, Y, Z by

$$V = W + 1 = \tfrac{1}{2}(1 - T^2 + X^2 + Y^2 + Z^2). \tag{9.2.8}$$

Every generator of $\mathscr{K}$ (set of points for which $T{:}V{:}W{:}X{:}Y{:}Z$ is constant and for which (9.2.5) is satisfied), unless it lies in the null hyperplane $V = W$, meets $\mathbb{M}$ in a unique point. The generators of $\mathscr{K}$ which lie in $V = W$ corresponds to points at infinity for $\mathbb{M}$. Now lines through the origin in $\mathbb{E}^6$ correspond to points in a projective 5-space $\mathbb{P}^5$, for which the five independent ratios $T{:}V{:}W{:}X{:}Y{:}Z$ serve as coordinates. The generators of $\mathscr{K}$ define the points of a *quadric* (i.e., a manifold given by the vanishing of a quadratic form) in $\mathbb{P}^5$, whose equation is (9.2.5), and which will be identified

* This is a higher-dimensional version of the situation depicted in Fig. 1-5, Volume 1, p. 13, where the Euclidean 2-plane is exhibited as a 'paraboloidal' section of the ordinary Lorentz null cone.

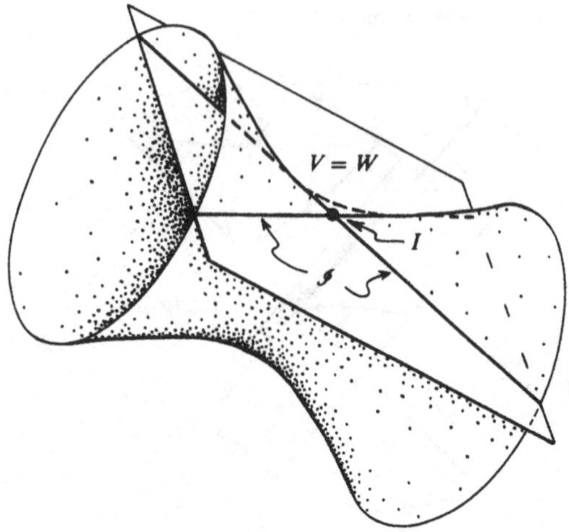

Fig. 9-7. In $\mathbb{P}^5$, $\mathbb{M}^\#$ is represented as a quadric hypersurface. The tangent hyperplane at I meets this quadric in $\mathscr{I}$.

with $\mathbb{M}^\#$. (See Fig. 9-7.) The points of this quadric not lying on $V = W$ correspond precisely to the points of $\mathbb{M}$, but now, in addition, there are those points of $\mathbb{M}^\#$ which lie on $V = W$, and which provide the required compactification of $\mathbb{M}$. (Being an algebraic subset of a projective space, $\mathbb{M}^\#$ must necessarily be compact.)

We shall not go into many of the details of the geometry involved here, but content ourselves with pointing out a few facts.* In the first place, $\mathbb{M}^\#$ has a well-defined conformal geometry everywhere. One way of seeing this is to observe that any two (local) hypersurface cross-sections of $\mathscr{K}$, which meet the same generators of $\mathscr{K}$, are mapped conformally to one another by the generators of $\mathscr{K}$. This can be established following an argument similar to one given in Chapter 1 (*cf.* Fig. 1-11, Volume 1, p. 38), or by showing that the metric of $\mathscr{K}$ can locally be put into the form

$$\mathrm{d}s^2 = q^2 a_{\alpha\beta}(x^\gamma)\mathrm{d}x^\alpha \mathrm{d}x^\beta + 0 \cdot \mathrm{d}q^2,$$

the generators being given by $x^\alpha = \text{constant}$, and the cross-sections by specifying q as a function of x^α. From this form of the metric, two such cross-sections are then obviously mapped conformally to one another. To put the metric in the required form we can select any variable, say W, as the 'q'

* See also Coxeter (1936), Dirac (1936*b*), Kuiper (1949), Hughston and Hurd (1983), Hurd (1985), Penrose (1967*a*, 1974).

coordinate and re-express (9.2.4) as

$$ds^2 = W^2\{d(T/W)^2 + d(V/W)^2 - d(X/W)^2 - d(Y/W)^2 - d(Z/W)^2\},$$

using (9.2.5). Then we eliminate one of the variables $T/W, \ldots, Z/W$ by expressing it in terms of the others, again using (9.2.5).

The space of generators of $\mathscr{K}$ thus acquires a conformal structure (as was the case with the light cone of the origin in ordinary Minkowski space, *cf.* Chapter 1 and §4.15), i.e., $\mathbb{M}^\#$ as a subspace of $\mathbb{P}^5$ has a canonically defined conformal structure of signature $(+\,-\,-\,-)$. The light cones of this conformal structure turn out to be the intersections of $\mathbb{M}^\#$ with projective 4-planes that touch $\mathbb{M}^\#$. These 4-planes meet $\mathbb{M}^\#$ in quadric cones which are generated by straight lines on $\mathbb{M}^\#$. The projective straight lines on $\mathbb{M}^\#$ (straight with respect to the projective space structure of $\mathbb{P}^5$) describe the light rays on $\mathbb{M}^\#$. The particular 4-plane whose equation is $V = W$ also *touches* $\mathbb{M}^\#$ – at the point whose equation is $T = X = Y = Z = (V - W) = 0$. This point is the point I (the identified points i^-, i^0, i^+ in our previous construction of compactified Minkowski space), and the remainder of the intersection of the 4-plane with $\mathbb{M}^\#$ is $\mathscr{I}$ (the identified surfaces $\mathscr{I}^+$, $\mathscr{I}^-$ of the previous construction).

O(2,4) and the conformal group

Now the pseudo-orthogonal group $O(2,4)$ of linear transformations in $\mathbb{E}^6$ that preserve (9.2.5), preserves $\mathscr{K}$ and therefore induces a transformation of $\mathbb{P}^5$ that sends $\mathbb{M}^\#$ to itself. Since it preserves linearity in $\mathbb{E}^6$, it sends projective straight lines in $\mathbb{M}^\#$ to other such lines, i.e. light rays, and hence it preserves the conformal structure of $\mathbb{M}^\#$. It is, in fact, the most general group with this property (*cf.* Penrose 1967*a*), and so it induces the conformal group $C(1,3)$ on $\mathbb{M}^\#$ (see Coxeter 1936, Dirac 1936*b*, Kuiper 1949, Penrose 1974). Furthermore, it has fifteen parameters, since the infinitesimal (pseudo-) orthogonal matrices are skew (with respect to the metric (9.2.4)) and therefore have fifteen linearly independent real components. However, $O(2,4)$ is not identical with $C(1,3)$, because the negative identity element of $O(2,4)$ reverses the orientation of each line through the origin of $\mathbb{E}^6$ while leaving $\mathbb{P}^5$ pointwise invariant. This and the identity element of $O(2,4)$ are the only two transformations in $O(2,4)$ which yield the identity element of $C(1,3)$, so the group homomorphism

$$O(2,4) \to C(1,3) \qquad (9.2.9)$$

is a 2–1 local isomorphism. It is also onto (i.e. surjective). In fact, since the negative identity in $O(2,4)$ belongs to the identity-connected component $O_+^\uparrow(2,4)$ (because $(T+\mathrm{i}V,\ W+\mathrm{i}X,\ Y+\mathrm{i}Z)\mapsto(\mathrm{e}^{\mathrm{i}\theta}(T+\mathrm{i}V),\ \mathrm{e}^{\mathrm{i}\theta}(W+\mathrm{i}X),\ \mathrm{e}^{\mathrm{i}\theta}(Y+\mathrm{i}Z))$ belongs to $O(2,4)$ for each $0\leqslant\theta\leqslant\pi$ and continuously connects the identity to the negative identity), we see that

$$O_+^\uparrow(2,4)\to C_+^\uparrow(1,3) \tag{9.2.10}$$

is also 2–1 and onto. (The '↑' refers to preservation of time sense and '+' to preservation of overall orientation.) The 2–1 nature of this map is rather analogous to that which relates spin transformations to restricted Lorentz transformations, but there is one essential difference: $O_+^\uparrow(2,4)$ is not the *universal* covering space of $C_+^\uparrow(1,3)$. It represents only a finite 'unwrapping' of $C_+^\uparrow(1,3)$, whereas, in fact, to pass to the universal covering space of $C_+^\uparrow(1,3)$ an infinite 'unwrapping' is required. We shall see in §9.4 that it is useful to pass also to a *four-fold* covering group of $C_+^\uparrow(1,3)$, namely the pseudo-unitary twistor group $SU(2,2)$, which is also a two-fold covering group of $O_+^\uparrow(2,4)$ (but, of course, not the universal covering group).

The Poincaré group is a subgroup of $C(1,3)$ and is characterized by leaving the hyperplane $V=W+1$ invariant as well as $\mathscr{K}$. For this entails that the space $\mathbb{M}$ is transformed to itself in a metric-preserving way, as required. The subgroup of $O(2,4)$ that arises in this way is in fact isomorphic (rather than 2–1 homomorphic) with the Poincaré group, because the negative identity of $O(2,4)$ does not preserve $V=W+1$. (The inverse image of the Poincaré group under the map (9.2.9) is, in fact, the subgroup of $O(2,4)$ preserving the *pair* of hyperplanes $V=W\pm 1$.) If we add the dilations to the Poincaré group elements, then the hyperplane $V=W+1$ is not invariant but the family of hyperplanes $V=W+$ constant is transformed into itself, only $V=W$ being invariant under the whole group. In terms of $\mathbb{P}^5$, this corresponds to the 4-plane that touches $\mathbb{M}^\#$ at I (see Fig. 9-7) being invariant, i.e., to $\mathscr{I}$ being transformed into itself. But for a *general* element of $O(2,4)$, this 4-plane is transformed to another 4-plane touching $\mathbb{M}^\#$, i.e., the light cone $\mathscr{I}$ is transformed to the complete light cone of some other point on $\mathbb{M}^\#$. This illustrates the fact that, from the point of view of the conformal structure of $\mathbb{M}^\#$ (as opposed to its metric structure), $\mathscr{I}$ is on an equal footing with any other light cone in $\mathbb{M}^\#$. Similarly, the point I (the vertex of $\mathscr{I}$) is on an equal footing with any other point of $\mathbb{M}^\#$.

We next examine the significance of the non-null (i.e. non-tangent) 4-planes in $\mathbb{P}^5$. The Minkowski coordinate hyperplane $t=0$ is, for example, represented in $\mathbb{P}^5$ as the intersection of $\mathbb{M}^\#$ with the 4-plane $T=0$. As we saw in the paragraph containing (9.2.2), a general element of $C(1,3)$ carries the hyperplane $t=0$ into a two-sheeted (within $\mathbb{M}$) spacelike 3-hyperboloid* in $\mathbb{M}$. The group $O(2,4)$ carries $T=0$ into other 4-planes in $\mathbb{P}^5$ which are 'spacelike' with respect to (9.2.5). These form a five-parameter system and intersect $\mathbb{M}^\#$ in 3-surfaces corresponding to the spacelike 3-hyperboloids (or spacelike 3-planes) in $\mathbb{M}$. Similarly, there are 4-planes in $\mathbb{P}^5$ (such as $X=0$) which are 'timelike' with respect to (9.2.5). They also form a 5-parameter system and they intersect $\mathbb{M}^\#$ in 3-surfaces corresponding to timelike one-sheeted (within $\mathbb{M}$) 3-hyperboloids, or, in special cases, timelike 3-planes in $\mathbb{M}$.

9.3 Complexified compactified Minkowski space and twistor geometry

In order to relate the above discussion to twistors, we consider the description given in (6.2.18) of a space–time point, with position vector r^a, in terms of a simple skew twistor $\mathrm{R}^{\alpha\beta}$. By (6.2.18) (*cf.* also (6.1.46), (6.1.47)), we have (with $\mathrm{R}^{\alpha\beta}=-\mathrm{R}^{\beta\alpha}$)

$$\mathrm{R}^{01}:\mathrm{R}^{02}:\mathrm{R}^{03}:\mathrm{R}^{12}:\mathrm{R}^{13}:\mathrm{R}^{23}=-\tfrac{1}{2}r_ar^a:-\mathrm{i}r^{01'}:\mathrm{i}r^{00'}:-\mathrm{i}r^{11'}:\mathrm{i}r^{10'}:1. \tag{9.3.1}$$

Now, referring to (9.2.8) and Fig. 9-6, we see that, choosing standard Minkowski coordinates

$$t=r^0,\quad x=r^1,\quad y=r^2,\quad z=r^3, \tag{9.3.2}$$

we can represent the Minkowski point (t, x, y, z) by the line through the origin $(\lambda T,\lambda V,\lambda W,\lambda X,\lambda Y,\lambda Z)$ in $\mathbb{E}^6$ with $\lambda\in\mathbb{R}$, where, for suitable λ, we have

$$\begin{aligned}&\lambda T=t,\quad \lambda X=x,\quad \lambda Y=y,\quad \lambda Z=z,\\ &\lambda V=\tfrac{1}{2}(1-t^2+x^2+y^2+z^2)=\lambda W+1.\end{aligned} \tag{9.3.3}$$

Thus

$$\lambda=(V-W)^{-1},$$

giving

$$t=T(V-W)^{-1},\quad x=X(V-W)^{-1},\quad y=Y(V-W)^{-1},\quad z=Z(V-W)^{-1} \tag{9.3.4}$$

whence from (9.2.5)

$$-r_ar^a=-t^2+x^2+y^2+z^2=(V+W)(V-W)^{-1}, \tag{9.3.5}$$

and also

$$T:V:W:X:Y:Z=t:\tfrac{1}{2}(1-r_ar^a):\tfrac{1}{2}(-1-r_ar^a):x:y:z. \tag{9.3.6}$$

Let us now compare (9.3.6) with (9.3.1). Using the standard vector-2-

* This is really a '3-sphere' – or perhaps one should say 'pseudosphere' – with respect to the Lorentzian metric (or pseudometric) of $\mathbb{M}$.

spinor correspondence (3.1.31) and the notation (9.3.2). we find

$$\mathsf{R}^{01}=\tfrac{1}{2}(V+W),\quad \mathsf{R}^{02}=\frac{1}{\sqrt{2}}(Y-\mathrm{i}X),\quad \mathsf{R}^{03}=\frac{\mathrm{i}}{\sqrt{2}}(T+Z),$$

$$\mathsf{R}^{12}=\frac{\mathrm{i}}{\sqrt{2}}(Z-T),\quad \mathsf{R}^{13}=\frac{1}{\sqrt{2}}(Y+\mathrm{i}X),\quad \mathsf{R}^{23}=V-W, \tag{9.3.7}$$

where the scale factor has been chosen to provide the standard twistor normalization (6.2.27), namely $\mathsf{R}^{23}=1$ when $V-W=1$. Inversely,

$$T=\frac{\mathrm{i}}{\sqrt{2}}(\mathsf{R}^{12}-\mathsf{R}^{03}),\quad V=\mathsf{R}^{01}+\tfrac{1}{2}\mathsf{R}^{23},\quad W=\mathsf{R}^{01}-\tfrac{1}{2}\mathsf{R}^{23},$$

$$X=\frac{\mathrm{i}}{\sqrt{2}}(\mathsf{R}^{02}-\mathsf{R}^{13}),\quad Y=\frac{1}{\sqrt{2}}(\mathsf{R}^{02}+\mathsf{R}^{13}),\quad Z=\frac{-\mathrm{i}}{\sqrt{2}}(\mathsf{R}^{03}+\mathsf{R}^{12}). \tag{9.3.8}$$

Note that, because $\mathsf{R}^{01}=\mathsf{R}_{23}$, $\mathsf{R}^{02}=-\mathsf{R}_{13}$, etc., and $\overline{\mathsf{R}^{01}}=\bar{\mathsf{R}}_{23}$, $\overline{\mathsf{R}^{02}}=\bar{\mathsf{R}}_{20}$, etc., the condition (6.2.31) that r^a be real asserts that

$$\overline{\mathsf{R}^{01}}=\mathsf{R}^{01},\quad \overline{\mathsf{R}^{02}}=\mathsf{R}^{13},\quad \overline{\mathsf{R}^{03}}=-\mathsf{R}^{03},\quad \overline{\mathsf{R}^{12}}=-\mathsf{R}^{12},\quad \overline{\mathsf{R}^{23}}=\mathsf{R}^{23}, \tag{9.3.9}$$

which is indeed satisfied by (9.3.7). It is unavoidable that the linear relations between the real coordinates (T, V, W, X, Y, Z) and the twistor components $\mathsf{R}^{\alpha\beta}$ should involve complex coefficients and that, consequently, the 'reality' condition on $\mathsf{R}^{\alpha\beta}$ be other than simply a statement of the reality of its components. This is because the equation of $\mathscr{K}$, in terms of $\mathsf{R}^{\alpha\beta}$, is the statement (*cf.* (6.2.23)) that $\mathsf{R}^{\alpha\beta}$ be simple:

$$\mathsf{R}^{\alpha\beta}\mathsf{R}_{\alpha\beta}=0 \tag{9.3.10}$$

whereas the quadratic form

$$\tfrac{1}{2}\mathsf{R}^{\alpha\beta}\mathsf{R}_{\alpha\beta}=\mathsf{R}^{01}\mathsf{R}^{23}-\mathsf{R}^{02}\mathsf{R}^{13}+\mathsf{R}^{03}\mathsf{R}^{12} \tag{9.3.11}$$

would have signature $(+++---)$ instead of the required $(++----)$ of (9.2.5), if the components $\mathsf{R}^{\alpha\beta}$ were to be all real. Twistors, after all, are essentially *complex* objects. To get a proper understanding of twistor geometry, it is therefore necessary to consider *complex* geometry and, in particular, the complexifications of the real spaces that we have been considering so far.

Let us denote by $\mathbb{CE}^6$, $\mathbb{C}\mathscr{K}$, $\mathbb{CM}$, $\mathbb{CP}^5$, $\mathbb{CM}^\#$, and $\mathbb{C}\mathscr{I}$, the complexifications of the respective spaces $\mathbb{E}^6$, $\mathscr{K}$, $\mathbb{M}$, $\mathbb{P}^5$, $\mathbb{M}^\#$, and $\mathscr{I}$. Thus, the coordinates T, V, W, X, Y, Z for $\mathbb{CE}^6$ are now complex variables taking values over the entire range $\mathbb{C}^6$. The original space $\mathbb{E}^6$ is a real six-dimensional subspace of $\mathbb{CE}^6$ (which itself is twelve-dimensional as a real manifold) defined by the vanishing of the imaginary parts of all the coordinates. The space $\mathbb{CE}^6$ possesses a complex-analytic metric, given formally by the same expression (9.2.4) as previously defined for $\mathbb{E}^6$. The complexified null cone $\mathbb{C}\mathscr{K}$ is then defined by the complex equation (9.2.5), and complexified Minkowski space $\mathbb{CM}$ by the intersection of $\mathbb{C}\mathscr{K}$ with the complex 5-plane $V-W=1$. The complex projective space $\mathbb{CP}^5$ is the space of complex lines through the origin of $\mathbb{CE}^6$, and the five independent complex ratios $T:V:W:X:Y:Z$ serve as coordinates for it. *Complexified compactified Minkowski space* $\mathbb{CM}^\#$ is the space of complex generators of $\mathbb{C}\mathscr{K}$, i.e., the locus of the complex equation (9.2.5) in $\mathbb{CP}^5$. Finally, $\mathbb{C}\mathscr{I}$ is defined as the intersection of $\mathbb{CM}^\#$ with the complex plane $V=W$ (but, strictly speaking, without the point I).

The Klein correspondence

We wish to explore the geometrical relation between $\mathbb{CM}^\#$ and projective twistor space* $\mathbb{PT}$, this being the space whose points are equivalence classes of proportional (non-zero) twistors Z^α. The space $\mathbb{PT}$ has the structure of a complex projective 3-space $\mathbb{CP}^3$ and can be coordinatized by the three independent complex ratios $\mathsf{Z}^0:\mathsf{Z}^1:\mathsf{Z}^2:\mathsf{Z}^3$. Now, as we have seen (in the paragraph containing (6.2.15)), complex linear 2-spaces in twistor space $\mathbb{T}$ (at least those for which not all the twistors have proportional projection parts) represent points in complexified Minkowski space $\mathbb{CM}$. We have also seen (in the paragraph containing (6.2.17)) that these 2-spaces can be represented as equivalence classes of proportional (non-zero) simple skew twistors $\mathsf{R}^{\alpha\beta}$. In terms of $\mathbb{PT}$, these 2-spaces are represented as *complex projective (straight) lines* (each a $\mathbb{CP}^1$) in $\mathbb{PT}$. So these lines in $\mathbb{PT}$ represent *points* of $\mathbb{CM}$. However, to obtain a *finite* point of $\mathbb{CM}$, we require that its representative line in $\mathbb{PT}$ does not meet the particular line I which is given by $\mathsf{R}^{\alpha\beta}=\mathsf{I}^{\alpha\beta}$. For the line I represents the system of twistors Z^α with vanishing projection part: so a line in $\mathbb{PT}$ meeting I represents a linear 2-space in $\mathbb{T}$ containing a twistor with vanishing projection part, whence the projection parts of *all* twistors in the 2-space are necessarily proportional. Yet such a line, being represented by a simple skew twistor $\mathsf{R}^{\alpha\beta}$, must correspond to a point of $\mathbb{CM}^\#$. It therefore corresponds to a point of $\mathbb{C}\mathscr{I}$, the

* We work exclusively with $\mathbb{T}^\bullet$ and $\mathbb{PT}^\bullet$ here, rather than with the dual spaces, but the raised dot (or $\alpha, \beta, \ldots$; *cf.* §§6.1, 6.10) will be omitted for notational convenience.

line I itself corresponding to the vertex I of $\mathscr{I}$. So we see that *the points of* $\mathbb{CM}^{\#}$ *represent the lines of* $\mathbb{PT}$. Such a correspondence, in which the lines of a projective 3-space are represented as the points of a quadric in projective 5-space, is known as the *Klein representation**. It is basic to the whole of twistor geometry.

Let us now examine the converse correspondence. We wish to see how a point Z of $\mathbb{PT}$ is to be represented in $\mathbb{CM}^{\#}$. Consider the lines in $\mathbb{PT}$ through Z. They constitute a 2-complex-parameter family (called a star of lines) which corresponds to a complex 2-surface Z in $\mathbb{CM}^{\#}$, and that 2-surface can be used to represent Z. To elucidate the nature of this 2-surface, we refer back to (6.2.2), which was the basis of the equation (6.2.15) whereby the position vector r^a of a point R in $\mathbb{CM}$ is defined from a linear 2-space of twistors. Equation (6.2.2) expresses *incidence*** between points of $\mathbb{CM}$ and twistors. Thus, we say that R is incident with the twistor $\mathrm{Z}^\alpha = (\omega^A, \pi_{A'})$ if

$$\omega^A = \mathrm{i} r^{AA'} \pi_{A'}, \tag{9.3.12}$$

$r^{AA'}$ being the position vector of R with respect to any point at which the spinor fields ω^A, $\pi_{A'}$ are to be evaluated. (For definiteness we shall consider all relevant equations to be evaluated at a fixed origin O.) If we hold $r^{AA'}$ fixed in (9.3.12) and allow $\mathrm{Z}^\alpha = (\omega^A, \pi_{A'})$ to vary, we get the linear 2-space in $\mathbb{T}$ that we considered above and which represents the point R in twistor terms. On the other hand, if we hold Z^α fixed and vary r^a, we obtain the required locus Z that represents Z in terms of $\mathbb{CM}$. Now provided that $\pi_{A'} \neq 0$, i.e., provided that $\mathrm{Z} \notin \mathrm{I}$, there will be at least one complex solution of (9.3.12) for r^a (as one can easily see by taking components). Call this solution $\underset{0}{r}^a$. Then the remaining solutions satisfy $(r^a - \underset{0}{r}^a)\pi_{A'} = 0$ and so, by (3.5.17), they are given by

$$r^a = \underset{0}{r}^a + \lambda^A \pi^{A'}, \tag{9.3.13}$$

where λ^A varies throughout $\mathfrak{S}^A[O]$. The complex vectors $\lambda^A\pi^{A'}$, at O, for

* Felix Klein (1870, 1926; *cf.*, for example, also Veblen and Young 1910) based his correspondence on the line coordinates of Julius Plücker (1865, 1868/9), these having been also obtained independently by Arthur Cayley (1860, 1869) (and which fell into the scheme put forward even earlier by Hermann Grassmann). Even more relevant to twistor theory was the observation by Sophus Lie in 1869 (*cf.* Lie and Scheffers 1896) that oriented spheres in (complex) Euclidean 3-space could be represented by lines in $\mathbb{CP}^3$, with (consistently oriented) contact between spheres represented by meeting lines in $\mathbb{CP}^3$. These spheres can be regarded as the intersections of light cones, of points in (complex) Minkowski space, with a constant time hypersurface. (We are grateful to Helmuth Urbantke for pointing out this work of Lie to us; *cf.* also Gindikin 1983, Penrose 1986.)

** The term 'incidence', generally, means that the dimension of the locus of intersection of the two spaces in question is larger than in the generic case.

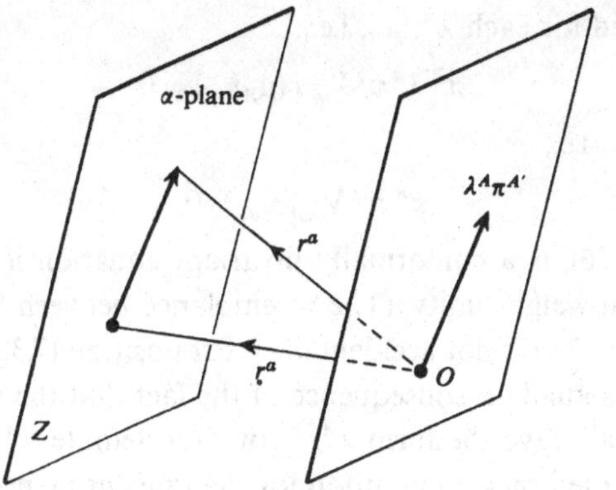

Fig. 9-8. An α-plane in $\mathbb{CM}$ is totally null. It has the property that the difference between the position vectors of any two of its points is a complex null vector $\lambda^A\pi^{A'}$, where $\pi^{A'}$ is fixed but λ^A varies. A β-plane is similar but the roles of λ^A and $\pi^{A'}$ are reversed.

fixed $\pi^{A'}$ and varying λ^A, constitute a two-complex-dimensional vector space, all of whose members are null and orthogonal to one another. Hence the space Z is a 2-plane in $\mathbb{CM}$ (see Fig. 9-8), all of whose tangent vectors are null and mutually orthogonal. The distance between any two points on Z is consequently zero, i.e., the induced metric on Z vanishes identically.

There are, in fact, two families of planes on $\mathbb{CM}$ with this property, the other being obtained when the roles of λ^A and $\pi^{A'}$ are interchanged, i.e., when (9.3.13) holds but λ^A is held fixed while $\pi^{A'}$ varies (compare (3.2.22) etc.). The planes of the first kind are called α-planes, while those of the second kind are called β-planes. (This terminology is consistent with that of §§6.2, 7.4.) What we have just shown is that a point $\mathbf{Z}$ of $\mathbb{PT}$ (with $\pi_{A'}\neq 0$) is represented by an α-plane in $\mathbb{CM}$. The restriction $\pi_{A'}\neq 0$ can be removed if we consider, instead, α-planes in $\mathbb{CM}^{\#}$. This follows easily from the general homogeneity of the space $\mathbb{CM}^{\#}$, and from the fact that the α-plane concept is conformally invariant – as we shall now show.

To this end, we observe that an α-plane is characterized by being a complex 2-surface, all of whose tangent vectors at any one point have the form $\lambda^A\pi^{A'}$ for some $\pi^{A'}$ (which is equivalent to saying that the tangent bivectors are self-dual and null, being of the form $\varepsilon^{AB}\pi^{A'}\pi^{B'}$, *cf.* (3.4.39)), and whose tangent spaces go into themselves under parallel propagation in tangent directions (i.e., the surfaces are flat). This last condition can be written as

$$\lambda^A\pi^{A'}\nabla_a(\mu_B\pi_{B'}) = \nu_B\pi_{B'} \qquad (9.3.14)$$

for some ν_B and for each λ^A, μ_B; i.e.,

$$\pi^{B'}\lambda^A\pi^{A'}\nabla_{AA'}(\mu_B\pi_{B'}) = 0$$

for each λ^A, μ_B, i.e.,

$$\pi^{B'}\pi^{A'}\nabla_{AA'}\pi_{B'} = 0, \tag{9.3.15}$$

which, by (7.4.56), is a conformally invariant equation if we take $\pi_{B'}$ to have conformal weight unity. (The resemblance between (9.3.15) and the SFR equation (7.3.1) is not accidental: *cf.* Proposition (7.3.18).) As it turns out, (9.3.15) is actually a consequence of the fact that the tangent vectors to the surface all have the form $\lambda^A\pi^{A'}$ (with a definite $\pi^{A'}$ at each point, because of the Lie bracket condition for the tangent planes to be surface-forming (*cf.* (7.3.17)):

$$\lambda^A\pi^{A'}\nabla_a(\mu^B\pi^{B'}) - \mu^A\pi^{A'}\nabla_a(\lambda^B\pi^{B'}) = \rho^B\pi^{B'}$$

for some ρ^B; transvection with $\pi_{B'}$ yields

$$\pi_{B'}\pi^{A'}\lambda^{[A}\mu^{B]}\nabla_a\pi^{B'} = 0,$$

from which (9.3.15) again follows*, with its consequence (9.3.14).

As for the β-planes, they arise in an exactly analogous way, with a 'dual' or $[{}^0_1]$-twistor $\mathsf{W}_\alpha = (\lambda_A, \mu^{A'})$ replacing the $[{}^1_0]$-twistor Z^α just considered, the roles of primed and unprimed spinor indices being now interchanged. The incidence relation (9.3.12) is replaced by

$$\mu^{A'} = -\mathrm{i}r^a\lambda_A, \tag{9.3.16}$$

whose solutions take the form (9.3.13), as before, but with λ^A remaining fixed and $\pi_{A'}$ varying. This gives us a β-plane as the description in $\mathbb{CM}$ or $\mathbb{CM}^\#$ of a projective dual twistor. But we can also represent a projective dual twistor as a complex projective 2-plane W (i.e. a $\mathbb{CP}^2$) in $\mathbb{PT}$; for W_α ($\neq 0$) is represented up to proportionality by the system W of points $\mathsf{Z} \in \mathbb{PT}$ for which

$$\mathsf{W}_\alpha \mathsf{Z}^\alpha = 0. \tag{9.3.17}$$

Equation (9.3.17) expresses *incidence* between a plane W and a point Z in $\mathbb{PT}$, or between a dual twistor W_α and a twistor Z^α. The incidence relation

* If instead of working in the specific space $\mathbb{CM}^\#$, which is conformally flat, we apply the same argument in a general curved complex-Riemannian 4-manifold, we obtain, upon further differentiation, the consistency requirement that $\Psi_{A'B'C'D'}\pi^{A'}\pi^{B'}\pi^{C'}\pi^{D'} = 0$ at each point of the surface (*cf.* the conjugate of (4.9.16)). The condition that (locally, at least) there should be as large a family of α-surfaces in the manifold as there is for $\mathbb{CM}^\#$ (i.e. 3-parameter) is that $\Psi_{A'B'C'D'} = 0$, i.e., that the conformal curvature be anti-self-dual, i.e., that the space be right-conformally *flat* (Penrose 1976*a*, Penrose and Ward 1980, Newman 1976, Hansen, Newman, Penrose and Tod 1978, Ko, Ludvigsen, Newman and Tod 1981, Plebański 1975).

(9.3.12) in purely twistor terms is

$$R^{[\alpha\beta}Z^{\gamma]} = 0 \quad \text{or, equivalently,} \quad R_{\alpha\beta}Z^{\beta} = 0. \qquad (9.3.18)$$

This expresses the fact that, in $\mathbb{PT}$, the point Z lies on the line R (represented by $R^{\alpha\beta}$). If $R_{\alpha\beta}Z^{\alpha} \neq 0$, this dual twistor represents the *plane joining* Z *to* R in $\mathbb{PT}$. Correspondingly, the incidence relation (9.3.16) can be written

$$R^{\alpha\beta}W_{\beta} = 0 \quad \text{or, equivalently,} \quad R_{[\alpha\beta}W_{\gamma]} = 0, \qquad (9.3.19)$$

and expresses the fact that, in $\mathbb{PT}$, the line R lies on the plane W. Similarly, if $R^{\alpha\beta}W_{\beta} \neq 0$, this twistor represents the *point of intersection* of R with W.

In terms of $\mathbb{CM}^{\#}$, the point Z, line R, and plane W are respectively represented, as we have seen, by an α-plane Z, a point R, and a β-plane W. Incidence between Z and R in $\mathbb{PT}$ translates to incidence between Z and R in $\mathbb{CM}^{\#}$: the point R lies on the α-plane Z. Incidence between R and W in $\mathbb{PT}$ translates to incidence between R and W in $\mathbb{CM}^{\#}$: the point R lies on the β-plane W. Incidence between Z and W in $\mathbb{PT}$ translates to incidence between Z and W in $\mathbb{CM}^{\#}$: the α-plane Z meets the β-plane W. This last relation can be seen either by referring back to (9.3.12), (9.3.16), and (9.3.17), and using suitable arguments to cover the cases when R is on $\mathscr{I}$ or I, or simply by observing that Z lying on W is necessary and sufficient for the existence of a line R which at the same time lies on W and passes through Z, i.e., for the existence of a point $R \in \mathbb{CM}^{\#}$ which at the same time lies on the β-plane W and on the α-plane Z. Note that when this occurs there is not only one line R lying in W and passing through Z but a whole 1-parameter family of such lines – referred to as a *plane pencil of lines.* Thus, in $\mathbb{CM}^{\#}$, there will be a curve of intersection of the β-plane W with the α-plane Z–evidently a *complex null geodesic* in the conformal metric of $\mathbb{CM}^{\#}$, since it is a null striaght line in $\mathbb{CM}^{\#}$ (because, for example, its parametric equation has the linear form $r^a = \underset{0}{r}^a + u\lambda^A\pi^{A'}$, $u \in \mathbb{C}$). Conversely, *every* complex null geodesic in $\mathbb{CM}^{\#}$ arises in this way, and so we have:

(9.3.20) PROPOSITION

Through every null geodesic in $\mathbb{CM}^{\#}$ *there passes a unique* α*-plane and a unique* β*-plane; when an* α*-plane and a* β*-plane intersect, they always intersect in a null geodesic.*

Any two distinct α-planes, on the other hand, always have a unique point of $\mathbb{CM}^{\#}$ in common – since distinct points on $\mathbb{PT}$ are joined by a unique line; similarly, any two distinct β-planes have a unique point of $\mathbb{CM}^{\#}$ in common – since distinct planes in $\mathbb{PT}$ have a unique line in common.

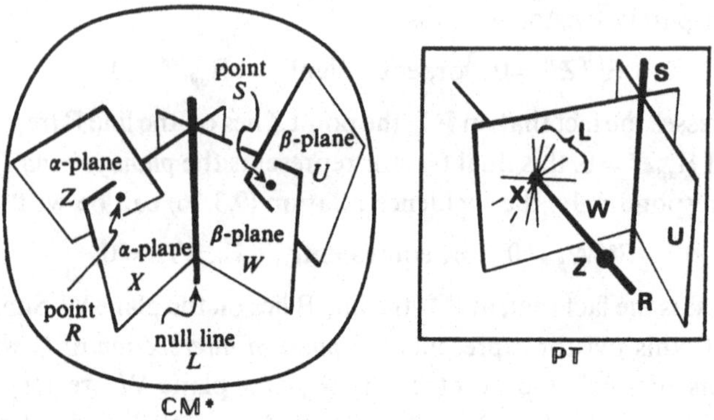

Fig. 9-9. The basic geometrical correspondence – the *Klein correspondence* – relating incidence properties of points, planes and lines in $\mathbb{PT}$ (a complex projective 3-space) with those of the corresponding α-planes, β-planes and points in $\mathbb{CM}^{\#}$ (a complex projective 4-quadric).

('Incidence' between two α-planes or between two β-planes would necessarily mean 'coincidence'.) Two lines in $\mathbb{PT}$ will not generally meet (i.e., they are 'skew'), but when they do meet, they lie in a plane. Meeting lines in $\mathbb{PT}$ thus correspond to *null-separated* points in $\mathbb{CM}^{\#}$. So we have:

(9.3.21) PROPOSITION

Points $P, Q \in \mathbb{CM}^{\#}$ are null-separated

⇔ P, Q lie on a common α-plane
⇔ P, Q lie on a common β-plane
⇔ the corresponding lines in $\mathbb{PT}$ meet.

Table (9.3.22)

$\mathbb{CM}^{\#}$		$\mathbb{PT}$
α-plane		point
on	⇔	on
point		line
on	⇔	on
β-plane		plane
null geodesic (= α-plane ∩ β-plane)		plane pencil (↔ point on plane)
null-separated points		meeting lines

The relation between the geometries of $\mathbb{CM}^{\#}$ and $\mathbb{PT}$ is summarized in Fig. 9.9 and Table (9.3.22).

Dual twistor correspondence

The above discussion has been carried out entirely in terms of $\mathbb{PT}^{\bullet}$ ($=\mathbb{PT}$ here) rather than the dual projective space $\mathbb{PT}_{\bullet}$. Had the latter space been used instead, the roles of α-planes and β-planes would have been interchanged. The modification to Table (9.3.22) that would be needed, if we keep the left-hand column unchanged but replace the right-hand one by a column referring to $\mathbb{PT}_{\bullet}$ rather than $\mathbb{PT}^{\bullet}$, is simply that the (upper) words 'point' and 'plane' must be interchanged and the curved arrows on the right must point upwards rather than downwards.

The geometrical relationship between $\mathbb{PT}_{\bullet}$ and $\mathbb{PT}^{\bullet}$ is a *duality* correspondence, where points correspond to planes, lines to lines and planes to points. The correspondence preserves the concept of *incidence*, but it reverses the direction of inclusion relations between spaces.

Reality structure

So far, in this correspondence, we have made no mention of the *reality* structure of $\mathbb{CM}^{\#}$ or of the notion of complex conjugation, to which we now turn. The complex conjugation operation $\mathscr{C}$ in $\mathbb{CM}^{\#}$ interchanges α-planes with β-planes; in $\mathbb{PT}$ ($=\mathbb{PT}^{\bullet}$) it is a duality operation within $\mathbb{PT}$ *itself* which interchanges points and planes, and which sends lines to other lines. The real points of $\mathbb{CM}^{\#}$ (i.e., the points of $\mathbb{M}^{\#}$) are those that are invariant under $\mathscr{C}$. They correspond to a family of lines in $\mathbb{PT}$ which are each left invariant by $\mathscr{C}$. According to the discussion of §6.2 (*cf.* (6.2.16)), these lines arise from linear 2-spaces consisting entirely of null twistors. Thus, the real points of $\mathbb{CM}^{\#}$ correspond to lines of $\mathbb{PT}$ that lie entirely in the subspace $\mathbb{PN}$ of *null projective twistors*. Twistor space $\mathbb{T}$ consists of $\{0\}$ and the three portions $\mathbb{T}^{+}$, $\mathbb{T}^{-}$, and $\mathbb{N}$, whose elements are those non-zero twistors Z^{α} for which $\mathrm{Z}^{\alpha}\bar{\mathrm{Z}}_{\alpha}$ is, respectively, positive, negative, or zero; so $\mathbb{PT}$ also consists of the three corresponding portions* $\mathbb{PT}^{+}$, $\mathbb{PT}^{-}$, and $\mathbb{PN}$. The special line I, representing the point I of $\mathbb{CM}^{\#}$, lies, of course, in $\mathbb{PN}$, since I is a 'real' point.

Any point $\mathrm{Z}\in\mathbb{PT}$ (corresponding to a twistor $\mathrm{Z}^{\alpha}\neq 0$) has a complex conjugate plane $\bar{\mathrm{Z}}\in\mathbb{PT}$ (corresponding to the dual twistor $\bar{\mathrm{Z}}_{\alpha}$) and,

* The topologies of these pieces are: $\mathbb{PT}^{+}\cong\mathbb{PT}^{-}\cong S^2\times\mathbb{R}^4$, $\mathbb{PN}\cong S^2\times S^3$, $\mathbb{T}^{+}\cong\mathbb{T}^{-}\cong S^3\times\mathbb{R}^5$, $\mathbb{N}\cong S^3\times S^3\times\mathbb{R}$.

conversely, any plane $\mathsf{W} \subset \mathbb{PT}$ has a unique complex conjugate point $\bar{\mathsf{W}} \in \mathbb{PT}$. The condition for a point Z of $\mathbb{PT}$ to lie on its complex conjugate plane $\bar{\mathsf{Z}}$ is $\mathsf{Z}^\alpha \bar{\mathsf{Z}}_\alpha = 0$, i.e., $\mathsf{Z} \in \mathbb{PN}$. Thus, the points of $\mathbb{PN}$ are those representing α-planes in $\mathbb{CM}^\#$ which meet their complex conjugate β-planes. These are the α-planes that *contain real points.* For if a point $P \in \mathbb{M}^\#$ lies on an α-plane Z, we see by applying conjugation that it also lies on the β-plane $\bar{Z}$. Conversely, if Z meets $\bar{Z}$, it must do so in a null geodesic which is necessarily 'real' (in the sense of being invariant under $\mathscr{C}$) and which therefore contains some real points. But such a 'real' null geodesic must also contain complex points (namely those with complex affine parameter values) and it has, altogether, the topology of S^2 (the Riemann sphere), its real points constituting a circle on this sphere. Recall from §6.2 that our original geometrical interpretation of a null twistor, up to proportionality, was as a null straight line in $\mathbb{M}$. We now see how that interpretation fits into the present geometrical description.

A non-null twistor Z^α corresponds to a point Z of $\mathbb{PT}^+$ or $\mathbb{PT}^-$. For its real interpretation in $\mathbb{M}^\#$ we can consider the intersection of the complex plane $\bar{\mathsf{Z}}$ with $\mathbb{PN}$, which is a three-real-dimensional set of points, each of which represents a null geodesic in $\mathbb{M}^\#$. This three-real-parameter family of null geodesics in $\mathbb{M}^\#$ is the Robinson congruence representing the twistor Z^α (up to proportionality), and this ties in with the discussion given in §6.2 of the geometrical interpretation of a non-null twistor and also with the discussion of §7.4; *cf.* Fig. 7-7, etc.). Note that the points of $\mathbb{PT}^+$ correspond to Robinson congruences which twist in a right-handed sense, while points of $\mathbb{PT}^-$ correspond to Robinson congruences twisting in a left-handed sense (*cf.* §6.2); furthermore, twistors in $\mathbb{T}^+$ and $\mathbb{T}^-$ describe, respectively, the angular-momentum structure of massless particles with right-handed or left-handed helicity (*cf.* §6.3).

The different kinds of complex point in $\mathbb{CM}^\#$

Thus far, we have examined the role of this reality structure only in relation to points in $\mathbb{PT}$. In $\mathbb{CM}^\#$ itself we can distinguish *six* different (conformally invariant) regions. A point $R \in \mathbb{CM}^\#$ can be represented by a complex position vector

$$r^a = u^a + iv^a \tag{9.3.23}$$

with respect to a real origin O, where u^a and v^a are real world-vectors. The spacelike/timelike and future/past properties of v^a are clearly invariant under real translations of the origin O. Less evident, though still true, is

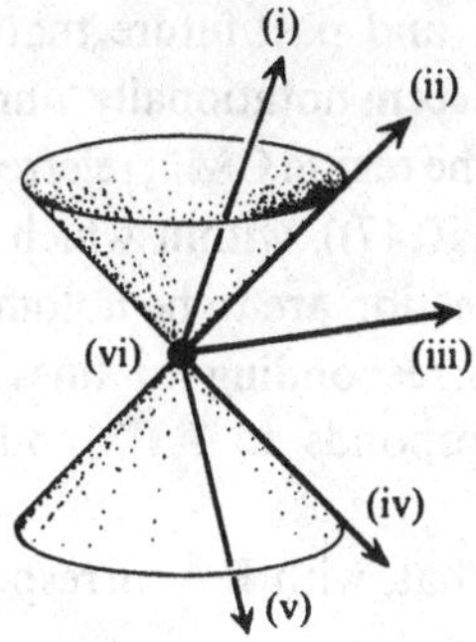

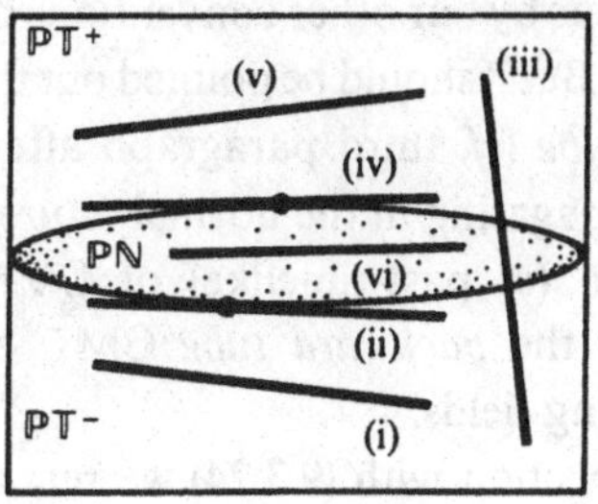

Fig. 9-10. The causal characterization of the imaginary part v^a of the position vector of a point R of $\mathbb{CM}$ has a (real) conformally invariant significance. This shows up in $\mathbb{PT}$ in terms of the way that the corresponding line R intersects the various regions $\mathbb{PT}^{\pm}$, $\mathbb{PN}$.

invariance under conformal rescaling, the effect on (9.3.23) being more complicated. This invariance is actually a consequence of the following proposition (see Fig. 9-10), which applies also when $R \in \mathbb{C}\mathscr{I}$ (since then we need only rescale so that the new $\mathbb{C}\mathscr{I}$ does not contain R).

(9.3.24) PROPOSITION

A point $R \in \mathbb{CM}^{\#}$ whose position vector in any standard Minkowski reference system has an imaginary part which is (i) *future-timelike,* (ii) *future-null,* (iii) *spacelike,* (iv) *past-null,* (v) *past-timelike,* (vi) *zero, corresponds, respectively, to a line* $\mathsf{R} \subset \mathbb{PT}$ *which* (i) *lies entirely in* $\mathbb{PT}^-$, (ii) *lies in* $\mathbb{PT}^-$ *except for touching* $\mathbb{PN}$ *at one point,* (iii) *meets all three of* $\mathbb{PT}^{\pm}$ *and* $\mathbb{PN}$, (iv) *lies in* $\mathbb{PT}^+$ *except for touching* $\mathbb{PN}$ *at one point,* (v) *lies entirely in* $\mathbb{PT}^+$, (vi) *lies entirely in* $\mathbb{PN}$.

Proof: Let $\mathsf{Z}^\alpha = (\omega^A, \pi_{A'})$ be incident with R, so that Z lies on the line $\mathsf{R} \subset \mathbb{PT}$; then we have

$$\omega^A = \mathrm{i} r^a \pi_{A'} = \mathrm{i} u^a \pi_{A'} - v^a \pi_{A'}.$$

Transvecting with $\bar{\pi}_A$ and taking the real part, we get

$$\tfrac{1}{2}\mathsf{Z}^\alpha \bar{\mathsf{Z}}_\alpha = -v^a(\bar{\pi}_A \pi_{A'}),$$

and the result follows easily by consideration of the various cases individually.

The association in (9.3.24) between $\mathbb{T}^{\pm}$ and past/future that has been forced on us by our other conventions may seem notationally 'unnatural' at first sight. But it should be pointed out that the region $\mathbb{CM}^+$, referred to as the *forward tube* (*cf.* third paragraph after (6.10.47)), within which quantum fields propagating in the normal *future* direction are to be holomorphic, is the region (v^a past-timelike) of $\mathbb{CM}^{\#}$ corresponding to lines in $\mathbb{PT}^+$. Similarly, the *backward tube* $\mathbb{CM}^-$ corresponds to $\mathbb{PT}^-$ and to past-propagating fields.

In connection with (9.3.24) we remark that, with $\mathrm{R}^{\alpha\beta}$ corresponding to $R \in \mathbb{CM}$ as above* and with (9.3.23),

$$\mathrm{R}^{\alpha\beta}\bar{\mathrm{R}}_{\alpha\beta} \gtrless 0 \quad according\ as \quad v^a v_a \gtrless 0, \tag{9.3.25}$$

and that

$$\mathrm{R}^{\alpha\beta}\mathrm{W}_\beta\bar{\mathrm{R}}_{\alpha\gamma}\bar{\mathrm{W}}^\gamma \textit{ is positive [negative] semidefinite in } \mathrm{W}_\beta \textit{ according as } v^a \textit{ is past [future-]causal.} \tag{9.3.26}$$

For the proof of (9.3.26), from which (9.3.25) follows, note that $\mathrm{R}^{\alpha\beta}\mathrm{W}_\beta$ represents the intersection of the line R with the plane W. Then use (9.3.24).

9.4 Twistor four-valuedness and the Grgin index

We have seen that the complex conformal geometry of $\mathbb{CM}^{\#}$ can, by the above correspondence, be elegantly described in terms of the complex projective geometry of $\mathbb{PT}$. For the real conformal geometry of $\mathbb{M}^{\#}$ we require, in addition to the complex projective geometry of $\mathbb{PT}$, the Hermitian $(++--)$ duality correspondence $\mathscr{C}$ that interchanges points and planes in $\mathbb{PT}$; equivalently, we need to know the location of the five-real-dimensional hypersurface $\mathbb{PN}$ in $\mathbb{PT}$. The symmetries of $\mathbb{PT}$ that preserve this structure induce conformal motions of $\mathbb{M}^{\#}$ to itself, and so correspond to elements of $O(2,4)$. These symmetries are obtained from linear transformations of $\mathbb{T}$. If they preserve not only $\mathbb{N}$ (i.e., the locus $\mathrm{Z}^\alpha\bar{\mathrm{Z}}_\alpha = 0$) but also the actual value of the $(++--)$ Hermitian form $\mathrm{Z}^\alpha\bar{\mathrm{Z}}_\alpha$, and are further normalized to have unit determinant, these transformations constitute the group $SU(2,2)$ – the twos coming from the $++$ and $--$ parts of the preserved Hermitian form. The normalization does not determine the linear transformation uniquely from its action on $\mathbb{PT}$ but allows a four-fold ambiguity, since scalar multiples of the identity by each of the four factors 1, i, -1, $-$i also give elements of $SU(2,2)$ yielding the identity on $\mathbb{M}^{\#}$. As we shall see shortly, each of these elements is connected

* We remark also that if $R \in \mathbb{CM}^{\#}$ and $\mathrm{R}^{\alpha\beta}\bar{\mathrm{R}}_{\alpha\beta} \neq 0$ then $R \in \mathbb{CM}$.

to the identity, so the ambiguity is essential. We thus have a 4–1 local isomorphism from $SU(2,2)$ onto $C^{\uparrow}_{+}(1,3)$. In fact, since our twistor construction gives the space on which $O^{\uparrow}_{+}(2,4)$ acts in terms of skew $[^{2}_{0}]$-twistors $R^{\alpha\beta}$, the (appropriately anti-symmetrized) tensor product of two elements of $SU(2,2)$ can be interpreted as an element of $O(2,4)$ (and, in fact, of $O^{\uparrow}_{+}(2,4)$). This gives a 2–1 local isomorphism from $SU(2,2)$ to $O^{\uparrow}_{+}(2,4)$ which, when combined with (9.2.10), shows that the 4–1 map from $SU(2,2)$ to $C^{\uparrow}_{+}(1,3)$ can be composed of the two 2–1 local isomorphisms (each onto):

$$SU(2,2) \to O^{\uparrow}_{+}(2,4) \to C^{\uparrow}_{+}(1,3).$$

Four-fold ambiguity in the space–time description of Z^α

This 4–1 relation between the twistor group $SU(2,2)$ and the space–time group $C^{\uparrow}_{+}(1,3)$ has a number of curious consequences. One of these is that any description of a twistor Z^α in space–time terms, which is *conformally invariant* (in the sense of being invariant under $C^{\uparrow}_{+}(1,3)$) must be quadruply ambiguous. That is to say, the descriptions of Z^α, iZ^α, $-Z^\alpha$, $-iZ^\alpha$ in terms of space–time geometry must be indistinguishable from one another. At first sight this seems paradoxical, since in Chapter 6 we defined a twistor in a conformally invariant way, as a solution ω^A of the twistor equation (6.1.1). Whereas it is clear from the discussion given in §1.5 that an essential *two-fold* (sign) ambiguity exists in the geometrical interpretation of ω^A, the nature of the ambiguity between ω^A and $i\omega^A$ is more subtle, and in fact involves $\mathscr{I}$. A continuous spatial rotation of $\mathbb{M}$ through 2π sends the particular solution ω^A of the twistor equation into another solution, namely $-\omega^A$, so it sends Z^α into $-Z^\alpha$. Since proper rotations are particular elements of $C^{\uparrow}_{+}(1,3)$, the closed loop in $C^{\uparrow}_{+}(1,3)$ which represents this active 2π-rotation of $\mathbb{M}$ corresponds to an open path in $SU(2,2)$ starting at the identity δ^α_β and ending at the negative identity $-\delta^\alpha_\beta$. Hence it carries Z^α into $-Z^\alpha$. The family of flag planes representing the ω^A field is, of course, sent into itself by this rotation, so the *sign* ambiguity in the geometrical interpretation of Z^α is something that we are already familiar with from the discussion of §1.5.

To discover the origin of the *four*-fold ambiguity, we must examine the *compactified* space $\mathbb{M}^\#$. An explicit path in $SU(2,2)$ which connects δ^α_β to $i\delta^\alpha_\beta$ can be obtained by setting

$$T^\alpha_\beta(\theta) = e^{i\theta}(\delta^\alpha_\beta - \bar{Q}_\beta Q^\alpha) + e^{-3i\theta}\bar{Q}_\beta Q^\alpha, \qquad (9.4.1)$$

where θ is real and Q^α is fixed, subject only to

$$\bar{Q}_\alpha Q^\alpha = 1.$$

One readily verifies that the transformations

$$Z^\alpha \mapsto T^\alpha_\beta(\theta) Z^\beta$$

indeed belong to $SU(2,2)$. (Preservation of $\bar{Z}_\alpha Z^\alpha$ is almost immediate; $\det T^\alpha_\beta(\theta) = 1$ follows from the eigenvalues being $e^{i\theta}$, $e^{i\theta}$, $e^{i\theta}$, $e^{-3i\theta}$.) In fact, these transformations constitute a one-parameter group:

$$T^\alpha_\beta(0) = \delta^\alpha_\beta, \quad T^\alpha_\beta(\theta) T^\beta_\gamma(\phi) = T^\alpha_\gamma(\theta + \phi).$$

The required path from δ^α_β to $i\delta^\alpha_\beta$ is given by θ varying from 0 to $\frac{1}{2}\pi$.

It is not hard to check that the transformations (9.4.1) in fact induce a conformal motion of $\mathbb{M}^\#$ along the rays of the Robinson congruence

$$\bar{Q}_\alpha Z^\alpha = 0.$$

(These are the integral curves of the null conformal Killing vector determined by the primary part of Q^α, *cf.* (6.3.19).) Such motions were discussed in §9.2 (after (9.2.1)) in order to clarify the topological structure of $\mathbb{M}^\#$. When we pass continuously from $\theta = 0$ to $\theta = \frac{1}{4}\pi$, $\mathbb{M}^\#$ slides over itself along these lines, which are all topologically $S^1$and, each point of $\mathbb{M}$ returns to its original position, having traversed one of these S^1s, and having intersected $\mathscr{I}$ once in the process. Thus to know how fields in $\mathbb{M}$ are affected by elements of $SU(2,2)$, and in particular by the transformation (9.4.1), we need to know how these fields extend across $\mathscr{I}$.

We shall see shortly that if, as before, we take ω^A as an ordinary spinor field of conformal weight zero (to ensure the conformal invariance of (6.1.1)), then although ω^A has a well-defined finite limit as $\mathscr{I}$ is approached from either side, these two limits differ by a factor i. In order that the description of ω^A in the vicinity of a point of $\mathscr{I}$ shall resemble its description elsewhere in $\mathbb{M}$, one therefore seems to be forced to continue ω^A across $\mathscr{I}$ in two distinct ways simultaneously, which differ by a factor i. So we find that the ω^A field must be coupled with $\pm i\omega^A$ in addition to $-\omega^A$ in the geometric description of Z^α. This provides a resolution of the apparent paradox concerning the four-fold ambiguity in the representation of Z^α. However, we shall also need to obtain a deeper understanding of what is really involved here.

The two spin structures on $\mathbb{M}^\#$

In order to discuss ω^A globally on $\mathbb{M}^\#$, or even to say what we *mean* by a spinor field ω^A on $\mathbb{M}^\#$, we must first specify a *spin structure* for $\mathbb{M}^\#$. This is not trivial, since $\mathbb{M}^\#$ is not simply-connected (see §1.5). The topology of $\mathbb{M}^\#$ being $S^3 \times S^1$ (*cf.* (9.2.1)), there is essentially just one 'unshrinkable' loop in

$\mathbb{M}^{\#}$, all others being continuously deformable to it or a multiple of it. We can take this loop to be a *null geodesic* or *ray* γ ($\cong S^1$) in $\mathbb{M}^{\#}$. (That no non-zero multiple of γ can be continuously deformed to a point should be clear from the discussion in the early part of §9.2.) According to §1.5, therefore, there will be precisely *two* spin structures for $\mathbb{M}^{\#}$, and we can distinguish between these by specifying how to carry a null flag continuously around γ so as to yield the original spin-vector rather than its negative.

It might be thought that there is an 'obvious' choice that can be made here, namely to deem the spin-vector unchanged if its flag is taken by parallel transport around γ. But this is to ignore the subtleties involved as $\mathscr{I}$ is crossed. The original Christoffel connection ∇_a, defined by the metric of $\mathbb{M}$, cannot be used to define parallelism across $\mathscr{I}$. One could use the Christoffel connection $\hat{\nabla}_a$ for some other metric to get across $\mathscr{I}$. Of course, this would involve some arbitrariness, but it is not this arbitrariness which is the heart of the problem. Let us consider a null flag *with its flagpole along* γ: this suffices, for if we have *one* non-zero spin-vector defined along γ we can complete that to a spin-frame along γ, and all such choices for completing the spin-frame are continuously deformable into one another. Now we saw in (7.1.20) that the condition for parallel propagation of the flag plane of o^A along the direction (γ) of its flagpole is that the spin-coefficient ε be real (with $\chi = 1$). But by (5.6.29), this property is preserved under conformal rescaling, so the propagation of flag planes along the flagpole direction is indeed *independent* of the choice of Christoffel connection $\hat{\nabla}_a$, i.e., independent of the choice of scaling for the metric $\hat{g}_{ab}$. (Clearly this argument is a local one and does not require a global definition of the spinor o^A.)

Thus we do have a natural conformally invariant way of carrying such null flags around γ. The subtlety referred to above arises because when we carry a flag once around γ in this way, the flag plane direction is *reversed* when we return to the starting point. This can be seen as follows.

Consider a scond ray γ' which is displaced infinitesimally from γ in such a way that the connecting vectors are orthogonal to the direction of γ, i.e. the rays γ and γ' are *abreast* (*cf.* §7.1). One possibility which meets this condition is that γ and γ' are neighbouring rays of a null hyperplane. It is clear that if in this case the flag plane of our null flag points from γ to γ' (i.e., is the half-plane of directions of all connecting vectors) then this flag plane is transported parallelly to itself along γ. Now suppose that, instead of being parallel to γ, γ' is a generator of a light cone with vertex at any given point P of γ. Conformally this is equivalent to the parallel case, where P is on $\mathscr{I}$ (*cf.* §§9.1, 9.2). Since parallel propagation of flag plane directions along γ is conformally invariant, it is just as well determined by γ' in the second case as in the first. Now at the vertex P of the cone the connecting vector between γ and γ' changes sign as we pass along γ (see Fig. 9-11). Since a null cone has only *one* vertex, this phenomenon is not repeated anywhere else on γ. It is therefore clear that parallel transport all around a loop γ in $\mathbb{M}^\#$ of a null flag, say that of γ^A with flagpole along γ, from a point just to the future of P to one just to the past of P, reverses the flag direction. It is up to us whether this reversal should be regarded as equivalent to a negative or positive rotation through π. And it is precisely this *choice* that gives rise to the aforementioned two possible spin structures for $\mathbb{M}^\#$. We define the *right-handed spin structure* for $\mathbb{M}^\#$ to be the one which decrees that the following closed motion of the flag of γ^A (i.e. 'flag path' of §1.5) shall also restore γ^A itself to its original value rather than to its negative: carry the flag plane of γ^A by parallel transport from a point P on γ once around γ in the future direction to a point just to the past of P and then apply a *right-handed* rotation through π about the flagpole direction in order to connect with the original location. The other spin structure is called the *left-handed* spin structure for $\mathbb{M}^\#$. Note that the results of these two possible motions of the null flag differ by a 2π-rotation, so they do indeed define *different* spin structures. And while the meaning of a 'right-handed' rotation – with respect to the *future*-pointing direction along γ^A – should be intuitively

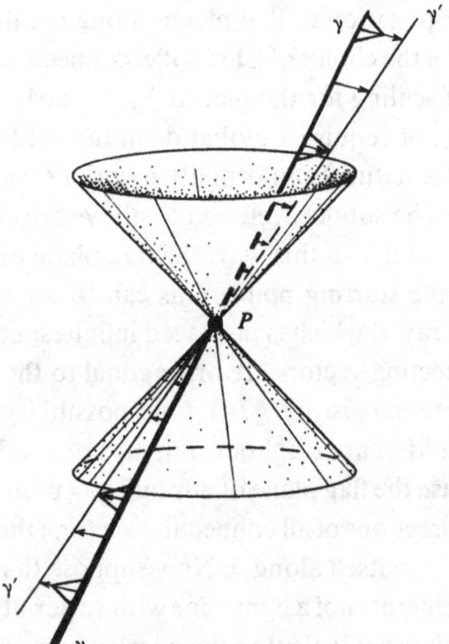

Fig. 9-11. Diagram illustrating the reversal of flag plane direction under parallel propagation as a ray γ is traversed, crossing $\mathscr{I}$ just once.

clear, it can be made precise by reference to the rotation of the tangent vector to the Riemann sphere S^+ which represents the direction of the flag plane. 'Right-handed' means anti-clockwise as viewed from outside the sphere. Equivalently we may think in terms of a right-handed rotation about the *spatial projection* of the future null direction along γ.

These two spin structures are on an equal footing, if we have no preference for one handedness over the other. In fact,

a space-reflection of $\mathbb{M}^{\#}$ *interchanges the two spin structures.* (9.4.2)

It may be recalled that space-reflections interchange the unprimed and primed spin-spaces. We can see the relation of this to the above choices of spin structure in the fact that for an unprimed spinor γ^A, a right-handed π-rotation about the flagpole effects $\gamma^A \mapsto \mathrm{i}\gamma^A$ and a left-handed π-rotation effects $\gamma^A \mapsto -\mathrm{i}\gamma^A$ (and in general right- and left-handed α-rotations effect $\gamma^A \mapsto \mathrm{e}^{\pm \mathrm{i}\alpha/2}\gamma^A$, respectively). On the other hand, for a primed spinor $\eta^{A'}$, the corresponding effects are, respectively, $\eta^{A'} \mapsto \mp \mathrm{i}\eta^{A'}$, $\eta^{A'} \mapsto \mathrm{e}^{\mp \mathrm{i}\alpha/2}\eta^{A'}$. Thus, if we adopt the right-handed spin structure for $\mathbb{M}^{\#}$ (as normally we shall), we find that parallel transport of γ^A in a future direction once around γ to its starting point results in $-\mathrm{i}$ times the original value of γ^A (since a π-rotation restores it). The same procedure applied to $\eta^{A'}$ results in $+\mathrm{i}$ times the original value. If, instead, we adopt the left-handed spin structure we find that the roles of $-\mathrm{i}$ and $+\mathrm{i}$ are reversed. (Compare Woodhouse 1980 for a somewhat different approach.)

It is of interest to note, however, that

a time-reflection of $\mathbb{M}^{\#}$ *leaves the two spin structures invariant,* (9.4.3)

even though it also interchanges primed and unprimed spinors. This can be seen as follows. If we reverse the time direction, this has *two* effects with regard to the motion that defines continuity for a spin-vector, say for the right-handed spin structure. In the first place, since all motions are now described with the time reversed, a positive π-rotation is thereby reversed into a negative one. But also the spatial direction about which we are measuring the rotation is reversed, since the future direction along γ is now spatially opposite to what it was. This causes a second reversal of the handedness, and consequently the right-handed spin structure is actually sent into itself by time-reversal, and the same is true of the left-handed one.

By combining the effects of separate space- and time-reflections, we obtain the result that *a space–time reflection of* $\mathbb{M}^{\#}$ *interchanges the two spin structures.* (9.4.4)

(We remark that space–time reflections do not, however, interchange primed and unprimed spinor indices.)

The behaviour of ω^A *at* $\mathscr{I}$

Now let us return to the discussion of the spinor field ω^A satisfying the twistor equation (6.1.1). We can choose γ to be one of the rays of the (possibly special) Robinson congruence that ω^A determines. Using the standard descriptions in $\mathbb{M}$ relative to an origin O on γ, we have, as in (6.1.10),

$$\omega^A = \mathring{\omega}^A - \mathrm{i}x^{AA'}\pi_{A'}, \tag{9.4.5}$$

where

$$x^{AA'} = u\gamma^A\bar{\gamma}^{A'} \quad (u \in \mathbb{R}) \tag{9.4.6}$$

along γ. Since γ belongs to the Robinson congruence, we can choose (assuming $\mathring{\omega}^A \neq 0$)

$$\gamma^A = \mathring{\omega}^A,$$

which is obviously parallelly propagated along γ. Then (9.4.5) gives

$$\omega^A = (1 - u(b + \mathrm{i}s))\gamma^A, \tag{9.4.7}$$

where

$$s = \mathrm{Re}\,(\mathring{\omega}^A\bar{\pi}_A), \quad b = \mathrm{Im}\,(\mathring{\omega}^A\bar{\pi}_A)$$

are real constants. Note that, by (6.1.74), s is the *helicity*

$$s = \tfrac{1}{2}\mathrm{Z}^\alpha\bar{\mathrm{Z}}_\alpha$$

of the twistor $\mathrm{Z}^\alpha = (\omega^A, \pi_{A'})$. In the Argand plane, the proportionality factor $(1 - u(b + \mathrm{i}s))$ of (9.4.7) executes a straight line as u varies from $-\infty$ to $+\infty$, its argument increasing or decreasing according as the helicity of Z^α is negative or positive, respectively (see Fig. 9-12*a*). In order to relate this to the geometry of the flag plane of ω^A it is better to consider the *square* of this factor, which executes a parabola with the origin as focus (see Fig. 9-12*b*). The argument of *this* point gives a direct realization of the amount of rotation of the flag plane of ω^A (*cf.* §3.2). We see that the limits of the flag plane direction are the same for $u \to -\infty$ and $u \to +\infty$, but the sign of the spinor ω^A in (9.4.7) gets reversed in the passage from one limit to the other.

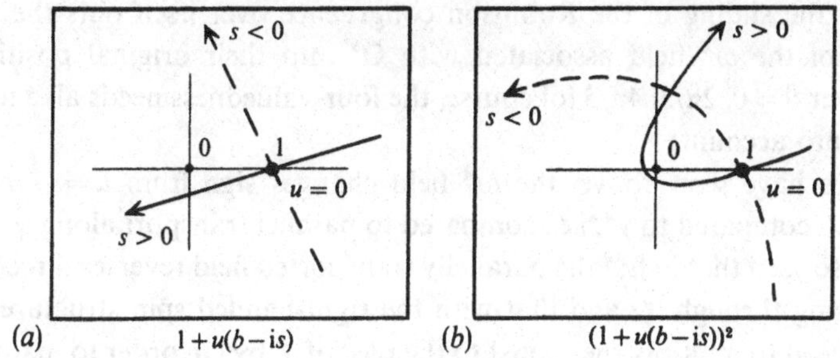

Fig. 9-12. Argand-plane diagrams showing how the flag plane direction of ω^A rotates, for a twistor of non-zero helicity, as a ray of the Robinson congruence is followed in the future direction. (The right-hand diagram gives the more direct realization, the flag plane being given by the direction of the moving point at the origin of the Argand plane.) The ray is oriented *towards* the viewer.

Note (e.g., from Fig. 9-12*b*) that the direction of rotation of the flag plane is *opposite* to the helicity of the twistor. We may contrast this with the rotation sense of neighbouring rays relative to γ in the Robinson congruence which, as we saw in §6.2, is in the *same* sense as the helicity. Moreover, the rate of rotation of the flag plane is (minus) *twice* that of the neighbouring rays. These facts can also be inferred from the spin-coefficient equations

$$\operatorname{Im}(\rho) = s, \quad 2\operatorname{Im}(\varepsilon) = -2s, \tag{9.4.8}$$

which are immediate consequences of (9.4.5) if we choose a spin-frame along γ with

$$o^A = \omega^A, \quad \iota^A = \bar{\pi}^A(\omega_A\bar{\pi}^A)^{-1},$$

and recall from the discussion of §7.1 that $\operatorname{Im}(\rho)$ and $2\operatorname{Im}(\varepsilon)$ measure, respectively, the rate of rotation of neighbouring rays of a congruence and the rate of rotation of the flag plane of o^A. (Note that despite the appearance of (9.4.8), these rotation rates are not constant, since they are scaled in relation to ω^A rather than γ^A. In fact, as is clear from (9.4.7), the 'absolute' rates are given by (9.4.8) scaled down by $|1 - u(b + is)|^{-2}$.) We may note the curious fact that for any one abreast neighbouring ray of the congruence, the flag plane of ω^A will point in the direction of the connecting vector at precisely three places on γ, a consequence of the rotation rates being as given above. (In an obvious notation, the equations $2\theta = -\phi, \theta - \phi = 2n\pi$ are solved by $\theta = 0, 2\pi/3, 4\pi/3$.) This is related to (9.4.1) which, if applied to the twistor Q^α itself, gives

$$\mathrm{Q}^\alpha \mapsto e^{-3i\theta}\mathrm{Q}^\alpha,$$

so that the sliding of the Robinson congruence over itself puts the flag planes of the ω^A field associated with Q^α into their original positions whenever $\theta = 0, 2\pi/3, 4\pi/3$ (of course, the four-valuedness needs also to be taken into account).

As we have seen above, the ω^A field changes sign from $u = -\infty$ to $u = +\infty$ compared to γ^A, i.e., compared to parallel transport along γ. We have also seen that in $\mathbb{M}^\#$ the parallelly transported field reverses direction on passing through $\mathscr{I}$, and that with the right-handed spin structure we would need to multiply the γ^A just to the past of $\mathscr{I}$ by i in order to match it with the γ^A just to the future of $\mathscr{I}$. To keep γ^A non-zero at $\mathscr{I}$ we rescale it: $\hat{\gamma}^A = \Omega^{-1}\gamma^A$, which corresponds to $\hat{\gamma}_A = \gamma_A$. For this is the scaling that preserves parallel propagation (*cf.* (7.1.18), and (5.6.25) which shows that $\varepsilon = 0$ is preserved when $w_0 = -1$). In the case of ω^A there is an extra *sign* difference, so with the right-handed spin structure we would have to multiply the value of ω^A just to the past of $\mathscr{I}$ by $-$i in order to get continuity with the value just to the future of $\mathscr{I}$.

Had we chosen a $[^0_1]$-twistor $W_\alpha = (\lambda_A, \mu^{A'})$ rather than a $[^1_0]$-twistor Z^α, we would correspondingly have obtained a field $\mu^{A'}$ whose flag plane just to the past of $\mathscr{I}$ would have to be rotated in the same sense as that of ω^A in order to give continuity with the value just to the future of $\mathscr{I}$ (still with the right-handed spin structure). But now this means that we must multiply $\mu^{A'}$ just to the past of $\mathscr{I}$ by $+$i in order to get continuity across $\mathscr{I}$. Of course, with the *left-handed* spin structure, the roles of these factors $+$i and $-$i would be reversed.

The twistor spin-bundles over $\mathbb{M}^\#$

The above rather elaborate geometric structure which is inherent in the geometry of the twistor equation, is also, as we shall shortly see (in the Grgin theorem (9.4.15)), present in the global solutions of the massless free-field equations – which at first sight seem unrelated to twistors. In all these cases the spinor fields involved are, by virtue of these geometric subtleties, strictly speaking not spinor fields in the ordinary sense but cross-sections of certain 'twisted' vector bundles (*cf.* §5.4), to whose discussion we now proceed.

The 'twist' that needs to be introduced is somewhat analogous to that of the Möbius band (which is illustrated in Fig. 5-3 on page 338 of Volume 1), and this is precisely the 'multiplication by $+$i' at $\mathscr{I}$ that we have just described. Thus if we are interested in the conformal invariance of the fields ω^A and $\mu^{A'}$, in the sense of invariance under $SU(2,2)$, we shall have to regard them as twisted fields in this sense.

In fact there is a direct 'twistor' way of defining the particular vector bundles that will be relevant (Eastwood, Penrose and Wells 1981). We shall refer to these as *twistor spin-bundles* over $\mathbb{M}^\#$ – and also over $\mathbb{CM}^\#$, since they extend in a natural way to the whole of $\mathbb{CM}^\#$. We begin by discussing the twistor bundle $\mathscr{S}_{A'}$ of twisted $[\begin{smallmatrix}0 & 0\\ 0 & 1\end{smallmatrix}]$-spinors over $\mathbb{CM}^\#$. Recall (from §§6.2, 9.3) that the points of $\mathbb{CM}^\#$ correspond uniquely to two-dimensional complex linear subspaces of twistor space $\mathbb{T}^\alpha$. Let X be such a linear subspace, corresponding to the point $X \in \mathbb{CM}^\#$. The various twistors incident with X are, on the one hand (in $\mathbb{T}^\alpha$), precisely the *points* of X, while on the other hand (in $\mathbb{CM}^\#$), each such twistor is uniquely defined by a choice of spinor $\pi_{A'}$ at the point X. In other words, the pair (X, Z^α), where Z^α is incident with X, has the interpretation in terms of $\mathbb{T}^\alpha$ as the pair

$$(\mathsf{X}, \mathsf{Z}^\alpha) \quad \text{with } \mathsf{Z}^\alpha \in \mathsf{X} \subset \mathbb{T}^\alpha,$$

and in terms of $\mathbb{CM}^\#$ as

$$(X, \pi_{A'}) \quad \text{with } \pi_{A'} \textit{ at } X \in \mathbb{CM}^\#.$$

The $\pi_{A'}$ spin-space at X becomes interpreted as the vector space X itself.

The point of view, then, is to *start* with $\mathbb{T}^\alpha$, which has a given complex-vector-space structure. Then we *define* $\mathbb{CM}^\#$ as the space of two-dimensional vector subspaces $\mathsf{X} \subset \mathbb{T}^\alpha$. From this the bundle $\mathscr{S}_{A'}$ comes out automatically as the space of pairs $(\mathsf{X}, \mathsf{Z}^\alpha)$ with $\mathsf{Z}^\alpha \in \mathsf{X}$, where varying Z^α and keeping X fixed gives us the fibres. Finally we interpret these fibres as the $\pi_{A'}$ spin-spaces $\mathfrak{S}_{A'}[X]$ at the various points of $\mathbb{CM}^\#$.

However, these are not quite spin-spaces in the ordinary sense, but *twisted* spin-spaces in the sense discussed above. In fact, they are the *duals* of the spin-spaces $\mathscr{S}^{A'}$ to which the earlier-discussed (twisted) $\mu^{A'}$-spinors (parts of twistors W_α) belong. To see this, consider any $[\begin{smallmatrix}0\\ 1\end{smallmatrix}]$-twistor W_α. In terms of $\mathbb{T}^\alpha$, we regard W_α as providing a linear map: $\mathbb{T}^\alpha \to \mathbb{C}$, while in terms of $\mathbb{CM}^\#$, W_α gives us a (twisted) $\mu^{A'}$ field. For each $X \in \mathbb{CM}^\#$ we get a particular $\mu^{A'}$, namely $\mu^{A'}[X] \in \mathfrak{S}^{A'}[X]$, and this corresponds, in $\mathbb{T}^\alpha$, to the restriction of the linear map W_α of $\mathbb{T}^\alpha$ to the subspace X. This restriction is a particular linear map: $\mathsf{X} \to \mathbb{C}$, i.e., an element of the dual space of the vector space X, which shows that X is canonically identifiable with $\mathfrak{S}_{A'}[X]$, the *dual* of the (conjugate) spin-space $\mathfrak{S}^{A'}[X]$ at X. As X varies over $\mathbb{M}^\#$, these spaces $\mathfrak{S}_{A'}[X]$ must be continuously related to one another with the appropriate twist, *dual to that of* $\mu^{A'}$, which characterizes $\mathscr{S}_{A'}$. This establishes our assertion. (Note that the fibres of the mutually *dual bundles* $\mathscr{S}_{A'}$ and $\mathscr{S}^{A'}$ are pointwise duals of each other. The $\mu^{A'}$ fields are cross-sections of $\mathscr{S}^{A'}$.)

Let us now consider the twistor bundle $\mathscr{S}_A$ of twisted $[\begin{smallmatrix}0 & 0\\ 1 & 0\end{smallmatrix}]$-spinors over $\mathbb{CM}^\#$. The simplest approach is just to regard $\mathscr{S}_A$ as the complex conjugate of the bundle $\mathscr{S}_{A'}$ defined above. This is equivalent to taking the points of

$\mathbb{CM}^\#$ as represented by two-dimensional linear subspaces in the dual twistor space $\mathbb{T}_\alpha$, since twistor complex conjugation interchanges $\mathbb{T}^\alpha$ and $\mathbb{T}_\alpha$. (In fact, regarding $\mathbb{T}_\alpha$ as the *dual* rather than the complex conjugate of $\mathbb{T}^\alpha$ is logically the more satisfactory here, since it leads to an entirely *holomorphic* construction. In certain contexts (e.g. Eastwood, Penrose and Wells 1981) it is important to keep all operations holomorphic, as far as possible, when one is concerned with the complex space $\mathbb{CM}^\#$, rather than $\mathbb{M}^\#$.) Thus, representing the point $X \in \mathbb{CM}^\#$ by the two-dimensional subspace $\mathsf{X}^* \subset \mathbb{T}_\alpha$, we define the bundle $\mathscr{S}_A$ over $\mathbb{CM}^\#$ to be the space of pairs $(\mathsf{X}^*, \mathsf{W}_\alpha)$ where $\mathsf{W}_\alpha \in \mathsf{X}^*$, so the fibre over X is simply the space X^*. Each such pair is equivalent to a pair (X, λ_A) and this serves to define what is meant by the 'twisted' spinor λ_A. The dual bundle $\mathscr{S}^A$ is then obtained by taking for its fibres the duals of the fibres of $\mathscr{S}_A$ in the usual way. Analogously to $\mu^{A'}$ and $\mathscr{S}^{A'}$, the solutions ω^A of the twistor equation (correctly 'twisted') are cross-sections of $\mathscr{S}^A$, namely those induced by the linear maps of the form $\mathsf{Z}^\alpha: \mathbb{T}_\alpha \to \mathbb{C}$.

Alternatively we can define $\mathscr{S}_A$ directly in terms of $\mathbb{T}^\alpha$ (which is more logical if $\mathbb{T}^\alpha$ is considered primary). Then we represent $x \in \mathbb{CM}^\#$ by the original two-dimensional subspace $\mathsf{X} \in \mathbb{T}^\alpha$. For each such X, consider the two-dimensional space of those linear maps $\mathsf{W}_\alpha: \mathbb{T}^\alpha \to \mathbb{C}$ which give zero at every point of X. These we define to constitute the fibres of $\mathscr{S}_A$, and this is easily seen to be equivalent to the definition given in the preceding paragraph. We can then define $\mathscr{S}^A$ as the dual of $\mathscr{S}_A$.

Having the bundles $\mathscr{S}_A, \mathscr{S}_{A'}, \mathscr{S}^A, \mathscr{S}^{A'}$ on $\mathbb{CM}^\#$, it is now a trivial matter to restrict them to $\mathbb{M}^\#$. Then we can define the spaces $\mathfrak{S}_A, \mathfrak{S}_{A'}, \mathfrak{S}^A, \mathfrak{S}^{A'}$ as smooth cross-sections of these respective bundles (of whatever degree of smoothness we may choose – say C^∞ for compatibility with our earlier work). If we are concerned with $\mathbb{CM}^\#$ we need holomorphic cross-sections, but then we must work locally since the general cross-sections are defined only over some open set of $\mathbb{CM}^\#$.

We can now follow the methods of §2.2 (and also §5.4, Volume 1) to define the elements of the general (twisted) $[^p_r\,{}^q_t]$-spinor space $\mathfrak{S}^{A\ldots F'}_{G\ldots N'}$, and we note that, in terms of ordinary spinor fields on $\mathbb{M}$, the jump across $\mathscr{I}$ with respect to the right-handed spin structure is given by the requirement that the field just to the past of $\mathscr{I}$ must be multiplied by a factor

$$\mathrm{i}^{r-p-t+q} \tag{9.4.9}$$

to achieve continuity with the field just to the future of $\mathscr{I}$. In particular, the jump for elements of $\mathfrak{S}_A$ must be opposite from that for elements of its dual $\mathfrak{S}^A$. (This is clear since, for example, a scalar product like $\kappa_A\omega^A$ must, at any point of $\mathbb{M}^{\#}$ (or $\mathbb{CM}^{\#}$), be an ordinary untwisted scalar, i.e., an element of $\mathbb{C}$.)

Although the above definition of the module $\mathfrak{S}^{\cdots}_{\cdots}$ does not provide us with spinor fields in the ordinary sense (unless $r - p - t + q$ is a multiple of 4), we may regard the 'twisted spin structure' from which it is derived as being more natural than either the right-handed or the left-handed spin structure on $\mathbb{M}^{\#}$. (We can, of course, express the 'jump' specified in (9.4.9) equally well in terms of the left-handed spin structure, by simply replacing i by $-$i in (9.4.9), or, equivalently, by changing the sign of the exponent.)

Conformal densities on $\mathbb{M}^{\#}$; the Grgin index

Note that (9.4.9) agrees with the behaviour of the parallelly propagated γ_A considered earlier in this section, provided we take the conformally invariant lower-index form. If we consider γ^A instead (or, for that matter, ω_A), which is a conformal *density* rather than a conformal invariant, we must incorporate an additional factor $(-1)^w$, for general conformal weight w, into the formula – where here w must be an *integer*. However, the interpretation of this needs some care. If we define a non-singular metric $\hat{g}_{ab}$ in some neighbourhood of a point of $\mathscr{I}$, conformal to g_{ab} in the standard way with

$$\hat{g}_{ab} = \Omega^2 g_{ab}, \tag{9.4.10}$$

using, say, the Ω factor of §9.1, we find that Ω is *negative* on one or the other side of $\mathscr{I}$. Thus the normal requirement of §5.6 that the conformal factor is to be everywhere positive has now to be relaxed. Quantities of *odd* conformal weight will change sign, in addition to being rescaled, in regions where we choose the *negative* sign for the square root of the Ω^2 of (9.4.10). If we follow the procedure of the beginning of §9.2 for constructing $\mathbb{M}^{\#}$, where the boundary hypersurfaces $\mathscr{I}^+$ and $\mathscr{I}^-$ are first added to $\mathbb{M}$ and then identified, we do not see negative factors Ω arising. But if we extend Ω smoothly across $\mathscr{I}^+$ into the future, or across $\mathscr{I}^-$ into the past, we enter regions of negative Ω. This is something of a nuisance, particularly since the interpretation of a spin-vector κ^A in terms of a null flag is achieved through the bivector

$$\kappa^A\kappa^B\varepsilon^{A'B'} + \varepsilon^{AB}\bar{\kappa}^{A'}\bar{\kappa}^{B'}$$

(*cf.* (3.2.9)), the ε-spinors all having *odd* conformal weight. In regions where $\Omega < 0$, therefore, we would seem to require the *reverse* association between κ^A and its flag plane from that given in §3.2 (and Chapter 1, Volume 1) (i.e., with the tangent vector to S^+ representing this flag plane pointing in the opposite direction). This would not enable us to avoid the problem of right- versus left-handed spin structure ambiguity as encountered above (spin structures being, in any case, entirely topological and essentially non-metric concepts; Milnor 1963), because with negative Ω we would have to 'go around' $\mathbb{M}^{\#}$ twice before returning to original values. For this reason, we adopt the 'gluing' procedure of the beginning of §9.2 (which takes $\Omega \geqslant 0$ throughout $\mathbb{M}^{\#}$) in making the following definition:

(9.4.11) DEFINITION

A $[{}^{p}_{r}\,{}^{q}_{t}]$*-spinor field in* $\mathbb{M}$ *which is a conformal density of (integer) weight w has* Grgin behaviour *at infinity if, when rescaled (with* $\Omega > 0$*), it extends smoothly across* $\mathscr{I}$ *when the field just to the past of* $\mathscr{I}$ *is multiplied by the factor* $(-\mathrm{i})^{p-r-q+t-2w}$*, using the right-handed spin structure.*

The integer (mod 4) $p-r-q+t-2w$ is called the *Grgin index* of the field (as suggested by N.M.J. Woodhouse). We note that the primary part of a $[{}^{1}_{0}]$-twistor has Grgin index $+1$ and that of a $[{}^{0}_{1}]$-twistor, -1. Hence, by taking tensor products, we arrive at:

(9.4.12) PROPOSITION

The Grgin index of the primary part of a $[{}^{p}_{q}]$*-twistor is* $p-q$ *(mod 4).*

Note also that, quite generally,

$$\text{if } \psi_{\dots}^{\dots} \text{ and } \chi_{\dots}^{\dots} \text{ have Grgin indices } a \text{ and } b, \\ \text{then } \psi_{\dots}^{\dots}\chi_{\dots}^{\dots} \text{ has Grgin index } a+b \text{ (mod 4).} \tag{9.4.13}$$

If desired, the above problems of interpretation in the case of non-zero conformal weight can be avoided if we replace the field $\phi_{G\dots N'}^{A\dots F'}$ by

$$\phi_{G\dots N'}^{A\dots F'}\varepsilon_{PQ}\dots\varepsilon_{TU} \quad \text{or} \quad \phi_{G\dots N'}^{A\dots F'}\varepsilon^{PQ}\dots\varepsilon^{TU} \tag{9.4.14}$$

in order to produce a field of zero conformal weight. The Grgin index remains unchanged since that of the ε-spinors is zero.*

* Note that the use of primed epsilons in place of some or all of the unprimed ones in (9.4.14) would make no difference. This is because (in this book) we have consistently assumed $\Omega = \tilde{\Omega}$ in our rescalings $\hat{\varepsilon}_{AB} = \Omega\varepsilon_{AB}$, $\hat{\varepsilon}_{A'B'} = \tilde{\Omega}\varepsilon_{A'B'}$; $\hat{g}_{ab} = \Omega\tilde{\Omega}g_{ab}$, whereas the more general choice of *independent* Ω and $\tilde{\Omega}$ could have been made. In the present context this corresponds to the fact that we are concerned with invariance under $SU(2,2)$ (or its complexification $SL(4,\mathbb{C})$) rather than the more general $U(2,2)$ (or $GL(4,\mathbb{C})$). The preservation of the twistor $\varepsilon_{\alpha\beta\gamma\delta}$ is what distinguishes these slightly more restricted transformations. We recall from (6.1.64) that the spinor parts of $\varepsilon_{\alpha\beta\gamma\delta}$ are terms like $\varepsilon^{A'B'}\varepsilon_{CD}$ which would scale with a factor $\tilde{\Omega}^{-1}\Omega$. So the preservation of $\varepsilon_{\alpha\beta\gamma\delta}$ entails $\tilde{\Omega} = \Omega$.

The Grgin theorem

The reason for the above terminology is the following remarkable result:

(9.4.15) THEOREM (Grgin 1966)

Any non-singular solution of the massless free-field equation (4.12.42) *on the Einstein cylinder* $\mathscr{E}$ *exhibits the behaviour* (9.4.11) *when* $\mathscr{E}$ *is identified in the standard way to become* $\mathbb{M}^{\#}$.

We shall presently give a proof. But first note that for a massless field of spin $\frac{1}{2}n$ with unprimed indices we have $r = n$, $w = -1$, and $p = q = t = 0$, so the Grgin index is $-n+2$, whereas for such a field with primed indices the Grgin index is $n+2$. Recall (*cf.* §5.7) that for fields of positive frequency the helicity is given by $s = -\frac{1}{2}n\hbar$ in the first case and by $s = \frac{1}{2}n\hbar$ in the second. So the Grgin index is $2s\hbar^{-1}+2$. In order to *avoid* the factor $(-\mathrm{i})^{\mp n+2}$ now occurring in (9.4.11) we would need to take a two-fold cover of $\mathbb{M}^{\#}$ for massless fields of even spin (e.g., for the conformally invariant massless scalar D'Alembert field), and a four-fold cover for half-odd integer spin. But for odd integer spin (e.g., for the Maxwell field) the space $\mathbb{M}^{\#}$ suffices.

To prove the Grgin theorem, we shall use an earlier lemma to reduce the case of general spin to that of zero spin. Suppose, in a conformally flat space–time $\mathscr{M}$, that $\lambda^{A\ldots D}$ is the primary part of a symmetric $[{}^{n}_{0}]$-twistor, so that (6.4.1) is satisfied, and suppose the symmetric n-index spinor $\phi_{A\ldots D}$ satisfies the massless field equation (4.12.42). Then, by (6.4.31) *et. seq.*

$$\phi = \phi_{A\ldots D}\lambda^{A\ldots D}$$

satisfies the conformally invariant wave equation (*cf.* 6.8.30))

$$(\Box + \tfrac{1}{6}R)\phi = 0.$$

Now suppose $\phi_{A\ldots D}$ is as above. For any point $Q \in \mathscr{E}$ the choices of $\lambda^{A\ldots D}$ satisfying (6.4.1) span the space $\mathfrak{S}^{(A\ldots D)}[Q]$, since a (symmetric) twistor has arbitrary (symmetric) primary spinor part at any one point. Let P be the point to which the future light cone of Q first re-converges, i.e., the first point to the future of Q which gets identified with Q when we make the identification to obtain $\mathbb{M}^{\#}$.

We know that $\lambda^{A\ldots D}$ has the correct Grgin behaviour, by (9.4.12). Because of the multiplicative property (9.4.13) and the space-spanning property of $\lambda^{\cdots}$ at each point, it is therefore sufficient to show that the Grgin property holds for the scalar ϕ. (The theorem for a *primed* massless field $\phi_{A'\ldots D'}$ will

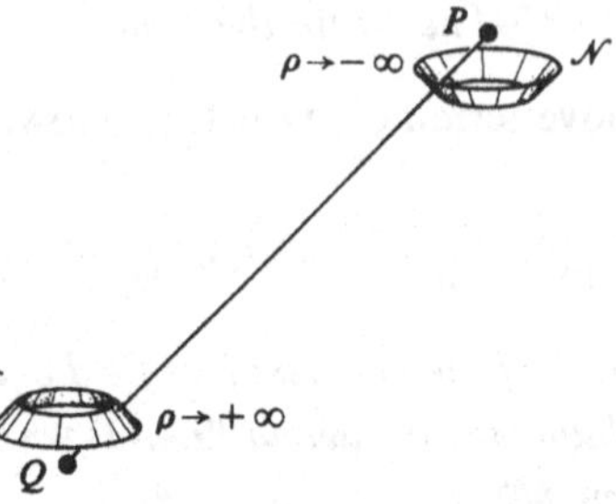

Fig. 9-13. An explanation of the Grgin phenomenon for massless scalar fields on the Einstein cylinder $\mathscr{E}$, in terms of the Kirchhoff–D'Adhémar integral formula (5.12.6). The rays through Q first converge again at P. The sign reversal between the field values at P and at Q can be attributed to the opposite sign for the convergence ρ as a null hypersurface $\mathcal{N}$, between P and Q, convergences on one or the other point.

then also follow from the corresponding complex conjugate argument.) The Grgin property for a scalar ϕ asserts that $\phi(P) = -\phi(Q)$ for each choice of Q. One way to see that this must hold is to note that every solution of the wave equation in $\mathbb{M}$ is composed of 'elementary' solutions having the form

$$t^{-1}\delta(|t| - (x^2 + y^2 + z^2)^{\frac{1}{2}}),$$

in standard coordinates, taken relative to various different origins. The coefficient of the δ-function has opposite signs on the future and past light cone and clearly the Grgin property is satisfied, and the theorem established. Alternatively, we may appeal to the Kirchhoff–D'Adhémar formula (5.12.6) which expresses $\phi(P)$ as an integral over any cross-section of the past light cone of P by some null hypersurface $\mathcal{N}$. As part of the proof of this formula we examined the limiting situation when the cross-section approaches P from the past (the value of the integral being independent of the cross-section). When we compare this limit with the corresponding expression obtained when the cross-section moves back along the past light cone of P to approach Q from the future (*cf.* Fig. 9-13), we find, since $\text{þ}_\varepsilon\phi = (D - \rho)\phi$ (taking $\varepsilon = 0$) is dominated by the $-\rho\phi$ term near P or Q, that the two limiting expressions differ only in the *sign* of the convergence ρ of the intersecting null hypersurface $\mathcal{N}$, and that in the first case it is the field value $\phi(P)$ which enters, and in the second case it is $\phi(Q)$. This again shows $\phi(Q) = -\phi(P)$, as required.

Non-Grgin fields

It should be remarked that the Grgin behaviour is a property of fields satisfying the field equations *globally* and it does not apply when the field has singularities. One familiar example of a *non*-Grgin field is the Coulomb field. Here the Grgin index is zero, so Grgin behaviour would entail that the field extends continuously across $\mathscr{I}$ in $\mathbb{M}^{\#}$. But it is easy to see directly that the field in fact changes sign at $\mathscr{I}$. For with a positive charge, the electric vector always points radially outwards, and so is associated with the outward pointing EPND (*cf.* §8.5, p. 255). But at $\mathscr{I}^+$ this is the EPND which points *across* $\mathscr{I}$ while at $\mathscr{I}^-$ it is the one that points tangentially to $\mathscr{I}$ (see Fig. 9-14). Thus the sign of the field must be opposite when $\mathscr{I}^+$ is matched to $\mathscr{I}^-$. Another way of seeing this is to consider, instead, the universal covering space $\mathscr{E}$ of $\mathbb{M}^{\#}$ and to extend the field analytically (without sign change at $\mathscr{I}$) to the whole of $\mathscr{E}$. Since the space–time $\mathscr{E}$ is spatially closed, its total charge must be zero (by e.g. (6.4.4) with $\mathscr{S} = \varnothing$). Therefore the image of the original charge line, which lies spatially at the antipodal S^3 point, must have the opposite (i.e., negative) charge to the original one. The continuation of the original Coulomb field across $\mathscr{I}$, therefore, must be a Coulomb field of the opposite charge, which makes the completed field have anti-Grgin behaviour.

In the linearized Schwarzschild solution we again get anti-Grgin behaviour for the field ϕ_{ABCD}. Here the Grgin index is 2, so ordinary spin-2 wave fields change sign at $\mathscr{I}$, whereas the linearized Schwarzschild field does not. However, we shall see in (9.6.40) that there is an extra Ω-factor in passing from ϕ_{ABCD} to the Weyl curvature Ψ_{ABCD}, so in the *full* theory we find the *opposite* behaviour, namely that there *is* a sign change in the mass as

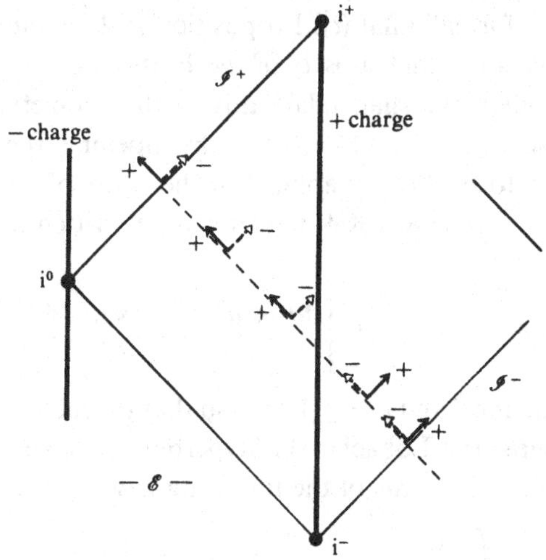

Fig. 9-14. The anti-Grgin behaviour of the Coulomb field. (The Grgin index is zero since this is a Maxwell field.)

the Schwarzschild solution is continued analytically across $\mathscr{I}$ (*cf.* §9.6, paragraph after that containing (9.6.7)).

Finally we may remark that there are indeed many non-singular solutions of the massless free-field equation over the whole of $\mathscr{E}$, for each integral or half-integral spin. For example, any appropriate initial data could be used on a spacelike S^3 cross-section of $\mathscr{E}$, and the field will evolve non-singularly to the whole of $\mathscr{E}$. The Grgin theorem tells us that the solutions have a curious kind of periodicity on $\mathscr{E}$. (This is a feature of the particular field equations being used here. For example, in 'massless $\lambda\phi^4$ theory', with field equation $(\square + \frac{1}{6}R)\phi = \lambda\phi^3$, no such periodicity would occur.) Explicit non-singular solutions on $\mathbb{M}^\#$ are easy to construct using twistor methods, e.g., the elementary states arising from the twistor function (6.10.48). The Grgin index $2s\hbar^{-1} + 2$ for massless fields is, as might be expected, determined by the homogeneity $-(2s\hbar^{-1} + 2)$ that appears in the twistor functions for massless fields of helicity s. But the matter will not be pursued further here.

9.5 Cosmological models and their twistors

Before examining the asymptotic structure of generally curved space-times, it will be of interest to consider that of the standard Friedmann–Robertson–Walker (FRW) cosmological models (*cf.* Rindler 1977). There is, by now, reasonably impressive observational evidence that the structure of the actual universe is very well approximated by such a model. The FRW models are all conformally flat (*cf.* Proposition (8.2.2)) and can therefore be represented as conformal subsets of the Einstein cylinder $\mathscr{E}$ (which is itself a FRW model). We shall relate this to the geometry and algebra that has been set up in §§9.1–9.3, thereby opening the way for the twistor formalism to be directly applied to the study of these models.

The metric of the general FRW model can be written in the standard form

$$\mathrm{d}s^2 = \mathrm{d}U^2 - [R(U)]^2\left\{\frac{\mathrm{d}q^2 + q^2(\mathrm{d}\theta^2 + \sin^2\theta\,\mathrm{d}\phi^2)}{(1 + \frac{1}{4}kq^2)^2}\right\}, \tag{9.5.1}$$

where U is 'cosmic time' and $k = \pm 1$ or 0, so that the metric in $\{\ \ \}$ is that of the unit 3-sphere, unit Lobachevski (hyperbolic) 3-space, or Euclidean 3-space, respectively. In terms of the following new coordinates

$$\tau = \int\frac{\mathrm{d}U}{R(U)}, \quad \rho = 2\tan^{-1}\tfrac{1}{2}q \quad (k = 1) \tag{9.5.2)(a}$$

or

$$\sigma = \int \frac{\mathrm{d}U}{R(U)}, \quad \mu = 2\tanh^{-1}\tfrac{1}{2}q \ (k = -1) \qquad (9.5.2)(b)$$

or

$$t = \int \frac{\mathrm{d}U}{R(U)}, \quad r = q \qquad (k = 0), \qquad (9.5.2)(c)$$

we obtain, respectively, the following alternative forms of these metrics:

$$\mathrm{d}s^2 = [S(\tau)]^2\{\mathrm{d}\tau^2 - \mathrm{d}\rho^2 - \sin^2\rho\,\mathrm{d}\omega^2\} \qquad (9.5.3)(a)$$

or

$$\mathrm{d}s^2 = [S(\sigma)]^2\{\mathrm{d}\sigma^2 - \mathrm{d}\mu^2 - \sinh^2\mu\,\mathrm{d}\omega^2\} \qquad (9.5.3)(b)$$

or

$$\mathrm{d}s^2 = [S(t)]^2\{\mathrm{d}t^2 - \mathrm{d}r^2 - r^2\mathrm{d}\omega^2\}, \qquad (9.5.3)(c)$$

where

$$\mathrm{d}\omega^2 = \mathrm{d}\theta^2 + \sin^2\theta\,\mathrm{d}\phi^2$$

and

$$R(U) = S(\tau), S(\sigma), S(t), \text{ respectively.}$$

The $\{\ \ \}$ in (9.5.3)(a) is the metric of the Einstein cylinder $\mathscr{E}$ as described in §9.1. We have already seen, in that section, how to relate $\mathscr{E}$ conformally to the metric of Minkowski space $\mathbb{M}$, which is the $\{\ \ \}$ in (c). With the obvious modifications we can also find corresponding formulae relating the metric of the *anti-Einstein universe* $\mathscr{A}$, which is the $\{\ \ \}$ in (b), to Minkowski space and hence find an explicit conformal map from $\mathscr{A}$ to $\mathscr{E}$. Collecting these results together, we obtain

$$(\cos\tau + \cos\rho)^{-2}\{\mathrm{d}\tau^2 - \mathrm{d}\rho^2 - \sin^2\rho\,\mathrm{d}\omega^2\}$$
$$= (\cosh\sigma + \cosh\mu)^{-2}\{\mathrm{d}\sigma^2 - \mathrm{d}\mu^2 - \sinh^2\mu\,\mathrm{d}\omega^2\} = \mathrm{d}t^2 - \mathrm{d}r^2 - r^2\mathrm{d}\omega^2, \qquad (9.5.4)$$

where

$$t = \frac{\sin\tau}{\cos\tau + \cos\rho} = \frac{\sinh\sigma}{\cosh\sigma + \cosh\mu}$$

$$r = \frac{\sin\rho}{\cos\tau + \cos\rho} = \frac{\sinh\mu}{\cosh\sigma + \cosh\mu}$$

$$\tan\tau = \frac{2t}{r^2 - t^2 + 1} = \frac{\sinh\sigma}{\cosh\mu}$$

$$\tan\rho = \frac{2r}{t^2 - r^2 + 1} = \frac{\sinh\mu}{\cosh\sigma}$$

(9.5.5 continued overleaf)

(continued)

$$\tanh\sigma = \frac{2t}{t^2 - r^2 + 1} = \frac{\sin\tau}{\cos\rho}$$

$$\tanh\mu = \frac{2r}{r^2 - t^2 + 1} = \frac{\sin\rho}{\cos\tau}. \tag{9.5.5}$$

The τ, ρ coordinates are the same as those used for the Einstein cylinder in §9.1 and relate to the Minkowski coordinates t, r by the equations of §9.1.

We saw in Figs. 9-1 and 9-2 the range of the variables p, q (and therefore τ, ρ) that correspond to the entire Minkowski space $\mathbb{M}$. In this range the conformal factor

$$\cos\tau + \cos\rho$$

from $\mathbb{M}$ to $\mathscr{E}$ is positive, the boundary being defined where this factor vanishes ($\rho \pm \tau = \pi$). In the same way that $\mathbb{M}$ is conformal only to a *portion* of $\mathscr{E}$, the anti-Einstein space $\mathscr{A}$ is also conformal only to a portion of $\mathscr{E}$. In this portion the conformal factor

$$(\cosh\sigma + \cosh\mu)^{-1}$$

from $\mathscr{A}$ to $\mathbb{M}$ is positive and vanishes on the boundary, defined by σ and μ becoming infinite:

$$r \pm t = 1, \quad \text{i.e.,} \quad \rho \pm \tau = \tfrac{1}{2}\pi.$$

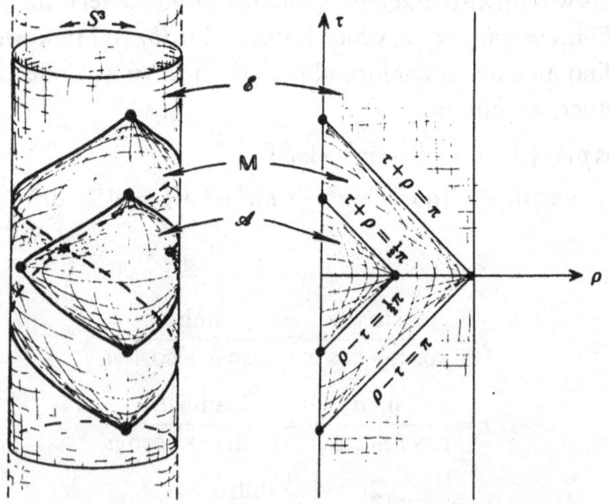

Fig. 9-15. The complete conformal spaces for various k-values: the Einstein cylinder $\mathscr{E}$ ($k = +1$), Minkowski space $\mathbb{M}$ ($k = 0$), the anti-Einstein space $\mathscr{A}$ ($k = -1$). The various conformal regions of $\mathscr{E}$ are indicated in terms of the ρ, τ coordinates for $\mathscr{E}$. The lines of constant time and of 'fundamental observer' world-lines are sketched in each case.

The nested regions $\mathscr{A} \subset \mathbb{M} \subset \mathscr{E}$ of this conformal mapping are illustrated in Fig. 9-15.

Cosmological horizons

For any particular cosmological model with a given function $R(U)$, only a certain range of the cosmic time variable U may be allowed. Then the model is conformal to some region of $\mathscr{E}$ bounded by hypersurfaces $\tau = \text{constant}$, $t = \text{constant}$, or $\sigma = \text{constant}$, as the case may be. Except when these constants are 'infinity' (in which case we have a boundary as already illustrated in Fig. 9-15), the bounding hypersurfaces are always spacelike and correspond to the existence of *particle horizons* or *event horizons* according as these boundaries are in the past or in the future. (Rindler 1956, Penrose 1964*b*, 1968*b*; *cf.* Hawking and Ellis 1973). The particle* [event] horizons are the boundaries of the TIFs [*or* TIPs] of maximally extended (idealized) galaxy world-line ($q, \theta, \phi = \text{constant}$: the 'fundamental observers'). The bounding hypersurfaces of the model can either represent *infinity* for the original cosmological space–time $\mathscr{M}$ (corresponding to $R(U) = \infty$, with $\int \mathrm{d}U/R(U)$ convergent, and a zero conformal factor in the passage from $\mathscr{M}$ to $\mathscr{E}$) or an infinite compression singularity (corresponding to $R(U) = 0$, with $\int \mathrm{d}U/R(U)$ convergent, and an infinite conformal factor in the passage from $\mathscr{M}$ to $\mathscr{E}$). A past singular boundary is referred to as the *big bang* and a future singular boundary as the *big crunch*. In all cases, whether the boundary represents a singularity or infinity, we may use the method described at the end of §9.1 (TIPs and TIFs) to provide an *intrinsic* definition of these boundary points in terms of constructions entirely within the original space–time $\mathscr{M}$. The relation between these spacelike boundaries and the physical properties of the horizons becomes especially transparent when viewed in this way (Penrose 1968*b*).

In Fig. 9-16 the regions corresponding to the three standard dust-filled Friedmann models with vanishing cosmological constant λ are illustrated. The 'height' of the cylinder in the case $k = +1$ is such that an observer following the $\rho, \theta, \phi = \text{constant}$ flow lines is just about to 'see' his creation event when he reaches the big crunch, the information having circled the universe exactly once. At the halfway point of maximum expansion he

* In Hawking and Ellis (1973), the term 'creation horizon' is preferred for such a TIF boundary.

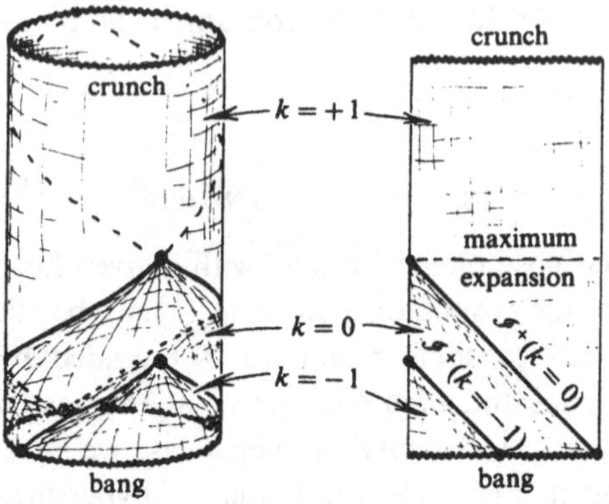

Fig. 9-16. The regions of Fig. 9-15 which correspond to the various dust-filled Friedmann models ($\lambda = 0$).

sees his antipodal point for the first time, so from that stage on all galaxies* are in view. (Proofs of these statements as well as of those to be given in the next paragraph follow readily from the standard equations as given, for example, in Rindler 1977, Wald 1984).

We consider a few other examples. The Tolman radiation-filled universes with $\lambda = 0$ (Tolman 1934: p. 413) are similar to the Friedmann cases, but for $k = +1$ the cylinder is only half as tall, so the observer only just sees the antipodal galaxy at the big crunch. The de Sitter universe (about which we shall have more to say presently) is conformally similar to the $k = +1$ Tolman universe, but the boundaries represent infinity rather than singularities. The Eddington–Lemaître model corresponds to a semi-infinite Einstein cylinder with future spacelike boundary representing infinity. The Lemaître model corresponds to a finite Einstein cylinder which can be made arbitrarily long, the past boundary being a singularity, the future boundary representing infinity. For models with $\lambda > 0$, infinity is always a spacelike boundary (see (9.6.18) below). 'Normal' FRW models with $\lambda < 0$ (which exclude, for example, the *static* anti-Einstein space $\mathscr{A}$ and the *empty* maximally extended anti-de Sitter space) do not have (temporal) infinities: they are always bounded by a spacelike bang and crunch. The (empty) Milne model is conformal to the whole of $\mathscr{A}$. (See Bondi 1960*a*, Hawking and Ellis 1973, Rindler 1977 and Wald 1984 for further details.) Anti-de Sitter space infinity is *timelike*: $\rho = \frac{1}{2}\pi$.

* Of course, as is the usual convention when discussing cosmology at this level, the term 'galaxy' refers to an idealized world-line originating at the big bang, and makes no reference to the time at which actual galaxies might first form.

Description in terms of $\mathbb{P}^5$

Let us now relate these correspondences to the discussion given in §9.3 where $\mathbb{M}^\#$ was regarded as a quadric hypersurface in $\mathbb{P}^5$. Directly translating (9.5.5) to the coordinates of that section, we find

$$\tan\tau = T/V,$$
$$\tanh\sigma = -T/W,$$
$$t = T/(V-W). \qquad (9.5.6)$$

Thus in each case the surfaces of constant cosmic time U are described by the intersections of a pencil of hyperplanes with the quadric $\mathbb{M}^\#$, namely:

$$T:V = \sin\tau:\cos\tau \qquad (k=1)$$
$$T:-W = \sinh\sigma:\cosh\sigma \qquad (k=-1)$$
$$T:V-W = t:1 \qquad (k=0). \qquad (9.5.7)$$

In the first case the (3-projective-dimensional) axis of the pencil ($T=V=0$) does not intersect the quadric $\mathbb{M}^\#$ (whose equation, we recall, is $T^2+V^2-W^2-X^2-Y^2-Z^2=0$), in the second case the axis ($T=W=0$) intersects $\mathbb{M}^\#$ in an S^2, and in the third case the axis ($T=0$, $V=W$) touches $\mathbb{M}^\#$ at one point (see Fig. 9-17.)

The de Sitter and anti-de Sitter models; $\mathsf{I}_{\alpha\beta}$ *and* $\mathsf{I}^{\alpha\beta}$

Worthy of some special attention among FRW models are de Sitter and anti-de Sitter space together with Minkowski space. For in these spaces (and only in these) there are extended symmetry groups so that the cosmic time slices $U=$ constant are not geometrically singled out. Indeed, it turns out (Schrödinger 1956, Rindler 1977) that descriptions of de Sitter space as FRW models can be given, for which $k=1$, -1, or 0 – though only in the case $k=1$ does the description hold globally. For Minkowski space we can have FRW models with $k=0$ or $k=-1$ (only the case $k=0$ being

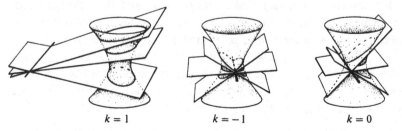

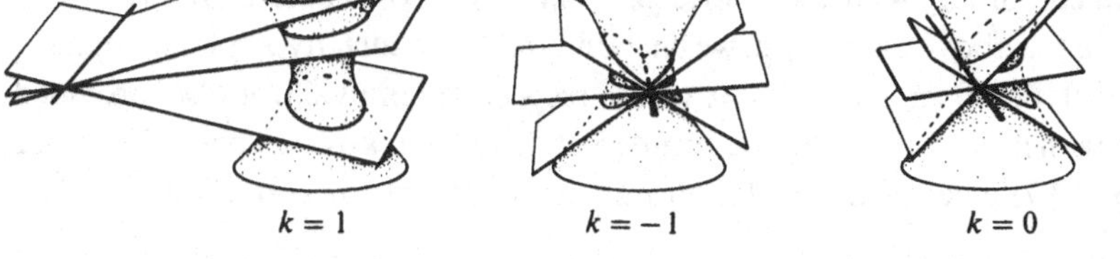

Fig. 9-17. The various constant-time slicings for $k=1$, $k=-1$, $k=0$, respectively, in terms of the projective quadric $\mathbb{P}\mathscr{K}$ in $\mathbb{P}^5$.

global). Anti-de Sitter space permits but one type of FRW model, and that not globally: $k = -1$.

One way to describe the complete de Sitter model is simply to use the T, V, W, X, Y, Z coordinates of §9.2, but instead of taking them as projective coordinates for $\mathbb{P}$, to *restrict* them—in $\mathbb{E}^6$ with metric (9.2.4)—to some hyperplane

$$T = Q, \tag{9.5.8}$$

where Q is a real constant. The equation (9.2.5) for $\mathbb{M}^\#$ then yields de Sitter space $\mathcal{M}$ as the 'pseudosphere'

$$V^2 - W^2 - X^2 - Y^2 - Z^2 = -Q^2,$$

of 'radius' Q. In $\mathbb{P}^5$, just the points given by $T = 0$ are now excluded from the model, so by removing the hyperplane $T = 0$ from $\mathbb{M}^\#$ one gets a space conformal to $\mathcal{M}$. This is very similar to the case of $\mathbb{M} \subset \mathbb{M}^\#$, the only difference being that, for $\mathbb{M}$, the removed hyperplane $V - W = 0$ in $\mathbb{P}^5$ touches $\mathbb{M}^\#$, the point of contact being I.

The case of anti-de Sitter space is also similar, but here we restrict the coordinates in $\mathbb{E}^6$ to a hyperplane.

$$W = Q, \tag{9.5.9}$$

so we get a pseudosphere of a different signature (still taking the metric (9.2.4) for $\mathbb{E}^6$):

$$T^2 + V^2 - X^2 - Y^2 - Z^2 = Q^2.$$

In $\mathbb{P}^5$, $\mathcal{M}$ is now given by the removal of the hyperplane $W = 0$ from $\mathbb{M}^\#$. (Strictly, anti-de Sitter space is the *universal covering space* of this $\mathcal{M}$, *cf.* Penrose 1968*b*, Hawking and Ellis 1973.)

A hyperplane $\mathcal{H}$ in $\mathbb{P}^5$ can be represented twistorially by a skew twistor $\mathsf{H}_{\alpha\beta}$ (which is 'real' in the normal twistor sense that its twistor complex conjugate $\bar{\mathsf{H}}^{\alpha\beta}$ and its dual $\mathsf{H}^{\alpha\beta}$ are equal). Of course we may also think of $\mathsf{H}^{\alpha\beta}$ as representing a *point* H in $\mathbb{P}^5$, but this point is simply the *pole* of $\mathcal{H}$ with respect to the quadric $\mathbb{M}^\#$. The point H and the hyperplane $\mathcal{H}$ represent equivalent information. (We recall that this is what 'raising the indices' of $\mathsf{H}_{\alpha\beta}$ with the metric $\frac{1}{2}\varepsilon^{\alpha\beta\gamma\delta}$ amounts to, geometrically.)

In the above discussion, we encountered three such hyperplanes, namely $T = 0$, $W = 0$, and $V - W = 0$, whose respective removal from $\mathbb{M}^\#$ provided us with models for de Sitter, anti-de Sitter, and Minkowski space. Reference to (9.3.7) gives us a translation into standard twistor coordinates:

$$T = \frac{\mathrm{i}}{\sqrt{2}}(\mathsf{R}^{12} - \mathsf{R}^{03}), \quad W = \mathsf{R}^{01} - \tfrac{1}{2}\mathsf{R}^{23}, \quad V - W = \mathsf{R}^{23}. \tag{9.5.10}$$

Thus we can define the region to be removed from $\mathbb{M}^\#$ (writing $\mathsf{I}_{\alpha\beta}$ for $\mathsf{H}_{\alpha\beta}$) as

$$\mathsf{I}_{\alpha\beta}\mathsf{R}^{\alpha\beta}=0, \tag{9.5.11}$$

where

$$\mathsf{I}_{\alpha\beta}=\frac{\mathrm{i}}{Q\sqrt{2}}\begin{bmatrix} 0&0&0&-1\\ 0&0&1&0\\ 0&-1&0&0\\ 1&0&0&0\end{bmatrix},\quad \frac{1}{Q}\begin{bmatrix} 0&1&0&0\\ -1&0&0&0\\ 0&0&0&-\tfrac{1}{2}\\ 0&0&\tfrac{1}{2}&0\end{bmatrix},$$

$$\begin{bmatrix} 0&0&0&0\\ 0&0&0&0\\ 0&0&0&1\\ 0&0&-1&0\end{bmatrix}, \tag{9.5.12}$$

in the de Sitter, anti-de Sitter, and Minkowski cases, respectively. The last expression for $\mathsf{I}_{\alpha\beta}$ is, of course, the standard one, (6.2.25), giving the Minkowski infinity twistor. But now we have corresponding *infinity twistors for de Sitter and anti-de Sitter space–times.* Moreover, the particular scalings given in (9.5.12) enable us to go further and define the actual *metric* of the respective space–times by imposing, in $\mathbb{E}^6$, the equation

$$\mathsf{I}_{\alpha\beta}\mathsf{R}^{\alpha\beta}=2 \tag{9.5.13}$$

(*cf.* (9.5.8), (9.5.9), (9.2.6), respectively, and (9.5.12); note that this is the standard normalization (6.2.27) in the Minkowski case).

We can interpret (9.5.13) as defining a subset (a hyperplane section) of the cone $\mathscr{K}$ (of Fig. 9-6), $\mathscr{K}$ having equation (9.2.5), i.e.,

$$2(T^2+V^2-W^2-X^2-Y^2-Z^2)\equiv\mathsf{R}^{\alpha\beta}\mathsf{R}_{\alpha\beta}\equiv\tfrac{1}{2}\varepsilon_{\alpha\beta\gamma\delta}\mathsf{R}^{\alpha\beta}\mathsf{R}^{\gamma\delta}=0, \tag{9.5.14}$$

in the space $\mathbb{E}^6$. The metric of $\mathbb{E}^6$ is (9.2.4), i.e.,

$$\begin{aligned} ds^2&=2\mathrm{d}\mathsf{R}^{01}\mathrm{d}\mathsf{R}^{23}+2\mathrm{d}\mathsf{R}^{02}\mathrm{d}\mathsf{R}^{31}+2\mathrm{d}^{03}\mathrm{d}\mathsf{R}^{12}\\ &=\tfrac{1}{4}\varepsilon_{\alpha\beta\gamma\delta}\mathrm{d}\mathsf{R}^{\alpha\beta}\mathrm{d}\mathsf{R}^{\gamma\delta}=\tfrac{1}{2}\mathrm{d}\mathsf{R}^{\alpha\beta}\mathrm{d}\mathsf{R}_{\alpha\beta}\\ &=\mathrm{d}T^2+\mathrm{d}V^2-\mathrm{d}W^2-\mathrm{d}X^2-\mathrm{d}Y^2-\mathrm{d}Z^2, \end{aligned} \tag{9.5.15}$$

and we have the standard reality condition $\bar{\mathsf{R}}_{\alpha\beta}=\tfrac{1}{2}\varepsilon_{\alpha\beta\gamma\delta}\mathsf{R}^{\gamma\delta}$. As we saw in §9.2, all sections of $\mathscr{K}$ are locally conformally identical (conformally flat), the selection of a particular metric from the entire conformal class being determined simply by specifying the section. In the cases of de Sitter, anti-de Sitter and Minkowski space, this section has the special feature of being given by a *linear* equation, namely (9.5.13), where (to preserve reality) $\mathsf{I}_{\alpha\beta}$ must be twistor-real,

$$\bar{\mathsf{I}}^{\alpha\beta}=\tfrac{1}{2}\varepsilon^{\alpha\beta\gamma\delta}\mathsf{I}_{\gamma\delta}=:\mathsf{I}^{\alpha\beta}, \tag{9.5.16}$$

and, without loss of generality, can be taken to be skew,

$$\mathsf{I}_{\alpha\beta} = -\mathsf{I}_{\beta\alpha}. \tag{9.5.17}$$

The de Sitter, anti-de Sitter and Minkowski cases are distinguished by

$$\mathsf{I}_{\alpha\beta}\mathsf{I}^{\alpha\beta} = \frac{2}{Q^2},\ -\frac{2}{Q^2},\ 0, \tag{9.5.18}$$

respectively, or equivalently by

$$\mathsf{I}_{\alpha\gamma}\mathsf{I}^{\beta\gamma} = \frac{1}{2Q^2}\delta^\beta_\alpha,\ -\frac{1}{2Q^2}\delta^\beta_\alpha,\ 0. \tag{9.5.19}$$

Having relations (9.5.16)–(9.5.18), we can dispense with explicit representations such as (9.5.12). In each case, the relevant symmetry group ([anti-] de Sitter group, Poincaré group) arises as the subgroup of $O(2,4)$ leaving $\mathsf{I}_{\alpha\beta}$ invariant (*cf.* penultimate paragraph of §9.2). The specially simple form of the metric in these cases also enables us to write down directly the actual geodetic distance (= time-interval) between points represented by $\mathsf{R}^{\alpha\beta}$, $\mathsf{P}^{\alpha\beta}$ as

$$Q\cosh^{-1}\left(1 - \frac{\mathsf{P}_{\alpha\beta}\mathsf{R}^{\alpha\beta}}{2Q^2}\right),\quad Q\cos^{-1}\left(1 + \frac{\mathsf{P}_{\alpha\beta}\mathsf{R}^{\alpha\beta}}{2Q^2}\right) \tag{9.5.20}$$

in the de Sitter and anti-de Sitter cases respectively. (This can be verified by reverting to a $T, V, \ldots, Z$ description.) These formulae may be compared with the corresponding Minkowski formula (6.2.30), which can be re-obtained from (9.5.20) by taking the limit $Q \to \infty$. There is also a version of (9.5.20) that corresponds to (6.2.26), which does not require the normalization (9.5.13) and so refers directly to descriptions in $\mathbb{RP}^5$. This can be written down simply by replacing the expressions in parentheses in (9.5.20) by

$$\left(1 - \frac{\mathsf{P}_{\alpha\beta}\mathsf{R}^{\alpha\beta}\mathsf{I}_{\gamma\delta}\mathsf{I}^{\gamma\delta}}{\mathsf{P}_{\rho\sigma}\mathsf{I}^{\rho\sigma}\mathsf{I}_{\tau\kappa}\mathsf{R}^{\tau\kappa}}\right). \tag{9.5.21}$$

The functions I and $\tilde{I}$ for conformally flat space–times

Let us now turn to a general conformally flat space–time $\mathscr{M}$ (compare also Hurd 1985). As before, we can define the scaling for the metric of $\mathscr{M}$ by specifying a section (but now not a hyperplane section) through the cone $\mathscr{K}$ in $\mathbb{E}^6$. Let us write the equation of this section as

$$\tilde{I}(\mathsf{R}^{\alpha\beta}) = 2, \tag{9.5.22}$$

where the function $\tilde{I}$ is homogeneous of degree 1 (so that in the above three cases, $\tilde{I}(\mathsf{R}^{\alpha\beta}) \equiv \mathsf{I}_{\alpha\beta}\mathsf{R}^{\alpha\beta}$), or, in terms of the dual twistor $\mathsf{R}_{\alpha\beta}$,

$$I(\mathsf{R}_{\alpha\beta}) = 2, \tag{9.5.23}$$

where I is also homogeneous of degree 1 and is defined by

$$I(\mathsf{R}_{\alpha\beta}) = \tilde{I}(\mathsf{R}^{\alpha\beta}) \tag{9.5.24}$$

(so in the above three cases, $I(\mathsf{R}_{\alpha\beta}) \equiv \mathsf{I}^{\alpha\beta}\mathsf{R}_{\alpha\beta}$). The reality of our section can be stated as $\bar{I} = \tilde{I}$, i.e.,

$$\overline{I(\mathsf{X}_{\alpha\beta})} = \tilde{I}(\bar{\mathsf{X}}^{\alpha\beta}). \tag{9.5.25}$$

In fact, we are not really concerned with the functions I and $\tilde{I}$ except at the places where $\mathsf{R}_{\alpha\beta}$ is *simple*. For this reason we shall prefer to consider them as functions of *pairs of univalent twistors* U_α, V_α or X^α, Y^α, where

$$\begin{aligned} I(\mathsf{U}_\alpha, \mathsf{V}_\alpha) &:= I(\mathsf{U}_\alpha\mathsf{V}_\beta - \mathsf{U}_\beta\mathsf{V}_\alpha), \\ \tilde{I}(\mathsf{X}^\alpha, \mathsf{Y}^\alpha) &:= I(\mathsf{X}^\alpha\mathsf{Y}^\beta - \mathsf{X}^\beta\mathsf{Y}^\alpha). \end{aligned} \tag{9.5.26}$$

Since I and $\tilde{I}$ depend on their arguments U_α, V_α or X^α, Y^α only through their skew products, and are homogeneous of degree 1 in each argument, we have

$$\begin{aligned} &\mathsf{U}_\alpha\frac{\partial I}{\partial \mathsf{V}_\alpha} = 0 = \mathsf{V}_\alpha\frac{\partial I}{\partial \mathsf{U}_\alpha}, \quad \mathsf{U}_\alpha\frac{\partial I}{\partial \mathsf{U}_\alpha} = I = \mathsf{V}_\alpha\frac{\partial I}{\partial \mathsf{V}_\alpha}, \\ &\mathsf{X}^\alpha\frac{\partial \tilde{I}}{\partial \mathsf{Y}^\alpha} = 0 = \mathsf{Y}^\alpha\frac{\partial \tilde{I}}{\partial \mathsf{X}^\alpha}, \quad \mathsf{X}^\alpha\frac{\partial \tilde{I}}{\partial \mathsf{X}^\alpha} = I = \mathsf{Y}^\alpha\frac{\partial \tilde{I}}{\partial \mathsf{Y}^\alpha} \end{aligned} \tag{9.5.27}$$

(see the extended 'footnote' on Young tableaux described in §3.3), or, equivalently,

$$\begin{aligned} I(\lambda\mathsf{U}_\alpha + \mu\mathsf{V}_\alpha, \rho\mathsf{U}_\alpha + \sigma\mathsf{V}_\alpha) &= (\lambda\sigma - \mu\rho)I(\mathsf{U}_\alpha, \mathsf{V}_\alpha), \\ \tilde{I}(\lambda\mathsf{X}^\alpha + \mu\mathsf{Y}^\alpha, \rho\mathsf{X}^\alpha + \sigma\mathsf{Y}^\alpha) &= (\lambda\sigma - \mu\rho)\tilde{I}(\mathsf{X}^\alpha, \mathsf{Y}^\alpha). \end{aligned} \tag{9.5.28}$$

Writing I and $\tilde{I}$ as functions of two twistor variables in this way enables us to give them another, and perhaps more significant, interpretation. The above relations, particularly (9.5.28), tell us that I and $\tilde{I}$ define skew bilinear forms on the linear spans of U_α, V_α and X^α, Y^α, respectively. These linear spans determine the point R of $\mathscr{M}$ defined by

$$\mathsf{R}_{\alpha\beta} = 2\mathsf{U}_{[\alpha}\mathsf{V}_{\beta]} \quad \text{or} \quad \mathsf{R}^{\alpha\beta} = 2\mathsf{X}^{[\alpha}\mathsf{Y}^{\beta]}$$

and the bilinear forms provide the spinors ε^{AB} and $\varepsilon^{A'B'}$, respectively, at R. For, using a local twistor description of $\mathsf{U}_\alpha, \ldots, \mathsf{Y}^\alpha$, we have, at R,

$$\begin{aligned} I(\mathsf{U}_\alpha, \mathsf{V}_\alpha) &= \varepsilon^{AB}\mathsf{U}_A\mathsf{V}_B \\ \tilde{I}(\mathsf{X}^\alpha, \mathsf{Y}^\alpha) &= \varepsilon^{A'B'}\mathsf{X}_{A'}\mathsf{Y}_{B'}. \end{aligned} \tag{9.5.29}$$

(These expressions can be obtained first in the Minkowski case from (6.2.25), and can then be rescaled to give the results for $\mathscr{M}$.)

The expressions (9.5.29) are useful in that they enable us to apply various twistor formulae directly to any conformally flat space $\mathcal{M}$. In particular, the massless free-field contour integrals of §6.10 can be applied in $\mathcal{M}$, the metric scaling for $\mathcal{M}$ entering only through the differential forms

$$\lambda_A \mathrm{d}\lambda^A = \varepsilon^{AB}\lambda_A \mathrm{d}\lambda_B, \quad \mathrm{d}\lambda_A \wedge \mathrm{d}\lambda^A = \varepsilon^{AB}\mathrm{d}\lambda_A \wedge \mathrm{d}\lambda_B$$
$$\pi_{A'}\mathrm{d}\pi^{A'} = \varepsilon^{A'B'}\pi_{A'}\mathrm{d}\pi_{B'}, \quad \mathrm{d}\pi_{A'} \wedge \mathrm{d}\pi^{A'} = \varepsilon^{A'B'}\mathrm{d}\pi_{A'} \wedge \mathrm{d}\pi_{B'} \qquad (9.5.30)$$

at each space–time point R (*cf.* (6.10.1), (6.10.3), (6.10.10)). Thus, the functions I and $\tilde{I}$ provide the required definitions of ε^{AB} and $\varepsilon^{A'B'}$ at each point R via the expressions (9.5.29).

FRW models; their bang and crunch twistors

Let us now return to the FRW models and determine I and $\tilde{I}$ explicitly for them. One method of doing this is to transform the metric $\mathrm{d}s^2$ of $\mathcal{M}$ to the Minkowski form, where we now write $\mathrm{d}\hat{s}^2 = \Omega^2 \mathrm{d}s^2$ for the *Minkowski* metric (9.5.4) and note that, in accordance with $\hat{\varepsilon}^{AB} = \Omega^{-1}\varepsilon^{AB}$, $\hat{\varepsilon}^{A'B'} = \Omega^{-1}\varepsilon^{A'B'}$, we require

$$\hat{I} = \Omega^{-1} I, \quad \hat{\tilde{I}} = \Omega^{-1}\tilde{I} \qquad (9.5.31)$$

with the Minkowski expressions

$$\hat{\tilde{I}}(\mathsf{R}^{\alpha\beta}) = 2\mathsf{R}^{23} = 2(V - W) = 2\mathsf{R}_{01} = \hat{I}(\mathsf{R}_{\alpha\beta}) \qquad (9.5.32)$$

(*cf.* (9.5.10), (9.5.12)). Reference to (9.5.3) and (9.5.4) provides

$$\Omega^{-1} = \begin{cases} (\cos\tau + \cos\rho)S(\tau) & (k=1) \\ (\cosh\sigma + \cosh\mu)S(\sigma) & (k=-1) \\ S(t) & (k=0), \end{cases} \qquad (9.5.33)$$

while from (9.5.5), (9.5.6) we obtain

$$\cos^2\rho = \frac{W^2}{V^2 + T^2}, \qquad \cos^2\tau = \frac{V^2}{V^2 + T^2}$$
$$\cosh^2\mu = \frac{V^2}{W^2 - T^2}, \quad \cosh^2\sigma = \frac{W^2}{W^2 - T^2}. \qquad (9.5.34)$$

Since in the relevant range of variables, for $k = 1$, $\cos\tau$ and V have the same sign while $\cos\rho$ and W have opposite signs, whereas for $k = -1$, $V > 0$ and $W < 0$, we finally obtain, using (9.5.6),

$$I(\mathsf{R}_{\alpha\beta}) = \tilde{I}(\mathsf{R}^{\alpha\beta}) = \begin{cases} 2(V^2 + T^2)^{\frac{1}{2}}/S(\tan^{-1}(T/V)) & (k=1) \\ 2(W^2 - T^2)^{\frac{1}{2}}/S(\tanh^{-1}(-T/W)) & (k=-1) \\ 2(V - W)/S(T/(V - W)) & (k=0). \end{cases} \qquad (9.5.35)$$

Note that $I = 0$ at $\mathscr{I}^{\pm}$, and $I = \infty$ at the singularities.

In these expressions, T, V and W are simply combinations of particular components of $\mathrm{R}^{\alpha\beta}$ (*cf.* (9.3.8)):

$$T=\frac{\mathrm{i}}{\sqrt{2}}(\mathrm{R}^{12}-\mathrm{R}^{03})=\frac{\mathrm{i}}{\sqrt{2}}(\mathrm{R}_{03}-\mathrm{R}_{12}) \quad V=\mathrm{R}^{01}+\tfrac{1}{2}\mathrm{R}^{23}=\mathrm{R}_{23}+\tfrac{1}{2}\mathrm{R}_{01}$$

$$W=\mathrm{R}^{01}-\tfrac{1}{2}\mathrm{R}^{23}=\mathrm{R}_{23}-\tfrac{1}{2}\mathrm{R}_{01} \quad V-W=\mathrm{R}^{23}=\mathrm{R}_{01}. \tag{9.5.36}$$

These components need have no absolute significance in themselves. They represent scalar products of $\mathrm{R}^{\alpha\beta}$ (or $\mathrm{R}_{\alpha\beta}$) with particular skew twistors $\mathrm{H}_{\alpha\beta}$ (or $\mathrm{H}^{\alpha\beta}$) corresponding to particular members of the pencil of hyperplanes in $\mathbb{P}^5$ (defined in (9.5.7)) whose intersections with $\mathbb{M}^{\#}$ provide the cosmic time slices $U=$ constant of $\mathcal{M}$. The geometry, in relation to $\mathbb{M}^{\#}$, of the two relevant hyperplanes (i.e., $T=0$ and $V=0$ when $k=1$, $T=0$ and $W=0$ when $k=-1$, and $T=0$ and $V-W=0$ when $k=0$) together with the explicit expressions (9.5.35) provides us with the required information about the metric on $\mathcal{M}$.

There is some arbitrariness, however, in the particular selections being made here of these hyperplanes from the pencil, which stems from the arbitrary constant involved in the original integrals (9.5.2). If the model contains a 'big bang' then we may choose to take the zero of the coordinates τ, σ, or t to represent the big bang, i.e., by (9.5.6),

$$T=0.$$

We can thus define a *bang twistor* $\mathrm{B}_{\alpha\beta}$ such that

$$\mathrm{B}_{\alpha\beta}\mathrm{R}^{\alpha\beta}=0 \quad \text{at the big bang,}$$

as well as

$$\mathrm{B}_{\alpha\beta}=-\mathrm{B}_{\beta\alpha}, \quad \mathrm{B}_{\alpha\beta}\mathrm{B}^{\alpha\beta}=4, \quad \mathrm{B}_{\alpha\gamma}\mathrm{B}^{\beta\gamma}=\delta_{\alpha}^{\beta}. \tag{9.5.37}$$

By the above coordinate conventions,

$$\mathrm{B}_{\alpha\beta}=\begin{bmatrix} & & & -\mathrm{i} \\ & & \mathrm{i} & \\ & -\mathrm{i} & & \\ \mathrm{i} & & & \end{bmatrix}, \tag{9.5.38}$$

so that

$$T=\frac{1}{\sqrt{8}}\mathrm{B}_{\alpha\beta}\mathrm{R}^{\alpha\beta}. \tag{9.5.39}$$

Similarly, if the model contains a big crunch, we can define a *crunch twistor* $\mathrm{C}_{\alpha\beta}$ for which $\mathrm{C}_{\alpha\beta}\mathrm{R}^{\alpha\beta}=0$ at the big crunch and which satisfies relations corresponding to (9.5.37). In the closed Friedmann dust-filled model with $\lambda=0$ we find, in fact, that $\mathrm{C}_{\alpha\beta}=\mathrm{B}_{\alpha\beta}$ (from the property mentioned earlier that light cones originating on the big bang refocus on the big crunch).

This also holds for the closed Tolman radiation-filled model, but in the Friedmann case there is the further degeneracy that the twistor representing the cosmic time of maximum expansion is also $(-)\mathsf{B}_{\alpha\beta}$.

If we wish to describe the structure of $\mathscr{M}$ by the selection of specific twistors (such as $\mathsf{I}_{\alpha\beta}$ in the case of $\mathbb{M}$ or the (anti-)de Sitter model), there is clearly some freedom of choice. There is some virtue in choosing, say, $\mathsf{B}_{\alpha\beta}$ and a corresponding twistor representing some other member of the pencil, e.g. that describing infinity in the cases where $\mathscr{M}$ possesses a $\mathscr{I}^+$. It makes little difference which two twistors are actually selected, since all other possible choices will be linear combinations of these. For uniformity and mathematical elegance it is convenient (except in the case $k=0$) to choose the two members of the pencil which *touch* $\mathbb{CM}^\#$, so the corresponding twistors are *simple*. In the case $k=1$ this entails selecting a pair of *complex conjugate* simple skew twistors $\mathsf{I}_{\alpha\beta}$ and $\bar{\mathsf{I}}_{\alpha\beta}$; in the case $k=-1$, a pair of distinct (twistor-)real simple skew twistors $\mathsf{I}_{\alpha\beta}$ and $\mathsf{J}_{\alpha\beta}$; and in the case $k=0$, one real simple skew twistor $\mathsf{I}_{\alpha\beta}$ and (say) a real non-simple skew twistor $\mathsf{B}_{\alpha\beta}$. We can normalize as follows:

$$\begin{aligned} &\mathsf{I}_{\alpha\beta}\bar{\mathsf{I}}^{\alpha\beta}=2 \quad (k=1)\\ &\mathsf{I}_{\alpha\beta}\mathsf{J}^{\alpha\beta}=2 \quad (k=-1)\\ &\mathsf{I}_{\alpha\beta}\mathsf{B}^{\alpha\beta}=0, \quad \mathsf{B}_{\alpha\beta}\mathsf{B}^{\alpha\beta}=4 \quad (k=0) \end{aligned} \tag{9.5.40}$$

(The reason for this normalization, in the cases $k=\pm 1$, is that $\mathsf{I}_{\alpha\gamma}\bar{\mathsf{I}}^{\beta\gamma}$ and $\bar{\mathsf{I}}_{\alpha\gamma}\mathsf{I}^{\beta\gamma}$, or $\mathsf{I}_{\alpha\gamma}\mathsf{J}^{\beta\gamma}$ and $\mathsf{J}_{\alpha\gamma}\mathsf{I}^{\beta\gamma}$, are then orthogonal idempotent projection operators which serve to decompose twistor space into two canonically determined spin-spaces. When $k=-1$, these are spin-spaces of the kind studied in these volumes but when $k=1$ their relation to complex conjugation is different.) In the cases $k=\pm 1$, we can also define

$$\mathsf{B}_{\alpha\beta}=\mathsf{I}_{\alpha\beta}+\bar{\mathsf{I}}_{\alpha\beta} \quad (k=1), \quad \mathsf{B}_{\alpha\beta}=\mathsf{I}_{\alpha\beta}+\mathsf{J}_{\alpha\beta} \quad (k=-1)$$

as the bang twistor (if there is a big bang), and this provides a $\mathsf{B}_{\alpha\beta}$ with precisely the same structure as that given in (9.5.40) for $k=0$ (though its relation to $\mathsf{I}_{\alpha\beta}$ is different).

In the normal (positive-density) expanding models we take $\mathsf{I}_{\alpha\beta}$ to represent $\mathscr{I}^+$ when $k=0,-1$ and $\lambda=0$:

$$\mathsf{I}_{\alpha\beta}\mathsf{R}^{\alpha\beta}=0 \quad \text{at } U=+\infty.$$

However, when $\lambda>0$, $\mathscr{I}^+$ is spacelike and does not have such a clear relation to $\mathsf{I}_{\alpha\beta}$. (In particular, this choice does not agree with that made earlier in the case of de Sitter space!) When $k=1$ and $\lambda=0$, $\mathsf{I}_{\alpha\beta}$ and $\bar{\mathsf{I}}_{\alpha\beta}$ represent 'virtual' (complex) infinities that can be reached only by complexifying the metric. In the expanding models with $k=-1$ and $\lambda=0$, $\mathsf{J}_{\alpha\beta}$ also represents a 'virtual' infinity, but this time it is the $\mathscr{I}^-$ of a hypothetical collapsing phase preceding the big bang. (Compare Fig. 9-15 with Fig. 9-16: $\mathsf{I}_{\alpha\beta}$ represents $\rho+\tau=\frac{1}{2}\pi$ while $\mathsf{J}_{\alpha\beta}$ represents $\rho-\tau=\frac{1}{2}\pi$. We note also that the *point* $\mathsf{I}^{\alpha\beta}$ is the vertex $\rho=0,\ \tau=\frac{1}{2}\pi$ and the *point* $\mathsf{J}^{\alpha\beta}$ is $\rho=0,\ \tau=-\frac{1}{2}\pi$.)

Explicit realizations of these twistors in terms of our coordinates, with

$$T-\mathrm{i}V=\frac{1}{\sqrt{2}}\mathsf{I}_{\alpha\beta}\mathsf{R}^{\alpha\beta},\quad T+\mathrm{i}V=\frac{1}{\sqrt{2}}\bar{\mathsf{I}}_{\alpha\beta}\mathsf{R}^{\alpha\beta}\quad (k=1)$$

$$T-W=\frac{1}{\sqrt{2}}\mathsf{I}_{\alpha\beta}\mathsf{R}^{\alpha\beta},\quad T+W=\frac{1}{\sqrt{2}}\mathsf{J}_{\alpha\beta}\mathsf{R}^{\alpha\beta}\quad (k=-1)$$

$$T=\frac{1}{\sqrt{8}}\mathsf{B}_{\alpha\beta}\mathsf{R}^{\alpha\beta},\qquad V-W=\tfrac{1}{2}\mathsf{I}_{\alpha\beta}\mathsf{R}^{\alpha\beta}\qquad (k=0),\tag{9.5.41}$$

are provided by

$$\mathsf{I}_{\alpha\beta}=\begin{bmatrix} & -\mathrm{i}/\sqrt{2} & & -\mathrm{i}/2\\ \mathrm{i}/\sqrt{2} & & \mathrm{i}/2 & \\ & -\mathrm{i}/2 & & -\mathrm{i}/\sqrt{8}\\ \mathrm{i}/2 & & \mathrm{i}/\sqrt{8} & \end{bmatrix}\quad (k=1)$$

$$\left.\begin{matrix}\mathsf{I}_{\alpha\beta}\\ \mathsf{J}_{\alpha\beta}\end{matrix}\right\}=\begin{bmatrix} & \mp 1/\sqrt{2} & & -\mathrm{i}/2\\ \pm 1/\sqrt{2} & & \mathrm{i}/2 & \\ & -\mathrm{i}/2 & & \pm 1/\sqrt{8}\\ \mathrm{i}/2 & & \mp 1/\sqrt{8} & \end{bmatrix}\quad (k=-1)\tag{9.5.42}$$

and, for $k=0$, by the expression for $\mathsf{B}_{\alpha\beta}$ given in (9.5.38) and the standard Minkowskian expression for $\mathsf{I}_{\alpha\beta}$ given in (9.5.12)(3). There is, however, no special merit in these explicit representations, since different coordinates for twistor space can be chosen which considerably simplify the form of (9.5.42). (Our standard coordinatization of twistor space was chosen to mesh with the Minkowskian spinor descriptions of Chapter 6, and have no special relevance here.) The essential properties required are all contained in (9.5.40), together with the stated simplicity and reality conditions; the one remaining 'invariant' choice we have made was to take the complex point represented by $\mathsf{I}^{\alpha\beta}$ in the *forward* tube (*cf.* just before (9.3.25)) so that the corresponding line in $\mathbb{PT}$ lies in $\mathbb{PT}^+$ rather than in $\mathbb{PT}^-$.

Explicit expressions for I

The explicit form of the function

$$I=I(\mathsf{R}_{\alpha\beta})=\tilde{I}(\mathsf{R}^{\alpha\beta})$$

is now obtained by substituting (9.5.41) into (9.5.35). We can then reconstruct the original 'universe radius' function $R(U)$ from the functional form of I (*cf.* (9.5.1)–(9.5.3)). In the cases $k = \pm 1$ some simplification can be achieved by introducing

$$a^2 = V + \mathrm{i}T = \frac{\mathrm{i}}{\sqrt{2}}\mathsf{I}_{\alpha\beta}\mathsf{R}^{\alpha\beta}, \qquad b^2 = V - \mathrm{i}T = \frac{-\mathrm{i}}{\sqrt{2}}\mathsf{T}_{\alpha\beta}\mathsf{R}^{\alpha\beta} \qquad (k=1)$$

$$a^2 = -W + T = \frac{1}{\sqrt{2}}\mathsf{I}_{\alpha\beta}\mathsf{R}^{\alpha\beta}, \quad b^2 = -W - T = \frac{1}{\sqrt{2}}\mathsf{J}_{\alpha\beta}\mathsf{R}^{\alpha\beta} \quad (k=-1) \tag{9.5.43}$$

so that a and b are conjugate complex if $k = 1$ and real if $k = -1$. We note that, by (9.5.7),

$$a^2{:}b^2 = V + \mathrm{i}T{:}V - \mathrm{i}T = \mathrm{e}^{\mathrm{i}\tau}{:}\mathrm{e}^{-\mathrm{i}\tau} \qquad (k=1)$$
$$a^2{:}b^2 = -W + T{:} -W - T = \mathrm{e}^{\sigma}{:}\mathrm{e}^{-\sigma} \qquad (k=-1)$$

so

$$\mathrm{e}^{\mathrm{i}\tau} = a/b, \quad \tau = -\mathrm{i}\log(a/b) \quad (k=1)$$
$$\mathrm{e}^{\sigma} = a/b, \quad \sigma = \log(a/b) \qquad (k=-1).$$

Substituting into (9.5.35), we get

$$I = \frac{2ab}{S(\varepsilon\log(a/b))}\begin{cases}\varepsilon = -\mathrm{i} & (k=1)\\ \varepsilon = 1 & (k=-1),\end{cases} \tag{9.5.44}$$

so, with (9.5.2), (9.5.3), we can complete the calculation of $R(U)$, simply noting that

$$R = 2abI^{-1}, \quad \mathrm{d}U = 2\varepsilon b^2 I^{-1}\mathrm{d}(a/b),$$

where the constant of integration can be fixed by setting $U = 0$ at the big bang, taken to be at $a = b$.

The case of the Friedmann dust-filled models with $\lambda = 0$ is particularly simple:

$$I = -\frac{8k}{C}\left(\frac{ab}{a-b}\right)^2. \tag{9.5.45}$$

This yields the familiar parametric forms

$$U = \tfrac{1}{2}C(\tau - \sin\tau), \quad R = \tfrac{1}{2}C(1 - \cos\tau) \quad (k=1)$$
$$U = \tfrac{1}{2}C(\sinh\sigma - \sigma), R = \tfrac{1}{2}C(\cosh\sigma - 1) \quad (k=-1),$$

where C is a constant (of dimensions of a density) which, in the case $k = 1$, defines the maximum value of R, given when $a = -b$.

The Tolman radiation-filled universes with $\lambda = 0$ are just as simple:

$$I = \frac{4}{\varepsilon C}\frac{a^2b^2}{(a^2 - b^2)}, \qquad (9.5.46)$$

yielding

$$U = C(1 - \cos\tau), \qquad R = C\sin\tau \quad (k = 1)$$
$$U = C(\cosh\sigma - 1), \qquad R = C\sinh\sigma \quad (k = -1).$$

Note that here (in the case $k = 1$) maximum expansion occurs when $a = ib = \mathrm{e}^{\mathrm{i}\pi/4}$, which is at a different ratio $T{:}V$ from that giving the big bang and big crunch, whereas in the corresponding dust-filled Friedmann case maximum expansion, big bang and big crunch all occur at the same ratio (namely $T = 0$) owing to the square roots involved in passing to a and b in (9.5.43), so these regions all coincide on $\mathbb{M}^\#$ as remarked earlier.

The perfect fluid models ($k = \pm 1$, $\lambda = 0$) for a polytrope of index γ (*cf.* Weinberg 1970) are only a little more complicated in the appearance of the form of I, which turns out to be proportional to

$$\left[\frac{ab}{(a^{3\gamma-2} - b^{3\gamma-2})^{1/(3\gamma-2)}}\right]^2$$

(We are grateful to K.P. Tod for this expression.)

9.6 Asymptotically simple space–times

Let us now turn to a study of general curved space–times having sufficiently 'nice' asymptotic properties to allow a smooth conformal boundary to be adjoined by a procedure similar to that followed in §9.1 for Minkowski space–time. It turns out, in fact, that stipulating the existence of such a conformal boundary, in the case of asymptotically flat space $\mathscr{M}$, represents a reasonable boundary condition – in the sense that it seems to be weak enough to allow for the presence of mass, momentum, angular momentum, and both incoming and outgoing freely varying radiation of the gravitational field, as well as of other massless fields; yet it is strong enough to allow precise mathematical results to be obtained from it, concerning the fall-off of radiation and the energy–momentum it carries.

Infinity for Schwarzschild's space–time

To begin our discussion, we examine conformal infinity of the Schwarzschild solution. The familiar form of the metric is

$$\mathrm{d}s^2 = (1 - 2m/r)\mathrm{d}t^2 - (1 - 2m/r)^{-1}\mathrm{d}r^2 - r^2(\mathrm{d}\theta^2 + \sin^2\theta\,\mathrm{d}\phi^2). \qquad (9.6.1)$$

Rather than attempt to obtain $\mathscr{I}^+$ and $\mathscr{I}^-$ simultaneously, as was done for Minkowski space, it is simpler to introduce a retarded time coordinate

$$u = t - r - 2m \log(r - 2m) \tag{9.6.2}$$

and an advanced time coordinate

$$v = t + r + 2m \log(r - 2m) \tag{9.6.3}$$

separately. In the first case the metric form becomes

$$ds^2 = (1 - 2m/r)du^2 + 2du dr - r^2(d\theta^2 + \sin^2\theta\, d\phi^2) \tag{9.6.4}$$

and in the second,

$$ds^2 = (1 - 2m/r)dv^2 - 2dv dr - r^2(d\theta^2 + \sin^2\theta\, d\phi^2). \tag{9.6.5}$$

In each case we can choose $\Omega = r^{-1} = w$, say. Then the 'unphysical' metric is

$$d\hat{s}^2 = \Omega^2 ds^2 = (w^2 - 2mw^3)du^2 - 2du dw - d\theta^2 - \sin^2\theta\, d\phi^2 \tag{9.6.6}$$

in the first case and

$$d\hat{s}^2 = (w^2 - 2mw^3)dv^2 + 2dv dw - d\theta^2 - \sin^2\theta\, d\phi^2 \tag{9.6.7}$$

in the second. The metrics (9.6.6) and (9.6.7) are manifestly regular (and analytic) on their respective hypersurfaces $w = 0$. (Clearly the determinants are non-zero at $w = 0$.) The physical space–time is given when $w > 0$ in (9.6.6) and we can extend the manifold to include the boundary hypersurface $\mathscr{I}^+$, given when $w = 0$. Similarly, in (9.6.7), the physical space–time corresponds to $w > 0$ and can be extended to include $\mathscr{I}^-$, given when $w = 0$. In fact, we could if we wanted extend the space–time *across* $w = 0$ to negative values of w, but this will not be done here. Only the boundary $\mathscr{I} = \mathscr{I}^- \cup \mathscr{I}^+$ will be adjoined to the space–time.

We may note here a difficulty that is encountered if we try to identify $\mathscr{I}^-$ with $\mathscr{I}^+$. If we do extend the region of definition of (9.6.6) to include negative values of w, and then make the replacement $w \mapsto -w$, we see that the metric has the form (9.6.7) (with u in place of v) but with a mass $-m$ in place of m. Thus, the extension across $\mathscr{I}$ involves a reversal of the sign of the mass. In fact, the derivative at $\mathscr{I}$ of the (conformal) curvature contains the information of the mass. (We shall find an explicit formula later, *cf.* (9.9.56).) It follows, therefore, that if we attempt to identify $\mathscr{I}^+$ with $\mathscr{I}^-$, and want the *same* sign of the (non-zero) mass to occur on the two sides, then there must be a discontinuity in the derivative of the curvature across $\mathscr{I}$ (so that the metric $d\hat{s}$ must fail to be C^3 at $\mathscr{I}$).

Accepting, then, that it is not reasonable to identify $\mathscr{I}^+$ with $\mathscr{I}^-$, we are led to a picture closely resembling that obtained in §9.1 for $\mathbb{M}$. The only essential difference occurs with the points i^-, i^0, i^+. It turns out that

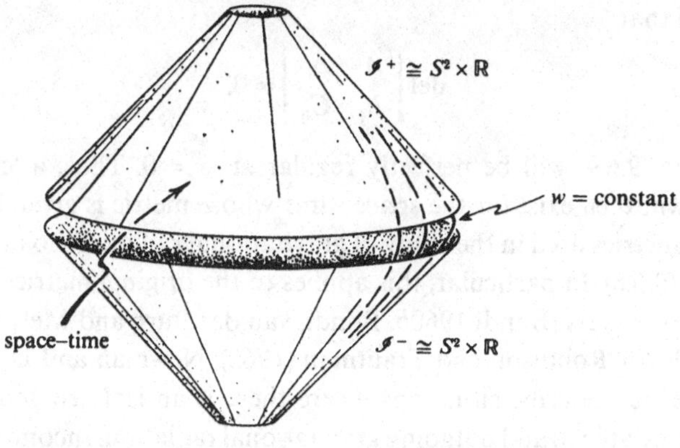

Fig. 9-18. Null infinity for the Schwarzschild space–time. Note that $w = 0$ corresponds both to $\mathscr{I}^+$ and to $\mathscr{I}^-$. The points $i^\pm$, i^0 are singular (divergent Weyl curvature) and have been deleted. This picture serves as a model for asymptotically flat spaces generally.

whenever mass is present, the point i^0, and normally also $i^\pm$, must, if adjoined to the manifold, be *singular* for the conformal geometry. (We shall not go into the argument here.) It is therefore reasonable not to attempt to include these points, in the general case, as part of the conformal infinity (and, as we saw earlier, even in Minkowski space the boundary surface at i^0, $i^\pm$ is not smooth). The picture, then, is as indicated in Fig. 9-18. We have two disjoint boundary hypersurfaces $\mathscr{I}^-$ and $\mathscr{I}^+$ each of which is a 'cylinder' with topology $S^2 \times \mathbb{R}$. It is clear from (9.6.6) and (9.6.7) that each of $\mathscr{I}^\pm$ is a *null* hypersurface (the induced metric at $w = 0$ being degenerate). These null hypersurfaces are generated by rays (given by θ, $\phi =$ constant, $w = 0$) whose tangents are *normals* to the hypersurfaces. These rays may be taken to be the '$\mathbb{R}$s' of the topological product $S^2 \times \mathbb{R}$.

Asymptotically simple space–times

We have obtained this structure explicitly in the case of the Schwarzschild metric. But it is clear that many other suitably asymptotically flat space–times will also give rise to such a structure. Let us start from a more general metric of the form

$$ds^2 = r^{-2}A\mathrm{d}r^2 + 2B_\mathbf{i}\mathrm{d}x^\mathbf{i}\mathrm{d}r + r^2C_{\mathbf{ij}}\mathrm{d}x^\mathbf{i}\mathrm{d}x^\mathbf{j}, \quad (\mathbf{i},\mathbf{j} = 1,2,3), \qquad (9.6.8)$$

the coordinates being r, x^1, x^2, x^3. Each of A, $B_\mathbf{i}$, $C_\mathbf{ij}$ is a suitably smooth function of $x^0 = r^{-1}, x^1, x^2, x^3$, also at $x^0 = 0$. Then setting $\Omega = r^{-1}$ we have

$$d\hat{s}^2 = \Omega^2 ds^2 = A\mathrm{d}x^0\mathrm{d}x^0 - 2B_\mathbf{i}\mathrm{d}x^\mathbf{i}\mathrm{d}x^0 + C_\mathbf{ij}\mathrm{d}x^\mathbf{i}\mathrm{d}x^\mathbf{j}. \qquad (9.6.9)$$

Provided that

$$\det\begin{bmatrix} A & B_{\mathrm{i}} \\ B_{\mathrm{j}} & C_{\mathrm{ij}} \end{bmatrix} \neq 0, \qquad (9.6.10)$$

the metric (9.6.9) will be perfectly regular at $x^0 = 0$. Thus, a 'conformal infinity' will then exist for the space–time whose metric is given by (9.6.8).

Many metrics used in the study of gravitational radiation do in fact have the form (9.6.8). In particular, this applies to the original metrics of Bondi and his coworkers (Bondi 1960*b*, Bondi, van der Burg and Metzner 1962), Sachs (1962*a*), Robinson and Trautman (1962), Newman and Unti (1962). These metrics describe situations where there is an isolated source (with asymptotic flatness) and outgoing gravitational radiation. Incoming gravitational radiation (of a suitably curtailed duration) may also be present. So may non-gravitational (e.g. electromagnetic or neutrino) massless radiation. In such situations, therefore, we expect a future-null conformal infinity $\mathscr{I}^+$ to exist. For the time-reversed situations we would expect $\mathscr{I}^-$ to exist. There should also be a wide class of 'physically reasonable' situations in which *both* $\mathscr{I}^+$ and $\mathscr{I}^-$ exist. It has, however, been argued occasionally (e.g. Bardeen and Press 1973) that the assumption of the existence of $\mathscr{I}^-$ may impose unnecessarily severe restrictions on the behaviour of the outgoing radiation in the infinite past. Indeed, examples may be constructed involving infinite wave trains in which either or both of $\mathscr{I}^\pm$ fail to exist. But whether such examples are regarded as 'physically reasonable' is often a matter of taste.* Asymptotic flatness is, in itself, a *mathematical* idealization, and so mathematical convenience and elegance constitute, in themselves, important criteria for selecting the appropriate idealization.

Asymptotically flat space–times of the type here considered constitute the most important subclass of those space–times which are termed '(weakly) asymptotically simple' (Penrose 1963, 1965). De Sitter space and certain space–times that are asymptotically de Sitter also come under this heading. The definition of an asymptotically simple space–time is as follows:

* There is, however, a class of physical situations that really ought to qualify as 'reasonable' regardless of one's tastes, but for which there exist unresolved questions as to the existence or regularity of $\mathscr{I}^-$. A typical example of this class consists of two gravitating bodies which come in from infinity (i^-) on approximately hyperbolic orbits, and then escape again to infinity (i^+) their encounter being accompanied by retarded gravitational radiation (Walker and Will 1979)

(9.6.11) DEFINITION

A space–time $\mathscr{M}$, with metric g_{ab}, is called k-asymptotically simple *if a C^{k+1} smooth manifold-with-boundary $\bar{\mathscr{M}}$ exists, with metric $\hat{g}_{ab}$, scalar field Ω and boundary $\mathscr{I} = \partial\bar{\mathscr{M}}$ such that:*

(a) $\mathscr{M} = \operatorname{int}\bar{\mathscr{M}}$

(b) $\hat{g}_{ab} = \Omega^2 g_{ab}$ *in* $\mathscr{M}$

(c) Ω *and* $\hat{g}_{ab}$ *are* C^k *smooth throughout* $\bar{\mathscr{M}}$

(d) $\Omega > 0$ *in* $\mathscr{M}$*; and* $\Omega = 0$, $\hat{\nabla}_a\Omega \neq 0$ *on* $\mathscr{I}$

(e) every null geodesic in $\mathscr{M}$ *acquires a past and future end-point on* $\mathscr{I}$.

Often the precise degree of differentiability* at $\mathscr{I}$ will not concern us ($k = 3$ will do for most purposes). Condition (*e*) is included to ensure that the *entire* null infinity is described by $\mathscr{I}$. (Without (*e*), any smooth space–time at all would satisfy the definition, simply with $\bar{\mathscr{M}} = \mathscr{M}$, $\Omega = 1$ and $\mathscr{I} = \varnothing$.) However, for some purposes (*e*) is rather too severe – for example, in the study of black holes; even outside the Schwarzschild horizon $r = 2m$ there are the circular (really, helical) null orbits at $r = 3m$ which, like the similar (but non-geodetic) null curve considered towards the end of §9.1, do not reach $\mathscr{I}$. To cover such situations, a weakened version of (*e*) would be needed. For example, a *weakly asymptotically simple* space–time $\mathscr{M}$ (Penrose 1968*b*) is one which possesses the conformal infinity of an asymptotically simple space–time but which may possess other 'infinities' as well; more precisely, for such an $\mathscr{M}$ there exists some asymptotically simple $\mathscr{M}'$ such that, for a neighbourhood $\mathscr{Q}'$ of $\mathscr{I}'$ in $\bar{\mathscr{M}}'$, the portion $\mathscr{Q}' \cap \mathscr{M}'$ is isometric with a subset of $\mathscr{M}$. As it stands, however, this condition may still not be considered quite satisfactory (*cf.* Geroch and Horowitz 1978), in effect because no assumption of 'physical reasonableness' has been placed on the auxiliary space $\mathscr{M}'$. Asymptotic simplicity is, in any case, a fruitful condition mainly when used in conjunction with Einstein's field equations (with 'reasonable' sources); and a reasonable mild strengthening of 'weak asymptotic simplicity' can be achieved when appropriate (but much weaker) restrictions of 'physical reasonableness' are placed not only on $\mathscr{M}$, but also on $\mathscr{M}'$ (for example, the rather minimal weak energy or null convergence condition on $\mathscr{M}'$, which requires that $R_{ab}l^a l^b \leqslant 0$ for any null vector l^a). One alternative approach that has been advocated (Geroch 1977, Geroch and Horowitz 1978) is to impose some extra conditions *directly* on

* Recall that 'C^k' means 'admitting continuous kth derivatives', where we also allow $k = \infty$ (derivatives of arbitrarily high order) and $k = \omega$ (real-analytic).

the structure of $\mathscr{I}$ – for example, the (strong) asymptotic Einstein condition (*cf.* (9.6.21) and after (9.6.37)) or the condition that the generators of a null $\mathscr{I}$ be infinitely long (*cf.* (9.8.1)) – but it is not *a priori* clear what those restrictions should be. No such conditions will be imposed here. It is remarkable how much detailed and useful asymptotic structure results merely when the apparently very unrestrictive assumption of (weak) asymptotic simplicity for $\mathscr{M}$ is supplemented by requiring Einstein's (vacuum) equations to hold in some neighbourhood of $\mathscr{I}$ in $\mathscr{M}$.*

For generality, let us allow the possibility of a cosmological term in the equations. Also, for the time being, we shall allow some massless matter fields in the neighbourhood of $\mathscr{I}$. (In the case of a massless scalar field we take the conformally invariant version (6.8.30) with energy tensor (6.8.36).) Then $T_a{}^a = 0$, and so, with Einstein's equations (4.6.32), we obtain

$$R = 4\lambda, \quad \text{i.e.,} \quad \lambda = 6\Lambda \tag{9.6.12}$$

near $\mathscr{I}$ (where 'near $\mathscr{I}$' means in $\mathscr{K} \cap \mathscr{M}$, for some neighbourhood $\mathscr{K}$ of $\mathscr{I}$ in $\bar{\mathscr{M}}$). Now, (6.8.22) gives

$$\mathrm{P}_{ab} = \hat{\mathrm{P}}_{ab} + \Omega^{-1}\hat{\nabla}_a\hat{\nabla}_b\Omega - \tfrac{1}{2}\Omega^{-2}\hat{g}_{ab}\hat{\nabla}^c\Omega\hat{\nabla}_c\Omega \tag{9.6.13}$$

near $\mathscr{I}$ (where, with regard to the last term, we recall that, according to the conventions of §5.6, 'hatted' quantities have their indices raised by $\hat{g}^{ab}$ and lowered by $\hat{g}_{ab}$). The quantity $\mathrm{P}_{ab} = \frac{1}{12}Rg_{ab} - \frac{1}{2}R_{ab}$ was introduced in (6.8.12). From (9.6.12),

$$\mathrm{P}_a{}^a = -\tfrac{2}{3}\lambda \tag{9.6.14}$$

near $\mathscr{I}$, whence, transvecting (9.6.13) with $g^{ab} = \Omega^2\hat{g}^{ab}$, we get

$$-\tfrac{2}{3}\lambda = \Omega^2\hat{\mathrm{P}}_a{}^a + \Omega\hat{\nabla}_a\hat{\nabla}^a\Omega - 2\hat{\nabla}^a\Omega\hat{\nabla}_a\Omega \tag{9.6.15}$$

near $\mathscr{I}$. *On* $\mathscr{I}$, $\Omega = 0$ and, by (9.6.11)(*c*) with $k \geqslant 2$, the 'hatted' derivatives of Ω and 'hatted' curvatures are all finite (continuous). So, putting

$$\hat{N}_a := -\nabla_a\Omega \tag{9.6.16}$$

(the minus sign for later convenience), we obtain

$$\hat{N}_a\hat{N}^a = \tfrac{1}{3}\lambda \quad on\ \mathscr{I}. \tag{9.6.17}$$

By (9.6.11)(*d*), $\hat{N}^a \neq 0$ on $\mathscr{I}$ and, being orthogonal to each locus $\Omega = \text{constant}$, $\hat{N}^a$ constitutes a normal to $\mathscr{I}$ at each of its points. Thus, assuming $k \geqslant 2$, equation (9.6.17) yields the following:

* When Einstein's vacuum field equations are assumed, condition (*d*) (3) in the definition (9.6.11) of asymptotic simplicity, namely the non-vanishing of the derivative of Ω on $\mathscr{I}$, is redundant. For it then follows from (9.6.21) below, or even more easily from (9.6.17) when the cosmological constant is non-zero.

(9.6.18) PROPOSITION

If the trace of the energy tensor vanishes near $\mathscr{I}$, then $\mathscr{I}$ is spacelike, timelike, or null according as λ is positive, negative, or zero.

This result also holds under somewhat weaker assumptions (Penrose 1965).

When $\mathscr{I}$ is spacelike or null, it consists naturally of two pieces $\mathscr{I}^+$ and $\mathscr{I}^-$. A point of $\mathscr{I}$ lies on $\mathscr{I}^+$ [*or* $\mathscr{I}^-$] if the interior of its past [future] light cone lies in $\mathscr{M}$. So the points of $\mathscr{I}^+$ [*or* $\mathscr{I}^-$] can be characterized in terms of structures within $\mathscr{M}$, namely as TIPs [*or* TIFs] as was done for Minkowski space in §9.1. In the timelike case ($\lambda < 0$), each point of $\mathscr{I}$ arises *both* as a TIP *and* as a TIF. Until recently,* the timelike case had seemed to have the least interest physically. Anti-de Sitter space (*cf.* §9.5) is an example, but it has never been seriously considered as a model for cosmology. The standard $\lambda < 0$ cosmologies are all expanding–collapsing models with singular beginnings and ends, in which no ray reaches infinity. Normal de Sitter space has a spacelike $\mathscr{I}^\pm$ and, as mentioned in §9.5, particle horizons and event horizons both consequently occur in this model. We have seen that horizons can also correspond to singular rather than regular boundary points, examples being the particle horizons in all the standard big bang models. Several such space–times can come under the heading of (9.6.11), but with conditions (*c*) and (*d*) suitably modified: the standard big bang is a spacelike boundary hypersurface '$\mathscr{I}^-$' with $\Omega = \infty$ (*cf.* §9.5).

The existence of two disconnected parts $\mathscr{I}^+$ and $\mathscr{I}^-$ to $\mathscr{I}$, when $\mathscr{I}$ is spacelike or null, allows us to refine our differentiability requirements for [weak] asymptotic simplicity. We say that $\mathscr{M}$ is [*weakly*] $\binom{k}{l}$*-asymptotically simple* if the conditions for [weak] k-asymptotic simplicity apply at $\mathscr{I}^+$ and those for [weak] l-asymptotic simplicity apply at $\mathscr{I}^-$. (In gravitational radiation problems one sometimes anticipates a difference in the differentiability properties at $\mathscr{I}^+$ and $\mathscr{I}^-$.) If we are concerned with the differentiability only at $\mathscr{I}^+$ we can refer to [weak] *future-k-asymptotic simplicity* (and correspondingly for $\mathscr{I}^-$).

The case when $\mathscr{I}$ is null is the most interesting, since it is relevant to the discussion of asymptotically flat space–times. Indeed, we have the following theorem (Penrose 1965, Geroch 1971):

* See Ashtekar and Magnon 1984, and references contained therein, for a discussion of this case.

(9.6.19) THEOREM

In any asymptotically simple space–time for which $\mathscr{I}$ is everywhere null, the topology of each of $\mathscr{I}^{\pm}$ is given by

$$\mathscr{I}^{+} \cong \mathscr{I}^{-} \cong S^2 \times \mathbb{R}$$

and the rays generating $\mathscr{I}^{\pm}$ can be taken to be the $\mathbb{R}$ factors.

Thus each of $\mathscr{I}^{\pm}$ contains S^2 null generators, and the topology is essentially identical with that for Minkowski space. We remark that two points of $\mathscr{I}^{+}$ [of $\mathscr{I}^{-}$] lie on the same generator if and only if one of the corresponding TIPs [TIFs] contains the other. The occurrence of this situation is what distinguishes an asymptotically simple space–time in which $\mathscr{I}$ is null from one in which $\mathscr{I}$ is spacelike.

Vacuum equations: asymptotic Einstein condition

In order to proceed further and obtain more detailed results concerning the structure of $\mathscr{I}$, we shall impose the condition that the Einstein vacuum equations hold near $\mathscr{I}$. In fact, if certain other more general sets of equations such as the Einstein–Maxwell equations hold near $\mathscr{I}$, then the consequences are almost the same, but the derivations can be considerably more involved (Penrose 1965). For this reason we restrict ourselves to the Einstein vacuum equations, when deriving properties of $\mathscr{I}$. When matter fields are present, we shall simply *assume* that these properties still hold for $\mathscr{I}$.

From (6.8.24) – which is the trace-free part of (9.6.13) – we have

$$\Phi_{ab} = \hat{\Phi}_{ab} + \Omega^{-1}\hat{\nabla}_{A'(A}\hat{\nabla}_{B)B'}\Omega. \qquad (9.6.20)$$

Einstein's vacuum equations, allowing for the possibility of a cosmological constant, take the form $\Phi_{ab} = 0$. Hence, multiplying (9.6.20) by Ω and noting that $\hat{\Phi}_{ab}$ must be continuous at $\mathscr{I}$ (assuming $k \geqslant 2$), we obtain the following important equation, which (whether or not the vacuum equations hold) we refer to as the *asymptotic Einstein condition*:

$$\hat{\nabla}_{A'(A}\hat{\nabla}_{B)B'}\Omega \approx 0, \quad \text{i.e.,} \quad \hat{\nabla}_a\hat{\nabla}_b\Omega \approx \tfrac{1}{4}\hat{g}_{ab}\hat{\nabla}_c\hat{\nabla}^c\Omega. \qquad (9.6.21)$$

Here we have begun to use the notation

$$A_{\dots}^{\dots} \approx B_{\dots}^{\dots} \qquad (9.6.22)$$

to denote the 'weak equality' that the tensor or spinor fields $A_{\dots}^{\dots}$, $B_{\dots}^{\dots}$ on $\bar{\mathscr{M}}$ are equal when restricted to $\mathscr{I}$, i.e., that $A_{\dots}^{\dots} - B_{\dots}^{\dots} = 0$ on $\mathscr{I}$. It must be borne in mind, when taking derivatives of a weak equation, that only the *tangential* derivatives can be relied upon to yield a new weak equation. Thus,

$$\hat{N}_{[a}\hat{\nabla}_{b]}A_{\cdots}^{\cdots} \approx \hat{N}_{[a}\hat{\nabla}_{b]}B_{\cdots}^{\cdots} \qquad (9.6.23)$$

is a valid deduction from (9.6.22), whereas $\hat{\nabla}_a A_{\cdots}^{\cdots} \approx \hat{\nabla}_a B_{\cdots}^{\cdots}$ would not be. We can henceforth drop the phrase 'near $\mathscr{I}$' whenever an ordinary equality sign is used ('strong equality'), it being assumed that *all* our calculations are performed in a suitable neighbourhood $\mathscr{K}$ of $\mathscr{I}$ in $\bar{\mathscr{M}}$. Clearly, it is always valid to take the covariant derivative of a strong equality between (C^1 smooth) quantities to derive a new strong equality. Note the obvious weak equality

$$\Omega \approx 0. \qquad (9.6.24)$$

Also we can now rewrite (9.6.17) as

$$\hat{N}_a \hat{N}^a \approx \tfrac{1}{3}\lambda. \qquad (9.6.25)$$

The asymptotic Einstein condition (9.6.21) can be written in the form

$$\hat{\nabla}_{A'(A}\hat{N}_{B)B'} \approx 0, \quad \text{i.e.,} \quad \hat{\nabla}_a \hat{N}_b \approx \tfrac{1}{4}\hat{g}_{ab}\hat{\nabla}_c \hat{N}^c, \qquad (9.6.26)$$

which tells us that the vectors $\hat{N}^a$ are *shear-free* (and also rotation-free, but this follows anyway from (9.6.16)). In the case of a null $\mathscr{I}$, we can put

$$\hat{N}^b \approx A\iota^B\bar{\iota}^{B'}, \qquad (9.6.27)$$

where A is a non-zero scalar which is positive on $\mathscr{I}^+$ and negative on $\mathscr{I}^-$, and then get, by transvecting the first of (9.6.26) with $\hat{\iota}^A\hat{\iota}^B$,

$$\hat{\sigma}' \approx 0, \qquad (9.6.28)$$

using the standard spin-coefficient notation of §4.5 (with 'hats'). (We can also get $\hat{\kappa}' \approx 0$, $\hat{\rho}' \approx \bar{\hat{\rho}}'$, but these merely restate the fact that $\mathscr{I}^{\pm}$ is a null hypersurface, *cf.* (7.1.58) *et seq*. Equation (9.6.28) is the shear-free condition for a null hypersurface. By the discussion of §7.1 (*cf.* Fig. 7-2) it tells us that

(9.6.29) PROPOSITION

If $R_{ab} = 0$ near $\mathscr{I}$, then any two cross-sections of $\mathscr{I}^{\pm}$ are mapped to one another conformally by the generators of $\mathscr{I}^{\pm}$.

By (9.6.19), these cross-sections are all topologically S^2, so the two S^2 factor spaces, whose points represent the various generators of $\mathscr{I}^+$ and $\mathscr{I}^-$, are naturally conformal spheres. We can invoke a classical theorem of Riemann to rescale their two metrics to that of a unit Euclidean 2-sphere, and assume, if we wish, that our original choice of Ω has been made to that effect.

Choosing spherical polar coordinates θ, ϕ for the Euclidean 2-sphere, or the equivalent complex stereographic coordinate (*cf.* (1.2.10))

$$\zeta = e^{i\phi} \cot \tfrac{1}{2}\theta, \tag{9.6.30}$$

the induced metric on $\mathscr{I}^+$ then becomes

$$\begin{aligned} dl^2 &= -d\hat{s}^2 = d\theta^2 + \sin^2\theta\, d\phi^2 + 0\cdot du^2 \\ &= \frac{4d\zeta d\bar{\zeta}}{(1+\zeta\bar{\zeta})^2} + 0\cdot du^2, \end{aligned} \tag{9.6.31}$$

where u is a retarded time coordinate, the corresponding form with an advanced time coordinate v in place of u holding on $\mathscr{I}^-$. Since the cross-sectional metric (9.6.31) is now constant along the generators, we now have $\hat{\rho}' = 0$ in addition to $\hat{\sigma}' = 0$ (see Fig. 7-2).

Vanishing of Weyl curvature of $\mathscr{I}$

An important further property of $\mathscr{I}$ is the following (for which we need $k \geqslant 3$):

(9.6.32) THEOREM

If $R_{ab} = \lambda g_{ab}$ near $\mathscr{I}$, then $\hat{C}_{abcd} \approx 0$.

Proof: Recall, first, the vacuum Bianchi identity in spinor form: $\nabla^{AA'}\Psi_{ABCD} = 0$ (*cf.* (4.10.9)). By (6.8.4) and (6.8.8) we have $\hat{\nabla}^{AA'}(\Omega^{-1}\hat{\Psi}_{ABCD}) = 0$ in $\mathscr{M}$, i.e. (*cf.* (7.3.42)),

$$\Omega\hat{\nabla}^{AA'}\hat{\Psi}_{ABCD} = \hat{\Psi}_{ABCD}\hat{\nabla}^{AA'}\Omega \tag{9.6.33}$$

which, by continuity, holds also on $\mathscr{I}$. Hence

$$\hat{\Psi}_{ABCD}\hat{N}^{AA'} \approx 0. \tag{9.6.34}$$

If $\lambda \neq 0$, the 'matrix' $\hat{N}^{AA'}$ is non-singular (*cf.* (9.6.25)) and can be inverted (in fact, $\hat{N}^{AA'}\cdot 6\lambda^{-1}\hat{N}_{A'E} = \varepsilon_E{}^A$), whence $\hat{\Psi}_{ABCD} \approx 0$ and the result follows.

The case $\lambda = 0$ is more difficult, and partly depends upon a global result requiring the topology (9.6.19). For (9.6.34), with (9.6.27), merely yields

$$\hat{\Psi}_{ABCD}\hat{\iota}^A \approx 0, \tag{9.6.35}$$

i.e.,

$$\hat{\Psi}_{ABCD} \approx \Psi\hat{\iota}_A\hat{\iota}_B\hat{\iota}_C\hat{\iota}_D \tag{9.6.36}$$

for some Ψ. So we differentiate (9.6.33) once more to obtain, using (9.6.24),

$$\hat{N}_{EE'}\hat{\nabla}^{AA'}\hat{\Psi}_{ABCD} \approx \hat{N}^{AA'}\hat{\nabla}_{EE'}\hat{\Psi}_{ABCD} + \hat{\Psi}_{ABCD}\hat{\nabla}_{EE'}\hat{N}^{AA'}.$$

Lowering A', symmetrizing over $A'E'$, and using the complex conjugate of (9.6.26)(1), we obtain, after applying the ε-identity (*cf.* (2.5.20)),

$$\hat{N}^A_{(A'}\hat{\nabla}_{E')A}\hat{\Psi}_{EBCD} \approx 0,$$

so by (9.6.27) and Proposition (3.5.15)

$$\hat{\iota}^A\hat{\nabla}_{AE'}\hat{\Psi}_{EBCD} \approx 0. \qquad (9.6.37)$$

On any spherical cross-section of $\mathscr{I}$ (taking $\hat{o}^A$ to have its flagpole orthogonal to it) we thus have an equation of the form $\eth'\Psi = 0$, where Ψ, being defined by (9.6.36), has spin-weight 2. Thus, by Proposition (4.15.59), $\Psi = 0$ on the sphere, whence the expression (9.6.36) vanishes on $\mathscr{I}$, and the proof is complete.

We refer to the condition that $\hat{\Psi}_{ABCD} \approx 0$, together with (9.6.26), as the *strong asymptotic Einstein condition* – irrespective of whether or not the vacuum equations hold near $\mathscr{I}$.

The vanishing of the Weyl curvature on $\mathscr{I}$ has a significant consequence, which follows from the following general result:

(9.6.38) LEMMA

Let $\mathscr{M}$ be (weakly) k-asymptotically simple and let $\mathscr{K}$ be a neighbourhood of $\mathscr{I}$ in $\hat{\mathscr{M}}$. Suppose $T^{\mathscr{A}} \in \underset{r}{\mathfrak{S}}^{\mathscr{A}}[\mathscr{K}]$ (with $r \leqslant k$) satisfies $T^{\mathscr{A}} \approx 0$. Then there exists a $U^{\mathscr{A}} \in \underset{r-1}{\mathfrak{S}}^{\mathscr{A}}[\mathscr{K}]$ such that $\Omega U^{\mathscr{A}} = T^{\mathscr{A}}$.

Here we make the *definition* that $\underset{r}{\mathfrak{S}}^{\mathscr{A}}$ denotes the module of spinor fields of index type $\mathscr{A}$ which are C^r-smooth – and we recall that '$[\mathscr{K}]$' means fields *restricted to the set* $\mathscr{K}$. This lemma tells us that it is legitimate to define $U^{\mathscr{A}} = \Omega^{-1}T^{\mathscr{A}}$ whenever a smooth $T^{\mathscr{A}}$ is given for which $T^{\mathscr{A}} \approx 0$. The proof of the lemma is immediate from the well-known result of analysis that if $f(x^0, x^1, \ldots, x^n)$ is a C^r function ($r > 0$), defined in some open subset $\mathscr{U}$ of $\mathbb{R}^{n+1}$, which vanishes at $x^0 = 0$, then $(x^0)^{-1}f(x^0, \ldots, x^n)$ defines a C^{r-1} function in $\mathscr{U}$, where $r = 1, 2, \ldots, \infty$, or ω (with $\infty - 1 = \infty$, $\omega - 1 = \omega$). To apply this result to a spinor field $T^{\mathscr{A}}$, we merely need to choose coordinates in $\mathscr{K}$ for which $x^0 = \Omega$, and a C^r basis, and then apply it to the components individually. This can be done provided $r \leqslant k$ (see (9.6.11)).

Now recall that in §6.8 we contrasted the conformal behaviour of the Weyl spinor Ψ_{ABCD} with that of a massless spin-2 field ϕ_{ABCD} (*cf.* (6.8.4), (6.8.6)). Since the vacuum Bianchi identities are simply the massless spin-2 field equations, it is legitimate to define a specific massless spin-2 field

$$\psi_{ABCD} = \Psi_{ABCD}. \qquad (9.6.39)$$

But when the metric is rescaled we would like to set

$$\hat{\psi}_{ABCD} = \Omega^{-1}\psi_{ABCD} = \Omega^{-1}\Psi_{ABCD} = \Omega^{-1}\hat{\Psi}_{ABCD}, \qquad (9.6.40)$$

since the massless field equations on ψ_{ABCD} are then preserved. From (9.6.32) and (9.6.38) we derive the important corollary:

(9.6.41) THEOREM

$\hat{\psi}_{ABCD}$ extends to a field continuous (C^{k-3}) at $\mathscr{I}$.

Note that, by (9.6.40), $\hat{\psi}_{ABCD}$ at $\mathscr{I}$ can be obtained from the *derivative* of the Weyl curvature at $\mathscr{I}$:

$$-\hat{\nabla}_{AA'}\hat{\Psi}_{BCDE} \approx \hat{N}_{AA'}\hat{\psi}_{BCDE}. \qquad (9.6.42)$$

We can think of ψ_{ABCD} as the *gravitational spin-2 field.* The various components of $\hat{\psi}_{ABCD}$ at $\mathscr{I}$ have a key significance for gravitational radiation theory. They can be interpreted relative to the physical metric g_{ab} of $\mathscr{M}$ in terms of the *Sachs peeling property* (Sachs 1961, 1962*a*, Penrose 1963, 1965, Newman and Penrose 1962). It is of interest to discuss this property somewhat more generally than just for the gravitational field, and this we shall proceed to do next.

9.7 Peeling properties

Suppose we have any field $\theta_{A\ldots HK'\ldots Q'}$ which is taken to be a conformal density of weight $-w$,

$$\hat{\theta}_{\ldots} = \Omega^{-w}\theta_{\ldots}, \qquad (9.7.1)$$

and for which $\hat{\theta}_{\ldots}$ is continuous C^h ($0 \leqslant h \leqslant k-1 \geqslant 2$) at some $P \in \mathscr{I}$. We shall assume (weak) k-asymptotic simplicity* and the asymptotic Einstein condition (9.6.21). Let γ be a complete ray in $\mathscr{M}$, having $P \in \mathscr{I}$ as one end-point (and not touching $\mathscr{I}$ at P – an unlikely eventuality which could, in any case, occur only in the unphysical situation of $\mathscr{I}$ being timelike and 'concave'). Choose the spin-frame (o^A, ι^A) to be parallelly propagated along γ with its flagpole

$$l^a = o^A o^{A'} \qquad (9.7.2)$$

tangent to γ (as in Chapter 7). Let r be an associated affine parameter on γ:

$$Dr = l^a \nabla_a r = 1. \qquad (9.7.3)$$

* We actually only need (weak) future- [*or* past-] k-asymptotic simplicity, if P lies in $\mathscr{I}^+$ [*or* $\mathscr{I}^-$].

Then if θ is any component of $\theta_{A\ldots Q'}$, having a total of q zeros (0 or 0′) among its indices, we shall find that

$$\theta = \sum_{i=w+q}^{w+q+h} \underset{i}{\theta} r^{-i} + o(r^{-w-q-h}), \tag{9.7.4}$$

where each $\underset{i}{\theta}$ is constant along γ.* We note especially that the leading term in this expansion is a multiple of $1/r^{w+q}$.

We shall establish this result presently, but first it is worth while to point out its special interpretation in the case of massless fields of general spin. Suppose that $\phi_{A\ldots L}$ has n symmetric indices, weight $w = -1$ (which – *cf.* (5.7.17) *et seq.* – makes the massless field equations conformally invariant), and that $\hat{\phi}_{A\ldots L}$ is continuous C^0 at P. Then the various components, $\phi_0 := \phi_{00\ldots0}, \phi_1 := \phi_{10\ldots0}, \ldots, \phi_n := \phi_{11\ldots1}$ behave as follows:

$$\phi_0 = \phi_0^0 r^{-n-1} + o(r^{-n-1}), \quad \phi_1 = \phi_1^0 r^{-n} + o(r^{-n}), \ldots,$$
$$\phi_n = \phi_n^0 r^{-1} + o(r^{-1}), \tag{9.7.5}$$

where the ϕ_i^0 are constant on γ. If we assume that $\hat{\phi}_{A\ldots L}$ is C^n at P, then, by (9.7.4), we can obtain

$$\phi_{A\ldots L} = \sum_{i=1}^{n} \underset{i}{\phi}_{A\ldots L} r^{-i} + o(r^{-n}), \tag{9.7.6}$$

where each $\underset{i}{\phi}_{A\ldots L}$ is parallelly propagated along γ and where

$$\underset{i}{\phi}_{A\ldots D\underbrace{0\ldots0}_{i}} \equiv \underset{i}{\phi}_{A\ldots DE_1E_2\ldots E_i} o^{E_1} o^{E_2} \ldots o^{E_i} = 0. \tag{9.7.7}$$

Referring back to Proposition (3.5.26), we find that the 'order r^{-i}' part $\underset{i}{\phi}_{A\ldots L}$ of the field $\phi_{A\ldots L}$ has at least (and, in general, exactly) $n-i+1$ principal null directions pointing along the direction of γ, i.e., along l^a (*cf.* Sachs 1961, 1962*a*, Newman and Penrose 1962, Penrose 1963, 1965). In particular, the r^{-1} part (or 'radiation field') is null. Thus we have the rough and intuitive picture (*cf.* Fig. 9-19) that as we move inwards along the null ray γ, and as the field becomes more and more influenced by the higher order terms, the PNDs 'peel off' one by one away from the radial direction. This is actually a slightly misleading picture, the exact form of the 'peeling' property being given by (9.7.5)–(9.7.7). The various leading terms ϕ_i^0 in (9.7.5) in fact provide the various components of $\hat{\phi}_{A\ldots L}$ at P. But the precise nature of this relation can be best described after we have given the proof of (9.7.4).

* The order symbol $o(r^{-n})$ stands for a quantity that, when multiplied by r^n, tends to zero for large $|r|$, i.e. at P. Similarly, $O(r^{-n})$ stands for a quantity that when multiplied by r^n remains bounded for large $|r|$.

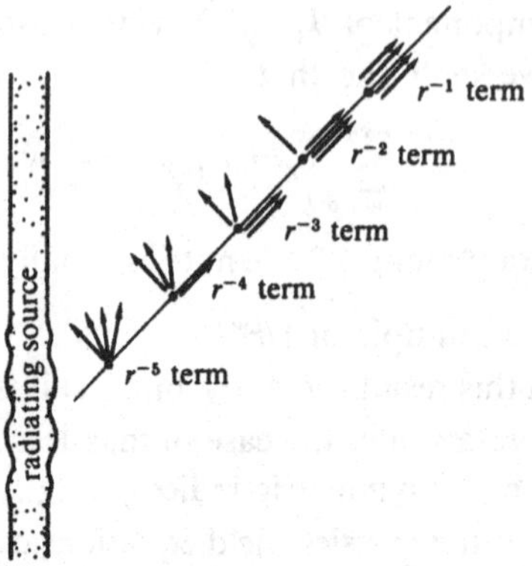

Fig. 9-19. The Sachs peeling property, illustrating the multiplicity of the radial PND of the Weyl curvature for the various terms in the expansion in negative powers of r along an outgoing ray

Parallelly propagated spin-frames

That proof depends on comparing parallel propagation of spin-frames along γ with respect to the two metrics g_{ab}, $\hat{g}_{ab}$. We note first that if we take

$$\hat{o}_A = o_A, \quad \hat{o}^A = \Omega^{-1} o^A, \quad \text{i.e.} \quad \hat{l}_a = l_a, \quad \hat{l}^a = \Omega^{-2} l^a \tag{9.7.8}$$

(*cf.* (9.7.2)), then the propagation equation $Do_A = 0$ is preserved: $\hat{D}\hat{o}_A = \Omega^{-2} l^b(\nabla_b o_A - \Upsilon_{AB'} o_B) = \Omega^{-2} D o_A$, by (5.6.15). Thus, completing $\hat{o}_A$ to a spin-frame $(\hat{o}^A, \hat{\iota}^A)$, we can arrange

$$\hat{D}\hat{o}^A = 0, \quad \hat{D}\hat{\iota}^A = 0, \tag{9.7.9}$$

where, to preserve the normalization with (9.7.8), we set

$$\hat{\iota}^A = \iota^A - \nu o^A \tag{9.7.10}$$

for some ν. We have, by (5.6.15),

$$0 = D\iota^A = D(\hat{\iota}^A + \nu o^A) = l^b(\hat{\nabla}_b \hat{\iota}^A - \varepsilon_B{}^A \Upsilon_{CB'} \hat{\iota}^C) + D(\nu o^A)$$
$$= 0 - o^A o^{B'} \hat{\iota}^C \Omega^{-1} \nabla_{CB'} \Omega + o^A D\nu + 0,$$

whence

$$\hat{D}\nu = \Omega^{-2}\eta, \tag{9.7.11}$$

where

$$\eta = \hat{\iota}^C \hat{o}^{B'} \hat{\nabla}_{CB'} \Omega. \tag{9.7.12}$$

Comparison between affine parameters

Next we compare affine parameters on γ. Take $\hat{r}$ to be an affine parameter on γ, with origin at P and associated with $\hat{l}^a$ (*cf.* (9.7.3)):

$$\hat{r} \approx 0, \quad \hat{D}\hat{r} = 1. \tag{9.7.13}$$

(The symbol $\approx$ now means equality at P; we ignore what happens at the other end of γ.) We have $\Omega \approx 0$ and $d\Omega/d\hat{r} \not\approx 0$, by (9.6.11)(*d*) and the assumed non-contact between γ and $\mathscr{I}$, hence

$$\frac{d\Omega}{d\hat{r}} \approx -A \tag{9.7.14}$$

for some non-zero A (constant along γ). Note that

$$-\hat{l}^a \hat{N}_a = \hat{l}^a \hat{\nabla}_a \Omega = \hat{D}\Omega \approx -A, \tag{9.7.15}$$

so this A coincides with the end-value of the A in (9.6.27), in the null case. From the C^k smoothness of Ω it follows that

$$\Omega = -A\hat{r} - A_2\hat{r}^2 - A_3\hat{r}^3 - \cdots - A_k\hat{r}^k + o(\hat{r}^k), \tag{9.7.16}$$

$A, A_2, \ldots$ being constant. But from the asymptotic Einstein condition (9.6.21) we get, since $\hat{l}^a$ is null and $\hat{D}\hat{l}^a = 0$,

$$\begin{aligned} 0 \approx \hat{l}^a \hat{l}^b \hat{\nabla}_a \hat{\nabla}_b \Omega &= \hat{l}^a \hat{\nabla}_a (\hat{l}^b \hat{\nabla}_b \Omega) \\ &= \hat{D}^2 \Omega, \end{aligned} \tag{9.7.17}$$

so in the expansion (9.7.16),

$$A_2 = 0. \tag{9.7.18}$$

Now

$$\frac{d\hat{r}}{dr} = D\hat{r} = \Omega^2 \hat{D}\hat{r} = \Omega^2, \tag{9.7.19}$$

whence

$$\begin{aligned} r &= \int \Omega^{-2}\, d\hat{r} \\ &= \int \hat{r}^{-2}\{A + A_3\hat{r}^2 + A_4\hat{r}^3 + \cdots + A_k\hat{r}^{k-1} + o(\hat{r}^{k-1})\}^{-2}\, d\hat{r} \\ &= \int A^{-2}\hat{r}^{-2}\{1 + B_2\hat{r}^2 + \cdots + B_{k-1}\hat{r}^{k-1} + o(\hat{r}^{k-1})\}\, d\hat{r} \\ &= -A^{-2}\hat{r}^{-1} + C_0 + C_1\hat{r} + \cdots + C_{k-2}\hat{r}^{k-2} + o(\hat{r}^{k-2}), \end{aligned} \tag{9.7.20}$$

by (9.7.16) and (9.7.18), all B and C coefficients being constant on γ, and C_0 being the constant of integration. Inverting (9.7.20) for large r yields

$$\hat{r} = -A^{-2}r^{-1} + D_2 r^{-2} + D_3 r^{-3} + \cdots + D_k r^{-k} + o(r^{-k}), \tag{9.7.21}$$

which, when substituted into (9.7.16), shows that Ω is of the form

$$\Omega = A^{-1}r^{-1} + E_2 r^{-2} + E_3 r^{-3} + \cdots + E_k r^{-k} + o(r^{-k}), \tag{9.7.22}$$

the D and E coefficients again being all constant on γ.

We observe that (9.7.22) justifies the statement made in §9.1 to the effect that the conformal factor behaves like the reciprocal of an affine parameter along any null geodesic. Here we have $r\Omega \to A^{-1}$, but (9.7.22) gives a considerably more detailed behaviour. (We can, of course, choose a scaling for r so that $A = 1$ if desired.) It is a remarkable fact that the asymptotic Einstein condition (9.6.21), which follows from Einstein's vacuum field equations, is just what is needed to ensure (9.7.18), the necessary condition for the elimination of logarithmic terms in (9.7.20), (9.7.21), and (9.7.22). Equally remarkably, the same equation (9.6.21) will eliminate the second possible source of logarithmic terms, this time in the comparison between the two spin-frames, to which we now proceed.

Comparison between the spin-frames

From (9.7.10) and (9.7.8) we have

$$\iota^A = \hat{\iota}^A + \nu\Omega\hat{o}^A. \tag{9.7.23}$$

We wish to choose ν so that $\nu\Omega \to 0$ at P, since then the two spin-frames will agree at P in the sense that $\hat{o}_A \approx o_A$, $\hat{\iota}^A \approx \iota^A$. This entails that η in (9.7.11) must vanish at P, otherwise integration of that equation would yield a ν which behaves like a non-zero multiple of Ω^{-1} at P. By (9.7.12), this means that $\hat{N}_{BB'}$ must be a linear combination of $\hat{\iota}_B\hat{\iota}_{B'} = \hat{n}_b$ and $\hat{o}_B\hat{o}_{B'} = \hat{l}_b$, i.e.,

$$\hat{N}^b \approx A\hat{n}^b + \tfrac{1}{6}\lambda A^{-1}\hat{l}^b, \tag{9.7.24}$$

the coefficient A being taken to agree with (9.7.15) (and (9.6.27)), the other coefficient being fixed by (9.6.25). When $\mathscr{I}$ is null, this means that we choose $\hat{n}^a$ to point along the future-null direction in $\mathscr{I}$. When $\mathscr{I}$ is non-null (and not touching γ, in the timelike case), we choose the future-null vector $\hat{n}^a$ to lie in the timelike 2-plane spanned by $\hat{l}^a$ and $\hat{N}^a$, and of course distinct from $\hat{l}^a$ (see Fig. 9-20).

With this choice, $\eta \approx 0$, and it seems at first sight that integration of (9.7.11) would yield a behaviour for ν at P like $\log \Omega$. Though this would give $\nu\Omega \to 0$ as required for (9.7.23), it would spoil the power series form of the higher terms in our expansions (9.7.4) of components. But it turns out that η actually vanishes to *second* order at P, so the logarithm is eliminated. For consider the difference between the two sides of (9.7.24). By (9.6.38), there is a C^{k-2} covector Q_b, defined along γ, such that

$$\Omega\hat{Q}_b = \hat{N}_b - A\hat{n}_b - \tfrac{1}{6}\lambda A^{-1}\hat{l}_b. \tag{9.7.25}$$

If we operate on this with $\hat{\nabla}_c$ and transvect the result with $\hat{o}^C\hat{o}^{B'}\hat{o}^{C'}$, we get,

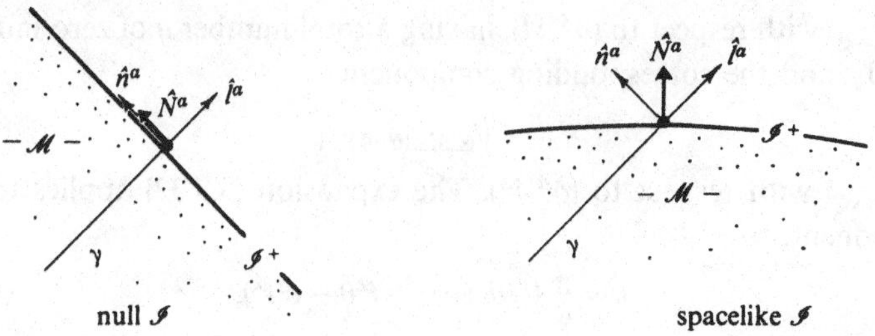

Fig. 9-20. When $\mathscr{I}^+$ is null, the direction of $\hat{n}^a$ at a point of $\mathscr{I}^+$ does not depend on the choice of γ, but this is not true when $\mathscr{I}^+$ is spacelike (or timelike) – the span of the null vectors $\hat{n}^a$ and $\hat{l}^a$ being such that it contains the normal $\hat{N}^a$ to $\mathscr{I}^+$. This has the implication that the radiation field concept (r^{-1} term) is less invariant when $\mathscr{I}^+$ is spacelike (or timelike) than it is when $\mathscr{I}^+$ is null.

using once more the asymptotic Einstein condition (9.6.21)(1) and (9.7.15),

$$-A\hat{Q}_{BB'}\hat{o}^{B'} \approx 0, \tag{9.7.26}$$

whence, by (9.6.38),

$$-\hat{Q}_{BB'}\hat{\iota}^B\hat{o}^{B'} = \Omega\mu \tag{9.7.27}$$

for some C^{k-3} scalar μ on γ. Substituting (9.7.27) and (9.7.25) into (9.7.12), we then obtain, as asserted above,

$$\eta = \Omega^2\mu. \tag{9.7.28}$$

Thus, by (9.7.11),

$$\nu = \int \mu \mathrm{d}\hat{r}, \tag{9.7.29}$$

which is regular (C^{k-2}) at P. From (9.7.23) and (9.7.22) we therefore have an expansion of the form

$$\iota^A = \hat{\iota}^A + \{r^{-1}\nu_1 + \cdots + r^{-k+1}\nu_{k-1} + o(r^{-k+1})\}\hat{o}^A, \tag{9.7.30}$$

the $\nu_1, \ldots, \nu_{k-1}$ being constant along γ, while from (9.7.8) we have

$$o^A = \Omega\hat{o}^A \tag{9.7.31}$$

Proof of the peeling property

We are now in a position to prove (9.7.4). Since the quantity $\hat{\theta}_{A\ldots HK'\ldots Q'}$ is assumed to be C^h at P, it is of the form

$$\hat{\theta}_{\ldots} = \underset{0}{\theta}_{\ldots} + \hat{r}\underset{1}{\theta}_{\ldots} + \cdots + \hat{r}^h\underset{h}{\theta}_{\ldots} + o(\hat{r}^h). \tag{9.7.32}$$

Consider a typical component

$$\theta = \theta_{0\ldots01\ldots10'\ldots0'1'\ldots1'} \tag{9.7.33}$$

of $\theta_{A\ldots Q'}$ with respect to (o^A, ι^A), having a total number q of zero indices (0 and 0′), and the corresponding component

$$\hat{\theta} = \hat{\theta}_{\hat{0}\ldots\hat{0}\hat{1}\ldots\hat{1}\hat{0}'\ldots\hat{0}'\hat{1}'\ldots\hat{1}'} \tag{9.7.34}$$

of $\hat{\theta}_{A\ldots Q'}$ with respect to $(\hat{o}^A, \hat{\iota}^A)$. The expansion (9.7.32) applies to each component, so

$$\hat{\theta} = \underset{0}{\hat{\theta}} + \hat{r}\underset{1}{\hat{\theta}} + \cdots + \hat{r}^h\underset{h}{\hat{\theta}} + o(\hat{r}^h), \tag{9.7.35}$$

with each $\underset{i}{\hat{\theta}}$ constant along γ. Now with $\theta_{A\ldots Q'}$ of conformal weight $-w$ (*cf.* (9.7.1)) we have

$$\theta = \Omega^w \hat{\theta}_{0\ldots 01\ldots 10'\ldots 0'1'\ldots 1'} \tag{9.7.36}$$

the components being obtained by transvection with o^A, ι^A and their complex conjugates. Substituting (9.7.30), (9.7.31), (9.7.22), and (9.7.35) successively into (9.7.36), we obtain (9.7.4) as required. This completes our somewhat lengthy argument.

Radiation fields

Note that the coefficient $\underset{w+q}{\theta}$ of the leading term in (9.7.4) can be explicitly identified at this stage:

$$\underset{w+q}{\theta} = A^{-w-q}\underset{w+q}{\hat{\theta}} \approx A^{-w-q}\hat{\theta}_{\hat{0}\ldots\hat{1}'}. \tag{9.7.37}$$

We may conveniently choose $A = 1$ (rescaling $r \mapsto Ar$, if necessary), whereupon the leading terms can be *directly* identified with the various components of the conformally rescaled field on $\mathscr{I}$. In the case of gravity, these components are $\psi_{0000}, \psi_{0001}, \ldots, \psi_{1111}$ and (with $A = 1$) they can be identified with the respective leading terms $\Psi_0^0, \Psi_1^0, \ldots, \Psi_4^0$ in the Weyl spinor expansions (*cf.* (9.7.5))

$$\Psi_0 = \Psi_0^0 r^{-5} + o(r^{-5}), \quad \Psi_1 = \Psi_1^0 r^{-4} + o(r^{-4}), \ldots,$$
$$\Psi_4 = \Psi_4^0 r^{-1} + o(r^{-1}) \tag{9.7.38}$$

(*cf.* the equations at the end of §9.8, below). In particular, therefore, Ψ_4^0, which may be thought of as describing the *gravitational radiation field*, can be identified with that component of $\hat{\psi}_{ABCD}$ on $\mathscr{I}$ which is totally contracted with the spinor $\hat{\iota}^A$. The situation is similar for electromagnetic radiation in the Einstein–Maxwell theory. Here again the asymptotic Einstein condition (9.6.21) is essentially a consequence of the field equations and asymptotic simplicity, and the rescaled field $\hat{\phi}_{AB}$ is finite on $\mathscr{I}$ (Penrose 1965). The peeling property thus holds also for the electromagnetic field, the radiation field being identified with the component of $\hat{\phi}_{AB}$ that is totally contracted with $\hat{\imath}^A$.

It should be pointed out, however, that when $\mathscr{I}$ is non-null, the radiation field concept is not very well defined. A glance at Fig. 9-20 will convince the reader that the direction of $\hat{n}^a$, and therefore of the spinor $\hat{\imath}^A$, is strongly dependent, in that case, upon the particular null geodesic γ that is chosen through the point P. Thus, on varying γ through P, the different components Ψ_i^0 get mingled with each other. Indeed, if we replace γ by a null geodesic through P whose tangent vector at P is $\hat{n}^a$, we find that the order of the terms $\Psi_0^0, \Psi_1^0, \ldots, \Psi_4^0$ is completely reversed! (The same holds for fields of other spins.) When $\mathscr{I}$ is null, on the other hand, the situation is much more satisfactory. It is true that the components Ψ_i^0 still get somewhat mingled with one another as γ is varied, but only in a comparatively mild way. Specifically, the 'radiation field' term Ψ_4^0 remains unaltered, Ψ_3^0 is modified only by the addition of a multiple of Ψ_4^0, Ψ_2^0 by the addition of multiples of Ψ_3^0 and Ψ_4^0, etc. This is because the spin-frame $(\hat{o}^A, \hat{\imath}^A)$ at P is replaced by $(\hat{o}^A + \omega\hat{\imath}^A, \hat{\imath}^A)$ for some ω. A corresponding property holds for the electromagnetic field.

When $\mathscr{I}$ is null, there is a useful and invariant description of radiation fields for each spin. The *outgoing* radiation may be thought of as described by $\hat{\phi}_{11\ldots1}$ on $\mathscr{I}^+$ and the *incoming* radiation by $\hat{\phi}_{11\ldots1}$ on $\mathscr{I}^-$ (taking the flagpole of ι^A tangential to $\mathscr{I}$ in both cases). Recall that these components, though used on a *finite* null hypersurface rather than on $\mathscr{I}^\pm$, provided the *null datum* (5.11.11) for the fields in question. This indicates that the radiation field (either incoming *or* outgoing, but not both) should provide appropriate initial (or final) data for the field. However, in some sense it is the $\hat{\text{þ}}'$ (or $\hat{n}^a\hat{\nabla}_a$) *derivative* of the null datum on $\mathscr{I}$ that could be expected more directly to influence the field in the interior space–time, since that is the quantity which enters into the generalized Kirchhoff–D'Adhémar integral formula (5.12.6) for flat-space massless fields. Of course, when conformal curvature or nonlinearities are present, this is not nearly so clear. Nevertheless, for a *scalar* field – in accordance with the Sommerfeld radiation condition (*cf.* Sommerfeld 1958), which in effect states that $\text{þ}'\hat{\phi} = 0$ on $\mathscr{I}^-$ for any retarded scalar field – this derivative indeed seems the more appropriate measure of the strength of the radiation field. The situation for higher spin is less clear-cut, because of the differing relations between the $\varphi_{11\ldots1}$ term and energy flow (*cf.* (9.10.13) *et seq.*, below).

We note that the peeling property arises in the present analysis because, in effect, an 'infinite boost' takes place in the limit, as P is approached along γ, in the comparison between a parallelly propagated frame (with respect to the physical metric) and a frame for which the conformally rescaled field is finite at $\mathscr{I}$. This serves to 'spread out' the different components of the field (at $\mathscr{I}$) so that they fall off with different powers of r. The actual finiteness of the rescaled field at $\mathscr{I}$ may be regarded as a 'probable' feature of conformally invariant fields since, in effect, such a field can propagate across $\mathscr{I}$ without 'noticing' its presence.

Finally it may be remarked that the work of Friedrich (1981*a*, *b*) has shown that (weak) future-asymptotic simplicity, in the case of vacuum space–times (with $\lambda = 0$) allows a full freedom for outgoing gravitational radiation. The status of combined regularity for $\mathscr{I}^+$ and $\mathscr{I}^-$ still remains unclear, however.

9.8 The BMS group and the structure of $\mathscr{I}^+$.

Minkowski space–time $\mathbb{M}$, and the cosmological models studied in §9.5, have interesting and useful groups of isometries. But for a general space–time $\mathcal{M}$, the isometry group is simply the identity and so provides no significant information. Yet symmetry groups have important roles to play in physics; in particular, the Poincaré group, describing the isometries of $\mathbb{M}$, plays a role in the standard definitions of energy–momentum and angular momentum. For this reason alone it would seem to be important to look for a generalization of the concept of isometry group that can apply in a useful way to suitable irregularly curved space–times.

The group (or pseudo-group) referred to as the 'general coordinate group' (or, equivalently, the 'diffeomorphism group') has, for historical reasons, frequently been invoked as a possible substitute for the Poincaré group for a general space–time $\mathcal{M}$. However, it is not really useful in this context, being much too 'large' and preserving only the differentiable structure of $\mathcal{M}$ rather than any of its more physically important properties. More significant is the concept of *asymptotic symmetry group*. This applies to any space–time $\mathcal{M}$ which suitably approaches, at infinity, either $\mathbb{M}$ or a suitable Friedmann–Robertson–Walker cosmological model. The idea is that by adjoining to $\mathcal{M}$ an appropriate conformal boundary (either $\mathscr{I}^+$ or $\mathscr{I}^-$ or the entire $\mathscr{I}$) we may obtain such asymptotic symmetries as conformal motions of the *boundary*, the boundary having a much better chance of having a meaningful symmetry group than $\mathcal{M}$ itself. (Clearly any isometry of $\mathcal{M}$ will show up as a conformal motion of the boundary, but conformal motions of the boundary need not extend into $\mathcal{M}$ in any meaningful way.)

Bondi parameters

We consider, here, only the case of a suitably asymptotically flat $\mathscr{M}$. We take $\mathscr{M}$ to be *future-3-asymptotically simple* (*cf.* §9.6), with $\mathscr{I}^+$ *null*, with the *strong asymptotic Einstein condition* ((9.6.21) together with $\Psi_{ABCD} \approx 0$) *holding on it*, and satisfying one further condition (Geroch and Horowitz 1978): the generators of $\mathscr{I}^+$ are to be *infinitely long*. (This additional condition was already alluded to earlier in this chapter, *cf.* after (9.6.11).) Explicitly, we have

(9.8.1) DEFINITION

A generator γ *of* $\mathscr{I}^+$ *is called* infinitely long *if a Bondi parameter on* γ *attains the full range* $(-\infty, \infty)$ *on* γ,

where

(9.8.2) DEFINITION

A future-increasing real parameter u on a generator γ *of* $\mathscr{I}^+$ *is called* a Bondi parameter *if, taking* n^a *tangential to the generators of* $\mathscr{I}^+$, *we have*

$$(\text{þ}' - 2\rho')\text{þ}'u = 0,$$

in the compacted spin-coefficient notation of §§4.5, 4.12.

The type of u is $\{0, 0\}$, so þ$'u$ has type $\{-1, -1\}$, whence, by (4.12.15), the equation in (9.8.2) stands for

$$(D' - \varepsilon' - \bar{\varepsilon}' - 2\rho')Du = 0. \tag{9.8.3}$$

If we also take u to have conformal weight 0, then, making use of the notation (5.6.33), we can rewrite the equation in (9.8.2) as

$$\text{þ}_c'^2 u = 0 \tag{9.8.4}$$

Equation (9.8.4) is manifestly conformally invariant (*cf.* (5.6.34)), so it follows that the concept of a Bondi parameter is independent of the conformal factor Ω (and, of course, of the choice of scaling for n^a). If we choose Ω so that $\rho' = 0$ (*cf.* after (9.6.31)), and also scale n^a to be parallelly propagated along γ, then we obtain (9.8.3) in the form

$$D^2 u = 0, \tag{9.8.5}$$

from which it is obvious that the different Bondi parameters on γ are related to one another by

$$u' = Gu + H, \tag{9.8.6}$$

where G, H are real constants with $G > 0$.

Note that in the discussion of (9.8.2) we have dropped the 'hats' on quantities defined at $\mathscr{I}^+$. This, indeed, will be our practice in the sequel, since most of the calculations will be concerned with quantities defined at $\mathscr{I}^+$. Thus, in particular (and in contrast to the two preceding sections), 'g_{ab}' will here denote an 'unphysical' metric, regular at $\mathscr{I}^+$ which is conformal to the physical one. When we have occasion to refer to 'physical' space-time quantities, a *tilde* will be used for those. Accordingly, our conformal rescaling is now

$$g_{ab} = \Omega^2 \tilde{g}_{ab}. \tag{9.8.7}$$

Our discussion in this section and the next is phrased, throughout, in terms of $\mathscr{I}^+$. It will be taken as obvious that our conclusions will also apply to $\mathscr{I}^-$ provided that the appropriate asymptotic conditions apply, instead, to $\mathscr{I}^-$. In physical problems one is, as a rule, more interested in the results at $\mathscr{I}^+$ than at $\mathscr{I}^-$. There are two reasons for this. In the first place, physical considerations lead one to be more concerned with retarded radiation, which shows up in the structure of $\mathscr{I}^+$, than with advanced radiation, which would show up in the structure of $\mathscr{I}^-$ if it were present. In the second place, it seems to be much less clear, in 'physically reasonable' situations, that the required asymptotic conditions should actually hold at $\mathscr{I}^-$, than that they should hold at $\mathscr{I}^+$ (*cf.*, for example, Walker and Will 1979, Porrill and Stewart 1981, Friedrich 1981*a*, *b*).

The Newman–Unti group

We have seen in §9.6 (*cf.* (9.6.31)) that with a suitable choice of Ω, (minus) the metric of $\mathscr{I}^+$ can be put into the form

$$\begin{aligned} dl^2 &= d\theta^2 + \sin^2\theta d\phi^2 + 0\cdot du^2 \\ &= 4d\zeta d\bar{\zeta}(1+\zeta\bar{\zeta})^{-2} + 0\cdot du^2 \end{aligned} \tag{9.8.8}$$

(with $\zeta = e^{i\phi}\cot\frac{1}{2}\theta$ as in (1.2.10)). It is clear (*cf.* (1.2.17)) that this (degenerate) metric for $\mathscr{I}^+$ is conformally preserved under the active point transformations

$$\zeta \mapsto \frac{a\zeta + b}{c\zeta + d}, \quad u \mapsto F(u, \zeta, \bar{\zeta}) \tag{9.8.9}$$

where a, b, c, d are complex constants, with $ad - bc = 1$, and where F is (appropriately) smooth on the whole of $\mathscr{I}^+ \cong \mathbb{R} \times S^2$ (so that $\zeta = \infty$ is adequately dealt with); also F must be monotonic increasing in u for each ζ, mapping the entire range for u for each ζ-generator to itself, and with non-vanishing u-derivative (so that the inverse transformation is also smooth). The transformations (9.8.9)(1) are the non-reflective conformal motions of the S^2-space of generators of $\mathscr{I}^+$ (the conformal structure being defined equivalently by any one of its cross-sections), while (9.8.9)(2), when (9.8.9)(1) is the identity ($a = d = 1$, $b = c = 0$), give the general non-reflective smooth motions of the generators to themselves. The group of transformations (9.8.9) is referred to as the *Newman–Unti (NU) group* – also as the *restricted NU group* (*cf.* Newman and Unti 1962), but we prefer to avoid consideration of the reflective transformations here, and take the term 'NU group' simply to refer to the identity-connected non-reflective transformations (9.8.9). We may thus regard the NU group as *the group of non-reflective motions of $\mathscr{I}^+$ preserving its intrinsic (degenerate) conformal metric.*

The $\mathscr{I}^+$ of $\mathbb{M}$ undergoes transformations of the form (9.8.9) whenever $\mathbb{M}$ is mapped to itself by a restricted Poincaré motion. This is so because the conformal structure of $\mathscr{I}^+$ is determined by the conformal structure of $\mathbb{M}$, and *that* is certainly preserved by Poincaré motions. Thus we may regard the restricted Poincaré group as a *subgroup* of the NU group. But the latter is clearly very much 'larger' than the former, being a function-space group (and therefore infinite-dimensional) rather than merely ten-dimensional.

The Bondi–Metzner–Sachs group

We can cut down the NU group in a natural way to a considerably smaller subgroup, by observing from (9.8.1), (9.8.2) that the Bondi parameters, as well as the intrinsic conformal metric (9.8.8), are determined by the conformal structure of $\mathscr{M}$. By (9.8.6), the preservation of Bondi parameters requires that the F of (9.8.9)(2) specializes to the form

$$F(u, \zeta, \bar{\zeta}) = uG(\zeta, \bar{\zeta}) + H(\zeta, \bar{\zeta}), \tag{9.8.10}$$

where we now demand that u be a Bondi parameter on each generator of $\mathscr{I}^+$ and use the assumption that these generators are infinitely long. This cuts down the freedom of F from a function of three real variables to two functions of two real variables.

It is possible to improve upon this by, in effect, eliminating the freedom in the function G in (9.8.10). For this we need to impose a further structure on $\mathscr{I}^+$ referred to as the *strong conformal geometry* of $\mathscr{I}^+$ (Penrose 1963, 1974).

(In fact, this terminology is slightly misleading because the structure in question involves certain aspects of the *metric* of $\mathscr{M}$ and not just its conformal geometry; but there is also a useful suggestiveness in the terminology, which will emerge shortly.) The group of motions of $\mathscr{I}^+$ to itself preserving this strong conformal geometry is referred to as the *Bondi–Metzner–Sachs* (*BMS*) *group* (or the *restricted* BMS group – but again we prefer to rule out all reflective transformations as part of our definition *cf.* Sachs 1962*b*). As we shall see presently, preservation of strong conformal geometry leads to G being restricted to the form

$$G(\zeta,\bar{\zeta})=\frac{1+\zeta\bar{\zeta}}{|a\zeta+b|^2+|c\zeta+d|^2}, \tag{9.8.11}$$

with a, b, c, d as in (9.8.9)(1).

There are several different ways of describing this additional structure for $\mathscr{I}^+$. Perhaps the most direct is to note that the particular null tangent (i.e. normal) vector N^a to $\mathscr{I}^+$, which is defined in terms of a given conformal factor Ω (satisfying the conditions of (9.6.11)) by

$$N_a=-\nabla_a\Omega \quad near\ \mathscr{I}^+ \tag{9.8.12}$$

(*cf.* (9.6.16)), will transform under a further conformal rescaling

$$g_{ab}\mapsto\Theta^2 g_{ab} \tag{9.8.13}$$

according to

$$N_a\mapsto\Theta N_a,\quad \text{i.e.,}\quad N^a\mapsto\Theta^{-1}N^a\ on\ \mathscr{I}^+. \tag{9.8.14}$$

Here Θ is taken to be smooth on $\bar{\mathscr{M}}$, and nowhere vanishing on $\mathscr{I}^+$; it represents the freedom of choice in Ω. Since g_{ab} is related to the physical metric $\tilde{g}_{ab}$ by $g_{ab}=\Omega^2\tilde{g}_{ab}$, the rescaling (9.8.13) accompanies any replacement

$$\Omega\mapsto\Theta\Omega \tag{9.8.15}$$

and the behaviour (9.8.14) follows at once from (9.8.12) (since the term involving a derivative of Θ vanishes at $\mathscr{I}^+$). The line-element dl of $\mathscr{I}^+$ (positive semi-definite and degenerate, *cf.* (9.8.8)) rescales according to

$$dl\mapsto\Theta\, dl, \tag{9.8.16}$$

whence the product

$$N^a dl \tag{9.8.17}$$

remains invariant:

$$N^a dl\mapsto N^a dl. \tag{9.8.18}$$

It is the invariant structure provided by (9.8.17) that can be taken to define the strong conformal geometry of $\mathscr{I}^+$.

The significance of (9.8.18) is that although no natural choice of parameter scale is defined on the generators of $\mathscr{I}^+$ and no natural metric is defined on its cross-sections, the *ratio* of the two is determined by the strong conformal geometry. Associated with any allowable choice of Ω we have, via the N^a defined by (9.8.12), a definite scaling for parameters u on the generators of $\mathscr{I}^+$, which is fixed by

$$N^a \nabla_a u = 1. \tag{9.8.19}$$

With this choice of scale, we have

$$du \mapsto \Theta\, du \tag{9.8.20}$$

under (9.8.15) (to compensate for (9.8.14)). Thus, by comparison with (9.8.16), we see that the ratio

$$du{:}dl \tag{9.8.21}$$

is independent of the choice of Ω.

We can use this invariance to define a concept of *null angle* between two tangent directions at a point P of $\mathscr{I}^+$, when their span contains the null tangent direction at P (neither of these given two directions being itself null). The angle in the ordinary sense, as defined by the (conformal) metric (9.8.8), is *zero* in these cases, so without some further structure assigned to $\mathscr{I}^+$, no meaning can be given to the notion of a null angle at one point of $\mathscr{I}^+$ being greater or smaller than another null angle at some other point of $\mathscr{I}^+$. But having the strong conformal geometry and with it the invariant ratio (9.8.21), we can numerically *define* the null angle $\nu\,(\in\mathbb{R})$ between two tangent directions at a point of $\mathscr{I}^+$ by

$$\nu = \frac{\delta u}{\delta l}, \tag{9.8.22}$$

where the (infinitesimal) increments δu, δl are as indicated in Fig. 9-21. Conversely, the induced conformal metric (9.8.8) of $\mathscr{I}^+$ supplemented by the concept of null angle, provides the strong conformal geometry of $\mathscr{I}^+$.

In fact, ν has a clear-cut meaning in terms of the geometry of $\mathscr{M}$, which we shall describe shortly. Our more immediate purpose is to relate the concept of Bondi parameter to the concepts we have just been discussing. First, recall the definition (9.6.27) of the quantity A on $\mathscr{I}^+$:

$$An^b :\approx N^b = -\nabla^b\Omega \tag{9.8.23}$$

(*cf.* (9.6.22)). We are, of course, at liberty to scale the *null-tetrad vector* n^b so that $A = 1$. However, here we shall *not* do this since we wish to take advantage of the compacted spin-coefficient formalism, and the making of such a choice (or making it too early) would cause difficulties (ambiguities)

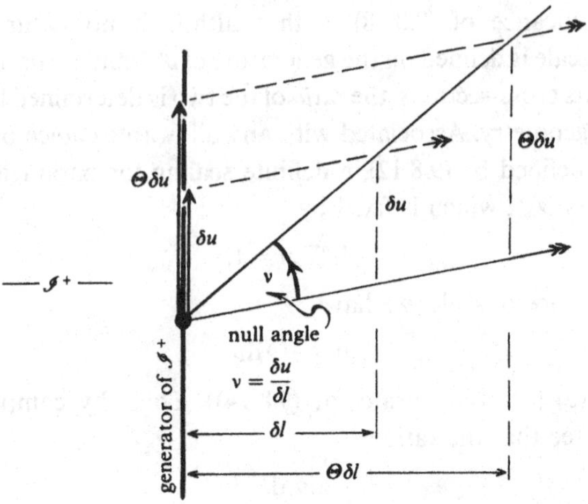

Fig. 9-21. A null angle ν on $\mathscr{I}^+$, given by $\nu = \delta u/\delta l$, is defined between a pair of directions on $\mathscr{I}^+$ whose span contains the null normal direction to $\mathscr{I}^+$. Under change of conformal factor for $\mathscr{I}^+$, null angles are invariant.

in the definitions of the derivative operators. Since n^b has type $\{-1, -1\}$ it follows that A is a real $\{1, 1\}$-scalar on $\mathscr{I}^+$. (We have $n^b = \iota^B\iota^{B'}$, and we choose o^B arbitrarily to complete the spin-frame.)

Next, recall the asymptotic Einstein condition (9.6.26)

$$\nabla_{B'(B}N_{C)C'} \approx 0. \tag{9.8.24}$$

For the dyad components of (9.8.24) that refer only to derivatives *tangential* to $\mathscr{I}^+$, we can substitute $A\iota_C\iota_{C'}$ for $N_{CC'}$ (*cf.* (9.6.23)). These components are then obtained by transvecting (9.8.24) successively with $\iota^B\iota^C$ and with $\iota^{B'}$. The first gives us a result,

$$\sigma' \approx 0 \approx \kappa', \tag{9.8.25}$$

which we have obtained previously (*cf.* (9.6.28)), while the second gives (using, say, (4.12.27) or (4.12.28))

$$(\text{þ}' + \rho')A \approx 0 \approx \eth A. \tag{9.8.26}$$

Note that, as a consequence, ρ' is real:

$$\rho' \approx \bar{\rho}', \tag{9.8.27}$$

but this we also know already, since $\mathscr{I}^+$ is a null hypersurface. Taking N_a to have conformal weight 1, in accordance with (9.8.14), we can write (9.8.26) in the conformally invariant form

$$\text{þ}'_c A \approx 0 \approx \eth_c A \tag{9.8.28}$$

(*cf.* (5.6.33); the choices of w_0 and w_1 are immaterial).

We also note, in passing, that if we specialize our choice of Ω so that the derivative *away* from $\mathscr{I}^+$ of the equation $A\iota_C\iota_{C'} = N_{CC'}$ also holds (i.e., so that each Ω = constant hypersurface 'near' $\mathscr{I}^+$ is also null), then we obtain, from the remaining components of (9.8.24),

$$\rho' \approx \tau' \approx 0 \approx \text{þ} A. \tag{9.8.29}$$

It is always possible to arrange this, but it would be of no particular advantage to us here. We do not generally adopt the full condition (9.8.29), though we shall see shortly that $\rho' \approx 0$ will be helpful for us.

Observe that the condition (9.8.19), which associates a particular choice of scaling for the parameter u on a generator of $\mathscr{I}^+$ with a particular conformal scale on $\mathscr{I}^+$, can be written

$$\text{þ}' u \approx A^{-1} \tag{9.8.30}$$

(or equivalently $D'u \approx A^{-1}$ since u has type $\{0,0\}$).

Once we have made a parameter specialization as in (9.8.30), we must be careful when using $\text{þ}'_c$ and $\eth_c$, etc., which cannot be applied to such a u that has no well-defined conformal weight. However, the standard compacted formalism is still unambiguous. We note that $(\text{þ}' - \rho')A^{-1} \approx 0$, by (9.8.26), so applying $\text{þ}' - \rho'$ to (9.8.30) we obtain $(\text{þ}' - \rho')\text{þ}'u \approx 0$. Comparison of this with the Definition (9.8.2) yields:

(9.8.31) PROPOSITION

The scaling (9.8.19) (*i.e.* (9.8.30)) *is compatible with u being a Bondi parameter on each generator of* $\mathscr{I}^+$ *iff* $\rho' = 0$ *on* $\mathscr{I}^+$, *i.e. iff the generators map the cross-sections of* $\mathscr{I}^+$ *to one another isometrically.*

(The fact that $\sigma' \approx \rho' \approx 0$ is the condition for these cross-sections to be mapped isometrically has been noted earlier, *cf.* after (9.6.31), and §7.1, Fig. 7-2.)

A smooth real function u on $\mathscr{I}^+$, which is a Bondi parameter on each generator, is called a *Bondi* (*retarded*) *time coordinate on* $\mathscr{I}^+$ if it satisfies not only the condition (9.8.30) (i.e. (9.8.19)), so that the generators of $\mathscr{I}^+$ map its cross-sections to one another isometrically, but so that the metric of these cross-sections is actually that of a *unit* 2-*sphere*. (The choice of u clearly fixes the metric on $\mathscr{I}^+$ by the invariance of the null angle ratio $du{:}dl$ of (9.8.21). The canonical selection of the value 'unity' for this null angle* gives a canonical proportionality between u-scalings on the generators and the cross-section metric.) As is implicit in the discussion of §1.2 (*cf.* also

* As we shall investigate presently, null angles actually have the dimension of 'length' or 'time', so this canonical selection involves a choice of units.

§4.15), there is a three-real-parameter freedom in the choice of such conformal scalings, corresponding to different selections of 'asymptotic time-axis' (see Fig. 1-11, Volume 1, p. 38, for a helpful illustration of this). The relation between the corresponding possible ζ-coordinates is given by (9.8.9)(1), yielding the restricted Lorentz transformations on the (anti-) celestial sphere. The conformal factor between two such scalings is given by the ratio of the T-coordinate of the null vector constructed from $\lambda^{\mathbf{A}} = (\zeta, 1)$ to that constructed from its transform $(a\zeta + b, c\zeta + d)$ under the spin-matrix of the transformation (9.8.9)(1). This is

$$\frac{1}{\sqrt{2}}(1 + \zeta\bar{\zeta}):\frac{1}{\sqrt{2}}(|a\zeta + b|^2 + |c\zeta + d|^2) \qquad (9.8.32)$$

We conclude that, to preserve the structure (9.8.21) as is required of a BMS transformation, the function G in (9.8.10) must indeed have the form (9.8.11) as we had previously asserted. With this restriction on the function F of (9.8.10), the general transformations of the BMS group are then given by (9.8.9).

We note the specializations that put the metric of $\mathscr{I}^+$ in the form (9.8.8):

$$\rho' \approx 0, \quad \Phi_{11} + \Lambda \approx \tfrac{1}{2}. \qquad (9.8.33)$$

The second relation follows from Proposition (4.14.21), from $\Psi_2 \approx 0$ by the strong asymptotic Einstein condition, and (9.8.25), the Gaussian curvature of a unit sphere being unity. It should be remarked, however, that (9.8.33) is simply a convenience: the BMS group, as defined here, is independent of such specialization.

Relation between strong conformal geometry and $|^{\alpha\beta}|_{\rho\sigma}$

We shall explore the significance of this group presently. But before doing so, it will be helpful to examine the strong conformal geometry of $\mathscr{I}^+$ from various other points of view. We first note that by squaring (9.8.18) to $N^aN^b\mathrm{d}l^2$ we may obtain this geometric structure in tensorial form. Here $\mathrm{d}l^2$ represents the (positive definite) intrinsic metric tensor $\gamma_{\Psi\Phi}$ for $\mathscr{I}^+$ – where capital Greek letters are now being used for the *abstract indices intrinsic to* $\mathscr{I}^+$. In (9.8.18) we used a 4-space-index on the vector N^a, which is more-or-less allowable for a contravariant index. But for *co*variant indices it is more important to maintain the notational distinction between 3-space and 4-space indices since the cotangent spaces of $\mathscr{I}^+$ are *factor* spaces of those of $\bar{\mathscr{M}}$, whereas the tangent spaces are *sub*spaces. (The difficulty arises because $\mathscr{I}^+$ is *null*, so orthogonal projectors, such as the $S_a{}^b$ of (4.14.6) cannot be naturally defined.) Writing N^a, also, with such intrinsic abstract indices, we have, for the tensor intrinsic to $\mathscr{I}^+$ that represents the square of (9.8.17), the expression

$$N^{\Gamma}N^{\Delta}\gamma_{\Psi\Phi}. \qquad (9.8.34)$$

In terms of 4-space quantities, we may interpret (9.8.34) as the abstract tensor

$$-N^cN^dg_{ef} \text{ modulo multiples of } N_f, N_e, \text{ i.e., } X^{cd}_eN_f, Y^{cd}_fN_e. \qquad (9.8.35)$$

The fact that N^a is orthogonal to all directions tangent to $\mathscr{I}^+$ finds expression in the equation

$$N^{\Psi}\gamma_{\Psi\Phi}=0 \qquad (9.8.36)$$

(which, in terms of (9.8.35), states the manifest result that N^ag_{ab} vanishes modulo multiples of N_b). Clearly we also have

$$\gamma_{\Psi\Phi}=\gamma_{\Phi\Psi}. \qquad (9.8.37)$$

The intrinsic tensor (9.8.34) – subject to (9.8.36), (9.8.37) – represents the strong conformal geometry of $\mathscr{I}^+$.

Taking the (degenerate) conformal metric of $\mathscr{I}^+$ as *given*, we may express the strong-conformal-geometry structure in certain other ways. For example, if $\mathscr{S}$ is the 2-form providing the measure of surface-area (as induced by g_{ab}) for 2-surfaces on $\mathscr{I}^+$, then

$$N^{\Gamma}N^{\Delta}\mathscr{S} \qquad (9.8.38)$$

also provides the required structure, or equivalently

$$N^{\Gamma}N^{\Delta}\sigma_{\Psi\Phi}, \qquad (9.8.39)$$

where $\sigma_{\Psi\Phi}=-\sigma_{\Phi\Psi}$ with $\mathscr{S}=\sigma_{\Psi\Phi}\mathrm{d}x^{\Psi}\wedge\mathrm{d}x^{\Phi}$ on $\mathscr{S}^+$; and $N^{\Psi}\sigma_{\Psi\Phi}=0$. In fact, we also have

$$\sigma_{\Gamma[\Delta}\sigma_{\Phi]\Psi}=\tfrac{1}{4}\gamma_{\Gamma[\Delta}\gamma_{\Phi]\Psi} \qquad (9.8.40)$$

(as is readily seen from a diagonal coordinate description). The *sign* of $\mathscr{S}$ determines the orientation of $\mathscr{S}^+$, this sign being ambiguous in (9.8.40). We take the orientation *and* time-orientation of $\mathscr{I}^+$ to be part of its given intrinsic structure.

Yet another way in which the strong conformal geometry of $\mathscr{S}^+$ arises is in relation to twistor theory. (This will have some significance in §9.9.) We recall that in the twistor description of $\mathbb{M}^\#$, the particular light cone which defines $\mathscr{I}^+$ is picked out by specifying, up to proportionality, the *infinity twistor* $\mathrm{I}^{\alpha\beta}$, or its dual $\mathrm{I}_{\alpha\beta}$ (*cf.* (6.2.25)). The choice of the actual scaling for $\mathrm{I}^{\alpha\beta}$ (or, rather, for $\mathrm{I}^{\alpha\beta}\mathrm{I}_{\rho\sigma}$) determines the strong conformal geometry of $\mathscr{I}^+$. We shall see that this, in fact, works equally well for $\mathscr{M}$, where we use a *local twistor* description of $\mathrm{I}^{\alpha\beta}$ (*cf.* §6.9).

If we suppose that $\mathcal{M}$ is empty near $\mathscr{I}^+$, then $\tilde{\mathrm{P}}_{ab}=0$ (*cf.* (6.8.12)), so the 'spinor part' description $(\tilde{\varepsilon}^{AB},0;0,0)$ of $\mathrm{I}^{\alpha\beta}$, with respect to the physical metric $\tilde{g}_{ab}$, is *constant* under local twistor transport (6.9.12). Because of the conformal invariance of local twistor transport, it follows that the description of $\mathrm{I}^{\alpha\beta}$ with respect to g_{ab} is also constant. By (6.9.6), this description is

$$\mathrm{I}^{\rho\sigma}=\begin{bmatrix}\tilde{\varepsilon}^{RS} & \mathrm{i}\Upsilon_{DS'}\tilde{\varepsilon}^{RD}\\ \mathrm{i}\Upsilon_{CR'}\tilde{\varepsilon}^{CS} & -\Upsilon_{CR'}\Upsilon_{DS'}\tilde{\varepsilon}^{CD}\end{bmatrix}=\begin{bmatrix}\Omega\varepsilon^{RS} & -\mathrm{i}N^{R}{}_{S'}\\ \mathrm{i}N_{R'}{}^{S} & (\Omega\Lambda+\tfrac{1}{4}\nabla_cN^c)\varepsilon_{R'S'}\end{bmatrix}$$

$$\approx A\begin{bmatrix}\mathrm{O} & -\mathrm{i}\iota^{R}\iota_{S'}\\ \mathrm{i}\iota_{R'}\iota^{S} & -\rho'\varepsilon_{R'S'}\end{bmatrix}. \tag{9.8.41}$$

Here we have not assumed $\tilde{R}_{ab}=0$, but merely (compare Proposition (9.6.18)) that the *physical scalar curvature is zero near* $\mathscr{I}^+$:

$$\tilde{\Lambda}=0, \tag{9.8.42}$$

which will be the case if, near $\mathscr{I}^+$, we have massless fields only. In the derivation of (9.8.41) we have used (6.8.21), (6.8.23), $\varepsilon^{RS}=\Omega^{-1}\tilde{\varepsilon}^{RS}$ (*cf.* (9.8.7)), and $\Upsilon_a=-\Omega^{-1}N_a$; also the relation

$$\nabla_cN^c\approx-4A\rho', \tag{9.8.43}$$

which follows from (9.8.24) (this latter showing that the index permutations of $\nabla_{BB'}N_{CC'}$ differ at most by signs) and the relation

$$o^{B}\iota^{B'}\iota^{C}o^{C'}\nabla_bN_c=A\rho', \tag{9.8.44}$$

which can be established by substituting (9.8.23), since the derivative in it is tangential.

It may be verified that the local twistor (9.8.41) is actually *constant* on $\mathscr{I}^+$ even in this more general case (9.8.42). The calculation is a straightforward application of the definitions given in §6.8, and is somewhat facilitated by making the specialization (9.8.29), since that entails

$$\iota^{B}\nabla_{BB'}N_c\approx 0. \tag{9.8.45}$$

Note that the significant spinor part of $\mathrm{I}^{\rho\sigma}$ at $\mathscr{I}^+$ is, by (9.8.41),

$$\mathrm{i}A\iota_{R'}\iota^{S} \tag{9.8.46}$$

(since ρ' can be scaled to zero). This is closely related to the strong conformal structure (9.8.34). For we can write (9.8.35) in spinor form as

$$-A^2\iota^{C}\iota^{C'}\iota^{D}\iota^{D'}\varepsilon_{EF}\varepsilon_{E'F'}\quad \textit{modulo expressions}\quad X^{cd}_{e}\iota_{F}\iota_{F'},\ Y^{cd}_{f}\iota_{E}\iota_{E'}. \tag{9.8.47}$$

The freedom in $X^{\cdot\cdot}_{\cdot}$ and $Y^{\cdot\cdot}_{\cdot}$ can be killed off precisely by transvecting with ι^F *or* $\iota^{F'}$ and with ι^E *or* $\iota^{E'}$. Clearly, if we transvect with two primed iotas, or with two unprimed iotas, the result is zero; so to obtain a non-trivial quantity we must take one of each. This leads to

$$A^2\iota^C\iota^{C'}\iota^D\iota^{D'}\iota_E\iota_{F'}$$

(and its complex conjugate), from which we must remove (say) $\iota^C\iota^{D'}$ in order to compensate for the extra $\iota^F\iota^{E'}$ that was introduced through the transvection. The result, which we write

$$(\mathrm{i}A\iota_{F'}\iota^D)(-\mathrm{i}A\iota_E\iota^{C'}), \qquad (9.8.48)$$

is of the form of an outer product of (9.8.46) with its complex conjugate. Thus the intrinsic tensor (9.8.34), on $\mathscr{I}^+$, in effect factorizes into two pieces (different from (9.8.17)), namely the (significant) spinor part of $\mathsf{I}^{\rho\sigma}$ and that of its complex conjugate $\mathsf{I}_{\rho\sigma}$. In this sense, the strong conformal metric of $\mathscr{I}^+$ effectively 'splits' into a product $\mathsf{I}^{\alpha\beta}\mathsf{I}_{\rho\sigma}$, reminiscent of the 'splitting' (3.1.9) of the space–time metric into $\varepsilon_{AB}\varepsilon_{A'B'}$.

The interpretation of null angles

As we mentioned earlier, the strong conformal geometry of $\mathscr{I}^+$ really arises as a *metric* property of $\mathscr{M}$ rather than as a conformal one. In order to see this most clearly, it is helpful to have a direct interpretation of it in terms of the geometry of $\mathscr{M}$. So let P be a point of $\mathscr{I}^+$ and consider two non-null tangent directions α and β to $\mathscr{I}^+$ at P. The significance of the two types of angle that can arise between α and β, in terms of space–time geometry, is best seen if we first consider, instead of α and β, their orthogonal complements (in terms of the space–time conformal structure at P). These are two hyperplane elements α^*, β^*, respectively, at P. Since α and β are spacelike, α^* and β^* must be timelike and so will have non-trivial intersections with the past null cone at P. These intersections are collections of null directions at P each of which extends to a *ray* in $\mathscr{M}$ – a generator of the past *light cone* $\mathscr{C}$ of P in $\mathscr{M}$. Thus we shall have an interpretation of α^*, β^*, and therefore of α, β, in terms of the ordinary space–time geometry of light rays (see Fig. 9-22).

Now recall from the discussion of §9.1 that a null hypersurface such as $\mathscr{C}$ has a spatial interpretation as an *asymptotically plane wave-front*. This property of asymptotic planarity entails that the geometry of a 2-space of (local) cross-section of $\mathscr{C}$ settles down to become that of a Euclidean plane, as the section proceeds into the future along $\mathscr{C}$. When $\mathscr{M}$ is Minkowski space $\mathbb{M}$, then $\mathscr{C}$ is necessarily a null hyperplane, having sections that are all intrinsically *exact Euclidean 2-planes*, and which are mapped isometrically to one another by the generators of $\mathscr{C}$. In the general case, we have a well-

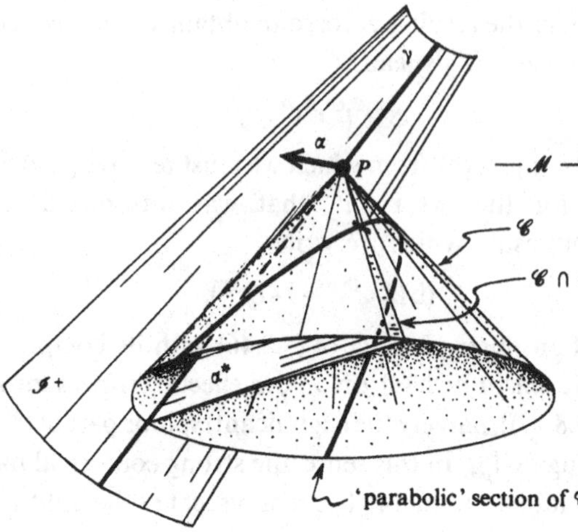

Fig. 9-22. A tangent direction α to $\mathscr{I}^+$ at P can be represented by its orthogonal complement α^*. This intersects a 'parabolic' section (with Euclidean 2-space structure $\mathbb{E}_p$), of the past cone $\mathscr{C}$ of P, in a 'straight line' in $\mathbb{E}_P$ – and this line can also be used to represent α.

defined *limiting* exact Euclidean plane $\mathbb{E}_P$, describing the geometry of the generators of $\mathscr{C}$, whose points correspond to the various (past) null directions at P, other than that of the one generator γ of $\mathscr{I}^+$ through P.

We may, in fact, think of $\mathbb{E}_P$ as the section, in the tangent space $\mathfrak{T}^\bullet[P]$ at P, of the past null cone at P by a null hyperplane parallel to γ. This 'parabolic' section has an intrinsic Euclidean 2-metric (*cf.* Fig. 9-22; compare Fig. 9-6 and, more appropriately, Fig. 1-5, Volume 1, p. 13), and we can arrange that it is scaled correctly by requiring that in terms of its induced metric g_{ab} ('unphysical'), and the associated N_a, the equation in $\mathfrak{T}^\bullet[P]$ of this section (with origin at P) is

$$x^a N_a + 1 = 0 = g_{ab} x^a x^b. \qquad (9.8.49)$$

The induced metric is then invariant under (9.8.13), (9.8.14).

We pass now to the orthogonal complements at P. For this, it is best to associate a point of $\mathbb{E}_P$ not just with a null direction at P, but with the 2-plane element spanned by this null direction and the null direction of γ. Indeed, each 2-plane element through the γ-direction (but not tangential to $\mathscr{I}^+$) contains just this one null direction, in addition to the null direction of γ, at P. The orthogonal complement of such a 2-plane element is another 2-plane element, but now *tangent* to $\mathscr{I}^+$ and *not* containing the γ-direction. Thus

the points of $\mathbb{E}_P$ *correspond to non-null tangent 2-plane elements to* $\mathscr{I}^+$ *at P.*

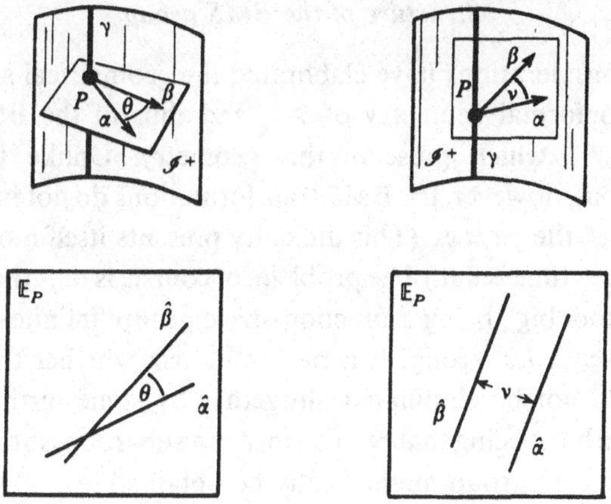

Fig. 9-23. In the representation of (spacelike) tangent directions to $\mathscr{I}^+$ by lines in $\mathbb{E}_P$, ordinary angles at $\mathscr{I}^+$ are directly equal to the corresponding angles between lines in $\mathbb{E}_P$. However null angles at $\mathscr{I}^+$ are represented as distances between parallel lines in $\mathbb{E}_P$.

Dually, within $\mathscr{I}^+$, a tangent *direction* α to $\mathscr{I}^+$ at P corresponds, in $\mathbb{E}_P$, to the dual (within $\mathbb{E}_P$) of a point, namely a Euclidean straight line (see Fig. 9-23). (Note that a tangent direction to $\mathscr{I}^+$ at P is the intersection of two tangent 2-planes to $\mathscr{I}^+$; a straight line in $\mathbb{E}_P$ is the join of two points in $\mathbb{E}_P$.) Thus we have:

non-null tangent directions to $\mathscr{I}^+$ at P correspond to straight lines in $\mathbb{E}_P$. (9.8.50)

Indeed, this is precisely the correspondence we obtained earlier whereby the directions α and β were respectively represented by the intersections of α^* and β^* with the past null cone of P. We now see, by (9.8.50), what these intersections correspond to: α and β are represented by straight lines, $\hat{\alpha}$ and $\hat{\beta}$, respectively, in $\mathbb{E}_P$. This enables us to visualize the two types of angle between α and β in familiar terms. In the general case there will be a non-zero angle θ between $\hat{\alpha}$ and $\hat{\beta}$. Straightforward geometry shows that θ is also the non-null angle between α and β. But θ can also be zero, in which case $\hat{\alpha}$ and $\hat{\beta}$ become parallel. Then a new measure of separation between $\hat{\alpha}$ and $\hat{\beta}$ arises, namely the Euclidean *distance* ν between them! Straightforward calculation shows that this ν is precisely the null angle (9.8.22) between α and β.

Structure of the BMS group

The above considerations have elaborated the geometrical significance of the strong conformal geometry of $\mathscr{I}^+$, and thus of the BMS group of motions of $\mathscr{I}^+$ which preserve this geometry. Unlike the Poincaré transformations, however, the BMS transformations do not in any obvious sense preserve* the *physics*. (This difficulty presents itself most manifestly when the space–time is flat.) The problem, of course, is that the BMS group is still much too 'big', being a function-space group (infinite-dimensional) and so not even a Lie group. It is natural to ask whether this functional freedom could not be eliminated altogether by some further geometric restriction, the hope being that by such means a subgroup isomorphic to the restricted Poincaré group might finally be obtained.

The source of the 'size' of the BMS group is the function H of (9.8.10) (*cf.* (9.8.11)). In the case of $\mathbb{M}$, restricted Poincaré motions induce transformations of $\mathscr{I}^+$, which can be written as (9.8.9), with (9.8.10) and (9.8.11) valid, but with H specialized to the form

$$H(\zeta,\bar{\zeta}) = \frac{H^{00'} - H^{10'}\zeta - H^{01'}\bar{\zeta} + H^{11'}\zeta\bar{\zeta}}{2^{-1/2}(1+\zeta\bar{\zeta})}, \qquad (9.8.51)$$

where $H^{AB'}$ is constant and Hermitian. To obtain this, u must be taken to be a special type of Bondi time coordinate, for which $u = 0$ is the intersection of some light cone in $\mathbb{M}$ with $\mathscr{I}^+$. We take the vertex of this light cone as origin O, and choose that unit future-timelike vector T^a at O which has the property that its associated scaling of the (anti-)celestial sphere as a unit sphere agrees with that which is determined by the Bondi time coordinate u (*cf.* after (9.8.31)). Then we find that u is simply the standard retarded time parameter of an inertial observer with origin O and time-axis T^a (see Fig. 9-24). That is to say, a given value u of the Bondi time on $\mathscr{I}^+$ is attained where the future light cone of the point with position vector uT^a in $\mathbb{M}$ meets $\mathscr{I}^+$. Adopting standard Minkowski coordinates in $\mathbb{M}$, one readily verifies that the particular BMS transformations (9.8.9) (with (9.8.10)) for which H takes the form (9.8.51) are given by the active (restricted) Poincaré motions (in terms of position vectors relative to O, which is taken as fixed):

$$x^{AA'} \mapsto S^A{}_B x^{BB'} \bar{S}^{A'}{}_{B'} + H^{BB'}, \qquad (9.8.52)$$

where $S^A{}_B$ and $H^{BB'}$ are the spinors whose standard components (*cf.*

* For the representation theory of the BMS group, see Sachs (1962*b*), Cantoni (1967), McCarthy (1976) and references contained therein.

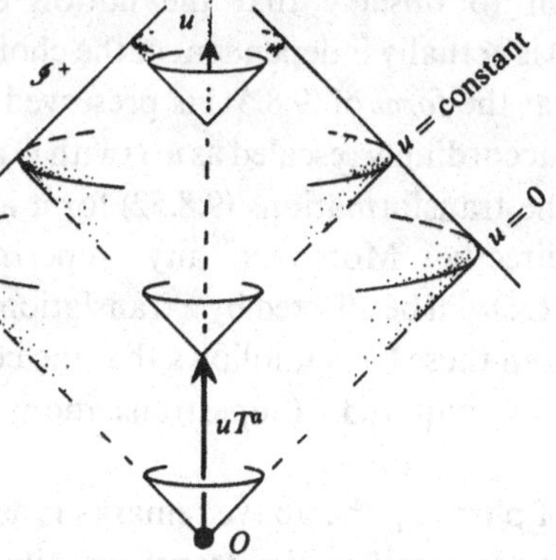

Fig. 9-24. A standard type of Bondi time coordinate for $\mathbb{M}$ is obtained by taking a timelike straight line in $\mathbb{M}$ and choosing u to be proper time along it, with u constant along the future light cones emanating from its points.

(3.3.31)) are the $H^{\mathbf{BB}'}$ of (9.8.51) and the

$$S^{\mathbf{A}}{}_{\mathbf{B}} := \begin{pmatrix} a & b \\ c & d \end{pmatrix} \tag{9.8.53}$$

of (9.8.9)(1), respectively. In fact, (9.8.51) can be expressed more 'invariantly' as

$$H(\lambda^A, \bar{\lambda}^{A'}) = \frac{H_a L^a}{T_b L^b}, \tag{9.8.54}$$

where

$$L^a = \lambda^A \bar{\lambda}^{A'}, \quad \lambda^{\mathbf{A}} \propto (\zeta, 1) \tag{9.8.55}$$

define the null direction at O (or at any other point of $\mathbb{M}$) in which the generator of $\mathscr{I}^+$ labelled by ζ lies.

Note that when (9.8.9)(1) is the identity (i.e. when (9.8.53) is the unit matrix), the transformations (9.8.52) are simply translations of $\mathbb{M}$. For this reason, whenever (9.8.9)(1) is the identity (so $G = 1$ in (9.8.11)) and H is of the form (9.8.51), the corresponding BMS transformations – even in the case of a *curved* $\mathscr{M}$ – are referred to as *translations*, and are said to constitute the 4-parameter *translation subgroup* $\mathscr{T}$ of the BMS group $\mathscr{B}$. The more general transformations for which (9.8.9)(1) is the identity (and so $G = 1$) but for which H is a general smooth function on the sphere, are called *supertranslations*, and they constitute an infinite-parameter subgroup $\mathscr{U}$ of $\mathscr{B}$. Thus we have

$$\mathscr{T} \subset \mathscr{U} \subset \mathscr{B}. \tag{9.8.56}$$

It is important to observe that the notion of translation (and of supertranslation) is actually independent of the choice of and of Bondi time coordinate u. That the *form* of (9.8.51) is preserved whenever ζ undergoes (9.8.9)(1) and u is accordingly rescaled as uG (with G as in (9.8.11)), is implicit in the fact that the transformations (9.8.52) form a group. (It can also be easily verified directly.) Moreover, any 'supertranslated' Bondi time coordinate $u + h(\zeta, \bar{\zeta})$ will be affected by a translation in just the same way as the original u. From these facts it follows that the concept of translation is BMS-invariant, as required. (Supertranslations will be dealt with presently.)

Another way of phrasing the above remarks is as follows. Let $\mathscr{R}$ denote the group of *Lorentz rotations* (i.e. transformations given by (9.8.9) and (9.8.11), with $H = 0$ in (9.8.10)), then

$$\boldsymbol{r}^{-1}\mathscr{T}\boldsymbol{r} = \mathscr{T} \quad \textit{for all } \boldsymbol{r} \in \mathscr{R}. \tag{9.8.57}$$

Moreover,

$$\boldsymbol{s}^{-1}\mathscr{T}\boldsymbol{s} = \mathscr{T} \quad \textit{for all } \boldsymbol{s} \in \mathscr{U}. \tag{9.8.58}$$

(Indeed, in the *second* case, $\mathscr{T}$ is preserved element-wise, since all supertranslations commute.) Now, by the form of (9.8.9), it is clear that every element of $\mathscr{B}$ has the form $\boldsymbol{sr}$ with $\boldsymbol{s} \in \mathscr{U}$ and $\boldsymbol{r} \in \mathscr{R}$, i.e. that

$$\mathscr{B} = \mathscr{U}\mathscr{R}. \tag{9.8.59}$$

Combining this with (9.8.57) and (9.8.58), we find that

$$\boldsymbol{b}^{-1}\mathscr{T}\boldsymbol{b} = \mathscr{T} \quad \textit{for all } \boldsymbol{b} \in \mathscr{B}, \tag{9.8.60}$$

i.e. $\mathscr{T}$ is a *normal subgroup* of $\mathscr{B}$. In the same way, $\mathscr{U}$ is also a normal subgroup of $\mathscr{B}$:

$$\boldsymbol{b}^{-1}\mathscr{U}\boldsymbol{b} = \mathscr{U} \quad \textit{for all } \boldsymbol{b} \in \mathscr{B}. \tag{9.8.61}$$

So the concept of supertranslation is also BMS-invariant. Moreover, there is the following result of Sachs (1962*b*) which we quote without proof:

(9.8.62) THEOREM

The translation subgroup of the BMS-group is its unique 4-parameter normal subgroup.

The invariance of the translation concept is essentially an instance of a phenomenon that was noted in §4.15. A general supertranslation is determined by an arbitrary (smooth) function H on the 2-sphere, and from the form of (9.8.51) – and with reference to formulae given at the end of §4.15 – we see that:

(9.8.63) PROPOSITION

The translations are precisely the supertranslations for which H is composed only of $j = 0$ and $j = 1$ *spherical harmonics.*

(Here the spin-weight is zero.) From table (4.15.60) we observe that this condition on H can be written as

$$ð^2 H = 0. \qquad (9.8.64)$$

We think of the $(\zeta, \bar{\zeta})$ 2-sphere as the space of generators of $\mathscr{I}^+$ (i.e. as the factor space of $\mathscr{I}^+$ by its generators). Under the action of $\mathscr{R}$ (or, indeed, of $\mathscr{B}$) the functions H behave as *conformally* weighted scalars of weight $w = 1$. This is because the 'scaling' du of the u-parameter has conformal weight $w = 1$, by (9.8.20)), though the u-parameters themselves are *not* conformally weighted objects. We recall from the discussion of §4.15 that for such scalars the $j = 0, j = 1$ parts transform among themselves under conformal motions of the sphere, while the higher j-value parts do not (i.e. they can pick up $j = 0$ or $j = 1$ parts under such motions). Thus while the concept of translation is Lorentz invariant, we have

(9.8.64) PROPOSITION

The property of a supertranslation that it be translation-free is not Lorentz invariant.

In this proposition we could substitute 'BMS-invariant' for 'Lorentz invariant' if desired. The Lorentz transformations referred to are simply the conformal motions (9.8.9) of the $(\zeta, \bar{\zeta})$-sphere. This (restricted) Lorentz group $\mathscr{L}$ therefore has a natural interpretation as a *factor* group of $\mathscr{B}$:

$$\mathscr{L} = \mathscr{B}/\mathscr{U}. \qquad (9.8.65)$$

On the other hand, the Lorentz group does not arise canonically as a *sub*group of $\mathscr{B}$. The subgroup $\mathscr{R}$ of $\mathscr{B}$ is isomorphic with $\mathscr{L}$, but it is far from being canonically singled out. For suppose that s is any element of $\mathscr{U}$, then the group

$$\mathscr{R}' = s^{-1}\mathscr{R}s \qquad (9.8.66)$$

will be another subgroup of $\mathscr{B}$, also isomorphic with the restricted Lorentz group, and – so far as the group structure of $\mathscr{B}$ is concerned – completely on an equal footing with $\mathscr{R}$. The distinguishing feature of $\mathscr{R}$ is that it consists of elements which leave a particular cross-section of $\mathscr{I}^+$ invariant. One refers to (smooth) cross-sections of $\mathscr{I}^+$ as *cuts*. The cut Γ left invariant by $\mathscr{R}$ is simply that defined by $u=0$ in the given coordinates. The supertranslation s^{-1} will carry Γ into some other cut $\Gamma' = s^{-1}\Gamma$, and we see from (9.8.66) that $\mathscr{R}'$ is the subgroup of $\mathscr{B}$ leaving Γ' invariant. Only when s is the identity element **1** will $\mathscr{R}'$ and $\mathscr{R}$ be the same, and as s ranges over the whole of $\mathscr{U}$, the invariant cut will range over all possible cuts.

Of course, it is not to be expected that the (restricted) Lorentz group should arise naturally as a subgroup. It does not even do so in relation to the ordinary (restricted) Poincaré group of $\mathbb{M}$, where also it arises naturally only as a factor group. (As a subgroup it depends on the choice of an arbitrary origin in $\mathbb{M}$.) However, in relation to $\mathscr{B}$ the situation is much 'worse' in that not even the restricted Poincaré group $\mathscr{P}$ itself arises naturally as a subgroup of $\mathscr{B}$ in general (or as a factor group, for that matter).

To appreciate this, let us consider first the case when $\mathscr{B}$ refers to the $\mathscr{I}^+$ of $\mathbb{M}$. We can regard $\mathscr{P}$ as the subgroup of $\mathscr{B}$ generated by the translations $\mathscr{T}$ and Lorentz rotations $\mathscr{R}$; indeed (compare (9.8.59)):

$$\mathscr{P} = \mathscr{T}\mathscr{R}. \tag{9.8.67}$$

While $\mathscr{R}$ is not canonically singled out within $\mathscr{P}$, the *family* of Lorentz rotation groups about the various different origins in $\mathbb{M}$ *is* so singled out. Any such Lorentz subgroup of $\mathscr{P}$ arises as the subgroup of $\mathscr{B}$ leaving invariant a certain type of cut, referred to as a *good cut*, which is the intersection with $\mathscr{I}^+$ of the light cone of some point in $\mathbb{M}$. These good cuts, which form a 4-parameter system, are obtained from one another by translations; but a supertranslation which is not a translation always takes a good cut into a *bad* cut (i.e. a cut which is *not* a good cut). The concept of a good cut is thus not BMS-invariant. Indeed, if the s of (9.8.66) is not a translation, then the $\mathscr{R}'$ it defines will be taken outside $\mathscr{P}$. Similarly, for such a supertranslation, the 10-parameter subgroup of $\mathscr{B}$,

$$\mathscr{P}' = s^{-1}\mathscr{P}s, \tag{9.8.68}$$

will be distinct from $\mathscr{P}$, though isomorphic with it. In fact, for a general s, the subgroups $\mathscr{P}$ and $\mathscr{P}'$ have only the translations $\mathscr{T}$ in common.

Shear structure of $\mathscr{I}^+$

This tells us that, so far as the group structure of $\mathscr{B}$ alone is concerned, there is no way to single out the restricted Poincaré group uniquely as a subgroup. (And, as we have seen, the problem is not to distinguish translations from supertranslations, but to say what we mean by a 'supertranslation-free' Lorentz rotation.) In the case of $\mathbb{M}$ we have some additional structure at $\mathscr{I}^+$, namely the notion of which cuts are to be labelled as 'good cuts', in relation to which a unique subgroup $\mathscr{P}$ may indeed be singled out. Thus it would seem that for us to obtain an appropriate unique analogue of $\mathscr{P}$ in the case of a *curved* $\mathscr{M}$, a corresponding notion of 'good cut' would be required.

There are, however, some serious difficulties with this. For several reasons it is not appropriate simply to take the intersections with $\mathscr{I}^+$ of actual light cones in $\mathscr{M}$ as good cuts. (For example, because of the appearance of caustics and crossing regions on the cone, such 'cuts' need not even be cross-sections of $\mathscr{I}^+$ – and even those that are do not generally transform among themselves according to a subgroup of $\mathscr{B}$.) A more reasonable plan is to use a definition which is 'local' on $\mathscr{I}^+$ (and hence entirely asymptotic in $\mathscr{M}$). In $\mathbb{M}$, the (future) light cones are singled out by the property that they diverge and have *vanishing shear*, $\sigma = 0$ (see §7.1). Thus in $\mathbb{M}$ we need only examine the shear at the cut itself to ensure that the cut is good. Taking the o-flagpoles to be orthogonal to the cut (as in the standard treatment of spacelike 2-surfaces given in §4.14), we may compute σ at the cut, and if we find $\sigma = 0$ over the whole cut we say that the cut is a good cut.

This procedure works not only for $\mathbb{M}$, but also for certain other space-times $\mathscr{M}$, notably those which are *stationary*. But there are fundamental difficulties when gravitational radiation is present. To see how these come about, we first develop some further formulae. To begin with, suppose that we have some smooth coordinate u on $\mathscr{I}^+$ which suitably increases up the generators, but which need not be a Bondi time parameter. Choose the o-flagpoles orthogonal to the cuts $u =$ constant, as above. Then we have (with u of type $\{0,0\}$)

$$\eth u \approx 0. \qquad (9.8.69)$$

Applying the commutator $\overline{(4.12.34)}'$ to u we get

$$(\text{þ}'\eth - \eth\text{þ}')u \approx \rho'\eth u - \tau\text{þ}'u,$$

by (9.8.25), and with (9.8.69) this reduces to

$$(\eth - \tau)\text{þ}'u \approx 0. \qquad (9.8.70)$$

Henceforth we shall adopt the specialization (9.8.33). Then the condition (9.8.30) for u to be a Bondi time coordinate (associated with the chosen scaling for $\mathscr{I}^+$) becomes $\text{þ}'u \approx A^{-1}$, and with (9.8.28) this further reduces (9.8.70) to

$$\tau \approx 0. \qquad (9.8.71)$$

(The argument is partly reversible: if (9.8.71) holds, (9.8.33) being assumed, then u is a *function* of this associated Bondi time parameter, so the latter and the former are constant on the same cuts.) We refer to the choice of o-flagpoles related in this way to a Bondi time parameter associated with the chosen scaling for $\mathscr{I}^+$ as a *Bondi system*. Thus we have:

(9.8.72) PROPOSITION

The condition for a Bondi system is $\tau \approx 0$.

Equation (4.12.32)(*e*) gives

$$\text{þ}'\sigma \approx \eth\tau - \tau^2 - \bar{\text{N}} \tag{9.8.73}$$

where

$$\text{N} \approx \Phi_{20} \tag{9.8.74}$$

is the Bondi–Sachs complex *news function*, whose important relation to gravitational energy-flux will be discussed in §9.9. In a Bondi system (9.8.73) becomes

$$\text{þ}'\sigma \approx -\bar{\text{N}}. \tag{9.8.75}$$

Recall that, as part of the strong asymptotic Einstein condition assumed for $\mathscr{M}$, we have

$$\Psi_{ABCD} \approx 0, \tag{9.8.76}$$

i.e.

$$\Psi_0 \approx \Psi_1 \approx \Psi_2 \approx \Psi_3 \approx \Psi_4 \approx 0. \tag{9.8.77}$$

Hence, from (4.12.32)(*a*)′ and (4.12.32)(*d*)′,

$$\Phi_{22} \approx 0 \approx \Phi_{21}. \tag{9.8.78}$$

Next we need the Bianchi identities at $\mathscr{I}^+$. Recalling that the spin-2 field

$$\psi_{PQRS} := \Omega^{-1}\Psi_{PQRS} \tag{9.8.79}$$

is C^1-smooth at $\mathscr{I}^+$ (with our initial assumptions on $\mathscr{M}$, *cf.* Theorem (9.6.41)), we have

$$A\iota_M\iota_{M'}\psi_{PQRS} \approx -\nabla_{MM'}\Psi_{PQRS} \tag{9.8.80}$$

(*cf.* (9.6.42)). With (9.8.77), this gives

$$A\psi_i \approx -\text{þ}\Psi_i \quad (i = 0, \ldots, 4) \tag{9.8.81}$$

(while, of course, $\eth\Psi_i \approx \eth'\Psi_i \approx \text{þ}'\Psi_i \approx 0$). Thus, by (4.12.39),

$$A\psi_4 \approx \text{þ}'\text{N}. \tag{9.8.82}$$

Also, taking the combination (4.12.38) + (4.12.41)′ + $\overline{(4.12.38)}$′, and noting that, by (9.8.33), $\eth'\Lambda + \eth'\Phi_{11} \approx 0$, we obtain

$$A\psi_3 \approx \eth \mathrm{N}. \tag{9.8.83}$$

We shall also be interested in a generalization of (9.8.83), obtained when we merely assume $\rho' \approx 0$ but not the other half of (9.8.33). Thus $\mathscr{I}^+$ is given a metric scaling so that all its cuts are mapped isometrically to one another by its generators, but they are not necessarily metric spheres. Then we have

$$A\psi_3 \approx \eth\Phi_{20} - \eth' K, \tag{9.8.84}$$

with

$$K \approx \Phi_{11} + \Lambda. \tag{9.8.85}$$

By (4.14.20), this *real* K is one-half the Gaussian curvature of the cuts. But for our immediate purposes we assume $K = \frac{1}{2}$, so (9.8.84) reduces to (9.8.83).

Next, subtracting (4.12.37) from its complex conjugate, we have

$$A\psi_2 - A\bar{\psi}_2 \approx \eth\Phi_{10} - \eth'\Phi_{01} + \sigma\mathrm{N} - \bar{\sigma}\bar{\mathrm{N}}, \tag{9.8.86}$$

while, from (4.12.32)(*d*),

$$\Phi_{01} \approx \eth\rho - \eth'\sigma. \tag{9.8.87}$$

In deriving (9.8.87) we used

$$\rho = \bar{\rho}, \tag{9.8.88}$$

which (*cf.* (4.14.2), (7.1.58)) expresses the fact that the 2-plane elements on $\mathscr{I}^+$ orthogonal to the *o*-flagpoles are *surface-forming* (namely, tangent to $u = \text{constant}$). Also, by (4.12.35) and the reality of K, $\eth'\eth\rho$ is real, so substituting (9.8.87) into (9.8.86) we get

$$A\psi_2 - \sigma\mathrm{N} + \eth'^2\sigma \approx A\bar{\psi}_2 - \bar{\sigma}\bar{\mathrm{N}} + \eth^2\bar{\sigma}, \tag{9.8.89}$$

a reality property which will have significance for us in §9.9.

In passing, we note that the form of the Bianchi identity given in (6.8.17),

$$\nabla^e C_{efgh} = -2\nabla_{[g}\mathrm{P}_{h]f}$$

with $\mathrm{P}_{ab} = \Phi_{ab} - \Lambda\varepsilon_{AB}\varepsilon_{A'B'}$, provides us with the equation

$$A\psi_{1FGH}\iota_{F'}\varepsilon_{G'H'} + A\iota_F\varepsilon_{GH}\bar{\psi}_{1'F'G'H'} \approx 2\nabla_{[g}\mathrm{P}_{h]f}, \tag{9.8.90}$$

from which various other expressions for the ψ_i may be directly obtained. (From it one can also rederive (9.8.82), (9.8.84), and (9.8.86), though not quite trivially; in this connection we observe that (9.8.78) and (9.8.85) can be combined into the relation $n^b\mathrm{P}_{ABB'A'} = Kn_a$.)

Of most immediate interest to us here are relations (9.8.82) and (9.8.75), which together imply that, *in a Bondi system*,

$$\text{þ}'^2\sigma \approx -A\bar{\psi}_4. \tag{9.8.91}$$

We recall from the discussion of §9.7 that ψ_4, on $\mathscr{I}^+$, measures the outgoing gravitational radiation field (i.e. the 'r^{-1}-part' of the physical Weyl curvature field). Thus, one implication of (9.8.91) is that *whenever outgoing gravitational radiation is present, we cannot preserve the 'good cut' condition* $\sigma = 0$ *throughout the Bondi system.* The essence of a Bondi system is that the cuts $u =$ constant are all time translations, with respect to a fixed 'time direction' (namely that provided by the given unit-sphere scaling of the cuts) of a given cut (say of $u = 0$). What we have just seen, in effect, is that the translation of a good cut is generally a bad cut. (The term 'translation' is used here in the sense that the second cut is a BMS-translation of the first while, however, the space–time itself – including $\mathscr{I}^+$ – is *not* moved.)

Thus in a system with outgoing gravitational radiation the 'good-cut structure' of $\mathscr{I}^+$ is different from what it is for $\mathbb{M}$, and we cannot use the method we described earlier for singling out a particular restricted Poincaré subgroup $\mathscr{P}$ of $\mathscr{B}$. A more appropriate 'good-cut structure' of $\mathscr{I}^+$ is a *shear structure* which assigns to any cut of $\mathscr{I}^+$ a $\{3, -1\}$ scalar function σ defined on it. This shear structure is one degree more 'extrinsic' than the strong conformal geometry of $\mathscr{I}^+$ (though what one calls 'intrinsic' or 'extrinsic', particularly for a null hypersurface, is to some extent a matter of convention – Penrose 1972*d*). For it refers to the shear, at $\mathscr{I}^+$, of null hypersurfaces in $\mathscr{M}$. We see from this discussion that the shear structure of the $\mathscr{I}^+$ of $\mathbb{M}$ differs from that of a general $\mathscr{M}$. In fact, the shear structure of $\mathscr{I}^+$ is *equivalent* (modulo at most two constants of integration on each generator of $\mathscr{I}^+$) to the information of the outgoing gravitational radiation field. The subgroup $\mathscr{P}$ of $\mathscr{B}$, in the case of $\mathbb{M}$, is the group preserving the shear structure of $\mathscr{I}^+$ in addition to its strong conformal geometry. In the case of a general $\mathscr{M}$, where ψ_4 has no symmetries on $\mathscr{I}^+$, the shear structure of $\mathscr{I}^+$ also has no symmetries. So we are back where we were at the beginning of this discussion: *the group of symmetries preserving the shear structure consists of the identity alone*, for a general $\mathscr{M}$.

The shear structure of $\mathscr{I}^+$ of course determines the good cuts, namely those for which $\sigma = 0$. However, in the general case, there need not be any good cuts at all! The essential reason for this is that σ is a *complex* quantity on the sphere whereas the freedom in choosing a cut is one *real* function defined on the sphere, namely the u-value where it intersects each generator. In the case of $\mathbb{M}$, the difference between the value $\sigma = \sigma_1$ on one cut and the value $\sigma = \sigma_2$ on a second cut, obtained from the first by the supertranslation $F(u, \zeta, \bar{\zeta}) = H(\zeta, \bar{\zeta})$ in (9.8.9)(2), is readily found to be (Sachs 1962*b*):

$$\sigma_2 \approx \sigma_1 + \eth^2 H \tag{9.8.92}$$

In $\mathbb{M}$, *all* cuts are supertranslations of some *good* cut ($\sigma = 0$), so the *reality* of H in (9.8.92) tells us that all cuts satisfy the condition called *purely electric* (Newman and Penrose 1966):

$$\eth'^2\sigma \approx \eth^2\bar{\sigma}, \tag{9.8.93}$$

and this reduces the effective freedom in the complex σ to a real function on the sphere. But in the case of a general $\mathscr{M}$, (9.8.92) is augmented by a term involving a u-integral of $\bar{\psi}_4$ (or of $\bar{\mathrm{N}}$) and condition (9.8.93) is then generally not satisfied.

$\mathscr{H}$-space and asymptotic twistor space

Various attempts at circumventing these several difficulties have been suggested, but none of them has led to a uniquely singled-out Poincaré subgroup of $\mathscr{B}$. The most noteworthy of these attempts is that due to Newman (Newman 1976; *cf.* also Aronson, Lind, Messmer and Newman 1971, Hansen, Newman, Penrose and Tod 1978, Ko, Ludvigsen, Newman and Tod 1981), which has led to the remarkable concept of *$\mathscr{H}$-space.** The key idea here is to allow $\mathscr{I}^+$ to become complexified, to $\mathbb{C}\mathscr{I}^+$, by allowing u to take complex values (and accordingly allowing ζ, $\bar{\zeta}$ to be replaced by independent complex parameters ζ, $\tilde{\zeta}$ – *cf.* the discussion early in §6.9), and then to define a *good cut* by the equation $\sigma = 0$, as before. The complexification of u removes the difficulty just referred to above, but it introduces the complication that the 'conjugate shear' $\tilde{\sigma}$ does *not* generally vanish. Assuming that the shear structure of $\mathscr{I}^+$ is adequately analytic (i.e. that it has a complexification which is sufficiently extensive), the points of the $\mathscr{H}$-space can be *defined* to be the good cuts in this sense. Recall that for $\mathbb{M}$, the real good cuts of $\mathscr{I}^+$ arise from actual light cones and therefore precisely correspond to the points of $\mathbb{M}$; in the same way, the complex good cuts of $\mathbb{C}\mathscr{I}^+$ correspond to the points of $\mathbb{CM}$. But for a general (adequately analytic) $\mathscr{M}$, the $\mathscr{H}$-space – with a remarkable definition of metric suggested by Newman – turns out to be a general holomorphic-Riemannian self-dual solution of the Einstein vacuum equations; and this holds whether or not the vacuum equations hold for $\mathscr{M}$ (*cf.* also p. 168).

There is a close relation between this construction and twistor theory. Recall the discussion of §7.3 (*cf.* (7.3.1) and Proposition (7.3.18)), in which we showed that the condition $\sigma = 0 = \kappa$ is an indication of the presence of totally null complex 2-surfaces ('dual twistor surfaces'). Here the geometry

* $\mathscr{H}$ stands for 'Heaven' – 'where the good cones (= Cohens) go'. The use of the tilde in this paragraph does *not* refer to the 'physical' metric $\tilde{g}_{ab}$.

is restricted to $\mathbb{C}\mathscr{I}^+$, so the '$\kappa = 0$' condition disappears and the dual twistor surfaces appear only as complex one-dimensional curves on $\mathbb{C}\mathscr{I}^+$. These curves turn out to be complex null geodesics on $\mathbb{C}\mathscr{I}^+$ (using the conformal metric of $\mathbb{C}\bar{\mathscr{M}}$), and are referred to as *twistor lines* (for $\tilde{\sigma} = 0$) or *dual twistor lines* (for $\sigma = 0$). The space of [dual] twistor lines on $\mathbb{C}\mathscr{I}^+$ provides a definition of the [dual] *projective asymptotic twistor space for* $\mathscr{M}$. In fact the [dual] twistor lines are precisely the α-curves [*or* β-curves] introduced in §7.4 for the construction of the projective *hypersurface* twistor space $\mathbb{P}\mathscr{T}^{\bullet}(\mathscr{I}^+)$ [*or* $\mathbb{P}\mathscr{T}_{\bullet}(\mathscr{I}^+)$]. Thus (projective) asymptotic twistors are particular examples of (projective) hypersurface twistors, the hypersurface being, in this instance, $\mathscr{I}^+$. (In the general case, the α-curves and β-curves are not complex null geodesics however; that is a special property of $\mathbb{C}\mathscr{I}^+$.)

Newman's good cuts of $\mathbb{C}\mathscr{I}^+$ turn out to be *ruled* by a one-complex-parameter family of dual twistor lines, providing a *holomorphic curve* in the dual asymptotic twistor space. In this way the $\mathscr{H}$-space construction arises as a close analogue of the construction of $\mathbb{CM}$ in terms of lines, in the standard [dual] projective twistor space picture (*cf.* §§6.10, 9.3), to which indeed it reduces when $\mathscr{M} = \mathbb{M}$. This procedure provides an example of (and is, in fact, the origin of) the so-called 'non-linear graviton' construction of all self-dual solutions of the Einstein vacuum equations (Penrose 1976*a*, Penrose and Ward 1980, Tod 1980, Atiyah, Hitchin and Singer 1978; *cf.* also Ward 1978, Tod and Ward 1979, Hitchin 1979), which in turn was the precursor of Ward's self-dual Yang–Mills construction, described in §6.10. But a detailed discussion of these matters would take us well beyond the scope of these volumes (*cf.* p. 168).

Relevance of BMS to momentum and angular momentum

Even $\mathscr{H}$-space, being in general without any symmetries, does not directly lead to a definition of a Poincaré 'symmetry group' for $\mathscr{M}$. Various other suggestions which have been made amount, in effect, to the following: Assume that the radiation ψ_4 falls off adequately fast, either into the future or into the past, along the generators of $\mathscr{I}^+$. Define limiting (say real) 'good cuts' of $\mathscr{I}^+$ in the appropriate asymptotic sense, as $u \to \infty(\mathrm{i}^+)$ or $u \to -\infty(\mathrm{i}^0)$. Then 'good cuts', generally, are defined by BMS translation from these limiting ones. This indeed provides an appropriate (restricted) Poincaré subgroup $\mathscr{P}$ of $\mathscr{B}$, though the ambiguity between the 'i^+-definition' and the 'i^0-definition' is worrysome. In general these two subgroups are not the same. (See Newman and Penrose 1966).

The seriousness of this ambiguity can be made graphic by the following

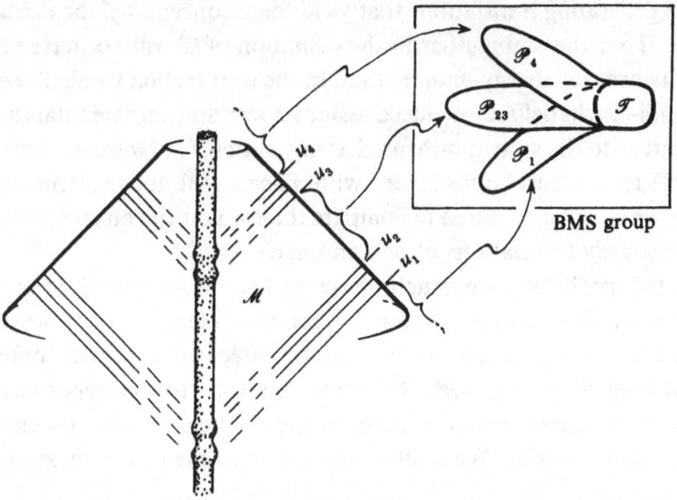

Fig. 9-25. An isolated gravitating system emits two bursts of radiation. It is (essentially) quiescent before, between and after the bursts. For each of the three quiescent periods a Poincaré subgroup of the BMS group is naturally defined in the asymptotic region, but these three Poincaré subgroups will generally have *only* their translation subgroup $\mathscr{T}$ in common.

consideration. Suppose we have an isolated physical system which is initially close enough to stationarity that the shear structure of its $\mathscr{I}^+$ (for $u < u_1$, say) agrees with that for $\mathbb{M}$ sufficiently closely so that, for the degree of approximation required, a unique restricted Poincaré subgroup $\mathscr{P}_1$ can be singled out as the subgroup of $\mathscr{B}$ preserving this shear structure.* Suppose that the system then radiates (retarded) gravitational radiation for a period, say for $u_1 < u < u_2$, after which there follows another period of quiescence, $u_2 < u < u_3$, so that the shear structure of $\mathscr{I}^+$ is once again adequately like that of $\mathbb{M}$ and a second Poincaré subgroup $\mathscr{P}_{23}$ is adequately singled out (see Fig. 9-25). Suppose the system then radiates once more ($u_3 < u < u_4$) before settling down for good. A third such Poincaré subgroup $\mathscr{P}_4$ (for $u_4 < u$) thus emerges, to a corresponding degree of approximation. In general, the subgroups $\mathscr{P}_1, \mathscr{P}_{23}, \mathscr{P}_4$ will all be distinct, and will have only the translation group $\mathscr{T}$ in common, being related to one another by conjugation by non-trivial supertranslations as in (9.8.68).

The group $\mathscr{T}$ is the one of relevance in providing the physical concepts of mass–energy and linear momentum, since in the case of $\mathbb{M}$ it is the Killing

* For a proof that in the *stationary* case the shear structure of the $\mathscr{I}^+$ of $\mathcal{M}$ is the same as for $\mathbb{M}$, see Sachs (1962*b*), Newman and Penrose (1968).

vectors generating translations that yield these concepts (*cf.* the discussion in §6.5). Thus the ambiguities in the definition of $\mathscr{P}$ will not have serious consequences for energy–momentum. In the next section we shall see how the Bondi–Sachs definition indeed assigns a very appropriate total energy–momentum to the system measured at any cut of $\mathscr{I}^+$. However, even here the BMS group must be contended with, since, as the above discussion has shown, we must be prepared to compare these quantities on cuts which are not necessarily translations of one another.

But the problems are much more serious with regard to *angular* momentum, since here generators of 'rotation' elements of $\mathscr{P}$ would be required. The concept of a 'supertranslation-free rotation' gets 'shifted' as we pass from $\mathscr{P}_1$ to $\mathscr{P}_{23}$ to $\mathscr{P}_4$. Thus it would seem that the very *concept* of angular momentum gets correspondingly 'shifted' by the presence of gravitational radiation. We shall return to this problem in the next section.

Asymptotic expansions for Einstein–Maxwell theory

In our analysis of the structure of $\mathscr{I}^+$ we have tried to avoid making unnecessary specializations, either in our choice of spin-frame and coordinates, or in that of the conformal factor. Thus we have kept our considerations general and have not prejudiced the way in which it may be helpful to make further specializations in particular calculations. For example, while in most discussions of problems involving outgoing gravitational radiation, *null* coordinates have been used (e.g. the $u=$ constant loci are extended inwards from $\mathscr{I}^+$ to become null hypersurfaces in $\mathscr{M}$), this is by no means essential; and for certain purposes *asymptotically* null (say spacelike) $u=$ constant hypersurfaces may be found preferable. All the arguments of this section would apply equally well.

Also, with regard to the conformal factor Ω we have merely specialized sufficiently to put the intrinsic metric of $\mathscr{I}^+$ into a particular form, but have left the choice of Ω otherwise free. In more detailed calculations it is often helpful to specialize further, and (as has been mentioned briefly around (9.8.29)) a convenient choice is to make the hypersurfaces $\Omega=$ constant null. For local calculations at $\mathscr{I}^+$ it is sometimes helpful even to 'flatten' $\mathscr{I}^+$ out completely by stereographically projecting one of its generators 'back to infinity' (*cf.* Penrose 1967*b*). Each 'cross-section' of $\mathscr{I}^+$ then acquires a Euclidean metric, and we can also arrange that

$$\nabla_{AA'}\iota^B \approx 0, \tag{9.8.94}$$

the ι-flag-planes being chosen to be parallel throughout $\mathscr{I}^+$. A specialization such as (9.8.94) would not have been helpful to us earlier, however, since it precludes the strict applicability of the compacted spin-coefficient formalism. Moreover, the discussion of global problems becomes more difficult with the topology change inherent in the deletion of a generator of $\mathscr{I}^+$.

To end this section we exhibit, without proof, the expressions for the leading terms in the asymptotic expansions of the physical metric, spin-coefficients, Maxwell field, and Weyl curvature, that result when the asymptotic structure for $\mathscr{I}^+$ obtained here is re-interpreted in terms of $\mathscr{M}$. For definiteness, we assume that the Einstein–Maxwell equations hold in $\mathscr{M}$, and this will allow the propagation inwards from $\mathscr{I}^+$ to be achieved in a well-defined manner. We select a Bondi time coordinate $x^1 = u$ on $\mathscr{I}^+$, and, in accordance with the remarks following Proposition (9.8.31), standard stereographic coordinates $x^3 = \zeta$, $x^4 = \bar{\zeta}$ to label the generators so that the metric of the spheres of cross-section of $\mathscr{I}^+$ takes the form (9.8.8). We propagate u inwards (uniquely, near $\mathscr{I}^+$) by requiring that u = constant are *null* hypersurfaces. These are the null hypersurfaces generated by the rays in $\mathscr{M}$ meeting the u = constant cuts of $\mathscr{I}^+$ orthogonally. Next we propagate ζ and $\bar{\zeta}$ inwards by requiring them to be constant along these rays. We take $x^2 = r$ to be an affine parameter on each ray, the *scaling* for r being chosen so that (specializing to $A = 1$) the physical metric component g^{12} is unity, and the *zero* of r being chosen so that the r^{-2}-term in the expansion of ρ vanishes.

In notation we are here reverting to the usage of earlier sections, where the physical metric is denoted by g_{ab}, not $\tilde{g}_{ab}$, and, correspondingly, 'tildes' will *not* now be used for physical spin-coefficients, curvature quantities, etc.

We also make a specific choice of spin-frame, and thus at last remove the invariance required for strict applicability of the compacted spin-coefficient formalism. Nevertheless we still adopt the ð notation as a shorthand and assume, in effect, the form (4.15.117) (the spin-weights of the relevant quantities being unambiguous). The choice of scaling for o^A is made so that

$$D = \frac{\partial}{\partial r} \tag{9.8.95}$$

along the generators (as in §9.7), and also so that the flag planes point along the ζ-lines of the spheres at $\mathscr{I}^+$ (i.e. Im ζ = constant, Re ζ increasing, at $\mathscr{I}^+$) and are parallelly propagated along the rays generating u = constant. (This is in agreement with the arrangement depicted in Fig. 4-6, Volume 1, p. 310.) The spin-frame is taken to be parallelly propagated along the rays. There remains a 'null rotation' freedom in the choice of ι^A, for each ray.

With these choices we have agreement with the work of Newman and coworkers (Newman and Unti 1962, Kozarzewski 1965, Newman and Penrose 1966, Newman and Tod 1980), except for minor variations in the choice of u, r coordinates, and the fact that $\eth$ replaces their $-2^{-\frac{1}{2}}\eth$. The equations below are adapted from the paper by Exton, Newman and Penrose (1969). Note that a dot denotes $\partial/\partial u$.

A translation to the (θ, ϕ)-system as depicted in Fig. 4-7, Volume 1 is also easy to achieve.

$$l^a = g_2{}^a, m^a = \omega g_2{}^a + \xi^{\mathrm{i}} g_{\mathrm{i}}{}^a, n^a = g_1{}^a + U g_2{}^a + X^{\mathrm{i}} g_{\mathrm{i}}{}^a \quad (\mathrm{i} = 3, 4)$$

$$g^{11} = g^{13} = g^{14} = 0, \; g^{12} = 1$$

$$g^{22} = 2(U - \omega\bar{\omega})$$

$$g^{2\mathrm{i}} = X^{\mathrm{i}} - (\xi^{\mathrm{i}}\bar{\omega} + \bar{\xi}^{\mathrm{i}}\omega) \quad (\mathrm{i} = 3, 4)$$

$$g^{\mathrm{ij}} = -(\xi^{\mathrm{i}}\bar{\xi}^{\mathrm{j}} + \bar{\xi}^{\mathrm{i}}\xi^{\mathrm{j}}) \quad (\mathrm{i}, \mathrm{j} = 3, 4)$$

$$\left.\begin{aligned} \Psi_0 &= \Psi_0^0 r^{-5} + \Psi_0^1 r^{-6} + O(r^{-7}) \\ \varphi_0 &= \varphi_0^0 r^{-3} + \varphi_0^1 r^{-4} + O(r^{-5}) \end{aligned}\right\} \text{ initial data on } u = 0$$

$$\Psi_1 = \Psi_1^0 r^{-4} + (6G\bar{\varphi}_1^0\varphi_0^0 - \eth'\Psi_0^0) r^{-5} + O(r^{-6})$$

$$\Psi_2 = \Psi_2^0 r^{-3} + (4G\varphi_1^0\bar{\varphi}_1^0 - \eth'\Psi_1^0) r^{-4} + O(r^{-5})$$

$$\Psi_3 = \Psi_3^0 r^{-2} + (2G\bar{\varphi}_1^0\varphi_2^0 - \eth'\Psi_2^0) r^{-3} + O(r^{-4})$$

$$\Psi_4 = \Psi_4^0 r^{-1} - \eth'\Psi_3^0 r^{-2} + O(r^{-3})$$

$$\varphi_1 = \varphi_1^0 r^{-2} - \eth'\varphi_0^0 r^{-3} + O(r^{-4})$$

$$\varphi_2 = \varphi_2^0 r^{-1} - \eth'\varphi_1^0 r^{-2} + O(r^{-3})$$

$$\rho = -r^{-1} - \sigma^0\bar{\sigma}^0 r^{-3} + O(r^{-5})$$

$$\sigma = \sigma^0 r^{-2} + (\bar{\sigma}^0\sigma^0\sigma^0 - \tfrac{1}{2}\Psi_0^0) r^{-4} + O(r^{-5})$$

$$\alpha = \alpha^0 r^{-1} + \bar{\sigma}^0\alpha^0 r^{-2} + \sigma^0\bar{\sigma}^0\alpha^0 r^{-3} + O(r^{-4})$$

$$\beta = -\bar{\alpha}^0 r^{-1} - \sigma^0\alpha^0 r^{-2} - (\sigma^0\bar{\sigma}^0\bar{\alpha}^0 + \tfrac{1}{2}\Psi_1^0) r^{-3} + O(r^{-4})$$

$$\tau = -\tfrac{1}{2}\Psi_1^0 r^{-3} + \tfrac{1}{3}(\tfrac{1}{2}\sigma^0\Psi_1^0 + \eth'\Psi_0^0 - 8G\varphi_0^0\bar{\varphi}_1^0) r^{-4} + O(r^{-5})$$

$$\lambda = -\sigma' = \dot{\bar{\sigma}}^0 r^{-1} + \bar{\sigma}^0 r^{-2} + (\sigma^0\bar{\sigma}\dot{\bar{\sigma}}^0 + \tfrac{1}{2}\bar{\sigma}^0\Psi_2^0 - G\bar{\varphi}_0^0\varphi_2^0) r^{-3} + O(r^{-4})$$

$$\begin{aligned} \mu = -\rho' &= -\tfrac{1}{2} r^{-1} - (\sigma^0\dot{\bar{\sigma}}^0 + \Psi_2^0) r^{-2} \\ &\quad - (\sigma^0\bar{\sigma}^0 + 2G\varphi_1^0\bar{\varphi}_1^0 - \tfrac{1}{2}\eth'\Psi_1^0) r^{-3} + O(r^{-4}) \end{aligned}$$

$$\gamma = -\tfrac{1}{2}\Psi_2^0 r^{-2} + \tfrac{1}{6}(2\eth'\Psi_1^0 + \alpha^0\Psi_1^0 - \bar{\alpha}^0\Psi_1^0 - 12G\varphi_1^0\bar{\varphi}_1^0) r^{-3} + O(r^{-4})$$

$$\nu = -\Psi_3^0 r^{-1} + (\tfrac{1}{2}\eth'\Psi_2^0 - 2G\bar{\varphi}_1^0\varphi_2^0) r^{-2} + O(r^{-3})$$

$$\begin{aligned} U &= -\tfrac{1}{2} - \tfrac{1}{2}(\Psi_2^0 + \bar{\Psi}_2^0) r^{-1} \\ &\quad + \tfrac{1}{6}(\eth'\Psi_1^0 + \eth\bar{\Psi}_1^0 - 12G\varphi_1^0\bar{\varphi}_1^0) r^{-2} + O(r^{-3}) \end{aligned}$$

$$X^3 = \overline{X^4} = \frac{1}{6\sqrt{2}}(1 + \zeta\bar{\zeta})\Psi_1^0 r^{-3} + O(r^{-4})$$

$$\xi^3 = \frac{-\zeta(1+\zeta\bar{\zeta})}{\bar{\zeta}\sqrt{2}}\sigma^0 r^{-2} + O(r^{-4})$$

$$\xi^4 = \frac{(1+\zeta\bar{\zeta})}{\sqrt{2}}(r^{-1} + \sigma^0\bar{\sigma}^0 r^{-3}) + O(r^{-4})$$

$$\omega = \eth'\sigma^0 r^{-1} - (\sigma^0\eth\bar{\sigma}^0 + \tfrac{1}{2}\Psi_1^0)r^{-2} + O(r^{-3})$$

where

$$\alpha^0 = \frac{\zeta}{\sqrt{8}}$$

$$\Psi_2^0 - \bar{\Psi}_2^0 = \eth'^2\sigma^0 - \eth^2\bar{\sigma}^0 + \bar{\sigma}^0\dot{\sigma}^0 - \dot{\bar{\sigma}}^0\sigma^0$$

$$\Psi_3^0 = -\eth\dot{\bar{\sigma}}^0$$

$$\Psi_4^0 = -\ddot{\bar{\sigma}}^0.$$

(Note that the last three of these equations correspond to our earlier equations (9.8.89), (9.8.83), (9.8.82), respectively, with (9.8.75).)

$$\Psi_0^0 - \eth\Psi_1^0 - 3\sigma^0\Psi_2^0 - 6G\varphi_0^0\bar{\varphi}_2^0 = 0$$

$$\Psi_0^1 + 4\eth'(\sigma^0\Psi_1^0) + \eth'\eth\Psi_0^0 - 8G\bar{\varphi}_1^0\eth\varphi_0^0 - 16G\sigma^0\varphi_1^0\bar{\varphi}_1^0 - 8G\varphi_0^1\bar{\varphi}_2^0 = 0$$

$$\dot{\Psi}_1^0 - \eth\Psi_2^0 - 2\sigma^0\Psi_3^0 - 4G\varphi_1^0\bar{\varphi}_2^0 = 0$$

$$\dot{\Psi}_2^0 - \eth\Psi_3^0 - \sigma^0\Psi_4^0 - 2G\varphi_2^0\bar{\varphi}_2^0 = 0$$

$$\dot{\varphi}_0^0 - \eth\varphi_1^0 - \sigma^0\varphi_2^0 = 0$$

$$\dot{\varphi}_0^1 + \eth'\eth\varphi_0^0 + 2\eth'(\sigma^0\varphi_1^0) = 0$$

$$\dot{\varphi}_1^0 - \eth\varphi_2^0 = 0$$

9.9 Energy–momentum and angular momentum

In §§6.4 and 6.5 we gave a discussion of energy–momentum and angular momentum in the weak-field linearized limit of general relativity. Our procedure was to use a $[^2_0]$-twistor to lower the spin of the field from two to one, so that the ten conservation laws in the gravitational case (namely those of energy–momentum and angular momentum) could be reduced to the form of the *one* in the electromagnetic case (namely of electric charge). Here we give a generalization of that procedure which applies to *full* general relativity (Penrose 1982; for earlier related procedures, *cf.* Synge 1960, Komar 1962, Streubel 1978). As we shall see, there are very severe difficulties standing in the way of obtaining as complete a set of conservation laws in the full theory as one has in its linearized limit. These are reflected, in our approach, in the peculiarities and limitations confronting the appropriate twistor concept in a curved space–time $\mathcal{M}$. The key to our procedure will, indeed, consist in providing a new kind of twistor, referred to as a 2-*surface* *twistor*, which achieves a form of spin-lowering even in a general curved $\mathcal{M}$. When applied at $\mathscr{I}^+$ our method yields the successful Bondi–Sachs definition of total 4-momentum.* It also provides a new concept of angular momentum which seems superior to those previously put forward. In addition, the procedure suggests a possible definition of an energy-momentum/angular-momentum complex which refers to the total matter and gravitational field surrounded by a *finite* closed 2-surface.

Linear theory

Recall that in our discussion in §§6.4, 6.5 we gave two integral descriptions, in $\mathbb{M}$, of the sources in linearized gravitational theory. In the first of these, a 2-form

$$\Theta = K^{*}_{i_1 i_2 ab} Q^{ab} \tag{9.9.1}$$

(*cf.* (6.5.45)) is integrated over a closed 2-surface $\mathscr{S}$, to obtain a measure of the sources surrounded by $\mathscr{S}$, while in the second, a 3-form

$$\mathrm{d}\Theta = \frac{16\pi G}{3} e_{i_1 i_2 i_3 a} E^{ab} \xi_b \tag{9.9.2}$$

(*cf.* (6.5.49)) is integrated over a region of 3-volume $\mathscr{V}$ to measure the total flux of source across $\mathscr{V}$. By the fundamental theorem of exterior calculus, the two methods agree, provided $\mathscr{V}$ is compact with boundary $\partial\mathscr{V} = \mathscr{S}$ (*cf.* (6.5.51)). Here K_{abcd} and E_{ab} are the linearized curvature and energy–momentum tensors, respectively. Also, as in (6.4.7),

$$Q^{ab} = \mathrm{i}\sigma^{AB}\varepsilon^{A'B'} - \mathrm{i}\bar{\sigma}^{A'B'}\varepsilon^{AB} \tag{9.9.3}$$

is a real skew tensor constructed from the primary part $\sigma^{AB}(=\sigma^{BA})$ of a *symmetric twistor* $\mathrm{S}^{\alpha\beta}$, so we have

$$\nabla^{(A}_{A'}\sigma^{BC)} = 0 \tag{9.9.4}$$

by (6.1.69). Equivalently, by (6.4.6), we have

$$\nabla^{(a}Q^{b)c} - \nabla^{(a}Q^{c)b} + g^{a[b}\nabla_d Q^{c]d} = 0, \tag{9.9.5}$$

Q^{ab} being related to the *Killing vector* ξ^a by

$$\begin{aligned}\xi^a &= \tfrac{1}{3}\nabla_b Q^{ba} \\ &= \tfrac{1}{3}(-\mathrm{i}\nabla^{A'}_{B}\sigma^{AB} + \mathrm{i}\nabla^{A}_{B'}\bar{\sigma}^{A'B'})\end{aligned} \tag{9.9.6}$$

(*cf.* (6.5.25), (6.5.40)). We recall also (*cf.* (6.5.15) and Figs. 6-6, 6-7) that,

* Bondi 1960*b*, Bondi, van der Burg and Metzner 1962, Sachs 1962*a*, *b*, Penrose 1963, 1964*b*, 1967*b*, Newman and Penrose 1968, Bonnor and Rotenberg 1966.

taking $\xi^{AA'}$ as the primary part of a (twistor-) Hermitian trace-free twistor $\mathrm{F}^{\alpha}{}_{\beta}$, the relation (9.9.6) assumes the form

$$\mathrm{F}^{\alpha}{}_{\beta} = \mathrm{S}^{\alpha\gamma}\mathrm{I}_{\gamma\beta} + \bar{\mathrm{S}}_{\beta\gamma}\mathrm{I}^{\gamma\alpha}. \tag{9.9.7}$$

Moreover, the Killing equation

$$\nabla^{(a}\xi^{b)} = 0 \tag{9.9.8}$$

is an automatic consequence of (9.9.4) (i.e. (9.9.5)) and (9.9.6), all *ten* linearly independent Killing vectors for $\mathbb{M}$ arising in this way. Since $\mathbb{T}^{(\alpha\beta)}$ is ten-complex-dimensional, there are ten *complex* linearly independent solutions of (9.9.4), and hence *twenty* real linearly independent solutions of (9.9.5). *Ten* of these yield the *zero* Killing vector in (9.9.6), the corresponding tensors Q^{ab} being those expressible in the form

$$Q^{ab} = e^{abcd}\nabla_d\gamma_c, \quad \text{i.e.} \quad \sigma^{AB} = \nabla^{(A}_{B'}\gamma^{B)B'} \tag{9.9.9}$$

(*cf.* (6.5.41), Figs. 6-6, 6-7), with γ^c a *conformal Killing vector.* The twistor expression for (9.9.9) is

$$\mathrm{S}^{\alpha\beta} = 2\mathrm{i}\mathrm{G}^{(\alpha}{}_{\rho}\mathrm{I}^{\beta)\rho}, \tag{9.9.10}$$

where $\gamma^{AB'}$ is the primary part of the (twistor-) Hermitian trace-free twistor $\mathrm{G}^{\alpha}{}_{\beta}$.

Difficulties in curved space–time

Our proposal is to carry out an analogous procedure in a general curved space–time $\mathcal{M}$, and then to specialize to the case when $\mathcal{M}$ is suitably asymptotically Minkowskian. We cannot merely take over equations (9.9.4) (or (9.9.5)), (9.9.6) and (9.9.8) directly into a curved-space setting. As we have remarked earlier (in §6.5), (9.9.8) has non-zero solutions only when $\mathcal{M}$ possesses (continuous) symmetries. Moreover, (9.9.4) is subject to severe algebraic consistency requirements with the Weyl curvature, similar to that encountered in (6.1.6) for the valence-$\left[\begin{smallmatrix}1\\0\end{smallmatrix}\right]$ twistor equation. In a general $\mathcal{M}$, (9.9.4) admits only the zero solution.

There is a good physical reason why the integral of an expression like (9.9.2), which vanishes whenever the local energy–momentum of the sources is zero, cannot yield a satisfactory expression for the total 4-momentum of a gravitating system. We know that the gravitational field itself must contribute to the total energy (sometimes negatively, as does the Newtonian potential energy of two mutually gravitating bodies, and sometimes positively, as with gravitational waves), yet there is no direct gravitational contribution to the energy–momentum tensor. That particular contribution may be viewed as residing in the nonlinearity of Einstein's field equations. Gravitational energy is, in an essential sense, a *non-local* quantity. It vanishes locally, but shows up in the total energy measure. One manifestation of this is the fact that the local covariant 'conservation law' $\nabla_a T^{ab} = 0$ does not integrate to give a total conserved four-momentum. It is as well that it should not, for otherwise we would have an energy–momentum concept to which the gravitational field could not contribute, contrary to physical experience.

As a replacement for (9.9.2) we shall therefore need an expression which contains, in addition to a term like the RHS of (9.9.2), another part which depends on non-local ingredients. In §9.10 we shall present a remarkable (positive-definite) expression due to Witten (1981) which appears to achieve what is needed for space–time regions that approach flatness suitably at infinity. A striking feature of this expression is that it depends essentially on *spinor* (spin $\frac{1}{2}$) ingredients. In this section, however, we shall be concerned with a satisfactory replacement for (9.9.1). It is remarkable that here, too, we shall find that it is necessary to use 'spin $\frac{1}{2}$' ingredients in the general case.

2-surface twistors

We desire an analogue of (9.9.4) at a given closed 2-surface $\mathscr{S}$. As we have just remarked, (9.9.4) itself does not generally have non-trivial solutions in $\mathscr{M}$. But again, this is physically desirable, because any such solution would yield $\mathrm{d}\Theta = 0$ in vacuum regions, and we would have a conservation law *without* gravity, in the sense that continuous deformations of $\mathscr{S}$ through such regions would not change the integral, regardless of the presence of a gravitational field. However we require σ^{AB} only to provide us with a *definition* of energy, etc. at $\mathscr{S}$, not with such a strong concept of conservation law. Thus we can try to consider only those components of (9.9.4) in which the derivative acts *tangentially*. Suppose that $\mathscr{S}$ is spacelike, and let us adopt the compacted spin-coefficient formalism, with spin-frame o^A, ι^A adapted to $\mathscr{S}$ in the standard way of §4.14. However, we find that only *two* terms, those obtained by transvecting (9.9.4) with

$$o_A o_B o_C \iota^{A'} \quad \text{and} \quad \iota_A \iota_B \iota_C o^{A'}, \tag{9.9.11}$$

involve only tangential derivatives, whereas σ^{AB} has *three* independent complex components. We therefore have an underdetermined system, with an infinite-dimensional space of solutions rather than the desired ten-dimensional one.

What is needed is something more subtle. Instead of defining the elements of the desired '$\mathbb{T}^{(\alpha\beta)}$' space directly, we think of that space as arising indirectly as the symmetric tensor product, with itself, of a valence-$[{}^1_0]$ 'twistor space' associated with $\mathscr{S}$. Thus, instead of looking for the tangential parts of (9.9.4), we look for the tangential parts of the original twistor equation (6.1.1):

$$\nabla^{(A}_{A'}\omega^{B)} = 0. \tag{9.9.12}$$

Transvecting with $\iota_A\iota_B o^{A'}$ and $o_A o_B \iota^{A'}$, we now have *two* tangential equations for *two* complex components, ω^0, ω^1, of respective types $\{-1,0\}$, $\{1,0\}$ (these equations having been earlier obtained as two of (4.12.46)); in the compacted spin-coefficient formalism they read

$$\eth'\omega^0 = \sigma'\omega^1, \quad \eth\omega^1 = \sigma\omega^0. \tag{9.9.13}$$

(Note: $\omega^0 = \omega_1 = \omega_A\iota^A$; $\omega^1 = -\omega_0 = -\omega_A o^A$.)

Any solution* $\{\omega^0, \omega^1\}$ of (9.9.13) (or the corresponding abstract-indexed ω^A) over the whole of the closed surface $\mathscr{S}$ is called a *2-surface twistor* on $\mathscr{S}$, of valence $[{}^1_0]$, and the space of such solutions is denoted by $\mathbb{T}^\alpha(\mathscr{S})$. It is in fact the case that, for an $\mathscr{S}$ having the topology of a 2-sphere, (9.9.13) will always have at least *four complex linearly independent solutions*. Moreover, in the generic case and in cases sufficiently close to the canonical situation when, as in §4.15, $\mathscr{S}$ arises as the compact intersection of two light cones in $\mathbb{M}$, (9.9.13) has *exactly* four independent solutions. Thus, in these 'normal' circumstances at least, $\mathbb{T}^\alpha(\mathscr{S})$ will be a four-complex-dimensional vector space. The index 'α' is then a four-dimensional abstract index and the standard rules and notations of §2.2 will apply.

The argument verifying the above statements is outside the scope of this book, but it may be outlined as follows: One can compute the Atiyah–Singer index (*cf.* Shanahan 1978, Gilkey 1974) for (9.9.13) by considering first the canonical situation, referred to above, of the intersection of two light cones in $\mathbb{M}$. Here the equations decouple to $\eth'\omega^0 = 0$, $\eth\omega^1 = 0$, with ω^0, ω^1 having respective spin-weights $-\frac{1}{2}, \frac{1}{2}$, so a glance at (4.15.60) tells us that each equation has two independent solutions, giving four in all. The *adjoint* equation has the form of the pair $\eth\lambda^0 = 0$, $\eth'\lambda^1 = 0$, where λ^0, λ^1 have respective spin-weights $-\frac{3}{2}, \frac{3}{2}$, and a glance at (4.15.60) (or Proposition (4.15.59)) now tells us that there is only the trivial solution. The index, being the difference between the dimensionalities of these solution spaces, is therefore $4 - 0 = 4$, this being an invariant under deformations of the

* We shall denote by { } quantities varying over $\mathscr{S}$. This is mainly to distinguish *locally* defined twistor expressions from the ones referring to elements of $\mathbb{T}^{\cdots}_{\cdots}(\mathscr{S})$ as a whole, such as in (9.9.25) later. When there is no confusion, this convention need not be adhered to.

differential equations. At worst, (9.9.13) can acquire additional solutions in particular circumstances, the adjoint equation simultaneously gaining the same dimensionality of solutions. This can happen only in exceptional cases (but such examples have been constructed by B.P. Jeffryes).

Instead of using (9.9.4) we now *define* the elements of $\mathbb{T}^{(\alpha\beta)}(\mathscr{S})$ as *symmetric tensor products of solutions of* (9.9.13), i.e.

$$\sigma^{AB} = \underset{1}{\omega}^{(A}\underset{2}{\omega}^{B)} + \cdots + \underset{2r-1}{\omega}{}^{(A}\underset{2r}{\omega}^{B)}, \tag{9.9.14}$$

where $\{\underset{i}{\omega}^0, \underset{i}{\omega}^1\}$, for $i = 1, \ldots, 2r$, are solutions of (9.9.13). In fact the general element of $\mathbb{T}^{(\alpha\beta)}(\mathscr{S})$ is already obtained when the sum (9.9.14) contains just *two* terms (i.e. $r = 2$).

We remark that, in $\mathbb{M}$, the space $\mathbb{T}^{\alpha}(\mathscr{S})$ can be *identified* with the standard twistor space $\mathbb{T}^{\alpha}$. (This is assuming that $\mathscr{S}$ is such that the 'normal' situation obtains, where $\mathbb{T}^{\alpha}(\mathscr{S})$ is four-dimensional.) For any solution of (9.9.12) is necessarily a solution of (9.9.13). Moreover, by (5.6.38), we can rewrite (9.9.13) in terms of the conformally invariant operators $\eth_c$, $\eth'_c$ (*cf.* (5.6.34)), so equations (9.9.13) are *conformally invariant* (taking ω^A to have conformal weight 0 and taking arbitrary weights for o^A, ι^A). Thus we may identify $\mathbb{T}^{\alpha}(\mathscr{S})$ with the standard $\mathbb{T}^{\alpha}$ also in *conformally flat* $\mathscr{M}$. In these cases, $\mathbb{T}^{(\alpha\beta)}(\mathscr{S})$ can similarly be identified with $\mathbb{T}^{(\alpha\beta)}$, and likewise for all the spaces $\mathbb{T}^{\alpha\ldots\gamma}_{\rho\ldots\tau}(\mathscr{S})$ constructed from $\mathbb{T}^{\alpha}(\mathscr{S})$ according to the usual prescriptions (*cf.* §2.2). However, when $\mathscr{S}$ and $\mathscr{M}$ are generic, the spaces of 2-surface twistors (of arbitrary valence) that we obtain provide us with some completely new objects of study.

Contorted 2-surfaces

The quantities σ and σ' which appear in (9.9.13) (or, rather, their real and imaginary parts) together with ρ and ρ' (which are already real, by (4.14.2) or by the arguments of §7.1) constitute an object normally referred to as the *extrinsic curvature* of $\mathscr{S}$. In (4.14.20) we considered a quantity K arising from the commutator of $\eth$ with $\eth'$ whose imaginary part is another type of extrinsic curvature, but involves derivatives within $\mathscr{S}$ of one order higher, namely two, like the intrinsic Gaussian curvature $K + \bar{K}$ (4.14.21). If we consider a second surface $\hat{\mathscr{S}}$, isometric with $\mathscr{S}$, which is embedded in a space–time $\hat{\mathscr{M}}$, and for which all these extrinsic curvature quantities are identical with those for $\mathscr{S}$, then the equations (9.9.13) are just the same for the two surfaces, and the solutions for one can be carried over directly to the solutions for the other. In the particular cases for which $\hat{\mathscr{M}}$ can be chosen to be conformally flat, this procedure is very useful because $\hat{\mathcal{M}}$ then admits a full complex four-dimensional family of solutions of (9.9.12) near $\mathcal{S}$, these being obtainable directly from the solutions (6.1.10) in $\mathbb{M}$ by conformal rescaling. The restriction of these solutions to $\hat{\mathcal{S}}$ can then be carried over to $\mathcal{S}$, so $\mathbb{T}^\alpha(\mathcal{S})$ is readily constructed. (This type of procedure was introduced and put to impressive use by Tod 1983*a*.) When such an embedding in conformally flat space–time exists, we say that $\mathcal{S}$ is *uncontorted*. Recall that Im(K), σ and σ' are conformal densities (5.6.28) and so are essentially unaffected in the passage from $\hat{\mathcal{M}}$ to $\mathbb{M}$. However that is not so for Re(K), ρ and ρ'. The condition that $\mathcal{S}$ be uncontorted is clearly conformally invariant.

In the more general case when such a conformally flat $\hat{\mathcal{M}}$ does not exist we call $\mathcal{S}$ *contorted*. Even in this case a similar construction is possible, but here the (conformally flat) embedding space is *complex*. The quantities σ and σ' carry over (perhaps rescaled) but their complex conjugates $\bar{\sigma}$ and $\bar{\sigma}'$ are replaced by new *unrelated* complex quantities. The restrictiveness of the condition on $\mathcal{S}$ that it be uncontorted is measured by three real equations per point of $\mathcal{S}$ (one of which is Im$(\Psi_2)=0$).

The quasi-local angular-momentum twistor

A possible definition of the 2-form $\mathbf{\Theta}$ on $\mathcal{S}$, applicable to full general relativity* (Penrose 1982), is obtained by substituting (9.9.14) into (9.9.3), and then using the resulting Q^{ab} together with the *full* curvature in (9.9.1):

$$\mathbf{\Theta} := R^{*}_{i_1 i_2 ab}Q^{ab} = R_{i_1 i_2 ab}{}^{*}Q^{ab}. \qquad (9.9.15)$$

There is indeed some impressive evidence (Tod 1983*a*) that in the case of uncontorted surfaces $\mathcal{S}$ this definition leads to a measure of mass–energy in general relativity which is in excellent (and somewhat remarkable) agreement with physical requirements. However there is now good reason to believe (as is strongly supported by work of Tod, R.M. Kelly and N.M.J. Woodhouse) that to cope with the general case of a *contorted* $\mathcal{S}$ one should not simply integrate (9.9.15) over $\mathcal{S}$ (in accordance with the standard procedure (4.3.24)), but an additional factor is needed.** Write

$$\mathbf{\Theta} = \mathbf{\Delta} + \bar{\mathbf{\Delta}},$$

* See Tod (1983*b*) for an analogous expression in Yang–Mills theory.

** *Added in proof*: Indeed, yet a further modification could also be considered. We can propose using a 'Tod form' (9.9.29) as on p.405, with the suggested replacements for $\pi_{0'}$, $\pi_{1'}$, but with the additional term $(\ldots)\eth v \eth' v$ *omitted*.

where (by (9.9.15), (9.9.3), (3.4.22))

$$\Delta = R_{i_1 i_2 AA'BB'}\sigma^{AB}\varepsilon^{A'B'}$$
$$= 2\sigma^{AB}(\Psi_{I_1 I_2 AB}\varepsilon_{I'_1 I'_2} + \Phi_{ABI'_1 I'_2}\varepsilon_{I_1 I_2}) + 4\Lambda\sigma_{I_1 I_2}\varepsilon_{I'_1 I'_2}, \qquad (9.9.16)$$

(*cf.* (4.6.38)). Rather than integrating Θ over $\mathscr{S}$ we integrate the 2-form

$$\Theta' := \eta\Delta + \bar{\eta}\bar{\Delta}$$

where η is a complex scalar quantity on $\mathscr{S}$ whose suggested definition will be given shortly. When $\mathscr{S}$ is uncontorted we shall have $\eta = 1$, and this will hold also for contorted cuts of $\mathscr{I}^+$. We obtain (with $\mathscr{S}$ denoting the surface-area element 2-form as in (4.14.65))

$$\oint_{\mathscr{S}} \Theta' = 2\,\mathrm{Re}\oint_{\mathscr{S}} \eta\Delta = -4\,\mathrm{Re}\oint \mathrm{i}\Delta_{01'10'}\eta\mathscr{S}$$
$$= 8\,\mathrm{Re}\oint \mathrm{i}\{\sigma^{AB}(\Psi_{01AB} - \Phi_{AB0'1'}) + 2\Lambda\sigma_{01}\}\eta\mathscr{S}$$
$$= -8\,\mathrm{Im}\oint\{\sigma^{00}(\Psi_1 - \Phi_{01}) + 2\sigma^{01}(\Psi_2 - \Phi_{11} - \Lambda)$$
$$+ \sigma^{11}(\Psi_3 - \Phi_{21})\}\eta\mathscr{S}, \qquad (9.9.17)$$

by (4.14.53), (4.14.66), with σ^{AB} given by (9.9.14).*

Recall now the duality between $\mathsf{S}^{\alpha\beta}$ and the angular momentum twistor $\mathsf{A}_{\alpha\beta}$ as discussed in §6.5. We have, equating (6.5.53) with (6.5.51) and adapting this to our present situation:

$$\mathrm{Re}(\mathsf{A}_{\alpha\beta}\mathsf{S}^{\alpha\beta}) = -\frac{1}{32\pi G}\oint\Theta', \qquad (9.9.18)$$

whence (from linearity)

$$\mathsf{A}_{\alpha\beta}\underset{1}{\mathsf{Z}^\alpha}\underset{2}{\mathsf{Z}^\beta} = -\frac{\mathrm{i}}{4\pi G}\oint\{(\Psi_1 - \Phi_{01})\underset{1}{\omega^0}\underset{2}{\omega^0} + (\Psi_2 - \Phi_{11} - \Lambda)(\underset{1}{\omega^0}\underset{2}{\omega^1} + \underset{1}{\omega^1}\underset{2}{\omega^0})$$
$$+ (\Psi_3 - \Phi_{21})\underset{1}{\omega^1}\underset{2}{\omega^1}\}\eta\mathscr{S}. \qquad (9.9.19)$$

Here $\underset{1}{\mathsf{Z}^\alpha}$ and $\underset{2}{\mathsf{Z}^\alpha}$ are arbitrary elements of $\mathbb{T}^\alpha(\mathscr{S})$, corresponding to solutions $\{\underset{1}{\omega^0}, \underset{1}{\omega^1}\}$, $\{\underset{2}{\omega^0}, \underset{2}{\omega^1}\}$ of (9.9.13). Equation (9.9.19) serves to define $\mathsf{A}_{\alpha\beta} \in \mathbb{T}_{(\alpha\beta)}(\mathscr{S})$ as the 'angular-momentum twistor' describing the total gravitational source surrounded by $\mathscr{S}$.

Note that, as defined by (9.9.19), $\mathsf{A}_{\alpha\beta}$ has ten complex components (since $\mathbb{T}_{(\alpha\beta)}(\mathscr{S})$ is ten-complex-dimensional). For its given physical interpretation

* With a cosmological constant λ present in Einstein's equations it seems reasonable to replace the middle term in the above integral by $2\sigma^{01}(\Psi_2 - \Phi_{11} - \Lambda + \frac{1}{8}\lambda)$.

to be completely satisfactory, however, we would anticipate that some Hermiticity property analogous to (6.3.12), namely

$$\overline{\mathsf{A}_{\alpha\gamma}\mathsf{I}^{\beta\gamma}} = \mathsf{A}_{\beta\gamma}\mathsf{I}^{\alpha\gamma}, \tag{9.9.20}$$

should reduce these ten complex components down to ten real ones. However, an appropriate formulation of (9.9.20) is, as yet, lacking. In order even to state what (9.9.20) would *mean*, two ingredients seem necessary. First, it appears that one would require an (involutory) operation of twistor *complex conjugation* for $\mathbb{T}^\alpha(\mathscr{S})$, carrying it to its dual space:

$$\mathbb{T}^\alpha(\mathscr{S}) \mapsto \mathbb{T}_\alpha(\mathscr{S}), \tag{9.9.21}$$

so that the resulting twistor 'norm' $\mathsf{Z}^\alpha\bar{\mathsf{Z}}_\alpha$ has signature $(+ + - -)$. And secondly, one apparently would need a (simple? twistor-real?) element

$$\mathsf{I}^{\alpha\beta} \in \mathbb{T}^{[\alpha\beta]}(\mathscr{S}) \tag{9.9.22}$$

with which to complete the relation (9.9.20).

One approach to these problems is to define a *local twistor field* $\{\omega^A, \pi_{A'}\}$ on $\mathscr{S}$ (see §6.9), for each $\mathsf{Z}^\alpha \in \mathbb{T}^\alpha(\mathscr{S})$, by setting

$$\pi_{0'} = \mathrm{i}\eth'\omega^1 - \mathrm{i}\rho\omega^0, \quad \pi_{1'} = \mathrm{i}\eth\omega^0 - \mathrm{i}\rho'\omega^1, \tag{9.9.23}$$

these, together with (9.9.13), being the tangential parts, in $\mathbb{M}$, of the equation (6.1.9). We easily check that the transformation formulae (6.9.6) for local twistors under conformal rescaling are satisfied. However, $\{\omega^A, \pi_{A'}\}$ will *not* normally be constant under local twistor transport (6.9.10), except when $\mathscr{M}$ is conformally flat. We now construct the *conformally invariant scalar field on* $\mathscr{S}$:

$$\begin{aligned}\{\mathsf{Z}^\alpha\bar{\mathsf{Z}}_\alpha\} &= \omega^A\bar{\pi}_A + \pi_{A'}\bar{\omega}^{A'} \\ &= \omega^0\bar{\pi}_0 + \omega^1\bar{\pi}_1 + \pi_{0'}\bar{\omega}^{0'} + \pi_{1'}\bar{\omega}^{1'} \\ &= \mathrm{i}(\bar{\omega}^{1'}\eth\omega^0 - \omega^0\eth\bar{\omega}^{1'} + \bar{\omega}^{0'}\eth'\omega^1 - \omega^1\eth'\bar{\omega}^{0'}) \\ &= 2\,\mathrm{Im}\,(\omega^0\eth\bar{\omega}^{1'} - \bar{\omega}^{1'}\eth\omega^0)\end{aligned} \tag{9.9.24}$$

since ρ and ρ' are real. Whenever the quantity (9.9.24) is *constant* over $\mathscr{S}$ this gives us a conformally invariant number which appropriately defines a Hermitian $(+ + - -)$ twistor norm $\mathsf{Z}^\alpha\bar{\mathsf{Z}}_\alpha$ for $\mathbb{T}^\alpha(\mathscr{S})$. However, in general, the expression $\{\mathsf{Z}^\alpha\bar{\mathsf{Z}}_\alpha\}$ is *not* constant over $\mathscr{S}$, the condition for such constancy, for all $\mathsf{Z}^\alpha \in \mathbb{T}^\alpha(\mathscr{S})$, being, in fact that $\mathscr{S}$ is uncontorted (see e.g., Jeffryes 1984*b*). In the contorted case one can try to get over this by averaging:

$$\mathsf{Z}^\alpha\bar{\mathsf{Z}}_\alpha := \left(\oint \mathscr{S}\right)^{-1} \oint \{\mathsf{Z}^\alpha\bar{\mathsf{Z}}_\alpha\}\mathscr{S}, \tag{9.9.25}$$

or perhaps alternatively:

$$\mathsf{Z}^\alpha \bar{\mathsf{Z}}_\alpha := \frac{1}{4\pi} \oint (K + \bar{K}) \{\mathsf{Z}^\alpha \bar{\mathsf{Z}}_\alpha\} \mathscr{S}, \tag{9.9.26}$$

where $K + \bar{K}$ is the Gaussian curvature of $\mathscr{S}$ (see (4.14.20), (4.14.21), (4.14.44)); or possibly some expression involving η should be used. This yields two (or more) suggestions for a product to achieve the mapping (9.9.21). But in neither case does one obtain conformal invariance.

The problems with (9.9.22) are, at first sight, similar. For example, we could define the scalar field over $\mathscr{S}$,

$$\{\mathsf{I}_{\alpha\beta} \underset{1}{\mathsf{Z}^\alpha} \underset{2}{\mathsf{Z}^\beta}\} = \underset{1}{\pi_{0'}} \underset{2}{\pi_{1'}} - \underset{1}{\pi_{1'}} \underset{2}{\pi_{0'}} \tag{9.9.27}$$

but again this is generally not constant (even when $\mathscr{S}$ is uncontorted). If we average this over $\mathscr{S}$ we are not now constrained by any considerations of conformal invariance. But the resulting $\mathsf{I}_{\alpha\beta} \in \mathbb{T}_{[\alpha\beta]}(\mathscr{S})$ generally turns out not to be simple (in the sense of (3.5.30), (3.5.35)). A non-simple $\mathsf{I}_{\alpha\beta}$ could still be used in (9.9.20); indeed, there are strong indications that generally it is necessary to use such a 'de Sitter' type $\mathsf{I}_{\alpha\beta}$. It is not yet known whether (9.9.20) holds with this $\mathsf{I}_{\alpha\beta}$ perhaps suitably modified by some term(s) in η.

It seems that a resolution of these difficulties must await further developments. As we shall see shortly the situation is much pleasanter at $\mathscr{I}^+$. Moreover, as has been shown by Shaw (1983*a*), in the limiting situation as $\mathscr{S}$ approaches spacelike infinity i^0, a good definition of $\mathsf{I}_{\alpha\beta}$ and of the norm exists, and (9.9.20) is satisfied provided that appropriate fall-off conditions are assumed for the curvature. Full agreement is then obtained with the Arnowitt, Deser and Misner (1961) definition of mass and the Ashtekar–Hansen (1978) definition of angular momentum. Before turning to $\mathscr{I}^+$, various remarks concerning a general $\mathscr{S}$ are worth making. We note first that in $\mathbb{M}$ there is a formula,

$$m^2 = -\tfrac{1}{2} \mathsf{A}_{\alpha\beta} \bar{\mathsf{A}}^{\alpha\beta} \tag{9.9.28}$$

for the squared *rest-mass m*, readily obtainable from (6.3.11), which requires only a twistor complex-conjugation operation and not a definition of $\mathsf{I}_{\alpha\beta}$. Thus we can use (9.9.25) or (9.9.26) to provide two tentative (alternative) definitions for a 'quasi-local rest-mass' in general relativity (i.e. for the rest-mass surrounded by a finite closed 2-surface). (Tod (1983*a*) has also suggested a modification of (9.9.28) in which the *determinant* of $\mathsf{A}_{\alpha\beta}$ is used rather than its norm.)

We take note, also, of an observation due to Tod (1983*a*), which holds when $\eta = 1$, namely that (9.9.19) can be re-expressed in a remarkably simple-looking way:

$$\mathsf{A}_{\alpha\beta}\underset{1}{\mathsf{Z}^{\alpha}}\underset{2}{\mathsf{Z}^{\beta}} = -\frac{\mathrm{i}}{4\pi G}\oint(\underset{1}{\pi_{0'}}\underset{2}{\pi_{1'}} + \underset{1}{\pi_{1'}}\underset{2}{\pi_{0'}})\mathscr{S}. \tag{9.9.29}$$

(The proof uses (9.9.13), (9.9.23), (4.12.32)(*d*), (4.12.35), and the parts-integration formula (4.14.71); it is straightforward if one starts from (9.9.29).) There is a tantalizing similarity between (9.9.29) and (9.9.27), suggesting that a relation such as (9.9.20) might be obtainable if the appropriate definitions could be found. The expression (9.9.29) will have importance for us at the end of this section. When $\eta \neq 1$, (9.9.29) becomes modified, with each $\pi_{0'}$ replaced by $\nu\pi_{0'} + \mathrm{i}\omega^1\eth'\nu$, each $\pi_{1'}$ replaced by $\nu\pi_{1'} + \mathrm{i}\omega^0\eth\nu$ and an additional term $(\underset{1}{\omega^1}\underset{2}{\omega^0} + \underset{1}{\omega^0}\underset{2}{\omega^1})\eth\nu\eth'\nu$ appearing, where $\nu^2 = \eta$.

In order to motivate the introduction of η into (9.9.19), and the definition of it that we shall give in a moment, it will be helpful to point out some results that have been obtained by Tod (1983*a*) in various uncontorted cases. We shall be concerned with the quasi-local rest-mass m as given by (9.9.28), which is unambiguously defined since in all uncontorted cases the norm (9.2.24) is constant over $\mathscr{S}$.

For the Schwarzschild space–time, the remarkable result is obtained that for $\mathscr{S}$ lying in any hypersurface $\mathscr{H}$ of spherical symmetry (defined, in the usual Schwarzschild coordinates, by t and r satisfying some fixed functional relation) m turns out to be precisely the Schwarzschild mass if $\mathscr{S}$ links the source (just once) and $m = 0$ if $\mathscr{S}$ does not link the source. (Note that $\mathscr{S}$ itself need not share the spherical symmetry of $\mathscr{H}$ and is just drawn arbitrarily within it. Such an $\mathscr{S}$ is necessarily uncontorted.)

Somewhat similar results are obtained in any vacuum space–time containing a conformally flat hypersurface $\mathscr{H}$ of *time-symmetry* – where $\mathscr{H}$ is allowed to contain 'sources' which can be either matter regions or 'wormholes' (*cf.* Misner, Thorne and Wheeler 1973). Again, provided $\mathscr{S} \subset \mathscr{H}$, the mass m depends only on the linking properties of $\mathscr{S}$ with the sources and not on its detailed location within $\mathscr{H}$. Denoting by m_j the value of m obtained when $\mathscr{S}$ links only the jth source (just once) and by m_{jk} the value obtained when $\mathscr{S}$ links just the jth and kth sources together (once each) we find the physically satisfying formula

$$m_{jk} < m_j + m_k.$$

This indicates that a gravitational potential energy contribution is already contained in the definitions (9.9.19), (9.9.28) (gravitational radiation contributions being necessarily zero* at $\mathscr{H}$). Indeed, in the limit of weak fields, the difference $m_j + m_k - m_{jk}$ turns out to have precisely the form of a Newtonian r^{-1}-potential energy term.**

[These results were obtained with the help of Tod's concept of 3-*surface twistor* which applies whenever there is a hypersurface $\mathscr{H}$, with normal t^a, with the property that (i) the *magnetic part*

$$H_{ac} = {}^*C_{abcd}t^bt^d$$

of the Weyl tensor *vanishes* on $\mathscr{H}$, and (ii) the magnetic part

$$^*(t^e\nabla_e C_{abcd})t^bt^d$$

of the normal derivative of the Weyl tensor also vanishes on $\mathscr{H}$. (This second condition can be restated in vacuum as the vanishing of the curl, on $\mathscr{H}$, of the *electric part* of C_{abcd}, which is defined like H_{ab} but without the dualization.) Such a hypersurface $\mathscr{H}$ is 'uncontorted' in the sense that it can be embedded in conformally flat space–time with the same intrinsic metric and extrinsic curvature at $\mathscr{H}$, and every 2-surface $\mathscr{S}$ within $\mathscr{H}$ is likewise uncontorted. Then $\mathscr{H}$ has the property that 3-*surface twistors* can be defined on it,† these being the solutions of those parts of (9.9.12) which refer to derivatives acting entirely within $\mathscr{H}$, namely

$$t^{A'(A}\nabla^B_{A'}\omega^{C)} = 0.$$

(In fact these are also hypersurface twistors for $\mathscr{H}$; *cf.* around (7.4.52), the equation (7.4.52) being in a certain sense *dual* to the above, with $o_C\omega^C$ constant along β-curves. Here $N^a = t^a$.)]

The quasi-local mass expressions can also be used when $\mathscr{S}$ lies in matter regions. In particular, for the spherically symmetric ('electrovac') Reissner–Nordstrom space–time, the quantity m does *not* now depend solely on the linking of $\mathscr{S}$ with the mass source, but varies in a way consistent with there being a direct contribution to the mass in agreement with the energy density

* Our definition does take gravitational radiation contributions into account when they are present, however (*cf.* below, in connection with the Bondi–Sachs mass).

** It is somewhat remarkable how this comes about in Tod's analysis. The angular-momentum twistor $A_{\alpha\beta}$ for the various sources turns out to be completely additive, the non-additiveness of the resulting scalars m arising from the nonlinearity of (9.9.28). Such inequalities would arise for rest-mass when 4-momenta are added, but here the inequality runs in the *reverse direction*. This comes about because the effective twistors $I_{\alpha\beta}$ get relatively 'distorted' for the different sources, and contributions arise from the parts of $A_{\alpha\beta}$ which would otherwise be always zero (*cf.* (6.3.11)).

† Tod also defines a modified concept of local twistor transport which is integrable within $\mathscr{H}$, for which the quantity $P_{...}$ appearing in (6.9.10) is augmented by a term involving the electric part of C_{abcd}. The 3-surface twistors are constant under this transport.

of the electromagnetic field.* The FRW models (*cf.* §9.5), being conformally flat are comparatively straightforward to handle. In particular, Tod showed that if $\mathscr{S}$ is a 2-sphere of rotational symmetry, then the value of m is equal to the measure of mass that would be surrounded by a sphere of equal area to $\mathscr{S}$ in Euclidean space, immersed in a fluid of density equal to that of the model. Thus, in particular, for a spatially closed model the value of m increases as the sphere moves outwards, reaching a maximum value for an 'equatorial' sphere, and then reduces back to zero again as the sphere shrinks to a point 'around the back'. This shows that the *total* mass of the entire model is zero, in agreement with expectations on other grounds (*cf.* Misner, Thorne and Wheeler 1973). The negative potential energy contributions exactly cancel those due to the matter. Indeed, it is a quite general feature of our construction that the total mass of *any* closed universe model must vanish. This is a limiting case of the more general property that the mass on one side of any finite $\mathscr{S}$ must be equal to that on the other – as follows obviously from the symmetry of (9.9.19).

Definition of η

It seems clear that the above (and other) highly satisfactory results for an uncontorted $\mathscr{S}$ ought to carry over to those cases when $\mathscr{S}$ is contorted. In particular, one would expect that, in the Schwarzschild space–time, if $\mathscr{S}$ does not link the source then one ought to obtain $m = 0$ whether or not $\mathscr{S}$ is contorted. However, K.P. Tod and R.M. Kelly have found, as part of an involved calculation, that if the factor η is omitted from the definition (9.9.19), then the physically unreasonable result $m^2 < 0$ would be obtained for certain small spheres not linking the Schwarzschild source.

These spheres are obtained by choosing a point P in $\mathscr{M}$ and taking cross-sections of the light cone of P which are defined at fixed affine distance u from P, normalized against a timelike vector T^a at P. Contorted sections arise when T^a does not lie in a (t,r)-plane of the ordinary Schwarzschild coordinates. Without the η-factor, a negative m^2 then arises at order u^5, but this is removed when an η-factor, defined according to the following prescription is included. This prescription had been suggested by a certain twistor contour integral argument we proposed, which shows that if the

* It may be remarked that this is more satisfactory than the result obtained using the Komar (1959) integral expression (*cf.* also Pirani 1962), according to which the contribution from the electromagnetic field is in error by a factor 2, by comparison with the gravitational contribution (which may be verified by passing to the weak-field limit).

Killing spinor (6.7.15) belongs to $\mathbb{T}^{(\alpha\beta)}(\mathscr{S})$, this η yields $m=0$ whenever $\mathscr{S}$ does not link the source in the Schwarzschild space–time.

To define η, form the (conformally invariant) determinant of any four linearly independent solutions, $\underset{1}{\omega}^{A}$, $\underset{2}{\omega}^{A}$, $\underset{3}{\omega}^{A}$, $\underset{4}{\omega}^{A}$ of (9.9.13) on $\mathscr{S}$:

$$Y=\begin{vmatrix} \underset{1}{\omega}^0 & \underset{2}{\omega}^0 & \underset{3}{\omega}^0 & \underset{4}{\omega}^0 \\ \underset{1}{\omega}^1 & \underset{2}{\omega}^1 & \underset{3}{\omega}^1 & \underset{4}{\omega}^1 \\ \eth\underset{1}{\omega}^0 & \eth\underset{2}{\omega}^0 & \eth\underset{3}{\omega}^0 & \eth\underset{4}{\omega}^0 \\ \eth'\underset{1}{\omega}^1 & \eth'\underset{2}{\omega}^1 & \eth'\underset{3}{\omega}^1 & \eth'\underset{4}{\omega}^1 \end{vmatrix} = \{\varepsilon_{\alpha\beta\gamma\delta}\underset{1}{Z}^{\alpha}\underset{2}{Z}^{\beta}\underset{3}{Z}^{\gamma}\underset{4}{Z}^{\delta}\} \qquad (9.9.30)$$

(see also Tod 1984, Jeffryes 1984*b*). When $\mathscr{S}$ is uncontorted this is constant over $\mathscr{S}$ (as follows by reverting to a local twistor description in the conformally flat embedding space $\mathscr{M}$), but in general Y will vary – though it normally does not vanish (assuming $\mathscr{S}$ is four-dimensional). Then:

choose η as some constant multiple of Y over $\mathscr{S}$.

There is, unfortunately, still some ambiguity in the precise choice of η. This could be eliminated (at the cost of breaking conformal invariance in the definition of η) by demanding, for example, that the average value of η over $\mathscr{S}$ be unity, or, perhaps more probably, that the average of its logarithm should be zero; but the status of such choices is unclear.

One significance of the quantity η is that it describes, locally on $\mathscr{S}$, the non-zero components of an alternating twistor

$$\varepsilon^{\alpha\beta\gamma\delta}\in\mathbb{T}^{[\alpha\beta\gamma\delta]}(\mathscr{S}).$$

The ambiguity in the choice of η for which $\eta = Y \times$ constant is equivalent to the freedom in such choice of $\varepsilon^{\alpha\beta\gamma\delta}$. The quantity $\eta\mathscr{S}$, which appears in (9.9.19), provides a surface-element 2-form for the *complex* conformally Minkowskian embedding space into which $\mathscr{S}$ can be put without changing σ or σ'. This 2-form is naturally defined up to the above ambiguity. The details of these matters cannot be discussed here, however, and we turn, for the rest of this section, to the case when $\mathscr{S}$ is at null infinity – a situation for which $\eta = 1$, as we shall see shortly.

$\mathrm{A}_{\alpha\beta}$ *at* $\mathscr{I}^+$

The conformal invariance of the definition of $\mathbb{T}^{\alpha}(\mathscr{S})$ is particularly useful when we come to apply our construction at a cut of $\mathscr{I}^+$ (*cf.* §9.8). We shall assume future-3-asymptotic simplicity and, in addition, that only massless fields are present near $\mathscr{I}^+$. Accordingly we take the physical energy tensor $\tilde{T}_{ab}$ to be traceless near $\mathscr{I}^+$ and (*cf.* (5.9.2), (6.7.34)) to scale as

$$T_{ab} = \Omega^{-2}\tilde{T}_{ab} \tag{9.9.31}$$

under

$$g_{ab} = \Omega^2 \tilde{g}_{ab}, \tag{9.9.32}$$

where we now denote the *physical* quantities with a tilde, as in §9.8. We suppose that the matter fields fall off at such a rate that T_{ab} is finite (and at least C^0) at $\mathscr{I}^+$, which is consistent with the peeling properties of §9.7 holding at $\mathscr{I}^+$. (If the fields $\phi_{...}$ are suitably regular at $\mathscr{I}^+$, then so is T_{ab}, by the assumed invariance (9.9.31) of the expressions for T_{ab} under (9.9.32), *cf.* (5.2.4), (5.8.3), (6.8.36).) Under these assumptions the strong asymptotic Einstein condition of §9.6 will hold.

Now consider the expression (9.9.19) (or (9.9.17)) written in terms of the physical quantities

$$\tilde{\Psi}_{ABCD} = \tilde{\psi}_{ABCD}, \quad \tilde{\Phi}_{ABC'D'} = 4\pi G \tilde{T}_{ABC'D'}, \quad \tilde{\Lambda} = 0, \tag{9.9.33}$$

but where, in each case, we substitute the RHS of (9.9.33). Adopting the scaling (9.9.31), together with

$$\psi_{ABCD} = \Omega^{-1}\tilde{\psi}_{ABCD} \tag{9.9.34}$$

(*cf.* (9.6.40)), and the spin-frame rescaling of §§9.6, 9.7,

$$o_A = \tilde{o}_A, \quad o^A = \Omega^{-1}\tilde{o}^A, \quad \iota_A = \Omega\tilde{\iota}_A, \quad \iota^A = \tilde{\iota}^A,$$

which allows the frame to remain finite at $\mathscr{I}^+$, we find

$$\begin{aligned}
&\tilde{\omega}^0 = \Omega^{-1}\omega^0, && \tilde{\omega}^1 = \omega^1; && \tilde{\mathscr{S}} = \Omega^{-2}\mathscr{S};\\
&\tilde{\psi}_1 = \Omega^4\psi_1, && \tilde{\psi}_2 = \Omega^3\psi_2, && \tilde{\psi}_3 = \Omega^2\psi_3;\\
&\tilde{T}_{000'1'} = \Omega^5 T_{000'1'}, && \tilde{T}_{010'1'} = \Omega^4 T_{010'1'}, && \tilde{T}_{110'1'} = \Omega^3 T_{110'1'}.
\end{aligned}$$

Hence (9.9.19) becomes, in terms of the 'unphysical' quantities,

$$\mathsf{A}_{\alpha\beta}\underset{1}{Z^\alpha}\underset{2}{Z^\beta} = -\frac{\mathrm{i}}{4\pi G}\oint\Big\{\psi_1\underset{1}{\omega^0}\underset{2}{\omega^0} + \psi_2\Big(\underset{1}{\omega^0}\underset{2}{\omega^1} + \underset{1}{\omega^1}\underset{2}{\omega^0}\Big) + \psi_3\underset{1}{\omega^1}\underset{2}{\omega^1}\Big\}\mathscr{S} + O(\Omega) \tag{9.9.35}$$

the contributions from T_{ab} being all included in the '$O(\Omega)$' and where we have anticipated $\eta = 1$ at $\mathscr{I}^+$, *cf.* below. If we take $\mathscr{S}$ to be a cut of $\mathscr{I}^+$, '$O(\Omega)$' disappears, and we are left with a definition of $\mathsf{A}_{\alpha\beta}$ entirely in terms of ψ_1, ψ_2, ψ_3 and the solutions of

$$\eth'\omega^0 = 0, \quad \eth\omega^1 = \sigma\omega^0 \tag{9.9.36}$$

on the cut (since $\sigma' = 0$ at $\mathscr{I}^+$). We choose Ω so that the intrinsic metric of $\mathscr{S}$ is a sphere.

The simplification of (9.9.36) over (9.9.13) is a considerable advantage to us. In the *first* place, it is clear that (9.9.36) *always* has exactly four independent solutions, so $\mathbb{T}^{\alpha}(\mathscr{S})$ is necessarily four-complex-dimensional. To see this, we refer to (4.15.60), recalling that ω^0 has spin-weight $-\frac{1}{2}$, and find that $\eth'\omega^0 = 0$ has just two independent solutions (linear combinations of ${}_{-\frac{1}{2}}Y_{\frac{1}{2},-\frac{1}{2}}$ and ${}_{-\frac{1}{2}}Y_{\frac{1}{2},\frac{1}{2}}$, *cf.* §4.15. Substituting each of these into $\eth\omega^1 = \sigma\omega^0$, and recalling that ω^1 and σ have respective spin-weights $\frac{1}{2}$ and 2, we again find, from (4.15.60), that for each ω^0 there are just two independent solutions for ω^1, making four in all, as required.

In the *second* place, we find that the η-factor discussed earlier can be chosen to be unity. To construct the relevant determinant (9.9.30) we need four linearly independent solutions of (9.9.36), two of which can have $\omega^0 = 0$ (i.e. $\eth\omega^1 = 0$), so (9.9.30) takes the form

$$Y = \begin{vmatrix} 0 & 0 & \underset{3}{\omega}{}^0 & \underset{4}{\omega}{}^0 \\ \underset{1}{\omega}{}^1 & \underset{2}{\omega}{}^1 & \underset{3}{\omega}{}^1 & \underset{4}{\omega}{}^1 \\ 0 & 0 & \eth\underset{3}{\omega}{}^0 & \eth\underset{4}{\omega}{}^0 \\ \eth'\underset{1}{\omega}{}^1 & \eth'\underset{2}{\omega}{}^1 & \eth'\underset{3}{\omega}{}^1 & \eth'\underset{4}{\omega}{}^1 \end{vmatrix} = -\begin{vmatrix} \underset{1}{\omega}{}^1 & \underset{2}{\omega}{}^1 \\ \eth'\underset{1}{\omega}{}^1 & \eth'\underset{2}{\omega}{}^1 \end{vmatrix} \times \begin{vmatrix} \underset{3}{\omega}{}^0 & \underset{4}{\omega}{}^0 \\ \eth\underset{3}{\omega}{}^0 & \eth\underset{4}{\omega}{}^0 \end{vmatrix}.$$

By the preceding discussion we can choose $\underset{1}{\omega}{}^1, \underset{2}{\omega}{}^1, \underset{3}{\omega}{}^0, \underset{4}{\omega}{}^0$ to be, say, ${}_{\frac{1}{2}}Y_{\frac{1}{2},-\frac{1}{2}}$, ${}_{\frac{1}{2}}Y_{\frac{1}{2},\frac{1}{2}}$, ${}_{-\frac{1}{2}}Y_{\frac{1}{2},-\frac{1}{2}}$, ${}_{-\frac{1}{2}}Y_{\frac{1}{2},\frac{1}{2}}$ respectively, which are explicit expressions (*cf.* §4.15) not involving σ, and so also are $\eth'\underset{1}{\omega}{}^1, \ldots, \eth\underset{4}{\omega}{}^0$. Thus the value of the determinant is the same quantity, *constant* over $\mathscr{S}$, as would be the case for Minkowski space. Thus we take $\eta = 1$.

In the *third* place, a unique $\mathsf{I}_{\alpha\beta} \in \mathbb{T}_{[\alpha\beta]}(\mathscr{S})$, with all the standard properties, arises (which is actually independent of the choice of cut in a well-defined sense). Recall the local twistor expression (9.8.41) for $\mathsf{I}^{\rho\sigma}$ on $\mathscr{I}^+$. Taking the complex conjugate and transvecting with $\underset{1}{\mathsf{Z}}{}^{\rho}\underset{2}{\mathsf{Z}}{}^{\sigma}$, we obtain the local twistor expression on $\mathscr{S}$:

$$\begin{aligned} \{\mathsf{I}_{\rho\sigma}\underset{1}{\mathsf{Z}}{}^{\rho}\underset{2}{\mathsf{Z}}{}^{\sigma}\} &= A(\mathrm{i}\underset{1}{\omega}{}^0\underset{2}{\pi}_{1'} - \mathrm{i}\underset{2}{\omega}{}^0\underset{1}{\pi}_{1'} - \rho'\underset{1}{\omega}{}^0\underset{2}{\omega}{}^1 + \rho'\underset{2}{\omega}{}^0\underset{1}{\omega}{}^1) \\ &= A(\underset{2}{\omega}{}^0\eth\underset{1}{\omega}{}^0 - \underset{1}{\omega}{}^0\eth\underset{2}{\omega}{}^0). \end{aligned} \tag{9.9.37}$$

Now we see from (4.15.60) that any scalar with spin-weight $-\frac{1}{2}$ satisfying $\eth'\omega^0 = 0$ must also satisfy

$$\eth^2\omega^0 = 0. \tag{9.9.38}$$

(This would not be true if $\mathscr{S}$ were not scaled to be a sphere.) Applying ð to the final expression in (9.9.37), we find that it is annihilated. But (9.9.37) has spin-weight 0, so by the discussion of §4.15 it must be *constant* over $\mathscr{S}$. This constant value serves to define $\mathsf{I}_{\rho\sigma}\underset{1}{\mathsf{Z}}^{\rho}\underset{2}{\mathsf{Z}}^{\sigma}\in\mathbb{C}$ as a skew bilinear function of $\underset{1}{\mathsf{Z}}^{\rho}$ and $\underset{2}{\mathsf{Z}}^{\sigma}$, and this yields our required

$$\mathsf{I}_{\rho\sigma}\in\mathbb{T}_{[\rho\sigma]}(\mathscr{S}). \tag{9.9.39}$$

In the *fourth* place, a definition of twistor complex conjugation (9.9.21) can be given which effectively satisfies all the necessary standard properties (involutory, $(++--)$ signature, defined without depending on an arbitrary choice of conformal factor Ω), and with respect to which $\mathsf{I}_{\alpha\beta}$ is twistor-real. The sought-for Hermiticity property (9.9.20) then turns out to hold. However, the definition is not quite direct and suffers from a certain awkwardness in its present form, as we shall shortly see. In the meantime, various relevant ideas need to be introduced.

Asymptotic spin-space

Note, to begin with, that any $\mathsf{Z}^{\alpha}\in\mathbb{T}^{\alpha}(\mathscr{S})$ for which $\omega^0\equiv 0$ on $\mathscr{S}$ must satisfy

$$\mathsf{I}_{\alpha\beta}\mathsf{Z}^{\alpha}=0 \tag{9.9.40}$$

since the ω^1 component does not appear in (9.9.37). Thus there is a two-dimensional subspace of $\mathbb{T}^{\alpha}(\mathscr{S})$ annihilated by $\mathsf{I}_{\alpha\beta}$ in the sense of (9.9.40). This is the space of $\{0,\omega^1\}$, where ω^1 satisfies

$$\eth\omega^1=0. \tag{9.9.41}$$

We refer to this space as *asymptotic spin-space* and denote it by $\mathsf{I}^{\alpha\beta}\mathbb{T}_{\beta}(\mathscr{S})$ when thinking of it as a subspace of $\mathbb{T}^{\alpha}(\mathscr{S})$, or by $\mathbb{S}^{A}(\mathscr{S})$ when thinking of $\{0,\omega^1\}$ as the two components of a spinor field ω^A on $\mathscr{S}$ which are the analogues of the restrictions to $\mathscr{S}$ of ω^A-fields that are *constant* in $\mathbb{M}$. These particular spinor fields on $\mathscr{S}$, satisfying (9.9.41), can be extended in a natural way to the whole of $\mathscr{I}^+$ by demanding that

$$\text{þ}'_c\omega^1:=(\text{þ}'+\rho')\omega^1=0. \tag{9.9.42}$$

So the concept of asymptotic spin-space refers to $\mathscr{I}^+$ as a whole and can, accordingly, be denoted by $\mathbb{S}^A(\mathscr{I}^+)$. (In fact, for $\omega^A\propto\iota^A$, (9.9.41) and (9.9.42) are the parts of the twistor equation (9.9.12) that are tangential to $\mathscr{I}^+$ (Bramson 1975*a*).) It is convenient to scale $\mathscr{I}^+$ so that $\rho'=0$ (as we did in §9.8, *cf.* (9.8.33)). Then (9.9.42) becomes

$$\text{þ}'\omega^1=0, \tag{9.9.43}$$

and we can directly identify all the $\mathbb{S}^A(\mathscr{S})$ with one another for different cuts $\mathscr{S}$. Having specialized to $\rho' = 0$, we may as well also choose the cuts of $\mathscr{I}^+$ all to have unit-sphere metrics. Thus *from now on we are adopting* (9.8.33).

Properties of the norm at $\mathscr{I}^+$

Our procedure for defining $\mathrm{Z}^\alpha \mathrm{Z}_\alpha$ is to construct the expression (9.9.24) as before, and to investigate its constancy over $\mathscr{S}$ by applying ðð′. In fact it will be helpful to use the slightly more general-looking (though essentially equivalent) 'polarized' form of (9.9.24):

$$\{\underset{0}{\mathrm{Z}}^\alpha \mathrm{Z}_\alpha\} = \mathrm{i}(\bar{\omega}^{1'} ð \underset{0}{\omega}^0 - \underset{0}{\omega}^0 ð \bar{\omega}^{1'} + \bar{\omega}^{0'} ð' \underset{0}{\omega}^1 - \underset{0}{\omega}^1 ð' \bar{\omega}^{0'}). \qquad (9.9.44)$$

We find, after a short calculation, that

$$ðð'\{\underset{0}{\mathrm{Z}}^\alpha \mathrm{Z}_\alpha\} = \underset{0}{\omega}^0 \bar{\omega}^{0'}(\mathrm{i}ð'^2\sigma - \mathrm{i}ð^2\bar{\sigma})$$
$$= 2\underset{0}{\omega}^0 \bar{\omega}^{0'} \,\mathrm{Im}\,(A\psi_2 - \sigma N), \qquad (9.9.45)$$

by (9.8.89). Note that this vanishes if the cut is purely electric (*cf.* (9.8.93)), as are all cuts of $\mathscr{I}^+$ for $\mathbb{M}$ (or for stationary $\mathscr{M}$). These are the *uncontorted* cuts of $\mathscr{I}^+$.

When (9.9.45) *does* vanish we see (*cf.* (4.15.60)) that $\{\underset{0}{\mathrm{Z}}^\alpha \mathrm{Z}_\alpha\}$ must indeed be constant over $\mathscr{S}$, and so a good definition of $\underset{0}{\mathrm{Z}}^\alpha \mathrm{Z}_\alpha$, and therefore of (9.9.21), is at hand. For, since $\{\underset{0}{\mathrm{Z}}^\alpha \mathrm{Z}_\alpha\}$ has conformal weight *zero* on $\mathscr{S}$, its *constancy* is invariant under conformal rescalings (9.8.13). However when (9.9.45) does *not* vanish (and this will be the general situation of a contorted $\mathscr{S}$), it will not be a conformally invariant operation to *average* $\{\underset{0}{\mathrm{Z}}^\alpha \mathrm{Z}_\alpha\}$ over $\mathscr{S}$, i.e. to extract its $j = 0$ part, since, as we have seen in the discussion of §4.15, this would only be invariant for a quantity of conformal weight -2; for all other weights the parts with higher j-values mix in with the $j = 0$ part under conformal transformations.

We shall see shortly that there is a (perhaps not entirely satisfactory) way of circumventing this problem. As a preliminary, we note that if either $\omega^0 = 0$ or $\underset{0}{\omega}^0 = 0$, then the RHS of (9.9.45) does indeed vanish. Thus we have an *unambiguous* meaning for the Hermitian scalar product $\underset{0}{\mathrm{Z}}^\alpha \mathrm{Z}_\alpha$ whenever *either* Z^α or $\underset{0}{\mathrm{Z}}^\alpha$ belongs to $\mathrm{I}^{\alpha\beta}\mathbb{T}_\beta(\mathscr{S})$. This implies that a certain map: $\mathbb{T}^\alpha(\mathscr{S}) \to \mathrm{I}^{\alpha\beta}\mathbb{T}_\beta(\mathscr{S})$, which we shall denote by

$$\mathrm{Z}^\alpha \mapsto \mathrm{Z}_\gamma \mathrm{I}^{\gamma\alpha}, \qquad (9.9.46)$$

is well-defined. It is characterized by the following property: For each $\underset{0}{Z^\alpha} \in \mathbb{T}^\alpha(\mathscr{S})$, if we construct from the left-hand member Z^α of (9.9.46) the well-defined quantity

$$\mathsf{I}_{\alpha\beta}\underset{0}{Z^\alpha Z^\beta}$$

(obtained from the constant expression (9.9.37)), we get the same result as when we construct from the right-hand member $\mathsf{Z}_\gamma \mathsf{I}^{\gamma\alpha}$ of (9.9.46) the well-defined scalar product

$$\underset{0}{Z^\beta}\overline{(\mathsf{Z}_\gamma \mathsf{I}^{\gamma\beta})}$$

(by substituting $\mathsf{Z}_\gamma \mathsf{I}^{\gamma\alpha}$ in the – now constant – expression (9.9.44)). Equating these expressions we obtain (9.9.46), and one easily verifies that this is in fact achieved by

$$\{\omega^0, \omega^1\} \mapsto \{0, -\mathrm{i}A\bar{\omega}^{0'}\}. \tag{9.9.47}$$

Hermiticity of $\mathsf{A}_{\alpha\beta}$ *at* $\mathscr{I}^+$; *Bondi–Sachs 4-momentum*

The reason for our interest in (9.9.46) is that it suffices for the formulation of the desired Hermiticity property (9.9.20). (The *full* definition of a norm on $\mathbb{T}(\mathscr{S})$ will be given shortly.) What is required for (9.9.20) is, in fact, that

$$Z^\alpha \mathsf{A}_{\alpha\beta}(\mathsf{Z}_\gamma \mathsf{I}^{\gamma\beta}) \in \mathbb{R} \quad \textit{for all } Z^\alpha \in \mathbb{T}^\alpha(\mathscr{S}) \tag{9.9.48}$$

Let us check that, with the definition (9.9.47), this is indeed the case. Substituting $\underset{1}{Z^\alpha} = Z^\alpha$ and $\underset{2}{Z^\beta} = \mathsf{Z}_\gamma \mathsf{I}^{\gamma\beta}$ into (9.9.35), and taking these as the LHS and RHS of (9.9.47), respectively, we obtain, for the expression in (9.9.48):

$$-\frac{1}{4\pi G}\oint\{\psi_2\omega^0\bar{\omega}^{0'} + \psi_3\omega^1\bar{\omega}^{0'}\}A\mathscr{S}. \tag{9.9.49}$$

Now, by (9.8.83),

$$\begin{aligned}\oint \psi_3\omega^1\bar{\omega}^{0'}A\mathscr{S} &= \oint \bar{\eth} N\omega^1\bar{\omega}^{0'}\mathscr{S}\\ &= -\oint N\bar{\eth}\omega^1\bar{\omega}^{0'}\mathscr{S}\\ &= -\oint \sigma N\omega^0\bar{\omega}^{0'}\mathscr{S}\end{aligned} \tag{9.9.50}$$

(using parts integration (4.14.71) and $\overline{(9.9.36)(1)}$ and (9.9.36)(2)). Substituting back into (9.9.49) then gives us

$$\mathsf{Z}^\alpha \mathsf{A}_{\alpha\beta}(\bar{\mathsf{Z}}_\gamma \mathsf{I}^{\gamma\beta}) = -\frac{1}{4\pi G}\oint\{A\psi_2 - \sigma N\}\omega^0\bar{\omega}^{0'}\mathscr{S}. \qquad (9.9.51)$$

The imaginary part of (9.9.51) can be re-expressed by means of (9.8.89) as

$$-\frac{1}{4\pi G}\operatorname{Im}\oint \eth^2\bar{\sigma}\ \omega^0\bar{\omega}^{0'}\mathscr{S},$$

which is seen to vanish by a double parts integration and use of (9.9.36) and (9.9.38). Thus (9.9.48) is indeed satisfied, and so the desired reduction from twenty to ten independent real components in $\mathsf{A}_{\alpha\beta}$ *is* achieved as in (9.9.20).

We refer to the flat-space descriptions of §§6.3, 6.4 and 6.5 for the meanings of these components. The vanishing of the projection spinor part in (6.3.11)(1) corresponds to the fact that had we taken *both* of $\underset{1}{\mathsf{Z}}^\alpha$ and $\underset{2}{\mathsf{Z}}^\alpha$ in (9.9.35) from $\mathsf{I}^{\alpha\beta}\mathbb{T}_\beta(\mathscr{S})$ we would have obtained zero (by an argument similar to (9.9.50)). The expression (9.9.51) itself is seen to refer to an off-diagonal part of $\mathsf{A}_{\alpha\beta}$ in (6.3.11), i.e. to the *total 4-momentum* surrounded by $\mathscr{S}$. Indeed, noting that, with standard flat-space twistor descriptions, the substitution of

$$\mathsf{S}^{\alpha\beta} = \bar{\mathsf{Z}}_\gamma \mathsf{I}^{\gamma(\alpha}\mathsf{Z}^{\beta)} \qquad (9.9.52)$$

into (9.9.7) yields

$$\mathsf{F}^\alpha{}_\beta = (\bar{\mathsf{Z}}_\gamma \mathsf{I}^{\gamma\alpha})(\mathsf{Z}^\delta \mathsf{I}_{\delta\beta}),$$

and recalling that the primary part of $\mathsf{F}^\alpha{}_\beta$ is the Killing vector $\xi^{AB'}$, we find that

$$\xi^{AB'} = \bar{\pi}^A\pi^{B'}, \qquad (9.9.53)$$

if Z^α is represented in the standard flat-space way as $(\omega^A, \pi_{A'})$. Then the LHS of (9.9.51) is, in this description,

$$p_a\xi^a = p_{BC'}\bar{\pi}^B\pi^{C'}, \qquad (9.9.54)$$

where p_a is the total 4-momentum described by $\mathsf{A}_{\alpha\beta}$. Thus (9.9.51) describes a null component of the 4-momentum surrounded by $\mathscr{S}$.

These descriptions make perfectly clear sense for $\mathscr{M}$, in terms of its asymptotic spin-space $\mathbb{S}^A(\mathscr{S})$ and the other spaces $\mathbb{S}^{A\ldots C}_{D'\ldots F'}(\mathscr{S})$ constructed from that in the standard way. Rather than just taking a null component, as in (9.9.54), it is more usual to take spacelike or timelike ones. This means taking elements of the *linear span* of the expressions $\omega^0\bar{\omega}^{0'}$ on $\mathscr{S}$, i.e. elements of $\mathbb{S}^a(\mathscr{S})$. Incorporating the factor A with the term $\omega^0\bar{\omega}^{0'}$ to remove the spin-frame dependence, we obtain the linear span of $A\omega^0\omega^{0'}$ as a space of spin-weight 0 conformal-weight 1 scalars W on $\mathscr{S}$. By (9.9.36),

(9.9.38), these satisfy

$$ð^2 W = 0, \quad ð'^2 W = 0, \tag{9.9.55}$$

and W is normally taken to be real. From (4.15.60) we see that any such W is composed only of $j=0$ and $j=1$ spherical harmonics. Thus, in the expression

$$-\frac{1}{4\pi G}\oint \{\psi_2 - \sigma N A^{-1}\} W \mathscr{S}, \tag{9.9.56}$$

the various such choices of W will provide the four components of energy–momentum. The choice $W=1$ provides the energy, with time-axis corresponding to our selection of unit-sphere metric on $\mathscr{S}$, and the $j=1$ choices (which in standard spherical polars would be $\sin\theta\cos\phi$, $\sin\theta\sin\phi$, and $\cos\theta$, respectively) provide the 3-momentum. There is no loss, at this stage, in taking $A=1$ since boost-weights are playing no essential role here. The resulting expression (9.9.56) is the total *Bondi–Sachs 4-momentum* surrounded by $\mathscr{S}$, which was originally arrived at by quite other means (Bondi 1960*b*, Bondi, van der Burg and Metzner 1962, Sachs 1962*a*).

Alternative expressions for Bondi–Sachs mass

There are various ways of re-expressing (9.9.56), one of which will be considered in the next section. But it is of some interest to point out another one of these here. Suppose we choose Ω so that the hypersurfaces $\Omega =$ constant are null. Then by (9.8.29) we have $\tau' \approx 0$. From (4.12.32)(*e*)′ we obtain

$$þ\sigma' \approx -\Phi_{20} \approx -\mathrm{N}, \tag{9.9.57}$$

by (9.8.74). Recall also that, by (9.8.81),

$$þ\psi_2 \approx -A\Psi_2. \tag{9.9.58}$$

These expressions tell us that we can interpret (9.9.56), with $W=1$, in the following way. Consider our cut of $\mathscr{I}^+$ to be a limiting member $\mathscr{S} = \mathscr{S}_0$ of a family of closed spacelike 2-surfaces $\mathscr{S}_\Omega$ obtained as the intersections with a fixed null hypersurface $\mathscr{N}$, of the null hypersurfaces $\Omega =$ constant. (Here $\mathscr{N}$ is the null hypersurface meeting $\mathscr{I}^+$ in $\mathscr{S}$.) Then we find that the derivative at $\Omega = 0$ with respect to $-\Omega$, of

$$\frac{1}{4\pi G}\oint_{\mathscr{S}_\Omega} (\Psi_2 - \sigma\sigma')\mathscr{S}_\Omega, \tag{9.9.59}$$

precisely measures the Bondi–Sachs energy ($W=1$). (This follows because $\partial/\partial(-\Omega)=A^{-1}D=A^{-1}\text{þ}$ here, and $\sigma'\approx 0$.) The integral in (9.9.59) was briefly considered in (4.14.41) and (4.14.45), where it was noted that it is both conformally invariant and real. We see now a significance of this reality in that it implies the reality of (9.9.56) and is thus intimately related to (9.9.48) and (9.9.20). We note also that (9.9.59) can be re-expressed, using (4.14.20) and (4.14.44), as

$$-\frac{1}{2G}+\frac{1}{4\pi G}\oint(\Phi_{11}+\Lambda-\rho\rho')\mathscr{S}_\Omega \tag{9.9.60}$$

and that, because of the conformal invariance of (9.9.59), the integral (9.9.60) can be written either in terms of the physical or unphysical curvature quantities. If we use the physical quantities, then, with the normal fall-off assumptions about the matter fields, the curvature terms disappear and we are left with

$$-\frac{1}{2G}-\frac{1}{4\pi G}\oint\tilde{\rho}\tilde{\rho}'\mathscr{S}_\Omega \tag{9.9.61}$$

Hence the rate at which this approaches zero at infinity also defines the Bondi–Sachs mass (*cf.* Hawking 1968).

A norm on $\mathbb{T}^\alpha(\mathscr{S})$ at $\mathscr{I}^+$

We shall return to the Bondi–Sachs 4-momentum $p_a(\mathscr{S})$ in the next section. Indeed, there we shall present an argument showing that if an appropriate inequality is satisfied by the physical energy tensor (the 'dominant energy condition') on a compact spacelike hypersurface spanned by $\mathscr{S}$, then $p_a(\mathscr{S})$ is future-timelike (or is zero if the space–time is flat all along $\mathscr{S}$). For the moment we assume this result and use it to circumvent the problem we had in defining an appropriate twistor norm for $\mathbb{T}^\alpha(\mathscr{S})$.

To this end, we assume that the space–time is not flat at $\mathscr{S}$ (otherwise there is in any case no problem), and that consequently (with the above assumption) a uniquely defined asymptotic time direction is singled out at $\mathscr{S}$, namely that of $p_a(\mathscr{S})$. We choose a conformal scale at $\mathscr{S}$ so that the sphere metric of $\mathscr{S}$ is the one associated with that particular direction, i.e. so that the choices $j=1$ for W all yield zero in (9.9.56). Then we define $\underset{0}{\mathbf{Z}^\alpha\mathbf{Z}_\alpha}$ by simply averaging (i.e. selecting the $j=1$ part in) (9.9.44). Though this is undoubtedly inelegant, it is perhaps not unreasonable, in view of the fact that the obstruction to the constancy of (9.9.44) involves the same quantity $A\psi_2-\sigma N$, in (9.9.45), as is now being used to fix the scale to overcome this very obstruction. However, in one case the imaginary part enters and in the other, the real part. A proper understanding of these matters is still lacking.

The exact sequence structure of $\mathbb{T}^\alpha(\mathscr{S})$ *at* $\mathscr{I}^+$

One reason that a full definition of twistor norm (i.e. of the complex conjugation operation (9.9.21) is needed is that otherwise we cannot properly interpret the remaining 'angular momentum' components of $\mathsf{A}_{\alpha\beta}$. What we have previously established, in effect, by obtaining the structure (9.9.39) and (9.9.46), is an analogue for $\mathbb{T}^\alpha(\mathscr{S})$ of the short exact sequence of (6.5.28):

$$0 \to \mathbb{S}^A(\mathscr{S}) \to \mathbb{T}^\alpha(\mathscr{S}) \to \mathbb{S}_{A'}(\mathscr{S}) \to 0, \qquad (9.9.62)$$

the second map being merely the inclusion $\mathsf{I}^{\alpha\beta}\mathbb{T}_\beta(\mathscr{S}) \subset \mathbb{T}^\alpha(\mathscr{S})$ and the third, the resulting factor space map (*cf.* (6.5.29)). We shall see that the structure (9.9.39), (9.9.46) will in fact tell us that $\mathbb{S}_{A'}(\mathscr{S})$, defined in this way, may be actually identified with the complex conjugate of the dual of $\mathbb{S}^A(\mathscr{S})$, in a natural way, as the notation implies. (The significance of (9.9.62) for angular momentum will emerge presently.)

To see that (9.9.39) and (9.9.46) do indeed have the above implication, we refer first to (9.9.46), as given by (9.9.47). Recalling that $\mathbb{S}^A(\mathscr{S})$ is defined by $\omega^0 = 0$, we see that (9.9.47) associates the factor space $\mathbb{T}^\alpha(\mathscr{S})/\mathbb{S}^A(\mathscr{S})$ (i.e. the space of ω^0 satisfying (9.9.36)(1)) with the complex conjugate $\mathbb{S}^{A'}(\mathscr{S})$ of $\mathbb{S}^A(\mathscr{S})$. But it is a little more natural to pass to the conjugate of the *dual* space $\mathbb{S}_A(\mathscr{S})$.

This can be done legitimately because (9.9.39) and (9.9.46) actually *also* provide a concept of

$$\varepsilon_{AB} \in \mathbb{S}_{[AB]}(\mathscr{S}), \qquad (9.9.63)$$

which, as we know, establishes an isomorphism between $\mathbb{S}^A(\mathscr{S})$ and its dual. To obtain (9.9.63), we first note that if $\underset{1}{\mathsf{Z}}^\alpha, \underset{2}{\mathsf{Z}}^\alpha \in \mathsf{I}^{\alpha\beta}\mathbb{T}_\beta(\mathscr{S})$, we can represent them, according to (9.9.46), as

$$\underset{j}{\mathsf{Z}}^\alpha = \underset{j}{\bar{\mathsf{X}}}_\gamma \mathsf{I}^{\gamma\alpha} \quad (j = 1, 2). \qquad (9.9.64)$$

Then, observing that according to standard flat-space spinor-part descriptions, as in $\mathbb{M}$, we would have

$$\underset{1}{\omega}^R \underset{2}{\omega}^S \varepsilon_{RS} = \overline{\underset{1}{\mathsf{X}}^\rho \underset{2}{\mathsf{X}}^\sigma \mathsf{I}_{\rho\sigma}}, \qquad (9.9.65)$$

where

$$\underset{j}{\mathsf{Z}}^\alpha = (\underset{j}{\omega}^A, 0), \quad \underset{j}{\mathsf{X}}^\alpha = (\underset{j}{\lambda}^A, -\underset{j}{\bar{\omega}}_{A'}) \quad (j = 1, 2), \qquad (9.9.66)$$

we use the RHS of (9.9.65) as providing a *definition* for (9.9.63). Reverting to descriptions on $\mathscr{S}$, and using (9.9.37), (9.9.47), we find for the value of the quantity (9.9.65), as applied to

$$\underset{j}{\mathsf{Z}}^\alpha = \{0, \underset{j}{\omega}^1\}, \quad \underset{j}{\mathsf{X}}^\alpha = \{-A^{-1}\mathrm{i}\underset{j}{\bar{\omega}}^{1'}, \underset{j}{\mu}\} \quad (j = 1, 2), \tag{9.9.67}$$

the expression

$$A^{-1}(\underset{1}{\omega}^1 \eth' \underset{2}{\omega}^1 - \underset{2}{\omega}^1 \eth' \underset{1}{\omega}^1), \tag{9.9.68}$$

which, like (9.9.37), is easily seen to be constant on $\mathscr{S}$. This expression (9.9.68) (= 9.9.65)) serves to define 'ε_{RS}' as a skew bilinear map

$$\varepsilon_{RS}: \mathbb{S}^R(\mathscr{S}) \times \mathbb{S}^S(\mathscr{S}) \to \mathbb{C}, \tag{9.9.69}$$

and so (9.9.63) is achieved, as desired.

The Minkowski space $\mathbb{M}(\mathscr{S})$ of origins

To understand the meaning of the structure imposed on $\mathbb{T}^\alpha(\mathscr{S})$ by (9.9.62) and (9.9.63), let us construct the space $\mathbb{CM}(\mathscr{S})$ whose points are the two-dimensional linear subspaces of $\mathbb{T}^\alpha(\mathscr{S})$ other than those which intersect $\mathsf{I}^{\alpha\beta}\mathbb{T}_\beta(\mathscr{S})$ non-trivially. That is to say, $\mathbb{CM}(\mathscr{S})$ is the '*complexified Minkowski space' associated with* $\mathbb{T}^\alpha(\mathscr{S})$. The points of $\mathbb{CM}(\mathscr{S})$ may be identified with *simple* elements $\mathsf{R}^{\alpha\beta} \in \mathbb{T}^{[\alpha\beta]}(\mathscr{S})$ normalized against $\mathsf{I}_{\alpha\beta}$:

$$\mathsf{R}^{\alpha\beta} = \mathsf{X}^\alpha\mathsf{Y}^\beta - \mathsf{Y}^\alpha\mathsf{X}^\beta, \quad \mathsf{R}^{\alpha\beta}\mathsf{I}_{\alpha\beta} = 2, \tag{9.9.70}$$

as in (6.2.17), (6.2.29). Then the concept of 'squared interval' between two such points (*cf.* (1.1.22) may be defined, as in (6.2.30), by the expression

$$-\tfrac{1}{2}\underset{1}{\mathsf{R}}^{\alpha\beta}\underset{2}{\mathsf{R}}^{\gamma\delta}\varepsilon_{\alpha\beta\gamma\delta}. \tag{9.9.71}$$

Here the element

$$\varepsilon_{\alpha\beta\gamma\delta} \in \mathbb{T}_{[\alpha\beta\gamma\delta]}(\mathscr{S}), \tag{9.9.72}$$

being skew, is fixed up to a scale factor which itself is determined by the equation

$$\mathsf{X}^\alpha\mathsf{Y}^\beta(\bar{\mathsf{X}}_\rho\mathsf{I}^{\rho\gamma})(\bar{\mathsf{Y}}_\sigma\mathsf{I}^{\sigma\delta})\varepsilon_{\alpha\beta\gamma\delta} = |\mathsf{X}^\alpha\mathsf{Y}^\beta\mathsf{I}_{\alpha\beta}|^2, \tag{9.9.73}$$

whose validity, in the standard $\mathbb{T}^\alpha$, is easily checked. The expression (9.9.71) provides $\mathbb{CM}(\mathscr{S})$ with a *flat complex metric* identical with that for $\mathbb{CM}$.

Moreover, $\mathbb{CM}(\mathscr{S})$ has a certain amount of 'reality structure' arising from the fact that its (constant) spin-space $\mathbb{S}^A(\mathscr{S})$ is a standard 'Lorentzian' one, with a complex conjugation operation mapping it to $\mathbb{S}^{A'}(\mathscr{S})$. The significance of the sequence (9.9.62) is that it shows how the two types of spin-space, unprimed and primed, arise in relation to the twistor-space structure.

The complex-conjugacy relation that these spin-spaces enjoy here, entails the reality structure for $\mathbb{CM}(\mathscr{S})$ that we mentioned. Another way of stating this reality concept is that a concept of *real elements* of the vector space $\mathbb{S}^a(\mathscr{S})$ exists. Thus, even though $\mathbb{CM}(\mathscr{S})$ is a complex space, the notion of a *real direction* in it is meaningful.

But what we need the *full* twistor complex conjugation for, is to define, via the relation

$$\bar{\mathsf{R}}_{\alpha\beta} = \tfrac{1}{2}\varepsilon_{\alpha\beta\gamma\delta}\mathsf{R}^{\gamma\delta},$$

a notion of *real point* in $\mathbb{CM}(\mathscr{S})$, i.e. to be able to pick out a canonical subspace

$$\mathbb{M}(\mathscr{S}) \subset \mathbb{CM}(\mathscr{S}), \tag{9.9.74}$$

with the structure of *real* Minkowski space. In relation to the angular-momentum concept, the significance of $\mathbb{M}(\mathscr{S})$ is that it supplies the space of 'origins' about which the angular momentum is to be defined. For, unlike 4-momentum, which belongs simply to $\mathbb{S}_a(\mathscr{S})$, angular momentum cannot be defined merely in relation to the asymptotic spin-spaces. Moreover, the space $\mathbb{CM}(\mathscr{S})$, by itself is not sufficient for this, i.e. for the physical interpretation of

$$\mathsf{A}_{\alpha\beta} \in \mathbb{T}_{\alpha\beta}(\mathscr{S}). \tag{9.9.75}$$

The reason is that a 'complex origin' can always be found, about which the *self-dual angular momentum* as described by $\mu^{A'B'}$ (where $2i\mu^{A'B'}$ is the primary spinor part of $\mathsf{A}_{\alpha\beta}$, *cf.* (6.3.11), (6.3.10)) *vanishes.*

To see this, we note how $\mu^{A'B'}$ varies as a function of a field point $X \in \mathbb{CM}(\mathscr{S})$ with (complex) position vector x^a relative to an origin $O \in \mathbb{CM}(\mathscr{S})$:

$$\mu^{A'B'} = \mathring{\mu}^{A'B'} + x^{A(A'}p_A^{B')} \tag{9.9.76}$$

(*cf.* the complex analogue of (6.1.51) for a twistor with two symmetric subscripts); here the 4-momentum p_a is constant (i.e. $p_a \in \mathbb{S}_a(\mathscr{S})$) and, as in §6.1, $\mathring{\mu}^{A'B'}$ $(\in \mathbb{S}^{(A'B')}(\mathscr{S}))$ is constant, agreeing with $\mu^{A'B'}$ at the point $X = O$. Now if we substitute

$$x^a = 2\mathring{\mu}^{A'B'}p_{B'}^{A}m^{-2} + \lambda p^a \tag{9.9.77}$$

for the x^a of (9.9.76), with $m^2 = p_c p^c$, we find that the resulting $\mu^{A'B'}$ vanishes. This tells us that the *complex-mass-centre world-line*, defined by the position vectors (9.9.77), as λ varies, has *vanishing self-dual angular momentum* about each of its points. (Note that p_a is assumed to be non-null in (9.9.77); as stated earlier, we take p_a in fact to be future-timelike.) Thus, the spinor-part description of $\mathsf{A}_{\alpha\beta}$ can be reduced to one where the angular-momentum part vanishes, by passing to a complex origin-point in $\mathbb{CM}(\mathscr{S})$ (*cf.* Newman 1975). Without the concept of 'real' points in $\mathbb{CM}(\mathscr{S})$ we would have no means of recognizing the magnitude of the *spin* of a system, this being a measure of the *minimum* total angular momentum of the system (in special relativity, say) as the *real* origin is varied.

We can examine this question in relation to the Pauli–Lubański spin-vector S_a of (6.3.5), in $\mathbb{M}$. With

$$M^{ab} = \bar{\mu}^{AB}\varepsilon^{A'B'} + \mu^{A'B'}\varepsilon^{AB} \tag{9.9.78}$$

as in (6.3.10), and

$$S_a = \tfrac{1}{2}e_{abcd}p^b M^{cd} \tag{9.9.79}$$

as in (6.3.5), we find, from (3.3.31), that

$$S_a = \mathrm{i}p_{A'}^{B}\bar{\mu}_{AB} - \mathrm{i}p_{A}^{B'}\mu_{A'B'}. \tag{9.9.80}$$

We recall that S_a is constant, so taking the origin $O \in \mathbb{M}$, we may substitute $\mathring{\mu}_{A'B'}$ for $\mu_{A'B'}$ in (9.9.80). Then comparison with (9.9.77) shows that the complex-mass-centre world-line is displaced into $\mathbb{CM}$, from $\mathbb{M}$, by an amount given by

$$\mathrm{Im}\,(x^a) = -m^{-2}S^a \tag{9.9.81}$$

(taking $\lambda \in \mathbb{R}$). Thus the spin of the system is, in effect, a measure of how far into the complex the complex mass centre has been displaced, and in order to know how far this is, we need to know where the 'real' part of the space is located.

It might be felt that these issues could be avoided simply by defining $\bar{\mathsf{A}}^{\alpha\beta}\ (=\overline{\mathsf{A}^{\alpha\beta}})$ alongside $\mathsf{A}_{\alpha\beta}$ and then constructing (9.9.80) from the pair of them. The difficulty here is that without the operation (9.9.21) of complex conjugation on $\mathbb{T}^{\alpha}(\mathscr{S})$ we would have no means of identifying the space $\mathbb{CM}(\mathscr{S})$, to which $\mathsf{A}_{\alpha\beta}$ refers, with its complex conjugate $\overline{\mathbb{CM}(\mathscr{S})}$, to which $\bar{\mathsf{A}}^{\alpha\beta}$ correspondingly refers, and so expressions like (9.9.80) could not be meaningfully constructed. (Indeed, without (9.9.21) it is not even legitimate to regard $\bar{\mathsf{A}}^{\alpha\beta}$ as an element of $\mathbb{T}^{\alpha\beta}(\mathscr{S})$.)

Angular momentum at $\mathscr{I}^+$

Let us next consider the form of the angular-momentum parts of the expression (9.9.35). We take $\underset{1}{\mathsf{Z}}^{\alpha} = \underset{2}{\mathsf{Z}}^{\alpha}(=\mathsf{Z}^{\alpha})$, for simplicity, and obtain:

$$\mathsf{A}_{\alpha\beta}\mathsf{Z}^\alpha\mathsf{Z}^\beta = -\frac{\mathrm{i}}{4\pi G}\oint\{\psi_1(\omega^0)^2 + 2\psi_2\omega^0\omega^1 + \psi_3(\omega^1)^2\}\mathscr{S}$$

$$= -\frac{\mathrm{i}}{4\pi G}\oint\{\psi_1(\omega^0)^2 + 2(\psi_2 - \sigma\mathrm{N}A^{-1})\omega^0\omega^1\}\mathscr{S}, \qquad (9.9.82)$$

where we have employed (9.8.83), (9.9.36)(1), and integration by parts. The coefficient $(\omega^0)^2$ of ψ_1 is a spin-weighted spherical harmonic with $s = -2$, $j = 1$; its linear span gives all such harmonics. In this particular respect, our expression (9.9.82) is identical with others that have been previously put forward. However, the coefficient $\omega^0\omega^1$ of the second term is something new, having a more complicated angular structure which depends on the detailed form of the solutions of (9.9.36)(2). (For earlier work, see Tamburino and Winicour 1966, Bramson 1975*b*, 1978, Lind, Messmer and Newman 1972, Prior 1977, Streubel 1978, Winicour 1968, 1980, Geroch and Winicour 1981.)

It is of some interest, for gaining insight into this, to consider a representation for the solutions of (9.9.36) which involves a 'potential' for σ. Since σ has spin-weight 2, it follows from the discussion of §4.15 (*cf.* (4.15.60)) that the equation

$$\eth^2\lambda = \sigma \qquad (9.9.83)$$

always has a solution on $\mathscr{S}$ and, moreover, that (9.9.83) is conformally invariant on the sphere (*cf.* (4.15.32)), where λ is taken as a conformal density of weight 1. The choice of λ (spin-weight 0) satisfying (9.9.83) is non-unique up to the addition of a part involving only $j = 0$ or $j = 1$ spherical harmonics (four-dimensional freedom). Also, λ can be chosen to be real if and only if σ is purely electric (*cf.* (9.8.92), (9.8.93)) (Newman and Penrose 1966) i.e. $\mathscr{S}$ uncontorted.

Now suppose ω satisfies (9.9.36)(i), $\eth'\omega^0 = 0$. We have seen that (9.9.38), $\eth^2\omega^0 = 0$, is a consequence. Thus we can solve (9.9.36)(2), $\eth\omega^1 = \sigma\omega^0$, by setting*

$$\omega^1 = \omega^0\eth\lambda - \lambda\eth\omega^0 + \xi, \qquad (9.9.84)$$

where

$$\eth\xi = 0, \qquad (9.9.85)$$

ξ having spin-weight $\frac{1}{2}$. So (9.9.85) has two solutions, (*cf.* (4.15.60)) providing the necessary freedom in ω^1, for each choice of ω^0. When (9.9.84) is

* Suggested by K.P. Tod.

substituted into (9.9.82), this freedom in ξ leads precisely to the kind of freedom that we would anticipate, whereby multiples of the 4-momentum are added into the angular momentum whenever the origin of angular momentum is moved (*cf.* (6.3.3), (9.9.51)). The angular dependence of the coefficient $\omega^0\omega^1$ in (9.9.82), in this sense, is a dependence on angular integrals of σ. This is an essential respect in which our expression (9.9.82) (Penrose 1982) differs from those that have been given previously (see also Dray and Streubel 1984).

The reason for this discrepancy is that a different point of view had been adopted in earlier work with regard to the 'origin' about which the angular momentum is considered to be taken. The cut $\mathscr{S}$ itself was previously thought of as providing the 'origin', just as for the $\mathscr{I}^+$ of $\mathbb{M}$ any good cut indeed corresponds to a well-defined point in $\mathbb{M}$, namely the vertex of the light cone intersecting $\mathscr{I}^+$ in the cut. However, for a bad cut, such an interpretation in $\mathbb{M}$ is unsatisfactory. If one applies any of these earlier definitions to a bad cut of $\mathscr{I}^+$ in the weak-field (linearized) limit of general relativity, one obtains incorrect results, whereas the present approach is specially designed so that the *correct* results are obtained in the weak-field limit. In our approach, the space of allowable (real) origins is the Minkowski space $\mathbb{M}(\mathscr{S})$ – which may be regarded, intuitively, as the 'best estimate' of where the flat space of origins should be when viewed simply from the neighbourhood of the particular cut of $\mathscr{I}^+$.

The Killing vector ξ^a which is associated with any particular choice of Q^{ab} according to (9.9.6), will actually correspond to a Killing vector in $\mathbb{M}(\mathscr{S})$. The cut $\mathscr{S}$ will itself correspond to a cut $\mathscr{S}(\mathscr{S})$ of the $\mathbb{C}\mathscr{I}^+(\mathscr{S})$ of $\mathbb{CM}(\mathscr{S})$ having the same σ as that of $\mathscr{S}$. Only when σ is purely electric (*cf.* (9.8.93)), so $\mathscr{S}$ is uncontorted, will $\mathscr{S}(\mathscr{S})$ be a cut of $\mathscr{I}^+(\mathscr{S})$. In that case, ξ^a will define a particular BMS generator for $\mathscr{I}^+(\mathscr{S})$ and therefore – since $\mathscr{I}^+(\mathscr{S})$ and $\mathscr{I}^+$ may be identified with one another at $\mathscr{S}(\mathscr{S}) = \mathscr{S}$ – a particular BMS generator for $\mathscr{I}^+$. When σ is not purely electric, we get a *complex* BMS generator. But *only* when $\mathscr{S}$ is a *good* cut, with ξ^a corresponding to a Lorentz rotation in $\mathbb{M}(\mathscr{S})$ preserving the cut $\mathscr{S}(\mathscr{S})$, can this BMS generator be represented as a vector field *tangential* to $\mathscr{S}$. (In this respect our approach differs essentially from those that have been given earlier.)

There is still an inherent difficulty, however, when one considers the time-evolution of the angular momentum of a system. When outgoing gravitational radiation is present, the space $\mathbb{M}(\mathscr{S})$, though remaining a standard Minkowski space as the cut $\mathscr{S}$ is moved, will 'shift', in a sense. Thus it is not yet clear whether the angular momentum concept that arises for each separate cut of $\mathscr{I}^+$ can be regarded as the 'same' or not. The difficulty is related to that discussed in §9.8 (*cf.* Fig. 9-25), where we saw that the natural choice of (restricted) Poincaré subgroup of the BMS group 'shifts' between bursts of gravitational radiation. But now we are in a better position in that the way in which this 'shift' takes place may be precisely followed as the cut $\mathscr{S}$ is moved along $\mathscr{I}^+$. (Strictly speaking, however, our Poincaré subgroup gets 'shifted into the complex', when σ is not purely electric – which can happen only while the space–time is non-stationary, *cf.* Sachs 1962*b*, Newman and Penrose 1968.)

It should be emphasized that this 'shifting' is entirely due to the presence of the radiation and is in no way affected by changes in the choice of cut during the periods of quiescence. The whole point of our approach, indeed, is that we have been able to compensate for 'bad' choices of cut (completely, in the case when the radiation field is weak) and thus largely to 'decouple' the concept of angular-momentum origin from the cut.

The problems that arise in attempting to identify the various spaces $\mathbb{M}(\mathscr{S})$ with one another are not yet properly understood. (It is a question of adequately identifying the twistor spaces $\mathbb{T}(\mathscr{S})$ with one another – or of propagating $\mathbb{T}(\mathscr{S})$ with time.) The freedom that one initially has in this identification is, to some extent, complementary to the freedom one has in selecting a Poincaré subgroup of the BMS group. In the exact sequence (9.9.62), if we regard the spin-spaces $\mathbb{S}^A(\mathscr{S})$ and $\mathbb{S}_{A'}(\mathscr{S})$ as *fixed* – and we have seen that this is legitimate (*cf.* our discussion around (9.9.42), (9.9.43)) – then the freedom lies in the way in which we fill in the middle term of the sequence. This corresponds to the freedom of making a complex translation in $\mathbb{CM}$, or of a real translation when we specify that the complex-conjugation structure of $\mathbb{T}(\mathscr{S})$ is also given. The precise propagation of the angular momentum concept would require the elimination of this freedom (*cf.* Shaw 1984*a*, *b*).

9.10 Bondi–Sachs mass loss and positivity

The propagation of Bondi–Sachs 4-momentum fortunately does not suffer from the uncertainties, referred to in the preceding two sections, that arise with the propagation of the angular momentum concept. The 4-momentum surrounded by any cut $\mathscr{S}$ of $\mathscr{I}^+$ belongs to the asymptotic (co-)vector space $\mathbb{S}_a(\mathscr{S})$ and, as we have seen (*cf.* around (9.9.42), (9.9.43)), these spaces can all be canonically identified with one another as the space $\mathbb{S}_a(\mathscr{I}^+)$. A detailed propagation of $\mathbb{T}^\alpha(\mathscr{S})$ is not required. In this section we give a proof of one important (and physically desirable) positivity property enjoyed by the Bondi–Sachs 4-momentum, and we outline a proof of a second. The first is the mass-loss formula* which shows, in effect, that the mass–energy carried by gravitational radiation is *positive*, its flux at $\mathscr{I}^+$ being measured by the squared norm $|\mathrm{N}|^2$ of the Bondi–Sachs complex news function N. The second property is the positivity of the Bondi–Sachs mass itself, given that the appropriate 'dominant energy condition' holds within $\mathscr{M}$. Our argument for the latter largely follows one given by Ludvigsen and Vickers (1982) (and also Horowitz and Perry 1982, Reula and Tod 1984) which is a development of a remarkable line of argument put forward by Witten (1981). The positivity of mass as measured at *spatial* infinity had previously been first proved by Schoen and Yau (1979*a*, *b*) and Witten's original argument had been concerned with finding an alternative proof for that case.

Bondi–Sachs mass-loss

We assume that $\mathscr{M}$ is future 4-asymptotically simple and that, as in §9.9 (just after (9.9.32)), the energy tensor when scaled as (9.9.31) remains regular at $\mathscr{I}^+$. As in §9.8 and the latter portions of §9.9, the *physical* quantities will be distinguished by a *tilde*. Now the physical Bianchi identity (4.10.12):

$$\tilde{\nabla}^P_{P'}\tilde{\Psi}_{PQRS} = 4\pi G\tilde{\nabla}^{Q'}_{(Q}\tilde{T}_{RS)P'Q'}, \qquad (9.10.1)$$

becomes, on substituting (9.9.33)(1) and conformally rescaling by (9.9.32),

$$\nabla^P_{P'}\psi_{PQRS} = 4\pi G\Omega\nabla^{Q'}_{(Q}T_{RS)P'Q'} + 12\pi G T_{P'Q'(RS}\nabla^{Q'}_{Q)}\Omega, \qquad (9.10.2)$$

by (9.9.31), (9.9.34), (6.8.4), (6.7.31), (5.6.15) and (5.6.14). Hence (*cf.* (9.8.23))

$$\nabla^P_{P'}\psi_{PQRS} \approx -12\pi G A\iota_{(Q}T_{RS)P'1'}. \qquad (9.10.3)$$

Expressing the 1111′, 0111′, 0011′, and 0001′ components of (9.10.3) in the compacted spin-coefficient formalism, we obtain, respectively (*cf.* (4.12.27))

$$þ'\psi_3 - (ð - \tau)\psi_4 \approx 0 \qquad (9.10.4)$$

$$þ'\psi_2 - (ð - 2\tau)\psi_3 - \sigma\psi_4 \approx 4\pi G A T_{111'1'} \qquad (9.10.5)$$

$$þ'\psi_1 - (ð - 3\tau)\psi_2 - 2\sigma\psi_3 \approx 8\pi G A T_{(01)1'1'} \qquad (9.10.6)$$

$$þ'\psi_0 - (ð - 4\tau)\psi_1 - 3\sigma\psi_2 \approx 12\pi G A T_{001'1'}, \qquad (9.10.7)$$

where we have scaled $\mathscr{I}^+$ so that $\rho' = 0$, as in §§9.8 and 9.9. (In fact, ρ' can easily be reinstated by replacing $þ'$ by the conformally invariant $þ_c$ of (5.6.33) in the above, the τ-terms being subsumed correspondingly into $ð$. Recall also that in a Bondi system we have $\tau = 0$, *cf.* (9.8.72).)

* Bondi 1960*b*, Bondi, van der Burg and Metzner 1962, Sachs 1962*a*, *b* Penrose 1963, 1964*b*, 1967*b*, Newman and Unti 1962, Newman and Penrose 1968.

We shall be directly concerned only with (9.10.5) here, but all these relations have some interest in their own right. For example, (9.10.4) is a consistency condition for the two formulae (9.8.82), (9.8.83) which relate ψ_4 and ψ_3, respectively, to the Bondi–Sachs news function N:

$$A\psi_4 \approx \text{þ}'\mathrm{N}, \quad A\psi_3 \approx \eth\mathrm{N} \tag{9.10.8}$$

((9.8.33) being assumed, and the commutator $\overline{(4.12.34)'}$ employed). Again, (9.10.6) would play a role in the time-evolution of angular momentum similar to that which will be played here by (9.10.5) in the time-evolution of 4-momentum. Finally, there is a connection between (9.10.7) and the Einstein (1918) formula for the energy loss due to changes in mass quadrupole moment of the system, the quantity ψ_0 being a measure of this quadrupole moment (Janis and Newman 1965). But we shall not pursue these matters here.

Our first purpose is to establish the Bondi–Sachs mass-loss formula, for which we adopt the form (9.9.56),

$$M = -\frac{1}{4\pi G}\oint\{\psi_2 - \sigma\mathrm{N}A^{-1}\}\mathscr{S}, \tag{9.10.9}$$

where, for simplicity, the weighting factor W has been put equal to unity. Thus M, in (9.10.9), describes that component of the Bondi–Sachs 4-momentum, surrounded by the cut $\mathscr{S}$, which refers to the time-direction defined by our particular choice of unit-sphere metric on $\mathscr{S}$. We shall refer to M simply as *the mass at* $\mathscr{S}$. We use the form (4.14.92) of the fundamental theorem of exterior calculus, applied to a region Σ of $\mathscr{I}^+$ bounded by two cuts $\mathscr{S}, \mathscr{S}'$, where $\mathscr{S}'$ lies entirely to the future of $\mathscr{S}$ along $\mathscr{I}^+$. The situation is indicated in Fig. 9-26 (which may be compared with Fig. 4-3 on p. 282 of Volume 1). The mass M at $\mathscr{S}$ can be interpreted as the total mass–energy (including the non-local gravitational contributions) intercepted by a compact spacelike 3-surface in $\mathscr{M}$ whose boundary lies entirely at $\mathscr{S}$. Thus, any outgoing radiation which intercepts this 3-surface will have its energy contribution included. The same holds for the mass M' at $\mathscr{S}'$. Now (4.14.92) (with (4.14.89)) states

$$\int_\Sigma\{\text{þ}'\mu_0 - (\eth - \tau)\mu_1\}A\mathscr{S}\wedge \mathrm{d}u = \oint_{\mathscr{S}'}\mu_0\mathscr{S} - \oint_{\mathscr{S}}\mu_0\mathscr{S}, \tag{9.10.10}$$

μ_0, μ_1 being weighted scalars of respective types $\{0,0\}$ and $\{-2,0\}$. Here we have put $\rho' = 0$ and $U = A^{-1}$, where U is given in (4.14.88). Comparing (4.14.88) with (9.8.30), we see that the differential in (9.10.10) must be

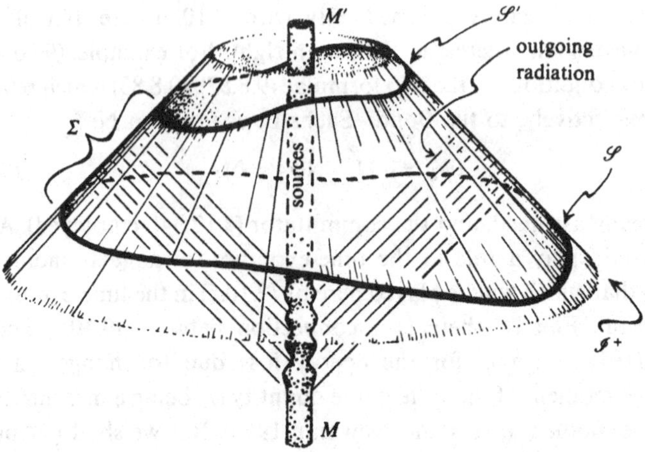

Fig. 9-26. To investigate mass–momentum loss by a radiating system we take the difference between the total Bondi–Sachs values for two arbitrary cuts $\mathscr{S}$, $\mathscr{S}'$ of $\mathscr{I}^+$ (where we take $\mathscr{S}'$ to lie entirely to the future of $\mathscr{S}$).

$A\mathscr{S} \wedge \mathrm{d}u$, by (4.14.89). We take

$$\mu_0 = \sigma \mathrm{N} A^{-1} - \psi_2 \quad \textit{and} \quad \mu_1 = \tau \mathrm{N} A^{-1} - \psi_3, \tag{9.10.11}$$

and compute, on Σ,

$$\begin{aligned} A\text{þ}'\mu_0 - A(\eth - \tau)\mu_1 &= \sigma\text{þ}'\mathrm{N} + \mathrm{N}(\text{þ}'\sigma - \eth\tau + \tau^2) \\ &\quad + A\{-\text{þ}'\psi_2 + (\eth - \tau)\psi_3\} - \tau\eth\mathrm{N} \\ &= \sigma A\psi_4 - \mathrm{N}\bar{\mathrm{N}} - A\sigma\psi_4 + A\tau\psi_3 - 4\pi G A^2 T_{111'1'} - \tau\eth\mathrm{N} \\ &= -\mathrm{N}\bar{\mathrm{N}} - 4\pi G A^2 T_{ab}n^a n^b, \end{aligned} \tag{9.10.12}$$

using (9.8.26), (9.8.73), (9.10.5), (9.10.8), and (9.8.28). Thus (9.10.10) tells us that

$$M - M' = \int_\Sigma \{A^2 T_{ab}n^a n^b + (4\pi G)^{-1}\mathrm{N}\bar{\mathrm{N}}\}\mathscr{S} \wedge \mathrm{d}u \geqslant 0. \tag{9.10.13}$$

The first term of the integrand measures the energy flux across $\mathscr{I}^+$ of the matter fields, and is non-negative for reasonable matter. (For example, in the case of an electromagnetic field we have $T_{111'1'} = (2\pi)^{-1}|\varphi_2|^2 \geqslant 0$, by (5.2.4).) Correspondingly, $(4\pi G)^{-1}\mathrm{N}\bar{\mathrm{N}}$ measures the flux across $\mathscr{I}^+$ of the gravitational energy, which establishes the important positive-definite property of the energy of gravitational radiation. (The original proofs applied only to cuts $\mathscr{S}$ and $\mathscr{S}'$ related by a *translation* of the BMS group. Our more general argument follows that of Penrose 1967*b*.)

Since our argument applies equally well for *any* choice of time-axis (i.e. any unit-sphere metric on $\mathscr{S}$), we have also established that the difference between the *4-momentum* at $\mathscr{S}$ and that at $\mathscr{S}'$ is necessarily a *future-causal vector* (*cf.* also Bonnor and Rotenberg 1966). If N is anywhere non-zero, this difference is *future-timelike.*

Non-locality of gravitational wave energy

It is important to note that even though we appear to have a definite *local* measure of the gravitational energy flux at $\mathscr{I}^+$, this is in fact not so. For the definition of N is *not* fixed locally at points of $\mathscr{I}^+$, but is fixed instead only by equations like (9.10.8)(1), which tells us that N is a *time-integral* of the gravitational radiation field, or (9.10.8)(2), which tells us that N is an *angular integral* of ψ_3, or (9.8.74), which gives N as a *Ricci tensor* component – dependent on a choice of Ω and characterized by the global condition that the metric of $\mathscr{S}$ be that of a unit sphere – or (9.8.73), or (9.8.75), which again require the global condition of a sphere metric for $\mathscr{S}$. Indeed, it can be shown that despite the conformal invariance

$$\mathrm{N} \mapsto \Theta^{-2}\mathrm{N} \tag{9.10.14}$$

enjoyed by N under conformal rescalings (9.8.13) which preserve the spherical metric of $\mathscr{S}$ (derived, say, from (9.10.8)(2)), there is no way to compute N at a point R of $\mathscr{I}^+$ merely from the geometry of $\mathscr{M}$ in a small neighbourhood of R. One can even arrange for $\mathscr{M}$ to be exactly flat in such a neighbourhood and yet have $\mathrm{N} \neq 0$ at R. This illustrates the essentially non-local character of gravitational energy.

NP-constants

As a brief digression, we remark that the calculation we performed in §5.12 to establish the validity of the generalized Kirchhoff–D'Adhémar integral (5.12.6), can also be performed when the light cone $\mathscr{C}$ becomes $\mathscr{I}^+$ and a result similar to (9.10.13) is obtained, in which, however, instead of a positive quantity on the RHS we get zero (Penrose 1967*b*, Newman and Penrose 1965, 1968). (This happens because of the vanishing of (5.12.17) in vacuum.) In place of the mass integral (9.10.9), we have, in this case, five linearly independent complex quantities referred to as *NP-constants*:

$$Q = \oint Y \text{þ}_c \psi_0 \mathscr{S}, \tag{9.10.15}$$

where Y is a type-$\{-5, -1\}$ spin-weighted spherical harmonic with $j = 2$.

In terms of the expansions of physical quantities given at the end of §9.8, $\text{þ}_c\psi_0$ becomes Ψ_0^1, the coefficient of r^{-6} in the expansion of Ψ_0. For their definition and proof of constancy we require just *one more* degree of differentiability at $\mathscr{I}^+$ than for 4-momentum, namely future 5-asymptotic simplicity. The proof follows exactly that given in Volume 1 for establishing (5.12.6). All the quantities which need to vanish do so because of properties established in §9.8 and §9.10 (vacuum assumed).

The constants (9.10.15) have some very remarkable and non-intuitive properties, which have led various authors to express disbelief in their reality (Bardeen and Press 1973, Misner, Thorne and Wheeler 1973). (However, the calculations at $\mathscr{I}^+$ are unambiguous, being not dependent on power series coordinate expansions, and requiring only a fairly reasonable degree of smoothness at $\mathscr{I}^+$.) In the first place, unlike 4-momentum, the quantities (9.10.15) are *exactly conserved in vacuum.* In the second place, there are corresponding quantities for massless fields of any other spin (e.g. three independent quantities $\oint Y\text{þ}_c\varphi_0\mathscr{S}$ in the electromagnetic case, where now Y has type $\{-3,-1\}$ with $j=1$). In the third place, modifications to (9.10.15) exist which are still exactly conserved even when certain sources are present at $\mathscr{I}^+$ (namely electromagnetic or neutrino sources to the gravitational field, Exton, Newman and Penrose 1969). In the fourth place, in *stationary space–times* the quantities (9.10.15) do not normally vanish, and they then describe certain curious origin-independent combinations of multipole moments of the general nature of

$$Q = \textit{mass} \times (\textit{complex quadrupole moment}) - (\textit{dipole moment} + \mathrm{i} \times \textit{angular momentum})^2 \qquad (9.10.16)$$

in the gravitational case, or a corresponding expression bilinear in gravitational and electromagnetic moments in the electromagnetic case.

This last property is perhaps the most striking, since it tells us that the integrals (9.10.15) have non-trivial content. In contrast, in the case of linearized gravity, the integrals (9.10.15) still exist and are exactly conserved, but no analogue of (9.10.16) occurs. Indeed, the quantities (9.10.15) always vanish for retarded fields in the linear theory, but this does not happen in the full theory. One implication of (9.10.16) for stationary gravitational fields, in conjunction with the exact conservation of (9.10.15), is that an exactly stationary system cannot radiate gravitationally and then return to *exact* stationarity if the multipole combination (9.10.16) is different after the radiation is emitted from what it was before.

It should be stressed, however, that while this may seem puzzling, it is not, in fact, unreasonable physically. The expected behaviour is roughly the following: We may start with an exactly stationary gravitating system in which the multipole-moment combination (9.10.16) has some prescribed value. The system then emits a burst of gravitational radiation, freely changing the multipole combination (9.10.16) to some other value. Owing to nonlinearity in Einstein's theory, this radiation then back-scatters to some extent, and from then on there is a significant contribution to (9.10.15) in which this back-scattered field enters *linearly*, and for this reason it can swamp the effect of the new moment combination (9.10.16). Though the back-scattered field might seem to be trivially small by comparison with the nearly stationary new multipole field, its continuing presence can remain sufficient to spoil the stationarity required for the applicability of (9.10.16).

Of course, this does not explain, in any physically comprehensible terms, why (9.10.15) should be conserved or what the physical status of (9.10.16) is. These are matters which remain decidedly mysterious and await further insights.

Witten's procedure for mass positivity

We finally turn to a question whose physical status is beyond dispute: the positivity of total mass in general relativity. The first complete argument for this was given in 1979 by Schoen and Yau (1979*a*, *b*). Their proof, however, referred to the mass defined at spatial infinity (the 'i^0' of §9.1), and not directly to the Bondi–Sachs mass (however, *cf.* Schoen and Yau 1982). The positivity of mass at *null* infinity is in fact an essentially stronger result than that concerning spatial infinity. If we assume the result (Ashtekar and Magnon-Ashtekar 1979) that the mass at spatial infinity is the past limit of the Bondi–Sachs mass taken as the cut $\mathscr{S}$ recedes into the past along $\mathscr{I}^+$ (i.e. approaches 'i^0') – though this is by no means obvious for the standard (ADM) mass expression of Arnowitt, Deser and Misner (1961) – then from the mass-loss formula (9.10.13) we would obtain positivity at spatial infinity as a consequence of positivity at null infinity. Moreover, the positivity at null infinity has a somewhat greater physical content since it tells us that a system cannot radiate more energy than its original mass, its resulting negative energy being not possible to estimate by a separate *spatial*-infinity measure, because the system perhaps continues to radiate for ever. (Of course, the most satisfactory positivity theorem would be one at the quasi-local level, but this is at present lacking.)

The key new ingredient, originally introduced by Witten (1981) to support an alternative positivity proof at spatial infinity, is to provide a means of extending inwards from infinity, along a spacelike hypersurface $\mathscr{H}$, a vector ξ^a which at infinity would approach the constant 'Killing vector' serving to define the relevant four-momentum component. This is done by taking ξ^a to be *null*, and thus of the form

$$\xi^a = \lambda^A \bar{\lambda}^{A'}, \tag{9.10.17}$$

over $\mathscr{H}$, where λ^A suitably approaches constancy at infinity. The spinor λ^A is subject to an elliptic differential equation on $\mathscr{H}$ – which we shall here refer to as the SW-equation – that serves to fix its value over $\mathscr{H}$ once the above asymptotic behaviour has been specified. A remarkable identity (found by Witten 1981 and also, for a different purpose, by Sen 1981) shows that a quantity constructed from λ^A (which actually becomes the asymptotic 4-momentum null-component $\xi^a p_a$ at infinity) is equal to a non-negative integral over $\mathscr{H}$ provided (i) that the local energy tensor satisfies the dominant energy condition (9.10.42) below, and (ii) that λ^A satisfies the SW-equation throughout $\mathscr{H}$.

Witten's original work was slightly corrected and made rigorous by a number of authors (Nester 1981, Parker and Taubes 1982), and subsequent modifications of the original argument were found so that it can also be applied at *null* infinity to obtain positivity for the Bondi–Sachs mass.* It seems unlikely that this argument has yet found its definitive form, but we shall indicate three approaches that appear to have been successful, namely those of Ludvigsen and Vickers (1982), Horowitz and Perry (1982), and Reula and Tod (1984), concentrating mainly on the first of these.

The common ingredient of these various approaches is the Sen–Witten identity, together with the general way in which it is used, as initiated by Witten. We consider a 2-form

$$\Xi = -\mathrm{i}\bar{\lambda}_{B'} \nabla_a \lambda_B \mathrm{d}x^a \wedge \mathrm{d}x^b, \tag{9.10.18}$$

where λ_A is chosen to be, in an appropriate sense, asymptotically constant. There are two properties of relevance that are claimed for this form. *First*, its integral over a closed 2-surface $\mathscr{S}$ has the limiting value

$$\frac{1}{4\pi G}\oint_{\mathscr{S}} \Xi \to p_a \lambda^A \bar{\lambda}^{A'} \tag{9.10.19}$$

as $\mathscr{S}$ recedes suitably to infinity, approaching a cut of $\mathscr{I}^+$ or else spacelike infinity, and where p_a is the relevant asymptotic measure of four-

* See also Walker (1982) for a survey of some of these results.

momentum. *Second*, if $\mathscr{H}$ is a compact spacelike (or locally achronal*) 3-surface with boundary $\partial\mathscr{H} = \mathscr{S}$, and if λ_A satisfies a suitable elliptic equation on $\mathscr{H}$ and the dominant energy condition holds throughout $\mathscr{H}$, then

$$\int_{\mathscr{H}} \mathrm{d}\Xi \geqslant 0, \tag{9.10.20}$$

with equality holding only if λ_A is constant over $\mathscr{H}$.

The usual choice for the equation to be satisfied by λ_A on $\mathscr{H}$ is the *SW-equation*:

$$D_{AA'}\lambda^A = 0 \tag{9.10.21}$$

(though there is actually some flexibility in the choice of equation (9.10.21), *cf.* Horowitz and Perry 1982), where

$$D_a = h_a{}^b \nabla_b, \tag{9.10.22}$$

with $h_a{}^b$ the orthogonal projector tangential to $\mathscr{H}$; thus

$$h_a{}^b = g_a{}^b - t_a t^b, \tag{9.10.23}$$

t^a being the unit normal to $\mathscr{H}$. Clearly

$$h_{[ab]} = 0, \quad h_a{}^b h_b{}^c = h_a{}^c, \quad t^a h_a{}^b = 0. \tag{9.10.24}$$

The following spinor expression for $h_a{}^b$ is sometimes useful

$$h_a{}^b = \tfrac{1}{2}\varepsilon_A{}^B \varepsilon_{A'}{}^{B'} - t_A^{B'} t_{A'}^B,$$

as is a consequent re-formulation for the SW-equation (9.10.21):

$$t_{(A}^{A'} \nabla_{B)A'} \lambda^B = 0. \tag{9.10.25}$$

(When working with spinors on spacelike 3-surfaces it can be convenient to normalize t^a by $t_a t^a = 2$ instead of as a unit normal. Then the operators $t_A^{A'}$ and $t_{A'}^A$ serve to convert primed into unprimed indices and vice versa, so that only one kind of index need be employed – compare the general discussion of spinors in the Appendix to this volume. It may be remarked that $D_{AA'}$, or rather $t_{(A}^{A'}\nabla_{B)A'}$, though not quite the Dirac operator intrinsic to the 3-surface, is closely related to it, *cf.* Sen 1981).

Relation to the Bondi–Sachs 4-momentum

Note that the reality of the integral (9.10.19) is manifest, because the imaginary part of (9.10.18) is the *curl*: $\tfrac{1}{2}\mathrm{d}(\lambda_A \bar{\lambda}_{A'}\, \mathrm{d}x^a)$. Indeed, it has been more

* The term 'achronal' is used here to allow the possibility that $\mathscr{H}$ may be null in places. This is needed for the Ludvigsen–Vickers argument.

usual to use the real part of Ξ (Nester 1981) in the integrals (9.10.19), (9.10.20), instead of the versions suggested here. Our reason for preferring the latter, apart from their greater simplicity, is that they appear to relate more directly to the expressions of §9.9, notably to (9.9.29).

To see this, and to derive the necessary connection with Bondi–Sachs 4-momentum, recall that in the discussion of §9.9 we obtained a null component of the 4-momentum p^a, measured at a cut $\mathscr{S}$ of $\mathscr{I}^+$, by choosing $\mathrm{Z}^\alpha \in \mathbb{T}^\alpha(\mathscr{S})$ and then substituting $\underset{1}{\mathrm{Z}}^\beta$ and $\underset{2}{\bar{\mathrm{Z}}}_\gamma \mathrm{I}^{\gamma\alpha}$ for Z^β and Z^α, respectively, in (9.9.19). Our plan here is to consider the '$\bar{\lambda}_{B'}$' of (9.10.18) as being, in effect, the projection part '$\pi_{B'}$' of Z^β; and the 'λ_B' (or, rather 'λ^B') of (9.10.18) as being, in effect, (minus) the primary part of $\bar{\mathrm{Z}}_\gamma \mathrm{I}^{\gamma\beta}$. To help with this picture we write $\pi_{B'}$ in place of $\bar{\lambda}_{B'}$ and $\tilde{\omega}_B\ (= -\bar{\pi}_B)$ in place of λ_B in (9.10.18). When Ξ is integrated over a spacelike 2-surface $\mathscr{S}$ (considered as a finite surface, for the moment) we find, in this notation, using (4.14.52), (4.14.53), (4.14.66), that

$$\begin{aligned}\oint_{\mathscr{S}} \Xi &= \mathrm{i}\oint_{\mathscr{S}} \pi_{B'}\nabla_a\tilde{\omega}_B\,\mathrm{d}x^a \wedge \mathrm{d}x^b \\ &= \oint_{\mathscr{S}} \{\pi_{0'}\varepsilon_1{}^B\nabla_{01'}\tilde{\omega}_B - \pi_{1'}\varepsilon_0{}^B\nabla_{10'}\tilde{\omega}_B\}\mathscr{S} \\ &= \oint_{\mathscr{S}} \{\pi_{0'}(\eth\tilde{\omega}^0 - \rho'\tilde{\omega}^1) + \pi_{1'}(\eth'\tilde{\omega}^1 - \rho\tilde{\omega}^0)\}\mathscr{S} \\ &= -\mathrm{i}\oint_{\mathscr{S}} \{\pi_{0'}\tilde{\pi}_{1'} + \pi_{1'}\tilde{\pi}_{0'}\}\mathscr{S}\end{aligned}$$

by (4.12.28), where we have adopted the notation (9.9.23) (with tildes). Note that (apart from the factor $(4\pi G)^{-1}$) this is formally identical with the Tod form (9.9.29) of the quasi-local '4-momentum' integral (9.9.48). When $\mathscr{S}$ is moved out to become a cut of $\mathscr{I}^+$ this correspondence becomes exact, provided that λ_A approaches constancy at infinity, appropriately. This requires that the conformally rescaled $\tilde{\omega}^A$ (taking the standard twistor scaling $\hat{\tilde{\omega}}^A = \tilde{\omega}^A$ of (6.1.2)) must actually satisfy the 2-surface twistor equations (9.9.36) on $\mathscr{S} \subset \mathscr{I}^+$, in the special case (9.9.41) corresponding to the elements of $\mathrm{I}^{\alpha\beta}\mathbb{T}_\beta(\mathscr{S})$, namely

$$\hat{\eth}\hat{\tilde{\omega}}^1 = 0, \quad \hat{\tilde{\omega}}^0 = 0. \tag{9.10.26}$$

Here, 'hatted' quantities refer to the unphysical metric, finite on $\mathscr{I}^+$, so we are again adopting (5.6.1), (5.6.2):

$$\hat{g}_{ab} = \Omega^2 g_{ab}, \quad \hat{\varepsilon}_{AB} = \Omega\varepsilon_{AB},$$

rather than (9.9.32).

This establishes, in effect, that the LHS of (9.10.19) indeed approaches the RHS as $\mathscr{S}$ becomes a cut of $\mathscr{I}^+$, where p_a is the Bondi–Sachs 4-momentum as described earlier in this section and in §9.9. For a completely rigorous discussion, some more attention would have to be paid to the limiting behaviour of λ_A, but we do not enter into the details of this here. (For a corresponding argument applied at spacelike infinity i^0, where agreement with the ADM mass is obtained, see Shaw 1983*a*, *b*.)

Ludvigsen–Vickers determination of λ_A

Such discussion of the limiting behaviour of λ_A also has a role to play in the positivity part of the argument (9.10.20). Different authors have adopted different approaches. For example, Ludvigsen and Vickers require that $\mathscr{H}$ is not actually spacelike all the way to $\mathscr{I}^+$, but instead consists of two pieces: an inner part, which is indeed spacelike, but bounded by a finite spacelike topological 2-sphere $\tilde{\mathscr{S}}$, the annular portion from $\tilde{\mathscr{S}}$ to the cut $\mathscr{S}$ of $\mathscr{I}^+$ being taken to be *null* (see Fig. 9-27). This necessitates a slight adaptation of the SW-equation so that it applies when $\mathscr{H}$ becomes null. For this one can use (9.10.25), with t^a replaced by the null vector $l^a = o^A o^{A'}$, and obtain

$$o_{(A}\nabla_{B)o'}\lambda^B = 0, \tag{9.10.27}$$

which takes the form

$$\text{þ}\lambda_0 = 0, \quad (\text{þ} - 2\rho)\lambda_1 = (2\eth' - \tau')\lambda_0 \tag{9.10.28}$$

in the compacted spin-coefficient formalism, where the flagpole direction of ι^A is chosen smoothly, to become tangent to $\mathscr{I}^+$ at $\mathscr{S}$. (We may take it

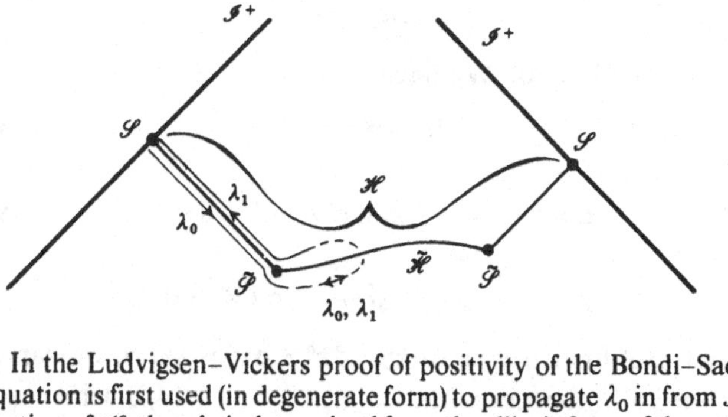

Fig. 9-27. In the Ludvigsen–Vickers proof of positivity of the Bondi–Sachs mass, the SW-equation is first used (in degenerate form) to propagate λ_0 in from $\mathscr{I}^+$ along the null portion of $\mathscr{H}$; then λ_1 is determined from the elliptic form of the equation on the spacelike region $\tilde{\mathscr{H}}$ and is finally propagated back to $\mathscr{I}^+$.

parallelly propagated along the rays as in §9.7, if desired, in which case $\tau' = 0$.)

We use (9.10.28)(1) to propagate λ_0 inwards from infinity, where $\lambda_0 = -\tilde{\omega}^1$ is subject to the required condition (9.10.26) at $\mathscr{S}$. When $\tilde{\mathscr{S}}$ is reached, λ_0 will take some smooth value, where we assume that $\mathscr{S}$ has been chosen sufficiently 'near' to $\mathscr{I}^+$ that no singularities (caustics or crossing regions) occur on the null portion of $\mathscr{H}$. The values of λ_0 on $\tilde{\mathscr{S}}$ are then taken as the *boundary values* for the elliptic SW-equation on the portion $\tilde{\mathscr{H}}$ (assumed to be smooth) of $\mathscr{H}$ bounded by $\tilde{\mathscr{S}}$. (This boundary value problem is well posed, *cf.* Atiyah, Patodi and Singer 1975.) A λ_A-field satisfying the SW-equation on $\tilde{\mathscr{H}}$ is obtained (unique if $\tilde{\mathscr{H}}$ has Euclidean topology), agreeing with λ_0 at $\tilde{\mathscr{S}}$, and this supplies the remaining component λ_1 at $\tilde{\mathscr{S}}$. This component is then propagated back to $\mathscr{S}$ along the null portion of $\mathscr{H}$ using (9.10.28)(2). It is not hard to verify, from the asymptotic form of the relevant quantities, that (9.10.26)(2) is then *necessarily satisfied* whatever values λ_1 has at $\tilde{\mathscr{S}}$.

In the Horowitz–Perry and Reula–Tod approaches, as in an earlier discussion by Tod and Horowitz (1982), $\mathscr{H}$ is taken to be spacelike everywhere, spanning the cut $\mathscr{S}$ of $\mathscr{I}^+$. But only Reula and Tod adopt (9.10.21) throughout. Horowitz and Perry use a modified version in which t^a is not taken normal to $\mathscr{H}$, whereas Horowitz and Tod adopt the Weyl neutrino equation $\nabla_{AA'}\lambda^A = 0$ in place of (9.10.21) for part of their argument. For all these approaches there is a considerable complication, not encountered with the Ludvigsen–Vickers argument, in that one needs some version of the boundary value problem for the SW-equation in which subtle fall-off conditions at infinity need to be considered. We do not enter into these matters here.

Positivity of dΞ

We next establish (9.10.20). We have

$$\mathrm{d}\Xi = \boldsymbol{\alpha} + \boldsymbol{\beta} \tag{9.10.29}$$

where

$$\boldsymbol{\alpha} = -\mathrm{i}\bar{\lambda}_{C'}\nabla_a\nabla_b\lambda_C\,\mathrm{d}x^a \wedge \mathrm{d}x^b \wedge \mathrm{d}x^c \tag{9.10.30}$$

and

$$\boldsymbol{\beta} = -\mathrm{i}(\nabla_a\bar{\lambda}_{C'})(\nabla_b\lambda_C)\,\mathrm{d}x^a \wedge \mathrm{d}x^b \wedge \mathrm{d}x^c. \tag{9.10.31}$$

Consider (9.10.30) first. Because $\mathrm{d}x^a \wedge \mathrm{d}x^b \wedge \mathrm{d}x^c$ is skew, we can replace $\nabla_a\nabla_b$ by (one-half) the commutator Δ_{ab} (*cf.* (4.2.14)), whereupon the derivative operators disappear, in favour of curvature terms. Since it is a little easier to work with duals, we write

$$X_a = \tfrac{1}{6}e_{abcd}\,dx^b \wedge dx^c \wedge dx^d \tag{9.10.32}$$

(*cf.* (3.4.30)), and obtain

$$\alpha = -\frac{i}{2}e^{abcd}\bar{\lambda}_{C'}\Delta_{ab}\lambda_C X_d. \tag{9.10.33}$$

By (3.4.22) and (4.9.1) this becomes

$$\begin{aligned}\alpha &= (\bar{\lambda}^{C'}\Box_{C'D'}\lambda_D - \bar{\lambda}_{D'}\Box_{CD}\lambda^C)X^d\\ &= (\Phi_{CDC'D'}\lambda^C\bar{\lambda}^{C'} + 3\Lambda\lambda_D\bar{\lambda}_{D'})X^d\\ &= 4\pi G T_{ab}\xi^a X^b,\end{aligned} \tag{9.10.34}$$

where we have used (4.9.11)(2), (4.9.17), (4.6.32), and (9.10.17).

Now consider (9.10.31), which we write as

$$\beta = i\bar{W}_{aC'}W_{bC}\,dx^a \wedge dx^b \wedge dx^c \tag{9.10.35}$$

with

$$W_{aB} = \nabla_a\lambda_B. \tag{9.10.36}$$

Again we consider duals; so, by (3.3.31),

$$\begin{aligned}\beta &= ie^{abcd}\bar{W}_{bC'}W_{dC}X_a\\ &= (\varepsilon^{AC}\varepsilon^{BD}\varepsilon^{A'D'}\varepsilon^{B'C'} - \varepsilon^{AD}\varepsilon^{BC}\varepsilon^{A'C'}\varepsilon^{B'D'})\bar{W}_{bC'}W_{dC}X_a\\ &= (\bar{W}_b{}^{B'}W^{BA'A} - \bar{W}_b{}^{A'}W^{AB'B})X_a\\ &= (\bar{U}_B W^{Ba} + U_{B'}\bar{W}^{B'a} - W_b{}^A\bar{W}^{bA'})X_a,\end{aligned} \tag{9.10.37}$$

where we have used the ε-identity (2.5.20) and where we have written

$$U_{B'} = W_{AB'}{}^A. \tag{9.10.38}$$

We are now in a position to consider the positivity of (9.10.20):

$$\int_{\mathcal{K}} d\Xi = \int\alpha + \int\beta \tag{9.10.39}$$

Here we can put

$$t^a\mathcal{H} \quad \textit{for} \quad X^a \tag{9.10.40}$$

where (assuming $\mathcal{H}$ is spacelike) t^a is the unit normal to $\mathcal{H}$ and $\mathcal{H}$ is its 3-volume element (where a suitable modification of this is needed for the portion of $\mathcal{H}$ which is null in the Ludvigsen–Vickers case but this is easily achieved). Then

$$\int_{\mathcal{H}}\alpha = 4\pi G\int T_{ab}\xi^a t^b\mathcal{H}, \tag{9.10.41}$$

which is always non-negative if the dominant energy condition (*cf.* Wald

1984, Hawking and Ellis 1973 and (5.2.9), Volume 1):

$$T_{ab}u^a v^b \geqslant 0 \quad \textit{for all future-causal} \quad u^a, v^a \tag{9.10.42}$$

holds. Moreover, the expression $e^{abcd}t_a$ when transvected by t_b or t_d (or, of course, t_c) gives zero, so

$$e^{abcd}t_a = h_p{}^b h_q{}^d e^{apcq}t_a, \tag{9.10.43}$$

by (9.10.23). It then follows from (9.10.40) that, on $\mathscr{H}$, we can make the replacement $\nabla_a \mapsto D_a$ and hence also the following:

$$W_{aB} \mapsto h_a{}^c W_{cB} = D_a \lambda_B, \quad U_{B'} \mapsto D_{AB'}\lambda^A \tag{9.10.44}$$

in the *algebraic identity* (9.10.37). The last term in (9.10.44) vanishes whenever the SW-equation holds, and thus, if we assume (9.10.21), we get

$$\int_{\mathscr{H}} \beta = -\int h^{ab}(D_a\lambda_C)(D_b\bar{\lambda}_{C'})t^c \mathscr{H}. \tag{9.10.45}$$

Now the intrinsic metric tensor 'h^{ab}' for $\mathscr{H}$ is *negative-definite* (for $\mathscr{H}$ spacelike), so we can express it as

$$h^{ab} = -x^a x^b - y^a y^b - z^a z^b,$$

with x^a, y^a, z^a real; and t^c, being timelike, can be expressed as

$$t^c = \frac{1}{\sqrt{2}}(o^C o^{C'} + \iota^C \iota^{C'}).$$

Consequently we find

$$h^{ab}(D_a\lambda_C)(D_b\bar{\lambda}_{C'})t^c \leqslant 0, \tag{9.10.46}$$

the left-hand side being a sum of negative terms, which are the squared moduli of the various components of

$$D_a\lambda_C. \tag{9.10.47}$$

Evidently, the equality in (9.10.46) holds only when (9.10.47) vanishes, i.e. only when λ_C is covariantly constant over the whole of $\mathscr{H}$.

This establishes the essential positivity of (9.10.39) and hence the required non-negativity property (9.10.20). Allowing λ_A to take different allowed values at infinity, the required non-negativity property – that the Bondi–Sachs 4-momentum p_a be future-causal – is thereby established. (By the constancy of λ_A whenever equality holds in (9.10.46), it is not hard to establish also the 'converse' result that $p_a = 0$, or indeed that p_a is null, *only* if the curvature vanishes throughout $\mathscr{H}$, *cf.* Witten 1981.)

Note that we obtain positivity for the total energy though, as we remarked earlier, gravitational potential energy is negative. Thus we see that it can never be *more* negative than the positive mass–energy that gives rise to it. In the Witten argument, there is a *positive total* gravitational energy arising from the integral of $\boldsymbol{\beta}$ given by (9.10.45). But this is manifestly *non-local* since it depends on the choice of solution of (9.10.21) that arises with the given boundary conditions. It is remarkable that the precise expression for this seems to require an essential description in terms of spinors.

Sparling's 3-form

The 2-spinor expression

$$\boldsymbol{\Omega} = \tfrac{1}{2}(\boldsymbol{\Xi} + \bar{\boldsymbol{\Xi}}) \tag{9.10.48}$$

(which can replace $\boldsymbol{\Xi}$ in (9.10.19) and (9.10.20)) was studied by Nester (1981) (Witten having originally used a Dirac 4-spinor formulation). Its exterior derivative is the same as that of $\boldsymbol{\Xi}$ itself (as we have seen), and is given by (9.10.29). By (9.10.34), we see that $\boldsymbol{\alpha}$ vanishes in vacuum – and, indeed, $\boldsymbol{\alpha}$ is a 3-form which, for varying λ_A, is essentially *equivalent* to the energy tensor T_{ab}. Thus, in vacuum we have

$$\mathrm{d}\boldsymbol{\Omega} = \mathrm{d}\boldsymbol{\Xi} = \boldsymbol{\beta} \tag{9.10.49}$$

and hence

$$\mathrm{d}\boldsymbol{\beta} = 0. \tag{9.10.50}$$

Indeed, Sparling (1983, 1984) has shown that the condition $\mathrm{d}\boldsymbol{\beta} = 0$ (to hold for all λ_A) is also *sufficient* for the vacuum equations to hold, so (9.10.50) provides a neat formulation of the Einstein vacuum equations. For this reason, $\boldsymbol{\beta}$ is frequently referred to as the *Sparling 3-form.*

It should be noted that the forms $\boldsymbol{\Omega}$, $\boldsymbol{\Xi}$, $\boldsymbol{\alpha}$ and $\boldsymbol{\beta}$ really refer to the (dual) *spin-vector bundle* $\mathscr{B}$ over $\mathscr{M}$. (See Fig. 1-15 of Volume 1, p. 48 for a graphic illustration of this bundle.) Thus, the vanishing of the Sparling form expresses the content of Einstein's field equations in terms of the spin-vector bundle. In fact Sparling also allows the presence of *torsion* in the connection ∇_a defining $\boldsymbol{\beta}$ in (9.10.31). Then $\mathrm{d}\boldsymbol{\beta} = 0$ expresses the vanishing of torsion *together* with the vacuum equations.

On $\mathscr{B}$ it is appropriate to use the notation

$$\boldsymbol{\Xi} = -\mathrm{i}\bar{\lambda}_{B'}\,\mathrm{d}\lambda_B \wedge \mathrm{d}x^b, \quad \boldsymbol{\beta} = -\mathrm{i}\mathrm{d}\bar{\lambda}_{C'} \wedge \mathrm{d}\lambda_C \wedge \mathrm{d}x^c \tag{9.10.51}$$

where, in accordance with the conventions of §4.3 we have, for any cross-section of $\mathscr{B}$ over a portion of $\mathscr{M}$ (i.e. spinor field λ_A)

$$\mathrm{d}\lambda_B = \nabla_{i_1}\lambda_B = \nabla_a\lambda_B\,\mathrm{d}x^a \tag{9.10.52}$$

and (though x^a itself does not exist)

$$\mathrm{d}x^a := g^a_{i_1}. \tag{9.10.53}$$

The latter is a reasonable notation (*cf.* also (4.2.55)) since, in local coordinates $x^{\mathbf{a}}$, the coordinate dual basis is

$$g^{\mathbf{a}} = g^{\mathbf{a}}_{i_1} = \mathrm{d}x^{\mathbf{a}}, \tag{9.10.54}$$

so (9.10.53) is simply the abstract-index version of (9.10.54). We do not have $\mathrm{d}^2 = 0$ when applied to λ_A, because of the abstract* index A. By (4.9.11), (4.6.34) and (4.2.31) we obtain instead (where, for generality we have included a torsion $T_{ab}{}^{c}$ as in (4.2.22))

$$\begin{aligned} 2\mathrm{d}^2\lambda_A + T_{i_1 i_2}{}^{b}\nabla_b\lambda_A &= 2\nabla_{[i_1}\nabla_{i_2]}\lambda_A \\ &= \{\varepsilon_{I'_1 I'_2}\square_{I_1 I_2} + \varepsilon_{I_1 I_2}\square_{I'_1 I'_2}\}\lambda_A \\ &= -\mathrm{X}_{I_1 I_2 A}{}^{B}\lambda_B\varepsilon_{I'_1 I'_2} - \varepsilon_{I_1 I_2}\Phi_A{}^{B}{}_{I'_1 I'_2}\lambda_B \\ &= -\Psi_{I_1 I_2 A}{}^{B}\lambda_B\varepsilon_{I'_1 I'_2} + 2\Lambda\varepsilon_{A(I_1}\lambda_{I_2)}\varepsilon_{I'_1 I'_2} - \varepsilon_{I_1 I_2}\Phi_A{}^{B}{}_{I'_1 I'_2}\lambda_B \end{aligned}$$

and

$$\mathrm{d}^2 x^a = -\tfrac{1}{2}T_{i_1 i_2}{}^{a} \tag{9.10.55}$$

(by (9.10.54), (4.2.22) and $\mathrm{d}^2 x^{\mathbf{a}} = 0$). From these relations we can readily obtain Sparling's result. (Note that $\mathrm{d}^2 x^a = 0$ iff the torsion vanishes.)

There are relations between the forms considered here and the forms $\boldsymbol{h}$ and $\boldsymbol{\Sigma}$ of the invariant contact structure of twistor theory considered in §7.4, these being forms defined on the space $\mathscr{N}$ of 'null twistors' in $\mathscr{M}$. In fact we can easily check that

$$\boldsymbol{\Sigma} = \mathrm{i}\boldsymbol{\Xi} - \mathrm{i}\bar{\boldsymbol{\Xi}} \tag{9.10.56}$$

where all forms now refer to $\mathscr{B}$ (which is a bundle over $\mathscr{N}$ as well as over $\mathscr{M}$, so that the forms can be pulled back to $\mathscr{B}$). Thus, by (9.10.56) and (9.10.48), $\boldsymbol{\Omega}$ and $-\frac{1}{2}\boldsymbol{\Sigma}$ are the real and imaginary parts, respectively, of $\boldsymbol{\Xi}$. However $\boldsymbol{\Xi}$ and the Sparling form $\boldsymbol{\beta}$ do not directly apply to the twistor space $\mathscr{N}$ and further work will be required to extract the full significance of these relations.

It is interesting, yet somewhat tantalizing to find so many such interconnections, relating energy–momentum, angular momentum, Einstein's field equations and twistor theory. The full role of twistor theory in this context remains problematical as of now. However the essentially spinorial nature of the Witten argument, and now also of the 2-surface twistor procedures of §9.9, provide convincing evidence of a hitherto unsuspected and deep role for spinor ideas in connection with energy and momentum. On the face of it this seems remarkable, since those physical

* Some people prefer to use a symbol other than 'd' when its use would entail $\mathrm{d}^2 \neq 0$. However, the usage adopted here is entirely logical in the context of the abstract-index formalism. See Volume 1, Chapters 2 and 4 for details.

quantities have always previously been described in terms of 'vectorial' or 'tensorial' (i.e. integral spin) entities, particularly those referring to translational motions of space–time. In general relativity such translational symmetries may be absent, and it appears that spinors are needed in order to reveal the deeper attributes of these fundamentally important physical quantities.

Appendix
Spinors in *n* dimensions

The spinors that we have been concerned with in these volumes are part of a general pattern. A spinor* concept exists for the group of proper rotations in *any* dimension *n*, and with *any* signature.

Whereas for the particular case $n = 4$ (and especially for the signature $+ - - -$ of space–time) spinors have an exceptional power and utility, they also have considerable importance for other *n*-values. However, since the spin-space dimension turns out to increase exponentially with *n*, one naturally finds that it is with comparatively small *n*-values that the utility of an entirely spinorial formalism is at its greatest. Moreoever, with *even* *n*-values the spinors can be broken down naturally into smaller objects – the *reduced* spinors, which are the most important ingredients of the theory. For $n = 4$, the reduced spinors are our familiar unprimed and primed spin-vectors, constituting two-dimensional spaces each. For $n = 6$, taking the relevant orthogonal group to be the $SO(2, 4)$ of §9.2, the reduced spinors are the (univalent) twistors and dual twistors – four-dimensional each. For $n = 2$, at a point of a spacelike 2-surface in space–time, they are the one-dimensional spin-weighted scalars, of spin-weight $\frac{1}{2}$ or $-\frac{1}{2}$ (like the ω^1 or ω^0 of §9.9). For general *even n*, the dimension of reduced spinors is $2^{\frac{1}{2}n-1}$, the combined space of (unreduced) spinors being of dimension $N = 2^{\frac{1}{2}n}$. For a general *odd n* there is no invariant reduction to simpler parts,** and the dimension of the spin-space is $N = 2^{\frac{1}{2}n-\frac{1}{2}}$. In each case, these spinors are two-valued *spinorial objects* (in the sense of §1.5, Volume 1) and thus change sign under rotation through 2π. The spin-group Spin(n) [or Spin(p, q)] is a two-fold cover of the corresponding proper† rotation group $SO(n)$ [or $SO(p, q)$ or $O^{\uparrow}_{+}(p, q)$].

* Unless otherwise stated, the term 'spinor', as used in this appendix, refers to a *univalent* object – a sort of 'spin-vector'.

** This is partly a question of the definition adopted. One could alternatively say that in the case of odd *n*, spin-space is of dimension $2^{\frac{1}{2}n+\frac{1}{2}}$ and reduces to two spaces each of dimension $2^{\frac{1}{2}n-\frac{1}{2}}$. But unlike the case of even *n*, only *one* of these spaces need be considered in the translation of vector–tensor expressions into spinor form.

† The corresponding notion for $O(n)$ [*or* $O(p, q)$] is referred to as Pin (n) [*or* Pin (p, q)]; *cf.* Porteous (1981).

In this appendix we present an outlined account of the basic algebraic and geometric ideas needed for the spinor theory of n dimensions. (For further details we refer the reader to the classic accounts of Cartan 1966 and of Brauer and Weyl 1935; *cf.* also Veblen 1933*a*, *b*, 1934, Chevalley 1954, Porteous 1981, Adams 1981, Lawson and Michelsohn 1986.)

The Clifford equation; γ-matrices

The usual starting point is the Clifford (–Dirac) equation

$$\gamma_{\mathbf{a}}\gamma_{\mathbf{b}} + \gamma_{\mathbf{b}}\gamma_{\mathbf{a}} = -2g_{\mathbf{ab}}\mathbf{I}, \tag{B.1}$$

where $\gamma_1, \ldots, \gamma_n$ are (complex) $N \times N$ matrices (and where N will turn out to be as given above), $g_{\mathbf{ab}}$ are the components of some (real or complex) n-dimensional non-degenerate symmetric tensor, and $\mathbf{I}$ is the $N \times N$ unit matrix. Somewhat more invariantly we can introduce n-dimensional abstract indices $a, b, c, \ldots$ and write

$$\gamma_a\gamma_b + \gamma_b\gamma_a = -2g_{ab}\mathbf{I}, \quad \text{i.e.} \quad \gamma_{(a}\gamma_{b)} = -g_{ab}\mathbf{I}. \tag{B.2}$$

The object γ_a may now be thought of as a matrix with abstract-indexed entries, each entry belonging to an n-dimensional vector space $\mathbb{V}_a$ (or perhaps module, in case we are concerned with spinor fields on an n-manifold) with non-degenerate symmetric metric tensor $g_{ab} \in \mathbb{V}_{(ab)}$. At this stage we shall assume that $\mathbb{V}_a$ is a complex* vector space (or module), so the question of signature for g_{ab} does not properly arise. Later on we shall briefly consider reality conditions and examine how the choice of signature can affect the structure of the spin-space.

For a properly invariant viewpoint, we should not think of γ_a explicitly as a matrix [or $\gamma_{\mathbf{a}}$ explicitly as matrices] but as an element [elements] of some abstract algebra, providing linear transformations of some abstract space: *spin-space*. Thus, if desired, we can introduce (say lower case Greek) abstract indices and write

$$\gamma_{a\rho}{}^{\sigma} \quad [\text{or } \gamma_{\mathbf{a}\rho}{}^{\sigma} = \gamma_{1\rho}{}^{\sigma}, \ldots, \gamma_{n\rho}{}^{\sigma}] \tag{B.3}$$

where the abstract indices ρ, σ refer to spin-space, which is a complex vector space $\mathbb{S}^\rho$ (or module $\mathbb{S}^\rho$, in the case of spinor fields on an n-manifold). We assume irreducibility for the algebra generated by $\gamma_{\mathbf{a}}$, and then it turns out that $\mathbb{S}^\rho$ has dimension $N = 2^{\frac{1}{2}n}$ (n even) or $N = 2^{\frac{1}{2}n - \frac{1}{2}}$ (n odd). (Matrices of

* The discussion is not substantially affected, however, if $\mathbb{V}_a$ is taken to be a vector space over any field closed under the taking of square roots and (with some reorganization) of any characteristic other than two.

this size $N \times N$ are the smallest that can satisfy (B.1)). The equation (B.2) can now be written

$$\gamma_{a\rho}{}^{\sigma}\gamma_{b\sigma}{}^{\tau} + \gamma_{b\rho}{}^{\sigma}\gamma_{a\sigma}{}^{\tau} = -2g_{ab}\delta_{\rho}^{\tau}. \tag{B.4}$$

If we choose some specific basis $\delta_1^{\rho}, \ldots, \delta_N^{\rho} = \delta_{\mathbf{p}}^{\rho}$ for $\mathbb{S}^{\rho}$, then, referring to this basis, we find an explicit realization of the matrix γ_a as

$$\gamma_a = (\gamma_{a\mathbf{p}}{}^{\boldsymbol{\sigma}}), \quad \gamma_{a\mathbf{p}}{}^{\boldsymbol{\sigma}} = \gamma_{a\rho}{}^{\sigma}\delta_{\mathbf{p}}^{\rho}\delta_{\sigma}^{\boldsymbol{\sigma}} \tag{B.5}$$

($\delta_{\sigma}^{\boldsymbol{\sigma}}$ being the dual basis of $\delta_{\mathbf{p}}^{\rho}$), and with a similar expression for the $\gamma_{\mathbf{a}}$. However, the notation γ_a or $\gamma_{\mathbf{a}}$ is not taken to be specific as to whether or not a basis for $\mathbb{S}^{\rho}$ has been selected, and can equally well apply when ρ and σ are abstract, or perhaps partially reduced to a direct sum of abstract indices ('block' abstract-indexed matrices).

Independently of $\delta_{\mathbf{p}}^{\rho}$, we may choose a basis $g_a{}^1, \ldots, g_a{}^n = g_a{}^{\mathbf{a}}$, for $\mathbb{V}_a$, with dual basis $g_{\mathbf{a}}{}^a$, which enables us to translate between the quantities

$$\gamma_1, \ldots, \gamma_n \tag{B.6}$$

of (B.1) and the more abstract γ_a of (B.2):

$$\gamma_{\mathbf{a}} = \gamma_a g_{\mathbf{a}}{}^a; \quad \gamma_a = \gamma_{\mathbf{a}} g_a{}^{\mathbf{a}} \tag{B.7}$$

It is usual to choose this basis so that $g_{\mathbf{ab}}$ is diagonal with entries ± 1. (Indeed, since $\mathbb{V}_a$ is complex we can choose all non-zero entries to be $+1$, but since we shall be interested in the real case later, it will be more helpful to allow for both signs at this stage.) Then (B.1) tells us that the different matrices (B.6) all anti-commute with one another and have squares equal to ∓ 1. The (complex) Clifford algebra $\mathfrak{C}$ (Clifford 1882) is the algebra over $\mathbb{C}$ generated by (i.e. complex polynomials in) the quantities (B.6) and subject to (B.1).

The element

$$\boldsymbol{\eta} = \gamma_1\gamma_2\cdots\gamma_n \tag{B.8}$$

of $\mathfrak{C}$ is of considerable interest. Note that if n is *even* this also anti-commutes with each of $\gamma_1, \ldots, \gamma_n$ while if n is *odd* it *commutes* with each of them. More invariantly we can write (B.8) as

$$\boldsymbol{\eta} = \frac{1}{n!}\varepsilon^{a_1 \ldots a_n}\gamma_{a_1}\ldots\gamma_{a_n} \tag{B.9}$$

or

$$\eta_{\rho}{}^{\sigma} = \frac{1}{n!}\varepsilon^{a_1 \ldots a_n}\gamma_{a_1\rho}{}^{\tau_1}\gamma_{a_2\tau_1}{}^{\tau_2}\cdots\gamma_{a_n\tau_{n-1}}{}^{\sigma}$$

where we choose an alternating tensor

$$\varepsilon^{a_1 \ldots a_n} \in \mathbb{V}^{[a_1 \ldots a_n]} \tag{B.10}$$

with $\varepsilon^{12\ldots n} = 1$, so that

$$\frac{1}{n!}\varepsilon^{a_1\ldots a_n}\varepsilon^{b_1\ldots b_n}g_{a_1b_1}\cdots g_{a_nb_n} = (-1)^u \tag{B.11}$$

where u is the number of negative elements of (the diagonal) $g_{\mathbf{ab}}$. Note that

$$\boldsymbol{\eta}^2 = (-1)^{\frac{1}{2}s(s+1)}\mathbf{I} \tag{B.12}$$

where $s = n - 2u$ is the 'signature' of $g_{\mathbf{ab}}$.

Consider the case: *n odd*, so $\boldsymbol{\eta}$ commutes with all elements of $\mathfrak{C}$. We assume *irreducibility** for our algebra, so Schur's lemma implies that $\boldsymbol{\eta}$ must, in this case, be represented by a multiple of $\mathbf{I}$:

$$\boldsymbol{\eta} = \pm\mathbf{I}\ or\ \pm\mathrm{i}\mathbf{I}\quad (\text{i.e. } \eta_\rho{}^\sigma = \pm\delta_\rho^\sigma\ or\ \pm\mathrm{i}\delta_\rho^\sigma)\ if\ n\ is\ odd \tag{B.13}$$

(*cf.* (B.12)). The real [imaginary] case occurs when $s \equiv 1 \pmod 4$ [*or* $s \equiv 3 \pmod 4$] (either sign in (B.13) being allowable).

But for *n even*, the anti-commutativity of $\boldsymbol{\eta}$ with each $\boldsymbol{\gamma}_{\mathbf{a}}$ implies that these can be represented as

$$\eta_\rho{}^\sigma\ or\ \mathrm{i}\eta_\rho{}^\sigma = \pm\begin{pmatrix}\delta_R^S & 0\\ 0 & -\delta_{R'}^{S'}\end{pmatrix},\quad\text{and}\quad \gamma_{a\rho}{}^\sigma = \begin{pmatrix}0 & \gamma_{aR}{}^{S'}\\ \gamma_{aR'}{}^S & 0\end{pmatrix} \tag{B.14}$$

where $\rho = R \oplus R'$, $\sigma = S \oplus S'$, the 'i' appearing when $s \equiv 2 \pmod 4$. The quantities $\gamma_{aR}{}^{S'}$ and $\gamma_{aR'}{}^S$ satisfy

$$\gamma_{(a|R|}{}^{S'}\gamma_{b)S'}{}^T = -g_{ab}\delta_R^T\quad\text{and}\quad \gamma_{(a|R'|}{}^{S}\gamma_{b)S}{}^{T'} = -g_{ab}\delta_{R'}^{T'} \tag{B.15}$$

by virtue of (B.4). The spin-space $\mathbb{S}^\rho$ splits into the direct sum

$$\mathbb{S}^\rho = \mathbb{S}^R \oplus \mathbb{S}^{R'} \tag{B.16}$$

where each *reduced spin-space* $\mathbb{S}^R$, $\mathbb{S}^{R'}$ has dimension $\frac{1}{2}N = 2^{\frac{1}{2}n-1}$.

The notation has been adopted, in (B.14), that any kernel symbol with abstract indices $\rho, \sigma, \ldots$ may have these indices replaced correspondingly by R or R', S or S', etc., and that the part of the quantity in question which is projected into the relevant reduced spin-space is thereby denoted. More explicitly, we have projection operators

$$\boldsymbol{\Pi} = \tfrac{1}{2}(\mathbf{I} - \mathrm{i}^{\frac{1}{2}s}\boldsymbol{\eta}),\quad \tilde{\boldsymbol{\Pi}} = \tfrac{1}{2}(\mathbf{I} + \mathrm{i}^{\frac{1}{2}s}\boldsymbol{\eta})\quad (n\ \text{even}), \tag{B.17}$$

so that

$$\Pi_\rho{}^\sigma = \mathrm{diag}\,(\delta_R^S, 0),\quad \tilde{\Pi}_\rho{}^\sigma = \mathrm{diag}\,(0, \delta_{R'}^{S'}) \tag{B.18}$$

(conventionally fixing the ambiguous sign in (B.14)), which achieve these projections to the reduced spin-spaces. We have

$$\boldsymbol{\Pi}^2 = \boldsymbol{\Pi},\quad \tilde{\boldsymbol{\Pi}}^2 = \tilde{\boldsymbol{\Pi}},\quad \boldsymbol{\Pi}\tilde{\boldsymbol{\Pi}} = \tilde{\boldsymbol{\Pi}}\boldsymbol{\Pi} = 0,\quad \boldsymbol{\Pi} + \tilde{\boldsymbol{\Pi}} = \mathbf{I}. \tag{B.19}$$

* See, for example Volume 1, p. 141.

Clifford algebra and forms

The entire Clifford algebra consists of (finite) sums of the form

$$A\mathbf{I} + B^a\gamma_a + C^{ab}\gamma_a\gamma_b + D^{abc}\gamma_a\gamma_b\gamma_c + \cdots \qquad \text{(B.20)}$$

However, because of (B.2), it turns out that we need consider only anti-symmetrized products

$$\gamma_{ab\ldots d} := \gamma_{[a}\gamma_b\cdots\gamma_{d]}. \qquad \text{(B.21)}$$

For example,

$$\gamma_a\gamma_b = \gamma_{[a}\gamma_{b]} - g_{ab}\mathbf{I},$$

$$\gamma_a\gamma_b\gamma_c = \gamma_{[a}\gamma_b\gamma_{c]} - g_{ab}\gamma_c + g_{ac}\gamma_b - g_{bc}\gamma_a, \qquad \text{(B.22)}$$

with corresponding (but more complicated) expressions holding for higher-order products. Thus, the coefficients in (B.20) may be assumed to be anti-symmetric:

$$C^{ab} = C^{[ab]}, \quad D^{abc} = D^{[abc]}, \ldots \qquad \text{(B.23)}$$

and the whole series (B.20) terminates at the $(n+1)$th term. The number of formally independent elements of the Clifford algebra is the total number of independent components A, $B^{\mathbf{a}}$, $C^{\mathbf{ab}}, \ldots$ (subject to (B.23)), i.e.

$$1 + n + \tfrac{1}{2}n(n-1) + \cdots + n + 1 = 2^n. \qquad \text{(B.24)}$$

Each element of the (formal) algebra may be viewed as a collection consisting of a 0-form, a 1-form, a 2-form, etc., often written formally as a (finite) sum

$$A + \mathbf{B} + \mathbf{C} + \mathbf{D} + \cdots \qquad \text{(B.25)}$$

where

$$\mathbf{B} = B_{i_1} = B^a g_{ai_1}, \quad \mathbf{C} = C_{i_1i_2}, \quad \mathbf{D} = D_{i_1i_2i_3}, \ldots \qquad \text{(B.26)}$$

(lowering indices, generally, with g_{ab} and adopting the notation of (4.3.10)). Clifford multiplication on expressions (B.25) (i.e. as induced by products of expressions (B.20) acts distributively (and associatively), where the Clifford product of a p-form with a q-form is a linear combination of terms each of which is obtained by first making a number of contractions on the outer product of the forms and then anti-symmetrizing the indices that remain. (The term with zero contractions is simply the exterior product (4.3.13) of the two forms.)

When n is *odd*, however, our assumption of irreducibility entails that not all formally distinct elements of the Clifford algebra are linearly independent, the $\boldsymbol{\eta}$ of (B.8) being proportional to the identity $\mathbf{I}$, as in (B.13). Indeed, we find that

$$\boldsymbol{\eta}\gamma_{a\ldots c} = \pm {}^*\gamma_{a\ldots c} \qquad \text{(B.27)}$$

where (compare Volume 1, p. 264)

$$^{*}\gamma_{a\ldots c} = \frac{1}{r!}\varepsilon_{a\ldots c}{}^{d\ldots f}\gamma_{d\ldots f} \tag{B.28}$$

$d,\ldots,f$ being r in number (indices on the quantity (B.10) being lowered with g_{ab}). Adopting (B.13), we find that each form (B.26) has to be equated to its dual (times ± 1 or $\pm$ i). With such an identification, the Clifford algebra becomes a *complete matrix algebra* of $N \times N$ matrices (over $\mathbb{C}$), where, as stated earlier, $N = (\frac{1}{2}2^n)^{\frac{1}{2}} = 2^{\frac{1}{2}n-\frac{1}{2}}$ (by (B.24)). The ambiguous sign in (B.13) gives us two inequivalent such representations, and the (unidentified) *formal* Clifford algebra becomes faithfully represented as the direct sum of these *two* complete matrix algebras when n is odd.

When n is *even*, there are no identifications, and the full Clifford algebra is faithfully represented as the complete algebra of $N \times N$ matrices (over $\mathbb{C}$), where $N = (2^n)^{\frac{1}{2}} = 2^{\frac{1}{2}n}$ (as stated earlier). However, it is useful to make the passage to reduced spinors, the quantities of the form $\theta_R{}^S$ [*or* $\theta_{R'}{}^{S'}$] arising as elements of $\mathfrak{C}$ which are annihilated both on the left and on the right by $\tilde{\Pi}$ [*or* Π] (*cf.* (B.18)), so that these are given by terms with an *even* number of γs; and those of the form $\theta_R{}^{S'}$ [*or* $\theta_{R'}{}^{S}$] arising as elements of $\mathfrak{C}$ which are annihilated on the left [right] by $\tilde{\Pi}$ and on the right [left] by Π, so these are given by terms with an *odd* number of γs.

The 2-valent ε-spinors

We next proceed to obtain analogues, in the general n-dimensional case, of the ε_{AB} and $\varepsilon_{A'B'}$ of standard Lorentzian 2-spinor theory. Define two elements of $\mathbb{V}^{\kappa\tau}_{\rho\sigma}$ by

$$^{(\pm)}E^{\kappa\tau}_{\rho\sigma} = \delta^{\kappa}_{\rho}\delta^{\tau}_{\sigma} \pm \frac{1}{1!}\gamma_{a\rho}{}^{\kappa}\gamma^{a}{}_{\sigma}{}^{\tau} + \frac{1}{2!}\gamma_{ab\rho}{}^{\kappa}\gamma^{ab}{}_{\sigma}{}^{\tau} \pm \frac{1}{3!}\gamma_{abc\rho}{}^{\kappa}\gamma^{abc}{}_{\sigma}{}^{\tau} + \cdots \tag{B.29}$$

these being finite sums since the non-zero quantities (B.21) appearing here are finite in number (indices raised with g^{ab}, inverse to g_{ab}). We note that when $g_{\mathrm{ab}} = \delta_{\mathrm{ab}}$ (positive-definite case $u = 0$) we can write (B.29) as

$$\begin{aligned} ^{(\pm)}\mathbf{E} = \mathbf{I}\otimes\mathbf{I} &\pm \sum_{\mathrm{i}=1}^{n} \gamma_{\mathrm{i}}\otimes\gamma_{\mathrm{i}} + \sum_{\mathrm{i}<\mathrm{j}} (\gamma_{\mathrm{i}}\gamma_{\mathrm{j}})\otimes(\gamma_{\mathrm{i}}\gamma_{\mathrm{j}}) \\ &\pm \sum_{\mathrm{i}<\mathrm{j}<\mathrm{k}} (\gamma_{\mathrm{i}}\gamma_{\mathrm{j}}\gamma_{\mathrm{k}})\otimes(\gamma_{\mathrm{i}}\gamma_{\mathrm{j}}\gamma_{\mathrm{k}}) + \cdots \end{aligned} \tag{B.30}$$

(whereas there would be appropriate sign changes in (B.30) if $u > 0$). In that part of the following discussion in which (B.30) is used, it will be adequate to assume that indeed $g_{\mathbf{ab}} = \delta_{\mathbf{ab}}$ holds. For we are not concerned with reality conditions at the moment, so a complex basis $\delta^a_{\mathbf{a}}$ for which this form holds will be allowable.

Observe that when n is *odd*, the terms in (B.30) (and therefore in (B.29)) simply repeat themselves in the reverse order after the middle point is reached, these terms either doubling the previous ones or cancelling them all out (*cf.* (B.13), (B.27)). In fact we have

$$^{(+)}\mathbf{E} = 0 \ (n \equiv 1 \ (\text{mod } 4)), \quad ^{(-)}\mathbf{E} = 0 \ (n \equiv 3 \ (\text{mod } 4)); \tag{B.31}$$

and $^{(\pm)}\mathbf{E} \neq 0$ otherwise. When n is *even*, neither $^{(+)}\mathbf{E}$ nor $^{(-)}\mathbf{E}$ will vanish.

Note the following important property, which follows from (B.30):

$$^{(\pm)}\mathbf{E}(\gamma_a \otimes \mathbf{I}) = \mp {}^{(\pm)}\mathbf{E}(\mathbf{I} \otimes \gamma_a) \tag{B.32a}$$

i.e.

$$^{(\pm)}E^{\lambda\tau}_{\rho\sigma}\gamma_{a\lambda}{}^{\kappa} = \mp {}^{(\pm)}E^{\kappa\lambda}_{\rho\sigma}\gamma_{a\lambda}{}^{\tau} \tag{B.32b}$$

Now consider $(^{(\pm)}\mathbf{E})^2$. By using (B.32), applied to the expansion (B.29) of the second $^{(\pm)}\mathbf{E}$, we obtain

$$^{(\pm)}E^{\lambda\mu}_{\rho\sigma}{}^{(\pm)}E^{\kappa\tau}_{\lambda\mu} = {}^{(\pm)}E^{\lambda\tau}_{\rho\sigma}P^{\kappa}_{\lambda}$$

where

$$P^{\kappa}_{\lambda} = \delta^{\kappa}_{\lambda} - \frac{1}{1!}\gamma_{a\lambda}{}^{\mu}\gamma^{a}{}_{\mu}{}^{\kappa} + \frac{1}{2!}\gamma_{ba\lambda}{}^{\mu}\gamma^{ab}{}_{\mu}{}^{\kappa}$$

$$-\frac{1}{3!}\gamma_{cba\lambda}{}^{\mu}\gamma^{abc}{}_{\mu}{}^{\kappa} + \cdots \tag{B.33}$$

But one finds, substituting numerical values for $a, b, \ldots$, that the successive terms in the above expansion of P^{κ}_{λ} are simply multiples by

$$1, n, \frac{1}{2!}n(n-1), \frac{1}{3!}n(n-1)(n-2), \ldots \tag{B.34}$$

of $\delta^{\kappa}_{\lambda}$, respectively. Summing these, we obtain

$$P^{\kappa}_{\lambda} = 2^n \delta^{\kappa}_{\lambda}, \tag{B.35}$$

whence

$$^{(\pm)}E^{\lambda\mu}_{\rho\sigma}{}^{(\pm)}E^{\kappa\tau}_{\lambda\mu} = 2^{n}{}^{(\pm)}E^{\kappa\tau}_{\rho\sigma}, \tag{B.36}$$

i.e.

$$2^{-n}{}^{(\pm)}\mathbf{E} \textit{ is idempotent.} \tag{B.37}$$

It follows that the matrix rank of the quantity in (B.37) is equal to its trace:

$$T = 2^{-n}{}^{(\pm)}E^{\rho\sigma}_{\rho\sigma}. \tag{B.38}$$

However the quantities

$$\gamma_{a\ldots c\rho}{}^{\rho} \tag{B.39}$$

which appear in (B.38) (*cf.* (B.29)) can be non-zero only when the number of indices $a, \ldots, c$ is either 0 or n. For otherwise (B.39) would define a non-zero skew tensor (or form) which is invariant under the appropriate orthogonal group – and this happens only for a scalar or n-form. Moreover, symmetry considerations show that when n is *even* (B.39) must vanish also in the case of n indices. (Permute the γs cyclicly.) When n is *odd*, the contraction of the last term in (B.33) must equal that of the first whenever ${}^{(\pm)}E_{\rho\sigma}^{\kappa\tau}$ is itself non-zero. Noting that $\delta_\rho^\rho = N$, we thus obtain for (B.38)

$$T = \begin{Bmatrix} 2^{-n} \cdot N^2 & (n \text{ even}) \\ 2^{-n} \cdot 2N^2 & (n \text{ odd}) \end{Bmatrix} = 1 \qquad \text{(B.40)}$$

in the cases when ${}^{(\pm)}\mathbf{E} \neq 0$ (*cf.* (B.31)).

The rank of (B.29) thus being unity (or zero), we can factorize it:

$$N^{-1(\pm)}E_{\rho\sigma}^{\kappa\tau} = {}^{(\pm)}\varepsilon_{\rho\sigma}{}^{(\pm)}\varepsilon^{\kappa\tau} \quad (n \text{ even}) \qquad \text{(B.41}a)$$

and (with (B.31))

$$\tfrac{1}{2}N^{-1(\pm)}E_{\rho\sigma}^{\kappa\tau} = \varepsilon_{\rho\sigma}\varepsilon^{\kappa\tau} \begin{cases} (n = 3, 7, 11, \ldots) \\ (n = 1, 5, 9, \ldots) \end{cases}. \qquad \text{(B.41}b)$$

Note that there is a choice of scalar multiplier involved in this splitting, which may be taken from one factor to the other. From (B.29), and using the stated properties of (B.39), we obtain

$${}^{(\pm)}E_{\rho\sigma}^{\rho\tau} = {}^{(\pm)}E_{\sigma\rho}^{\tau\rho} = \begin{cases} N\delta_\sigma^\tau & (n \text{ even}) \\ 2N\delta_\sigma^\tau \text{ or } 0 & (n \text{ odd}) \end{cases} \qquad \text{(B.42)}$$

so that

$$\varepsilon_{\rho\sigma}\varepsilon^{\rho\tau} = \delta_\sigma^\tau = \varepsilon_{\sigma\rho}\varepsilon^{\tau\rho} \qquad (n \text{ odd}), \quad \text{(B.43}a)$$

$${}^{(+)}\varepsilon_{\rho\sigma}{}^{(+)}\varepsilon^{\rho\tau} = {}^{(+)}\varepsilon_{\sigma\rho}{}^{(+)}\varepsilon^{\tau\rho} = {}^{(-)}\varepsilon_{\rho\sigma}{}^{(-)}\varepsilon^{\rho\tau} = {}^{(-)}\varepsilon_{\sigma\rho}{}^{(-)}\varepsilon^{\tau\rho} = \delta_\sigma^\tau \quad (n \text{ even}). \quad \text{(B.43}b)$$

Now consider the diagonal contraction ${}^{(\pm)}E_{\lambda\rho}^{\rho\kappa}$ applied to (B.29). We obtain an expression like (B.33), but where, instead, the signs of the successive terms are

$$+, \pm, -, \mp, +, \pm, \cdots$$

and we have, by (B.34)

$${}^{(\pm)}E_{\lambda\rho}^{\rho\kappa} = \delta_\lambda^\kappa\left(1 \mp n - \frac{1}{2!}n(n-1) \pm \frac{1}{3!}n(n-1)(n-2) + \cdots\right).$$

The sum of the terms in the bracket for *even* n is either $2^{\frac{1}{2}n}$ or $-2^{\frac{1}{2}n}$, and for *odd* n it is either $2^{\frac{1}{2}n+\frac{1}{2}}$ or 0. Thus the contraction of (B.41a, b) over κ, σ yields either δ_ρ^τ or $-\delta_\rho^\tau$. The arrangement of signs has a periodicity (mod 8) and we find

$$ {}^{(+)}\varepsilon_{\rho\sigma}{}^{(+)}\varepsilon^{\sigma\tau} = \pm\,\delta_\rho^\tau \quad \begin{cases} (n \equiv 0, 6 \pmod 8) \\ (n \equiv 2, 4 \pmod 8), \end{cases} \tag{B.44a} $$

$$ {}^{(-)}\varepsilon_{\rho\sigma}{}^{(-)}\varepsilon^{\sigma\tau} = \pm\,\delta_\rho^\tau \quad \begin{cases} (n \equiv 0, 2 \pmod 8) \\ (n \equiv 4, 6 \pmod 8) \end{cases} \tag{B.44b} $$

and

$$ \varepsilon_{\rho\sigma}\varepsilon^{\sigma\tau} = \pm\,\delta_\rho^\tau \quad \begin{cases} (n \equiv 1, 7 \pmod 8) \\ (n \equiv 3, 5 \pmod 8). \end{cases} \tag{B.44c} $$

Comparing (B.44) with (B.43) we obtain the symmetry properties given in table (B.45). In the cases of *odd n*, the *non-zero* ${}^{(\pm)}\varepsilon_{\rho\sigma}$, ${}^{(\pm)}\varepsilon^{\rho\sigma}$ stand for $\varepsilon_{\rho\sigma}$, $\varepsilon^{\rho\sigma}$. Thus, $\varepsilon_{\rho\sigma}$ and $\varepsilon^{\rho\sigma}$ are *skew* if $n \equiv 3$, 5 (mod 8) and *symmetric* if $n \equiv 1$, 7 (mod 8).

${}^{(+)}\varepsilon_{\rho\sigma}, {}^{(+)}\varepsilon^{\rho\sigma}$	${}^{(-)}\varepsilon_{\rho\sigma}, {}^{(-)}\varepsilon^{\rho\sigma}$	n (mod 8)
symmetric	symmetric	0
zero	symmetric	1
skew	symmetric	2
skew	zero	3
skew	skew	4
zero	'skew	5
symmetric	skew	6
symmetric	zero	7

(B.45)

When n is *even*, we can reduce these quantities further, using the $\Pi_\rho{}^\sigma$ and $\tilde{\Pi}_\rho{}^\sigma$ of (B.17). First, observe that

$$ {}^{(\pm)}E^{\kappa\lambda}_{\rho\sigma}\eta_\lambda{}^\tau = \eta_\sigma{}^{\lambda(\mp)}E^{\kappa\tau}_{\rho\lambda} \quad (n \text{ even}). \tag{B.46} $$

It follows from this and (B.41*a*) that ${}^{(\pm)}\varepsilon^{\kappa\lambda}\eta_\lambda{}^\tau$ is a multiple of ${}^{(\mp)}\varepsilon^{\kappa\tau}$ and $\eta_\sigma{}^{\lambda(\mp)}\varepsilon_{\rho\lambda}$ is a multiple of ${}^{(\pm)}\varepsilon_{\rho\sigma}$. It is convenient to choose the arbitrary factor in this definition (B.41*a*) of the εs so that

$$ {}^{(\pm)}\varepsilon^{\kappa\lambda}\eta_\lambda{}^\tau = -\,\mathrm{i}^{-\frac{1}{2}s(\mp)}\varepsilon^{\kappa\tau}, \quad {}^{(\pm)}\varepsilon_{\rho\lambda}\eta_\sigma{}^\lambda = -\,\mathrm{i}^{-\frac{1}{2}s(\mp)}\varepsilon_{\rho\sigma} \tag{B.47} $$

and then define (*cf.* (B.17))

$$ \begin{aligned} \varepsilon^{\rho\sigma} &:= {}^{(+)}\varepsilon^{\rho\lambda}\Pi_\lambda{}^\sigma = \tfrac{1}{2}({}^{(+)}\varepsilon^{\rho\sigma} + {}^{(-)}\varepsilon^{\rho\sigma}) \\ \varepsilon_{\rho\sigma} &:= {}^{(+)}\varepsilon_{\rho\lambda}\Pi_\sigma{}^\lambda = \tfrac{1}{2}({}^{(+)}\varepsilon_{\rho\sigma} + {}^{(-)}\varepsilon_{\rho\sigma}) \\ \tilde{\varepsilon}^{\rho\sigma} &:= {}^{(+)}\varepsilon^{\rho\lambda}\tilde{\Pi}_\lambda{}^\sigma = \tfrac{1}{2}({}^{(+)}\varepsilon^{\rho\sigma} - {}^{(-)}\varepsilon^{\rho\sigma}) \\ \tilde{\varepsilon}_{\rho\sigma} &:= {}^{(+)}\varepsilon_{\rho\lambda}\tilde{\Pi}_\sigma{}^\lambda = \tfrac{1}{2}({}^{(+)}\varepsilon_{\rho\sigma} - {}^{(-)}\varepsilon_{\rho\sigma}). \end{aligned} \tag{B.48} $$

Recall that Π and $\tilde{\Pi}$ are the projection operators to the reduced spin-spaces (*cf.* (B.16), (B.18), (B.19)). Writing $\rho = R \oplus R'$, $\sigma = S \oplus S'$, etc. as before, we can write each of $\varepsilon^{\rho\sigma}, \ldots, \tilde{\varepsilon}_{\rho\sigma}$ in terms of its four reduced parts:

$$\varepsilon^{\rho\sigma} = \begin{pmatrix} \varepsilon^{RS} & \varepsilon^{RS'} \\ \varepsilon^{R'S} & \varepsilon^{R'S'} \end{pmatrix}, \ldots, \tilde{\varepsilon}_{\rho\sigma} = \begin{pmatrix} \tilde{\varepsilon}_{RS} & \tilde{\varepsilon}_{RS'} \\ \tilde{\varepsilon}_{R'S} & \tilde{\varepsilon}_{R'S'} \end{pmatrix}. \tag{B.49}$$

For each (even) value of n, precisely *one* of the four entries in each matrix (B.49) is non-vanishing, depending upon the value of n (mod 4):

$$\left.\begin{aligned} \varepsilon^{\rho\sigma} &= \begin{pmatrix} \varepsilon^{RS} & 0 \\ 0 & 0 \end{pmatrix}, & \varepsilon_{\rho\sigma} &= \begin{pmatrix} \varepsilon_{RS} & 0 \\ 0 & 0 \end{pmatrix} \\ \tilde{\varepsilon}^{\rho\sigma} &= \begin{pmatrix} 0 & 0 \\ 0 & \tilde{\varepsilon}^{R'S'} \end{pmatrix}, & \tilde{\varepsilon}_{\rho\sigma} &= \begin{pmatrix} 0 & 0 \\ 0 & \tilde{\varepsilon}_{R'S'} \end{pmatrix} \end{aligned}\right\} (n \equiv 0 \pmod 4), \tag{B.50a}$$

whereas

$$\left.\begin{aligned} \varepsilon^{\rho\sigma} &= \begin{pmatrix} 0 & 0 \\ \varepsilon^{R'S} & 0 \end{pmatrix}, & \varepsilon_{\rho\sigma} &= \begin{pmatrix} 0 & 0 \\ \varepsilon_{R'S} & 0 \end{pmatrix} \\ \tilde{\varepsilon}^{\rho\sigma} &= \begin{pmatrix} 0 & \tilde{\varepsilon}^{RS'} \\ 0 & 0 \end{pmatrix}, & \tilde{\varepsilon}_{\rho\sigma} &= \begin{pmatrix} 0 & \tilde{\varepsilon}_{RS'} \\ 0 & 0 \end{pmatrix} \end{aligned}\right\} (n \equiv 2 \pmod 4)). \tag{B.50b}$$

We have, when $\frac{1}{2}n$ is even,

$$\varepsilon^{RT}\varepsilon_{ST} = \varepsilon^{TR}\varepsilon_{TS} = \delta_S^R, \quad \tilde{\varepsilon}^{R'T'}\tilde{\varepsilon}_{S'T'} = \tilde{\varepsilon}^{T'R'}\tilde{\varepsilon}_{T'S'} = \delta_{S'}^{R'} \tag{B.51a}$$

and when $\frac{1}{2}n$ is odd,

$$\varepsilon^{R'T}\varepsilon_{S'T} = \tilde{\varepsilon}^{TR'}\tilde{\varepsilon}_{TS'} = \delta_{S'}^{R'}, \quad \tilde{\varepsilon}^{RT'}\tilde{\varepsilon}_{ST'} = \varepsilon^{T'R}\varepsilon_{T'S} = \delta_S^R. \tag{B.51b}$$

Moreover, from (B.45) we find

$$\varepsilon^{RS}, \varepsilon_{RS}, \tilde{\varepsilon}^{R'S'}, \tilde{\varepsilon}_{R'S'} \text{ are } \begin{cases} \text{symmetric} & (n \equiv 0 \pmod 8)) \\ \text{skew} & (n \equiv 4 \pmod 8)) \end{cases} \tag{B.52a}$$

and

$$\tilde{\varepsilon}^{RS'} = \pm\, \varepsilon^{S'R}, \tilde{\varepsilon}_{RS'} = \pm\, \varepsilon_{S'R} \begin{cases} (n \equiv 6 \pmod 8)) \\ (n \equiv 2 \pmod 8)). \end{cases} \tag{B.52b}$$

We note that when $n \equiv 0$ (mod 4) the spaces of primed and unprimed reduced spinors each possess an ε-object which can be used to raise and lower reduced spinor indices* (as is the case for standard Lorentzian 2-spinors). Thus, a canonical isomorphism exists, in these cases, between each of the spaces $\mathbb{S}^A$, $\mathbb{S}^{A'}$ and their respective duals $\mathbb{S}_{A'}$, $\mathbb{S}_{A'}$. On the other hand, when $n \equiv 2$ (mod 4), the ε-object establishes an isomorphism between $\mathbb{S}^A$ and the dual $\mathbb{S}_{A'}$ of $\mathbb{S}^{A'}$ (and therefore also between $\mathbb{S}^{A'}$ and $\mathbb{S}_A$) and can

* For definiteness, we make the convention that when ε and $\tilde{\varepsilon}$ are skew, both primed and unprimed indices are raised and lowered according to (2.5.14), (2.5.15).

therefore be used to eliminate the primed indices (say) from all expressions (as has been our implicit procedure in the case of twistor theory). Note also that for *odd n*, the ε-object establishes an isomorphism between $\mathbb{S}^\rho$ and its dual $\mathbb{S}_\rho$ and therefore can be used for raising and lowering indices in these cases also.

Translation of tensors to spinors

Implicit in the foregoing discussion is a procedure for translating tensorial objects (elements of the various $\mathbb{V}^{d\dots f}_{a\dots c}$) into spinorial form, using the quantities $\gamma_{a\rho}{}^\sigma$, or their various raised, lowered or reduced versions (*cf.* (B.14)):

$$\gamma_a{}^{\rho\sigma}, \gamma^a{}_{\rho\sigma}, \gamma_a{}^{RS'}, \gamma_a{}^{RS}, \text{ etc.} \tag{B.53}$$

The use of these quantities could, if one desired it, eliminate all tensor indices in favour of spinor ones (and *cf.* (B.55) below, with $r = 1$). For tensors possessing (sets of) anti-symmetrical indices the procedure is somewhat more economical since the quantities

$$\gamma_{a\dots c}{}^{\rho\sigma}, \gamma^{a\dots c}{}_{\rho\sigma}, \gamma_{a\dots c}{}^{RS'}, \gamma_{a\dots c}{}^{RS}, \text{ etc.} \tag{B.54}$$

(*cf.* (B.21)) can be used directly to translate a whole block of such indices into merely a *pair* of spinor indices. Moreover, it follows from our discussion that the quantities (B.54) have special properties. Taking $a,\dots,c$ to be r in number, we find that for odd n

$$\gamma_{a\dots c}{}^{\rho\sigma} \textit{ is } \begin{cases} \text{symmetric in } \rho,\sigma & \text{if } n-2r \equiv 1,7 \pmod 8 \\ \text{skew in } \rho,\sigma & \text{if } n-2r \equiv 3,5 \pmod 8), \end{cases} \tag{B.55a}$$

while for *even n* we have

$$\gamma_{a\dots c}{}^{RS} \text{ and } \gamma_{a\dots c}{}^{R'S'} \text{ are } \begin{cases} \text{symmetric in } RS, R'S', & \text{if } n-2r \equiv 0 \pmod 8 \\ \text{skew in } RS, R'S', & \text{if } n-2r \equiv 4 \pmod 8 \end{cases} \tag{B.55b}$$

and

$$\gamma_{a\dots c}{}^{RS'} = \pm\gamma_{a\dots c}{}^{S'R} \begin{cases} n+2r \equiv 6 \pmod 8 \\ n+2r \equiv 2 \pmod 8 \end{cases} \tag{B.55c}$$

where the reduced parts of $\gamma_{a\dots c}{}^{\rho\sigma}$ with different primed/unprimed index structures from these all *vanish*. The same statements (B.55) also apply when all the indices are in reverse upper/lower position.

Property (B.55*a*) follows by repeated application of (B.32) applied to (B.21) together with (B.41*b*), where the index-raising convention

$$\psi_{\mathscr{A}}{}^\rho = \varepsilon^{\rho\lambda}\psi_{\mathscr{A}\lambda}, \quad \psi_{\mathscr{A}\rho} = \varepsilon_{\lambda\rho}\psi_{\mathscr{A}}{}^\lambda \tag{B.56}$$

is used (*cf.* (B.43*a*) and compare (2.5.14), (2.5.15)). Properties (B.55*b*, *c*) follow from

$$\tilde{\varepsilon}^{\rho\lambda}\gamma_{a\lambda}{}^{\sigma} = -\varepsilon^{\lambda\sigma}\gamma_{a\lambda}{}^{\rho}, \quad \varepsilon^{\rho\lambda}\gamma_{a\lambda}{}^{\sigma} = -\tilde{\varepsilon}^{\lambda\sigma}\gamma_{a\lambda}{}^{\rho}$$

(*cf.* (B.32), (B.41*b*), (B.48), applied repeatedly to (B.21)). The index-raising convention for the unprimed and primed capital indices follows the same ordering as in (B.56) (even though when $n \equiv 2 \pmod 4$ primed indices are converted to unprimed ones and unprimed to primed).

Translation of spinors to tensors; pure spinors

We remark that this translation procedure for converting tensorial quantities into spinor terms is also, in a sense, reversible: it allows a translation of any spinor ξ^{ρ} (n odd) or reduced spinor ξ^{R}, or $\xi^{R'}$ (n even) into tensor terms (up to sign). For the various quantities

$$\xi^{\rho}\xi^{\sigma}\gamma^{a\ldots c}{}_{\rho\sigma} \quad (n \text{ odd}) \tag{B.57a}$$

or

$$\xi^{R}\xi^{S}\gamma^{a\ldots c}{}_{RS} \quad (n \text{ even}) \tag{B.57b}$$

or

$$\xi^{R'}\xi^{S'}\gamma^{a\ldots c}{}_{R'S'} \quad (n \text{ even}) \tag{B.57c}$$

for differing numbers of indices $a, \ldots, c$, will, in each case, together serve to define the spinor uniquely up to sign. Note that the relevant tensors (B.57) are those for which $\rho\sigma$, RS or $R'S'$ are symmetric, i.e. by (B.55), for which the number r of indices $a, \ldots, c$ satisfies

$$n - 2r \equiv 0, 1, 7 \pmod 8. \tag{B.58}$$

Of particular interest is the case when

$$r = \tfrac{1}{2}n \pm \tfrac{1}{2} \quad (n \text{ odd}) \tag{B.59a}$$

or

$$r = \tfrac{1}{2}n \quad (n \text{ even}). \tag{B.59b}$$

If the expressions (B.57) all *vanish* for each value of r apart from the one (or, two) given by (B.59), then $\boldsymbol{\xi}$ is called a *pure spinor*. (We use the term 'pure spinor' to imply 'reduced', when n is even.) We note that, by (B.58), the spinor $\boldsymbol{\xi}$ is necessarily pure if $n < 7$.

A special significance of the pure spinors lies in the fact that the (non-zero) skew tensors (B.57) are then necessarily *simple*, which, by proposition (3.5.30), is the condition

$$\xi^{R}\xi^{S}\gamma^{[a_1\ldots a_r}{}_{RS}\gamma^{a_{r+1}]a_{r+2}\ldots a_{2r}}{}_{TK}\xi^{T}\xi^{K} = 0 \tag{B.60}$$

or a similar version for $\xi^{R'}$, when $n\ (=2r)$ is even, and a similar (pair of) expressions when $n\ (=2r \pm 1)$ is odd. We omit the proof of (B.60) here but note that it has the interesting geometric consequence that any pure spinor can be represented, up to proportionality, by a $\tfrac{1}{2}n$-plane through the origin in the vector space $\mathbb{V}^a$ (n even), or by a pair consisting of a $(\frac{1}{2}n-\frac{1}{2})$-plane and a $(\frac{1}{2}n+\frac{1}{2})$-plane through the origin in $\mathbb{V}^a$ (n odd). In fact in the *odd* case the two are orthogonal complements of one another (because of (B.27), (B.13)) and so only the $(\frac{1}{2}n-\frac{1}{2})$-plane need be considered. For similar reasons, in the *even* case the $\frac{1}{2}n$-plane is the orthogonal complement of *itself*, i.e. it is *(anti-) self-dual.* Indeed, this (anti-) self-dual property does not depend on the simplicity (B.60), but holds merely by virtue of the fact that it is only the quantity $\gamma^{a\ldots c}{}_{RS}$ with entirely *unprimed* spinor indices which is involved and not the primed quantity $\gamma^{a\ldots c}{}_{R'S'}$. More generally, *any* element of $\mathbb{V}^{[a\ldots c]}(r=\frac{1}{2}n)$ has the form

$$\alpha^{RS}\gamma^{a\ldots c}{}_{RS}+\tilde{\alpha}^{R'S'}\gamma^{a\ldots c}{}_{R'S'}, \tag{B.61}$$

where $\alpha^{RS}\in\mathbb{S}^{(RS)}$, $\tilde{\alpha}^{R'S'}\in\mathbb{S}^{(R'S')}$. It is self-dual or anti-self-dual if and only if one of the two terms in (B.61) is zero.

This (anti-) self-duality property, together with (B.60) entails, when n is *even*, that

$$g_{ad}(\xi^R\xi^S\gamma^{a\ldots c}{}_{RS})(\xi^T\xi^K\gamma^{d\ldots f}{}_{TK})=0, \tag{B.62}$$

$a\ldots c$ and $d\ldots f$ each being $\frac{1}{2}n$ in number; and the primed version of (B.62) also holds. When n is *odd*, we correspondingly have

$$g_{ad}(\xi^\rho\xi^\sigma\gamma^{a\ldots c}{}_{\rho\sigma})(\xi^\tau\xi^\kappa\gamma^{d\ldots f}{}_{\tau\kappa})=0, \tag{B.63}$$

when $a\ldots c$ and $d\ldots f$ are each $\frac{1}{2}n-\frac{1}{2}$ in number, and *also* when one of these sets of indices is $\frac{1}{2}n-\frac{1}{2}$ in number and the other, $\frac{1}{2}n+\frac{1}{2}$ in number – this last condition being equivalent to the 'simplicity' condition (*cf.* (B.60), with Greek in place of capital Latin indices) because the two bracketed terms of (B.63) are then duals of one another. (The proofs of these facts are similar to that of (B.60) and are omitted here.)

Geometry of pure spinors

These properties have a very direct geometrical interpretation. This is best described in terms of the $(n-1)$-dimensional (complex) *projective space* $\mathbb{PV}$ associated with $\mathbb{V}^a$ (i.e. the space of one-dimensional linear subspaces of $\mathbb{V}^a$). The (non-zero) *null* vectors $v^a\in\mathbb{V}^a$ (i.e. $g_{ab}v^av^b=0$), defining the null cone in $\mathbb{V}^a$, provide the points of a non-singular quadric $(n-2)$-surface $\mathcal{Q}$ in $\mathbb{PV}$. The properties described above tell us that any pure spinor determines a *projective* $\frac{1}{2}(n-3)$*-plane on* $\mathcal{Q}$ if n is *odd* and a *projective* $\frac{1}{2}(n-2)$*-plane on* $\mathcal{Q}$ if n is *even*. According to the theory of (complex projective) quadric surfaces (*cf.* for example, Hodge and Pedoe (1952), Porteous (1981), the maximum dimension of a linear projective space lying on a non-singular $(n-1)$-

quadric is indeed $\frac{1}{2}(n-3)$ if n is *odd*, and such $\frac{1}{2}(n-3)$-planes form a $\frac{1}{8}(n^2-1)$-dimensional family; and when n is *even*, this maximum dimension is indeed $\frac{1}{2}(n-2)$, these $\frac{1}{2}(n-2)$-planes forming *two* disjoint $\frac{1}{8}n(n-2)$-dimensional families.* These two families, in the even case, correspond to the unprimed and primed pure spinors, respectively, and a $\frac{1}{2}(n-2)$-plane of the first family is frequently called an α-plane and one of the second family a β-plane. (This is consistent with the terminology of §9.3, when $n=6$ and $\mathcal{Q}=\mathbb{CM}^{\#}$.) Let us refer to the $\frac{1}{2}(n-3)$-planes on $\mathcal{Q}$ as γ-planes. These planes are determined by the projective pure spinors (i.e. non-zero pure spinors up to proportionality) in each case, so we have:

For even n, the projective pure spinors, unprimed and primed, are in natural 1–1 *correspondence with the α-planes and β-planes on $\mathcal{Q}$, respectively.* (B.64*a*)

For odd n, the projective pure spinors are in natural 1–1 *correspondence with the γ-planes on $\mathcal{Q}$.* (B.64*b*)

The purity condition; structure of spin-space

We see from this that the dimension d_n of the space of pure spinors is $\frac{1}{8}n(n-2)+1$ in the even case and $\frac{1}{8}(n^2-1)+1$ in the odd case, which values may be compared with the dimensions $\frac{1}{2}N=2^{\frac{1}{2}n-1}$ and $N=2^{\frac{1}{2}n-\frac{1}{2}}$, respectively, of (reduced) spinors *not* necessarily satisfying the purity condition. The first few values of these dimensions are listed in table (B.65):

n	1, 2	3, 4	5, 6	7, 8	9, 10	11, 12	13, 14	...
d_n	1	2	4	7	11	16	22	...
$N, \frac{1}{2}N$	1	2	4	8	16	32	64	...

(B.65)

* In the even case, with $n=2r$, the equation of $\mathcal{Q}$ can always be put into the form $x^1y_1 + \cdots + x^ry_r = 0$. Then the generic $(r-1)$-plane of *one* family can be expressed: $y_i = S_{ij}x^j$ where S_{ij} is *skew* $r \times r$, giving $\frac{1}{8}n(n-2)$ independent components. If an *even* number of the x^i are interchanged with their corresponding y_i then this form can be re-obtained, in general. But if an *odd* number are interchanged, an $(r-1)$-plane of the *opposite* family is obtained. In the case of *odd n*, with $n=2r+1$, the equation of $\mathcal{Q}$ can be put in the form $x^iy_i = z^2$. The generic $(r-1)$-plane can then be expressed $z = T_ix^i$, $y_i = S_{ij}x^j + T_iz$, giving $\frac{1}{8}(n^2-1)$ independent parameters.

From this we see that whereas purity represents just a single condition where this first appears at $n=7$, 8, the number of conditions increases rapidly with n thereafter. For $n=7$ the purity condition is

$$\varepsilon_{\rho\sigma}\xi^{\rho}\xi^{\sigma}=0 \tag{B.66}$$

and, for $n=8$ we have

$$\varepsilon_{RS}\xi^{R}\xi^{S}=0 \quad \text{or} \quad \tilde{\varepsilon}_{R'S'}\zeta^{R'}\zeta^{S'}=0. \tag{B.67}$$

Using the notation

$$^{(r)}G^{\kappa\tau}_{\rho\sigma}=\frac{1}{r!}\gamma_{a\ldots c\rho}{}^{\kappa}\gamma^{a\ldots c}{}_{\sigma}{}^{\tau} \tag{B.68}$$

(where $a,\ldots,c$ are r in number) for the $(r+1)$st term in (B.29), we find that for $n=9$, 11, 13, the purity condition is the pair:

$$^{(r)}G^{\kappa\tau}_{\rho\sigma}\xi^{\rho}\xi^{\sigma}=0 \quad \begin{cases} r=\frac{1}{2}(n-9) \\ r=\frac{1}{2}(n-7) \end{cases} \tag{B.69}$$

whereas for $n=15$ we also need

$$^{(0)}G^{\kappa\tau}_{\rho\sigma}\xi^{\rho}\xi^{\sigma}=0. \tag{B.70}$$

Generally, for *odd n*, the needed conditions are those like (B.69), (B.70) for which $^{(r)}G^{\kappa\tau}_{\rho\sigma}$ is symmetric in $\rho\sigma$ (i.e. $n-2r\equiv\pm 1)\ (\mathrm{mod}\,8))$ and for which $0\leqslant 2r<n-1$.

For *even n* we use

$$^{(r)}G^{KT}_{RS},\ ^{(r)}G^{K'T'}_{R'S'}(r\ \text{even})\ \text{and}\ ^{(r)}G^{K'T'}_{RS},\ ^{(r)}G^{KT}_{R'S'}(r\ \text{odd}) \tag{B.71}$$

the purity conditions being a set of relations of the form

$$^{(r)}G^{KT}_{RS}\xi^{R}\xi^{S}=0 \quad \text{or} \quad ^{(r)}G^{K'T'}_{RS}\xi^{R}\xi^{S}=0 \tag{B.72a}$$

in the unprimed case, and

$$^{(r)}G^{K'T'}_{R'S'}\zeta^{R'}\zeta^{S'}=0 \quad \text{or} \quad ^{(r)}G^{KT}_{R'S'}\zeta^{R'}\zeta^{S'}=0 \tag{B.72b}$$

in the primed case. The needed r-values are those for which

$$n-2r\equiv 0\ (\mathrm{mod}\,8), \quad 0\leqslant 2r<n. \tag{B.73}$$

It is clear from all this that for large n, the structure of the various spin-spaces can get very complicated. For this kind of reason, and not simply because of the exponentially large spin-space dimension, it would be hard to view spinor algebra as providing a practical *alternative* to tensor algebra, as it clearly is when $n=4$ (or other small values). Nevertheless, spinors are important in all dimensions (e.g. for the Atiyah–Singer index theorem, *cf.* Shanahan (1978)) and can yield very significant insights. In principle (rather than practice), one can carry out all tensor procedures entirely spinorially. But the structure possessed by the (reduced) spin-spaces is, for large n, much more complicated than that of the original vector space. In essence, this structure is defined by the object

$$G_{\rho\sigma}^{\kappa\tau} := g^{ab}\gamma_{a\rho}{}^{\kappa}\gamma_{b\sigma}{}^{\tau} = {}^{(1)}G_{\rho\sigma}^{\kappa\tau} \quad (n \text{ odd}) \tag{B.74}$$

or by the corresponding pair of objects

$$G_{RS}^{K'T'},\ G_{R'S'}^{KT} \quad (n \text{ even}) \tag{B.75}$$

since (because of relations like (B.22)) all quantities (B.68) (or their reduced parts) can be expressed in terms of products like

$$G_{\rho\sigma}^{\lambda_1\mu_1}G_{\lambda_1\mu_1}^{\lambda_2\mu_2}\ldots G_{\lambda_k\mu_k}^{\kappa\tau}.$$

The ε-quantities are also needed, since their individual scalings are not determined by the Gs, although the εs are already fixed up to proportionality by them. There are also various identical relations satisfied by the Gs implicit in

$${}^{(r)}G_{\rho\sigma}^{\kappa\tau} = 0 \quad (r > n)$$

and

$${}^{(r)}G_{\rho\sigma}^{\kappa\tau} = \pm\, {}^{(n-r)}G_{\rho\sigma}^{\kappa\tau} \quad (n \text{ odd}).$$

This structure for $\mathbb{S}^\rho$, or for $\mathbb{S}^R$ and $\mathbb{S}^{R'}$, is clearly far more complicated (when $r > 8$, say) than the original structure for $\mathbb{V}^a$ that g_{ab} (and choice of $\varepsilon^{a_1\ldots a_n}$) define.

Inductive construction of spin-space

One good way of building up the spin-spaces for larger n is *inductively*, assuming that the structure for $n-1$ is already understood. This is a standard procedure for building up explicit representations for the γ-matrices. If $n-1$ is *even*, then (with $\boldsymbol{\eta}$ for $n-1$ dimensions)

$$\boldsymbol{\gamma}_1, \ldots, \boldsymbol{\gamma}_{n-1}, \boldsymbol{\eta}$$

all anti-commute, as we have seen, so we directly obtain a representation for n-dimensions from that of $(n-1)$-dimensions by taking $\boldsymbol{\gamma}_n$ to be a multiple of $\boldsymbol{\eta}$ (*cf.* (B.8) with $n-1$ for n, (B.12)). If $n-1$ is *odd*, then the *reduced* γs for n dimensions can be represented as faithful copies of the γs for $(n-1)$ dimensions, and the complete algebra for n dimensions is obtained from the block matrix description (B.14). (Note that each $\gamma_{\mathbf{a}\rho}{}^{\sigma}$, for $n-1$, gets represented as $\gamma_{\mathbf{a}R}{}^{S'}$ *and* as $\gamma_{\mathbf{a}R'}{}^{S}$, for n. Products are now only allowed which *alternate* between these two types, the sums of such products having to be all of the same type.)

We thus have a direct way of passing from spinors for $n-1$ dimensions to those for n dimensions. In the even-to-odd case, the new $\mathbb{S}^{\rho}$ is represented as the old $\mathbb{S}^{R} \oplus \mathbb{S}^{R'}$ and in the odd-to-even case the new $\mathbb{S}^{R}$ and $\mathbb{S}^{R'}$ are each represented as copies of the old $\mathbb{S}^{\rho}$. However, from the abstract point of view, there is an arbitrariness in this procedure since it depends on the choice of the basis element $g_a{}^n$ for $\mathbb{V}_a$ (see (B.7); n is not an abstract index!). Writing

$$u_a = g_a{}^n \in \mathbb{V}_a \tag{B.76}$$

we see that the only restriction on u_a is that it be a unit vector:

$$u_a u^a = \pm 1 \tag{B.77}$$

since the remaining basis element $g_a{}^1, \ldots, g_a{}^{n-1}$ can be reconstructed if desired, once u^a has been chosen. (A modified procedure can also be adopted if u_a is chosen to be null – and, in effect, this is what *is* adopted in the standard description of a twistor as a pair of 2-spinors, *cf.* §6.1. But we do not pursue the general discussion of this here.)

The quantity u_a is a covector for $\mathbb{V}^a$ and thus defines an $(n-1)$-dimensional subspace (hyperplane) $\mathbb{U}^a \subset \mathbb{V}^a$ with normal u^a. We thus obtain the spinors for $\mathbb{V}^a$ by building them up from those for $\mathbb{U}^a$. To pass back down to the spinors for $\mathbb{U}^a$ from those for $\mathbb{V}^a$, we introduce the quantity*

$$u_{\rho}{}^{\sigma} = 2^{-\frac{1}{2}} u^a \gamma_{a\rho}{}^{\sigma} \quad (n \text{ odd}) \tag{B.78a}$$

or, when n is *even*

$$u_R{}^{S'} = 2^{-\frac{1}{2}} u^a \gamma_{aR}{}^{S'}, \quad u_{R'}{}^{S} = 2^{-\frac{1}{2}} u^a \gamma_{aR'}{}^{S} \quad (n \text{ even}) \tag{B.78b}$$

When n is *odd* we can use (for the two signs in (B.77))

$$\tfrac{1}{2}(\delta_{\rho}^{\sigma} \pm 2\mathrm{i}u_{\rho}{}^{\sigma}), \quad \tfrac{1}{2}(\delta_{\rho}^{\sigma} \pm 2u_{\rho}{}^{\sigma}) \tag{B.79}$$

for the projection operators $\Pi_{\rho}{}^{\sigma}$, $\tilde{\Pi}_{\rho}{}^{\sigma}$ of (B.18) that are needed for the breakdown of the spinors for $\mathbb{V}^a$ to the two sets of reduced spinors for $\mathbb{U}^a$. (Note that $u_{\rho}{}^{\tau} u_{\tau}{}^{\sigma} = -\frac{1}{2}\delta_{\rho}{}^{\sigma} u_a u^a$.) When n is *even*, the quantities (B.78*b*) serve to translate between primed and unprimed indices. Thus the spinors for $\mathbb{U}^a$ constitute the thereby identified *reduced* spin-spaces for $\mathbb{V}^a$.

It is instructive to examine the geometry in $\mathbb{PV}$ that is involved in this procedure (Fig. B-1). The hyperplane $\mathbb{U}^a$ gives us a projective $(n-2)$-plane $\mathbb{PU}$ in $\mathbb{PV}$ which intersects $\mathcal{Q}$ in some $(n-3)$-quadric $\mathcal{Q}'$. Since we assume here that $u_a u^a \neq 0$, $\mathbb{PU}$ does not touch $\mathcal{Q}$, so $\mathcal{Q}'$ is consequently a non-singular quadric. Consider first the case when n is *even* (Fig. B-1(*a*)). We

* The factor $2^{-\frac{1}{2}}$, which looks awkward here, is introduced only to obtain agreement with the standard 2-spinor translation conventions.

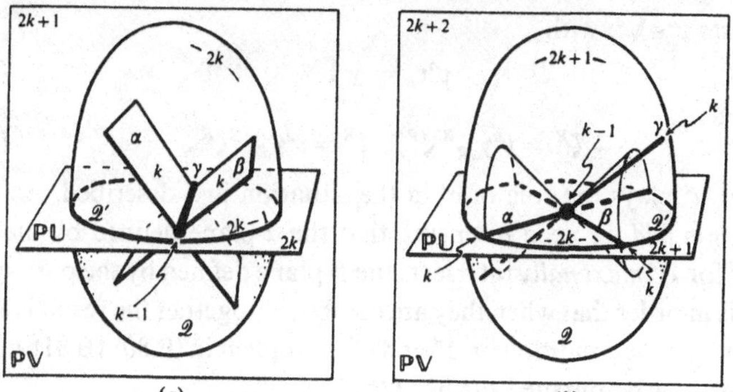

(*a*)
even $n = 2k+2$

(*b*)
odd $n = 2k+3$

Fig. B-1. The pure spinors for n dimensions can be built up inductively from those for $n-1$ dimensions by examining how the α, β and γ spaces for an $(n-2)$-quadric $\mathcal{Q}$ are related to those for a general hyperplane section – an $(n-3)$-quadric $\mathcal{Q}'$.

write $n = 2k+2$, so $\mathcal{Q}$ is a $2k$-quadric and $\mathcal{Q}'$ a $(2k-1)$-quadric. The γ-planes of $\mathcal{Q}'$ are $(k-1)$-dimensional and form a $\frac{1}{2}k(k+1)$-dimensional system. The α- and β-planes of $\mathcal{Q}$ are in (1–1) relation to their γ-plane intersections with $\mathcal{Q}'$. Consider next the case *n odd* (Fig. B-1(*b*)). We now write $n = 2k+3$, so $\mathcal{Q}$ is a $(2k+1)$-quadric and $\mathcal{Q}'$ a $2k$-quadric. The γ-planes of the *larger* quadric are k-dimensional, forming a $\frac{1}{2}(k^2+3k+2)$-dimensional system. The general member of this system meets $\mathcal{Q}'$ in a $(k-1)$-plane, through which passes a unique α-plane of $\mathcal{Q}'$ and a unique β-plane of $\mathcal{Q}'$. (In particular cases the γ-plane of $\mathcal{Q}$ actually lies in $\mathcal{Q}'$ and *becomes* either an α-plane or a β-plane of $\mathcal{Q}'$.) The α-planes and β-planes of $\mathcal{Q}'$ form $\frac{1}{2}k(k+1)$-dimensional systems.

The relationship between this geometry and the foregoing spinor discussion is that this concerns the *pure* spinors for $\mathbb{V}^a$ as they relate to the pure spinors for $\mathbb{U}^a$. The general (reduced) spinors constitute the *linear span* of the pure spinors. We note that when *n* is *even*, any pure spinor ξ^P for $\mathbb{U}^a$ directly determines a pair of pure spinors ξ^R and $\xi^{R'}$ for $\mathbb{V}^a$, the γ-plane for $\mathcal{Q}'$ directly providing an α-plane and a β-plane for $\mathcal{Q}$. However, the α-plane and the β-plane are specially related to one another, not just in that their intersection lies entirely on $\mathcal{Q}'$, but also because their intersection has the maximal dimension* $k-1$. Two pure spinors ξ^A and $\zeta^{R'}$ for $\mathbb{V}^a$ correspond to an α-plane and a β-plane on $\mathcal{Q}$ which intersect maximally if and only if

* For a non-singular $2k$-quadric, the dimension of intersection of an α-plane with a β-plane can take the values $k-1, k-3, k-5, \ldots, -1$ (k even) or 0 (k odd) while the dimension of intersection of two α-planes or two β-planes can be $k, k-2, k-4, \ldots, 0$ (k even) or -1 (k odd). For a non-singular $(2k+1)$-quadric, two γ-planes can intersect with dimension $k, k-1, \ldots, 0, -1$. (Dimension -1 means vacuous.) In each case, the generic situation gives the smallest dimension of intersection.

there *exists* $v^a \in \mathbb{V}^a$, with

$$v^a v_a = \pm 1 \tag{B.80}$$

and

$$\mp \zeta^{R'} = v^a \gamma_{aR}{}^{R'} \xi^R, \quad \xi^R = v^a \gamma_{aR'}{}^R \zeta^{R'} \tag{B.81}$$

Note that u^a plays the role of v^a in the situation just described.

When n is *odd* we need to ensure that the α-plane defined by the pure spinor ξ^R for $\mathbb{U}^a$ *maximally* intersects the β-plane defined by the pure spinor $\zeta^{R'}$. This is in order that when they are combined together the result is a *pure* spinor for $\mathbb{V}^a$. The condition on ξ^R and $\zeta^{R'}$ is implicit in (B.80), (B.81), but the matter will not be pursued further here.

We can also reduce the spin-space for $\mathbb{V}^a$ by considering a subspace $\mathbb{U}^a$ of some arbitrary dimension r, and its $(n-r)$-dimensional orthogonal complement $\mathbb{W}^a$ (where $\mathbb{U}^a$ is non-null so that its induced metric, and consequently also that of $\mathbb{W}^a$, is non-degenerate). Thus we have a *direct sum*

$$\mathbb{V}^a = \mathbb{U}^a \oplus \mathbb{W}^a \tag{B.82}$$

(though, strictly, we should perhaps introduce different labelling a', a'', say, where $a = a' \oplus a''$). The spinors for $\mathbb{V}^a$ then turn out to be, in an appropriate sense, the *direct product* (tensor product) of those for $\mathbb{U}^a$ and for $\mathbb{W}^a$. For this to work systematically it is more satisfactory to consider that in the case of *odd* dimension the spin-space consists of *two 'copies'* of the spin-spaces for odd dimension that we have been considering here (*cf.* second footnote on p. 440), where we now allow *each* of the two signs in (B.13) to occur, one for each 'copy'*. Then if both r and $n-r$ are odd, so that n is even, the $2^{\frac{1}{2}n}$-dimensional total spin-space for $\mathbb{V}^a$ arises as a direct sum of its two $2^{\frac{1}{2}n-1}$-dimensional reduced spin-spaces each of which is regarded as a direct product of one of the $2^{\frac{1}{2}r-\frac{1}{2}}$-dimensional 'copies' of the spin-space for $\mathbb{U}^a$ and one of the two 'copies' of the $2^{\frac{1}{2}(n-r)-\frac{1}{2}}$-dimensional spin-space for $\mathbb{W}^a$. If one of r and $n-r$ is even and the other is odd, the products work out naturally whether or not 'copies' are used. If both r and $n-r$ are even the question of such 'copies' does not arise. However, we note that each *reduced* spin-space for $\mathbb{V}^a$ arises as direct sum of the products of reduced spaces for $\mathbb{U}^a$ and $\mathbb{W}^a$ in the general form:

$$\begin{aligned} \text{unprimed space} &= (\text{unprimed} \otimes \text{unprimed}) \oplus (\text{primed} \otimes \text{primed}) \\ \text{primed space} &= (\text{unprimed} \otimes \text{primed}) \oplus (\text{primed} \otimes \text{unprimed}). \end{aligned} \tag{B.83}$$

* Strictly speaking these two spin-spaces are not quite the same as each other, and become interchanged under improper transformations.

Reality; complex conjugation

The foregoing discussion has been carried out assuming that $\mathbb{V}^a$ is a *complex* space, but for the most part this has played no critical role (i.e. except for the discussion of linear subspaces on $\mathscr{Q}$), since we have allowed for an arbitrary choice of 'signature' for $g_{\mathbf{ab}}$. Many applications of spinor ideas do in fact require $\mathbb{V}^a$ to be real. For our present purposes we shall still think of $\mathbb{V}^a$ as a complex space but allow it to possess a 'reality structure' defined by an involutory complex-conjugation operator $\mathscr{C}$. The required real space is then the (real) subspace of $\mathbb{V}^a$ whose elements are invariant under $\mathscr{C}$. The 'signature' s of $g_{\mathbf{ab}}$ now acquires an invariant significance in relation to $\mathscr{C}$, where we now demand that the basis $g_{\mathbf{a}}{}^{a}$ consist of *real* elements of $\mathbb{V}^a$.

When n is *even*, it turns out that $\mathscr{C}$ can be extended to apply to spinors so that:

$$\textit{each of } \mathbb{S}^A, \mathbb{S}^{A'} \textit{ is invariant under } \mathscr{C} \textit{ if } \tfrac{1}{2}s \textit{ is even;} \qquad \text{(B.84}a\text{)}$$

$$\mathbb{S}^A \textit{ and } \mathbb{S}^{A'} \textit{ are interchanged by } \mathscr{C} \textit{ if } \tfrac{1}{2}s \textit{ is odd.} \qquad \text{(B.84}b\text{)}$$

This is clearly consistent with the result for ordinary Lorentzian 2-spinors, and also for twistors, when we bear in mind that our twistor conventions have been to eliminate all primed indices in favour of unprimed ones in the reverse position, which is legitimate (*cf.* (B.52*b*) *et seq.*) since here $n = 6$. It is also consistent with the fact that spin-weight $\frac{1}{2}$ functions on a spacelike 2-surface in space–time become spin-weight $-\frac{1}{2}$ functions under $\mathscr{C}$ ($s = -2$), whereas for a timelike 2-surface the boost-weight $\frac{1}{2}$ and boost-weight $-\frac{1}{2}$ functions are each invariant under $\mathscr{C}$ ($s = 0$).

In more detail, the value (mod 8) of the signature s has relevance to the question of the existence of a non-trivial subspace, in $\mathbb{S}^\rho$, of *real spinors* (or 'Majorana' spinors, in the context of physics). We find that for some signatures the extension of $\mathscr{C}$ to spinors cannot be involutory, but merely satisfies* $\mathscr{C}^4 = 1$. There are some matters of convention that further complicate things. Moreover, in the case of odd n, we see from (B.13) *et seq.* that if $s \equiv 3 \pmod 4$, the choice of factor relating $\boldsymbol{\eta}$ to the identity is *reversed* under $\mathscr{C}$. This has the implication that $\mathbb{S}^\rho$ itself is not, strictly speaking, invariant under $\mathscr{C}$, but is interchanged with its other 'copy' in these cases. We shall ignore these last two complications, however, and we can make the following general comments (for which we gratefully acknowledge the assistance of F. Reese Harvey).

When $s \equiv 3, 4, 5 \pmod 8$ we have $\mathscr{C}^2 = -1$ when acting on $\mathbb{S}^\rho$. These are

* This evades the problem mentioned on p. 107 of Volume 1 in relation to the action of $\mathscr{C}$ as it applies to spinors for $SO(3)$ (and $SO(4)$. Here $\mathscr{C}^2\xi^A = -\xi^A$, so one cannot sensibly define a 'real part' of ξ^A as $\frac{1}{2}(\xi^A + \mathscr{C}\xi^A)$ since this is not real.

the cases when $\mathbb{S}^\rho$ acquires a *quaternionic* structure and there are *no* real spinors ($\neq 0$). (The spinors for $SO(3)$ and $SO(4)$ are examples.) When $s \equiv 2$, 6 (mod 8) (*cf.* B.84*b*), each of $\mathbb{S}^R$, $\mathbb{S}^{R'}$ is naturally a *complex* space, with *no* real spinors ($\neq 0$), but the combined space $\mathbb{S}^\rho$ can be said to have real ('Majorana') spinors (using suitable conventions – though the usual ones require $s \equiv 6$ (mod 8)). When $s \equiv 0, 1, 7$ (mod 8) we have a *full system of real* spinors (i.e. real spinors whose complex span is the full relevant spin-space. When, in particular, $s = 0, \pm 1$, this last property is related to the maximum dimensionality of *real* projective subspaces on $\mathcal{Q}$, which is

$$\tfrac{1}{2}(n - |s|) - 1. \tag{B.85}$$

Note that $\mathcal{Q}$ has no real points in the positive-definite case, but that there are real α-planes and β-planes when $s = 0$ and real γ-planes when $s = \pm 1$.

Some cases of physical interest; triality

We have discussed Lorentzian 2-spinors, twistors and functions of spin-weight $\pm\frac{1}{2}$ on a spacelike 2-surface, as examples of our general procedure in the cases $n = 4, 6$ and 2, respectively. Little has been said directly of *Dirac* 4-*spinors*, which are the *unreduced* spinors for $n = 4$, $s = -2$ (or for $n = 4$, $s = 2$). These do, of course, fall under our general scheme and, indeed, have tended to be much more directly used in the literature than the reduced Lorentzian 2-spinors that have formed the central topic of these volumes. Our neglect of Dirac 4-spinor notation is to some extent a matter of practicality, the reduced spinors having, in this case, an exceedingly simple form that is much the easier and more powerful to work with. The elimination of all tensor indices is easily achieved in the 2-spinor formalism, and the γ-matrices, together with their various somewhat complicated identities (trace formulae, etc.) simply evaporate. They are given explicitly (*cf.* Volume 1, p. 221) by

$$\gamma_{a\rho}{}^{\sigma} = \sqrt{2}\begin{pmatrix} 0 & \varepsilon_{AR}\varepsilon_{A'}{}^{S'} \\ \varepsilon_{A'R'}\varepsilon_{A}{}^{S} & 0 \end{pmatrix}, \quad \eta_{\rho}{}^{\sigma} = \begin{pmatrix} -\mathrm{i}\varepsilon_{R}{}^{S} & 0 \\ 0 & \mathrm{i}\varepsilon_{R'}{}^{S'} \end{pmatrix} \tag{B.86}$$

where $\boldsymbol{\eta}$ is usually written $\boldsymbol{\gamma}_5$; and moreover we have

$$\gamma_{ab\rho}{}^{\sigma} = 2\begin{pmatrix} \varepsilon_{A'B'}\varepsilon_{R(A}\varepsilon_{B)}{}^{S} & 0 \\ 0 & \varepsilon_{AB}\varepsilon_{R'(A'}\varepsilon_{B')}{}^{S'} \end{pmatrix}. \tag{B.87}$$

As a general rule, calculations tend to be considerably easier in the 2-spinor formalism than with the use of γ-matrices, especially when there is a proliferation of expressions involving $\mathbf{I} \pm \mathrm{i}\boldsymbol{\gamma}_5$. (For a striking example of such simplifications in the case of quantum chromodynamics, see Farrar and Neri (1983).)

Spinors have found value in relativity theory also when $n = 3$, since these apply directly to a spacelike hypersurface ($s = -3$) or to a timelike hypersurface ($s = -1$). Taking the normal to this hypersurface to be $u^a = u^{AA'}$ (normalized so that $u_a u^a = \pm 2$), we can apply our earlier discussion (*cf.* (B.80), (B.81)) and use $u_A{}^{A'}$ and $u_{A'}{}^{A}$ to convert all primed indices to unprimed form. This provides a useful calculus, and one that has been exploited by various authors (*cf.* Sommers (1980), Shaw (1983*b*)). (The procedure is relevant also to our discussion of the Witten argument for mass-positivity; see §9.10.)

A particularly interesting situation which seems to have some relevance in physics is given when $n = 8$ and where either $s = 8$ or $s = 0$), since it then turns out that the three spaces $\mathbb{V}^a$, $\mathbb{S}^A$ and $\mathbb{S}^{A'}$ are, remarkably, all on an *equal footing*. The quantities ε_{AB} and $\varepsilon_{A'B'}$ are both symmetric (*cf.* (B.52*a*)) and are on an equal footing with g_{ab}. Thus, we may regard $\mathbb{S}^{A'}$ and $\mathbb{V}^a$ as the 'unprimed' and 'primed' reduced spin-spaces for $\mathbb{S}^A$, with metric ε_{AB}; or $\mathbb{V}^a$ and $\mathbb{S}^A$ as the 'unprimed' and 'primed' spin-spaces for $\mathbb{S}^{A'}$. This curious additional symmetry is referred to as the *triality* principle for $SO(8, \mathbb{C})$, $SO(8)$ or $SO(4, 4)$. Note that, by (B.67), the *pure* spinors of $\mathbb{S}^A$, $\mathbb{S}^{A'}$ correspond, under this symmetry, to the *null* vectors of $\mathbb{V}^a$. In terms of $\mathscr{Q}$, this means that the family of α-planes on $\mathscr{Q}$, the family of β-planes on $\mathscr{Q}$ and the family of points on $\mathscr{Q}$ are all on an equal footing with one another. The *incidence* property that a point lie on an α-plane corresponds, under this symmetry, to the incidence property that an α-plane meet a β-plane maximally, i.e. in a 2-space. In the case of $SO(8)$, these α-planes, β-planes and points are all imaginary but the symmetry between them still exists and shows up as a symmetry between the *non*-null (non-pure) elements. (See Adams 1981, Chevalley 1954, for more details.)

The quantity

$$\gamma_{aAA'} \tag{B.88}$$

now plays a symmetrical role between the three types of index. It satisfies an identity

$$\gamma_{(a}{}^{AA'}\gamma_{b)A}{}^{B'} = g_{ab}\tilde{\varepsilon}^{A'B'} \tag{B.89}$$

and also others obtained by permuting the index types. There is a relation to the algebra of octonians (Cayley numbers). This is obtained by choosing a pair of fixed unit elements of two of the three spaces, say

$$k^a \in \mathbb{V}^a, \quad m^{A'} \in \mathbb{S}^{A'}, \quad k^a k_a = 1 = m^{A'} m_{A'},$$

defining a third by

$$l^A = \gamma_a{}^A{}_{A'}\, k^a m^{A'}$$

and then using

$$\gamma_{aAA'}k^a,\ \gamma_{aAA'}l^A,\ \gamma_{aAA'}m^{A'}$$

and the metric quantities g^{ab}, ε^{AB}, $\tilde{\varepsilon}^{A'B'}$, to translate between all the different types of index. Applying this translation to (B.88) we obtain an object γ_{abc} for which $\gamma_{ab}{}^c$ provides the required Cayley multiplication on $\mathbb{V}^a$. The identity (B.89) then provides all the algebraic properties that are needed (i.e. $\gamma_{(ab)}{}^d\gamma_{dc}{}^e = \gamma_{(a|d|}{}^e\gamma_{b)c}{}^d$ which gives the 'alternative' law $(AA)B = A(AB)$).

It is perhaps noteworthy, in relation to all this, that there is some relevance of $SO(8,\mathbb{C})$ and $SO(4,4)$ to twistor theory. We consider pairs each consisting of a $[^1_0]$ and a $[^0_1]$ twistor:

$$(\mathrm{Z}^\alpha, \mathrm{W}_\beta). \tag{B.90}$$

The metric on the space of these objects is to be defined by

$$(\mathrm{Z}^\alpha, \mathrm{W}_\beta) \mapsto 2\mathrm{Z}^\alpha\mathrm{W}_\alpha \tag{B.91}$$

and the complex conjugation,

$$\mathscr{C}:(\mathrm{Z}^\alpha, \mathrm{W}_\alpha) \mapsto (\bar{\mathrm{W}}^\alpha, \bar{\mathrm{Z}}_\alpha). \tag{B.92}$$

The signature of $\mathrm{Z}^\alpha\bar{\mathrm{Z}}_\alpha + \mathrm{W}_\alpha\bar{\mathrm{W}}^\alpha$ is $(+\,+\,+\,+\,-\,-\,-\,-)$, (i.e. $s = 0$, $n = 8$) and the above discussion applies. (Note that 'nullity' here is the *ambitwistor* relation $\mathrm{Z}^\alpha\mathrm{W}_\alpha = 0$, *cf.* footnote on p. 164.)

Spinor fields; twistors for *n* dimensions

Recall that at the outset we allowed for the possibility that $\mathbb{V}^a$ might be a (suitable) module, rather than a vector space, and might describe say vector fields on some (pseudo-) Riemannian n-manifold. The corresponding $\mathbb{S}^\rho$ (or $\mathbb{S}^R$ and $\mathbb{S}^{R'}$) would then describe spinor fields. The quantity $\gamma_{a\rho}{}^\sigma$ (or $\gamma_{aR}{}^{S'}$ and $\gamma_{aR'}{}^S$) then serves to translate tangent vector fields to spinor form. Thus, we have analogues of the Dirac equation or Dirac–Weyl equation

$$\gamma_{a\rho}{}^\sigma\nabla^a\psi^\rho = \hbar^{-1}m\psi^\sigma \quad \text{or} \quad \gamma_{aR}{}^{S'}\nabla^a\psi^R = 0, \tag{B.93}$$

for each n. There are also analogues of local twistors and local twistor transport (*cf.* Cartan 1923, and §6.9; and, more explicitly, unpublished work by G.A.J. Sparling).

The question of what one should regard as the twistor theory proper, for n dimensions, has various interpretations. One method of procedure is to think of the hyper-Minkowski space, with dimension $p + q$ and metric signature $p-q$ as being compactified, as in §9.2, and the group $SO(p+1, q+1)$ as acting on this compactification as its conformal group. Thus we have a situation, with dimension $p+q+2$ and signature $p-q$, to which we can apply the foregoing theory. The twistors for this hyper-Minkowski space may be taken to be the spinors for $SO(p+1, q+1)$.

Alternatively (but essentially equivalently) one may adopt an n-dimensional version of the twistor equation (6.1.1):

$$\gamma_{(a|R|}{}^{S'}\nabla_{b)}\omega^R = \frac{1}{n} g_{ab}\gamma^c{}_R{}^{S'}\nabla_c\omega^R \quad (n \text{ even}) \tag{B.94a}$$

or

$$\gamma_{(a|\rho|}{}^{\sigma}\nabla_{b)}\omega^\rho = \frac{1}{n} g_{ab}\gamma^c{}_\rho{}^{\sigma}\nabla_c\omega^\rho \quad (n \text{ odd}) \tag{B.94b}$$

(i.e. $\nabla_{(a}\boldsymbol{\varpi}\gamma_{b)} = n^{-1}g_{ab}\nabla_c\boldsymbol{\varpi}\gamma^c$ in the second case) whose general solutions, in hyper-Minkowski space, take the form

$$\omega^R = \Omega^R + x^a\gamma_{aS'}{}^R\Pi^{S'} \quad (n \text{ even}) \tag{B.95a}$$

or

$$\omega^\rho = \Omega^\rho + x^a\gamma_{a\sigma}{}^\rho\Pi^\sigma \quad (n \text{ odd}) \tag{B.95b}$$

where $\boldsymbol{\Omega}$ and $\boldsymbol{\Pi}$ are constant. The equation (B.94*b*) was introduced by Wess and Zumino (1974) in the context of supersymmetry theory. Here $n=4$, but (B.94*b*) is equivalent to two copies of (B.94*a*), one unprimed and one primed, these being taken as complex conjugates of one another by the imposition of a 'Majorana' (i.e. reality) condition on ω^ρ, namely $\omega^{R'} = \overline{\omega^R}$.

These ideas are sometimes useful for solving differential equations in an analogous way to that provided by the contour integral expressions of §6.10. An elegant example of this procedure has been suggested by L.P. Hughston. Let $f(\mathsf{Z}^\alpha, \mathsf{W}_\alpha)$ be a function of two twistors Z^α, W_α which is holomorphic in a suitable region and has overall homogeneity degree -4 (not necessarily homogeneous in Z^α and W_α separately). The $\mathsf{X}_{\alpha\beta}$ is to be *skew* and is taken to label the points in a six-dimensional space whose metric is given by

$$\varepsilon^{\alpha\beta\gamma\delta}\mathrm{d}\mathsf{X}_{\alpha\beta}\mathrm{d}\mathsf{X}_{\gamma\delta}. \tag{B.96}$$

Then the result of the contour integral

$$\phi(\mathsf{X}_{\alpha\beta}) = \oint f(\mathsf{Z}^\alpha, \mathsf{X}_{\alpha\beta}\mathsf{Z}^\beta)\varepsilon_{\alpha\beta\gamma\delta}\mathsf{Z}^\alpha \mathrm{d}\mathsf{Z}^\beta \wedge \mathrm{d}\mathsf{Z}^\gamma \wedge \mathrm{d}\mathsf{Z}^\delta \tag{B.97}$$

satisfies the (complex) Laplace or wave equation

$$\varepsilon_{\alpha\beta\gamma\delta}\frac{\partial^2\phi}{\partial\mathsf{X}_{\alpha\beta}\partial\mathsf{X}_{\gamma\delta}} = 0. \tag{B.98}$$

This is related to the aforementioned twistor relationship to $SO(8, \mathbb{C})$. (Other examples of analogous expressions have recently been provided by R.S. Ward and by M.F. Atiyah; see also Murray 1984 for what appears to be a very general such procedure. Other generalizations of twistor theory have also been suggested; see, for example, Bryant 1985; Eastwood 1985*b*.)

References

Abbott, L.F. and Deser, S. (1982). Stability of gravity with a cosmological constant, *Nucl. Phys.* **B195**, 76–96.

Adams, J.F. (1981). Spin (8), triality, F_4 and all that, in *Superspace and Supergravity*, ed. S.W. Hawking and M. Roček (Cambridge University Press, Cambridge).

Arnol'd, V.I. (1978). *Mathematical Methods of Classical Mechanics* (Springer, New York).

Arnowitt, R., Deser, S. and Misner, C.W. (1961). Coordinate invariance and energy expressions in general relativity, *Phys. Rev.* **122**, 997–1006.

Aronson, B., Lind, R., Messmer, J. and Newman, E.T. (1971). A note on asymptotically flat spaces, *J. Math. Phys.* **12**, 2462–7.

Ashtekar, A. (1980). Asymptotic structure of the gravitational field at spatial infinity, in *General Relativity and Gravitation, One Hundred Years after the Birth of Albert Einstein*, ed. A. Held (Plenum, New York).

Ashtekar, A. and Hansen, R.O. (1978). A unified treatment of null and spatial infinity in general relativity. I. Universal structure, asymptotic symmetries, and conserved quantities at spatial infinity, *J. Math. Phys.* **19**, 1542–66.

Ashtekar, A. and Magnon, A. (1984). Asymptotically anti-de Sitter spacetimes, *Class. Quant. Grav.* **1**, L39–44.

Ashtekar, A. and Magnon-Ashtekar, A. (1979). Energy–momentum in general relativity, *Phys. Rev. Lett.* **43**, 181–4.

Ashtekar, A. and Streubel, M. (1981). Symplectic geometry of radiative mode and conserved quantities at null infinity, *Proc. Roy. Soc. London* **A376**, 585–607.

Atiyah, M.F. (1979). *Geometry of Yang–Mills Fields* (Lezioni Fermiane, Scuola Normale Superiore, Pisa).

Atiyah, M.F., Hitchin, N.J., Drinfeld, V.G. and Manin, Yu. I. (1978). Construction of instantons, *Phys. Lett.* **65A**, 185–7.

Atiyah, M.F., Hitchin, N.J. and Singer, I.M. (1978). Self-duality in four-dimensional Riemannian geometry, *Proc. Roy. Soc. London* **A362**, 425–61.

Atiyah, M.F., Patodi, V.K. and Singer, I.M. (1975). Spectral asymmetry and Riemannian geometry I., *Math. Proc. Camb. Phil. Soc.* **77**, 43–69; II., *ibid.* **78**, 405–32; III., *ibid.* **79**, 71–99.

Atiyah, M.F. and Ward, R.S. (1977). Instantons and algebraic geometry, *Comm. Math. Phys.* **55**, 111–24.

Bach, R. (1921). Zur Weylschen Relativitätstheorie, *Math. Z.* **9**, 110–35.

Bailey, T.N. (1985). Twistors and fields with sources on worldlines, *Proc. Roy. Soc. London* **A397**, 143–55.

Bailey, T.N., Ehrenpreis, L. and Wells, R.O., Jr. (1982). Weak solutions of the massless field equations, *Proc. Roy. Soc. London* **A384**, 403–25.

Bardeen, J.M. and Press, W.H. (1973). Radiation fields in the Schwarzschild background, *J. Math. Phys.* **14**, 7–19.

Bateman, H. (1904). The solution of partial differential equations by means of definite integrals. *Proc. Lond. Math. Soc.* (2) **1**, 451–8.

Bateman, H. (1910). The transformation of the electrodynamical equations, *Proc. Lond. Math. Soc.* (2) **8**, 223–64.

Bateman, H. (1944). *Partial Differential Equations of Mathematical Physics* (Dover, New York).

Bel, L. (1959). Introduction d'un tenseur du quatrième ordre, *Comptes Rend.* **248**, 1297–300.

Bôcher, M. (1914). The infinite regions of various geometries, *Bull. Amer. Math. Soc.* **20**, 185–200.

Bondi, H. (1960*a*). *Cosmology* (Cambridge University Press, Cambridge).

Bondi, H. (1960*b*). Gravitational waves in general relativity, *Nature*, London **186**, 535.

Bondi, H., van der Burg, M.G.J. and Metzner, A.W.K. (1962). Gravitational waves in general relativity. VII. Waves from axi-symmetric isolated systems. *Proc. Roy. Soc. London* **A269**, 21–52.

Bonnor, W.B. and Rotenberg, M.A. (1966). Gravitational waves from isolated sources, *Proc. Roy. Soc. London* **A289**, 247–74.

Boyer, C.P., Finley, J.D., III and Plebański, J.F. (1980). Complex general relativity, $\mathfrak{H}$ and $\mathfrak{H}\mathfrak{H}$ spaces – a survey of one approach, in *General Relativity and Gravitation, One Hundred Years after the Birth of Albert Einstein*, ed. A. Held (Plenum, New York).

Bramson, B.D. (1975*a*). The alinement of frames of reference at null infinity for asymptotically flat Einstein–Maxwell manifolds, *Proc. Roy. Soc. London* **A341**, 451–61.

Bramson, B.D. (1975*b*). Relativistic angular momentum for asymptotically flat Einstein–Maxwell manifolds, *Proc. Roy. Soc. London* **A341**, 463–90.

Bramson, B.D. (1978). The invariance of spin, *Proc. Roy. Soc. London* **A364**, 383–92.

Brauer, R. and Weyl, H. (1935). Spinors in *n* dimensions, *Amer. J. Math.* **57**, 425–49.

Bryant, R.L. (1985). Lie groups and twistor spaces, *Duke Math. J.* **52**, 223–61.

Buchdahl, N.P. (1982). Applications of Several Complex Variables to Twistor Theory, Oxford University D. Phil. thesis.

Buchdahl, N.P. (1983). On the relative de Rham sequence, *Proc. Amer. Math. Soc.* **87**, 363–6.

Buchdahl, N.P. (1985). Analysis on analytic spaces and non-self-dual Yang–Mills fields, *Trans. Amer. Math. Soc.* **288**, 431–69.

Callan, C.G., Coleman, S. and Jackiw, R. (1970). A new improved energy–momentum tensor, *Ann. Phys.* **59**, 42–73.

Cantoni, V. (1967). Reduction of some representations of the generalized Bondi–Metzner group, *J. Math. Phys.* **8**, 1700–6.

Cartan, É. (1914). Groupes réels simples, finis et continus, *Ann. Sci. Ec. Norm. Sup.* **31**, 263–315.

Cartan, É. (1922*a*). Sur les équations de la gravitation d'Einstein, *J. Math. Pures et Appl.* **1**, 141–203.

Cartan, É. (1922*b*). Sur les espaces conformes généralisés et l'univers optique, *Comptes Rend.* **174**, 734–6.

Cartan, É. (1923). Les espaces à connexion conforme, *Ann. Soc. Po. Math.* **2**, 171–221.

Cartan, É. (1932). Sur la géométrie pseudo-conforme des hypersurfaces de l'espace de deux variables complexes, *Annali Mat.* **11**, 17–90.

Cartan, É. (1966). *The Theory of Spinors* (Hermann, Paris).

Carter, B. (1968a). Hamilton–Jacobi and Schrödinger separable solutions of Einstein's equations, *Comm. Math. Phys.* **10**, 280–310.

Carter, B. (1968*b*). Global structure of the Kerr family of gravitational fields, *Phys. Rev.* **174**, 1559–71.

Carter, B. (1970). The commutation property of a stationary, axisymmetric system, *Comm. Math. Phys.* **17**, 233–8.

Carter, B. and McLenaghan, R.G. (1979). Generalized total angular momentum operator for the Dirac equation in curved space–time, *Phys. Rev.* **D19**, 1093–7.

Cayley, A. (1860). On a new analytic representation of curves in space, *Quart. J. Pure and Appl. Math.* **3**, 225–36.

Cayley, A. (1869). On the six co-ordinates of a line, *Trans. Camb. Phil. Soc.* **11** (2), 290–323.

Chandrasekhar, S. (1979). An introduction to the theory of the Kerr metric and its perturbations, in *General Relativity: An Einstein Centenary Survey*, ed. S.W. Hawking and W. Israel (Cambridge University Press, Cambridge).

Chandrasekhar, S. (1983). *The Mathematical Theory of Black Holes* (Oxford University Press, Oxford).

Chern, S.S. (1979). *Complex Manifolds Without Potential Theory* (Springer-Verlag, New York).

Chevalley, C. (1954). *The Algebraic Theory of Spinors* (Columbia University Press, New York).

Churchill, R.V. (1932). Canonical forms for symmetric linear vector functions in pseudo-euclidean space, *Trans. Amer. Math. Soc.* **34**, 784–94.

Clifford, W.K. (1882). A preliminary sketch of biquaternions, *Mathematical Papers* (London), p. 181.

Connors, P.A. and Stark, R.F. (1977). Observable gravitational effects on polarised radiation coming from near a black hole, *Nature* **269**, 128–9.

Cox, D. and Flaherty, E.J., Jr. (1976). A conventional proof of Kerr's theorem, *Comm. Math. Phys.* **47**, 75–9.

Coxeter, H.S.M. (1936). The representation of conformal space on a quadric, *Ann. Math.* **37**, 416–26.

Crampin, M.J. and Pirani, F.A.E. (1971). Twistors, symplectic structures and Lagrange's identity, in *Relativity and Gravitation*, ed. Ch. G. Kuper and A. Peres (Gordon and Breach, London).

Curtis, G.E. (1978*a*). Twistors and linearized Einstein theory on plane-fronted impulsive wave backgrounds, *Gen. Rel. Grav.* **9**, 987–97.

Curtis, G.E. (1978*b*). Twistors and multipole moments, *Proc. Roy. Soc. London* **A359**, 133–49.

Curtis, W.D., Lerner, D.E. and Miller, F.R. (1978). Complex *pp* waves and the non-linear graviton construction, *J. Math. Phys.* **19**, 2024–7.

Curtis, W.D., Lerner, D.E. and Miller, F.R. (1979). Some remarks on the nonlinear graviton, *Gen. Rel. Grav.* **10**, 557–65.

Dietz, W. and Rudiger, R. (1980). Space-times admitting Killing–Yano tensors I, *Proc. Roy. Soc. London* **A375**, 361–78.

Dietz, W. and Rudiger, R. (1981). Space-times admitting Killing–Yano tensors II, *Proc. Roy. Soc. London* **A381**, 315–22.

Dighton, K. (1972). The Theory of Local Twistors and its Applications, Ph.D. thesis, Birkbeck College, University of London.

Dighton, K. (1974). An introduction to the theory of local twistors, *Int. J. Theor. Phys.* **11**, 31–43.
Dirac, P.A.M. (1928). The quantum theory of the electron, *Proc. Roy. Soc. London* **A117**, 610–24.
Dirac, P.A.M. (1931). Quantised singularities in the electromagnetic field, *Proc. Roy. Soc. London* **A133**, 60–72.
Dirac, P.A.M. (1936*a*). Relativistic wave equations, *Proc. Roy. Soc. London* **A155**, 447–59.
Dirac, P.A.M. (1936*b*). Wave equations in conformal space, *Ann. of Math.* **37**, 429–42.
Dodson, C.T.J. and Poston, T. (1977). *Tensor Geometry* (Pitman, London).
Doebner, H.D. and Palev, T.D. (eds.) (1982). *Twistor Geometry and Non-Linear Systems*, Proc. Primorsko 1980, Lecture Notes in Mathematics No. 970 (Springer-Verlag, Berlin).
Dray, T. (1985). The relationship between monopole harmonics and spin-weighted spherical harmonics, *J. Math. Phys.* **26**, 1030–3.
Dray, T. and Streubel, M. (1984). Angular momentum at null infinity, *Class. Quant. Grav.* **1**, 15–26.
du Plessis, J.C. (1970). Polynomial conformal tensors, *Proc. Camb. Phil. Soc.* **68**, 329–44.
Eastwood, M.G. (1985*a*). A duality for homogeneous bundles on twistor space, *J. Lond. Math. Soc.*, **31**, 349–56.
Eastwood, M.G. (1985*b*). Supersymmetry, twistors and the Yang–Mills equations, *Trans. Amer. Math. Soc.* (in press).
Eastwood, M.G. and Ginsberg, M.L. (1981). Duality in twistor theory, *Duke Math. J.* **48**, 177–96.
Eastwood, M.G., Penrose, R. and Wells, R.O., Jr. (1981). Cohomology and massless fields, *Comm. Math. Phys.* **78**, 305–51.
Eastwood, M.G. and Singer, M. (1985). A conformally invariant Maxwell gauge, *Phys. Lett.* (in press).
Eddington, A.S. (1924). *The Mathematical Theory of Relativity* (Cambridge University Press, Cambridge).
Einstein, A. (1918). Uber Gravitationswellen, *Sitz. Ber. Preuss. Akad. Wiss.* 154–67.
Exton, A.R., Newman, E.T. and Penrose, R. (1969). Conserved quantities in the Einstein–Maxwell theory, *J. Math. Phys.* **10**, 1566–70.
Farrar, G. and Neri, F. (1983). How to calculate 35,640 $O(\alpha^5)$ Feynman diagrams in less than an hour, *Phys. Lett.* **130B**, 109–14.
Fefferman, C. (1976). Monge–Ampère equations, the Bergman kernel and geometry of pseudo-convex domains, *Ann. of Math.* **103**, 395–416.
Field, M. (1982). *Several Complex Variables and Complex Manifolds*, Lond. Math. Soc. Lecture Notes no. 65, Vol. 1 (Cambridge University Press, Cambridge), p. 140.
Fierz, M. (1944). Zur Theorie magnetisch gelandener Teilchen, *Helv. Phys. Acta* **17**, 27–34.
Flaherty, E.J., Jr. (1976). *Hermitian and Kahlerian Geometry in Relativity*, Lecture Notes in Physics No. 46 (Springer-Verlag, Berlin).
Flaherty, E.J., Jr. (1980). Complex variables in relativity, in *General Relativity and Gravitation, One Hundred Years after the Birth of Albert Einstein*, ed. A. Held (Plenum, New York).
Flaherty, F.J. (ed.) (1984). *Asymptotic Behavior of Mass and Spacetime Geometry*, Proc. Corvalis, Oregon 1983 (Springer-Verlag, Berlin).

Floyd, R. (1973). The Dynamics of Kerr Fields, Ph.D. thesis, University of London.

Folland, G.B. and Kohn, J.J. (1972). *The Neumann Problem for the Cauchy–Riemann Complex*, Annals of Math. Studies No. 75 (Princeton University Press, Princeton).

Friedrich, H. (1981*a*). On the regular and the asymptotic characteristic initial value problem for Einstein's vacuum field equations, *Proc. Roy. Soc. London* **A375**, 169–84.

Friedrich, H. (1981*b*). The analytic characteristic initial value problem for Einstein's vacuum field equations as an initial value problem for a first-order quasilinear symmetric hyperbolic system. *Proc. Roy. Soc. London* **A378**, 401–21.

Frolov, V.P. (1979). The Newman–Penrose method in the theory of general relativity, in *Problems in the General Theory of Relativity and Theory of Group Representations*, ed. N.G. Basov (Plenum, New York).

Gårding, L. (1945). Relativistic wave equations for zero rest-mass, *Proc. Camb. Phil. Soc.* **41**, 49–56.

Geroch, R.P. (1971). Space-time structure from a global view point, in *General Relativity and Cosmology*, Proc. of Int. Sch. in Phys. 'Enrico Fermi', Course XLVII, ed. R.K. Sachs (Academic Press, New York), pp. 71–103.

Geroch, R.P. (1977). Asymptotic structure of space-time, in *Asymptotic Structure of Space-Time*, ed. F.P. Esposito and L. Witten (Plenum, New York).

Geroch, R.P., Held, A. and Penrose, R. (1973). A space-time calculus based on pairs of null directions. *J. Math. Phys.* **14**, 874–81.

Geroch, R.P. and Horowitz, G.T. (1978). Asymptotically simple does not imply asymptotically Minkowskian, *Phys. Rev. Lett.* **40**, 203–6.

Geroch, R.P., Kronheimer, E.H. and Penrose, R. (1972). Ideal points of space-times, *Proc. Roy. Soc. London* **A327**, 545–67.

Geroch, R.P. and Winicour, J. (1981). Linkages in general relativity, *J. Math. Phys.* **22**, 803–12.

Gibbons, G.W. and Hawking, S.W. (1979). Classification of gravitational instanton symmetries, *Comm. Math. Phys.* **66**, 291–310.

Gilkey, P.B. (1974). *The Index Theorem and the Heat Equation* (Publish or Perish, Inc., Boston, Mass.).

Gindikin, S.G. (1982). Integral geometry and twistors, in *Twistor Geometry and Non-Linear Systems*, ed. H.D. Doebner and T.D. Palev (Springer-Verlag, Berlin), pp. 2–42.

Gindikin, S.G. (1983). The complex universe of Roger Penrose, *Math. Intelligencer*, **5**, No. 1, 27–35.

Ginsberg, M.L. (1983). Scattering theory and the geometry of multi-twistor spaces, *Trans. Amer. Math. Soc.* **276**, 789–815.

Godement, R. (1964). *Théorie des Faisceaux* (Hermann, Paris).

Goldberg, J.N., Macfarlane, A.J., Newman, E.T., Rohrlich, F. and Sudarshan, E.C.G. (1967). Spin-*s* spherical harmonics and ð. *J. Math. Phys.* **8**, 2155–61.

Goldberg, J.N. and Sachs, R.K. (1962). A theorem on Petrov types, *Acta Phys. Polon.*, Suppl. **22**, 13–23.

Grace, J.H. and Young, A. (1903). *The Algebra of Invariants* (Cambridge University Press, Cambridge).

Grgin, E. (1966). A global technique for the study of spinor fields, Ph.D. thesis, Syracuse University, Syracuse, New York.

Griffiths, P. and Harris, J. (1978). *Principles of Algebraic Geometry* (John Wiley, New York).

Grothendieck, A. and Dieudonné, J. (1961). *Eléments de Géométrie Algébrique* (Pub. Math. IHES, Paris), Ch. 3, § 1.1.
Gunning, R.C. (1967). *Lectures on Vector Bundles over Riemann Surfaces*, Mathematical Notes No. 6 (Princeton University Press, Princeton).
Gunning, R.C. and Rossi, H. (1965). *Analytic Functions of Several Complex Variables* (Prentice-Hall, Englewood Cliffs, NJ).
Hansen, R.O. and Ludvigsen, M. (1977). A new $\mathscr{H}$-space formalism, *Gen. Rel. Grav.* **8**, 761–86.
Hansen, R.O. and Newman, E.T. (1975). A complex Minkowski space approach to twistors, *Gen. Rel. Grav.* **6**, 361–85.
Hansen, R.O., Newman, E.T., Penrose, R. and Tod, K.P. (1978). The metric and curvature properties of $\mathscr{H}$-space, *Proc. Roy. Soc. London* **A363**, 445–68.
Hawking, S.W. (1968). Gravitational radiation in an expanding universe, *J. Math. Phys.* **9**, 598–604.
Hawking, S.W. (1977). Gravitational instantons, *Phys. Lett.* **60A**, 81–3.
Hawking, S.W. and Ellis, G.F.R. (1973). *The Large Scale Structure of Space-Time* (Cambridge University Press, Cambridge).
Hepner, W.A. (1962). The inhomogeneous Lorentz group and the conformal group, *Nuovo Cim.* **26**, 351–68.
Hermann, R. (1968). *Differential Geometry and the Calculus of Variations* (Academic Press, New York), p. 71.
Hicks, N.J. (1965). *Notes on Differential Geometry* (Van Nostrand, Princeton).
Higgins, P.J. (1974). *An Introduction to Topological Groups* (Cambridge University Press, Cambridge).
Hirzebruch, F. (1962). *New Topological Methods in Algebraic Geometry* (Springer, Berlin).
Hitchin, N.J. (1979). Polygons and gravitons, *Math. Proc. Camb. Phil. Soc.* **85**, 465–76.
Hitchin, N.J. (1982). Complex manifolds and Einstein's equations, in *Twistor Geometry and Non-Linear Systems*, ed. H.D. Doebner and T.D. Palev (Springer-Verlag, Berlin), pp. 73–99.
Hodge, W.V.D. and Pedoe, D. (1952). *Methods of Algebraic Geometry*, Vol. 2 (Cambridge University Press, Cambridge).
Hodges, A.P. (1982). Twistor diagrams, *Physica* **114A**, 157–75.
Hodges, A.P. (1985*a*). A twistor approach to the regularisation of divergences, *Proc. Roy. Soc. London* **A397**, 341–74.
Hodges, A.P. (1985*b*). Mass eigenstates in twistor theory, *Proc. Roy. Soc. London* **A397**, 375–96.
Hodges, A.P. and Huggett, S. (1980). Twistor diagrams, *Surveys in High Energy Physics* **1**, 333–53.
Hopf, H. (1935). Uber die Abbildungen von Sphären auf Sphären neidrigerer Dimension, *Fund. Math.* **25**, 427–40.
Hörmander, L. (1966). *An Introduction to Complex Analysis in Several variables* (Van Nostrand-Reinhold, New York).
Horowitz, G.T. and Perry, M.J. (1982). Gravitational energy cannot become negative, *Phys. Rev. Lett.* **48**, 371–4.
Huggett, S. and Tod, K.P. (1985). *Introduction to Twistor Theory*, London Math. Soc. Lecture Notes Series (Cambridge University Press, Cambridge).
Hughston, L.P. (1979). *Twistors and Particles*, Lecture Notes in Physics No. 97 (Springer-Verlag, Berlin).
Hughston, L.P. (1980). The twistor particle programme, *Surveys in High Energy Physics* **1**, 313–32.

Hughston, L.P. (1986). Applications of $SO(8)$ spinors, in *Gravitation and Geometry* (I. Robinson Festschrift volume), eds. W. Rindler and A. Trautman (Bibliopolis, Naples).

Hughston, L.P. and Hurd, T.R. (1981). A cohomological description of massive fields, *Proc. Roy. Soc. London* **A378** 141–54.

Hughston, L.P. and Hurd, T.R. (1983). A CP^5 calculus for space–time fields, *Phys. Rep.* **100**, 273–326.

Hughston, L.P., Penrose, R., Sommers, P. and Walker, M. (1972). On a quadratic first integral for the charged particle orbits in the charged Kerr solution, *Comm. Math. Phys.* **27**, 303–8.

Hughston, L.P. and Sommers, P. (1973). Spacetimes with Killing tensors, *Comm. Math. Phys.* **32**, 147–52.

Hughston, L.P. and Ward, R.S. (eds.) (1979). *Advances in Twistor Theory*, Research Notes in Mathematics No. 37 (Pitman, San Francisco).

Hurd, T.R. (1985). The projective geometry of simple cosmological models, *Proc. Roy. Soc. London* **A397**, 233–43.

Isenberg, J., Yasskin, P.B. and Green, P.S. (1978). Non-self-dual gauge fields, *Phys. Lett.* **78B**, 462–4.

Jacobowitz, H. and Trèves, F. (1982). Non-realizable CR structures, *Invent. Math.* **66**, 231–49.

Janis, A.I. and Newman, E.T. (1965). Structure of gravitational sources, *J. Math. Phys.* **6**, 902–14.

Jeffryes, B.P. (1984*a*). Space-times with two-index Killing spinors, *Proc. Roy. Soc. London* **A392**, 323–41.

Jeffryes, B.P. (1984*b*). Two-surface twistors and conformal embedding, in *Asymptotic Behavior of Mass and Spacetime Geometry*, Proc. Corvalis, Oregon 1983, ed. F.J. Flaherty (Springer-Verlag, Berlin).

Jordan, P., Ehlers, J. and Sachs, R.K. (1961). Beiträge sur Theorie der reinen Gravitationsstrahlung, *Akad. Wiss. Lit. Mainz, Abhandl. Math. Nat. K1.* 1960 no. 1.

Kerr, R.P. (1963). Gravitational field of a spinning mass as an example of algebraically special metrics, *Phys. Rev. Lett.* **11**, 237–8.

Kerr, R.P. and Schild, A. (1965). Some algebraically degenerate solutions of Einstein's gravitational field equations, *Proc. Symp. Appl. Math.* **17**, 199–209.

Kerr, R.P. and Schild, A. (1967). A new class of vacuum solutions of the Einstein field equations, in *Atti del convegno sulla relatività generale* (Firenze), p. 222.

Klein, F. (1870). Zur Theorie der Liniercomplexe des ersten und zweiten Grades, *Math. Ann.* **2**, 198–226.

Klein, F. (1892). *Autographierte Vorlessungen über nicht-euklidische Geometrie*, Vol II (Göttingen), p. 245.

Klein, F. (1926). *Vorlesungen über höhere Geometrie* (Springer-Verlag, Berlin), pp. 80, 262.

Ko, M., Ludvigsen, M., Newman, E.T. and Tod, K.P. (1981). The theory of $\mathscr{H}$-space, *Phys. Repts.* **71**, 51–139.

Ko, M., Newman, E.T. and Penrose, R. (1977). The Kähler structure of asymptotic twistor space, *J. Math. Phys.* **18**, 58–64.

Kobayashi, S. and Nomizu, K. (1963). *Foundations of Differential Geometry*, Vol. 1 and 2 (Interscience, London).

Komar, A.B. (1959). Covariant conservation laws in general relativity, *Phys. Rev.* **113**, 934–6.

Komar, A.B. (1962). Generators of coordinate transformations in the Penrose

formalism of general relativity, *Phys. Rev.* **127**, 955–9.
Komar, A.B. (1964). Commutators on characteristic surfaces, *Phys. Rev.* (2) **134**, B 1430–40.
Kozameh, C.N., Newman, E.T. and Porter, J.R. (1984). Maxwell's equations, linear gravity and twistors, *Found. Phys.* **14**, 1061–81.
Kozameh, C.N., Newman, E.T. and Tod, K.P. (1985). Conformal Einstein spaces, *Gen. Rel. Grav.* **17**, 343–52.
Kozarzewski, B. (1965). Asymptotic properties of the electromagnetic and gravitational fields, *Acta. Phys. Polon.* **27**, 775–81.
Kramer, D., Stephani, H., MacCallum, M. and Herlt, E. (1980). *Exact Solutions of Einstein's Field Equations* (VEB Deutscher Verlag der Wissenschaften, Berlin and Cambridge University Press, Cambridge).
Kristian, J. and Sachs, R.K. (1966). Observations in cosmology, *Astrophys. J.* **143**, 379–99.
Kuiper, N.H. (1949). On conformally-flat spaces in the large, *Ann. Math.* **50**, 916–24.
Kundt, W. and Thompson, A. (1962). Le tenseur de Weyl et une congruence associée de géodésiques isotropes sans distorsion, *C.R. Acad. Sci. Paris* **254**, 4257–9.
Lang, S. (1972). *Differentiable manifolds* (Addison Wesley, Reading, Mass.).
Law, P. (1985). Twistor theory and the Einstein's equations, *Proc. Roy. Soc. London*, (in press)
Lawson, H.B. Jr. and Michelsohn, M.L. (1986) *Spin Geometry* (to appear)
LeBrun, C.R. (1982). $\mathscr{H}$-space with a cosmological constant, *Proc. Roy. Soc. London* **A380**, 171–85.
LeBrun, C.R. (1983). Spaces of complex null geodesics in complex Riemannian geometry, *Trans. Amer. Math. Soc.* **278**, 209–31.
LeBrun, C.R. (1984). Twistor CR-manifolds and three-dimensional conformal geometry, *Trans. Amer. Math. Soc.* **284**, 601–16.
LeBrun, C.R. (1985). Ambitwistors and Einstein's equations, *Class. and Quantum Grav.* (in press).
Lerner, D.E. (1977). Twistors and induced representations of *SU* (2,2), *J. Math. Phys.* **18**, 1812–17.
Lerner, D.E. and Clarke, C.J.S. (1977). Some global properties of massless free fields, *Comm. Math. Phys.* **55**, 179–82.
Lerner, D.E. and Sommers, P.D. (eds.) (1979). *Complex Manifold Techniques in Theoretical Physics*, Research Notes in Mathematics No. 32 (Pitman, San Francisco).
Lewy, H. (1956). On the local character of a solution of an atypical linear differential equation in three variables and a related theorem for regular functions of two complex variables, *Ann. of Math.* (2) **64**, 514–22.
Lewy, H. (1957). An example of a smooth linear partial differential equation without solution, *Ann. of Math.* (2) **66**, 155–8.
Lichnerowicz, A. (1958). Sur les ondes et radiations gravitationelles, *C.R. Acad. Sci. Paris* **246**, 893–6.
Lie, S. (1872). Über Complexe, unbesondere Linien- und Kugelcomplexe mit Anwendung auf die Theorie partieller Differentialgleichungen, *Math. Ann.* **5**, 145–265.
Lie, S. and Scheffers, G. (1896). *Geometrie der Beruhrungstransformationen I* (Leipzig).
Lind, R.W., Messmer, J. and Newman, E.T. (1972). Equations of motion for the sources of asymptotically flat spaces, *J. Math. Phys.* **13**, 1884–91.

Lind, R.W. and Newman, E.T. (1974). Complexification of the algebraically special gravitational fields, *J. Math. Phys.* **15**, 1103–12.
Ludvigsen, M. and Vickers, J.A.G. (1981). The positivity of the Bondi mass, *J. Phys.* **A14**, L389–91.
Ludvigsen, M. and Vickers, J.A.G. (1982). A simple proof of the positivity of the Bondi mass, *J. Phys.* **A15**, L67–70.
Ludwig, G. (1969). Classification of electromagnetic and gravitational fields, *Amer. J. Phys.* **37**, 1225–38.
Ludwig, G. (1976). On asymptotically flat space–times, *Gen. Rel. Grav.* **7**, 293–311.
Ludwig, G. (1981). On asymptotic flatness, *Gen. Rel. Grav.* **13**, 291–7.
Ludwig, G. and Scanlan, G. (1971). Classification of the Ricci tensor, *Comm. Math. Phys.* **20**, 291–300.
Lugo, G. (1982). Structure of asymptotic twistor space, *J. Math. Phys.* **23** (2) 276–82.
McCarthy, P.J. (1976). The Bondi–Metzner–Sachs group in the nuclear topology, *Proc. Roy. Soc. London* **A343**, 489–523.
McLennan, J.A., Jr. (1956). Conformal invariance and conservation laws for relativistic wave equations for zero rest mass, *Nuovo Cim*, **3**, 1360–79.
Manin, Yu. I. (1982). Gauge field and cohomology of analytic sheaves, in *Twistor Geometry and Non-Linear Systems*, ed. H.D. Doebner and T.D. Palev (Springer-Verlag, Berlin), pp. 43–52.
Mariot, L. (1954). Le champ électromagnétique singulier, *C.R. Acad. Sci. Paris* **238**, 2055–6.
Maunder, C.R.F. (1980). *Algebraic Topology* (Cambridge University Press, Cambridge).
Merkulov, S.A. (1984). A conformally invariant theory of gravitation and electromagnetism, *Class. Quantum Grav.* **1**, 349–54.
Milnor, J. (1963). Spin structure on manifolds, *Enseign. Math.* **9**, 198–203.
Misner, C.W., Thorne, K. and Wheeler, J.A. (1973). *Gravitation* (Freeman, San Francisco).
Morrow, J. and Kodaira, K. (1971). *Complex Manifolds* (Holt, Rinehart and Winston, New York).
Murai, Y. (1953). On the group of transformations in six-dimensional space, *Prog. Theor. Phys.* **9**, 147–68.
Murai, Y. (1954). On the group of transformations in six-dimensional space, II – conformal group in physics, *Prog. Theor. Phys.* **11**, 441–8.
Murai, Y. (1958). New wave equations for elementary particles, *Nucl. Phys.* **6**, 489–503.
Murray, M.K. (1984). A twistor correspondence for homogeneous polynomial differential operators, Research Report No. 47; Reprint, Mathematical Sciences Research Centre, Australian National University, Canberra.
Nester, J.M. (1981). A new gravitational energy expression with a simple positivity proof, *Phys. Lett*, **83A**, 241–2.
Newlander, A. and Nirenberg, L. (1957). Complex analytic coordinates in almost complex manifolds, *Ann. of Math.* (2) **65**, 391–404.
Newman, E.T. (1975). Complex space–time and some curious consequences, in *Quantum Theory and the Structures of Time and Space*, ed. L. Castell, M. Drieschner and C.F. von Weizsäcker (Carl Hanser Verlag, München).
Newman, E.T. (1976). Heaven and its properties. *Gen. Rel. Grav.* **7**, 107–11.
Newman, E.T. and Penrose, R. (1962). An approach to gravitational radiation by a method of spin coefficients, *J. Math. Phys.* **3**, 896–902 (Errata **4** (1963) 998).

Newman, E.T. and Penrose, R. (1965). 10 exact gravitationally-conserved quantities, *Phys. Rev. Lett.* **15**, 231–3.
Newman, E.T. and Penrose, R. (1966). A note on the Bondi–Metzner–Sachs group. *J. Math. Phys.* **7**, 863–70.
Newman, E.T. and Penrose, R. (1968). New conservation laws for zero rest-mass fields in asymptotically flat space-time, *Proc. Roy. Soc. London* **A305**, 175–204.
Newman, E.T., Porter, J.R. and Tod, K.P. (1978). Twistor surfaces and right-flat spaces, *Gen. Rel. Grav.* **9**, 1129–42.
Newman, E., Tamburino, L. and Unti, T. (1963). Empty-space generalization of the Schwarzschild metric, *J. Math. Phys.* **4**, 915–23.
Newman, E.T. and Tod, K.P. (1980). Asymptotically flat space-times, in *General Relativity and Gravitation, One Hundred Years after the Birth of Albert Einstein*, ed. A. Held (Plenum, New York).
Newman, E.T. and Unti, T.W.J. (1962). Behavior of asymptotically flat empty space, *J. Math. Phys.* **3**, 891–901.
Newman, E.T. and Winicour, J. (1974). A curiosity concerning angular momentum, *J. Math. Phys.* **15**, 1113–5.
Nirenberg, L. (1973). *Lectures on Linear Partial Differential Equations*, CBMS Regional Conf. Ser. in Math., No. 17 (Amer. Math. Soc., Providence, R.I.).
Nirenberg, L. (1974). On a question of Hans Lewy, *Russian Math. Surveys* **29**, 251–62.
Palais, R.S. (1965). *Seminar on the Atiyah–Singer index theorem*, Ann. of Math. Study No. 57 (Princeton University Press, Princeton).
Parker, T. and Taubes, C.H. (1982). On Witten's proof of the positive energy theorem, *Comm. Math. Phys.* **84**, 223–38.
Payne, W.T. (1952). Elementary spinor theory, *Amer. J. Phys.* **20**, 253–62.
Penrose, R. (1959). The apparent shape of a relativistically moving sphere, *Proc. Camb. Phil. Soc.* **55**, 137–9.
Penrose, R. (1960). A spinor approach to general relativity, *Ann. Phys.* **10**, 171–201.
Penrose, R. (1962). General relativity in spinor form, in *Les Théories Relativistes de la Gravitation*, ed. A. Lichnerowicz and M.A. Tonnelat (CNRS, Paris), pp. 429–32.
Penrose, R. (1963). Asymptotic properties of fields and space–times. *Phys. Rev. Lett.* **10**, 66–8.
Penrose, R. (1964*a*). The light cone at infinity, in *Conférence Internationale sur les Théories Relativistes de la Gravitation*, ed. L. Infeld (Gauthier-Villars, Paris and PWN, Warsaw).
Penrose, R. (1964*b*). Conformal approach to infinity, in *Relativity, Groups and Topology*: the 1963 Les Houches Lectures, ed. B.S. DeWitt and C.M. DeWitt. (Gordon and Breach, New York).
Penrose, R. (1965). Zero rest-mass fields including gravitation: asymptotic behaviour, *Proc. Roy. Soc. London* **A284**, 159–203.
Penrose, R. (1966). General-relativistic energy flux and elementary optics, in *Perspectives in Geometry and Relativity*, ed. B. Hoffmann (Indiana University Press, Bloomington), pp. 259–74.
Penrose, R. (1967*a*). Twistor algebra, *J. Math. Phys.* **8**, 345–66.
Penrose, R. (1967*b*). Conserved quantities and conformal structure in general relativity, in *Relativity Theory and Astrophysics*, Lectures in Applied Mathematics Vol. 8, ed. J. Ehlers (American Mathematical Society, Providence).

Penrose, R. (1968*a*). Twistor quantization and curved space–time, *Int. J. Theor. Phys.* **1**, 61–99.
Penrose, R. (1968*b*). Structure of space-time, in *Battelle Rencontres: 1967 Lectures in Mathematics and Physics*, ed. C.M. DeWitt and J.A. Wheeler (Benjamin, New York).
Penrose, R. (1969). Solutions of the zero rest-mass equations, *J. Math. Phys.* **10**, 38–9.
Penrose, R. (1972*a*). Spinor classification of energy tensors, in *Gravitatsya*, A.Z. Petrov Festschrift volume (Naukdumka, Kiev).
Penrose, R. (1972*b*). *Techniques of Differential Topology in Relativity*, CBMS Regional Conf. Ser. in Appl. Math., No. 7 (S.I.A.M., Philadelphia).
Penrose, R. (1972*c*). On the nature of quantum geometry, in *Magic without Magic: J.A. Wheeler Festschrift*, ed. J.R. Klauder (Freeman, San Francisco).
Penrose, R. (1972*d*). The geometry of impulsive gravitational waves, in *General Relativity, Papers in Honour of J.L. Synge*, ed. L. O'Raifeartaigh (Clarendon Press, Oxford), pp. 101–15.
Penrose, R. (1973). Naked singularities, *Ann. N.Y. Acad. Sci.*, **224**, 125–34.
Penrose, R. (1974). Relativistic symmetry groups, in *Group Theory in Non-Linear Problems*, ed. A.O. Barut (Reidel, Dordrecht).
Penrose, R. (1975*a*). Twistor theory: its aims and achievements, in *Quantum Gravity, an Oxford Symposium*, eds. C.J. Isham, R. Penrose and D.W. Sciama (Oxford University Press, Oxford).
Penrose, R. (1975*b*). Twistors and particles: an outline, in *Quantum Theory and the Structures of Time and Space*, eds. L. Castell, M. Drieschner and C.F. von Weizsäcker (Carl Hanser Verlag, Munich).
Penrose, R. (1976*a*). Nonlinear gravitons and curved twistor theory, *Gen. Rel. Grav.* **7**, 31–52; The non-linear graviton, *ibid.*, 171–6.
Penrose, R. (1976*b*). Any space–time has a plane wave as a limit, in *Differential Geometry and Relativity*, eds. M. Cahen and M. Flato (Reidel, Dordrecht).
Penrose, R. (1979*a*). A googly graviton, in *Advances in Twistor Theory*, ed. L.P. Hughston and R.S. Ward (Pitman, San Francisco); On the twistor description of massless fields, in *Complex Manifold Techniques in Theoretical Physics*, ed. D.E. Lerner and P.D. Sommers (Pitman, San Francisco).
Penrose, R. (1979*b*). The topology of ridge systems, *Ann. Hum. Genet. London* **42**, 435–44.
Penrose, R. (1980*a*). A brief outline of twistor theory, in *Cosmology and Gravitation; Spin Torsion, Rotation and Supergravity*, ed. P.G. Bergmann and V. de Sabbata (Plenum, New York).
Penrose, R. (1980*b*). A brief introduction to twistors, *Surveys in High Energy Physics*, **1**, 267–88.
Penrose, R. (1982). Quasi-local mass and angular momentum in general relativity, *Proc. Roy. Soc. London* **A381**, 53–63.
Penrose, R. (1983*a*). Spinors and torsion in general relativity, *Foundations of Physics*, **13**, 325–39.
Penrose, R. (1983*b*). Physical space–time and nonrealizable CR-structures, *Proc. Symp. Pure Math.* **39**, 401–22.
Penrose, R. (1984*a*). Integrals for general-relativistic sources: a development from Maxwell's electromagnetic theory, in *Maxwell Symposium Volume*, ed. M.S. Berger (North-Holland, Amsterdam).
Penrose, R. (1984*b*). Mass and angular momentum at the quasi-local level in general relativity, in *Asymptotic Behavior of Mass and Spacetime Geometry*, Proc. Corvalis, Oregon 1983, ed. F.J. Flaherty (Springer-Verlag, Berlin).

Penrose, R. (1986). On the origins of twistor theory, in *Gravitation and Geometry* (I. Robinson festschrift volume), eds. W. Rindler and A. Trautman (Bibliopolis, Naples).

Penrose, R. and MacCallum, M.A.H. (1972). Twistor theory: an approach to the quantization of fields and space–time, *Phys. Repts.* **6C**, 241–315.

Penrose, R. and Sparling, G.A.J. (1979). A note on the *n*-twistor internal symmetry group, in *Advances in Twistor Theory*, Research Notes in Mathematics No. 37, eds. L.P. Hughston and R.S. Ward (Pitman, San Francisco).

Penrose, R. and Ward, R.S. (1980). Twistors for flat and curved space–time, in *General Relativity and Gravitation, One Hundred Years after the Birth of Albert Einstein*, ed. A. Held (Plenum, New York).

Perjés, Z. (1975). Twistor variables in relativistic mechanics, *Phys. Rev.* **D11**, 2031–41.

Perjés, Z. (1977). Perspectives of Penrose theory in particle physics, *Rept. Math. Phys.* **12**, 193–211.

Perjés, Z. (1982). Introduction to twistor particle theory, in *Twistor Geometry and Non-Linear Systems*, ed. H.D. Doebner and T.D. Palev (Springer-Verlag, Berlin), pp. 53–72.

Perjés, Z. and Sparling, G.A.J. (1979). The twistor structure of hadrons, in *Advances in Twistor Theory*, ed. L.P. Hughston and R.S. Ward (Pitman, San Francisco).

Petrov, A.Z. (1954). Classification of spaces defined by gravitational fields, *Uch. Zap. Kazan Gos. Univ.* (Sci. Not. Kazan State Univ.), **114**, 55–69. English translation: Tran. No. 29, Jet Propulsion Lab. California Inst. Tech. Pasadena 1963.

Petrov, A.Z. (1966). *New methods in General Relativity* (in Russian) (Nauka, Moscow). English edition of Petrov's book: *Einstein Spaces* (Pergamon Press, Oxford, 1969).

Pirani, F.A.E. (1957). Invariant formulation of gravitational radiation theory, *Phys. Rev.* **105**, 1089–99.

Pirani, F.A.E. (1959). Gravitational waves in general relativity IV. The gravitational field of a fast moving particle, *Proc. Roy. Soc. London*, **A252**, 96–101.

Pirani, F.A.E. (1962). Gauss's theorem and gravitational energy in *Les Théories Relativistes de la Gravitation*, eds. M.A. Lichnerowicz and M.A. Tonnelat (C.N.R.S., Paris), pp. 85–91.

Plebański, J.F. (1964). The algebraic structure of the tensor of matter, *Acta Phys. Polon.* **26**, 963–1020.

Plebański, J.F. (1975). Some solutions of complex Einstein equations, *J. Math. Phys.* **16**, 2395–402.

Plebański, J.F. and Robinson, I. (1977). The complex vacuum metric with minimally degenerated conformal curvature, in *Asymptotic Structure of Space-Time*, ed. F.P. Esposito and L. Witten (Plenum, New York).

Plebański, J.F. and Schild, A. (1976). Complex relativity and double KS metrics. *Nuovo Cim.* **B35**, 35–53.

Plücker, J. (1865). On a new geometry of space, *Phil. Trans. Roy. Soc. London* **155**, 725–91.

Plücker, J. (1868). *Neue Geometrie des Raumes*, Vol. 1 (Leipzig).

Plücker, J. (1869). *Neue Geometrie des Raumes*, Vol. 2, ed. F. Klein (Leipzig).

Porrill, J. and Stewart, J.M. (1981). Electromagnetic and gravitational fields in a Schwarzschild space–time, *Proc. Roy. Soc. London* **A376**, 451–63.

Porteous, I.R. (1981). *Topological Geometry* (Cambridge University Press, Cambridge).
Porter, J.R. (1982). The nonlinear graviton: superposition of plane waves, *Gen. Rel. Grav.* **14**, 1023–33.
Press, W.H. and Bardeen, J.M. (1971). Non-conservation of the Newman–Penrose conserved quantities, *Phys. Rev. Lett.* **27**, 1303–6.
Prior, C.R. (1977). Angular momentum in general relativity I. Definition and asymptotic behaviour, *Proc. Roy. Soc. London* **A354**, 379–405.
Qadir, A. (1978). Penrose graphs, *Phys. Repts.* **39C**, 131–67.
Qadir, A. (1980). Field equations in twistors, *J. Math. Phys.* **21**, 514–20.
Reula, O. and Tod, K.P. (1984). Positivity of the Bondi energy, *J. Math. Phys.* **25**, 1004–8.
Rindler, W. (1956). Visual horizons in world-models, *Mon. Not. R. Astr. Soc.* London **116**, 662–7.
Rindler, W. (1977). *Essential Relativity* (Springer-Verlag, New York).
Rindler, W. (1982). *Introduction to Special Relativity* (Clarendon Press, Oxford).
Robinson, I. (1961). Null electromagnetic fields, *J. Math. Phys.* **2**, 290–1.
Robinson, I. and Schild, A. (1963). Generalization of a theorem by Goldberg and Sachs, *J. Math. Phys.* **4**, 484–9.
Robinson, I. and Trautman, A. (1962). Some spherical gravitational waves in general relativity. *Proc. Roy. Soc. London* **A265**, 463–73.
Sachs, R.K. (1961). Gravitational waves in general relativity VI. The outgoing radiation condition, *Proc. Roy. Soc. London* **A264**, 309–38.
Sachs, R.K. (1962*a*). Gravitational waves in general relativity VIII. Waves in asymptotically flat space–time, *Proc. Roy. Soc. London* **A270**, 103–26.
Sachs, R.K. (1962*b*). Asymptotic symmetries in gravitational theory. *Phys. Rev.* **128**, 2851–64.
Sachs, R.K. (1962*c*). Distance and the asymptotic behavior of waves in general relativity, in *Recent Developments in General Relativity* (Pergamon Press, Oxford and PWN, Warsaw).
Sachs, R.K. and Bergmann, P.G. (1958). Structure of particles in linearized gravitational theory. *Phys. Rev.* **112**, 674–80.
Schoen, R. and Yau, S.T. (1979*a*). Positivity of the total mass of a general space–time, *Phys. Rev. Lett.* **43**, 1457–9.
Schoen, R. and Yau, S.T. (1979*b*). On the proof of the positive mass conjecture in general relativity, *Comm. Math. Phys.* **65**, 45–76.
Schoen, R. and Yau, S.T. (1982). Proof that the Bondi mass is positive, *Phys. Rev. Lett.* **48**, 369–71.
Schouten, J.A. (1954). *Ricci-Calculus,* 2nd edition (Springer, Berlin).
Schrödinger, E. (1956). *Expanding Universes* (Cambridge University Press, Cambridge).
Seifert, H.J. (1971). The causal boundary of space–times, *Gen. Rel. Grav.* **1**, 247–59.
Sen, A. (1981). On the existence of neutrino 'zero-modes' in vacuum spacetimes, *J. Math. Phys.* **22**, 1781–6.
Shanahan, P. (1978). *The Atiyah–Singer Index Theorem,* Lecture Notes in Mathematics, No. 638 (Springer-Verlag, Berlin).
Shaw, W.T. (1983*a*). Twistor theory and the energy-momentum and angular momentum of the gravitational field at spatial infinity, *Proc. Roy. Soc. London* **A390**, 191–215.
Shaw, W.T. (1983*b*). Spinor fields at spacelike infinity, *Gen. Rel. Gray,* **15**, 1163–89.

Shaw, W.T. (1984*a*). Twistors, asymptotic symmetries and conservation laws at null and spatial infinity, in *Asymptotic Behavior of Mass and Spacetime Geometry*, Proc. Corvalis, Oregon 1983, ed. F.J. Flaherty (Springer-Verlag, Berlin).

Shaw, W.T. (1984*b*). Symplectic geometry of null infinity and two-surface twistors, *Class. Quant. Grav.* **1**, L33–7.

Sommerfeld, A. (1958). *Partielle Differentielgleichungen der Physik* (Akademische Verlagsgesellschaft, Leipzig).

Sommers, P.D. (1973). On Killing tensors and constants of motion, *J. Math. Phys.* **14**, 787–90.

Sommers, P.D. (1976). Properties of shear-free congruences of null geodesics, *Proc. Roy. Soc. London*, **A349**, 309–18.

Sommers, P. (1977). Type N vacuum space–times as special functions in C^2, *Gen. Rel. Grav.* **8**, 855–63.

Sommers, P. (1980). Space spinors, *J. Math. Phys.* **21**, 2567–71.

Sparling, G.A.J. (1975). Homology and twistor theory, in *Quantum Gravity*, eds. C.J. Isham, R. Penrose and D.W. Sciama (Oxford University Press, Oxford).

Sparling, G.A.J. (1981). Theory of massive particles I: Algebraic structure, *Phil. Trans. Roy. Soc. London* **301**, 27–74.

Sparling, G.A.J. (1983). Differential ideals and the Einstein vacuum equations, (Pittsburg preprint).

Sparling, G.A.J. (1984). Twistors, spinors and the Einstein vacuum equations, (Pittsburg preprint).

Sparling, G.A.J. (1985). Twistor theory and the characterization of Fefferman's conformal structures, (Pittsburg preprint, to appear).

Streubel, M. (1978). 'Conserved' quantities for isolated gravitational systems, *Gen. Rel. Grav.* **9**, 551–61.

Strooker, J.R. (1978). *Introduction to Categories, Homological Algebra and Sheaf Cohomology* (Cambridge University Press, Cambridge).

Study, E. (1903). *Jahresbericht der Deutschen Mathematikervereinigung*, Vol. XI, p. 319.

Synge, J.L. (1955). *Relativity: The Special Theory* (North-Holland, Amsterdam).

Synge, J.L. (1960). *Relativity: The General Theory* (North-Holland, Amsterdam).

Szekeres, P. (1963). Spaces conformal to a class of spaces in general relativity, *Proc. Roy. Soc. London* **A274**, 206–12.

Szekeres, P. (1967). Conformal tensors, *Proc. Roy. Soc. London* **A304**, 113–22.

Tafel, J. (1985). On the Robinson theorem and shearfree geodesic null congruences, *Lett. Math. Phys.* (to appear).

Tamburino, L.A. and Winicour, J.H. (1966). Gravitational fields in finite and conformal Bondi frames, *Phys. Rev.* **150**, 1039–53.

Tamm, Ig. (1931). Die verallgemeinrten Kugelfunktionen und die Wellenfunktionen eines Elektrons im Felde eines Magnetpoles, *Z. Phys.* **71**, 141–50.

Terrell, J. (1959). Invisibility of the Lorentz contraction. *Phys. Rev.* **116**, 1041–5.

Thorpe, J.A. (1979). *Elementary Topics in Differential Geometry* (Springer-Verlag, New York).

Tod, K.P. (1980). Curved twistor spaces and $\mathscr{H}$-space. *Surveys in High Energy Physics* **1**, 299–312.

Tod, K.P. (1983*a*). Some examples of Penrose's quasi-local mass construction, *Proc. Roy. Soc. London* **A388**, 457–77.

Tod, K.P. (1983*b*). Quasi-local charges in Yang–Mills theory, *Proc. Roy. Soc. London* **A389**, 369–77.

Tod, K.P. (1984). Three-surface twistors and conformal embedding, *Gen. Rel. Grav.* **16**, 435–43.

Tod, K.P. and Horowitz, G.T. (1982). A relation between local and total energy in general relativity, *Comm. Math. Phys.* **85**, 429–47.

Tod, K.P. and Perjés, A. (1976). Two examples of massive scattering using twistor theory, *Gen. Rel. Grav.* **7**, 903–13.

Tod, K.P. and Ward, R.S. (1979). Self-dual metrics with self-dual Killing vectors, *Proc. Roy. Soc. London* **A368**, 411–27.

Todd, J.A. (1947). *Projective and analytic geometry* (Pitman, London).

Tolman, R.C. (1934). *Relativity, Thermodynamics, and Cosmology* (Oxford University Press, Oxford).

Turnbull, H.W. and Aitken, A.C. (1948). *An Introduction to the Theory of Canonical Matrices* (Blackie, London).

Veblen, O. (1933*a*). Geometry of two-component spinors. *Proc. Nat. Acad. Sci.* **19**, 462–74.

Veblen, O. (1933*b*). Geometry of four-component spinors. *Proc. Nat. Acad. Sci.* **19**, 503–17.

Veblen, O. (1934). Spinors, *Science* **80**, 415–19.

Veblen, O. and Taub, A.H. (1934). Projective differentiation of spinors, *Proc. Nat. Acad. Sci.* **20**, 85–92.

Veblen, O. and Young, J.W. (1910). *Projective Geometry*, Vol. I (Ginn, Boston), p. 333.

Veblen, O. and Young, J.W. (1918). *Projective Geometry*, Vol. II (Ginn, Boston), p. 374.

Wald, R.M. (1984). *General Relativity* (University of Chicago Press, Chicago).

Walker, M. (1982). On the positivity of total gravitational energy at retarded times, in *The 1982 Les Houches Summer School on Gravitational Radiation*, ed. R. Ruffini.

Walker, M. and Penrose, R. (1970). On quadratic first integrals of the geodesic equations for type {22} spacetimes, *Comm. Math. Phys.* **18**, 265–74.

Walker, M. and Will, C.M. (1979). Relativistic Kepler problem, I: Behavior in the distant past of orbits with gravitational radiation damping, *Phys. Rev.* **D19**, 3483–94; II. Asymptotic behavior of the field in the infinite past, *ibid.* 3495–508. (Erratum: *Phys. Rev.* **D20**, 3437).

Walker, R.J. (1950). *Algebraic Curves* (Princeton University Press, Princeton, N.J.).

Ward, R.S. (1977). On self-dual gauge fields, *Phys. Lett.* **61A**, 81–2.

Ward, R.S. (1978). A class of self-dual solutions of Einstein's equations, *Proc. Roy. Soc.* **A363**, 289–95.

Ward, R.S. (1980*a*). Self-dual gauge fields. *Surveys in High Energy Physics*, **1**, 289–97.

Ward, R.S. (1980*b*). Self-dual space–times with cosmological constant, *Comm. Math. Phys.* **78**, 1–17.

Ward, R.S. and Wells, R.O., Jr. (1985). *Twistor Geometry* (Cambridge University Press, Cambridge).

Weinberg, S. (1970). *Gravitation and Cosmology: Principles and Applications of the General Theory of Relativity* (Wiley, New York).

Wells, R.O. Jr. (1979). Complex manifolds and mathematical physics, *Bull.* (*New Series*) *Amer. Math. Soc.* **1**, 296–336.

Wells, R.O. Jr. (1980). *Differential Analysis on Complex Manifolds* (Springer-Verlag, New York).

Wells, R.O. Jr. (1981). Hyperfunction solutions of the zero-rest-mass field equations, *Comm. Math. Phys.* **78**, 567–600.

Wells, R.O. Jr. (1982). *Complex Geometry in Mathematical physics* (Presses de l'Université de Montréal, Montréal).
Wess, J. and Zumino, B. (1974). Supergauge transformations in four dimensions, *Nucl. Phys.*, **70**, 39–50.
Whittaker, E.T. (1903). On the partial differential equations of mathematical physics, *Math. Ann.* **57**, 333–55.
Willmore, T.J. and Hitchin, N. (eds.) (1984). *Global Riemannian Geometry* (Ellis Horword, Chichester; Wiley, New York).
Winicour, J. (1968). Some total invariants of asymptotically flat space–times, *J. Math. Phys.* **9**, 861–7.
Winicour, J. (1980). Angular momentum in general relativity, in *General Relativity and Gravitation*, Vol. 2, ed. A. Held (Plenum, New York).
Witten, L. (1959). Invariants of general relativity and the classification of spaces, *Phys. Rev.* **113**, 357–62.
Witten, E. (1978). An interpretation of classical Yang–Mills theory, *Phys. Lett.* **77B**, 394–8.
Witten, E. (1981). A new proof of the positive energy theorem, *Com. Math. Phys.* **80**, 381–402.
Woodhouse, N.M.J. (1977). The real geometry of complex space–times, *Int. J. Theor. Phys.* **16**, 663–70.
Woodhouse, N.M.J. (1979). Twistor theory and geometric quantization, in *Group Theoretical Methods in Physics*, Springer Lecture Notes in Physics No. 50, eds. A. Janner, T. Janssen and N. Boon (Springer-Verlag, Berlin), pp. 149–63.
Woodhouse, N.M.J. (1980). *Geometric Quantization* (Clarendon Press, Oxford).
Woodhouse, N.M.J. (1983). On self-dual gauge fields arising from twistor theory, *Phys. Lett.* **94A**, 269–70.
Woodhouse, N.M.J. (1985). Real methods in twistor theory *Class. Quantum Grav.* **2**, 257–91.
Wu. T.T. and Yang, C.N. (1976). Dirac monopole without strings: monopole harmonics, *Nucl. Phys.* **B107**, 365–80.
Zharinov, V.V. and Sergeev, A.G. (1983). *Twistors and Gauge Fields* (in Russian) (MIR, Moscow).

Subject and author index

Italics denote main references; page numbers in parentheses refer to footnotes; a dash indicates continuation for more than two pages.

Index of symbols

Symbols are listed in order of appearence in text; page numbers in italics denote main references; page numbers in parentheses refer to footnotes; A dash indicates continuation for more than two pages.